S0-AEN-072

NATIVE
New Zealand
FLOWERING
PLANTS

NATIVE

New Zealand
FLOWERING
PLANTS

J. T. Salmon

REED

Dedicated to all those fine people who work
daily to preserve our environment and who
care deeply for the earth and all its living
things.

Cover photographs (left to right): kowhai flowers (fig. 593): flowers of
rata vine (fig. 572); drupes of the karaka tree (fig. 50).

The verso of the half title page shows heketara flowering in scrub
(fig. 231); the title page shows flowers of akepiro (fig. 226). Page
vii shows a southern rata in flower, Haast River, January.

Published by Reed Books, a division of Reed Publishing (NZ)
Ltd, 39 Rawene Road, Birkenhead, Auckland. Associated companies,
branches and representatives throughout the world.

This book is copyright. Except for the purpose of fair reviewing, no part
of this publication may be reproduced or transmitted in any form or by
any means, electronic or mechanical, including photocopying, recording,
or any information storage and retrieval system, without permission in
writing from the publisher. Infringers of copyright render themselves
liable to prosecution.

ISBN 0 7900 0223 X

© 1991 John T. Salmon

The author asserts his moral rights in the work.

First published 1991
This edition 1999

Printed in Hong Kong

CONTENTS

PREFACE

This book started out as a revision of my previous book *New Zealand Flowers and Plants in Colour*, published in 1963 and since reprinted eight times. First, the idea was to make it a 'field guide' but it soon became clear that a more comprehensive production was desirable and, indeed, possible. This new volume represents about forty years of exploring the New Zealand wilderness, whenever opportunity presented itself, photographing our native plants — at first by myself, but later with my wife and sons. The wild places and their plants became the centres of family holidays and the photographs gathered the backbone for future writings. In all this I have received help from many people who have been acknowledged in my earlier books, but for this one I have been wonderfully supported by my wife, Pam, not only when we went into the wilderness areas but most especially during the writing and preparation of the book. She has toiled unsparingly as a proof-reader and as a critic, offering many useful suggestions as we went along, and for all of this I am really grateful.

J. T. Salmon
Taupo
December 1990

INTRODUCTION

Within the small zone encompassing the islands of New Zealand, early European explorers found a flora that proved to be one of the most remarkable on earth. In the lowlands or in the mountains, by coast, stream, river or lakeside, plants of all kinds grew in unlimited profusion. Though the number of plant species found in our islands proved ultimately to be less than was at first predicted, few other regions have possessed such a high proportion of plant species not found anywhere else. Among flowering plants alone, 75 percent of the species proved peculiar to New Zealand, and few other regions could show within so small an area such an infinite variety of habitat, from seaside to alpine situations.

Into this near pristine environment with its lovely world of plants, about 150 years ago, came Europeans with their axes and saws, animals and fire. The early settlers saw the forest as a hindrance to farming, grass, crops and wealth. So the frontiers of the 'bush', as the forests were called, were pushed back towards and into the mountains, while introduced game animals increased the havoc being wrought upon the land.

From the earliest European pioneering days New Zealand society developed a spirit of indifference towards our native plants, which has changed for the better only within the last thirty years. Meanwhile many New Zealand native plants have been cultivated and appreciated in such other countries of the world as Australia, California, Britain and Spain.

I hope that this book will help to enhance the public appreciation of our natural heritage of native plants, many of which are unique, too many of which are now threatened with extinction, and almost all of which are beautiful to behold.

This book of pictures is designed to show the flowers and fruits of the native plants of New Zealand most likely to be met with when exploring the countryside, and to help interested people to identify our native plants. It moves from seaside plants through scrub and forests to the open spaces of the high mountains, grouping the photos according to the habitats in which the plants grow. However, though many plants can grow in more than one type of habitat, each is shown only once, usually in the habitat in which it is commonly found. The short text accompanying each picture should, with the picture, provide sufficient detail to accurately identify each plant.

The flowering times given in the legends and the months in which each photograph was taken (as recorded in the captions) are a general guide only, as flowering times can vary by as much as three to four weeks from place to place, depending on latitude, elevation, situation and seasonal climate.

To identify a particular plant, look first in that section of the book that corresponds to the natural habitat in which you were when you found it. A comprehensive index at the end of the book will assist in tracking down plant names. A list of reference works is appended for those readers who wish to learn more about our native flora.

NORTH ISLAND

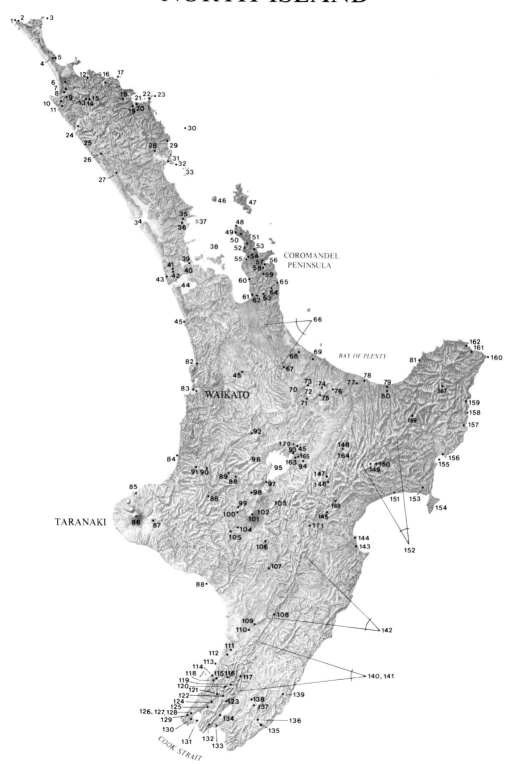

•o

1 •2 •3

•5
4

12 16 17
6
7 18
8 9 13 15 22 23
10 14 21
11 19 20

•30

24
28 29
25
31
26 32
33

27

34 35 37
36

38 46 •47

48
49 51
50 53
52
55 54 57 56 COROMANDEL
58 PENINSULA
39 59
41 40 60 65
42
43 62 64
44 61 63

45 66

68 69 BAY OF PLENTY 81
82 67
45 73 74 77 78 79
70 72 76 80
83 WAIKATO 71 75

92
170
93 45 148
96 163 165 164
84 95 94
91 90 147
89 146
88
85 98
99 103 168
100 102 145
86 101 171
87 104
105 106

107

88

162
161
160

167

159
158
157

169

156
155

150
149

151 153
154
152

144
143

108
109 142
110

111
112
113
114 140, 141
118 115 116
119 117
120 121
122 138 139
124 123 137
125
126, 127 128
129 136
130 134 135
131 132
133

COOK STRAIT

TARANAKI

SEASIDE PLANTS

Seaside plants are those found growing on the rocks, cliffs, sand-hills and sandy beaches, marshes and swamps within close proximity to the sea. They are all hardy plants adapted to withstand strong winds, shifting sands and frequent drenching with salty spray. Let's look first at those among the rocks and cliffs.

1   Pohutukawa flowers, Coromandel Peninsula (January)

2   The hairy seeds of pohutukawa, Karaka Bay, Wellington (July)

1–2   Pohutukawa/New Zealand Christmas tree, *Metrosideros excelsa*, grows naturally around Auckland, the coasts of Northland, Coromandel Peninsula, Bay of Plenty and East Cape as a medium-sized, rounded tree or a massive, often twisted and gnarled, large, spreading tree. During December and January it covers itself with masses of deep crimson to blood-red flowers (fig. 1). The conspicuous part of the flower is the mass of red stamens that surrounds a calyx funnel filled with nectar, which is much sought after by birds. The conspicuous hairy seed capsules (fig. 2) split open during June–July to release many seeds but seedlings are seldom found abundantly.   MYRTACEAE

3   Kermadec pohutukawa, *Metrosideros kermadecensis*, is similar to the familiar pohutukawa, *M. excelsa*, but is a smaller tree with shorter, oval-shaped leaves, 2.5 cm long by 10–20 mm wide. Flowers occur throughout most of the year and, though native to the Kermadec Islands, the tree is grown extensively in parks and gardens in New Zealand.   MYRTACEAE

3   Kermadec Island pohutukawa flowers, showing also the typical rounded leaves of this species (January)

4 ***Pyrrosia serpens*** is a peculiar spore-bearing plant found commonly in coastal rocky places but it also grows as an epiphyte on trees in lowland and montane forests throughout New Zealand, the Kermadec and Chatham Islands.

POLYPODIACEAE

4 Plant of *Pyrrosia serpens,* showing spores, Wellington Harbour (January)

5–7 **Taupata/angiangi/naupata,** *Coprosma repens,* grows along our coasts, south from North Cape to Marlborough, in rocky or sandy situations and, with its shining, bright green leaves, is one of the best known and most beautiful of our coastal shrubs. Flowers occur during October and November (fig. 5, female; fig.6, male), with the 6 mm long drupes (fig. 7) colouring in March and April.

RUBIACEAE

7 Taupata berries, Wellington (January)

8 **Woollyhead,** *Craspedia uniflora* var. *maritima,* grows in coastal rock crevices and grassy places around Cook Strait and near Ocean Beach at Oamaru.   ASTERACEAE

5 Taupata, female flowers, Wellington (October)

6 Taupata, male flowers, Wellington (October)

8 Woollyhead in flower, Cook Strait (November)

9   Horokaka in flower, Cook Strait (December)

10   Thick-leaved porcupine plant showing berries, leaves and stems, Wellington hills (February)

9   Horokaka/Maori ice plant, *Disphyma australe,* is found along the coasts of New Zealand as well as the Kermadec and Chatham Islands, trailing on rocky cliffs, coastal banks and gravels. The flowers, 2.5 cm across, occur from October to March.

AIZOACEAE

10   Thick-leaved porcupine plant, *Melicytus crassifolius,* occurs along our coasts as large, flat, springy cushions, up to 20 cm thick, closely adpressed to rocky surfaces. The flowers are only 3 mm across and appear from September to January, and the berries, 6 mm across, are ripe from October to March.

VIOLACEAE

11   Wild celery, *Apium australe,* occurs all along New Zealand coasts in rocky places and flowers from December to January. It is also found on the Kermadec, Three Kings and Chatham Islands.

APIACEAE

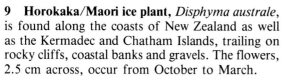

11   Wild celery in flower, East Wairarapa coast (December)

12 Wharanui showing leaves and
a flower-spike, southern
Wairarapa coast (August)

12 Wharanui, *Peperomia urvilleana*, is a succulent, prostrate and branching herb up to 30 cm high, found on rocky coasts near the sea throughout the North Island and the northern tip of the South Island. The leaves are about 4 cm long by 2 cm wide, with petioles up to 5 mm long; the flower-spike is up to 5 cm long. The plant flowers and fruits throughout the year. PIPERACEAE

13 Pigweed, *Rhagodia triandra*, grows as large mats often hanging over rocky faces beside the sea, from North Cape south to about Timaru in the east and Jackson Bay in the west, and also on the Kermadec Islands. Minute flowers occur from September to November, and the berries, 5 mm across, ripen from December to March.

CHENOPODIACEAE

13 Pigweed with berries, Wellington coast
(December)

14–15 Large-leaved porcupine plant, *Melicytus obovatus*, is an erect, spreading shrub up to 3 m high, with hairy young branchlets. It grows in rocky coastal places throughout both the North and South Islands. The thick adult leaves are simple, 4 cm long by 15 mm wide (fig. 14), but juvenile leaves are toothed and lobed. Small solitary flowers occur during November and December; the berries (fig. 15), 3–5 mm across, ripen from March to December.
VIOLACEAE

14 Large-leaved porcupine plant,
showing leaves and berries,
Cook Strait (December)

15 Large-leaved porcupine plant
heavy with fruit, Cook Strait
(December)

17   Creeping fuchsia with berries, Wellington (March)

16   Creeping fuchsia with flowers, Wellington (November)

16-17   Creeping fuchsia, *Fuchsia procumbens*, forms a sprawling or trailing shrub found along North Island coasts from North Cape to about Thames in stony, rocky, sandy and gravelly places above normal high-tide mark. Flowers, about 18 mm long, with blue pollen (fig. 16) occur from October to January, and the berries (fig. 17) ripen from January through March.   ONAGRACEAE

18   Cook Strait Islands senecio in flower, Wellington Harbour (January)

18   Cook Strait Islands senecio, *Senecio sterquilinus*, is a plant up to about 1 m high, rather similar to *S. lautus*. It is found growing in rocky places and sea-bird nesting areas on the islands of Cook Strait only.   ASTERACEAE

19   Shore groundsel, *Senecio lautus*, is a much-branched, herbaceous plant that grows commonly in rocky places, sometimes on sand dunes, near the sea but can be found also in inland rocky places up to 1,500 m altitude. Flowers 10–25 mm across occur during November and December.   ASTERACEAE

19   The shore groundsel in flower, Karaka Bay, Wellington (November)

20-21 *Colobanthus muelleri* is one of 15 species of *Colobanthus* found in New Zealand. Others occur in South America, Australia, Tasmania and some subantarctic islands. *C. muelleri* is found along the coasts of the North Island and the Chatham Islands in gravelly places, forming flat rosettes 10–20 mm across with leaves 10–15 mm long (fig. 20). Flowers and fruit capsules each 4–6 mm long (fig. 21) appear from September till April.

CARYOPHYLLACEAE

22 Pohuehue with flowers and seeds, Makara coast (April)

22 Pohuehue, *Muehlenbeckia complexa*, grows as a dense, tangled mass, several metres across and up to 60 cm high, all along our rocky coasts as well as inland in coastal and montane forests, where it can cover low shrubs with impunity. It flowers profusely from October to June, producing black seeds sitting in the white, cup-like tepals.   POLYGONACEAE

20 *Colobanthus muelleri* plant with fruiting capsules, Makara coast (November)

21 Fruiting and flowering capsules of *C. muelleri*, Makara coast (November)

23 Sprawling pohuehue, *Muehlenbeckia ephedrioides*, forms a much-branching, sprawling shrub, which bears small flowers from November to June. It is found all over the North and South Islands in coastal scrub and in rocky, sandy and gravelly places from sea-level to 1,000 m. It is similar to *M. complexa* except that the rigid, wiry stems have only a few narrow leaves 5–25 mm long.

POLYGONACEAE

23 Stem of sprawling pohuehue showing seeds, Wairarapa coast (February)

Turning to seaside plants that grow in sandy and gravelly places, we find:

24 Flowers of mutton bird sedge, male above, female below, Stewart Island (October)

24 Mutton bird sedge, *Carex trifida*, has a characteristic spike with the long, dark brown male flower above the shorter, lighter-coloured female flower. This sedge is found on coastal cliffs and rocky places as well as in inland swamps south of Timaru, on the subantarctic islands, the Chatham Islands and in Chile. It forms dense clumps about 60 cm across and has keeled leaves, 10 mm wide and 1–2 m long, with rough or scabrous margins.   CYPERACEAE

25 Pingao in flower, Tautuku Beach (December)

25 Pingao, *Desmoschoenus spiralis*, is a sedge found covering large areas of sandhills along the New Zealand coasts. Flowers and seed-heads are produced during December and January.
                                   CYPERACEAE

26 Sand-dune coprosma, *Coprosma acerosa*, forms a low-growing, cushiony mass of interlacing branches with narrow leaves, 1–5 mm wide and 12 mm long. Tiny axillary flowers are followed by drupes, 7 mm long, which turn pale blue during March and April.
                                     RUBIACEAE

26 Sand-dune coprosma plant with drupes, Tautuku Beach (January)

27 Sand-dune pin cushion, *Cotula trailii*, with its conspicuous dark, hairy flower-stems, is found sparsely among sand dunes, on sandy beach terraces and on salt marshes from Cape Foulwind south to Stewart Island. Flowers occur from December till February. ASTERACEAE

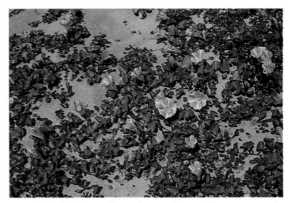

29 Nihinihi plant flowering and spreading over sand at Tautuku Beach (January)

27 Sand-dune pin cushion showing flowers and leaves, Stewart Island (December)

28 Spiny rolling grass, *Spinifex hirsutus*, occurs as a sand-binding plant that stabilises sand dunes along the coasts of the North Island and the northern parts of the South Island. Not indigenous, it is also found in Australia, Tasmania, some Pacific islands and India. GRAMINEAE

30 Nihinihi flowers, Tautuku Beach (January)

29–30 Nihinihi/sand convolvulus, *Calystegia soldanella*, is found throughout New Zealand, growing on sand dunes, sandy beaches (fig. 29) by sea or lake shore, gravelly riverbanks and occasionally on rocky places. Flowers (fig. 30), 2.5–7.5 cm across, occur in profusion from December to March.
CONVOLVULACEAE

28 Spiny rolling grass with flowers and seed-heads, Warehou Bay (December)

Seaside plants that grow in sandy and swampy places or salty marshes include:

31 These plants of the **seaside daisy,** *Celmisia major* var. *brevis*, shown here in flower, were found at Wharariki Beach, near Farewell Spit, growing in the sand and on rocky ledges close to high-water mark, where they are frequently drenched by salt spray. They are similar to *C. major* var. *brevis* from Mt Taranaki but are much smaller.

ASTERACEAE

32   Triangular-stemmed sedge in flower, Farewell Spit (January)

31   Seaside daisy plants in flower on Wharariki Beach, near Farewell Spit (January)

33   Triangular-stemmed sedge growing near Farewell Spit (January)

32–33 **Triangular-stemmed sedge,** *Scirpus americanus*, is commonly found in damp, sandy places, sometimes covering extensive areas, along the New Zealand coasts from the Coromandel Peninsula southwards but is absent from Westland and Fiordland. It flowers (fig. 32) during November–February.

CYPERACEAE

34   Sand sedge flower-head, South Wairarapa coast (November)

34 **Sand sedge,** *Carex pumila*, is a coarse, tufted sedge that grows along our coasts on the seaward slopes of sandy and gravelly beaches. It is not indigenous and is found in many places round the Pacific Basin.

CYPERACEAE

35 Maori musk, *Mimulus repens*, is found through-out New Zealand, growing in coastal salt marshes or shallow swamps near the coast. The musk-scented flowers, which occur from October till January, when first open are a deep mauve colour as shown but, after a few days, they fade to white.

SCROPHULARIACEAE

35 Maori musk in flower, Lake Wairarapa outlet (December)

36 Remuremu, *Selliera radicans*, is a perennial that grows on mud-flats, coastal, damp, sandy and rocky places or inland along stream, pond or lake margins. Flowers, usually white but sometimes pale blue, occur freely from November to April.

GOODENIACEAE

36 Remuremu plant in flower, Karaka Bay, Wellington (February)

37 Flowers of maakoako, Wellington Harbour (April)

37 Maakoako, *Samolus repens*, is a creeping peren-nial that grows in salty marshes and rocky places beside the sea throughout New Zealand. Flowers arise at the branch tips as well as from the axils of the thick leaves and occur from December till April.

PRIMULACEAE

38 Yellow buttons, *Cotula coronopifolia*, is a creeping plant found along the edges of coastal muddy swamps, in damp sand-dune hollows and along damp lowland streamsides. The flowers, about 8 mm across, arise from October to January, but the seeds do not ripen until the following November.

ASTERACEAE

38 Yellow buttons plant in flower, Wellington (October)

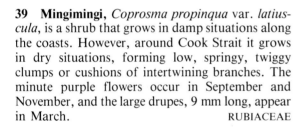

39   Mingimingi plant showing leaves and drupes, Cook Strait (March)

40   Southern salt horn plant growing on rock shelf at Karaka Bay, Wellington (February)

39   Mingimingi, *Coprosma propinqua* var. *latiuscula*, is a shrub that grows in damp situations along the coasts. However, around Cook Strait it grows in dry situations, forming low, springy, twiggy clumps or cushions of intertwining branches. The minute purple flowers occur in September and November, and the large drupes, 9 mm long, appear in March.   RUBIACEAE

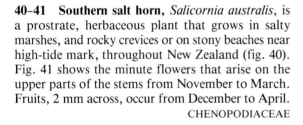

41   Close view of southern salt horn flowers, Karaka Bay, Wellington (January)

40–41   Southern salt horn, *Salicornia australis*, is a prostrate, herbaceous plant that grows in salty marshes, and rocky crevices or on stony beaches near high-tide mark, throughout New Zealand (fig. 40). Fig. 41 shows the minute flowers that arise on the upper parts of the stems from November to March. Fruits, 2 mm across, occur from December to April.   CHENOPODIACEAE

42   Mangrove flower and flower-bud showing pollen, Whangaparaoa (April)

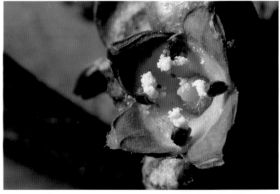

43   Male mangrove flower showing anthers ready to discharge pollen, Whangaparaoa (April)

44 Mangrove flowers showing (left) perfect anthers
and (right) anthers after pollen has been
discharged, Whangaparaoa (April)

45 A perfect mangrove flower showing female
stigma, Whangaparaoa (April)

46 A mature mangrove tree with
seeds near Paihia (February)

42–48 Manawa/New Zealand mangrove, *Avicennia
marina* var. *resenifera*, is a medium-size tree or shrub
that grows, except at low tide, with its roots covered
by salt water (fig. 46). Because of this, structures
called aerial roots grow upwards (fig. 48) to allow
the main roots to breathe air when uncovered by the
tides. Flowers are 6–7 mm across; the female flower
(fig. 45), with its broad pistil, arises as 4–8-flowered
clusters from February to April, but the fruit (fig.
47) takes until the following January to ripen.

AVICENNIACEAE

47 Mangrove seeds, Russell (December)

48 Aerial roots exposed at low tide. These allow the
main root system of the mangrove to breathe,
Tapatupoto Estuary, Northland (January)

COASTAL PLANTS

Trees, shrubs and herbaceous plants found growing along the coastal strips and hills, around river mouths and estuaries, on cliffs and rocky places or in bogs and swamps, near the coast but not beside the sea, are our coastal plants. Some unusual forms of these can range far inland as well and high into the mountains.

49   Flowers of the karaka tree, Karaka Bay, Wellington (October)

50   Drupes of the karaka tree, Karaka Bay, Wellington (March)

51   Close-up of karaka flowers, Karaka Bay, Wellington (October)

49–51 Karaka, *Corynocarpus laevigatus*, is a large tree, up to 16 m high, with large, dark green, very glossy leaves. It grows as isolated specimens or in small groves in coastal districts of the North Island, and in the South Island as far south as Banks Peninsula in the east and Jackson Bay in the west. The flowers are borne on stiff panicles (fig. 49), up to 22 cm long, from August to November, with the individual flowers (fig. 51) each about 5 mm across. The drupes (fig. 50) ripen from January through April and are 2.5–4.3 cm long.

CORYNOCARPACEAE

52 Rauhuia, *Linum monogynum*, is a bushy herb found throughout New Zealand. The flowers are up to 3.2 cm across and appear from October to December, though in warmer districts they may continue to April.   LINACEAE

52   Rauhuia flowers, Karaka Bay, Wellington (March)

53 Tauhinu spray with flowers, Makara coast (April)

53 Tauhinu/cottonwood, *Cassinia leptophylla*, grows plentifully in coastal areas from East Cape to Nelson, bearing clusters of flower-heads, up to 2 cm across, from November to January and again during March and April. ASTERACEAE

56 Flowers of northern koromiko, Manukau Heads (April)

54–55 Napuka, *Hebe speciosa*, is a lovely evergreen shrub, having either purple (fig. 55) or magenta (fig. 54) flowers. Both colour forms are found growing naturally on western coastal cliffs of the North Island from Hokianga to Tongaporutu and Cook Strait, and in Pelorus Sound; it is now often grown in private gardens. The flower-spikes, up to 5 cm long, arise from January through winter till October.
SCROPHULARIACEAE

54 Napuka with magenta flowers, Cook Strait (October)

55 Napuka with purple flowers, Cook Strait (February)

56 Northern koromiko, *Hebe obtusata*, with its pale lilac-coloured flower-spikes, up to 6 cm long, is found along the coast from Manukau Heads to Muriwai Beach. It flowers from January till June.
SCROPHULARIACEAE

59 Rengarenga in flower, Waikanae (December)

59 Rengarenga, *Arthropodium cirratum*, is a handsome lily found growing in dry, rocky, coastal regions of the North Island, Nelson and Marlborough. The plant flowers abundantly during November and December. ASPHODELACEAE

60 Green clematis, *Clematis hookeriana*, grows in dry places along both the north and south shores of Cook Strait. Sprawling over rocks and shrubs, it produces its green or pale yellow, sweet-scented flowers, about 15 mm across, from November till January. RANUNCULACEAE

57 Spray of flowers of coastal tree daisy, Karaka Bay, Wellington (April)

57–58 Coastal tree daisy, *Olearia solandri*, grows on coastal hills and rocky cliffs near the sea, up to 500 m altitude, throughout the North Island and in coastal areas of Marlborough and Nelson in the South Island. The flowers (fig. 57), up to 10 mm across, appear from February through to April, and the fluffy seeds (fig. 58) remain until October. ASTERACEAE

58 Seeds of coastal tree daisy, Karaka Bay, Wellington (September)

60 Flower spray of green clematis, Cook Strait (November)

61 Flowers and leaves of Castlepoint groundsel, Castlepoint (December)

62 Cook Strait groundsel in flower, Cook Strait (December)

61 Castlepoint groundsel, *Brachyglottis compacta,* grows naturally only on limestone cliffs near Castlepoint and flowers profusely from December to February. ASTERACEAE

62 Cook Strait groundsel, *Brachyglottis greyi,* is a handsome, grey-leaved shrub restricted in the wild state to rocky places on the north shores of Cook Strait and northwards to the Pahaoa River. The bright yellow corymbs of flowers, each flower about 2.5–3 cm across, occur from December till March. ASTERACEAE

63 Wharariki flowers, Karaka Bay, Wellington (November)

64 Vittadinia plant in flower, south Wairarapa (November)

63 Mountain flax/wharariki, *Phormium cookianum,* is as common on the coasts as it is in the mountains and can be found on rocks close to the water or, more often, on steep cliffs and rocky promontories. The flowers appear from November to January, and the seeds ripen from February to March. PHORMIACEAE

64 Vittadinia, *Vittadinia australis,* is a small, low-growing, bushy plant found throughout New Zealand, growing in rocky or grassy places in coastal regions up to 900 m altitude. Flowers, about 10 mm across, occur from November till March. ASTERACEAE

65 Rangiora tree in full flower,
    Karaka Bay, Wellington
    (September)

65–67 Rangiora, *Brachyglottis repanda*, is found in coastal scrub and lowland forest, from sea-level to 800 m, throughout the North Island and in Marlborough and Nelson. Fig. 65 shows a tree in full bloom. The large, sweet-scented, much-branched panicles of flowers (fig. 66) occur from August till November; the panicles are terminal on the branches and made up of many single florets (fig. 67).

ASTERACEAE

68 Pinatoro branch with flowers,
    Cass (November)

66 Flower-head of rangiora, Kaitoke (November)

69 Pinatoro plant in flower, Cass (November)

68–69 Pinatoro/wharengarara/common New Zealand daphne, *Pimelia prostrata*, is a prostrate or erect shrub (fig. 69) found throughout the country in rocky crags from sea-level to 1,600 m in the mountains. The sweet-scented flowers (fig. 68), about 6 mm across, occur from October till March. The drupes are either white or pinkish.

THYMELAEACEAE

67 Close-up of rangiora flowers, Kaitoke (November)

71 North Cape hibiscus flower, Ahipara Bay
(February)

71 North Cape hibiscus, *Hibiscus diversifolius*, is
found in sandy situations from North Cape to about
Ahipara Bay and Cape Brett. The perennial plant
reaches to 2 m high and the flowers, up to 8 cm
across, occur from November to March.

MALVACEAE

70 Perennial hibiscus flowers, Hicks Bay (October)

70 Perennial hibiscus, *Hibiscus trionum*, produces
flowers up to 5 cm across from October till March
and is found in sheltered coastal positions round
Northland, and on Mayor and Great Barrier Islands.

MALVACEAE

72 Coastal cutty grass, *Mariscus ustulatus*, is a
sedge found in damp situations in most coastal
regions. It flowers during November, with the seed-
heads ripening through December and January.

CYPERACEAE

72 Coastal cutty grass with seed-heads, south
Wairarapa (January)

73 New Zealand iris plant in flower, Lake Pounui
(November)

74 Seeds of New Zealand iris, Lake Pounui (May)

73–74 New Zealand iris, *Libertia ixioides*, is found
everywhere in both coastal and inland regions and
along lowland scrub and forest margins. Flowers
(fig. 73) appear during October and November, and
the berries (fig. 74) are ripe from January through
to December.   IRIDACEAE

76 Patotara with berries, Cupola Basin (April)

75 Patotara flowers, Sugarloaf, Cass, 1,000 m
(November)

75–76 Patotara, *Leucopogon fraseri*, is a low,
shrubby plant found on coastal dunes, in dry, rocky
places, dry riverbeds and lowland or subalpine grasslands. Flowers (fig. 75) occur abundantly from
September to December, and the berries (fig. 76)
ripen during February and March.

EPACRIDACEAE

77 Kopata, *Geum urbanum* var. *strictum*, grows in open country from sea-level to 900 m altitude. The flowers, 2 cm across, occur from November to January and the fruit from February to March.

ROSACEAE

79   Close-up of some flowers of pink tree broom, Woodside Gorge (December)

77   Kopata flower and seed, Hinakura (February)

78 Giant-flowered broom, *Carmichaelia williamsii*, is a much-branched shrub that bears 2.5 cm long flowers during February and March. It is found along the Bay of Plenty coast to East Cape and on the Poor Knights, Little Barrier and Aldermen Islands.   FABACEAE

80   Pink tree broom in full flower at the Woodside Gorge (December)

78   Flowers of giant-flowered broom, East Cape (March)

79-80 Pink tree broom, *Notospartium glabrescens*, when in flower (fig. 80), is one of New Zealand's most spectacular plants. Found only in Marlborough, it grows as a shrub or a small tree, up to 10 m high, on river terraces and in rocky situations from sea-level to 1,200 m. The slender, pendulous branchlets are compressed, and the flowers (fig. 79) occur as open racemes, up to 5 cm long, during December and January.   FABACEAE

81   Kaka beak flowers, Lake
     Waikaremoana (November)

82   White flowers of kaka beak,
     var. *albus*, Hinakura (October)

83   Flowers of a horticultural
     hybrid kaka beak, Hinakura
     (October)

81–83   Kaka beak/red kowhai/kowhai ngutu-kaka,
Clianthus puniceus, is a spreading shrub found wild
in the inlets of the Bay of Islands, on the coast near
Thames and at Lake Waikaremoana. The vivid red
to pale pink flowers, each about 8 cm long (fig. 81),
occur in abundance from October to December. A
white variety, *Clianthus puniceus* var. *albus* (fig. 82),
and hybrid forms (fig. 83) are also known.

FABACEAE

84   Wild lobelia, *Lobelia anceps*, is a low, herbaceous plant found in coastal and lowland areas
over most of New Zealand. Flowers, 6 mm long,
arise from November to March, blue at first but
quickly fading to pale blue or white.

LOBELIACEAE

84   Wild lobelia plant in flower, Lake Pounui
     (February)

85 Taurepo, yellow-flowered form, Hinakura (November)

86 Taurepo, red-flowered form, Hinakura (November)

85–86 Taurepo/waiu-atua/New Zealand gloxinia, *Rhabdothamnus solandri*, is a small shrub found in dry, shady, shingly gullies, along forest margins and streamsides throughout the North Island. Flowers, commonly red (fig. 86) but also yellow (fig. 85), are 18–25 mm long and occur almost all the year round.

GESNERIACEAE

87 Leafless sedge/wiwi, *Scirpus nodosus*, occurs in sandy, swampy areas on the coast and inland and around seepages on coastal hillsides.

CYPERACEAE

88 Kaikoura rock daisy, *Pachystegia insignis*, forms a low, stiff shrub found on coastal cliffs along the Kaikoura coast and in rocky places in river valleys near the coast throughout Marlborough. The flowers, up to 7.5 cm across, appear from December to February.

ASTERACEAE

87 Leafless sedge with flowers, Lake Pounui (February)

88 Kaikoura rock daisy plant in flower, Kaikoura coast (December)

89–90 Whau/cork tree, *Entelea arborescens*, is a shrub or a small, spreading tree with very large, soft, heart-shaped leaves, 15–25 cm long and 15–20 cm wide. The flowers, 18–25 mm across, arise as erect, flat or drooping clusters in the leaf axils from September to December (fig. 89) and the seeds form in spiney fruit capsules (fig. 90) from November to January. Whau is the only New Zealand tree to produce fruits with spines and these can be up to 2.5 cm long. The wood rivals balsawood as one of the lightest woods in the world. Whau occurs from North Cape south to Nelson at the bases of western coastal cliffs and on the edges of lowland forests; it is common around the Mokau River mouth.

TILIACEAE

90   Ripening seeds of whau, Hukutaia Domain (January)

89   The flowers of whau, Mokau River estuary (November)

91 Limestone harebell, *Wahlenbergia matthewsii*, is a tall herb, up to 30 cm high, that grows only on limestone rocky places between the Waima and Clarence Rivers. The flowers, 2–3 cm across, are distinctly blue when first opened but quickly fade to white.   CAMPANULACEAE

91   Limestone harebell, near Woodside Gorge (December)

92   Tawapou, berries and leaves, Kaitaia (April)

92–94   Tawapou, *Planchonella novo-zelandica*, is a small tree found along the east coast of the North Island from North Cape to Tolaga Bay, including the adjacent offshore islands, and to the Manukau Harbour in the west. The axillary flowers (fig. 93) occur in January and are very small, being 3–6 mm across, but the berries are large, up to 2.5 cm long. The red, yellow and orange berries (fig. 92) colour up from March through to May.   SAPOTACEAE

93   A flower of tawapou, Piha (January)

94   Spray of tawapou showing axillary flowers and a green berry from last year's flowering, Piha (January)

95 Maori jasmine in flower, Barton's Bush, Upper Hutt (November)

95-98 Maori jasmine/kaiku/kaiwhiria, *Parsonsia heterophylla*, found throughout New Zealand in coastal scrub and forest and along lowland forest margins, is a stronger, tougher vine (fig. 95) than akakiore (opposite), producing large, white (fig. 96) or yellow (fig. 97), peculiarly scented flowers, 8 mm across, from September to March, followed by seed-pods (fig. 98), 15 cm long, from February onwards.

APOCYNACEAE

96 Close-up of flowers of Maori jasmine, Stokes Valley (October)

97 Yellow-flowered Maori jasmine, Hinakura (November)

98 Seed-pods of Maori jasmine, Hinakura (February)

99-100 Akakiore/small Maori jasmine/pink jasmine, *Parsonsia capsularis*, is a slender climber (fig. 99) with small, pink, cream or white, fragrant flowers about 4 mm long (fig. 100). Found in coastal and lowland forests, it flowers from September till February. APOCYNACEAE

99   Akakiore with pinkish-coloured flowers, Hinakura (November)

100   Close-up of akakiore flowers, Hinakura (November)

101   Lowland lycopodium with strobili, Central Plateau (February)

102   Traill's daisy flower with leaves, Stewart Island (November)

101   Lowland lycopodium, *Lycopodium scariosum*, is a creeping plant found in lowland and subalpine forest or open scrub from the Bay of Plenty southwards, also occurring in Victoria and Tasmania. The strobili arise from the tips of ascending branches and are 2.5–5 cm long. LYCOPODIACEAE

102   Traill's daisy/tupare, *Olearia traillii*, is a handsome hybrid shrub or a small tree, 3–4 m high, the result of a cross between *O. angustifolia* and *O. colensoi*. The plant grows around the coasts of Stewart Island, flowering from November to January, and the rays of the flower-head may be violet or white. ASTERACEAE

103–106   Karo/turpentine tree, *Pittosporum crassi-folium*, forms a shrub or a small tree, up to 9 m high, with thick, tough, leathery and wavy leaves, 5–10 cm long and up to 2.5 cm wide. It is found naturally along coastal forest margins and streamsides from North Cape to Poverty Bay and is extensively cultivated in gardens, often as a hedge plant. The scented flowers (figs 103–104) are 12 mm across and occur from September to December, with the seed capsules, 2–3 cm across (fig. 105) ripening during the following August through December. Cultivars with variegated leaves have been produced (fig. 106). On a calm evening when the tree is in full flower the rich, sweet scent from the flowers fills the air.   PITTOSPORACEAE

104   Close-up of flower of variegated form of karo, Otari (September)

103   Flowers and leaves of karo, Karaka Bay, Wellington (September)

106   Horticultural variegated form of karo in flower, Otari (October)

105   Karo seed capsules opened showing ripe black seeds, Otari (May)

107   *Pittosporum huttonianum* is somewhat similar to *P. crassifolium* but a larger tree, with leaves 12 cm long by 5 cm wide. It grows only on the Coromandel Peninsula and the Great and Little Barrier Islands, and is distinguished by the very hairy branchlets and leaf petioles as shown in fig. 107.

PITTOSPORACEAE

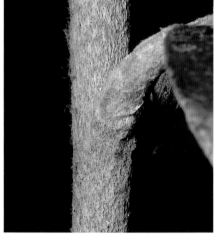

107   Hairy branchlet and leaf petiole typical of *Pittosporum huttonianum*, Hukutaia Domain, Opotiki (April)

108   Heart-leaved kohuhu branchlet with flowers and leaves, Kaitaia (November)

109   Close-up of flowers of the heart-leaved kohuhu, Otari (October)

110   Flower spray of *Hebe elliptica*, Wharariki Beach, Farewell Spit (January)

111   Flowers, close up, of *H. elliptica*, Wharariki Beach (January)

108–109   **Heart-leaved kohuhu,** *Pittosporum obcordatum*, is a narrow, erect, upward-branching, small tree, 3–4 m high, found growing wild only near Kaitaia and the Wairoa River. It is now sometimes grown in parks and gardens. The leaves are 4–8 mm long and 2–7 mm wide (fig. 108) and the flowers are borne along the branches, arising as groups from the leaf axils (fig. 109).   PITTOSPORACEAE

110–111   *Hebe elliptica* forms a much-branched shrub, about 2 m high, found along the west coast from about New Plymouth southwards and along the east coast of Otago. The lateral flower racemes (fig. 110) have 4–14 flowers (fig. 111), which occur from November to March. The leaves, 2–4 cm long by 6–16 mm wide, are somewhat fleshy and keeled.   SCROPHULARIACEAE

112   The flowers of manatu, Trentham
(October)

113   Manatu tree in full flower,
Waiorongomai (October)

112–116 Manatu/lowland lacebark, *Plagianthus
regius*, is a tree up to 15 m high, with soft, coarsely
toothed leaves, 10–15 cm long, many of which fall
in the autumn. The tree has a juvenile form with
interlacing branches that form a dense bush. Small
flowers, each 3–4 mm across (fig. 114), appear in
great profusion (fig. 113) as dense clusters (fig. 112)
in October and November, and the downy seed
capsules are ripe in January and February. Manatu
is found in coastal areas and marginally in lowland
forest throughout New Zealand.   MALVACEAE

115   Close-up of seeds of manatu, Lake Pounui
(March)

114   Close-up of manatu flowers, Trentham (October)

116   Close-up of manatu flowers, Trentham (October)

117-118 Ngaio, *Myoporum laetum*, forms a rounded tree up to 8 m high, found in coastal places and lowland forests from North Cape to Dunedin and on the Kermadec and Chatham Islands. The alternate, soft, fleshy, gland-dotted leaves, 4–10 cm, long and 10–30 mm wide, are on petioles up to 3 cm long (fig. 117). Flowers (fig. 118) about 10 mm across appear in the axils of the leaves during November and December and often again during April and May. The drupes (fig. 117) ripen three months later in each case.   MYOPORACEAE

117   Ngaio berries, Karaka Bay, Wellington (March)

118   Close-up of ngaio flower, Karaka Bay, Wellington (December)

119   Prostrate ngaio, leaves and flowers, Kawhia (March)

119-120   Prostrate ngaio, *Myoporum debile*, forms a low-growing shrub often covering large areas in scrub and exposed places between Raglan and Kawhia. Fig. 119 shows a spray with flowers and typical long, narrow leaves, 4–8 cm long by 5–10 mm wide, with toothed margins towards the apex. Flowers (fig. 120), about 5 mm across, occur during March and April.   MYOPORACEAE

120   Close-up of prostrate ngaio flowers, Kawhia (March)

121–125 Coastal maire, *Nestegis apetala,* is a shrub or small tree with spreading, tortuous branches and a furrowed bark. It is found on rocky headlands around Whangarei Heads, the Bay of Islands, the Hen and Chickens, Fanal, Cuvier, Poor Knights, and Great and Little Barrier Islands. The glossy, leathery, elliptic leaves are 4.5–12.5 cm long and 15–60 mm wide (fig. 121). The flowers are without petals and arise as racemes of up to 21 flowers from the leaf axils or direct from branches. Male racemes are shown in figs 122 and 125; female racemes in figs 123 and 124. Both flowers appear from late January into February, and the lovely drupes (fig. 121) are ripe by the following November through December.   OLEACEAE

121   Drupes and leaves of coastal maire, Oke Bay (December)

122   Raceme of male flowers of coastal maire, Oke Bay (January)

123   Raceme of female flowers of coastal maire, Oke Bay (January)

124   Close-up of female flowers of coastal maire, Oke Bay (January)

125   Close-up of male flowers of coastal maire, Oke Bay (January)

126   Parapara flowers and leaves,
Warkworth (February)

127   Close-up of flower of parapara,
Warkworth (January)

126–128   Parapara, *Pisonia brunoniana,* is the sole
representative in New Zealand of a group of subtropical plants that produce sticky fruits that can
capture and kill quite large birds and sometimes
small reptiles. Parapara is a shrub or a small tree,
up to 6 m high, with large, glossy, thick leaves,
10–40 cm long by 5–15 cm wide. Many-flowered
panicles (fig. 126), with each flower up to 10 mm
long (fig. 127), occur in December through January,
and the sticky fruits (fig. 128) turn black and become
very sticky during February and March when they
ripen. Small birds can be caught and held by the
sticky mass until they die. The plant then feeds upon
them and the seeds may germinate on the birds'
remains, the seedlings later falling to the ground to
grow.   NYCTAGINACEAE

129   Hooker's daisy in flower,
Stewart Island (October)

128   The sticky fruits of parapara, Warkworth
(February)

129   Hooker's daisy, *Celmisia hookeri,* is a tufted
herb found in coastal to montane grassland or scrub
of north-east Otago and Stewart Island. The leathery
leaves, 20–50 cm long by 4–8 cm wide, are smooth
above but have a dense, white, appressed tomentum below. Flowers, 2–5 cm across, occur from
November to January.   ASTERACEAE

130   Houhere flowers and leaves,
       Karaka Bay, Wellington
       (March)

131   Houhi ongaonga spray with
       seeds and leaves, Karaka
       Bay, Wellington (March)

130   Lacebark/houhere, *Hoheria populnea*, takes
the name lacebark from its stringy, interlaced bark
characteristic of the genus *Hoheria*. It forms a small,
erect tree up to 10 m high, with large, shining,
coarsely serrated leaves, 14 cm long by 6 cm wide
(fig. 130). The flowers, 2.5 cm across (fig. 130),
occur as 5–10-flowered cymes from February
through May. A form having flowers with purple
stamens is known as var. *osbornei* and was discov-
ered on Great Barrier Island in 1910. Houhere grows
wild only between North Cape and the Waikato in
the west and Bay of Plenty in the east; it is the only
lacebark that grows wild between Kaitaia and North
Cape It has become one of the most popular native
plants in cultivation and a variegated form has been
developed.   MALVACEAE

132   Houhi ongaonga flowers and leaves, Mt
       Holdsworth (March)

131–133   Houhi ongaonga, *Hoheria sextylosa*, is a
small tree, to 6 m high, with deeply incised, serrated
leaves, 10–30 cm long by 10–25 mm wide. Flowers,
2 cm across, (figs 132–133), occur in 2–5-flowered
cymes during March and April, and seeds typical of
lacebarks (fig. 131) are produced from March
through June.   MALVACEAE

133   Close-up of flower of houhi ongaonga, Karaka
       Bay, Wellington (March)

134 Narrow-leaved lacebark/houhere/ribbon-wood, *Hoheria angustifolia*, is an attractive-looking, slender tree, to 10 m high, which matures from a straggling shrub with slender, interlacing branches bearing a few scattered leaves. The coarsely toothed adult leaves are up to 7 cm long by 10 mm wide on petioles 5 mm long. Flowers, about 15 mm across, occur singly or in 2–5-flowered cymes from December to March. Found in coastal and lowland forests from Taranaki to Southland.   MALVACEAE

136   Wharangi flowers, close up, Karaka Bay, Wellington (September)

134   Narrow-leaved lacebark spray showing leaves and flowers, Peel Forest (January)

137   Wharangi seed-pods and black seeds, Karaka Bay, Wellington (September)

135   Wharangi leaves with developing flower-buds, Karaka Bay, Wellington (February)

135–137 Wharangi/koheriki, *Melicope ternata*, is a spreading and branching shrub or a small tree, up to 6 m high, found along coastal and lowland forest margins and on sheltered, rocky places near the sea from North Cape to Nelson and Kaikoura. The leaves (fig. 135) are trifoliate, wavy, 7–10 cm long by 3–4 cm wide, on petioles 2–3 cm long. Flowers (fig. 136) arise as paired, three-branched cymes from September to November, the single flowers each 10 mm across. The black seeds (fig. 137) appear from September to October.   RUTACEAE

138 Poataniwha flowers and typical leaf, Trentham (October)

139 Close-up of male flowers of poataniwha, Barton's Bush, Trentham (November)

140 Stewart Island daisy in flower (December)

138-139 Poataniwha, *Melicope simplex*, is a twiggy shrub about 3 m high found in coastal situations throughout New Zealand up to 600 m altitude. Young plants have trifoliate leaves, up to 10 mm long on petioles 2 cm long, but mature plants have simple, rhomboid-leaves, up to 2 cm long by 2 cm wide, on petioles 5 mm long (fig. 138). Flowers (female, fig. 138; male, fig. 139) occur from September through November, and the seeds, similar to those of *M. ternata* but smaller, occur from December through to April.   RUTACEAE

140 Stewart Island daisy, *Celmisia rigida*, is a tufted herb with rigid, leathery leaves, 10–15 cm long by 5–7 cm wide, found on coastal cliffs of Stewart Island. Flowers, 4–5 cm across, occur during November and December.   ASTERACEAE

141 White harebell plant in flower, Charleston (February)

141 White harebell, *Wahlenbergia congesta*, is a creeping herb forming dense patches to 25 cm across on lowland grassy banks in sand-dune country from Cape Foulwind to Charleston. A varietal form is found on Mt Taranaki.   CAMPANULACEAE

143 **Chatham Island forget-me-not**, *Myosotidium hortensia*, is a handsome perennial with large, glossy, kidney-shaped leaves, 25–40 cm wide, each with a stout petiole, which, with the leaf, can be 1 m long. Flowers are in dense cymes, 10–15 cm across, on stalks up to 60 cm high, from September to November. Occurring naturally only on the Chatham Islands, it is now a very highly prized garden plant.

BORAGINACEAE

143   Chatham Island forget-me-not plant in flower, Otari (November)

142   Pink broom in full flower, Clarence River (December)

142 **Pink broom**, *Notospartium carmichaeliae*, grows as a shrub or a small tree, up to 10 m high, with drooping, leafless branches and compressed branchlets that become smothered with small, crowded racemes of pink-flushed and pink-veined flowers, each flower about 8 mm long, during December and January.   FABACEAE

PLANTS OF LOWLAND SCRUB

Scrub is an association of shrubby and herbaceous plants and occasional small trees found in coastal regions; sometimes it is open but is more often dense and perhaps impenetrable. Scrub forms strange and peculiar associations of plants almost unique to New Zealand. It is generally encountered in gullies or covering dry, windswept hillsides, dry coastal terraces or river terraces and plateaus. It can be the forerunner of forest, providing a sheltered, nourishing seed-bed for larger trees. Many plants found in scrub also occur in forest or along forest margins, and the species associations can vary considerably, even within small areas, or be particularly locally defined. Lowland scrub can extend inland for 1–2 km from the coast and up to elevations of 700 m.

144–146 Golden tainui/kumarahou/gum-diggers' soap, *Pomaderris kumaraho*, is a rounded shrub bearing large, golden yellow corymbs of flowers (fig. 144) in profusion during September and October. It is found growing on poor soils from North Cape to Tauranga on the east coast and to Kawhia on the west. Leaves 6 cm long by 3 cm wide are soft, with depressed veins and star-like hairs beneath. A somewhat similar, more open shrub with leaves having fine tomentum beneath and the flower-heads smaller and more open is found round Warkworth to Thames and is known as *P. hamiltonii* (fig. 146).
RHAMNACEAE

144 Flower-heads of golden tainui, from my Waikanae garden (September)

145 Close-up of golden tainui flowers, Waikanae (September)

146 Spray of *Pomaderris hamiltonii* showing leaves and flower-head, Waipahihi Botanical Reserve, Taupo (September)

147-148 Wrinkled-leaved pomaderris, *Pomaderris rugosa*, is an erect shrub found growing on poor soils around Northland, the Coromandel Peninsula and Mayor Island. It is distinguished from the golden tainui by its wrinkled leaves (fig. 148) and longer flower-heads (fig. 147), and from tainui by the clothing of close, stellate hairs on the undersides of the leaves and the less-prominent leaf veins (fig. 148).

RHAMNACEAE

149 Flower-head of tainui, Kawhia (November)

147 Flower-head of wrinkled-leaved pomaderris, Coromandel Peninsula (November)

150 Spray of tainui showing prominent leaf veins below, Kawhia (November)

149-150 Tainui, *Pomaderris apetala*, is an erect, branching shrub up to 4 m high, with wrinkled, crenulate-margined, prominent-veined leaves, 7 cm long and 3 cm wide, hairy below with occasional hairs above (fig. 150). The greenish flowers (fig. 149), without petals, occur as lateral and terminal clusters on the branchlets from November into January. Tainui is an Australian shrub that grows in New Zealand naturally from Kawhia to Waitara and has become popular and widespread in gardens.

RHAMNACEAE

148 Undersurfaces of leaves of wrinkled-leaved pomaderris showing stellate hairy surface, Coromandel Peninsula (October)

151 Dwarf pomaderris/tauhinu, *Pomaderris phylic-*
ifolia, is a low-growing, heath-like plant that smothers
itself with corymbs of yellow flowers during October
and November. The small, thick leaves, with recurved
margins, are 10 mm long by 3 mm wide and densely
hairy below. There are two varieties, var. *ericifolia,*
with flower corymbs laterally along the branches,
and var. *polifolia,* (fig. 151), which has flower
corymbs both along the branches and at their tips.
RHAMNACEAE

151 Flowers of dwarf pomaderris var. *polifolia,*
Whangarei Heads (October)

152 Poroporo flowers, Karaka Bay (October)

154 Spray of manuka flowers,
Lake Pounui (December)

153 Poroporo berries, Karaka Bay (January)

152–153 Poroporo, *Solanum aviculare,* forms a
branching shrub up to 3 m high. Along with the very
similar *S. laciniatum* (fig. 272, p. 68), the flowers
(fig. 152), are up to 3.5 cm across and occur from
September till April. The berries (fig. 153) ripen from
November onwards. It is found growing in frost-free
areas from Auckland to Dunedin and on the Kermadec Islands.   SOLANACEAE

155 Close-up of manuka flower, Lake Pounui
(November)

156 Manuka seeds, Woodside Gorge
(September)

157 Flower spray of horticultural manuka var.
'Ruby Glow', Otari (November)

158 Close-up of flower of manuka var. 'Martinii',
Otari (September)

159 Flowers of manuka var. *keatleyi*, Otari
(December)

154–159 Manuka, *Leptospermum scoparium,* is probably one of the best-known New Zealand plants, growing as a shrub or a small tree throughout both the North and South Islands. Manuka flowers (figs 154–155) normally are about 12 mm across but the variety *keatleyi* (fig. 159) has flowers 2 cm across. Flowers occur from September till February, in the wild usually white (fig. 155), but coloured wild forms have occurred, such as var. *keatleyi,* and from these several fine horticultural forms with single or double flowers have been developed; var. 'Ruby Glow' (fig. 157) and var. 'Martinii' (fig. 158) are good examples. The characteristic seeds of manuka are shown in fig. 156; they ripen during April and May but persist on the tree until the following year. The flowers of the related tree **kanuka,** *Leptospermum ericoides,* are borne much more densely, and the stamens spread more than they do in manuka flowers.

MYRTACEAE

160   Close-up of flowers of kanuka, Taupo (November)

160–163 Kanuka/teatree, *Leptospermum ericoides,* is a tall shrub or a small tree, up to 15 m high, which grows throughout the North and South Islands from sea-level to 900 m altitude. The acute, narrow, aromatic leaves are 12–15 mm long by 2 mm wide. The small flowers (fig. 160), about 5 mm across, clothe the branches in great profusion from September into February (figs 162–163), and the seeds, similar to those of manuka, remain on the tree until the following year.   MYRTACEAE

161   A kanuka tree in full flower, Lake Rotoiti, Nelson (January)

162   Spray of kanuka flowers showing density of flowering, Taupo (January)

163   Spray of kanuka flowers from Hinakura (December)

164–165 Red-fruited karamu, *Coprosma rham-noides*, is one of the small-leaved, twiggy coprosmas, with interlacing branches forming tight bushes (fig. 164). It is found growing in lowland and subalpine scrub and forests all over New Zealand. The drupes (fig. 165) appear in early November as small, red berries, 3–4 mm in diameter, which turn dark crimson or black as they ripen.   RUBIACEAE

164 Red-fruited karamu bushes growing on windswept hillside, Wellington (February)

165 Ripe red-fruited karamu drupes, Pahaoa Valley (December)

166 Karamu drupes, Opepe Bush (March)

167 Male flowers of karamu, Karaka Bay (October)

166–168 Karamu, *Coprosma lucida*, is a small tree, 3–6 m high, found in scrub, along forest margins and in forests throughout the North, South and Stewart Islands from sea-level to 1,060 m altitude. One of the larger-leaved coprosmas, its glossy, leathery leaves, 12–19 cm long by 3–4 cm wide, are dark green above but paler below, with very prominent domatia on the lower surface. The flowers (fig. 167, male flowers; fig. 168, a spectacular group of female flowers) appear from September into October, while the drupes (fig. 166), some 12 mm long, mature during the following winter to ripen by February and March, some 14 months after the flowering.
         RUBIACEAE

168 A magnificent display of karamu female flowers, Karaka Bay (October)

169–171 Kakaramu, *Coprosma robusta*, is probably the most widespread coprosma species found throughout New Zealand in scrub and forests. Kakaramu grows as a shrub or a small, spreading tree up to 6 m high, with stout, hairless branches and large, leathery leaves (fig. 169), 12 cm long and 5 cm wide. Flowers (fig. 169), occur from September till November. The brilliant red drupes are ripe by March and smother the tree with colour well into the winter (figs 170–171).   RUBIACEAE

171   Kakaramu drupes, close up, Waipunga Gorge (April)

169   Male flowers of kakaramu, Taupo (October)

172–173 Stinkwood/hupiro, *Coprosma foetidissima*, is a shrub or small tree (fig. 173) found growing in scrub, along forest margins and in forests from sea-level to 1,360 m throughout New Zealand. The quite large, solitary flowers, about 10 mm long, occur during September and October, and the drupes, 7–10 mm long, arise singly along the branches and ripen during May and June (fig. 172). When crushed, or even brushed against, the plant emits a vile odour like that of many rotten eggs. With its fresh, rather fleshy looking foliage and brilliant red, single drupes, stinkwood, however, looks quite striking.   RUBIACEAE

170   Dense clusters of drupes clothe a branch of kakaramu, Taupo (March)

172   Stinkwood drupe and leaves close up, Mt Holdsworth (June)

173 Stinkwood tree on forest margin,
Mt Ruapehu (January)

174 Hairy coprosma leaves
showing upper sides, Ure
River Gorge (January)

174–176 Hairy coprosma, *Coprosma crassifolia,* grows as a shrub or a small tree, with spreading, stiff, interlacing branches, hairy when young. It is found in scrub, tussocklands, river terraces, rocky places and forests from sea-level to 400 m southwards from Mangonui through both islands. The leaves (figs 174, upper surfaces; 175, lower surfaces showing hairy margins and tomentum) are small, thick and rounded, with the petioles and leaf margins very hairy. The white to cream-coloured drupes (fig. 176), 5–6 mm long, are ripe in March and April.

RUBIACEAE

176 Twigs of hairy coprosma with leaves and drupes,
Ure River Gorge (March)

175 Hairy coprosma leaves
showing lower surfaces, Ure
River Gorge (January)

177 *Coprosma macrocarpa* grows as a shrub or a small tree, 5–10 m high, and is found on the Three Kings Islands, and from North Cape to Kaipara Harbour. The wavy, leathery leaves are 9–13 cm long by 4–8 cm wide and have a prominent mid-vein. The drupes, orange-red when ripe, are 10–25 mm long and ripen during December and January.

RUBIACEAE

177 *Coprosma macrocarpa*, branch showing leaves and immature drupes, Warkworth (December)

178 **Maruru/hairy buttercup,** *Ranunculus hirtus,* is a tall, very hairy-stemmed and hairy-leaved buttercup, up to 60 cm high, found in scrub, often in rocky places, and in forest throughout New Zealand. The flowers are 15 mm across and occur from September till February.

RANUNCULACEAE

178 Maruru flowers, Mt Egmont (December)

180 Flowers and flower-buds of *Cassytha paniculata*, Cape Reinga (January)

179 *Cassytha paniculata* plant sprawling over low shrubs, Cape Reinga (January)

179–180 *Cassytha paniculata* grows parasitically on herbs and low shrubs, forming a tangled mass of stems that sprawls over everything nearby (fig. 179). It is found in scrubland from North Cape south to Ahipara and Mangonui. The minute, sessile flowers arise in succulent perianth tubes (fig. 180), and these and the fruits occur from November through till April.

LAURACEAE

181   Tutu flowers, Taupo (January)

181–185   Tutu, *Coriaria arborea*, is a shrub or small tree, up to 8 m high, that is very poisonous to humans and other animals. In spring the sap and in autumn the seeds contain a poisonous glucoside 'tutin', which makes tutu New Zealand's most poisonous native plant. One milligram of tutin is sufficient to severely upset a healthy person. Tutu occurs all over New Zealand and the Chatham Islands, from sea-level to 1,060 m, in scrublands, along forest margins, in gullies and on river terraces and is recorded as being one of the first plants to appear after a forest fire. The stems are four-sided and the opposite leaves are 5–10 cm long by 4–5 cm wide (fig. 185). The flowers (fig. 181) occur as pendulous racemes, up to 30 cm long, from September till February (fig. 182 shows a male flower; fig. 183 a female flower), and the berries (fig. 184) ripen from November through April.   CORIARIACEAE

182   Close-up of male tutu flower,
       Taupo (October)

183   Close-up of female tutu flower, Karaka Bay
       (November)

184   The ripe poisonous berries of
       tutu, Hanmer (January)

185   A tutu leaf, Waipahihi
       Botanical Reserve (February)

186   Spray of korokio flowers, Waikanae (November)

187   Close-up of korokio flower,
Waikanae (November)

188   Leaves and yellow berries of korokio, Waikanae
(April)

189   Red berries of korokio, close up, Waikanae
(April)

186–190   Korokio, *Corokia cotoneaster*, is a much-branched shrub, up to 3 m high, with stiff, often interlacing branches, found in scrub and on dry river flats and rocky places throughout New Zealand and on the Three Kings Islands. The leaves (fig. 186) are small, 2–15 mm long by 2–10 mm wide, on flattened petioles. Flowers (figs 186–187) appear in profusion from September through December, and the drupes, which may be either yellow or red (figs 188–189), ripen from January through May. A related plant, *C. buddleioides*, which is found in the northern parts of the North Island and which has large leaves similar to those of *C. macrocarpa* (page 67), sometimes hybridises with *C. cotoneaster* to produce a hybrid form known as var. *cheesemanii* (fig. 190).
CORNACEAE

190   Korokio var. *cheesemanii* in full berry, Otari
(August)

191 Bracken fern, *Pteridium aquilinum* var. *excelsum*, can grow luxuriantly to a height of 4 m, often covering large areas as pure stands in scrub, and along forest margins throughout New Zealand. During December through January the young uncurling fronds can be quite a spectacular sight.
PTERIDACEAE

191   Bracken fern, Taupo (January)

192 The hairy nertera, *Nertera dichondraefolia*, forms patches 40 cm across in dry scrub, lowland forests, and sometimes in grasslands from Coromandel Ranges south to Stewart Island. Recognised by its hairy stems and leaves, it flowers from October till February and carries these berries from December to May.   RUBIACEAE

192   The hairy nertera with drupes, Otira Gorge (April)

193   Powhiwhi flowers, Akatarawa (December)

193–194 Powhiwhi/New Zealand convolvulus, *Calystegia tuguriorum*, and **pohue/rauparaha,** *Calystegia sepium*, are both creeping plants, with powhiwhi usually creeping higher than pohue. Powhiwhi has its leaves on petioles 4 cm long and flowers from November to February, while pohue has leaf petioles 10 cm long and flowers from October to March.   CONVOLVULACEAE

194   Pohue flowers, Taupo (February)

195   Common hairy nertera, *Nertera setulosa,* is a very hairy to setose plant, forming patches 30 cm across in scrub, exposed open places and along forest margins from North Cape to Stewart Island. It flowers from November to February, with red drupes from January through May.   RUBIACEAE

195   Common hairy nertera plant with flower, near Opotiki (November)

196   Ciliated nertera, *Nertera ciliata,* forms patches 20 cm across in scrub and lowland to montane forests from Arthur's Pass southwards. The heart-shaped leaves have occasional cilia-like hairs. Flowers and drupes occur from October till March.   RUBIACEAE

199   Native broom, *Carmichaelia aligera,* is found commonly in scrub and along forest margins throughout the northern half of the North Island. Fig. 199 shows stems bearing some ripened seed-pods, which appear during March and April but often remain attached to the plant until November. FABACEAE

196   Ciliated nertera plant with flowers and drupes, Mt Egmont (November)

197   Tutukiwi flowers, Akatarawa (December)

197   Tutukiwi/hooded orchid/elf's hood, *Pterostylis banksii,* is a ground orchid found throughout New Zealand in shady places in lowland and sub-alpine scrub and forests. This orchid flowers from September through December, with the flowers, 5–7.5 cm long including the tails, on stems 30–45 cm high.   ORCHIDACEAE

198   Grassy hooded orchid, *Pterostylis graminea,* is a ground orchid, usually about 20 cm high, with grass-like leaves. It flowers from September to January and is found in similar situations to tutukiwi.   ORCHIDACEAE

198   Grassy hooded orchids in flower, Hinakura (December)

199 Native broom seeds in seed-pods, Whangarei (September)

201 Thick-leaved tree daisy, *Olearia pachyphylla*, has thick, leathery, smooth leaves, 7–12.5 cm long by 5–6.5 cm wide. This olearia is found naturally from the Bay of Plenty southwards in lowland to montane scrub and forests. Flowers occur in large corymbs from October through February, and the fruits ripen from November till March.

ASTERACEAE

201 Spray of thick-leaved tree daisy with flowers, Hukutaia Domain (January)

200 Tamingi spray in flower, Whangarei (October)

202 Koromiko taranga, spray with flowers and leaves, Rimutaka Hill (February)

200 Tamingi, *Epacris pauciflora*, forms an upright, slender shrub, up to 2 m in height, with flowers, 8 mm across, tightly packed around the stems from October through April. It is found in lowland scrub, and boggy places from North Cape south to Marlborough. EPACRIDACEAE

202 Koromiko taranga, *Hebe parviflora*, showing flowers and leaves. It is a shrub or a small tree, much-branched with twiggy branches and narrow leaves 2.5–6.5 cm long. Flowers occur on stalks 5–10 cm long from December through March. The plant is found from Auckland south throughout both islands from sea-level to 600 m altitude.

SCROPHULARIACEAE

203-206 Mingimingi, *Leucopogon fasciculata,* is
an open-branched shrub or tree to 6 m high, found
from the Three Kings Islands to Canterbury in
lowland scrub or forest, and in rocky places from
sea-level up to 1,150 m. The narrow, lanceolate
leaves, 12–25 mm long by 2–4 mm wide, are
spreading and sharp pointed (fig. 206). Flowers occur
as drooping racemes (fig. 203) or spikes, each with
6–12 small flowers from August through to December
(fig. 204). The fruits (fig. 205) ripen from September
to April or May.   EPACRIDACEAE

205 Mingimingi (*L. fasciculata*)
with ripening berries, Taupo
(February)

203 Branchlet of mingimingi
(*Leucopogon fasciculata*)
with terminal raceme of
flowers, Taupo (October)

206 The leaves of mingimingi, (*L. fasciculata*),
showing the veins and the hairy stem, Taupo
(February)

204 Close-up of mingimingi (*L. fasciculata*) flowers
showing the intricate, beautiful structural detail,
Taupo (October)

207-211 Mingimingi, *Cyathodes juniperina,* is an
open-branching shrub to 5 m high, with narrow pungent leaves, 6–20 mm long by about 1 mm wide, with
pale, white-striped lower surfaces (fig. 209). The
flowers, unlike those of *L. fasciculata,* occur singly
along the branches (figs 207–208) from August till
December, and the fruits ripen from October
through March, occurring in varying shades from
white through pink to red (figs 210–211), and
persisting on the branches till July. *C. juniperina* is
found in similar situations to *L. fasciculata* but also
occurs further south to Stewart Island.
                                         EPACRIDACEAE

209 Undersides of leaves of *C. juniperina*, showing typical white stripes of this species, Volcanic Plateau (October)

207 Mingimingi *(C. juniperina)*, flowers around tip of branchlet, Volcanic Plateau (October)

210 Sprays of mingimingi (*C. juniperina*) with berries, Volcanic Plateau (March)

208 Mingimingi (*C. juniperina*), showing flowers along the branchlet, Volcanic Plateau (October)

211 Mingimingi (*C. juniperina*) with white berries, Lake Pounui (February)

212 *Hebe ligustrifolia* is a rather openly branched shrub, found from Dargaville northwards in scrub and along forest margins. The leaves are up to 5 cm long, and the flowers, occurring during December and January, are about 10 cm long.

SCROPHULARIACEAE

212   Spray of leaves and flowers of *Hebe ligustrifolia*, near Waipoua (December)

214 **Chatham Island tree daisy,** *Olearia chathamica,* has finely serrate leaves, 2.5–8 cm long, and flowers 3–4.5 cm across, produced in abundance from October till February. It is found in both dry and damp scrub on the Chatham Islands.

ASTERACEAE

214   Flowers of Chatham Island tree daisy, Otari (November)

215–216 **Akiraho/yellow akeake,** *Olearia panic-ulata,* is a small tree up to 6 m high, readily recognised by its crinkly leaves (fig. 216), 3–10 cm long by 2–4 cm wide, and its sweet-scented flowers (figs 215–216), which occur during March and April. It is found in scrub from East Cape to Greymouth and Oamaru.   ASTERACEAE

213 **Koromiko/koromuka/willow koromiko,** *Hebe stricta,* is a slender, much-branched shrub, up to 4 m high, with grey bark and willow-like, sessile leaves, 10–15 cm long. The flowers on stalks up to 20 cm long occur from December till March. The shrub is found throughout the North Island from sea-level to 1,100 m. A variety *atkinsonii* is found in coastal Marlborough. A similar species, *H. salicifolia*, is found throughout the South Island.

SCROPHULARIACEAE

213   Koromiko flowers, Rimutaka Hill (February)

215   Close-up of flowers of akiraho, Karaka Bay (March)

217–218 Twiggy tree daisy, *Olearia virgata*, is a twiggy shrub or a small tree with four-angled branchlets (fig. 218) and small leaves arising in widely separated, opposite fascicles of two to four leaves. Flowers (fig. 217), about 9 mm across, occur in abundance from October till January. *O. virgata* is found in damp or boggy ground in scrub from Thames south to Stewart Island.

ASTERACEAE

218 Branchlet of twiggy tree daisy var. *implicata* from the Port Hills, Christchurch (January)

217 Twiggy tree daisy flowers, close up, Otari (November)

219 Teteaweka plant with flowers grown at Happy Valley, Wellington, by Mrs Natusch (November)

219–220 Teteaweka, *Olearia angustifolia,* is a shrub or a small tree, up to 6 m high, found in coastal scrub along the coasts of Southland, Foveaux Strait and the headlands of Stewart Island. The thick, narrow-lanceolate leaves, up to 15 cm long and 2 cm wide, have crenate dentate margins and a thick, soft, white tomentum below (fig. 219). The large, scented flower-heads, 3.5–5 cm across, occur from October through January (figs 219–220).   ASTERACEAE

216 Akiraho flowers, Karaka Bay (March)

220 Close-up of flower from plant in fig. 219.

221   Spray of fragrant tree daisy with flowers, Otari (November)

222   Close-up of flowers of fragrant tree daisy showing also the ridged, four-angled branchlet, Otari (November)

221-222   **Fragrant tree daisy,** *Olearia fragrantissima,* forms an erect, much-branched shrub or small tree, up to 5 m high, with zig-zagging branches bearing variable, more or less elliptic leaves 7–30 mm long by 5–10 mm wide (fig. 221). The delightfully fragrant flowers (fig. 222) occur in dense clusters about 2 cm across from October through February. *O. fragrantissima* is found in east coast lowland scrub from Banks Peninsula southwards.   ASTERACEAE

223-224   **Deciduous tree daisy,** *Olearia hectori,* found in scrub from sea-level to 900 m around Taihape and from the Clarence River to Southland, is a shrub or a small tree up to 5 m high and has slender, rounded, slightly grooved, smooth branchlets. The narrow to broad, obovate leaves, 2.5 cm long by about 10–20 mm wide, are in fascicles of 2–4 on short branchlets (fig. 224) Flowers occur from October through December as drooping fascicles on silky pedicels, each with 2–5 heads about 5 mm across (fig. 223).   ASTERACEAE

223   Flowers, close up, of deciduous tree daisy, Otari (November)

224   Leaf spray of deciduous tree daisy, Otari (November)

225-226 Akepiro/tanguru, *Olearia furfuracea,*
occurs as a shrub or small tree 5 m high from sea-
level to 900 m along forest margins, streamsides and
in scrub from North Cape to the Southern Ruahine
mountains. The branchlets are angled, grooved and
pubescent, with the thick, leathery, wavy-margined
leaves, 7–13 cm long by 5–6.5 cm wide, on petioles
2.5 cm long, having a buff-coloured, satiny tomentum
below (fig. 225). Flowers are sweet scented and occur
as large, flat corymbs from November to February
(fig. 226).   ASTERACEAE

226   Flowers of akepiro, Taupo
      (December)

227-228 Common tree daisy, *Olearia arborescens,*
is a shrub 4 m high with angled branchlets and thin,
leathery leaves, 2–6 cm long by 2–4 cm wide, with
a thin, satiny tomentum below and a 2 cm long
petiole (fig. 228). Flowers (fig. 227) are in large, flat
corymbs, 25 cm across, produced in profusion from
October through January. This is one of the most
common and beautiful of the *Olearia* species and is
found from East Cape to Stewart Island.

ASTERACEAE

225   Leaf spray of akepiro, Taupo
      (December)

228   Leaves of common tree
      daisy, Whangamoa Saddle
      (November)

227   Flower-head of common tree daisy, Whangamoa
      Saddle (November)

229–230 Akeake, *Olearia avicenniaefolia,* is a shrub or small tree 6 m high, with narrow, elliptic, slightly wavy, thick leaves, 5–10 cm long by 3–5 cm wide, shining above but with a soft, white or fawn-coloured tomentum below (fig. 229). The flowers are in large flat corymbs on long stalks and occur from January through April (fig. 230). The shrub is found from sea-level to 900 m in scrub throughout the South Island and Stewart Island.

ASTERACEAE

230   Flowers of akeake, Waipunga Gorge (April)

229   Leaves of akeake with flower-buds, Lewis Pass (December)

231–233 Heketara/forest tree daisy, *Olearia rani,* is a shrub or a small tree 7 m high found in scrub and lowland forest, along forest margins, river and stream banks, in forest clearings and in second-growth forest from North Cape to Nelson and Marl-borough (fig. 231, in Otaki Gorge). It is one of the most beautiful of our tree daisies. The characteristic and unmistakable leathery, dentate leaves, 5–15 cm long by 5 cm wide, (fig. 232), on petioles 4 cm long, are, like the branchlets and petioles, covered below with a dense whitish or brownish, soft tomentum. The flowers (fig. 233) appear from August through November as large sprays up to 18 cm across, the individual ray florets 10–20 mm across.

ASTERACEAE

231   Heketara flowering in scrub, Otaki Gorge (October)

232   Heketara leaves, Stokes Valley (October)

234 Scented tree daisy *Olearia odorata*, is an erect, spreading shrub, up to 4 m high, with divaricating branches and rounded leaves, 10–25 mm long by 6-15 mm wide, which arise in opposite fascicles. Small, strongly sweet-scented flowers, 8–12 mm across, occur in fascicles along the branches from December to February, and the shrub is found from Kaikoura southwards in montane and subalpine scrublands.   ASTERACEAE

234   Branches of scented tree daisy with flowers and leaves, Otari (February)

235–237 Inanga/grass tree, *Dracophyllum longi-folium*, is the most widespread grass tree, being found in lowland and subalpine scrub and forest from East Cape southwards through the North, South and Stewart Islands, the Auckland Islands and the Chatham Islands. It grows as a slender, erect tree, 12 m high, with spreading branchlets, which have the stiff leaves crowded towards their tips. The leaves are 10–25 cm long by 3–5 mm wide, tapering to a long, acuminate tip (fig. 236). The shape and form of the basal leaf sheath (fig. 237) is very useful in identifying species of *Dracophyllum*. Flowers in 6–15-flowered racemes (Fig. 235) appear terminally on the branchlets during November and December.

EPACRIDACEAE

235   Inanga flowers, Mt Dobson (December)

233   Heketara flowers, close up, Kaitoke (November)

236   Inanga leaves, Otari (April)

237   Inanga leaves, showing basal leaf sheaths, Otari (April)

238–240 *Dracophyllum lessonianum* is a small tree, 10 m high, found in scrub from North Cape to Kaitaia often among manuka. Fig. 238 shows a tree, fig. 239 the leaves, 6–10 cm long by 1–1.5 mm wide, and fig. 240 shows the flower racemes, each about 3 cm long, which arise singly or in clusters terminally on short, lateral branches.   EPACRIDACEAE

240   Flowers, close up, of *D. lessonianum*, Kaitaia (April)

238   A small grass-tree, *Dracophyllum lessonianum*, Kaitaia (April)

241–242 *Dracophyllum sinclairii* is a slender tree, 3–6 m high, with close-set leaves, 3.5–12.5 cm long by 4–8 mm wide, which have finely serrulate margins and long, acuminate apices (figs 241–242). Flowers in 4–8-flowered racemes arise terminally and subterminally (fig. 241) on lateral branches during April and May. The tree is found from North Cape south to Kawhia.   EPACRIDACEAE

239   Leaves and flower-buds of *D. lessonianum*, Kaitaia (April)

241   Tip of branchlet of *Dracophyllum sinclairii* with flowers, Kaitaia (May)

243-245 *Dracophyllum viride* is a small tree, up to 5 m high, with slender, ascending, leafy branches bearing long (juvenile) and short (adult) leaves. It is found in scrub only between North Cape and Kaitaia. Adult leaves, 5–7 cm long by 5–6.5 mm wide, tend to spread and have a distinct ciliated shoulder at the top of the sheath (fig. 244). Juvenile leaves occur on both young and old plants and are 2–3 times longer than adult leaves (fig. 243). Flower racemes, each with 5–6 flowers, arise just below leaf tufts on the branchlets (fig. 245) during March and April.   EPACRIDACEAE

244   The ciliated shoulder on the leaf of
       D. viride, Herekino Saddle, Kaitaia
       (April)

243   Juvenile and adult foliage of the grass-tree,
       Dracophyllum viride, Herekino Saddle, Kaitaia
       (April)

242   The shoulder in the leaf of
       D. sinclairii, Kaitaia (April)

245   Growing apex of grass-tree,
       D. viride, showing adult leaves below
       with flowers and juvenile leaves
       above, Herekino Saddle, Kaitaia (April)

246–247 Lesser caladenia, *Caladenia carnea* var. *minor*, is a hairy ground orchid, 5–16 cm high, usually with a single, narrow leaf, found throughout the North Island, sometimes concentrated in large numbers in dry places under open scrub canopy and on dry, open clay hills. The flowers, 1–2 cm across, occur from October through January, varying in colour from bluish white (fig. 246) through pinks (fig. 247) to purple.   ORCHIDACEAE

248 Giant sedge, *Gahnia xanthocarpa*, has densely tufted stems reaching to 4 m in height and leaves 2 cm wide, and is the largest sedge found in New Zealand. It forms large clumps on the edges of bogs, in scrub, along forest margins and inside forests, especially in Northland, but also discontinuously throughout both North and South Islands. The flower panicles, 60 cm to 1.5 m high, appear from December through to March.   CYPERACEAE

246   The lesser caladenia, greenish white form, Hinakura (December)

248   Giant sedge plants in flower, Lake Pounui (December)

247   The lesser caladenia, pink form, Hinakura (December)

249 *Teucridium parviflorum* is a much-branched, twiggy shrub, sometimes forming thickets up to 1.5 m high or sprawling over rocks in lowland scrub and forest throughout New Zealand. The branchlets are four-sided and hairy when young; the small leaves, 4–15 mm long, are on petioles almost as long as themselves. Solitary axillary flowers, 8 mm across, occur from October through January.

VERBENACEAE

249   Flowers of *Teucridium parviflorum*, Hinakura (December)

250 Papataniwhaniwha, *Lagenophora pumila*, is found along lowland forest margins and in scrub or grasslands from Rotorua southwards. This is one of five species of *Lagenophora* found in New Zealand and has leaves 2–6 cm long.   ASTERACEAE

250 Papataniwhaniwha plant in flower, Waihohonu Stream, Tongariro National Park (January)

252 Spray of ongaonga with female flowers and showing stinging hairs, Hinakura (November)

251–254 Ongaonga/tree nettle, *Urtica ferox*, is a soft-wooded shrub or small tree, up to 3 m high, with many intertwining branches. The leaves, branchlets and branches all bear stout stinging hairs that can inflict a painful sting, and, if stung many times, a person can lose co-ordination of muscle movement for upwards of three days; deaths from the stings are recorded. Ongaonga is found in scrub and along forest margins, forming thickets from sea-level to 600 m. Leaves are 8–12 cm long by 3–5 cm wide, and fig. 253 shows a leaf underside with stinging hairs and female flowers. Branchlets, leaves and female flowers are shown in fig. 252; male flowers are shown close up in fig. 251 and females flowers close up in fig. 254. Flowers are in spikes about 8 cm long and occur from November through March.

URTICACEAE

253 Underside of ongaonga leaf, showing stinging hairs and female flowers, Hinakura (November)

251 Male flowers of ongaonga, Hinakura (November)

254 Close-up view of ongaonga female flowers, Hinakura (November)

255–256 Rohutu, *Neomyrtus pedunculata,* is a shrub or small tree up to 6 m high, with four-sided branches that are hairy when young. It occurs in lowland scrub and along forest margins all over New Zealand from sea-level to 1,050 m. The leaves are of variable shape, obovate-oblong and 15–20 mm long by 10–15 mm wide or obovate and 6–15 mm long by 4–10 mm wide, all dotted with glands, thick and leathery with rolled or thickened margins. Solitary flowers, 5 mm across (fig. 256), arise from leaf axils from December through to April, and the berries (fig. 255) are ripe from February through May.

MYRTACEAE

255   Branch of rohutu with berries, Ponatahi (April)

256   Flowers and leaf of rohutu, Makarora Valley (January)

257   Flowers and leaves of *Pittosporum lineare,* ,Kaimanawa Mountains (October)

257 *Pittosporum lineare* is a much-branched, divaricating shrub about 3 m high found in lowland scrub and along forest margins throughout the North Island, Nelson and Marlborough. The fascicled leaves, 10–15 mm long by 1–3 mm wide, are paler below. The small black flowers, about 5 mm across, arise singly and mostly terminally on branchlets during October through January.

PITTOSPORACEAE

258–259 *Neopanax anomalum* is a shrub, 3 m high, with densely divaricating branches. The branchlets have fine bristle-like setae and small, rounded leaves, 10–20 mm long by 10–15 mm wide, on petioles 5 mm long. Minute flowers (fig. 258) in 2–10-flowered umbels occur from November to February, with berries ripe during March and April (fig. 259).   MYRTACEAE

258   Flowers of *Neopanax anomalum,* Rahu Saddle (December)

260-263 Horopito/pepper tree, *Pseudowintera colorata*, is commonly a shrub but sometimes a small tree, up to 10 m high, noted for its reddish coloured, aromatic leaves, 2–8 cm long by 10–30 mm wide (fig. 260), with pungent taste. It is common in lowland scrub and forests as well as in alpine regions up to 1,200 m throughout New Zealand. Aromatic flowers (figs 261–262) occur singly or in fascicles along the branches, each on a pedicel 10 mm long, from October to March, and the dark reddish black fruits (fig. 263) are ripe from December to June.

WINTERACEAE

260   Foliage of horopito, Hihitahi State Forest (April)

261   Branchlets of horopito with flowers and showing typical blue-coloured leaf undersurfaces, Opepe Bush (October)

262   Close-up of horopito flowers, Waipunga Gorge (October)

259   *N. anomalum* showing branchlets with leaves and berries, Rahu Saddle (April)

263   Spray of horopito with berries, Waipunga Gorge (April)

264 Northern cassinia plant in
flower, North Cape
(January)

265 Northern cassinia flower-
head, close up, Cape Reinga
(January)

264–265 Northern cassinia, *Cassinia amoena*, is a
low shrub, about 80 cm high (fig. 264), with thick,
fleshy leaves that are covered below with a dense
white tomentum. It is found only in scrub on cliff
faces near North Cape, and flowers (fig. 265) occur
in corymbs, about 3 cm across, during January and
February.   ASTERACEAE

266 Totorowhiti/grass tree/turpentine shrub,
Dracophyllum strictum, is found in damp lowland
and subalpine scrub from Thames south to Nelson
and Marlborough from sea-level to 800 m. The
leaves, 3.5–10 cm long by 7–12 mm wide, have finely
serrulate-crenulate margins. Flowers in terminal
panicles, 5–10 cm long, occur during April and May.
                                        EPACRIDACEAE

266   Totorowhiti flower spray, Nelson (April)

267 Niniao, *Helichrysum glomeratum*, is a much-
branched shrub about 3 m high with interlacing
branches and rounded leaves, 10–25 mm long, on
flattened petioles about 5 mm long. It is found in
lowland scrub, along forest margins and sometimes
in rocky places throughout New Zealand. The small,
ball-like clusters of flowers, about 5 mm across,
occur from November to January.   ASTERACEAE

267   Niniao flower-heads, Lake Pounui (December)

268 Variable coprosma, *Coprosma polymorpha,* forms a spreading shrub with divaricating branches, pubescent branchlets and variable leaves, 10–25 mm long by 2–6 mm wide, varying from ovate to lanceolate to linear in shape. Drupes ripen during April and May, and the shrub is found in lowland scrub and forest throughout New Zealand, though often local in occurrence. ASTERACEAE

268 Branchlet of variable coprosma with drupes and leaves, Rahu Saddle (April)

269 Ure Valley tree daisy, *Olearia coriacea,* is a 3 m high shrub with very thick, rounded, leathery leaves having a dense brownish white tomentum below, and is found in scrub along the Seaward Kaikoura Range. Flowers occur singly during March and April. ASTERACEAE

269 Flowering stem of Ure Valley tree daisy, Upper Ure River (March)

270 Shrubby kohuhu, *Pittosporum rigidum,* is a densely branched, rigid shrub, to 3 m high, with small, thick, leathery leaves, 8–10 mm long by 5–8 mm wide, found occasionally in scrub but more commonly from montane to subalpine regions from the Kaimanawa to the Southwest Nelson mountains. Small axillary flowers occur from September to February. PITTOSPORACEAE

270 Branchlets of shrubby kohuhu with flowers and leaves, Kaimanawa Mountains (October)

271 Hokotaka, *Corokia macrocarpa,* forms a shrub, up to 6 m high, with narrow, leathery leaves, 4–15 cm long. Flowers, 10 mm across, occur in axillary racemes from October to December; the drupe, ripe during April and May, is dark red. It is found along forest margins and in scrub in the Chatham Islands. CORNACEAE

271 Hokotaka flowers, Otari (October)

272 Poroporo, *Solanum laciniatum,* is similar to
S. aviculare (p. 40) except that *S. laciniatum* has distinctly purplish blue leaves and stems. It is found
mainly in the North Island along forest margins and
in scrub.   SOLANACEAE

273 North Cape hebe, *Hebe macrocarpa* var.
brevifolia, is found only around the North Cape as
a stiffly branched shrub, about 2 m high, with
spreading leaves, 6–12 cm long by 10–30 mm wide,
and bearing these lovely flowers almost all the year
round.   SCROPHULARIACEAE

272   Poroporo flowers, Waikanae
       (November)

273   North Cape hebe flower-
       spike, North Cape (October)

274   Houpara leaves, Coromandel Peninsula
       (December)

274–275 Houpara, *Pseudopanax lessonii,* is one of
three species of *Pseudopanax,* all of which have
thick, leathery, unifoliolate or 3–5 foliolate leaves
on long petioles. The leaves of *P. lessonii,* with their
petioles 5–15 cm long, are shown in fig. 274, and
the flower, which occurs from December to February,
is shown in fig. 275. It is a shrub or small tree, up
to 6 m high, and grows in lowland scrub and forest
from the Three Kings Islands to Poverty Bay.
                                          ARALIACEAE

275   Houpara flower, close up, Coromandel
       Peninsula (January)

276-279 Toothed lancewood, *Pseudopanax ferox,* forms a small tree up to 5 m high, with a slender, 'roped' trunk (fig. 279) and long, toothed, upward-spreading, adult leaves, 5-15 cm long by 10-20 mm wide (fig. 277); leaves of juveniles are up to 50 cm long, and drooping (fig. 278). Flowers and fruits occur from January till April (fig. 276), and the tree is found in lowland scrub and forest from Mangonui southwards in the North and South Islands but is rather rare in occurrence.   ARALIACEAE

276   Toothed lancewood berries, Otari (April)

277   Toothed lancewood, showing juvenile pendant leaves and adult ascending leaves, Otari (November)

278   Toothed lancewood leaves, Otari (November)

279   The 'roped' trunk of the toothed lancewood, Otari (September)

280 *Pseudopanax discolor* leaf
from Kauaeranga Valley
(November)

281 *P. discolor* leaf from near
Thames (November)

280–281 ***Pseudopanax discolor*** is a shrub or small tree, up to 5 m high, that grows in lowland scrub and forest from Mangonui to about Thames and the Manukau Harbour. Trifoliolate and five-foliolate leaves on their petioles, 2–8 cm long, are shown in figs 280 and 281 respectively.   ARALIACEAE

282 ***Pseudopanax gilliesii*** is a shrub or tree up to 5 m high, with slender branches and unifoliolate or trifoliolate, and sometimes irregular-lobed, leaves, 4–8 cm long on long, slender petioles up to twice as long as the leaves (fig. 282). It is found growing in scrub only near Whangaroa and on Little Barrier Island.   ARALIACEAE

282 *Pseudopanax gilliesii* leaf spray showing the very long leaf petioles, Whangaroa (October)

PLANTS OF BOGS AND SWAMPS

Plants that grow in these situations are tolerant of continuous wet conditions around their roots. They are found in swampy places in both lowland and alpine regions and often occur round the seepages of springs in bush-clad gullies or on mountainsides. Bog conditions of only a few square metres in area often provide ideal places in which these plants can thrive.

283 Blue swamp orchids in flower, Te Anau (January)

283 Blue swamp orchid, *Thelymitra venosa*, is a lovely ground orchid, 20–50 cm high, found in swamps throughout New Zealand. Flowers appear during December and January. ORCHIDACEAE

284 Raupo, *Typha muelleri*, is a bullrush, up to 2.7 m high, found in swamps and marshes throughout New Zealand. The leaves are up to 2.5 cm wide, and flowers occur from December through March. TYPHACEAE

284 Raupo in flower on edge of Lake Pounui (January)

285 Oliver's dracophyllum, *Dracophyllum oliveri*, is a species of *Dracophyllum* that grows to 1 m high and is found in swamps near Te Anau. Flower racemes occur along the branches from November to January. EPACRIDACEAE

285 Flowers and flower-buds of Oliver's dracophyllum, Lake Manapouri (November)

286 Toetoe, *Cortaderia toetoe,* is New Zealand's largest endemic grass, found throughout the country in lowland swampy places and along riverbanks, and on sandhills. The plume-like flowers, up to 3 m high, occur from November through March.

GRAMINEAE

286   Toetoe plant in flower, Lake Pounui (December)

287   Flax plant in flower, Waiorongomai (December)

287–288 Flax/harakeke, *Phormium tenax,* is one of New Zealand's most handsome plants. It is common in lowland swamps throughout the country and on the Kermadec and Chatham Islands. Leaves are up to 3 m long by 5–12.5 cm wide and the flowers stalks reach 5 m high; the flowers, each about 2.5 cm long, appear along the stalks from November to January. Several cultivars with richly coloured leaves have been developed (fig. 288).

PHORMIACEAE

289   Sweet-scented grass, Hinakura (February)

288   A bronze-leaved form of flax, Taupo (August)

289 Sweet-scented grass, *Hierochloe redolens,* is an erect, strongly sweet-scented grass, up to 1 m high, common around boggy places throughout New Zealand. The flower panicles, 35 cm long, appear from December through February.   GRAMINEAE

290-291 Soft herb, *Myriophyllum pedunculatum*, is a soft aquatic herb with creeping, rooting stems producing simple, upright branches up to 10 cm high (fig. 290). Fig. 291 shows the flowering patch made by this tiny plant growing amongst a gunnera. Flowers, 4 mm across, occur from October to February, with female flowers at the branch apex and male flowers just below (fig. 290). Found in peaty, wet places from Mangonui southwards to Stewart and Chatham Islands.   HALORAGACEAE

291   Soft herb growing among solitary gunnera and flowering, Boulder Lake (February)

290   Soft herb plants in flower, Boulder Lake (February)

292 Kuwawa, *Scirpus lacustris*, is a leafless sedge, up to 2 m high, found along swamp and lake margins throughout New Zealand. Flower-heads, 5-10 cm long, occur from November to February, and the fruits ripen by May.   CYPERACEAE

292   Kuwawa in fruit, Lake Pounui (May)

293 Wi, *Juncus pallidus*, is a tall, dense, tufted herb, 1.75 m high, found in lowland swampy places throughout New Zealand. Large cymes of flowers, each 3 mm across, occur from December to February.   JUNCACEAE

293   Flowers of wi, near Rotorua (February)

294 Cutty-grass, *Carex geminata,* is a robust sedge, 50 cm–1.3 m high, with flat, scabrid-margined leaves, 5–12 mm wide, which can cut the skin if grasped. It is found throughout New Zealand on the edges of swamps, along streamsides and in damp places in the forests. The flower-spikes, up to 10 cm long, appear from November to February.

CYPERACEAE

294   Cutty-grass with flower-spikes, Mt Tauhara (January)

295 Dense nertera, *Nertera balfouriana,* forms dense patches of a creeping plant up to 25 cm across in damp or boggy ground, often among sphagnum moss, from the Kaimanawa Range southwards, at elevations between 600 m and 1,000 m. The leaves and flowers are small but the brilliant drupes, 7–9 mm long, are ripe during February and March.

RUBIACEAE

295   Dense nertera plant with drupes, Volcanic Plateau (February)

296 Alpine cushion, *Donatia novae-zealandiae,* is a small, firm, tufted plant, which forms hard cushions in alpine bogs from the Tararua Range southwards to Stewart Island, where it descends to sea-level. It flowers profusely during January and February, with small flowers 8–10 cm across.

DONATIACEAE

296   Alpine cushion plant in flower, Arthur's Pass (January)

297 Prostrate grass tree, *Dracophyllum muscoides,* is a creeping grass tree with stout, dark, brownish black stems, up to 30 cm long, found around sub-alpine bogs, damp grasslands and herbfields of the South Island from Lake Ohau southwards. Branchlets are clothed with thick, leathery, imbricating leaves about 6 mm long by 2.5 mm wide; the flowers appear during January and February.

EPACRIDACEAE

297   Prostrate grass tree with flowers, Key Summit (January)

298   Peat swamp orchids in flower,
Moana Tua Tua Bog (September)

299   Alpine sundew, *Droscera arcturi,* is an insecti-vorous plant found in bogs and swamps from the Volcanic Plateau to Stewart Island. The narrow, strap-like leaves with their petioles are 5–12 cm long and bear sticky hairs that, as in all sundews, entangle insects, which the plant consumes to obtain its nitrogen. Flowers 16 mm across on stalks 15 cm high occur from November to January. DROSCERACEAE

299   Alpine sundew plants in flower, Key Summit (January)

298   Peat swamp orchid, *Corybas unguiculatus,* is one of the rarest New Zealand plants, found only in wet, peaty swamps of the northern North Island. Flowers appear in mid-July, at first as horizontal buds, which continue growing in size and length of stalk until September or early October, when they open. Unless pollinated, flowers collapse and wither within ten days. Flowers are 2 cm long on stalks 10–20 mm high.   ORCHIDACEAE

300   Spathulate sundew, *Droscera spathulata,* is a small insectivorous plant, usually occurring in groups in lowland and subalpine bogs and swamps through-out New Zealand. The glandular leaf petiole, 10 mm long, widens to a spoon-shaped lamina, 5 mm long, bearing stalked, glandular hairs. Flower-stalks, each carrying racemes of up to 15 white or pink-coloured flowers, occur from November till January.
                                    DROSCERACEAE

300   Spathulate sundews, Key Summit (January)

301   Scented sundew, *Droscera binata,* has upright leaves, 15 cm long by 2 mm wide, bearing glandular hairs and arising in twos and threes from petioles up to 35 cm long. Flower-stalks up to 50 cm high bear white flowers, 2 cm across, in cymes from November to February. Scented sundew is found in lowland and montane bogs and swamps throughout New Zealand.   DROSCERACEAE

301   Scented sundews, Lake Te Anau (January)

302 Hollow-stemmed sedge with
     flower-spikelets, Lake
     Pounui (January)

303 Wahu showing spathulate tip
     of leaf, Renata Peak, Tararua
     Mountains (November)

302   Hollow-stemmed sedge, *Eleocharis sphacelata*,
is a sedge with hollow, tubular stems up to 1 m high,
found throughout New Zealand in wet swamps, and
along the margins of lakes from sea-level to 800 m.
The terminal flower-spikelets, white when fresh, are
up to 5 cm long.   CYPERACEAE

303–304   Wahu, *Droscera stenopetala*, is a variable
sundew, normally with narrow, spathulate leaves,
2 cm long by 5–6 mm wide on stout petioles 5 cm
long by 8 mm wide (fig. 304). The leaves (fig. 303)
have long, sticky, glandular hairs. Solitary white
flowers, 2 cm across on peduncles 20 cm long, occur
from November to January. This wahu is found in
montane and subalpine bogs and swamps from the
Ruahine Range south to Stewart Island, coming
down to sea-level in the far south.

DROSCERACEAE

304 Wahu plant, Renata Peak, Tararua Mountains
     (November)

305   Moss daisy, *Abrotanella caespitosa*, is a rather
moss-like plant, with thick, leathery leaves,
10–15 mm long by 1–1.5 mm wide, which form
rounded matted patches, up to 10 cm across, along
the edges of subalpine bogs and seepages in grasslands and herbfields from the Ruahine Range south
to northern Fiordland. Minute flowers occur during
January and February.   ASTERACEAE

305 Moss daisy with flower-buds, Renata Peak,
     Tararua Mountains (November)

306–308 Niggerhead, *Carex secta,* forms a large sedge up to 1.5 m high whose matted roots and decaying leaves form huge, broad pillars, 1–1.5 m high (fig. 306), which lift the sedge up to form the conspicuous objects in swamps called 'niggerheads' (fig. 308). The leaves are 1–2 m long by 3–4 mm wide, flat above, keeled below, with sharp cutting margins. Flower panicles (fig. 307), up to 1 m in length, occur from December to January. Niggerheads are found throughout New Zealand in swamps, bogs, along stream banks and in other damp places.

CYPERACEAE

306   Niggerhead in flower, Lake Lyndon (January)

307   Flower panicles of niggerhead, Lake Lyndon (January)

308   Niggerheads in bog near Garston (January)

309   Horizontal orchid, *Lyperanthus antarcticus,* is an unusual orchid, with its flower held horizontally, almost at right angles to the stem. The leaves are 2.5–7 cm long and the flowers, which occur from November to March, 8–14 mm long. It is found in subalpine bogs and swamps from the Tararua Range to Stewart Island.   ORCHIDACEAE

309   Horizontal orchids flowering, Key Summit (January)

310 Bog daisy, *Celmisia graminifolia,* is a small daisy found throughout the country on the edges of montane and subalpine bogs and swamps and in damp places in alpine grasslands, herbfields and fell-fields. The pointed leaves, 5–20 cm long by 4–5 mm wide, are clothed below with dense felt; the flowers, 10–15 mm across, occur during December and January.   ASTERACEAE

312 Native oxalis, *Oxalis lactea,* is one of three species of *Oxalis* found near bogs, swamps and damp places alongside streams in lowland and subalpine regions throughout New Zealand. It spreads by creeping rhizomes, and the patches of three-lobed leaves can reach 20 cm across, with the flowers, 12–20 mm across, occurring from October to March.

OXALIDACEAE

310   Bog daisies in flower, Key Summit (January)

312   Native oxalis in flower, Opotiki (November)

311 Small alpine swamp sedge, *Carex flaviformis,* is a short-leaved sedge found in swamps in mountain regions of the South Island between 500 and 1,200 m altitude. Tufted flower-heads arise from December to February.   CYPERACEAE

313 Turf-forming astelia, *Astelia linearis,* and *A. subulata* are two species of turf-forming *Astelia* found in alpine wet and boggy places in the South Island mountains from the Paparoa Range to the Longwoods. The leaves are 5–10 cm long by 2–6 mm wide, and the plant forms extensive patches.

ASPHODELACEAE

311   Small alpine swamp sedge with seed-heads, Key Summit (January)

313   Turf-forming astelia with seeds, Upper Stillwater (December)

PLANTS OF STREAMSIDES, DAMP OR SHADY PLACES

Plants that grow in these situations prefer cool, moist, but not wet, root runs. They are found from lowland to alpine regions, growing on banks, either shaded by trees and shrubs or open and facing away from the hot noonday sun. They also grow on the drier, but still damp, ground above the edges of bogs and swamps and in the cool alluvium of river and streamsides. They are found, as well, in rocky places kept damp by seepages and rivulets.

314 Creeping gunnera, *Gunnera prorepens*, is a creeping plant, forming patches 40 cm or more across in lowland and subalpine damp or boggy places throughout New Zealand, and is often found in sphagnum moss growing along the edges of bogs. The leaves are about 3 cm long and the flowers appear on 6 cm high stalks from September to January; the drupes, 3–4 mm long, ripen from February on.   HALORAGACEAE

314   Creeping gunnera plant with ripe drupes, Outerere Stream, Volcanic Plateau (February)

315   Red-fruited gunnera plant, Travers Valley (April)

315   Red-fruited gunnera, *Gunnera dentata*, is a creeping plant, forming mats up to 60 cm across along subalpine streamsides and in damp places throughout New Zealand; sometimes several plants may coalesce to form one large mat 2–3 m across. The dentate leaves are 8–15 mm long on petioles up to 5 cm long; the flowers appear in November and December, and the pendulous drupes (fig. 315), occur as open clusters on elongated stalks during March and April.   HALORAGACEAE

316–317 Solitary gunnera, *Gunnera monoica,* is a creeping mat plant found up to 1,000 m altitude on damp, shaded clay banks or damp, stony river-beds, from about Thames southwards. The dentate-serrate leaves, 10–15 mm long, are on hairy petioles up to 4 cm long (fig. 317). The flowers (fig. 316) appear during October and November on erect stalks up to 7 cm high, with male flowers occupying the upper three-quarters of the stalk and female flowers the lower quarter. Ripe drupes appear from December to February.   HALORAGACEAE

317 Plant of solitary gunnera with immature drupes, Boulder Lake (February)

316 Solitary gunnera with flower-stalks, Travers Valley (November)

318 Swamp musk, *Mazus radicans,* is a creeping perennial herb, found between 200 m and 1,200 m, in damp places and along the edges of bogs or swamps from the Ruahine Range to Otago. The narrow, obovate leaves, with their petioles, are hairy and 2–5 cm long. The flowers occur from November to March. A similar plant, *M. pumilio,* has longer, obovate-spathulate leaves that are less hairy.   SCROPHULARIACEAE

318 Swamp musk in flower, Travers Valley (December)

319 Purple bladderwort, *Utricularia monanthos,* is a plant that drops its leaves before it flowers and has small bladders on its roots. These bladders, 1.5–2.5 mm wide, have opening lids and trap microscopic water animals, which are digested to provide nitrogen for the plant. The flowers are about 10 mm across on stems 10 cm high and occur during January. Found in bogs and swamps from the Kaimanawa Range southwards.

LENTIBULARIACEAE

319 Purple bladderwort flowers, Key Summit (January)

320 *Carex dissita* forms a handsome, tall, tufted sedge, up to 1 m high, found in abundance in damp areas of lowland and subalpine forests throughout New Zealand. The broad, flat, dark green leaves are deeply grooved, and the flower-heads appear from September to January. CYPERACEAE

320 *Carex dissita* leaves with flower- and seed-heads, near Taupo (January)

321 Parataniwha, *Elatostema rugosum,* is a fleshy, decumbent and spreading plant found in shaded, damp places, especially by streamsides, in lowland forests from about North Cape to the Tararua Ranges. The attractive, alternate, coarsely serrate and rather rough leaves are 8–25 cm long by 2.5–6 cm wide. URTICACEAE

321 A parataniwha plant displays its colourful leaves, Hukutaia Domain (April)

322 Puatea, *Gnaphalium keriense,* is a small, hanging or prostrate everlasting daisy, up to 24 cm high, with spreading branches and sessile leaves, 4–7 cm long; those towards the branch tips are shorter. Flowers arise as flat corymbs, each flower about 1 cm across, from September to February. Found from Northland to North Canterbury in lowland and subalpine regions along streamsides and shaded banks, often on shady roadside banks. ASTERACEAE

322 Puatea in flower, near Taihape (November)

323 Maori calceolaria, *Jovellana sinclairii,* found from the East Cape southwards in damp places, is a small herb, up to 30 cm high, with upright, downy stems and thin, opposite leaves about 8 cm long. Flowers occur as branched panicles from October till February. SCROPHULARIACEAE

323 Maori calceolaria flowers, Pahaoa River Valley (December)

324   Red spider orchid, *Corybas macranthus*, is a small, low-growing orchid with thin leaves, 5 cm long, found throughout New Zealand from sea-level to 700 m on clay banks and logs and among moss on logs in damp, shaded places. The spider-like flowers occur during October and November. When the seeds have set, the stalks elongate to about 10 cm so that the seeds may be borne away by wind.

ORCHIDACEAE

325   Close-up of flowers of cudweed, Haast (November)

325   Cudweed, *Gnaphalium hookeri*, is a trailing everlasting daisy with alternate leaves, 5–10 cm long, and flowers, 15 mm across, occurring in corymbs 7.5–10 cm across, from November to January. Found on damp, shady banks in lowland and subalpine regions from Taupo southwards.

ASTERACEAE

324   Red spider orchids on a bank near Hinakura (November)

326–328   Lantern berry/puwatawata, *Luzuriaga parviflora*, is a delicate creeping lily found throughout New Zealand in lowland forests, growing on damp, moss-covered banks and moss-covered tree trunks and logs. Leaves, 10–27 mm long by 3–6 mm wide, are alternate, and flowers, 15–35 mm across, occur from November to February, with berries appearing from January to March.

LILIACEAE

327   Lantern berry leaves and flower, Mt Ngamoko (January)

326   The flower of lantern berry, Mt Ngamoko (January)

328   Lantern berry, spray with berry, Mt Ngamoko (January)

329 Yellow-leaved sedge, *Carex coriacea*, forms a small, yellow-green-leaved sedge, 50–100 cm high, found in damp grassland, swampy seepages or swampy river flats from sea-level to 1,200 m. Leaves are longer than the culms; flowers occur during January, and fig. 329 shows a plant with male flowers on the tip of a spike and female flowers below. The shoots of this sedge die back completely during winter.   CYPERACEAE

329   Yellow-leaved sedge with
       flower- and seed-heads,
       Lake Te Anau (January)

332 Scented broom, *Carmichaelia odorata*, is a branching shrub, up to 3 m high, with drooping, compressed, grooved and striated branchlets up to 20 cm long. Scented flowers occur as 15–20-flowered racemes during October and November and, often again, during February and March. Found along shady banks, streamsides and forest margins from Lake Waikaremoana southwards.   FABACEAE

332   Scented broom in flower, my garden at
       Waikanae (October)

330–331 Tree broom, *Chordospartium stevensonii*, is a small, leafless canopy tree, up to 8 m high, with pendulous branchlets that bear 9 cm long racemes of mauve flowers in profusion from November to January. The tree is found on isolated alluvial river and stream terraces in the vicinity of the Clarence, Awatere and Wairau rivers.   FABACEAE

330   Tree broom flowers, Jordan
       River (December)

331   Flowers of tree broom, close
       up, Jordan River (December)

334   Plant of yellow snow marguerite in full flower, Homer Saddle (January)

333   Flowers of yellow snow marguerite, Arthur's Pass (January)

333–334   Yellow snow marguerite, *Dolichoglottis lyallii,* forms a plant 50 cm high with slender stems terminating in corymbs of brilliant yellow flowers from December to February (fig. 334), each flower 4–5 cm across (fig. 333). The daffodil-like leaves are up to 25 cm long by 10 mm wide, and the plant is found in damp rocky places, and damp situations in grasslands, herbfields and fellfields along the Southern Alps, and in Fiordland and Stewart Island.   ASTERACEAE

335   Flowers of snow marguerite, Arthur's Pass (January)

335–336   Snow marguerite, *Dolichoglottis scorzoneroides,* forms a plant reaching 60 cm high with stout stems and broad, pointed leaves up to 20 cm long by 2–3 cm wide (fig. 336). The snow marguerite bears broad corymbs of white flowers from December to February, each flower 4–6 cm across (fig. 335). It is found in similar situations to *D. lyallii,* and these two species readily hybridise with one another, producing plants of varying form and flower colour.   ASTERACEAE

336   Plant of snow marguerite in flower, Homer Saddle (January)

337 Flowers and leaves of large-
flowered pink broom,
Milford Sound (December)

337–338 Large-flowered pink broom, *Carmichaelia (Thomsoniella) grandiflora* is a branching, spreading shrub up to 2 m high, bare in winter but leafy in spring and summer, and bearing 5–10-flowered racemes of fragrant flowers (fig. 337), each 6–8 mm long, during December and January. The seed-pods shown in fig. 338 illustrate the typical apical, beak-like structure of native broom seed-pods. Found along streamsides and in shaded places in the South Island west of the Southern Alps.

FABACEAE

338 Seeds of large-flowered pink broom, Milford Sound (January)

339 Leafy broom, *Carmichaelia angustata*, forms a shrub or small tree reaching 2 m high, with grooved, flattened branches and occasional 3–7-foliolate leaves. Flowers in 10–40-flowered racemes occur from February to April. Found west of the Southern Alps along streamsides and montane forest margins.

FABACEAE

339 Leaf spray of leafy broom, Otari (April)

340　Stems and flowers of South
　　Island broom, Milford Track
　　(April)

341　Flowers of South Island
　　broom, Milford Track
　　(April)

340–341 South Island broom, *Carmichaelia arborea*, forms a tree up to 5 m high with ascending branches (fig. 340) and closely set, compressed, striated branchlets. Flowers in 3–5-flowered racemes (fig. 341) occur from February to April. Found only in the South Island, west of the Southern Alps, along lowland streamsides and forest margins, in scrub and on alluvial swampy and boggy ground. FABACEAE

342 Common nertera, *Nertera depressa*, forms patches up to 30 cm across on damp, shady banks, in forest and scrub, on the edges of bogs and in damp grasslands and herbfields in lowland and subalpine regions throughout New Zealand. Minute flowers occur from November to February, and red drupes ripen from January to May. RUBIACEAE

343–344 Broad-leaved sedge, *Vincentia sinclairii*, is a tall, leafy sedge, with broad leaves up to 1.3 m long, found on shaded cliffs or banks and along stream and lakesides from North Cape to the Kaimanawa Mountains. The flowers (fig. 343) occur from October to January, followed by the large, conspicuous seed-heads (fig. 344). CYPERACEAE

343　Broad-leaved sedge with
　　flower-head, Waipunga
　　Gorge (October)

342　Common nertera with drupes, Hollyford Valley
　　(April)

345–346 **New Zealand violet,** *Viola cunninghamii,* is a small, tufted herb (fig. 345), up to 15 cm high, found in damp places in lowland and subalpine regions throughout the country. Flowers (fig. 346), 10–20 mm across, on stalks up to 10 cm long, occur from October to January. The leaves, 3 cm long, are on petioles 1–10 cm long.   VIOLACEAE

346   Close view of New Zealand violet, *V. cunninghamii,* flowers, Dun Mountain (November)

345   New Zealand violet, *Viola cunninghamii,* plant in flower, Thomas River (December)

347   **New Zealand violet,** *Viola lyallii,* forms small plants about 15 cm high with stems that creep only a short distance before turning upwards at their tips. The cordate leaves, 3 cm long, are on petioles up to 10 cm long and the flowers, 10–20 mm across, are without any scent and occur from October to January. Found in damp, shady places from sea-level to 1,200 m throughout New Zealand. VIOLACEAE

344   Broad-leaved sedge with seed-heads, Waitahanui Stream, Taupo (December)

347   New Zealand violet, *Viola lyallii,* flower sprays with leaves, Wharite Peak (December)

348 Thistle-leaved senecio, *Senecio solandri* var. *rufiglandulosus*, is a leafy shrub up to 1 m high, with deeply lobed, dentate leaves 20 cm long. Flowers, 2 cm across, occur in corymbs from November to February. Found along damp, shaded banks and streamsides in lowland and subalpine regions from Mt Taranaki southwards.   ASTERACEAE

350   Mountain cotula plant in flower, Cupola Basin (December)

348   Thistle-leaved senecio plant in flower, Mt Egmont (January)

350 Mountain cotula, *Cotula pyrethrifolia*, is a very aromatic herb, with creeping and rooting stems bearing pinnatifid leaves. Flowers are 8–20 mm across and occur on long stems from October to January. Found throughout New Zealand, between 800 and 2,000 m, along streamsides, and in damp gravel, grassland and herbfields.   ASTERACEAE

349 Native pin cushion, *Cotula squalida*, forms a creeping plant, up to 40 cm across, with serrated, fern-like leaves, 2.5–5 cm long. Flowers, like small pin-cushions, 6 mm across, occur in profusion from December to February. Found in damp grassland and rocky situations in lowland and alpine regions all over New Zealand.   ASTERACEAE

351   Feathery-leaved cotula plant in flower, Waitahanui Stream (February)

349   Native pin cushion plant in flower, Lewis Pass (December)

351 Feathery-leaved cotula, *Cotula minor* is a creeping, rooting plant with silky-hairy stems up to 40 cm long bearing many thin, pinnate and pinnatisect leaves, 10–50 mm long by 5–10 mm wide. Small flowers, 2–5 mm across, occur from December to February. Found in lowland to montane regions on wet ground along the margins of swamps, streamsides, shaded grassy places and the edges of sandy tidal estuaries.   ASTERACEAE

352 Leafy forstera, *Forstera bidwillii* var. *densifolia*, is a herb with densely leafy stems, the leaves 12 mm long by 6 mm wide on stems 20 cm long. Flowers, 8–15 mm across, occur on long stems as 1–3-flowered clusters from December to March. *F. bidwillii* is found in subalpine and alpine regions, growing on damp rock faces or in damp grasslands and herbfields. The variety *densifolia* grows only on Mt Taranaki.   STYLIDIACEAE

354   Flowers of panakenake, Mt Taranaki (December)

352   Leafy forstera plant in flower, Mt Taranaki (January)

355   Panakenake plant with berries, Arthur's Pass (April)

353 Slender forstera, *Forstera tenella*, is a smooth-leaved, herbaceous plant found in alpine regions from the Ruahine Range southwards. It grows in damp grasslands, bogs and herbfields, and flowers, up to 10 mm across, occur from December to February.   STYLIDIACEAE

353   Flowers of the slender forstera, Lewis Pass (January)

354–355 Panakenake/creeping pratia, *Pratia angulata,* is a slender, creeping herb, forming mats up to 1 m across and found in damp, sheltered places up to 1,500 m altitude. The flowers (fig. 354), 8–16 mm across, occur in profusion from October to March. The berries, 8–12 mm long (fig. 355), ripen from February to April.   LOBELIACEAE

356   Chatham Island pratia, *Pratia arenaria,* forms patches 1 m across in sandy, damp places throughout the Chatham Islands, in south-east Otago, and on the Antipodes Islands. The leaves are 10–15 mm long, the flowers 10–12 mm across and the berries 7–10 mm long. Flowers occur from December to March and ripe berries from February to May.

LOBELIACEAE

356   Chatham Island pratia plant in flower, Otari (February)

357   Close-up of creeping ourisia flowers, Travers Valley (December)

357–358   Creeping ourisia, *Ourisia caespitosa* var. *caespitosa* (fig. 357), is a mat-forming plant, with leaves 10 mm long by 5 mm wide found in damp, rocky places from the Ruahine Range southwards. Fig. 358 shows the variety *gracilis,* which has shorter leaves, 6 mm long by 3 mm wide, and which is found in damp places only in the mountains of Canterbury and Otago. Flowers of *gracilis* occur mainly in pairs while those of *caespitosa* occur mostly singly.

SCROPHULARIACEAE

358   Flowering plant of creeping ourisia var. *gracilis,* Homer Cirque (December)

359   Sand ranunculus, *Ranunculus acaulis,* is a low-growing, creeping plant with tufts of fleshy, 3-foliolate to deeply three-lobed leaves, up to 2 cm long, on petioles 2.5–5 cm long. Flowers, 6–9 mm across, occur from September through November, and the plant is found in damp, sandy places along the coasts or along damp, sandy shores of inland lakes.

RANUNCULACEAE

359   The sand ranunculus in flower, Opotiki coast (November)

360-361 North Island mountain foxglove, *Ourisia macrophylla*, is a perennial herb up to 60 cm high, which forms colonies from a creeping rhizome. The broad, dentate leaves are up to 15 cm long, and the whorls of flowers (fig. 361), with each flower about 2 cm across, occur from October to February. Found only in the mountains in damp, shaded places in subalpine scrub and herbfields and along shaded banks and streamsides. Several varieties are known from different localities (fig. 360).

SCROPHULARIACEAE

362 Hairy ourisia, *Ourisia sessilifolia* var. *simpsonii*, forms a small plant with hairy leaves formed as rosettes along a creeping stem. Flowers in pairs, each 10-15 mm across, occur from December to February on hairy stems 5-15 cm high. Found in the South Island mountains in damp grassland and wet fellfields up to 2,000 m altitude.

SCROPHULARIACEAE

362 Hairy ourisia plants in flower, Cupola Basin (December)

360 North Island mountain foxglove plants, var. *meadii*, in flower, Mangakino (October)

361 Whorls of flowers of North Island mountain foxglove, Mt Taranaki (December)

363 Hairy ourisia, *Ourisia sessilifolia* var. *splendida*, is a similar plant to *O. sessilifolia* var. *simpsonii* but with the leaves and stems much more heavily clothed with long, soft hairs. Flowers, 15-20 mm across, on hairy stalks up to 15 cm long, occur from December to February. Found in damp situations in the mountains from north-west Nelson to Fiordland.

SCROPHULARIACEAE

363 Hairy ourisia, var. *splendida*, in flower, Wapiti Lake, Fiordland (December)

364 South Island mountain foxglove plant in flower, Temple Basin (January)

364 South Island mountain foxglove, *Ourisia macrocarpa* var. *calycina*, is a similar plant to *O. macrophylla* and is found in damp scrub and herbfields and along streamsides in the South Island mountains from Nelson to Fiordland. Flower stalks up to 70 cm high bear flowers from October to January.
SCROPHULARIACEAE

365 *Potentilla anserinoides* is a woody, prostrate plant with masses of pinnate leaves, up to 15 cm long, each having a single terminal pinna. The plant is found in lowland to montane damp grasslands and on the edges of bogs from Thames southwards. Flowers occur from September to February.
ROSACEAE

365 *Potentilla anserinoides* plant with flower, Waipahihi Botanical Reserve (February)

366 Small New Zealand gentian, *Gentiana grisebachii*, is an annual with weak, ascending stems, 7–20 cm across, and thin, spathulate leaves, 15–20 mm long by 8–10 mm wide. Solitary or paired flowers about 12 mm long occur in January and February. Found in damp places in herbfields, subalpine scrub and grasslands throughout New Zealand.
GENTIANACEAE

366 Flower of small New Zealand gentian, Mt Ruapehu (February)

367 Maori hypericum plant in flower, The Wilderness (January)

367 Maori hypericum, *Hypericum japonicum*, is a procumbent, much-branched plant, forming patches up to 20 cm across and bearing small yellow flowers, singly, during November and December. Found in open damp places, along edges of tarns and slow-flowing seepages throughout New Zealand.
HYPERICACEAE

368   Wet rock hebe, *Parahebe linifolia*, is a branching shrub, forming a mat up to 40 cm across in wet rocky places and alongside streams between 800 and 1,400 m in the South Island mountains. The thick, sessile, narrow leaves are 8–20 mm long by 1.5–4 mm wide, and the flowers appear in 2–4-flowered racemes during December and January.

SCROPHULARIACEAE

368   Wet rock hebe plant in flower at Arthur's Pass (December)

369   Small-leaved wet rock hebe, *Parahebe lyallii*, forms a prostrate, spreading shrub, carpeting wet rocks and streamsides from 800 to 1,400 m in the mountains from the Ruahine Range to Fiordland. The thick, fleshy leaves are 5–10 mm long by 4–8 mm wide, and the racemes of flowers appear on long stalks, 8 cm long, from November to March.

SCROPHULARIACEAE

369   Spray of small-leaved wet rock hebe with flowers, Boulder Lake (January)

370   Flower spray and leaves of streamside hebe, Tararua Ranges (December)

370   Streamside hebe, *Parahebe catarractae*, is a low, open-branched shrub with serrated leaves, 10–40 mm long by 5–20 mm wide, found in sub-alpine regions throughout New Zealand along damp streamsides, cliff faces and wet, rocky places. Flowers, 8–12 mm across, are borne profusely as few- or many-flowered racemes on long stalks from November to April.

SCROPHULARIACEAE

371   Tangle herb with flowers and seeds, Lake Rotoiti, Nelson (December)

371   Tangle herb, *Rumex flexulosus*, is a peculiar tangled plant found throughout New Zealand in damp, rocky places and damp grasslands. The leaves may reach 30 cm in length. Tiny flowers appear in clusters from November to March, and fruits about 1 mm long ripen from January to April. The stems are grooved, and both stems and leaves are always brown in colour.

POLYGONACEAE

372 **Ranunculus godleyanus** grows to 60 cm high, with rounded crenate leaves up to 18 cm long. Yellow flowers, 2–5 cm across, occur on long stalks from January to March, and the plant is found in wet, subalpine, rocky places in the South Island from about Mt Travers to the Ben Ohau Range.
RANUNCULACEAE

374 Creeping lily plants in flower, The Wilderness (December)

372 *Ranunculus godleyanus* plant, Waterfall Valley, Cass River (December)

373 **Star herb,** *Libertia pulchella*, is a small, dainty plant about 12 cm high, with grass-like leaves having the lower surface duller than the upper. The flowers, 15 mm across, occur from November to February, and the rounded, smooth seeds are yellow when ripe.
IRIDACEAE

375 Close-up of creeping lily flower, The Wilderness (December)

373 Star herb plant in flower, Wharite Peak (December)

374–375 **Creeping lily,** *Herpolirion novae-zelandiae*, is a creeping plant with grass-like leaves, making patches up to 1 m across. The flowers, 2–3 cm across, may be white, pale blue or pale mauve and occur during January and February. This is one of the smallest lilies in the world; it is found in damp places and along the edges of bogs in lowland and subalpine regions, up to 1,250 m, throughout New Zealand.
LILIACEAE

376   Small feathery-leaved buttercup, *Ranunculus gracilipes*, with its finely pinnate or divided leaves on hairy petioles 2–12 cm long, is a delicate little herb, about 15 cm high, found in damp grasslands and herbfields alongside streams in the South Island from Mt Travers southwards. The flowers, 2 cm across, occur singly from November to December.
RANUNCULACEAE

376   Small feathery-leaved
buttercup plants in flower,
Cupola Basin (December)

377   False buttercup, *Schizeilema haastii*, looks like a buttercup but on closer examination will show tiny umbelliferous flowers. The plant is found alongside water seepages from steep rock faces and among rocks in fellfields between 1,000 and 1,400 m from the Ruahine Range southwards.   APIACEAE

378   Forest floor lily, *Arthropodium candidum*, grows as small colonies on the forest floor and other shaded places throughout New Zealand. The leaves are 10–30 cm long by 3–10 mm wide, and the flowers, about 10 mm across, occur from November to January.   ASPHODELACEAE

378   Forest floor lily in flower,
Wharite Peak (December)

379   Common willow herb, *Epilobium nerterioides*, is widespread in lowland and montane regions throughout New Zealand, growing in open, damp places and stream beds. It forms patches up to 20 cm across; the fleshy leaves are crowded, and the flowers, 2–3 mm across, occur from September to April.   ONAGRACEAE

377   False buttercup with flowers, Homer Cirque
(January)

379   Common willow herb in flower, Lake Tekapo
(October)

380–382 Mountain astelia, *Astelia nervosa*, is a tussock-like plant (fig. 380) with large, silvery leaves, 50 cm to 1.5–2 m long by 2–4 cm wide, covered by scales that become ruffed up to form a white fur. This lily can form large colonies covering extensive areas on mountainsides all over New Zealand in damp grasslands, herbfields and fellfields between 700 and 1,400 m. The flowers (fig. 382) are strongly sweet scented and appear from November to January, and the berries, about 8 mm across (fig. 381), ripen from March to May.   ASPHODELACEAE

380   Mountain astelia plant growing on Mt Arthur (January)

381   Mountain astelia berries, Mt Holdsworth (May)

382   Mountain astelia flowers, Dun Mountain (November)

383 Kahaka/fragrant astelia, *Astelia fragrans*, is a tufted plant, growing either singly or in colonies. Leaves, 50 cm to 2–2.5 m long by 2.5–7.5 cm wide, are stiff and ascending in the lower half but flaccid in the upper half, usually with a strong red-coloured costa on either side of the midrib. Flowers occur in October and November, with berries ripening from December to May (fig. 383).   ASPHODELACEAE

383   Kahaka showing berries and red costa on each side of leaf midrib, Mt Hector, Tararua Range (March)

384 Streamside tree daisy in full
flower, Arthur's Pass (December)

385 Close-up of flowers of
streamside tree daisy,
Arthur's Pass (December)

384–385 Streamside tree daisy, *Olearia cheesemanii*, is a branching shrub, up to 4 m high, found along lowland and montane streamsides near forest margins, from Auckland south to about Arthur's Pass. The leaves, 5–9 cm long by 2–3 cm wide, are pale buff coloured below with a fine tomentum. Corymbs of flowers, up to 15 cm across, occur in profusion from October to January.

ASTERACEAE

386 Large white-flowered daisy plant in full flower,
Takaka Hill (January)

387 Close-up of large white-flowered
daisy, Takaka Hill (January)

386–387 Large white-flowered daisy, *Brachyglottis hectori*, is a shrub or small tree, up to 4 m high, with stout, brittle, spreading branchlets bearing oblanceolate leaves, 10–25 cm long by 4–12 cm wide, on 4 cm long petioles. Flowers are in large corymbs (fig. 386), each flower up to 5 cm across (fig. 387), and occur in profusion during December and January. Found along lowland to montane streamsides and forest margins from north-west Nelson to just south of Westport.   ASTERACEAE

388   Yellow eyebright, *Euphrasia cockayniana,* is a low, succulent herb, 5–10 cm high, which produces bright yellow flowers, 12–14 mm long, either singly or in pairs, towards the tips of the branches from December to March. Found in damp herbfields and boggy places from the Paparoa Range to Arthur's Pass.   SCROPHULARIACEAE

389–390   Tararua eyebright, *Euphrasia drucei,* is a perennial herb (fig. 389) with erect branches, 5 cm high, and close-set, sessile leaves, 2–10 mm long by 1–5 mm wide, having thickened margins rolled slightly outwards. Flowers, 10–15 mm across (fig. 390), occur in great abundance from December to February. This eyebright is found from the Ruahine Range south to Fiordland in damp or boggy places in alpine herbfields, fellfields and grasslands.

SCROPHULARIACEAE

388   Yellow eyebright in flower, Arthur's Pass (January)

389   Tararua eyebright in full flower, Mt Holdsworth (February)

390   Flowers of tararua eyebright, close up, Mt Holdsworth (February)

391   Creeping eyebright flowers, close up, Mt Holdsworth (December)

391   Creeping eyebright, *Euphrasia revoluta,* forms a creeping plant, up to 10 cm high, often found in subalpine and alpine damp herbfields, tussock grasslands and on the edges of bogs from the Ruahine Range southwards. Large flowers, 10–15 mm across, occur on slender stalks from December to March.

SCROPHULARIACEAE

392 New Zealand eyebright, *Euphrasia cuneata*, is a herb up to 60 cm high, with smooth, wedge-shaped leaves, 5–15 mm long, with slightly thickened, flat margins. Flowers, 15–20 mm long, occur in profusion from January to March. Found in damp herbfields, fellfields, subalpine scrub, damp, rocky places and along streamsides up to 1,550 m altitude, throughout the North Island and as far south as North Canterbury. SCROPHULARIACEAE

394 Nelson eyebright growing among *Celmisia sessiliflora* on Mt Robert, Nelson (January)

392 New Zealand eyebright in flower, Mt Holdsworth (February)

393 Lesser New Zealand eyebright, *Euphrasia zelandica*, is a sparingly branched, rather succulent, hairy herb, 5–20 cm high. The sessile, hairy leaves, 4–10 mm long by 2–6 mm wide, have thickened, recurved margins and tend to cluster as rosettes at the branch tips. Flowers, about 10 mm long, occur during January and February. Found in damp places in herbfields and fellfields from Mt Hikurangi southwards. A similar but smaller eyebright, without hairs on the leaves, *E. australis*, is found throughout the Fiordland mountains. SCROPHULARIACEAE

394 Nelson eyebright, *Euphrasia townsonii*, is a tufted herb with slender stems rooting at the nodes and with erect branches about 15 cm high. The leaves, 4–15 mm long by 2–7 mm wide, are sessile; the flowers, 15–20 mm long, occur singly towards the tips of the branches from December to February or March. Found on the mountains of north-west Nelson and the Paparoa Range.

SCROPHULARIACEAE

395 New Zealand portulaca plant with flowers, Gertrude Cirque, Homer (January)

395 New Zealand portulaca, *Claytonia australasica*, is a succulent, spreading herb, forming patches up to 15 cm across in wet subalpine grasslands and herbfields and along streamsides from the Central Volcanic Plateau southwards. Flowers, 2 cm across, either solitary or paired, occur from November to January. PORTULACACEAE

393 Lesser New Zealand eyebright in flower on Mt Ruapehu (January)

396   Green cushion daisy plant in flower, Cupola Basin (January)

397   Crenate-leaved cotula, *Cotula potentillina*, forms a creeping herb, with pinnate leaves, 4–8 cm long, and with the pinnae irregularly crenate along their margins. Flowers, 8–10 mm across, occur from November to January, and the plant is found in damp situations on the Chatham Islands.

ASTERACEAE

397   Crenate-leaved cotula with flowers, Otari (November)

398–399   Makaka/marsh ribbonwood, *Plagianthus divaricatus*, is usually a much-branching, deciduous shrub, or a small tree up to 4m high, but sometimes it becomes prostrate. Leaves on juvenile plants are 3 cm long by 5 mm wide; on adult plants 15 mm by 2 mm. Flowers (fig. 398), 5 mm across, are axillary and solitary with a strong sweet scent and occur from September to November, and the fruit, about 5 mm across, matures from December to March. Found usually alongside salty swamps or damp, gravelly places in coastal regions throughout New Zealand.

MALVACEAE

396   Green cushion daisy, *Celmisia bellidioides*, is a creeping, branching, rooting daisy, forming mats up to 1 m across on wet rocks or gravel through which water is slowly flowing. It is found from 600 to 1,600 m in the South Island mountains. The fleshy, slightly spathulate leaves are 8–12 mm long by 3–4 mm wide, and the flowers, up to 2 cm across, occur during December and January.

ASTERACEAE

398   Makaka, close-up of flowers, Colville (October)

399·   Makaka spray, showing leaves and berries, Colville (December)

400 **Cardamine,** *Cardamine debilis,* is one of the six cardamines found throughout New Zealand. The leaves are compound. The flowers, 4–5 mm across, occur from November to January, and the plant is found in damp situations from the coast to subalpine herbfields. CRUCIFERAE

402 Flowers and flower-buds of leathery-leaved mountain hebe, The Wilderness (December)

400 Cardamine in full flower, Temple Basin (January)

402 **Leathery-leaved mountain hebe,** *Hebe odora,* is a shrub to 1.5 m high, found in damp places in herbfields and fellfields throughout the New Zealand mountains. The leaves, 10–30 mm long, are imbricate, stiff, concave, leathery and tend to be uniform. Flowers in lateral spikes occur from October to March. SCROPHULARIACEAE

401 **Mountain koromiko,** *Hebe subalpina,* forms a densely branched shrub, up to 2 m high, with pubescent branchlets and narrow, pointed leaves, 2.5 cm long by 5–8 mm wide. Found in damp montane to subalpine regions of Westland and Canterbury. The simple lateral flowers occur in profusion during December and January, and each is about 2.5 cm long. SCROPHULARIACEAE

403 Hairy forget-me-not flowers and leaves, Key Summit (January)

401 Mountain koromiko in full flower, Arthur's Pass (December)

403 **Hairy forget-me-not,** *Myosotis forsteri,* has very hairy leaves, flower-stalks and buds; the leaves, 15–40 mm long by 10–30 mm wide, are on stout hairy petioles. Flowers, 2–6 mm across, occur from October till April. Found along streamsides and in forest from the Urewera southwards.
BORAGINACEAE

404 Early winter orchids in flower, Hinakura (September)

405 Plants of early winter orchid, Hinakura (September)

404–405 Early winter orchid, *Acianthus reniformis* var. *oblonga*, is a small ground orchid that grows as small clumps on damp banks (fig. 404) and lowland grassland or in scrub and among moss throughout the North Island but only occasionally in the South Island. Flowers, 8 mm across, occur during July and August, and the basal sessile leaf is 10–40 mm long (fig. 405).   ORCHIDACEAE

406 Tufted grass cushion with fruiting capsules, Kaimanawa Mountains (November)

406–407 Tufted grass cushion, *Colobanthus apetalus*, is a loosely tufted cushion plant (fig. 406) with grass-like, pointed leaves, 10–25 mm long. Flowers up to 6 mm across occur during December and January, and the seed capsules mature during March and April, later they open to display round, black seeds (fig. 407).   CARYOPHYLLACEAE

407 Close-up of fruiting capsule of tufted grass cushion, Kaimanawa Mountains (November)

408–409 Dense sedge, *Uncinia uncinata*, forms clumps (fig. 408) up to 50 cm high in shaded, open places in lowland and subalpine forest up to 900 m altitude, throughout New Zealand. It is common in clearings and alongside walking tracks. Flowers occur from November to February as dense spikes, up to 15 cm long, which are shorter than the leaves (fig. 409). CYPERACEAE

408 Plant of dense sedge, Tauhara Mountain (January)

409 Flower-heads of dense sedge, Tauhara Mountain (January)

410 Potts' forget-me-not, flower and leaf, Norman Potts garden, Opotiki (November)

410 Potts' forget-me-not, *Myosotis petiolata* var. *pottsiana*, is an open-branched plant with rosettes of leaves, 15–25 mm long by 10–17 mm wide, and flowers 9–12 mm across, occurring from November to February. Found along the Otara River near Opotiki. BORAGINACEAE

411 Bush rice grass, *Microlaena avenacea*, is a grass, often quite abundant in damp, shaded places in lowland and subalpine forests throughout New Zealand. Plants reach a height of 1.3 m, and the graceful flower panicles appear during December and January. GRAMINEAE

411 Bush rice grass flower panicles, Tauhara Mountain (January)

412   Scrambling broom, stem with
       typical seeds, Otari (February)

412   Scrambling broom, *Carmichaelia (Kirkiella)*
kirkii, is a semi-prostrate plant, which scrambles
over the ground with long, thin, interlacing stems
that bear leaves only during spring and summer.
Flowers occur as open racemes followed by the
stout, straight-beaked seed-pods shown in fig. 412.
Found on lowland wet river terraces and streamsides from Canterbury southwards.   FABACEAE

416   Flowers and leaves of Traver's hebe, French Pass
       (January)

416   Traver's hebe, *Hebe traversii,* is a shrub, up
to 2 m high, with slender branches and narrow,
spreading leaves, about 2.5 cm long by 4–7 mm
wide, found along damp streamsides and banks in
montane to subalpine regions from Marlborough
to mid-Canterbury. Flowers arise laterally from
December to March.   SCROPHULARIACEAE

413–415   Koru, *Pratia physalloides,* is an erect-
branching herb, 1 m high, woody towards its base
and with large, soft, conspicuously veined and
toothed leaves, up to 20 cm long by 10 cm wide (fig.
414). The odd-shaped flowers (fig. 413) on hairy
peduncles occur as terminal racemes from October
till May, and the berries, 10–15 mm long, are ripe
from April on (fig. 415). Found in shaded places
along coastal and lowland forest margins and
streamsides in the northern part of the North Island
and the adjacent offshore islands.   LOBELIACEAE

413   Koru flowers, Waipahihi Botanical Reserve
       (February)

414   The striking leaves of koru, Waipahihi Botanical
       Reserve (February)

415   Spray of koru with berries, Waipahihi Botanical
       Reserve (May)

PLANTS OF FORESTS

The forests of New Zealand are evergreen, made up of massive timber trees forming the upper or primary canopy and, below them, lesser trees, which form secondary canopies. Below these again is a thick undergrowth of shrubs, lianes, ferns, mosses, lichens, liverworts and numerous epiphytic and parasitic plants. Many of these trees produce spectacular flowers and fruits, and this, along with the association within the forest of many and diverse species, imparts a richness and extravagance to the forest that is more typical of subtropical than of temperate regions. In places, such as on Mt Taranaki and the Westland forests, this assemblage of plants is a typical rain forest, and in its primeval condition formed a dense, in places almost impenetrable, barrier extending from sea-level to 1,200 m in the warmer north but ascending to 900 m in the cooler south. Known to New Zealand people as 'the bush' these mixed forests were dominated by different species in different localities, and few extensive stands of any single species, other than beech, were known. Since European settlement of the country began, most of these forests have disappeared and, today, the original primeval New Zealand forests can only be seen in some National Parks and special reserves such as those at Pureora, Whirinaki and South Westland.

Forest trees, like other plants, have preferences for differing habitats; some prefer growing where they have cool, damp root runs, others prefer drier places, while others prefer warmer or cooler climates and yet others thrive near the sea while their opposites thrive high in the mountains. Some grow best on limestone-based soils while others do not. Steep, shaded banks, bogs and swamps can occur within forests as well as in open country and, accordingly, as we move into the forest arena, we look first at those trees that prefer damp or wet situations in which to grow.

DAMP AND WET PLACES

417   Kahikatea forest exposure, Lake Rotoroa (April)

417–420   Kahikatea/white pine, *Dacrycarpus dacry-dioides*, is New Zealand's tallest tree; specimens growing in the estuary of the Kauaeranga River, near what is now Thames, were measured by Cook, on the *Endeavour*, at over 200 feet high. Some small stands of such trees still exist in South Westland. Kahikatea prefers growing in wet swamps but can also grow on dry land and even on dry hillsides, and is found naturally in wet and damp areas throughout New Zealand forests from sea-level to 600 m. Fig. 417 shows a stand of kahikatea on the shore of Lake Rotoroa; fig. 418 shows kahikatea foliage and the black seeds in their red receptacles, while fig. 419 shows the seed and receptacle close up. Fig. 420 shows the ripe male cones of kahikatea.

PODOCARPACEAE

418   Mature seeds of kahikatea, Lake Pounui (April)

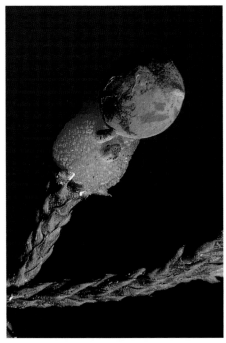

419   Close-up of a mature kahikatea seed, showing the mature bluish-coloured seed on top of the orange-red-coloured receptacle, Lake Pounui (April)

420   Mature male cones of kahikatea, Lake Pounui (October)

421–422   **Kawaka**, *Libocedrus plumosa*, is a tree up to 25 m high with connate, sheathing, compressed leaves, 2.5–5 mm long, arranged in four rows on the branchlets (fig. 421). It is found in damp lowland forests from Mangonui south to about Opotiki, and from Collingwood to Westhaven. Male cones (figs 421–422) are produced sparingly during September and October.   CUPRESSACEAE

421   Spray of kawaka showing male cones, Opotiki (September)

422   Close-up of male cones of kawaka, Opotiki (September)

423–426 Pahautea/mountain cedar, *Libocedrus*
bidwillii, is a tree up to 20 m high found in damp
places in montane and subalpine forests from the
Coromandel Peninsula south to the forests of Fiordland. The appressed, triangular, pointed leaves are
2 mm long, and the male cones appear in profusion
(fig. 423) during September and October (fig. 424).
Female cones (fig. 426) appear in October, and seeds
(fig. 425) are ripe by November.

CUPRESSACEAE

423   Spray of pahautea bearing mature male cones,
       Tongariro National Park (September)

424   Close-up of male cones of pahautea,
       Tongariro National Park
       (September)

426   Pahautea showing leaves and newly formed
       female cones, Otari (October)

425   Close-up of a ripe seed of pahautea,
       Otari (November). Kawaka has a
       similar seed.

427   Maire tawaki, ripe berries, Lake Pounui
(October)

427–429   Maire tawaki, *Syzygium maire*, forms a
tree up to 15 m high, with a smooth, white bark,
four-sided branchlets and sinuate, opposite leaves,
4–5 cm long by 10–15 mm wide, on petioles
5–10 mm long (fig. 428). Flowers (figs 428–429),
about 12 mm across, occur from January to March,
and the berries (fig. 427) ripen from March to July.
Found throughout both islands in lowland to montane, boggy and swampy forests.   MYRTACEAE

429   Close-up of flower of maire tawaki, Lake Pounui
(March)

428   Opening flower-head of maire tawaki,
Lake Pounui (March)

431 Flowers of mountain lacebark, *Hoheria glabrata*, Rahu Saddle (January)

430 Flowers of mountain lacebark, *Hoheria lyallii*, Peel Forest (January)

430–431 Mountain lacebarks, *Hoheria lyallii* (fig. 430) and *Hoheria glabrata* (fig. 431), are both deciduous shrubs or small trees, 6–10 m high, found in wetter places in subalpine forest and scrub in the South Island, *lyallii* on the east side, *glabrata* on the west side of the Southern Alps. The thin, delicate leaves, 2–7 cm long by 2–6 cm wide in *lyallii*, are usually smaller in *glabrata*, and the branchlets, leaves and stems are more heavily clothed with stellate hairs in *lyallii*. Flowers, 2–4 cm across, occur in profusion from November to February.   MALVACEAE

432 Hunangamoho with flowers, Hollyford Valley (January)

432–433 Hunangamoho, *Chionochloa conspicua*, is a large tussock, up to 2 m high, found throughout New Zealand in lowland and subalpine forests and natural forest clearings, especially near streams and seepages. The leaves are strongly nerved, 45 cm–1.2 m long by 6–8 mm wide, flat and often hairy along the margins. Handsome flower panicles, about 45 cm high, occur from November to January.

GRAMINEAE

433 Flower panicle of hunangamoho, Hollyford Valley (January)

434–438 Pukatea, *Laurelia novae-zelandiae,* is a large, aromatic tree, up to 35 m high, with a 2 m diameter trunk buttressed around its base. Leaves are glossy, serrated, thick and leathery, 4–8 cm long by 2.5–5 cm wide, on petioles 10 mm long. Flowers are unusual, occurring as perfect, with both male and female parts in the one flower (fig. 436); male flowers (fig. 434) have stamens (red) and female flowers have stigmas but no stamens (fig. 435); all as well have yellow, triangulate structures called staminodes. Seeds (figs 437–438) are contained in elongated, pear-shaped seed-cases. Pukatea is found in lowland swampy forest, creek beds and damp gullies throughout the North Island and south to Fiordland. MONIMIACEAE

434   Male flowers of pukatea, Kaitoke (November)

435   Close-up of female pukatea flowers, Kaitoke (November)

437   Pukatea seed-cases opened releasing wind-borne seeds, Waikanae (May)

436   Close-up of perfect pukatea male flower, Kaitoke (November)

438   Pukatea spray showing unopened and opened seed-cases, Waikanae (May)

439–440 Cabbage tree/ti kouka, *Cordyline aus-tralis*, grows to a height of 12–20 m with an unbranched trunk topped by a mass of leaves and flowers. Cabbage tree is a collective name for five species, of which ti kouka is the most common, being found in damp situations along forest margins and open spaces, especially alongside bogs and swamps, throughout the country from sea-level to 600 m altitude. Isolated specimens often occur on hillsides near seepages. The long leaves are are up to 1 m long by 3–6 cm wide. Sweet-scented flowers, each 2 cm across, arise on dense panicles up to 1.5 m long, above the leaf canopy, from September to December followed in January–February by whitish berries speckled with blue.   ASPHODELACEAE

441   Toi tree with flowers, Mt Taranaki (December)

439   Cabbage trees heavy with flowers, Hinakura (November)

441–445 Toi/broad-leaved cabbage tree/mountain cabbage tree, *Cordyline indivisa,* is similar to ti kouka but has longer, broader leaves, 1–2 m long by 10–15 mm wide, and grows only in the wetter mountains from the Hunua and Coromandel Ranges to Fiordland. Sweet-scented flowers occur from November to January as tightly packed panicles (fig. 445) in a flower-head up to 1.6 m long by 30 cm wide, which arises beneath the leaf canopy (fig. 442), the individual flowers up to 2 cm across (fig. 444). This flower-head is followed by a mass of black berries, each about 6 mm in diameter, during April and May (fig. 443).   ASPHODELACEAE

440   Close-up of cabbage tree flowers, Lake Pounui (November)

446 Dwarf cabbage tree, *Cordyline pumilio,* seldom reaches 2 m high and is sometimes stemless, when it can be mistaken for a tussock or a sedge. The stiff leaves are up to 60 cm long by 10–20 mm wide. Small, very sweet-scented flowers, 5 mm across, arise, widely spaced, along the panicle, which is up to 60 cm long, during December and January. Found in the North Island only in damp places in open lowland forest and scrub. ASPHODELACEAE

442 The pendant flower-head of toi, Mt Taranaki (December)

443 The pendant fruiting head of ripe berries of toi, Mt Taranaki (May)

444 Close-up of an individual toi flower, Mt Taranaki (December)

446 Dwarf cabbage tree plant, Waipahihi Botanical Reserve (October)

445 A section of flower-head of toi, Mt Taranaki (December)

447–450 Yellow pine, *Halocarpus biformis,* in forest forms a small tree, 10 m high, but on exposed mountainsides it forms a tight, rounded, cypress-like shrub or small tree, sometimes only 1 m high. It occurs from the central Volcanic Plateau south to Stewart Island, from sea-level to 1,400 m, usually in damp places. Adult leaves are 2 mm long, densely imbricate, appressed and keeled; juvenile leaves and leaves of reversion shoots are 10–20 mm long and spreading (fig. 450). Male cones (fig. 448) are produced in profusion during October, shedding pollen during January (fig. 447). Small female cones appear in December and, after pollination, grow slowly to reach maturity fourteen months later (fig. 449).

PODOCARPACEAE

447   Yellow pine, foliage with opened male cones, Mt Ruapehu (February)

451   Yellow silver pine, young and adult foliage, Taita (October)

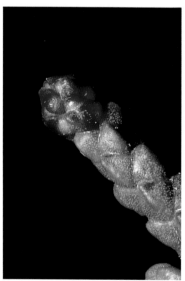

448   Yellow pine, mature male cone, Mt Ruapehu (December)

449   Yellow pine, mature seed sitting on yellow aril, Mt Ruapehu (February)

450   Juvenile foliage above with adult foliage below of yellow pine, Mt Ruapehu (December)

452 Mature female cone of yellow silver pine, Coromandel Peninsula, (April)

453 Male cones of yellow silver pine, Taita (October)

454 Close-up of male cones of yellow silver pine, Taita (October)

451–454 Yellow silver pine, *Lepidothamnus intermedius*, is a spreading tree up to 15 m high, found in wet places in lowland, montane and subalpine forests from the Coromandel south to Stewart Island; in the latter it is the principal tree in swamp forests. The imbricate adult leaves are 1.5–3 mm long; juvenile leaves are 9–15 mm long and spreading (fig. 451). Male cones (fig. 454) mature in profusion from October to December (fig. 453), and female cones are mature by April (fig. 452).

PODOCARPACEAE

455–456 Silver pine, *Lagarostrobos colensoi*, is a somewhat spreading tree, up to 15 m high, found from Mangonui to Mt Ruapehu and, in the South Island, throughout Westland in shady places with a rich soil and a wet climate. Leaves of juveniles, 6–12 mm long, are spreading; leaves of adults, 1–2.5 mm long, are imbricate, appressed and keeled. Male cones (fig. 455) occur abundantly from September to November; female cones are borne sparingly and take about eighteen months to mature (fig. 456), usually in November, but at irregular intervals.

PODOCARPACEAE

455 Silver pine, male cone sheding pollen, Mt Ruapehu (November)

456 Mature female cone of silver pine, Mt Ruapehu (March)

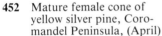

457–459 Kotukutuku/tree fuchsia/konini, *Fuchsia excorticata,* forms a deciduous shrub or a small tree, up to 12 m high, with a red, papery bark that peels readily to reveal a smooth, satiny, greenish inner layer. Leaves, 3–10 cm long by 15–30 mm wide, are pale green or pale silvery below on petioles 10–40 mm long. Flowers, notable for their blue-coloured pollen, begin to arise in August, in some districts, and become abundant during October and November, often persisting through to February. Berries (fig. 459) occur from September to February. Found throughout New Zealand and the Auckland Islands in lowland and montane forests, especially along forest margins and streamsides where the soil is damp.   ONAGRACEAE

458 A kotukutuku flower, close up, Wellington (October)

457 Kotukutuku branch with flowers displaying typical blue-coloured pollen, Lewis Pass (January)

460 Sprawling fuchsia, *Fuchsia colensoi,* is a low, sprawling, woody shrub with long straggling branches bearing variable-shaped leaves and numerous flowers very similar to those of *F. excorticata.* The berries, however, are narrower and more cylindrical in shape than those of *F. excorticata.*

ONAGRACEAE

459 Kotukutuku spray with berries, Kau Kau, Wellington (January)

460 Sprawling fuchsia with flowers and berries, Waikanae (December)

461–462 Round-leaved coprosma, *Coprosma rotundifolia*, is a slender shrub or small tree, up to 5 m high, with rounded leaves (fig. 461), up to 25 mm long, and entangled branches. It is found throughout New Zealand in damp lowland and montane forests or in scrub and along riverbanks. Slender, drooping flowers (fig. 462) arise during September and October, and the drupes (fig. 461), which ripen during February and March, occur singly or in twos and threes along the branches.

RUBIACEAE

463–464 Great Barrier tree daisy, *Olearia allomii*, is a shrub about 1 m high with very thick, leathery leaves, 2.5–5 cm long (fig. 464). It is found only on the Great Barrier Island, in damp forest from sea-level to 700 m. Sweet-scented flowers (fig. 463), 15 mm across, occur during October and November.

ASTERACEAE

463 Great Barrier tree daisy, close-up of flowers, Otari (October)

461 Round-leaved coprosma drupe and leaves, Pelorus Bridge (January)

464 Great Barrier tree daisy, spray with flowers and leaves, Otari (October)

462 Branchlet of round-leaved coprosma with male flowers, Silverstream (October)

COASTAL FOREST

Several trees grow in coastal areas, mostly as isolated groves or specimens, but in its primeval state New Zealand's coasts, in many places, were clothed by coastal forests. Today, these are represented only by scattered reserves such as that near Paraparaumu, which preserves a remnant of the once-dominant coastal kohekohe forests.

465 Panicles of flowers sprouting from the trunk of a kohekohe tree, Huntleigh Park, Wellington (May)

465–468 Kohekohe, *Dysoxylum spectabile,* is a tree, up to 15 m high, that grows in damp areas of lowland coastal forests from North Cape to Nelson and once formed extensive coastal forests in restricted areas. From March till June the tree produces spectacular, long, drooping panicles of greenish white, waxy flowers (figs 466–467), sprouting directly from the trunk and branches (fig. 465). These are followed from April to September by panicles of green capsules, 2.5 cm across, which split to reveal bright orange or red arils (fig. 468) that carry the seeds.

MELIACEAE

466 Portion of a flower panicle of a kohekohe tree, Huntleigh Park, Wellington (May)

468 Seed capsules of kohekohe opening to reveal the red arils, Waikanae (May)

467 Close-up of a kohekohe flower, Waikanae (June)

469–472   Nikau, *Rhopalostylis sapida*, is a palm reaching to 10 m in height, found in lowland and coastal forests throughout the North Island and, in the South Island, as far south as Banks Peninsula and Greymouth. The rings on the trunk are leaf scars from feather-like leaves up to 3 m long and 2 m wide. The flower panicle shown in fig. 469 has the upper bud sheath behind it; this usually falls within minutes of the opening of the flower. To the left in fig. 469 are the green berries from the previous year's flower, and below are the brownish seeds from two years before. Flower-spikes are about 30 cm long and arise from December to February. Fig. 471 shows male flowers closely packed on the flower-spike, with occasional tiny female buds between them. There are three male flowers to each single female, and fig. 470 shows a male flower, close up, with the tiny female flower beside it to the right and another female to the left, beneath the tips of the two anthers. Female flowers do not fully open until after the male flowers have fallen off. The orange to red berries are ripe twelve months later (fig. 472).

PALMAE

470   Close-up of nikau flowers: male flowers with anthers, female flowers are the tiny bud-like structures beside the two male flowers, Hukutaia Domain (December)

471   A section of a panicle of nikau flowers, females visible between male flowers, Hukutaia Domain (December)

469   The flower panicle (above), last year's seeds (middle) and the year before's seed remain on a nikau palm, Lake Pounui (February). Note the flower-sheath behind the flower panicle.

472   Ripe nikau berries, Waiorongomai (April)

473   Ti ngahere trees, Wharariki
       Beach (January)

475   Young ti ngahere tree in flower, Kauaeranga
       Valley (December)

473–477   Ti ngahere/forest cabbage tree, *Cordyline banksii*, unlike other cabbage trees, usually has several stems, up to 4 m high (fig. 473), and tends to be more bushy. It grows, preferably in damp situations, in lowland to montane forest, in coastal forests, or along forest margins from North Cape to North Westland. Fragrant white flowers in rather open panicles (fig. 474), 1–2 m long, occur from November to January, and whitish berries with blue markings (fig. 477) follow in February, persisting till April. Quite small plants, only 1 m high, will flower (fig. 475).   AGAVACEAE

476   Close-up of flower of ti
       ngahere tree, Kauaeranga
       Valley (December)

474   Flower panicle of ti ngahere tree, Lake Pounui
       (November)

477   Ti ngahere berries, Lake Pounui (February)

478 Akeake seeds, Ure River Valley (March)

479 Spray of akeake with male flowers, Otari (October)

480 Close-up of male flowers of akeake, Otari (October)

478–482 **Akeake,** *Dodonaea viscosa*, is a shrub or small tree, up to 6 m high, with narrow leaves up to 10 cm long (figs 478–479); the branches and branchlets are usually sticky. The very small male flowers (fig. 479) occur from September to January, and the unusual-looking fruits (fig. 478) from November to March. It is found from North Cape to Banks Peninsula in the east and Greymouth in the west. The thin, reddish-coloured bark peels in thin flakes, and the sexes occur on separate trees (fig. 480, male flowers; fig. 481, female flowers). A variety, *purpurea*, with copper-coloured leaves and copper-coloured fruits (fig. 482), is known and is often grown in gardens.   SAPINDACEAE

482 Seeds of the purple form of akeake, Taupo (December)

481 Female flowers of akeake, Otari (October)

LOWLAND AND MONTANE FORESTS

Lowland forests are inland from coastal forests and usually situated on rolling or flat country. They are rich in species and included much of the original primeval podocarp dominant forests of New Zealand. As the land rises and slopes become steeper, we move into montane forests of mixed podocarp and other species, which extend towards the upper subalpine and alpine forests that consist, in the main, of beech trees.

483–484 Coastal tree daisy, *Olearia albida*, is a shrub up to 5 m high, with grooved branchlets clothed with a loose, white tomentum. The leathery leaves, 7–10 cm long by 2.5–3.5 cm wide, are on petioles 2 cm long (fig. 483). Flowers (fig. 484) appear from January to May, and the plant is found in coastal forests from North Cape to East Cape and Kawhia.   ASTERACEAE

483   Coastal tree daisy, spray with flower-heads, Seatoun, Wellington (April)

485   Kohuhu spray with flowers, Karaka Bay (September)

485 Kohuhu, *Pittosporum tenuifolium*, is a small tree, up to 8 m high, found commonly throughout New Zealand, except in Westland, in lowland and coastal forests. The thin, glossy, wavy leaves are 3–7 cm long by 10–20 mm wide. Flowers, 12 mm across, are produced in abundance from September to November and, in the evening, fill the air with a strong, sweet fragrance.   PITTOSPORACEAE

484   Close-up of flowers of coastal tree daisy, Seatoun, Wellington (April)

486–488 Kauri, *Agathis australis,* is New Zealand's
most massive tree, reaching a height of 30–60 m with
a trunk diameter of 3–7 m. Found in lowland forests
in Northland, the Coromandel and south to Maketu
in the east and Kawhia in the west. Male and female
cones occur on the same tree (fig. 486); the male
(fig. 487) appears during August and September, the
female (fig. 488) during September and October. The
thick, parallel-veined adult leaves are 2–3.5 cm long,
those of juvenile trees are 5–10 cm long by 5–12 mm
wide. The male cone is mature and releases pollen
when one year old; the female opens for pollination,
also when about twelve months old, but does not
open to release seeds until another two years have
passed.   ARAUCARIACEAE

487 Kauri branchlet with mature male cone, Waipoua
(November)

486 Kauri branch with two immature male cones and
one female, Waipoua (October)

488 Mature female kauri cones, Waipoua (September)

489 Tarairi flowers and leaves,
       Whangarei (December)

491 Tarairi drupes, Whangarei
       (May)

489–491 Tarairi, *Beilschmiedia tarairi*, forms a medium-sized tree, up to 22 m high, found in lowland forests, especially kauri forests, in Northland and as far south as Kawhia and Tauranga. The prominently veined, glossy, leathery leaves, 4–15 cm long by 3–6 cm wide, on petioles 10–15 mm long, have the veins below and the petioles clothed with a thick brown to golden-coloured wool (fig. 489). Small apetalous flowers (fig. 490), each 5 mm across, occur in panicles up to 10 cm long from September to December (fig. 489). The purple-bloomed drupes (fig. 491), up to 3.5 cm long, ripen during the following April and May.   LAURACEAE

492 Branch of wavy-leaved
       coprosma heavy with drupes,
       Opepe Bush (May)

492 Wavy-leaved coprosma, *Coprosma tenuifolia*, is a shrub or small tree to 5 m high, with wavy-margined leaves, 4–10 cm long. The drupes, 7–8 mm long, ripen in April and May. The plant occurs in lowland to montane forests and in subalpine scrub from Te Aroha Mountain south to the Ruahine Range; it is very common on Mt Taranaki.
                                        RUBIACEAE

490 Tarairi flowers, close up, Colville (December)

493 Kanono/raurekau, *Coprosma australis,* is a shrub or small tree to 6 m high, found in lowland and montane forests and along forest margins or in scrub throughout the country. The leaves, 10–20 cm long by 5–10 cm wide, are on petioles 2–5 cm long. Flowers occur from April to June, often along with the drupes of the previous season, which are 9 mm long and take some twelve months to ripen.

RUBIACEAE

494–496 Tawa, *Beilschmiedia tawa,* is a tree to 25 m high, with leaves 5–10 cm long by 10–20 mm wide on petioles 10 mm long. Panicles, about 8 cm long, arising from the axils of the leaves and carrying tiny flowers (fig. 494), 2–3 mm across (fig. 495), appear from September to December. The drupe, 2–3 cm long (fig. 496), appears from October to February. Tawa is found in lowland and montane forests from North Cape to Nelson and Marlborough.

LAURACEAE

494   Tawa spray with flower panicles, Barton's Bush, Silverstream (December)

493   Kanono showing leaves and drupes, Opepe Bush (May)

495   Tawa flower, Barton's Bush, Silverstream (December)

496   Tawa drupe, Wilton's Bush, Wellington (March)

497–499 Makamaka, *Ackama rosaefolia*, is a small bushy tree, up to 12 m high, with pinnate, serrated leaves having a terminal pinna (fig. 497). Found in lowland forest and along forest margins and streamsides from Mangonui to about Dargaville. The flowers arise in much-branched panicles (fig. 497), up to 15 cm long, with each flower 3 mm across (fig. 498), from August till November, and the fruits (fig. 499) ripen from January to March.

CUNONIACEAE

498   Close-up of makamaka flowers, Otari (October)

497   Makamaka flowers and leaves, Otari (October)

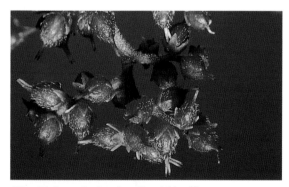

499   Makamaka berries, Otari (April)

500–501 Putaputaweta, *Carpodetus serratus*, is usually a small tree up to 10 m high, with flat, rather fan-like branches bearing leaves 4–6 cm long (fig. 501). White, star-like flowers, 5–6 mm across (fig. 500), occur as broad panicles (fig. 501) from November to March. The fruit is a black capsule, 4–6 mm across, that ripens from March through May. Putaputaweta is found throughout New Zealand in lowland and montane forests and along forest margins and streamsides and in natural clearings.   ESCALLONIACEAE

501   Spray of putaputaweta with flowers and leaves, Lake Pounui (December)

500   Putaputaweta flowers, close up, Lake Pounui (December)

502   Rimu foliage with ripe female
      cones, Mangamuka (February)

503   Spray of rimu with male
      cones, Bushy Park (November)

502–503 Rimu/red pine, *Dacrydium cupressinum*, is a tall forest tree, up to 60 m high, found in lowland and montane forests throughout the North, South and Stewart Islands. With its characteristic feather-like foliage (fig. 502), rimu is one of New Zealand's finest timber trees. Female cones with their red arils (fig. 502) appear at irregular intervals over several years, usually about April through May. The same applies to the male cones (fig. 503), which appear late November to December in a year when they occur.   PODOCARPACEAE

505   Hutu branch with flowers, Otari (September)

504   Hutu leaves, Hukutaia
      Domain (June)

504–505 Hutu, *Ascarina lucida*, is a closely branched, small, highly aromatic tree, up to 8 m high, found in lowland and montane forests from Hokianga Harbour southwards to Stewart Island but more common in the South Island. The branchlets are purple coloured (fig. 504), the leaves, 2–7 cm long by 15–35 mm wide, are serrated (fig. 504) and on petioles about 10 mm long. Flowers as spikes occur from September to November (fig. 505), followed by small, whitish berries from October to January.   CHLORANTHACEAE

506 Pigeonwood spray with male
flowers, Waiorongomai (October)

507 Close-up of male pigeonwood flower,
Waiorongomai (November)

506–509 Pigeonwood/porokaiwhiri, *Hedycarya
arborea*, is an aromatic tree, up to 12 m high, with
thick, tough, wavy-margined leaves, 5–12 cm long
by 2.5–5 cm wide, on petioles 2 cm long. Racemes
of lemon-scented flowers (fig. 506), each flower
about 10 mm across, occur from September to
December. The sexes are on separate trees, with the
male (fig. 507) and the female (fig. 508) quite
different. The drupes (fig. 509) take twelve months
to mature and are ripe from October to March of
the following year. Pigeonwood is found in damp
situations in lowland and montane forests from
North Cape to Banks Peninsula.   MONIMIACEAE

508 Female pigeonwood flowers, Lake Pounui
(December)

509 Pigeonwood drupes, Otaki Forks (March)

510 Mingimingi branchlets with leaves and drupes,
Ure River Gorge (March)

510 Mingimingi, *Coprosma propinqua*, is a small-
leaved shrub or small tree, 3–6 m high, found
throughout New Zealand in lowland forest, along
forest margins and streambanks, in scrub, gravelly
places and along the edges of bogs and swamps.
Flowers occur during October and November, and
the white or blue drupes are mature from March to
May.   RUBIACEAE

511 Titoki spray showing leaves and flowers, Barton's Bush (November)

512 Winged seed capsules of titoki opening to show black seeds sitting on red arils, Barton's Bush (December)

513 Close-up of titoki flower, Barton's Bush (November)

511–513 **Titoki,** *Alectryon excelsus*, is a spreading tree, up to 10 m high, with pinnate leaves (fig. 511), 10–40 cm long. It is found in lowland forests, growing on alluvial soils from near North Cape south to Banks Peninsula. Apetalous flowers (fig. 513) arise as panicles (fig. 511) during October and November, but the seeds that follow (fig. 512) take one year to ripen so that flowers and seeds can occur together on the tree. SAPINDACEAE

514 *Coprosma rubra* is a shrub about 2–3 m high with stems and leaf petioles clad with soft hairs and the leaves, 10–15 mm long by 10–12 mm wide, are strongly reticulated with ciliated margins when young. Fig. 514 shows the elongate yellow drupe, which occurs singly, but sometimes in groups of two, and the female flower. This *Coprosma* grows in lowland forest and scrub from the Pahaoa River in the Wairarapa south to about Banks Peninsula.
RUBIACEAE

514 *Coprosma rubra* branchlet with leaves and drupe, Otari (September)

515–517   Matai/black pine, *Prumnopitys taxifolia,*
forms a robust forest tree, up to 25 m high, with a
broad crown and narrow, elongate leaves, 5–10 mm
long by 1–2 mm wide. Male cones (figs 516–517)
arise on spikes about 5 cm long during October and
November, and the fruit ripens during the following
year to a black drupaceous seed with a purplish
bloom (fig. 515). Matai is a common tree in lowland
forests of both the North and South Islands but is
rare on Stewart Island.   PODOCARPACEAE

515   Matai seeds, Hinakura (February)

516   Male cones of Matai, Lake Pounui (November)

517   Close-up of mature matai male cones, Lake
Pounui (November)

518   Miro/brown pine, *Prumnopitys ferruginea,*
forms a round-headed tree up to 25 m high, with
sickle-shaped leaves, 10–25 mm long by 1–2 mm
wide, that are in one plane. Male cones similar to
those of *P. taxifolia* occur during October and
November, and the brilliant seeds (fig. 518) mature
one year later. Miro is a common tree in lowland
forests throughout New Zealand.

PODOCARPACEAE

518   Miro seeds, Huntleigh Park, Wellington (April)

520   Immature totara seed on its aril, Woodside Gorge (September)

519   Totara spray with cones, Hinakura (April)

519–521   Totara, *Podocarpus totara,* is a tall forest tree, up to 30 m high, with thick, stringy, furrowed bark and narrow, leathery, pungent leaves, up to 3 cm long by 3–4 mm wide. Male cones (fig. 521) are abundant during October and November, each 15 mm long, and produced singly or four together on a short peduncle. The seeds sit in red, fleshy arils (fig. 520). Totara grows throughout New Zealand in lowland and montane forests, often as small pure stands.   PODOCARPACEAE

521   Ripe male cones of totara, Akatarawa Saddle (November)

522   Montane totara/Hall's totara, *Podocarpus cunninghamii,* is a similar but smaller tree than *P. totara,* but differs in having thin, papery bark and subsessile leaves. Male cones and seeds are similar, with the receptacle in *P. cunninghamii* tending to be more elongate. Montane totara is found in lowland, montane and subalpine forests throughout New Zealand.   PODOCARPACEAE

522   Ripe female cone of montane totara, Moawhango (April)

523–526 Five finger/pouahou, *Pseudopanax arboreum*, is a much-branching shrub or small tree, up to 8 m high, with compound leaves, each of 5–7 serrate-dentate-margined leaflets, 10–15 cm long (fig. 523). Five finger flowers profusely from June to August (fig. 523), the flowers in umbels, each flower 6 mm across (fig. 524). Sexes are on separate trees (figs 524–525). The rounded flower-buds are shown in fig. 526 but the dark brown to black seeds that occur from August to February are noticeably flattened. Five finger is common in lowland and montane forests and lowland scrub throughout New Zealand.   ARALIACEAE

525   Close-up of female flowers of five finger, Taupo (August)

523   Five finger plant in flower, Taupo (August)

526   Flower-buds of five finger are easily confused with the dark, flattened, similarly arranged seeds that occur from September onwards, Karaka Bay (April)

524   Close-up of male flower of five finger, Taupo (June)

527   Haumakaroa, *Pseudopanax simplex*, forms a shrub or low, open-branched tree, 8 m high, in lowland forests from the Coromandel Ranges south to Stewart Island and on the Auckland Islands. Seedling plants have divided leaves that quickly change to 3–5 foliolate, then to the acuminate, sharply serrate, leathery, trifoliate and unifoliolate leaves, 5–10 cm long, with petioles 3–8 cm long, of adult plants. Umbels of 5–15 greenish-coloured flowers arise from June through to March.   ARALIACEAE

527   Haumakaroa male flowers, close up, Taurewa Intake (January)

528–530 Lancewood/horoeka, *Pseudopanax crassifolium*, is a small, round-headed tree, 15 m high, with very thick, long, narrow, serrated leaves (fig. 530). The tree passes through distinct seedling and juvenile stages that bear little resemblance to the adult stage. Lancewood occurs throughout the lowland and montane forests of New Zealand, and in lowland scrub. Flowers arise in large terminal umbels (fig. 528) from January to April, and the fruits, large bunches of berries (fig. 529), each 5 mm across, ripen during the following twelve months.

ARALIACEAE

530 Spray of lancewood, showing leaf undersides and green berries, Lake Pounui (March)

528 Umbel of lancewood flowers, Waikanae (February)

531 Raukawa spray showing leaves, flowers, flower-buds and immature seeds, Mt Kaitarakihi (January)

529 Lancewood with ripe seeds, Lake Pounui (March)

531–532 Raukawa, *Pseudopanax edgerleyi*, is a tree 10 m high, with aromatic, green, shining, 1–3-foliolate leaves, paler below, and each 7–15 cm long by 3–5 cm wide. Greenish-coloured flowers (fig. 532) in small, 10–15-flowered, axillary umbels (fig. 531) arise from November to March; both the buds and seeds are rounded (fig. 532). Raukawa is found in lowland forests from about Mangonui southwards to Stewart Island.   ARALIACEAE

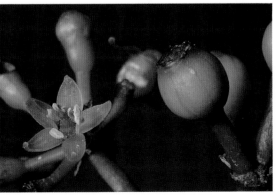

532 Close-up of flower, flower-bud and seed of raukawa, Mt Kaitarakihi (January)

533 Mapau showing ripe berries and leaves, Karaka Bay (April)

533–535 Mapau, *Myrsine australis*, is a shrub or small tree to 6 m high, found in lowland forests throughout New Zealand. The wavy-margined, leathery leaves (fig. 533) are 3–6 cm long by 15–25 mm wide and are on strong petioles 5 mm long. Flowers (male, fig. 534; female fig. 535) are 1.5–2.5 mm across and arise along the stems from December to April, followed in October through to February by clusters of berries (fig. 533).

MYRSINACEAE

535 Spray of mapau with female flowers, Whangarei (January)

534 Mapau male flowers shedding pollen, Karaka Bay (December)

536–537 Tawari, *Ixerba brexioides*, forms a canopy tree 6–16 m high, found from Mangonui to Mamaku, which smothers itself with creamy white flowers (fig. 536) in November and December. The narrow, leathery, serrated leaves (fig. 537), 6–16 cm long by 10–40 mm wide, are on petioles 2 mm long. Figure 537 shows a panicle of flower-buds, a seed capsule, and the unusual seeds revealed when the capsules open from January to April.   ESCALLIONIACEAE

536 The striking flowers of tawari set among terminal leaves, Rotorua (December)

538-539 Rewarewa/New Zealand honeysuckle, *Knightia excelsa,* is an upright branching tree, 30 m high, with long, thick, leathery, coarsely serrated leaves, 10-20 cm long by 2.5-4 cm wide. Brilliant flowers, in racemes up to 10 cm long (fig. 538), arise on the branchlets from October to December, and the seeds (fig. 539) mature by the following June. Rewarewa is found in lowland forests from North Cape to the Marlborough Sounds. PROTEACEAE

538 Rewarewa flower, Ngaio Gorge, Wellington (November)

540 Supplejack/karewao/pirita, *Ripogonum scandens,* is often met with as a mass of twining, twisting stems blocking progress through the forest. Supplejack is a climbing liane that, up in the light of the forest canopy, produces leaves, 5-16 cm long by 2-6 cm wide, and tiny flowers on branching stalks followed by clusters of berry-like fruits (fig. 540).

SMILACACEAE

539 Rewarewa seeds, Otari (June)

537 Tawari spray showing seeds of this season and the flower-buds for next season, Mamaku Forest (May)

540 Supplejack spray with leaves and berries, Lake Pounui (October)

541 Flowers and leaves of toru,
Otari (October)

542 Close-up of toru flowers,
Otari (October)

541–543 Toru, *Toronia toru,* forms an erect tree,
12 m high, with narrow, alternate or whorled, thick,
leathery leaves (fig. 541), 16–20 cm long by 8–15 mm
wide. Fragrant flowers in 6–15-flowered axillary
racemes (figs 541–542) occur from October to
February and are followed by fruits clustered along
the stems from December to March (fig. 543). Toru
is found in lowland and montane forests from
Mangonui to about Tolaga Bay.   PROTEACEAE

544–545 Puataua, *Clematis forsteri,* is a small vine
found in lowland forest, along forest margins and
in lowland scrub throughout the North Island.
Sweet-scented flowers (fig. 545), 2.5–4 cm across,
occur from September to November, followed by
clusters of downy seeds from November to February
(fig. 544).   RANUNCULACEAE

543 Toru berries arise from December to March,
but persist throughout April and May, Taupo
(April)

544 Puataua seeds, Hinakura (January)

545 Puataua leaves and flowers, Hinakura (October)

548 Toro spray showing leaves and flowers clustered round the branch, Mt Holdsworth (November)

550 Toro branch with berries, Mt Holdsworth (December)

549 Close-up of flowers of toro, Mt Holdsworth (November)

546–547 Bush clematis/puawhanganga, *Clematis paniculata,* is a strong, woody liane that climbs to the tree tops to produce, in springtime, great clusters of white flowers (fig. 546), each flower softly fragrant and up to 10 cm across. Found throughout New Zealand in forest and along forest margins and in scrub.   RANUNCULACEAE

548–550 Toro, *Myrsine salicina,* is a small canopy tree to 8 m high, with long, smooth, narrow, leathery leaves, 7–18 cm long by 2–3 cm wide. Flowers occur from August to January as dense-flowered fascicles along the stems (fig. 548), with each flower (fig. 549) about 3 mm across, followed from September to May by clusters of berry-like fruits (fig. 550). Found in lowland and montane forests from North Cape to about Greymouth.   MYRSINACEAE

546 Bush clematis in flower, Renata Track, Tararua Range (November)

547 Close-up of flowers of bush clematis, Renata Track, Tararua Range (November)

The genus *Pittosporum* in New Zealand

Altogether there are 26 species of *Pittosporum* known from New Zealand; all are shrubs or small trees found in lowland and montane forests and lowland scrub, either throughout the country or confined to restricted localities as with *P. dallii*, found only in north-west Nelson. Some species have flowers borne singly, others are in panicles or umbels, and they may be monoecious or dioecious, and many are strongly sweet scented, especially in the evenings. Their leaves can vary from 5 mm to 15 cm in length, and from pinnatifid to elliptic in shape; some are aromatic. Most species flower between September and December, with seeds following from December onwards. A selection of these plants is shown here; others are already figured on pages 28, 29, 64, 67 and 122.   PITTOSPORACEAE

551–553 Tarata/lemonwood, *Pittosporum eugenioides*, is found throughout both North and South Islands. Berries (fig. 553) go black when mature.

551   Tarata, umbels of male flowers, Wilton's Bush, Wellington (October)

553   Tarata berries, Wilton's Bush, Wellington (March)

552   Tarata, female flowers, close up, Otari (October)

554–555 Black mapou, *Pittosporum tenuifolium* subsp. *colensoi*, is found from Opotiki southwards, and grows to 10 m high with leaves 4–10 cm by 2–5 cm wide.

554   Black mapou flowers, close-up, Erua (November)

556 Tawhirikaro/perching kohuhu, *Pittosporum cornifolium*, is found throughout both islands on rocks and also growing as an epiphyte on rata and other trees. It grows to 2 m high with young hairy branchlets and thin leathery leaves up to 7 cm long and 3 cm wide.

557 Golden-leaved kohuhu, *Pittosporum ellipticum*, is found from Mangonui to the Coromandel Ranges, and has golden tomentum covering young leaves, leaf undersides, petioles and peduncles.

555 Black mapou, flowers and leaves,
Kaimanawa Mountains
(September)

557 Golden-leaved kohuhu, flowers and leaves
clustered round branch tip, Piha (September).
Note the golden tomentum on the young leaves.

558 **_Pittosporum anomalum_** is a dwarf divaricating
shrub found around Waitomo, the Volcanic Plateau
and Arthur's Pass. Leaves are pinnatifid, 5–10 mm
long and aromatic. The flowers, 6–8 mm across, are
mostly solitary and terminal on branchlets.

559 **Haekaro,** _Pittosporum umbellatum_, is found
from North Cape to Gisborne, as a tree to 8 m high.
Flowers occur in many-flowered umbels from
September to January. The leathery leaves are
5–10 cm long by 2 cm wide.

558 _Pittosporum anomalum_
branch with flowers and
leaves, Otari (September)

556 Tawhirikaro, flowers and leaves, Wilton's Bush
(September)

559 Haekaro flowers, Ponatahi (September)

560 Ralph's kohuhu, *Pittosporum ralphii*, is found along streamsides and forest margins from Thames to Wanganui. Leaves are 7–12 cm long, densely clothed below with buff-coloured tomentum.

561 Dall's pittosporum, *Pittosporum dallii*, is found only near Boulder Lake on the Cobb Ridge and in isolated spots in north-west Nelson. It has coarsely serrate leaves up to 10 cm long by 3 cm wide. Sweet-scented flowers occur from October to January.

561   Dall's pittosporum flowers, Cobb Ridge (November)

560   Ralph's kohuhu flowers, Hukutaia Domain (November)

562–563 Monoao, *Halocarpus kirkii*, is a handsome canopy tree, 25 m high, not unlike a kauri when seen in the distance, and easily recognised by its juvenile foliage. Found occasionally in lowland forest to montane forests from Hokianga Harbour to the southern Coromandel Ranges and on Great Barrier Island. The female cone (fig. 562) is sometimes twinned and matures by April; the male cone (fig. 563) is mature in December.

PODOCARPACEAE

562   Monoao, female cone, Mt Kaitarakihi (April)

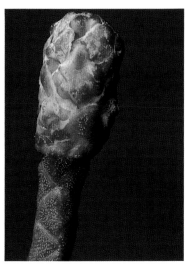

563   Male cone of monoao close up, Otari (December)

564 Kawakawa female flowers, close up, Waikanae (November)

565 Ripe fruit of kawakawa, Karaka Bay (February)

566 Kawakawa, male spike shedding pollen, Karaka Bay (September)

564–566 Kawakawa/pepper tree, *Macropiper excelsum*, is an aromatic shrub or small tree found in the undergrowth of lowland forests from North Cape to Westland and on the Chatham Islands. The zigzag branches, swollen at the nodes, bear broad leaves, 5–10 cm long by 6–12 cm across; the flowers and fruits are on spikes, 2.5–7.5 cm long, that arise all through the year. **PIPERACEAE**

567 Hangehange flowers, close up, Wilton's Bush (October)

568 Green berries of hangehange, Akatarawa Saddle (February)

567–568 Hangehange/whangewhange/Maori privet, *Geniostoma ligustrifolium*, forms a bushy shrub or small tree, with brittle branches and shining pointed leaves, 7–9 cm long by 3–4 cm wide on petioles 10 mm long. Flowers (fig. 567), each about 5 mm across, arise in axillary cymes from September to November, and the berries, produced in profusion, persist from November to March or even as late as June (fig. 568). **LOGANIACEAE**

569–571 Karapapa, *Alseuosmia pusilla* (figs 570–571) and **Northern karapapa,** *Alseuosmia banksii* (fig. 569), are both shrubs up to 2 m high found in undergrowth of lowland and montane forests from North Cape to Marlborough, generally in dense shaded areas. The leaves are of variable shape, 6–20 cm long, crenulated or toothed. The very strongly sweet-scented, trumpet-shaped flowers (fig. 570), 4 cm long, arise on drooping stalks beneath the leaves from April to November, and the berries (up to 9 mm long) ripen during February and March.   ALSEUOSMIACEAE

569   Northern karapapa flowers and leaves, near Waipoua (April)

570   Karapapa flowers and leaves, Mt Holdsworth (June)

571   Karapapa berries, Mt Holdsworth (February)

572   Flowers of rata vine, New Plymouth (May)

572   Rata vine, *Metrosideros fulgens*, is a stout-stemmed liane found in lowland forests from the Three Kings Islands south to Westland. The thick leaves, 3.5–6 cm long by 10–25 mm wide, are on stout petioles, and the flowers occur in terminal cymes from February to June.   MYRTACEAE

573   Large-leaved white rata, *Metrosideros albiflora*, is one of several white-flowering rata vines, with flowers occurring from November to March. *M. albiflora* has smooth, leathery leaves, 3.5–9 cm long by 2–3.5 cm wide, and is found in lowland forests, mostly kauri forests, in Northland.
                                                    MYRTACEAE

574 Hairy-leaved white rata, showing leaves and flowers, Kaitoke Gorge (November)

574 Hairy-leaved white rata, *Metrosideros colensoi*, is a slender stemmed liane with leaves 15–20 mm long by 7–10 mm wide, the young leaves very hairy. It is found in lowland and coastal forests from Mangonui to Marlborough. Flowers arise as terminal and lateral small cymes from November to January. MYRTACEAE

575 Small-leaved white rata, French Pass (January)

575 Small-leaved white rata, *Metrosideros perforata*, is usually a slender liane, but if exposed without support, this plant becomes a bushy shrub with entangled branches. Found in lowland forests, coastal forests and along forest margins from the Three Kings Islands to Banks Peninsula. The close-set, leathery leaves are 6–12 mm long by 5–9 mm wide. Flowers occur in profusion from January to March; they are normally white but pinkish and yellowish forms do occur. MYRTACEAE

576 Small white rata spray with leaves and flowers, Lake Pounui (November)

573 Large-leaved white rata flowers, Mt Kaitarakihi (November)

576 Small white rata, *Metrosideros diffusa*, is a slender liane found in lowland forests from Northland to Nelson and Marlborough. The subsessile leaves, 7–15 mm long by 3–8 mm wide, are hairy when young, and the white or pinkish flowers occur from October till January. MYRTACEAE

577   Shrubby rata flower, Canaan Track, Abel
Tasman National Park (October)

577   Shrubby rata, *Metrosideros parkinsonii*, is a
straggly shrub to 7 m high, with four-sided
branchlets bearing leathery leaves, 3.5–5 cm long by
15–20 mm wide. The large, rounded flowers are produced in abundance along the branches, below the
leaves, from October to January. Found in lowland
and subalpine forests of the North Island, Great
Barrier Island and in the mountains of the Nelson
region.   MYRTACEAE

578–579   Rata/northern rata, *Metrosideros
robusta*, forms a large tree, to 30 m high, with four-
sided branchlets bearing leathery leaves, 2.5–5 cm
long by 1.5–2 cm wide, and large flowers (fig. 579)
produced in profusion on the canopy (fig. 578) from
November to January. Rata is found in lowland and
subalpine forests from the Three Kings Islands to
Nelson.   MYRTACEAE

578   Northern rata in flower,
Lake Waikaremoana
(January)

580   Flower of yellow-flowered
southern rata, Denniston
(January)

579   Flowers, close up, of northern rata, Pahiatua
(December)

580–581   Southern rata, *Metrosideros umbellata*,
is a tree to 15 m high, with rounded branches
bearing narrow, thick, silky leaves, 5–7.5 cm long.
Flowers (fig. 581) occur in profusion all over the
tree from November to January. Yellow-flowered
forms of this tree (fig. 580) occur sparingly at Otira
and in Westland.   MYRTACEAE

582–584 **Broadleaf/kapuka/papauma,** *Griselinia littoralis*, and **Puka,** *Griselinia lucida*, have similar leaves and flowers, but whereas kapuka forms a round-headed tree to 15m high, puka starts life as an epiphyte on another tree, sending roots down to the ground to finally establish it as a separate tree. Both are found in lowland forests to subalpine scrub throughout New Zealand. The thick, leathery, glossy leaves of kapuka are normally equal sided at their bases (fig. 582), but in puka they are very lopsided. The tiny flowers, each 2–4 mm across (fig. 584), occur in panicles from October to December. The fruits (fig. 583), each up to 10 mm long, are black when mature and remain on the tree from December to August.   CORNACEAE

583   Puka spray showing berries and leaf shape, Woodside Gorge (April)

582   Kapuka spray showing flowers and leaves, Canaan Track, Abel Tasman National Park (November)

584   Kapuka flowers, close up, Renata Track, Tararua Range (November)

581   Spray of southern rata with flowers and leaves, Tautuku Beach (December)

585-588 Kaikomako, *Pennantia corymbosa*, is a small tree to 10 m high, which passes through a shrubby juvenile stage and has thick, leathery, sinuate-margined leaves, 5–10 cm long by 1–4 cm wide, on 10 mm long petioles (fig. 588). Found in lowland forests throughout New Zealand. Strongly scented flowers (fig. 586, male; fig. 587, female) occur in panicles (fig. 585), 4–8 cm long, from November to February. The drupes (fig. 588) ripen from February to May.   ICACINACEAE

585 Kaikomako showing panicles of flowers at branch tips, Lake Pounui (December)

586 Close-up of male kaikomako flowers, Lake Pounui (December)

588 Kaikomako drupes at the branch tips, Woodside Gorge (March)

587 Close-up of female kaikomako flower, Lake Pounui (December)

589-593 Kowhai, *Sophora microphylla, Sophora tetraptera, Sophora prostrata*, are all trees varying from 2 m to 12 m in height. The kowhai flower is New Zealand's national flower blooming from July to October, starting earlier in the north, later in the south and at higher altitudes. *S. microphylla* (fig. 589) occurs naturally throughout New Zealand, *S. tetraptera* (fig. 590) in the North Island from East Cape to the Wairarapa, both in lowland open forest, along forest margins and river banks,and in open damp and rocky places; *S. prostrata* (fig. 593) is found only in the South Island, mainly in open rocky places. All have feathery leaves, *S. microphylla* with 20–40 pairs of leaflets, *S. tetraptera* with 10–20 pairs and *S. prostrata* with 4–5 pairs.   FABACEAE

589 Kowhai flowers, *Sophora microphylla*, Peel Forest (October)

590 Kowhai flowers, *Sophora tetraptera*, Taupo (October)

591 Kowhai flowers, *S. tetraptera*, massed along a branch, Ponatahi (September)

592 Seeds of kowhai, *S. microphylla*, Woodside Gorge (January)

593 Kowhai flowers, *Sophora prostrata*, Woodside Gorge (September)

594-596 Rohutu, *Lophomyrtus obcordata,* is a spreading shrub to 5 m high, with rounded leaves, 5-10 mm across. Found in lowland forests from Mangonui southwards through both islands. The flowers (figs 594 & 596), about 6 mm across, occur from December to February, and the berries (fig. 595) ripen from April through May.

MYRTACEAE

595   Rohutu berries, Rotorua (May)

594   Rohutu flowers and leaves, Rotorua (December)

596   Rohutu, close-up of flower and leaves, Barton's Bush (December)

597   Pate flowers and leaves at branch tip, Akatarawa Saddle (February)

597-599 Pate/patete/five finger, *Schefflera digitata,* is a shrub or small tree to 8 m high, bearing compound leaves on petioles up to 25 cm long, and each made up of 3-9 thin leaflets, up to 20 cm long, with serrated margins. Found in lowland forests throughout New Zealand, usually in dampish places. The flowers, each about 7 mm across (fig. 599), arise in large, drooping panicles (fig. 597) below the leaves from January to March, and the berries (fig. 598) are coloured by March or April of the following year.   ARALIACEAE

600 Scented clematis, *Clematis foetida,* is a liane with strongly lemon-scented flowers produced in profusion from September to November. The plant sprawls over shrubs along lowland forest margins throughout New Zealand. The leaves are three foliolate with the leaflets sinuate or wavy margined.

RANUNCULACEAE

598   Pate berries, Lake Pounui (April)

601   Hinau, *Elaeocarpus hookerianus*, is a branching canopy tree, 15 m high, with leathery leaves, 10–12 cm long by 10–30 mm wide, having a silky tomentum on their undersides. Flowers similar to pokaka occur in great profusion from October to February, and the drupes, 8–12 mm long, mature from December to March but persist on the tree till May. Hinau is found in lowland forests from North Cape south to about Tuatapere.

ELAEOCARPACEAE

601   Hinau flowers, close up, Otari (November)

599   Close-up of pate flower, Akatarawa Saddle (February)

602   Westland quintinia, *Quintinia acutifolia*, is a bushy tree, to 12 m high, with wavy-margined leaves, 6–16 cm long by 3–5 cm wide, on 2 cm long petioles. Flowers, each about 6 mm across, occur in racemes during October and November. Found in lowland forests on both Little and Great Barrier Islands, around Coromandel Ranges and National Park to Taranaki and from Collingwood to Hokitika. Seeds ripen during December and January.

ESCALLONIACEAE

600   Scented clematis flowers with leaves, Hinakura (October)

602   Westland quintinia showing leaves and flowers, Lake Kanieri (November)

603 Tawherowhero flowers, close up, Kauaeranga Valley (December)

603–605 Tawherowhero, *Quintinia serrata*, is a small, open-branching tree to 9 m high with narrow, often blotchy, coarsely serrate leaves, 6–12.5 cm long by 10–25 mm wide, usually wavy margined and on petioles 2 cm long. Found in lowland forests from Mangonui south to about Taumarunui. Flowers (fig. 603), in racemes (fig. 605), occur from October to December, with seeds (fig. 604) maturing from January to March. ESCALLONIACEAE

604 Tawherowhero, newly set seeds, Kauaeranga Valley (December)

605 Tawherowhero spray with leaves and flower racemes, Kauaeranga Valley (December)

606 Northern forest hebe, *Hebe diosmifolia*, is a much-branched shrub, 1–6 m high, found in lowland forests and scrub in Northland. Leaves are 10–30 mm long by 3–6 mm wide, and the flowers arise in large corymbose heads, either white or lavender coloured, each flower about 8 mm across, during September and October. SCROPHULARIACEAE

606 Northern forest hebe, flowers and leaves, Mangamuka Gorge (October)

607 Pokaka flowers, close up, Hinakura (November)

608 Pokaka branches heavy with
flowers, Hinakura (November)

609 Pokaka branch with drupes
and leaves, Hinakura
(February)

607–609 Pokaka, *Elaeocarpus dentatus*, is a
branching, rounded canopy tree, up to 12 m high,
with leathery, serrate-margined leaves, 3–11 cm long
by 10–30 mm wide. Flowers (fig. 607), each about
10 mm across, in drooping 8–12-flowered racemes,
are borne in great profusion (fig. 608) from October
to January. The purple-coloured drupe (fig. 609),
8 mm long, is mature from November to March.
Found in lowland forests from Mangonui south to
Stewart Island.   ELAEOCARPACEAE

610–612 Mahoe/whitey-wood, *Melicytus rami-
florus*, is a small, branching tree, to 10 m high, with
serrated leaves, 5–15 cm long and 3–5 cm wide.
Flowers, 3–4 mm across, arise on 10 mm long
pedicels as fascicles from the leaf axils or directly
from the branchlets (figs 610 & 612). The violet berries, 4–5 mm long, form clusters along the branches
(fig. 611) from November to March. Mahoe is found
in lowland and montane forests throughout New
Zealand and the Kermadec Islands.   VIOLACEAE

610 Mahoe flowers cluster thickly around a branch,
Karaka Bay (December)

611 Mahoe berries on branches with leaves, Karaka
Bay (February)

612 Mahoe flowers, close up, Karaka Bay
(December)

613 Large-leaved whitey-wood
with berries, Waipoua Forest
(March)

614 Spray of large-leaved whitey-
wood with flowers,
Waitakere Ranges (March)

615 Mahoe-wao showing branch
heavy with berries and leaves,
Waipunga Gorge (April)

613–614 Large-leaved whitey-wood, *Melicytus
macrophyllus,* is a small tree to 6 m tall, with large,
leathery, serrated leaves, up to 20 cm long and 10 cm
wide, on petioles 2 cm long. Flowers (fig. 614),
6–7 mm across, and berries (fig. 613) are more sparse
than in mahoe and occur round the same periods.
Found in lowland to montane forests from Mangonui
to Opotiki.   VIOLACEAE

615–619 Mahoe-wao, *Melicytus lanceolatus,* is a
slender shrub or a small tree, 5–6 m high, with
narrow, serrate leaves, 5–16 cm long by 5–30 mm
wide. Flowers (fig. 616) arise as in mahoe from June
to November, each about 5 mm across, and vary in
colour from a rich yellow to a dark blackish brown
(figs 617–619). Berries (fig. 615) are similar to those
of mahoe and ripen from July to February, and can
persist to the next flowering time. Found in lowland
and montane forests and along forest margins from
Cape Brett southwards throughout New Zealand.
                                               VIOLACEAE

616 Mahoe-wao, branch with
dense flowers, Hauhungaroa
Range (September)

617 Flowers of mahoe-wao, close up, showing colour
variation, Mt Ruapehu (October)

618   Close-up of yellow flower of mahoe-wao, Otari
(September)

619   Close-up of flowers of mahoe-wao showing
stalks and colour variation, Mt Ruapehu
(October)

622   Dwarf bush nettle in flower, Wilton's Bush,
Wellington (January)

622   Dwarf bush nettle, *Urtica incisa,* is a small
herb, about 45 cm high, found in lowland and montane forests, along forest margins and in shaded open
places throughout New Zealand. Leaves, up to 5 cm
long, are on petioles up to 7 cm long, and bear
occasional stinging hairs. Flowers, shown here, occur
from September to February.   URTICACEAE

620   Dianella/turutu/blue-berry/ink-berry, *Dianella*
nigra, forms a lily-like plant growing to about 1 m
high found in moist and dry forest and on dry,
forested or scrub covered hillsides throughout New
Zealand. Leaves are 25–60 cm, but sometimes up to
100 cm, long, by 10–15 mm wide. The plant is noted
for its magnificent blue berries that arise from
December till March from small inconspicuous
whitish flowers with yellow stamens.

PHORMIACEAE

620   Dianella plant with berries, Lake Pounui
(February)

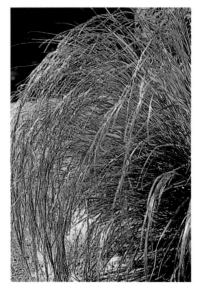

621   Bamboo grass in flower,
Otari (January)

621   Bamboo grass, *Oryzopsis rigida,* is a graceful
forest grass to 1 m high, found in lowland forests
and along forest margins throughout New Zealand.
Flowers occur in profusion during January.

GRAMINEAE

623–626 Turepo/milk tree, *Paratrophis micro-*
phylla, forms a small tree, to 12 m high, found in
lowland forests throughout New Zealand. Minute
male flowers, 1–2 mm across, occur as spikes
10–20 mm long from October to February (fig. 624).
Fig. 625 shows a minute female flower, about 2 mm
across, which occurs on dense spikes up to 3 cm long
from August to October. The flower-spikes of turepo
and of towai are often affected by a fungal disease,
which produces flower-like growths as shown in fig.
626. The leaves (fig. 623), 8–25 mm long by
5–12 mm wide, are crenate margined and on stout
petioles 5 mm long. A white milky sap exudes from
cut branches, leaves or twigs. From November to
March the tree is adorned by brilliant red drupes (fig.
623).   MORACEAE

625   Female flowers of turepo, Waiorongomai
       (November)

623   Turepo branch with drupes and leaves,
       Waiorongomai (February)

626   Fungus disease affecting flowers of turepo,
       Wairongomai (November)

627 Towai/large-leaved milk tree, *Paratrophis*
banksii, forms a spreading tree, to 12 m high, found
in lowland forests from Mangonui south to the Marlborough Sounds. Except for the large crenate-
margined, distinctly veined leaves, 3.5–8.5 cm long
by 2–3.5 cm wide, on stout 10 mm long petioles,
towai is very similar to turepo. Fig. 627 shows the
brilliant drupes occuring from February to April.
                                         MORACEAE

624   Male flowers of turepo, Waiorongomai
       (November)

627   Towai drupes, Waiorongomai (February)

628 Puka flower-head, Waikanae (May)

629 Puka flowers, close up, Waikanae (May)

630 Ripening puka fruits, Lake Pounui (March)

628–630 Puka, *Meryta sinclairii,* forms a handsome, round-headed tree, up to 8 m high, with large, glossy, thick, leathery leaves, 30–50 cm long by 15–25 cm wide on stout petioles 25–35 cm long. Flowers (fig. 628) arise as terminal panicles to 50 cm long from February to May. The succulent fruits are black when they mature 11–12 months later. Found on the Hen and Chickens and Three Kings Islands but now cultivated and grown in parks and gardens, sometimes also as street trees. ARALIACEAE

631–632 Raukumara, *Brachyglottis perdicioides*, is a shrub to 2 m high found in lowland forests from East Cape to Mahia Peninsula. Thin coarsely serrated leaves are up to 5 cm long, and the strongly scented flowers (fig. 631), each to 10 mm across, occur from November to January. A hybrid formed by crossing with *Brachyglottis hectori* is grown in gardens and known as *Brachyglottis* 'Alfred Atkinson' (fig. 632). ASTERACEAE

631 Raukumara in flower, Mahia (November)

632 *Brachyglottis* 'Alfred Atkinson' in flower, Otari (December)

633   Maire spray with drupes, Hinakura (October)

634   Female flowers of maire, Hinakura (September)

633–636 Maire/black maire, *Nestegis cunning-hamii,* is a forest canopy tree to 20 m high, with narrow, leathery, rough-to-touch, willow-like leaves on stout petioles about 10 mm long. Minute flowers (fig. 636 shows a female) occur as racemes (fig. 634), 10–25 mm long, from September to December, and red or yellow drupes (figs 633 & 635) are mature from September to October of the following year. Maire is found in lowland forests from North Cape to Marlborough and Nelson.   OLEACEAE

635   Yellow drupes of maire, close up, Mt Ruapehu (November)

636   Close-up of female flower of maire, Hinakura (September)

637   White maire drupes, Waipunga Gorge (March)

638   White maire flowers, Waipunga Gorge (December)

639-640 Bush pohuehue, *Muehlenbeckia australis,* is a scrambling vine that can completely cover trees, shrubs or rock faces in lowland forests throughout New Zealand. Leaves are 3–10 cm long, and the small flowers, each about 4 mm across, and the seeds (fig. 640) appear in panicles almost the year round.

POLYGONACEAE

639   Bush pohuehue in flower, Lake Pounui (January)

640   Bush pohuehue with seeds, Lake Pounui (April)

642   Close-up of mairehau flower, Kauaeranga Valley (December)

641   Mairehau flower-head and leaves, Rotorua (December)

637-638 White maire, *Nestegis lanceolata,* is a similar tree to black maire but smaller, to 15 m high, with smooth, leathery leaves, glossy above, 5–12 cm long by 10–35 mm wide. Minute flowers (fig. 638) occur as racemes about 2 cm long from November to January, and the drupes (fig. 637), which are more attenuated than those of black maire, mature about one year later.   OLEACEAE

641-642 Mairehau, *Phebalium nudum,* is an aromatic, branching shrub or small tree, to 3 m high, found in lowland forests from Mangonui to about Waihi. The narrow, finely crenulate, gland-dotted, leathery leaves, 2.5–4.5 cm long by 5–10 mm wide, are on short, twisted petioles. Flowers arise as corymbs (fig. 641), 5 cm across, with each flower 5–10 mm across (fig. 642), from October to December.   RUTACEAE

643–645 Tanekaha/celery pine, *Phyllocladus tri-chomanoides,* is a forest tree, to 20 m high, with the cladodes, which replace the leaves, arranged pinnately in two rows on rachides that arise in whorls along the branches. The cladodes, shaped like a celery leaf, are 10–25 mm long. Male catkins arise in bunches of 5–10 at the tips of the branches (fig. 645), and female cones arise arranged around the outsides of the cladodes (fig. 643), both from October to January. The seeds (fig. 644) take at least six months to mature. Found from North Cape to Nelson and Marlborough.   PODOCARPACEAE

643   Tanekaha, female cones at tip of branch, last season in outer ring, present season in inner group, Otari (October)

645   Mature male cones of tanekaha, Otari (October)

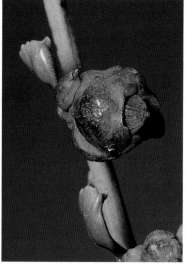

644   Mature female cone of tanekaha, Otari (April)

646   Kohurangi flowers and leaves, Mangamuka Gorge (January)

646 Kohurangi/kohuhurangi, *Urostemon kirkii,* grows either as an epiphyte or on the ground as a shrub or small tree, to 3 m high, in lowland and montane forests, more commonly from North Cape to the Coromandel Ranges but also sparingly throughout the North Island. The soft, fleshy, variable-shaped leaves, 4–10 cm long by 2–4 cm wide, are on petioles 10 mm long. Flowers, each about 5 cm across, occur in corymbs up to 30 cm across during December and January.

ASTERACEAE

647–650 Toatoa, *Phyllocladus glaucus,* is a tapering tree to 15 m high, with large, wedge-shaped, leathery cladodes (fig. 647), 4–6 cm long by 2–4 cm wide. Male cones (figs 648–649) arise in clusters on stout stalks at the branch tips (fig. 649), and the females (fig. 647) arise near the bases of the rachides during November and December. The seeds (fig. 650) mature some five to six months later. Found in lowland and montane forests from Mangonui south to a line from Taumarunui to Wairoa.

PODOCARPACEAE

647 Toatoa showing cladodes and ripe seeds, Waipahihi Botanical Reserve (May)

648 Toatoa, mature male cone with young leaf, Otari (November)

649 Toatoa spray with male cones, Otari (November)

651 Rautini with terminal panicle of flowers and rosettes of leaves, Otari (December)

650 Toatoa mature female cones showing protruding seeds, Taupo (May)

651 Rautini, *Senecio huntii*, is a shrub to 6 m high, with its branchlets marked by leaf scars. The long, narrow leaves, 5–10 cm long, occur mainly as rosettes of 20–30 leaves around the tips of the upturned branchlets. Flowers, each about 2 cm across, arise as dense terminal panicles, 12–18 cm high, from December to February. Found only on the Chatham Islands, in forests or on the edges of drier bogs, but now grown in parks and gardens.

ASTERACEAE

652 Black beech heavy with male flowers, Rimutaka Hill (October)

652 Black beech/tawhairauriki, *Nothofagus solandri*, forms a spreading tree, to 25 m high, with tough, leathery leaves, 10–15 mm long by 5–10 mm wide, clothed below with a dense whitish tomentum. In a good season male flowers are produced in profusion (fig. 652) from September to December. Black beech is found in lowland and montane forests from the Rotorua district south to Canterbury and Westland. FAGACEAE

653 Mountain beech spray with male flowers, Volcanic Plateau (December)

654 Mountain beech seeds, Mt Ruapehu (February)

655 *Cyttaria gunnii* fungus on silver beech, Mt Robert Track (January)

655 *Cyttaria gunnii*, sometimes mistaken for a flower, is a fungus disease of beech trees, especially silver beech.

653–654 Mountain beech/tawhairauriki, *Nothofagus solandri* var. *cliffortioides*, is a smaller tree with more pointed leaves (fig. 654) than black beech, their margins rolled under and the lower surfaces pubescent. Male flowers (fig. 653) occur from November to January; female flowers are small and inconspicuous. Seeds (fig. 654) occur from February to April. Found in montane and subalpine forests and subalpine scrub from the Central Volcanic Plateau south but absent from Mt Taranaki and the Tararua Ranges. FAGACEAE

656-658 Silver beech/tawhai, *Nothofagus men-ziesii*, is a tall tree, to 30 m high, with its branches arranged as if in tiers, and with round to oval, thick, leathery, crenate leaves (fig. 656), 6–15 mm long by 5–15 mm wide. Found in lowland and montane forests from Thames southwards, excluding Mt Taranaki. It can occur in subalpine scrub as a shrub or stunted small tree. Male flowers (fig. 656) and female flowers (fig. 657) occur from November to January, and the seeds, typical of beech trees, (fig. 658) mature from January to March.

FAGACEAE

656 Silver beech spray with male flowers, Tararua Forest Park (November)

657 Silver beech, female flowers, Eglinton Valley (December)

659-660 Hard beech/tawhairaunui, *Nothofagus truncata*, is a tall tree, to 30 m high, often with flanged buttresses to the trunk. Leaves (fig. 659) are 2.5–3.5 cm long by 2 cm wide, thick, leathery, smooth and coarsely serrated with a truncate apex. Flowers occur profusely (fig. 659) from September to December, and male flowers are often brilliant (fig. 660) as in *N. solandri*. Found in lowland and montane forests from Mangonui south to Kaikoura and Greymouth but absent from Mt Taranaki.

FAGACEAE

658 Silver beech seeds, Lewis Pass (January)

659 Hard beech spray, showing male flowers and leaves, Kaitoke (October)

660 Close-up of male flowers of hard beech, Woodside Gorge (January)

661   Red beech male flower, close up, Tararua Forest
Park (November)

662   Male flowers and leaves of red beech, Hinakura
(October)

661–663   Red beech/tawhairaunui, *Nothofagus
fusca,* is a tall tree to 30 m high, often buttressed,
with thin, leathery, coarsely and deeply serrate
leaves, 2–4 cm long by 15–25 mm wide, on 4 mm
long petioles (fig. 662). Flowers occur from September to December, male flowers (fig. 661), often
brilliant red, carry huge quantities of pollen (fig.
663). Found in lowland and montane forests from
Thames to Southland but absent from Mt Taranaki.
FAGACEAE

663   Red beech male flowers shedding pollen, Renata
Ridge, Tararua Range (November)

664   Maire spray with leaves and
racemes of flowers, Mt
Meredith, Kaitaia (October)

665   Maire, pale coloured male
flowers close up, Mt
Meredith, Kaitaia (October)

666   Red-coloured female flowers
of maire, Rewa Reserve,
Kaitaia (October)

667 Fruits of maire, *M. salicifolia*, Rewa Reserve, Kaitaia (October)

664-667 Maire, *Mida salicifolia*, forms a slender tree, to 6 m high, which occurs locally in lowland forests from North Cape southwards but more commonly in the north. The leaves, 5-12 cm long by 3-10 mm wide, are entire, somewhat glossy, and usually arise alternately (fig. 664). Flowers, greenish or reddish tinged (fig. 665, male flowers; fig. 666, female flowers), occur from September to November, and the fruits mature from October to February.
SANTALACEAE

668-671 Kamahi, *Weinmannia racemosa*, is a spreading tree, to 25 m high, with coarsely serrate adult leaves, 3-10 cm long by 2-4 cm wide; juvenile trees have compound leaves with three leaflets and can bear flowers. Young leaves (fig. 670) are a rich red as they unfurl. Flowers on adult trees occur plentifully as racemes, to 12 cm long (fig. 668), from November to January, and the preceding reddish buds are very attractive (fig. 671). Found in lowland and montane forests from about Thames southwards.
CUNONIACEAE

670 Young leaves of kamahi, Tararua Forest Park (November)

668 Kamahi flowers, Akatarawa (November)

669 Close-up of kamahi flowers, Akatarawa (November)

671 Flower-buds of kamahi, Kauaeranga Valley (December)

672-674 Tawhero, *Weinmannia sylvicola*, is a small tree, to 15 m high, with persistent juvenile foliage of long, 5-10-foliolate leaves (fig. 673), and juvenile trees flower freely (fig. 673). Adult leaves are simple or 3-5 foliolate, leathery and serrate. Flowers (fig. 674) occur profusely as racemes, 8-12 cm long, from September to January. Tawhero is found in lowland forests from Mangonui south to the Rotorua district.   CUNONIACEAE

673 Adult form of tawhero showing terminal leaves and flowers, Kauaeranga Valley (January)

672 Juvenile form of tawhero showing terminal leaves and flowers, Kauaeranga Valley (January)

674 A small tree of tawhero in full flower, Mangamuka Gorge (January)

675 Kohia vine with fruits and leaves, Hinakura (May)

675-676 Kohia/New Zealand passion vine, *Tetrapathea tetrandra*, is a soft-wooded liane, reaching to 10 m high with alternate, entire, glossy, leathery leaves, 5-10 cm long by 2-3 cm wide, on petioles 2 cm long. Flowers (fig. 676) occur from October to December, and fruits (fig. 675) mature from March to May. Found in lowland forests, often along forest margins, from North Cape to Canterbury and Westland.   PASSIFLORACEAE

676 Kohia vine with flowers, Hinakura (November)

677–679 Bush lawyer/tataramoa, *Rubus cissoides,* is a liane, reaching to 15 m high, with interlacing, prickly branchlets and elliptic, serrate leaves in threes (fig. 677). Flowers arise as panicles up to 60 cm long (fig. 677), with male flowers (fig. 678) and female flowers (fig. 679) on separate lianes from September to November. Found in lowland and montane forests throughout New Zealand.   ROSACEAE

678 Bush lawyer, male flower and flower-buds, close up, Tararua Forest Park (November)

677 Bush lawyer in flower, Tararua Forest Park (November)

679 Bush lawyer, female flower, close up, Tararua Forest Park (November)

680 Tangled prickly branchlets of leafless lawyer, Woodside Gorge (January)

681 Fruits of leafless lawyer, Upper Ure River (March)

680–681 Leafless lawyer, *Rubus squarrosus,* is a liane that forms tangled masses of more or less leafless, prickly branchlets, the prickles recurved and yellow (fig. 680). It is found in lowland and montane forests and open, rocky places near forests throughout New Zealand. Flowers occur in panicles or racemes to 20 cm long from September to November, and the fruits (fig. 681), typical of lawyer fruits, mature from November to April. ROSACEAE

682   Puriri flowers and leaves, Wellington (April)

682–685   Puriri, *Vitex lucens*, is a massive spreading tree, to 20 m high, with four-angled branchlets and leaves made up of 3–5 glossy, undulate, leathery leaflets, each 5–12.5 cm long by 3–5 cm wide, with conspicuous veins, the whole on stout petioles to 10 cm long. Flowers arise in panicles of 4–15 each, and these and the drupes, about 2 cm in diameter, occur all the year round, with the most profuse flowering from June to October. Puriri grows naturally in lowland and coastal forests from North Cape to about Mahia and New Plymouth but is now extensively grown in parks and as a street tree.

VERBENACEAE

683   Puriri flowers, close up, Wellington (April)

684   Puriri berries, Wellington (February)

685   A puriri leaf, Wellington (March)

686–687 Tecomanthe, *Tecomanthe speciosa,* is a woody liane found naturally only on the Three Kings Islands but now grown in parks and gardens in frost-free districts. Flowers (fig. 687) occur from May to August, and fig. 686 shows the glossy, 3–5-foliolate leaves. BIGNONIACEAE

686 Tecomanthe leaves, Hukutaia Domain (April)

688 Oro-oro male flowers after shedding pollen, Waipunga Gorge (October)

687 Tecomanthe flowers, Waikanae garden (June)

688–689 Oro-oro/narrow-leaved maire, *Nestegis montana,* is a small, much-branching, round-headed tree, 10–15 m high, found in lowland, montane and subalpine forests from Mangonui to Nelson and Marlborough. The very narrow-lanceolate leaves, 3.5–9 cm long by 6–9 mm wide, are leathery, slightly glossy, with a raised midrib. Slender racemes of flowers (fig. 688, male flowers) arise along the branches (fig. 689) from October to January, and the fruits ripen from December to March. OLEACEAE

689 Oro-oro spray, showing narrow leaves and flower racemes, Waipunga Gorge (October)

PERCHING AND PARASITIC PLANTS

690–691 Kiekie/tawhara (edible bracts)/**ureure** (fruit), *Freycinetia banksii*, is a shrub that climbs on standing trees and fallen logs, with aerial roots that cling into fissures and crevices in the host bark. It sometimes grows on rocky ground or near water and is found from North Cape to Westland. The leaves, 1.5 m long by 2.5 cm wide, are spirally arranged, with the flowers (fig. 691) and fruits (fig. 690) hidden among the bases of the apical leaves. Flowers occur from September to November and fruits from January to February, ripening by May.   PANDANACEAE

691   Kiekie flowers, Lake Pounui (October)

690   Kiekie fruits set among whorls of leaves, Lake Pounui (April)

693   Kowharawhara female flower panicle, Mt Holdsworth (February)

692   Kowharawhara male flower panicle, Mt Holdsworth (February)

692–694 Kowharawhara/perching lily, *Astelia solandri*, is an epiphyte, mostly growing on branches and trunks of forest trees but occasionally on the ground. The drooping leaves, 1–2 m long by 2–3.5 cm wide, have three subequal nerves on either side of the midrib. Flowers in drooping panicles 15–40 cm long, males (fig. 692) yellowish white or maroon coloured, females (fig. 693–694) yellowish white to greenish, occur from October to June, with fruits present all the year round. ASPHODELACEAE

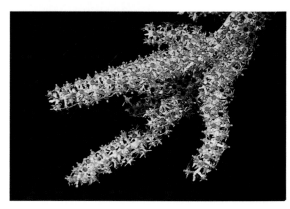

694   Close-up of female flowers of kowharawhara, Mt Holdsworth (February)

695-698 Kahakaha/perching lily, *Collospermum hastatum*, forms huge clumps on trees (fig. 695) or on rocks, and is generally similar to *A. solandri*. The leaves are 60 cm-1.7 m long by 3-7 cm wide, and the panicles of flowers, 15-30 cm long, appear from January to March, each extending out from a fan of leaves (fig. 696, female). Male flower panicles are similar, and male flowers are shown close up in fig. 698. The fruits (fig. 697) ripen from March till August, ultimately turning red.   ASPHODELACEAE

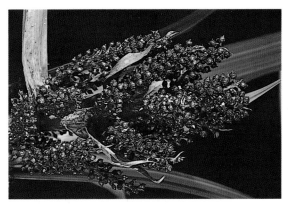

696 Female flower panicle of kahakaha, Lake Pounui (November)

695 Kahakaha plant on tree trunk, Waiorongomai (December)

697 Kahakaha fruits, Lake Pounui (October)

698 Close-up of male flowers of kahakaka, Lake Pounui (October)

699 Wharawhara/shore astelia, *Astelia banksii*, is found on the forest floor of lowland and coastal forests of the North Island. It is similar to *A. solandri* in general appearance, with the leaves displaying several nerves on either side of the midrib and ascending for the lower half then drooping in the upper half. Flowers occur in panicles, up to 50 cm long, from March to June, with fruits (fig. 699), black when mature, being present the year round.   ASPHODELACEAE

699 Wharawhara with panicle of ripe fruits, Lake Pounui (January)

700–701 **Tree orchid,** *Dendrobium cunninghamii,* is found throughout New Zealand on trees in lowland forests and scrub and occasionally on rocks. The long, thin, wiry, drooping stems bear narrow leaves, 3–5 cm long and 3 mm wide, and the strongly sweet-scented flowers, 2–2.5 cm across, arise in groups of 1–56 during December and January (fig. 701). The plant shown in fig. 700, growing on a cabbage tree in the south Wairarapa, was the largest *Dendrobium cunninghamii* I have ever seen, being almost 2 m high, but unfortunately it was destroyed during the 'Wahine storm' of 1965.

ORCHIDACEAE

700   Tree orchid in flower, Waiorongomai (December)

701   Tree orchid flowers, close up, Waiorongomai (December)

702 **Hanging tree orchid/peka-a-waka,** *Earina mucronata,* has drooping stems, 30 cm to 1 m long, bearing narrow, leathery, finely ribbed leaves up to 7.5 cm long. Sweet-scented flowers, about 6 mm across, occur as long drooping sprays during October and November. Found in lowland forests throughout New Zealand, preferably in shaded, somewhat damp places.

ORCHIDACEAE

702   Hanging tree orchid showing flowers and leaves, Hinakura (November)

703   Flowers of raupeka, Lake Pounui (February)

704-705 Red mistletoe/pirirangi, *Elytranthe tetra-petala,* is a bushy, branching, parasitic shrub, up to 1 m high and 2 m across (fig. 704), found growing on beech and *Quintinia* in lowland and montane forests from Dargaville southwards. The red flowers and buds (fig. 705) make a brilliant display from October to January.   LORANTHACEAE

706   Korukoru spray with flowers and leaves, Boulder Lake (December)

706 Korukoru, *Elytranthe colensoi,* is a smaller parasitic plant found on *Nothofagus* in the South Island, where it is more common, and on pohutukawa and *Pittosporum* in the North Island. The flowers occur from November to February and differ from those of *E. tetrapetala* by having their petals curled back on themselves.

LORANTHACEAE

704   Red mistletoe mass on a beech tree, Mt Ruapehu (January)

705   Red mistletoe flowers, close up, Mt Ruapehu (January)

707   Common mistletoe spray with berries, Ure River Gorge (April)

703 Raupeka/Easter orchid, *Earina autumnalis* is a stout-stemmed, mostly erect, but occasionally drooping, orchid, found on trees and also on the ground in lowland forest throughout New Zealand. It has broader leaves than peka-a-waka and reaches 15–45 cm high. Sweet-scented flowers, 8 mm across, are produced at the tips of the stems from February to April.   ORCHIDACEAE

707 Common mistletoe, *Loranthus micranthus,* is a bushy shrub attached to its host by a ball-like mass and found on many species of trees and shrubs in lowland forests and scrub throughout New Zealand. The small, greenish flowers are about 2.5 mm long and occur from October to December, being replaced from December to April by the conspicuous yellow berries shown in fig. 707.

LORANTHACEAE

PLANTS OF DRY RIVER BEDS

Many plants grow in these situations, forming mats and bushes that provide humus upon which, in time, other plants may grow.

708 Scabweeds The stone-strewn dry river and stream beds with their adjacent stony flats are normally covered by an assemblage of dense, mat-forming plants called scabweeds. These plants are often very beautiful and impart a splash of colour (fig. 708) to an otherwise drab, grey environment. They are also found in alpine dry, rocky places and screes. Except where indicated, they all belong to the family Asteraceae.

708   Scabweeds, Bealey River bed (January)

709 Mat daisy, *Raoulia parkii*, is a densely matted, prostrate, creeping and rooting daisy, found in South Island dry, rocky places. Flowers, 3–4 mm across, are produced in abundance during December and January.

709   Mat daisy in flower, Bealey River bed (January)

710 Golden scabweed, *Raoulia australis*, is found from the Tararua Range to Otago from sea-level to 1,000 m; it flowers in December and January.

711   Mossy scabweed with flowers, Bealey River bed (January)

711 Mossy scabweed, *Scleranthus uniflorus*, is a moss-like perennial herb found throughout the South Island in grasslands as well as in dry, rocky places. Tiny flowers occur from November to January.

CARYOPHYLLACEAE

713 Common scabweed, *Raoulia hookeri*, forms dense mats, up to 1 m across, varying in colour from green to grey, and is found in riverbeds from East Cape southwards. Flowers, 5–7 mm across, arise during December and January.

712 Tutahuna/green scabweed, *Raoulia tenuicaulis*, forms dense, silvery green to bright green mats, up to 1 m across, in gravel riverbeds and herbfields throughout New Zealand. The leaves are thick, 5 mm long by 2 mm wide, tapering to a pointed apex. White flowers, 6 mm across, arise from November to January, followed by silky seeds.

710   Golden scabweed in flower, Bealey River bed (January)

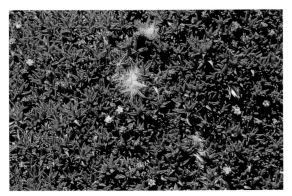

712   Tutahuna with flowers and seeds, Mt Taranaki (November)

713   Common scabweed in flower, Bealey River bed (January)

714   Plateau scabweed, *Raoulia hookeri* var. *albosericea,* is a greyish coloured scabweed, 60–70 cm across, flowering during December–January at the tips of erect branches with closely imbricate leaves. It grows on the Volcanic Plateau, Tararua Range and Ruahine Range.

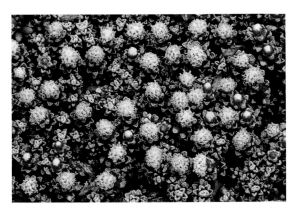

714   Plateau scabweed in flower, Mt Ruapehu (December)

715   Green mat daisy, *Raoulia haastii,* forms dense, green cushions, up to 1 m across and 30 cm high, and is found on the eastern side of the Southern Alps on riverbeds and river flats.

715   Green mat daisy, Bealey River bed (January)

716–717   Wild Irishman/matagouri/tumatakuru, *Discaria toumatou,* is a much-branched, stiff, spiny shrub (fig. 716), to 5 m high, found in dry riverbeds, open rocky places and sand dunes from coastal to subalpine regions throughout New Zealand. Flowers (fig. 717), 3–5 mm across, occur from October to January, followed by rounded berries from December to March.   RHAMNACEAE

716   Matagouri spray showing small leaves, flowers and large thorns, Thomas River (December)

717   Close-up of matagouri flowers, Otari (October)

PLANTS OF DRY SCREES, ROCKY AND STONY PLACES

Many unusual and beautiful plants grow in dry, rocky or stony places in New Zealand, from about sea-level to high in the mountains. Often being exposed in precipitous situations, these are mostly very hardy plants.

718–719 **Leafless clematis,** *Clematis afoliata,* forms large masses of tangled stems, up to 1 m high, on open rocky ground from Hawke's Bay to Southland from sea-level to 1,000 m. Flowers, 1–2 cm across (fig. 718), in fascicles of 2–5 flowers each, occur from September to October, and the seed (fig. 719) sets during November and December.

RANUNCULACEAE

718   Leafless clematis in flower, Woodside Gorge (October)

719   Leafless clematis seeds, Otari (November)

721–722 **Creeping lawyer,** *Rubus parvus,* is a prostrate, creeping, rooting bramble with characteristic red bark and leaves 2.5–9 cm long. Flowers, 2–5 cm across, occur from September to November, and the juicy, edible fruits are ripe from November to April. Found in Nelson and Westland on stony ground and the stony banks of river valleys from about sea-level to 900 m.   ROSACEAE

720   Creeping pohuehue seed in capsule, Haast Pass (January)

720 **Creeping pohuehue,** *Muehlenbeckia axillaris,* is a prostrate, spreading, tangled shrub, forming patches, up to 1 m across, in subalpine rocky places, riverbeds and grasslands throughout New Zealand. Flowers, 3–4 mm across, occur singly and in pairs from November to April, filling the evening air with a rich, sweet scent. The black seeds, each about 3 mm long, sit in a white cup and are mature from December onwards.   POLYGONACEAE

721   Creeping lawyer, showing leaves and flower, Boulder Lake (November)

723   Porcupine plant with berries, Otari (February)

722   Creeping lawyer with fruits, Boulder Lake (February)

724   Close-up of flowers and leaves of porcupine plant, Otari (September)

723–725 Porcupine plant, *Melicytus alpinus,* is a stiff shrub, to 60 cm high and 1 m across, with rigid, interlacing branches bearing many lenticels and each terminating in a spine. Leaves are 6–18 mm long, and the tiny bell-shaped flowers occur in November, hanging in thousands along the branches. White berries, 5 mm across, ripen during February. This curious plant occurs east of the Southern Alps in exposed rocky places between 600 and 1,300 m.

VIOLACEAE

725   Porcupine plant, showing interlacing branches and flowers, Otari (October)

726   Leafless porcupine plant in flower, Otari (November)

726   Leafless porcupine plant, *Melicytus angusti-folius*, is similar to *M. alpinus* but grows to 1.75 m high and is without leaves. The tiny flowers occur in thousands from August to October, and the berry, 3–4 mm long, with blue blotches on white, is ripe from September to January.   VIOLACEAE

727   Pink hebe flowers, Mt Terako (November)

727   Pink hebe, *Hebe raoulii*, forms a straggly shrub sprawling over dry rocks among the hills of Nelson, Marlborough and Canterbury from 200 to 900 m. Flowers occur abundantly from October through to March.   SCROPHULARIACEAE

729   Lake Tekapo willow herb, *Epilobium ros-tratum*, is a small, spreading herb with erect stems, 4–15 cm high, found along streambeds, on rocky outcrops and in grasslands from Arthur's Pass to Lake Wakatipu, being particularly common on the mountains around Lake Tekapo. Flowers, 3–5 mm across, occur from December to February.

The genus *Epilobium*

The genus *Epilobium* contains some 200 species worldwide, of which 50 are endemic to New Zealand. Many of the New Zealand species grow in dry, rocky places, including screes and riverbeds, and a few of our common species are shown here.

ONAGRACEAE

728   Glossy willow herb in flower, Cupola Basin (January)

728   Glossy willow herb, *Epilobium glabellum*, is abundant and common in mountain regions (100–1,800 m) from East Cape southwards, in riverbeds, stony subalpine places, grasslands and herbfields. Flowers, 5–6 mm across, occur from December to February.

729   Lake Tekapo willow herb in flower, Lake Tekapo (December)

730 Scree epilobium, *Epilobium pychnostachyum,* is a small, woody-stemmed herb with stems decumbent at the base, then erect, to 25 cm long. Leaves, 10–20 mm long by 2–4 mm wide, tinged with orange-red, are coarsely dentate. Flowers, 8–9 mm across, occur in the upper leaf axils from October to January. Found on rock screes in the Ruahine Range, North Island, and from Marlborough to Lake Wakatipu in the South Island.

730 The scree epilobium in flower, Fog Peak (January)

731 Woody willow herb with flowers, Homer Cirque (January)

731 Woody willow herb, *Epilobium novae-zelandiae,* is a woody, much-branched herb, up to 25 cm high with denticulate leaves to 15 mm long, found from Mangonui southwards in subalpine riverbeds, grasslands and open stony ground. Flowers and fruits occur from November to February.

732 Red-leaved willow herb, *Epilobium crassum,* is a woody, creeping herb, rooting at the nodes; the thick, fleshy, leathery leaves, 3–4 cm long by 9–14 mm wide, are entire and basally dull green strongly tinged with red. Flowers and fruits occur from November to February, and seeds can persist on the plant until April. The plant is found from the Nelson mountains to Lake Wakatipu in subalpine rocky places, screes and herbfields.

733 Large-flowered mat daisy plant in flower, Homer Saddle (January)

734 Close-up of flowers of large-flowered mat daisy, Renata Trig, Tararua Range (November)

733–734 Large-flowered mat daisy, *Raoulia grandiflora,* is a spreading shrub forming cushions or mats (fig. 733), up to 15 cm across, on rocks in exposed places between 900 and 2,000 m from East Cape southwards. Flowers (fig. 734), 8–16 mm across, occur during December and January.

ASTERACEAE

732 Red-leaved willow herb with seeds, Cupola Basin (April)

735 Mountain sandalwood; the black and the red spines are leaves, the small, dark-coloured, grape-like clusters are male flowers, Jack's Pass (January)

736 Mountain sandalwood with fruits, Boulder Lake (March)

737 Mountain sandalwood, showing female flowers, Jack's Pass (January)

735–737 Mountain sandalwood, *Exocarpus bidwillii,* is a rigid, much-branched, spreading shrub, to 60 cm high, found in subalpine rocky places and open areas from the Nelson mountains south throughout the South Island. In fig. 735 the broad, triangular, black spines on the stems are leaves, the small grape-like clusters behind the leaves on the short lateral spikes off the main stems are the developing male flowers; female flowers are below the depressions in the spikes (fig. 737). Flowers occur during January and February; the nut-like fruits (fig. 736) appear from January to April.

SANTALACEAE

The genus *Helichrysum*

The genus *Helichrysum* contains several species with appressed, more or less triangular leaves, found in dry, rocky places among the mountains of the South Island; some of these are illustrated here.

ASTERACEAE

738–739 Yellow-flowered helichrysum, *Helichrysum microphyllum,* has flower-heads (fig. 738), 5 mm across, from November to March. It is a shrub (fig. 739), to 50 cm high, found in the mountains from Nelson to North Canterbury, between 500 and 1,300 m.

738 Spray of yellow-flowered helichrysum with flowers, Waipahihi Botanical Reserve (February)

739   Yellow-flowered helichrysum plant in flower, Mt Cupola (January)

740 Marlborough helichrysum, *Helichrysum coralloides,* is a shrub to 60 cm high, with leaves densely surrounded by soft, woolly hairs; flowers occur only occasionally in summer. Found on the Kaikoura Mountains.

740   Marlborough helichrysum, showing the very woolly stems, Otari (March)

741–742 Common helichrysum, *Helichrysum selago,* is a shrub to 30 cm high, flowering profusely during summer (fig. 742), the flower-heads (fig. 741) 6–7 mm across. Found in rocky places throughout the South Island mountains.

743 Hairy helichrysum, *Helichrysum plumeum,* forms a shrub, to 60 cm high completely covered by tangled, whitish hairs. Solitary terminal flowers occur during summer months. Found among the rocky mountains of West Canterbury and the Hunter Hills.

741   Common helichrysum flowers close up, Otari (November)

742   Common helichrysum plant in flower, Otari (November)

743   Hairy helichrysum showing hairy stems, Otari (March)

744   Everlasting daisy plant in flower, Dun Mountain
(November)

745   Close-up of flowers of everlasting daisy, Tauhara
Mountain (December)

744–745 Everlasting daisy, *Helichrysum belli-dioides*, is a prostrate, creeping and rooting shrub (fig. 744), with leaves, 5–6 mm long by 3–4 mm wide, clothed below with a soft, woolly tomentum. Flowers (fig. 745), 2–3 cm across, are produced on tall, woolly stems in great profusion from October to February. Found in lowland to subalpine rocky places, grasslands and open scrub throughout New Zealand.

Mountain hebes

Plants belonging to the genus *Hebe* are common in rocky places throughout the New Zealand mountains and a selection of these are illustrated here; all belong to the family SCROPHULARIACEAE.

746 Black-barked mountain hebe, *Hebe decum-bens*, forms a decumbent shrub with shining, purplish black or dark brown bark and spreading, simple, slightly concave leaves, 12–20 mm long by 5–10 mm wide, with reddish-tinged margins. Flowers occur from November to February, and the plant is found among mountains from Nelson to Canterbury on rocky ledges at around 1,000–1,400 m.

747 Large-flowered hebe, *Hebe macrantha*, is a straggling, erect shrub, to 60 cm high, with thick, leathery leaves, 15–30 mm long, and flowers, about 18 mm across, occurring from December to February, Found between 800 and 1,600 m, in the mountains from Nelson to North Canterbury, on steep rock faces and in alpine scrub.

747   Large-flowered hebe flowers
with leaves, Wilberforce
River Gorge (January)

750 Thick-leaved hebe, *Hebe buchananii*, is a shrub, to 20 cm high, with spreading, slightly dished dull bluish-tinged, thick, leathery leaves, 3–7 mm long. Flowers occur from November to March, and the plant is found from the Godley Valley southwards, in the drier mountain areas.

746   Black-barked mountain hebe in flower, Mt Peel,
Nelson (November)

748-749 Mt Arthur hebe, *Hebe albicans*, is a low, spreading shrub, 1 m high, with long, spreading, bluntly pointed, bluish green leaves (fig. 748), 15-30 mm long and 8-15 mm wide, found flowering from December to April (fig. 749) in rocky places among the mountains of Nelson. A similar species with blunter leaves, *H. amplexicaulis*, is found among the mountains of Canterbury.

751 Nelson mountain hebe plant with flower, Dun Mountain (December)

748 Mt Arthur hebe plant in flower, Mt Arthur (December)

751 Nelson mountain hebe, *Hebe gibbsii*, is a sparingly branched shrub, to 30 cm high, with thick, red-edged, hairy-margined leaves, 10-18 mm long. Sessile flowers occur as terminal spikes from December to February. Found on the Nelson mountains, Ben Nevis and Mt Rintoul.

749 Mt Arthur hebe flowers and leaves, close up, Mt Peel, Nelson (December)

752 Dish-leaved hebe spray, showing leaves and flowers, Cass (December)

750 Thick-leaved hebe spray with flowers, Lindis Pass (December)

752 Dish-leaved hebe, *Hebe treadwellii*, is a low, sprawling shrub with green, deeply concave, thick leaves 10-15 mm long and 5-10 mm wide, flowering during December and January and found in rocky places from the Victoria Range to Mt Cook.

753–754 Spiny whipcord, *Hebe ciliolata*, is a stiff shrub (fig. 754), to 30 cm high, with appressed leaves, found in rocky places among the Nelson and Canterbury mountains to 2,400 m. Terminal, sessile flowers (fig. 753) occur from November till January.

753 Spiny whipcord flowers, close up, Porter's Pass (November)

754 Spiny whipcord plant in flower, Porter's Pass (November)

756 Trailing whipcord, *Hebe haastii*, is a sprawling shrub with woody stems ascending at their tips, bearing imbricate, thick and leathery leaves, 6–13 mm long, with ciliated margins. Terminal flowers, each about 8 mm across, occur in terminal spikes from October to February, and the plant is found in rocky places to 2,200 m in the mountains from Nelson to Otago.

755 Spray of Colenso's hebe, showing flowers and leaves, Taruarau River (October)

755 Colenso's hebe, *Hebe colensoi*, is a spreading, low-growing, North Island hebe, found in the head-waters of the Taruarau River on the Ruahine Range and in the Kaweka Range south to the Moawhango River. The leaves are blue, about 3 cm long and 1 cm wide. Flowers occur from August to November, both laterally and terminally on the branches.

756 Trailing whipcord stem, showing flowers and leaves, Mt Terako (September)

757-758 Scree hebe, *Hebe epacridea*, is a low-growing, spreading shrub (fig. 757) with ascending stems bearing recurved leaves and terminal flowers (fig. 758). Flowers occur from November to February, and the plant is found on open rocky scree slopes among the mountains of Nelson, Marlborough and Canterbury.

757   The scree hebe plant in flower, Waterfall Valley, Cass River, Lake Tekapo (December)

758   Stem of scree hebe showing leaves and flowers, Waterfall Valley, Cass River, Lake Tekapo (December)

760 New Zealand lilac, *Hebe hulkeana*, occurs naturally in dry, rocky places, to 900 m, over the northern half of the South Island; now extensively cultivated in parks and gardens. Branchlets are clothed with fine hairs, and the elliptic leaves, 7–10 cm long by 2–3 cm wide, have serrate or dentate margins. Flowers occur in profusion in crowded terminal spikes from October till December.

759   Canterbury whipcord spray, *Hebe cheesemanii* showing leaves and flowers, Waterfall Valley, Cass River, Lake Tekapo (December)

759 Canterbury whipcords, *Hebe cheesemanii*, *Hebe tetrasticha*, are both shrubs, 20–30 cm high, with erect, leafy stems, 1–2.5 mm across, stouter in *tetrasticha*. Leaves in *tetrasticha* are longer than broad, shorter than broad in *cheesemanii*. Sessile flowers, in 1–3 pairs, in *tetrasticha* and 1–2 pairs in *cheesemanii* (fig. 759), occur from November to January. Both species are found in dry, rocky places in the mountains of Canterbury.

760   New Zealand lilac flowers, Otari (November)

761 Spray of blue-flowered mountain hebe with flowers, Upper Rangitata (December)

761 Blue-flowered mountain hebe, *Hebe pimeleoides*, is an erect, branching shrub, to 45 cm high, with narrow, lanceolate, bluish leaves, 5–15 mm long by 2–6 mm wide, found in the drier rocky places of the mountains from Marlborough southwards. Flowers, varying from bluish white to deep blue, occur in pairs on lateral hairy stalks from November to March.

Vegetable sheep

Vegetable sheep is the common name given to peculiar cushion-like daisy plants found in rocky places in subalpine and alpine regions of the New Zealand mountains, since these plants appear, in the distance, not unlike a flock of sheep. Two genera, *Haastia* and *Raoulia*, are involved.   ASTERACEAE

762 Tufted haastia, *Haastia sinclairii*, is a woolly, erect or decumbent shrub, sometimes forming mats. The leaves, 3.5 cm long by 15 mm wide, are covered by a thick, white, appressed tomentum; but plants in Fiordland can have a buff-coloured tomentum, and these are known as variety *fulvida*. Flowers, 2–3 cm across, occur from December to January. Found in subalpine and alpine screes, rocky places and fellfields on the eastern slopes of the mountains from Nelson to Fiordland.

762 Tufted haastia in flower, Homer Saddle (January)

763 Red-flowered vegetable sheep, *Raoulia rubra*, is a daisy forming hard cushions up to 25 cm across and 15 cm high; found on the Tararua Range in the North Island and on the Paparoa, Haupiri and Mt Arthur Ranges in the South Island, on screes and in rocky places. Flowers, 2–3 mm across, occur during January and February.

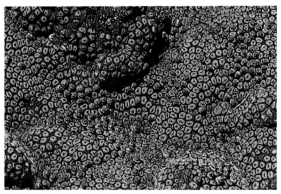

763 Red-flowered vegetable sheep with flowers, Mt Holdsworth (February)

764 Silvery vegetable sheep, Mt Torlesse (November)

766 Giant vegetable sheep, close-up showing flowers, Cupola Basin, 1,710 m altitude (April)

767 Giant vegetable sheep; the seated figure gives a comparison of size, Cupola Basin, 1,710 m altitude (April)

764–765 Silvery vegetable sheep, *Raoulia mammillaris*, is a small plant, 50 cm across, with compacted silvery leaves found on rocks on Mts Torlesse (fig. 764), St Bernard, Hutt and Somers and the Craigieburn Range. Flowers, 2 mm across (fig. 765), occur from December to February, followed by downy seeds.

765 Silvery vegetable sheep, showing flowers, Fog Peak (December)

766–767 Giant vegetable sheep, *Haastia pulvinaris*, forms huge flat cushions, up to 2 m or more across (fig. 767, measured 8 square metres), and is found throughout the mountains of Nelson and Marlborough, at around 1,700 m altitude, on loose screes and rocky places, fellfields and the edges of more stable screes where the rocks are large. Flowers (fig. 766) occur during January and February, occasionally as late as April, followed by downy seeds.

768 Common vegetable sheep growing on rocks, Gunsight Pass, Cupola Basin, 1,900 m (December)

769 A large plant of common vegetable sheep, Cupola Basin, 1,700 m (April)

768–769 Common vegetable sheep, *Raoulia eximia*, forms tight, rounded cushions (fig. 769), up to 2 m across and 20–60 cm high. Found in dry, rocky places from 1,600 m to 2,700 m throughout the mountains of Nelson, Marlborough and Canterbury (fig. 768). The woolly leaves are tightly packed and overlapping, while the tiny flowers, 3 mm across, appear in January and February.

770   North Island edelweiss plant in flower, Mt Holdsworth (February)

770 North Island edelweiss, *Leucogenes leontopodium*, is a small, branching, woody plant with decumbent stems that have turned-up tips and sessile leaves, 8–20 mm long by 4–5 mm wide, entirely covered with silvery white wool. Flower-heads, 2.5 cm across, terminal on each stem, occur from November till March, and the plant, photographed on Mt Holdsworth, is found in subalpine and alpine rocky places, herbfields and fellfields.
ASTERACEAE

771   Black daisy plant in flower, scree on Fog Peak (January)

771 Black daisy, *Cotula atrata*, is a creeping, scree-inhabiting plant, found flowering during January and February between 1,250 and 2,000 m among the mountains of Marlborough and Canterbury. The flower-heads, 2 cm across, may be either black or brown.   ASTERACEAE

772 South Island edelweiss, *Leucogenes grandiceps*, is similar to *L. leontopodium* and is found in flower from November to February in rocky and stony places, up to 1,600 m, throughout the South Island mountains.   ASTERACEAE

773–774 Mountain cress, *Notothlaspi australe*, is a fleshy plant with grey-black, hairy leaves (fig. 773), 1–5 cm long, forming rosettes to 12 cm across, each rosette with many fragrant flowers (fig. 774), each 10 mm across, during December. Found on fairly stable alpine screes, between 1,600 and 1,800 m, among the mountains of Nelson and Marlborough.
CRUCIFERAE

773   Mountain cress plant on scree, Cupola Basin, 1,720 m altitude (April)

774   Mountain cress in flower, Cupola Basin, 1,600 m (December)

772   South Island edelweiss plant in flower, Homer Saddle (February)

775 **Barbless bidibidi,** *Acaena glabra,* is a creeping, prostrate, hairless plant, with branches, 50 cm long, rising at their tips and bearing leaves up to 5 cm long. Flower-heads, 2 cm across, arise from December to February. Found on screes, riverbeds and stony places throughout the eastern South Island.

ROSACEAE

776 Scarlet bidibidi plant with seed-heads, Volcanic Plateau, (January)

775 Barbless bidibidi plant with seed-heads, Mt Torlesse (December)

776-777 **Scarlet bidibidi,** *Acaena microphylla,* is a prostrate, spreading herb, forming bright red patches, to 75 cm across, when in flower from December to February. Found in stony stream beds, herbfields and grasslands of the Central Volcanic Plateau. ROSACEAE

778 Penwiper plant, showing rosulate form, Sugarloaf, Cass (November)

777 Scarlet bidibidi plant with two flowers and masses of seed-heads, Volcanic Plateau (January)

778-779 **Penwiper plant,** *Notothlaspi rosulatum,* is a fleshy herb, forming rosettes (fig. 778), 8-10 cm across and to 25 cm high when in flower. It is found on partially stabilised screes on the eastern side of the mountains from Marlborough to South Canterbury (fig. 779). Highly fragrant flowers occur during December and January. CRUCIFERAE

779 Penwiper plant in flower and with seeds round the base of the flower-head, scree, Porter's Pass (January)

780 Fleshy lobelia, *Lobelia roughii*, is a spreading, hairless, fleshy herb found on screes and rock outcrops, between 900 and 1,800 m, in the mountains from Nelson to Otago. Flowers occur from October to February.   LOBELIACEAE

780   Fleshy lobelia in flower, scree on Fog Peak (January)

781 Grey-leaved succulent, *Lignocarpa carnosula*, forms a succulent plant, to 15 cm high, found on loose screes of Mt Torlesse and other mountains from Marlborough to Canterbury between 1,200 and 1,400 m. Tiny flowers, 2–3 mm across, occur from November to February.   APIACEAE

783 Snowy woollyhead, *Craspedia incana*, is a soft plant, covered all over with a silvery white wool, found on screes and rocks of the dry mountains of Canterbury. Leaves, 5–10 cm long by 2–3 cm wide, are arranged in rosettes, and the flower-heads, 2–3 cm wide, occur during January and February.   ASTERACEAE

781   Grey-leaved succulent in flower, scree on Fog Peak (January)

782 Mountain chickweed, *Stellaria roughii*, is a branching, succulent herb, with leaves 2 cm long by 3 mm wide, found on screes, between 1,000 and 2,000 m among the South Island mountains. Flowers, 2 cm across, having sepals longer than the petals, occur from December to February.   CARYOPHYLLACEAE

782   Mountain chickweed in flower, scree on Fog Peak (January)

783   Snowy woollyhead plant, showing dense wool on leaves and two opening flowers, scree on Porter's Pass (January)

784 Hairy pimelea, *Pimelea traversii*, forms a rather dense shrub, up to 60 cm high, found in rocky and stony montane to subalpine places throughout the South Island. Each individual flower of the flower-head has its perianth densely clothed with long silky hairs. The flowers are always pale pink and the thick leaves are faintly keeled.   THYMELAEACEAE

785 New Zealand chickweed, *Stellaria gracilenta*, is a stiff, erect herb, 10 cm high, with thick, awl-shaped leaves, 3–6 mm long, found in dry rocky places, fellfields and grasslands to 1,500 m from Mt Hikurangi to Fiordland. Flowers, 10–12 mm across, occur from October to March.

CARYOPHYLLACEAE

786 Rock cushion plant in flower among stones on Mt Lucretia (January)

785 New Zealand chickweed flowers, Cass (November)

787 Rock cushion, close-up to show flowers, Mt Robert (January)

786–787 Rock cushion, *Phyllachne colensoi*, forms hard cushions or mats to 40 cm across (fig. 786) in exposed rocky places in herbfields and fellfields to 1,850 m, from Mt Hikurangi south to Stewart Island. Tiny white flowers, with their anthers extending far beyond the lip of the corolla (fig. 787), occur during January and February.

STYLIDIACEAE

788 Arthur's Pass forget-me-not, *Myosotis explanata*, is a rosette-shaped herb, to 30 cm high, with leaves to 7 cm long, covered all over by soft, white hairs. Flowers occur as terminal cymes from December to February. Found in rocky places between 900 and 1,400 m, in the vicinity of Arthur's Pass.

BORAGINACEAE

784 Spray of hairy pimelea showing flower-heads, leaves and the very hairy basal parts of the flowers, Lake Lyndon (January)

788 Arthur's Pass forget-me-not plant in flower, Arthur's Pass (December)

789   Colenso's forget-me-not, *Myosotis colensoi*, is a creeping, prostrate perennial with lanceolate leaves, 2–3 cm long by 5–10 mm wide, more hairy on the upper surfaces. Flowers, 8 mm across, occur singly or in clusters from November through December, and the plant is found on limestone rocks around Broken River.   BORAGINACEAE

790–791   Small forget-me-not, *Myosotis monroi*, is a prostrate, rosette-forming plant with hairy leaves, found in rocky places on Dun Mountain, Nelson. Flowers occur as terminal cymes from late November to February.   BORAGINACEAE

789   Colenso's forget-me-not plant in flower, Castle Hill (November)

790   Small forget-me-not flowers, Dun Mountain (November)

791   Plant of small forget-me-not with flower-stalk, Dun Mountain (November)

Mountain ranunculi

Mountain ranunculi occur widely throughout the New Zealand mountains, with many species growing in rocky and stony places while others are found in grasslands, herbfields and fellfields. They are all conspicuous plants, and some of those found in rocky situations are depicted here.   RANUNCULACEAE

792   Haast's buttercup in flower, scree in Waterfall Valley, Lake Tekapo (December)

792   Haast's buttercup, *Ranunculus haastii*, has deeply divided leaves, to 15 cm long by 10 cm wide, and large flowers, 2–4 cm across, produced from November to January. *R. haastii* is found on screes in the mountains of Nelson and Canterbury.

793   Large feathery-leaved buttercup, *Ranunculus sericophyllus*, is found in the higher alpine rocky places and fellfields of the wetter regions from Lewis Pass to Northern Fiordland. Pictured here from above Wapiti Lake in Fiordland, the species is characterised by its deeply divided, hairy leaves on petioles 2–12 cm long, and the flowers, 2.5–4 cm across, produced from December to February.

794 Large white-flowered buttercup, *Ranunculus buchananii*, is found in rocky clefts between 1,500 and 2,300 m in the mountains from Lake Wakatipu to Lake Hauroko. The deeply divided leaves, to 15 cm long, are not hairy, and the flowers, 3–7 cm across, occur during December and January.

795–796 Korikori, *Ranunculus insignis*, is a branching hairy herb, to 90 cm high, found in rock crevices, herbfields and alpine grasslands from East Cape to the Kaikoura Mountains. Leaves, 10–16 cm wide, are thick and leathery. Flowers, 2–5 cm across, occur in profusion from November to February. Pictured in fig. 795 is the variety *glabratus* from Mt Ruapehu, while fig. 796 shows the parent form from Mt Holdsworth.

794 Large white-flowered buttercup in flower above Wapiti Lake, Fiordland, 1,600 m altitude (December)

795 Korikori plant in flower, Mt Ruapehu (January)

797 Snow buttercup, *Ranunculus nivicolus*, is an erect, hairy buttercup, producing flowering stems to 80 cm high, with flowers 3–5 cm across from September to December. Found on scoria slopes, in scrub and herbfields between 1,200 and 1,850 m, on Mt Taranaki, the central volcanoes, and the Kaweka and Raukumara Ranges.

796 Korikori flowers close up, Mt Holdsworth (December)

793 Large feathery-leaved buttercup in flower above Wapiti Lake, Fiordland, at 1,650 m (December)

797 Snow buttercup plant in flower, Mt Taranaki (September)

798   Blue-leaved ranunculus, *Ranunculus crithmifolius,* is a bluish black-leaved herb, to 80 cm high, found on screes above the Wairau River Gorge at about 2,100 m. Narrow-petalled flowers, about 2 cm across, occur during December and January.

798   Blue-leaved buttercup plant in flower, scree on Mt Dobson (December)

799   Small variable buttercup flowers and leaves, Porter's Pass (November)

799   Small variable buttercup, *Ranunculus enysii,* is a rosette-forming, hairless buttercup, with leaves either divided or lobed, 2–10 cm long by 10–70 mm wide, on grooved petioles to 10 cm long. Flowers, 15–30 mm across, occur during October and November. Found in rocky clefts or rocky sheltered places in tussock and scrub along the eastern mountains from North Canterbury to southwest Otago.

800   Haast's carrot plant in flower, Arthur's Pass (January)

800   Haast's carrot, *Anisotome haastii,* is up to 60 cm high, with purple stemmed, 2–3-pinnate, carrot-like leaves, 15–25 cm long by 6–12 cm wide, on stout petioles 8–12 cm long. Flowers occur from October till February, and the plant is found throughout the South Island mountains in rocky places and fellfields, between 500 and 1,600 m, mainly in western wetter regions.   APIACEAE

801   Snowball Spaniard, *Aciphylla congesta,* is a soft-leaved *Aciphylla* that forms compact masses of rosettes alongside cracks and water trickles that cross rocky outcrops and basins, between 1,300 and 1,600 m, in the mountains of west Otago and Fiordland. The flowers occur as white rounded umbels, up to 12 cm across, during December and January and appear like balls of snow on the rocks.

APIACEAE

801   Snowball Spaniard plant in flower above Wapiti Lake, Fiordland (December)

Mountain daisies

Mountain daisies belong to the genus *Celmisia* and are among the most common plants in the New Zealand mountains. They grow in herb- and fellfields, on grasslands and screes or other stony places, and a selection of those found in rock situations are illustrated here. ASTERACEAE

802 Fiordland rock daisy, *Celmisia inaccessa*, forms large mats, up to 1 m across, with lush, vivid green leaves, 2–6 cm long by 10–20 mm wide. Flowers, 5 cm across, occur during December and January, and the plant grows on limestone outcrops in the Wapiti Lake to Barrier Peaks area between the Doon and Stillwater Rivers.

803 White daisy flowers and leaves, Mt Ruapehu (October)

802 Fiordland rock daisy plant in flower, Wapiti Lake, Fiordland (December)

803 White daisy, *Celmisia incana*, is found in exposed rocky places, herbfields, fellfields and grasslands from the Coromandel Ranges to Otago. The thick, leathery leaves and flower stalks are densely clothed with appressed white hairs. Flowers, 2–3.5 cm across, occur from October to January.

804 Trailing celmisia, *Celmisia ramulosa*, has stiff, erect branches bearing thick, overlapping leaves with recurved margins. Flowers, 2–2.5 cm across, occur on woolly stalks during November and December. The plant is found trailing over rocks at about 1,400 m in the mountains of Otago, Southland and Fiordland.

805 Dusky Sound daisy plant in flower, Stillwater Basin, Fiordland (December)

804 Trailing celmisia in flower above Wapiti Lake, Fiordland (December)

805 Dusky Sound daisy, *Celmisia holosericea*, is found in rocky coastal to subalpine places throughout Fiordland. The shining leaves, 15–30 cm long by 4–6 cm wide, have silvery hairs on the lower surfaces, and the flowers, 5–7 cm across, occur in December and January.

806 Hector's daisy in flower and sprawling over rocks above Wapiti Lake, Fiordland (December)

807 Close-up of flowers and leaf rosettes of Hector's daisy, Homer Saddle, Fiordland (December)

806–807 Hector's daisy, *Celmisia hectori*, forms patches, to 1 m across (fig. 806), in rocky places or over rocks in herb- and fellfields from Arthur's Pass to Fiordland. The branches are clothed with leafy remains, and the hairy leaves are in rosettes, each of which bears a single flower (fig. 807), 2–2.5 cm across, during December and January.

808 Mountain rock daisy, *Celmisia walkeri*, forms large mats over rocks and rock-clefts in alpine regions southwards from the Spencer Mountains. The woody, 2 m long branches bear sticky leaves, up to 5 cm long by 5 mm wide, and the flowers, 2–4 cm across, occur from October to January.

809 Crag-loving daisy, *Celmisia philocremna*, forms a cushion, to 70 cm across, of thick, overlapping, succulent-like leaves 1.5–2.5 cm long by 5 mm wide, in crevices and on rock ledges between 900 and 1,800 m, in the Eyre Mountains. Hairy flower-stalks bear flowers, to 3 cm across, in December and January.

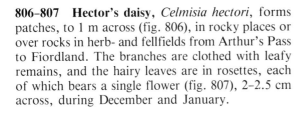

808 Mountain rock daisy plant in flower, Lake Ohau (November)

809 Crag-loving daisy plant with flower, photographed in the garden of Mr J. Anderson, Albury (December)

810 Fiordland mountain daisy, *Celmisia verbasci-folia*, is a large, tufted daisy, with thick, smooth, shining, leathery leaves, 15–25 cm long by 2.5–5 cm wide, having distinct veins above and velvety hairs below. Flowers, 2–2.5 cm across, occur on hairy stalks, 30–40 cm high, from December to February. Found in rocky places and herbfields among the mountains of West Otago and Fiordland.

811–812 Spiny-leaved daisy, *Celmisia brevifolia*, forms large, loose, sprawling clumps, the woody branches clothed with leaf remains. The thick leaves, 10–15 mm long by 6–9 mm wide, have their margins bearing widely separated, small spines. Flowers, 2–3 cm across, occur on short, sticky stalks from October to January. Found in rocky places and herb-fields, between 1,400 and 1,840 m, among the mountains of Canterbury and Otago.

810 Fiordland mountain daisy plant with flower, Wapiti Lake (December)

811 Spiny-leaved daisy in flower, Old Man Range (October)

812 Spiny-leaved daisy, close-up to show spines on leaves, Old Man Range (October)

813 Bristly carrot plant with flowers, Temple Basin (January)

813 Bristly carrot, *Anisotome pilifera*, grows to 60 cm high with pinnate leaves, 10–30 cm long by 5–10 cm wide, on stout petioles to 10 cm long. Flowers, as compound umbels, arise from October to March, and the plant grows in rocky places in the South Island mountains, at about 1,850 m, from north-west Nelson southwards. APIACEAE

814 Broad-leaved carrot, *Gingidium montanum,* is a stout aromatic herb, to 50 cm high, leaves radical, each with 5–7 pairs of serrate, sessile leaflets on 40 cm long, purplish petioles; terminal leaflets are usually 3–5 lobed. Flowers in umbels, 10 cm across, occur from October to January. Found in rocky and gravelly places and grasslands in coastal to subalpine regions from the central volcanoes southwards.

APIACEAE

815 Thread-like carrot with flowers and seeds, Mt Dobson (December)

814 Broad-leaved carrot plant with flowers, Te Anau (January)

815–817 Thread-like carrot, *Gingidium filifolium,* is a slender herb, to 30 cm high, with grooved, slender stems branching near their tops (fig. 815). Flowers are in umbels, 5–10 cm across, with male flowers (fig. 816) and females (fig. 817) in the same umbel from November to February. Found east of the Alps on scree and debris slopes from Dun Mountain to the Ben Ohau Range.   APIACEAE

818 Prostrate dwarf broom, *Carmichaelia enysii,* forms patches to 10 cm wide and 5 cm high on stony river terraces and grasslands at around 1,400 m on the eastern slopes of the Southern Alps from Arthur's Pass to Fiordland. Flowers occur from November to January.   FABACEAE

817 Thread-like carrot, close-up to show female flowers, Dun Mountain (November)

818 Prostrate dwarf broom, plants with flowers, Kurow (November)

819 Kopoti plant with flower, Dun Mountain (November)

816   Thread-like carrot, close-up of male flowers,
       Dun Mountain (November)

821   Cushion plant with flowers, Fog Peak (January)

819–820 Kopoti, *Anisotome aromatica,* grows
throughout New Zealand in alpine and subalpine
rocky places, grasslands, herb- and fellfields. Flowers
occur as umbels from October till February, and
plants can be 50 cm high at lower altitudes but only
10 cm in high alpine regions.   APIACEAE

821 Cushion plant, *Pygmea pulvinaris* and *P.
ciliolata,* both form a soft, moss-like, dense cushion,
up to 10 cm across and about 4 cm high, in subalpine
and alpine rocky and shingly places from Nelson to
Canterbury. The tiny leaves, 2.5–4 mm long by
1 mm wide, are sparingly clothed with long, bristle-
like hairs. Flowers, 5–6 mm across, occur in
December and January.   SCROPHULARIACEAE

820   Kopoti plant in full flower, Mt Ruapehu
       (January)

822   Variable cushion plant *Pygmea ciliolata,* in
       flower, Stillwater, Fiordland (December)

822   Variable cushion plant, *Pygmea ciliolata,* is
similar to *P. pulvinaris* but rather variable in size and
rigidity, with the young leaves ciliated along their
margins. Found in bare rocky places and fellfields
from north-west Nelson to Fiordland. Fig. 822 shows
the variety *fiordensis,* from the Stillwater, in flower
in December.   SCROPHULARIACEAE

823   Auckland Island forget-me-not, *Myosotis
capitata,* grows on the Auckland Islands, between
sea-level and 600 m, on rocky clefts and ledges. The
oblong to spathulate, hairy leaves, 3–12 cm long, are
in rosettes. Flowers occur on hairy stems from
November to February.   BORAGINACEAE

823   Auckland Island forget-me-not plant in flower,
       Auckland Islands (February)

824   Chatham Island geranium, *Geranium traversii,* forms prostrate stems to 60 cm long with upturned tips. Both leaves and stems are densely clothed with silvery hairs, and the flowers, 2–2.5 cm across, are produced in abundance from October to February.

GERANIACEAE

825   Poor Knights lily plant with flowers, from a plant grown at Seatoun, Wellington, by the late Dr W. R. B. Oliver (November)

824   Chatham Island geranium flower and leaves, Otari (October)

825   Poor Knights lily/raupo-taranga, *Xeronema callistemon,* grows on the Poor Knights and Hen Islands only, in sunny, rocky situations between sea-level and 300 m. Leaves are 60–100 cm long by 3–5 cm wide, and the flowers appear on stout stalks, 25–30 cm long, from October to December.

LILIACEAE

826   Dusky Sound pimelea spray with flowers, Stillwater, Fiordland (December)

827   Northern snowberry in full flower, Mt Ruapehu (January)

826   Dusky Sound pimelea, *Pimelea gnidia,* is a much-branched shrub, to 1 m high, with dark red-brown bark and shining, oblong-lanceolate leaves, 1.5–2 cm long by 4–6 mm wide, on short petioles. Flowers with densely hairy perianth occur during December and January. Found in subalpine, stony scrub or herbfields from Coromandel Ranges to Fiordland.

THYMELAEACEAE

827   Northern snowberry, *Gaultheria colensoi,* is a sprawling, bushy shrub, to 60 cm high, with thick, generally crenulate margined leaves, 8–12 mm long by 5–9 mm wide, on petioles 1 mm long. Flowers, 2–3 mm long, occur in profusion from November to January. The berries, 3 mm across, ripen during February and March. Found in rocky places and grasslands among the North Island mountains from Mt Hikurangi southwards.

ERICACEAE

ALPINE AND SUBALPINE SCRUB

As one moves upwards out of the forest towards the open slopes of the New Zealand mountains there is usually a distinct zone of shrubs and small trees with some tussocks and herbaceous plants that forms a transition from the forest to the subalpine tussocklands, herbfields and fellfields. This zone constitutes the subalpine and alpine scrub.

828 **Springy coprosma,** *Coprosma colensoi,* is a slender, springy shrub, to 2 m high, with narrow, leathery leaves, 15–20 mm long by 5–8 mm wide, on hairy petioles 2–3 mm long. Flowers arise from September till December, and the elongated drupes are ripe by the following May–June. Found in subalpine scrub and forest from the Coromandel Ranges south to Stewart Island.   RUBIACEAE

829 **Sprawling coprosma,** *Coprosma cheesemanii,* is a sprawling, prostrate shrub with divaricating, pubescent branches, often forming a tight compact bush. Leaves are 8–11 mm long by 1–2 mm wide, and the drupes, 6–7 mm across, ripen during February and March. Found in subalpine scrub, tussock and grasslands, and sometimes in forest, fellfields and alpine bogs from East Cape south to Stewart Island.   RUBIACEAE

828   Springy coprosma, showing leaves and drupes, Mt Holdsworth (June)

829   Sprawling coprosma showing spray heavy with drupes, Volcanic Plateau (February)

830   Musk tree daisy leaves and flowers, Mt Huxley (December)

830 **Musk tree daisy,** *Olearia moschata,* is a much-branched shrub, to 4 m high, producing a strong, musk-like scent, especially during hot weather. The close-set leaves, 8–15 mm long by 5 mm wide, are glabrous or nearly so above, but have an appressed white tomentum below. Flowers occur from November to March and the plant is found in subalpine scrub around 1,500 m altitude from the Lewis Pass southwards. Hybridisation with other species occurs, and the form illustrated may be a hybrid.

ASTERACEAE

831–832 Rough-leaved tree daisy, *Olearia lacunosa,* is a shrub or small tree, up to 5 m high, with its branchlets, leaf petioles, leaf undersides and flower-stalks all clothed with a thick, brown or rust-coloured wool. Leaves, 7.5–17 cm long by 8–25 mm wide, have prominent midribs and veins imparting an alveolate appearance to their undersides (fig. 832). Flowers (fig. 831), 10 mm across, in corymbs, occur from November to February. Found in subalpine scrub and forest from the Tararua Ranges south to the mountains of Nelson and Marlborough.

ASTERACEAE

833–834 Hakeke, *Olearia ilicifolia,* is a musk-scented shrub or tree, to 5 m high, found in subalpine scrub or forest from the Bay of Plenty southwards. The strongly dentate leaves, 5–10 cm long by 10–20 mm wide, on petioles up to 2 cm long, have an appressed yellow tomentum on their undersides and are narrower in North Island plants than in South Island plants. Flowers, each about 15 mm across (fig. 834), arise in large corymbs from November till January. Hakeke often grows luxuriantly along stream banks and tracks or around clearings, and quite large specimens grow in the upper Hollyford Valley.   ASTERACEAE

831 Rough-leaved tree daisy flowers, close up, Mt Arthur (January)

833 Hakeke showing leaves and corymbs of flowers, Ruahine Range (December)

832 Rough-leaved tree daisy leaves, Mt Arthur (January)

834 Hakeke, close-up of flower, West Taupo (December)

835–836 Pigmy pine, *Lepidothamnus laxifolius,* is the smallest pine in the world, often fruiting in the juvenile stage, and forms a prostrate, trailing, creeping plant, making dense mats in subalpine and alpine scrub, herbfields and fellfields from the central volcanoes southwards. Spreading juvenile leaves are up to 12 mm long while adult leaves are 1–2 mm long, appressed to the stems. Male strobili (fig. 835) appear during November and December, with fruits (fig. 836) following through March and April.

PODOCARPACEAE

836 Pigmy pine plant with ripe seeds, Arthur's Pass (April)

835 Spray of pigmy pine with male cones, Arthur's Pass (November)

837–838 Hard-leaved tree daisy, *Olearia nummularifolia* and variety *cymbifolia* are shrubs, to 3 m high, with sticky branchlets clothed with white or yellowish, star-shaped hairs during the young or immature stage. The leaves are thick and leathery, 5–10 mm long by 4–6 mm wide in *nummularifolia* (fig. 837), and 6–14 mm long with margins recurved to the midrib in *cymbifolia* (fig. 838), giving the appearance of a swollen or fat leaf. Flowers, 3–5 mm across, occur from November to April except for *cymbifolia*, in which flowers seldom occur after January. The species is found in subalpine and alpine scrub from the Volcanic Plateau southwards, but the variety *cymbifolia* is found only throughout the South Island.

ASTERACEAE

837 Hard-leaved tree daisy, *Olearia nummularifolia*, spray with leaves and flowers, Mt Ruapehu (February)

838 Hard-leaved tree daisy, var. *cymbifolia*, spray showing leaves and flowers, Mt Terako (December)

839   Ruapehu hebe spray with flowers, Mt Ruapehu
(December)

839   Ruapehu hebe, *Hebe venustula*, is an erect
hebe found from Mt Hikurangi to Mt Ruapehu
among alpine scrub. Flowers occur in profusion as
racemes, 3–4 cm long, from December to February.
SCROPHULARIACEAE

840   Canterbury hebe in full flower, Arthur's Pass
(December)

840   Canterbury hebe, *Hebe canterburiensis*, forms
an erect shrub, to 1 m high, with narrow leaves,
7–17 mm long. It becomes smothered with white
flowers during December and January. Found in
subalpine scrub or tussock grassland on the Tararua
Ranges and the mountains of Nelson, Marlborough
and Canterbury.   SCROPHULARIACEAE

841   Western mountain koromiko. *Hebe gracillima*,
is a rather loosely branched shrub, to 2 m high, with
finely pubescent branchlets and narrow, lanceolate,
spreading leaves, 4 cm long by 6 mm wide. Flower
spikes occur laterally, from January to April, and
are longer than the leaves. Found in damp areas of
subalpine scrub, mainly on the western slopes, from
the Nelson mountains south to Arthur's Pass.
SCROPHULARIACEAE

841   Western mountain koromiko
in flower, Cobb Valley
(January)

842–843   Tupare/leatherwood, *Olearia colensoi*,
forms a closely branched shrub, up to 3 m high, with
the branches clothed in a fawn-coloured wool.
Leaves are thick, leathery and serrated, 8–20 cm long
by 3–6 cm wide, with a white, appressed tomentum
below. Flowers, 2–3 cm across, occur from
December to January. Tupare is found in subalpine
scrub from Mt Hikurangi southwards to Stewart
Island.   ASTERACEAE

842   Tupare in flower, Wharite Peak (December)

844 Variable coprosma, *Coprosma pseudocuneata,* is a shrub to 3 m high of variable form and habit, being reduced to 5–6 cm high in exposed situations. Found in subalpine scrub, forest, grasslands, herbfields and bogs from Mt Hikurangi to Southland. The thick, leathery leaves are 15–20 mm long by 2–6 mm wide, and the drupes, 5–6 mm long, are produced in profusion from February to May and may persist on the plants in alpine situations until the following December. RUBIACEAE

845 Mountain lycopodium with fruiting spikes, Mt Holdsworth (January)

845 Mountain lycopodium, *Lycopodium fastigiatum,* is a creeping plant found all over New Zealand in subalpine scrub and herbfields; the creeping stems extend for up to 2 m with ascending branches to 30 cm high. The fruiting bodies are 2.5–5 cm long, occurring singly at the tips of the branches from December through to April. LYCOPODIACEAE

844 Variable coprosma sprays with drupes, Temple Basin (April)

846 Club-moss whipcord stems and flowers, Garvie Mountains (December)

843 Tupare flowers close up, Wharite Peak (December)

846 Club-moss whipcord, *Hebe lycopodioides,* is a stiff, erect, branching shrub, to 1 m high, with four-angled branchlets clothed by thick, appressed, triangular leaves, 1.5–2 mm long, with ciliated margins and each terminating with an apical cusp. Flowers occur during December and January.

SCROPHULARIACEAE

847   Shrub groundsel, spray with
flowers and leaves, Mt
Holdsworth (January)

847   Shrub groundsel, *Brachyglottis elaeagnifolia,*
is a shrub or small tree, to 3 m high, readily recognised by its grooved branches and the brown to dark
brown, appressed hairs that clothe the branchlets,
leaf petioles, flower stalks and lower leaf surfaces.
The thick, leathery leaves, 10–12.5 cm long by
7.5 cm wide, have conspicuous veins, and the
flowers, each 8 mm across, occur in panicles to
15 cm long during January and February.
ASTERACEAE

848   Kaikoura shrub groundsel, *Brachyglottis
monroi,* is a shrub, to 1 m high, with serrated leaves,
2–4 cm long by 15 mm wide, found in subalpine
scrub among the Kaikoura Ranges and the mountains of Nelson. Flowers arise from December to
March.   ASTERACEAE

849   South Island shrub groundsel, *Brachyglottis
revoluta,* forms a compact shrub, to 3 m high, with
ribbed leaves, 3–6 cm long by 2–3 cm wide, white
tomentum below and grooved petioles 10–20 mm
long. Flowers, 2 cm across, occur from January to
March, terminally, on ascending stalks up to 10 cm
long. Found in subalpine scrub and fellfields in
western Otago and Fiordland.   ASTERACEAE

849   South Island shrub groundsel,
showing flowers and leaves,
Homer Cirque (January)

850   Nelson mountain groundsel, *Brachyglottis laxi-
folia,* forms a laxly branched shrub to 1 m high with
grey-green leaves. It smothers itself with flowers,
2 cm across, from December to February. Leaves are
up to 6 cm long and 2 cm wide, their undersides and
the branchlets densely clothed with white hairs.
Found among the mountains of Nelson and Marlborough in subalpine scrub.   ASTERACEAE

848   Kaikoura shrub groundsel, plant with flowers,
Otari (December)

850   Nelson mountain groundsel flowers and leaves,
Marlborough mountains (December)

851 Thick-leaved shrub groundsel, spray of flowers and leaves, Mt Holdsworth (February)

851 Thick-leaved shrub groundsel, *Brachyglottis bidwillii*, is a tightly branched shrub to 1 m high, with thick, leathery, elliptic leaves, 2–2.5 cm long and 10–15 mm wide, both leaves below and branchlets densely clothed with soft, white to fawn-coloured hairs. Flowers, 15 mm across, occur in panicles from December to March, and the plant is found in subalpine scrub and fellfields from Mt Hikurangi southwards. ASTERACEAE

852 Reticulate coprosma, *Coprosma serrulata*, is easily recognised by its markedly reticulate, thick, leathery leaves, 4–7 cm long by 2.5–4 cm wide, and the white bark that falls in flakes. Found in subalpine scrub, forest, grasslands and herbfields throughout the South Island. The drupes, 7–8 mm long, make a fine display in the autumn. RUBIACEAE

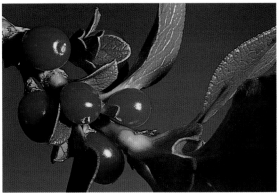

852 Spray of the reticulate coprosma showing leaves and drupes, Arthur's Pass (April)

853–854 Coromandel groundsel, *Brachyglottis myrianthos*, is a sparingly branched shrub, up to 4 m high, with membraneous, very coarsely serrated leaves (fig. 854), 7–18 cm long, found on the Coromandel Ranges in subalpine scrub on valley sides. Flowers (fig. 853), up to 10 mm across, occur in large panicles (fig. 854) from November to January. ASTERACEAE

854 Spray of Coromandel groundsel with leaf and flowers, Kauaeranga Valley, Billy Goat Track (December)

853 Coromandel groundsel flower, close up, Kauaeranga Valley, Billy Goat Track (December)

855   Mountain wineberry flowers,
Volcanic Plateau (November)

855–856   Mountain wineberry, *Aristotelia fruticosa*, forms a much-branched, tangled, rather rigid shrub, to 2 m high, with ovate to ovate-oblong leaves (fig. 856), 5–7 mm long by 4–5 mm wide. Flowers (fig. 855), varying in colour from white to red, occur singly or in 3–6-flowered cymes from October to December, and the berries (fig. 856), 3–4 mm across, ripen from November to April. Found in subalpine scrub and forest, and fellfields, throughout New Zealand.   ELAEOCARPACEAE

856   Mountain wineberry spray with berries, showing growth habit, Outerere Stream, Volcanic Plateau (December)

857   Needle-leaved mountain coprosma, *Coprosma rugosa*, found in lowland, montane and subalpine grasslands, scrublands and forest margins from Mt Hikurangi southwards, forms an erect rather rigid shrub, to 3 m high, often with divaricating branches clothed with reddish bark. The thick, pointed, narrow leaves are 10–14 mm long by 1–1.5 mm wide. Flowers appear in October and November, with the translucent drupes, 6–8 mm long, following from February to April and sometimes persisting on the plant through the winter.   RUBIACEAE

857   Needle-leaved mountain coprosma with drupes, Arthur's Pass (April)

858   Mountain five finger, *Pseudopanax colensoi*, is a shrub, up to 5 m high, found in subalpine scrub and forest throughout New Zealand. Strongly sweet-scented male and female flowers occur from May to October, and the rather flat, rounded, black seeds occur in large bunches, ripening by about the following March. Leaves are 3–5 foliolate with the thick, leathery, sessile leaflets serrate on the upper halves.   ARALIACEAE

858   Mountain five finger with male flowers and flower-buds, Mt Taranaki (May)

859 Red mountain heath in flower, Cleddau Canyon (January)

860 Pepper tree flower, Mt Holdsworth (October)

859 Red mountain heath, *Archeria traversii,* is a shrub, 2–5 m high, with many erect or spreading branches. The leathery leaves are 7–12 mm long by 1.5–3 mm wide, Flowers occur as terminal racemes, 10–25 mm long, from December to February. The plant is found in subalpine scrub and forest throughout the South and Stewart Islands but is local in occurrence. EPACRIDACEAE

860–861 Pepper tree, *Pseudowintera axillaris,* is a shrub or small tree, up to 8 m high, found in subalpine scrub and forest and in lowland forests throughout New Zealand. The leaves, 6–10 cm long by 3–6 cm wide, are shining green on both surfaces, highly aromatic and pungent to taste. Flowers (fig. 860), 10 mm across, occur as axillary fascicles from September to December, and the fruits (fig. 861) ripen to red from October till January, persisting on the plant through to June. WINTERACEAE

861 Spray of pepper tree with berries, Maungutukutuku (April)

862 Varnished koromiko, *Hebe vernicosa,* is a semi-erect to prostrate shrub found in subalpine scrub and fellfields throughout the north-west Nelson and Marlborough regions. The leaves, 15 mm long, have a varnished appearance and their petioles twist so that, though they arise opposite and alternate, they do not immediately appear to do so. Flower-heads, 5 cm long, arise laterally towards the branch tips during January and February. This hebe also grows, to 1 m high, in the beech forests of the Nelson Lakes region. SCROPHULARIACEAE

862 Varnished koromiko spray with flower panicle, Mt Robert track (January)

863   Tararua hebe plant in flower, Mt Holdsworth (February)

863   Tararua hebe, *Hebe evenosa*, is a spreading, branching hebe, to 1 m high, found in the Tararua Ranges. The smooth, thick leaves are 15–20 mm long; the flowers occur as crowded lateral spikes around the branch tips during January and February.
SCROPHULARIACEAE

864   Cypress-like hebe branches with flowers, Marlborough mountains (December)

864   Cypress-like hebe, *Hebe cupressoides*, forms a slender, densely branched, round shrub, to 2 m high, found in subalpine scrub and on river flats east of the Southern Alps from Marlborough to Otago. Flowers occur during December and January.
SCROPHULARIACEAE

865   Ochreous whipcord, *Hebe ochracea*, is an erect whipcord *Hebe*, to 30 cm high, with glossy green branches, ochreous towards their tips. Flowers occur as terminal spikes during December and January. Found in subalpine scrub of west Nelson mountains.
SCROPHULARIACEAE

866   Dish-leaved hebe, *Hebe pinguifolia*, is one of a complex of species with dish-shaped leaves occurring in subalpine-alpine scrub on the eastern side of the drier mountains from Nelson to Canterbury. Crowded, sessile flowers occur during December and January. Other species belonging to the complex are *H. buchananii* and *H. treadwellii*, both found in similar places.   SCROPHULARIACEAE

866   Dish-leaved hebe, *Hebe pinguifolia*, showing leaves and flowers, Jack's Pass (January)

865   Ochreous whipcord branches with flowers, Mt Peel, Nelson (December)

867 North-west Nelson hebe, *Hebe coarctata*, is a spreading shrub, to 1 m high, with arching branches tending to be decumbent, found along the edges of beech forest in the Nelson Lakes area and on the Brunner Range. Flowers occur in December and January. SCROPHULARIACEAE

867 Spray of north-west Nelson hebe with flowers, Mt Robert (January)

Snowberries

Low shrubs belonging to the genera *Gaultheria* and *Pernettya* are found throughout the New Zealand mountains from subalpine scrub to the higher alpine fellfields. All species bear flowers and edible fruits called snowberries in profusion during spring, summer and autumn. ERICACEAE

868–869 Snowberry, *Gaultheria antipoda*, is 30 cm to 2 m high, with branchlets clothed with a fine, silvery down mingled with black or yellow bristles. Thick, leathery leaves, 5–15 mm long by 3–15 mm wide, have conspicuous veins and bluntly serrated margins. Flowers (fig. 869) occur from October to February, and the snowberries (fig. 868) ripen from December to April. Found all through New Zealand in subalpine scrub, fellfields and rocky places.

868 Snowberry plant with berries, Volcanic Plateau (February)

870 Tararua snowberry plant in full flower, Mt Holdsworth (December)

871 Tararua snowberry flowers close up, Mt Holdsworth (December)

870–871 Tararua snowberry, *Gaultheria subcorymbosa*, forms a branching shrub, to 1 m high, which smothers itself with flowers (fig. 870). It is found in the Ruahine and Tararua Ranges and the mountains of Nelson. Leaves, 15–20 mm long by 5–7 mm wide, are thick and finely serrate-crenulate. Flowers occur as terminal and subterminal racemes from November till March, followed by snowberries from January till May.

869 Close-up of snowberry flowers and leaves, Taupo (October)

872   Weeping matipo flowers close up, Upper Travers Valley (August)

873   Weeping matipo branches with leaves and berries, Upper Travers Valley (April)

872–873   **Weeping matipo,** *Myrsine divaricata*, is a shrub or small tree, to 3 m high, found in subalpine scrub and forests, preferably where the ground is moist, throughout New Zealand. The obovate leaves, 5–15 mm long by 5–10 mm wide, are emarginate (fig. 873); flowers, 2–3 mm across, occur along the branches (fig. 872) from June till November, and the berries, 4–5 mm across, ripen from August to April.
MYRSINACEAE

874–876   **Mountain tutu,** *Coriaria pteridioides*, forms a small shrub, 60 cm high, with square, slender, hairy branchlets and narrow, lanceolate leaves, 15–25 mm long by 2–4 mm wide, each with a prominent pair of lateral veins (fig. 874). Flowers (fig. 875) occur as racemes to 5 cm long from October to February, and the black berries (fig. 876) are ripe between November and April. Found in the North Island only in subalpine scrub on the Central Volcanic Plateau and on Mt Taranaki.
CORIARIACEAE

874   Spray of mountain tutu with flowers, Volcanic Plateau (December)

875   Flowers of mountain tutu, close up, Mt Dobson (December)

876   Ripe berries of mountain tutu, Volcanic Plateau (February)

877 Dense tutu spray with flowers, Arthur's Pass (December)

878 Ripe berries of dense tutu, Haast Pass (January)

877–878 Dense tutu, *Coriaria angustissima*, forms large patches, to 50 cm high, spreading by branching rhizomes. The erect stems bear narrow leaves, 7–10 mm long (fig. 877), and racemes of flowers (fig. 877), 3–5 cm long, from November to February. The shining, intensely black berries follow the flowers till May. Found in subalpine scrub and rocky places, mainly on the west side of the mountains.

CORIARIACEAE

879 Feathery tutu, *Coriaria plumosa*, is a shrub with narrow, feathery-looking leaves, 6–10 cm long by 1.5–3 mm wide, found from Mts Hikurangi and Taranaki southwards to Stewart Island in subalpine scrub and grasslands, and along lowland and subalpine stream banks. Flowers occur from October till February and are followed by crimson-black berries from November till March. CORIARIACEAE

880–881 Mountain totara, *Podocarpus nivalis*, is a much-branching, prostrate, spreading shrub, to 1.3 m high, with leathery, closely set, thick-margined, spirally arranged, rigid leaves, 5–15 mm long by 2–4 mm wide. Found in subalpine scrub and along subalpine forest margins from the Coromandel Ranges southwards but growing in lowland forests in Westland. Male cones (fig. 880) occur during November and December; female seeds (fig. 881), produced on separate plants, ripen from March to May. PODOCARPACEAE

880 Mountain totara spray, showing male cones, Arthur's Pass (December)

879 Feathery tutu with flowers, Mt Belle, Homer Tunnel (December)

881 Mountain totara spray with seeds on red arils, Mt Ruapehu (February)

882–885   Mountain toatoa, *Phyllocladus asplenii-folius* var. *alpinus*, is a shrub or a small tree, up to 9 m high, in which leaves are replaced by modified branchlets called cladodes. These are 2.5–6 cm long, thick, leathery in texture and with thickened margins. Flowers are produced from October till January, the male as catkins, 6–8 mm long, in clusters at the tips of the branches (fig. 882), and the female as clusters imbedded in swollen carpidia along the edges of cladodes (fig. 883). The seeds are black, nut-like and held in a white cupule (figs 884–885). Found in subalpine scrub and forest from the southern Coromandel Ranges southwards; also found in lowland forest in south Westland.   PODOCARPACEAE

886–888   Bog pine/mountain pine, *Halocarpis bidwillii*, is a spreading shrub or small tree, up to 3.5 m high and 6 m across, with 1–2 mm long, leathery, thick, appressed leaves. Found in subalpine scrub from the Central Volcanic Plateau southwards and in lowland forest in south Westland and Stewart Island. Male (fig. 887) and female (figs 886 & 888) cones occur from November till March.

PODOCARPACEAE

884   Mountain toatoa seeds, Mt Ruapehu (February)

882   Mountain toatoa leaves and male cones, Cupola Basin (December)

883   Mountain toatoa branches with female flowers, Mt Ruapehu (November)

885   Mountain toatoa spray heavy with seeds, Mt Ruapehu (February)

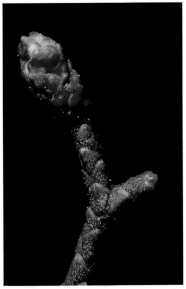

886 Bog pine spray with seeds, Lewis Pass (March)

887 Male cone of bog pine, Lewis Pass (November)

888 Female cone of bog pine with the seed on its white aril, Lewis Pass (March)

889 Mountain cottonwood, *Cassinia vauvilliersii,* is a shrub, to 3 m high, with grooved branches clothed with a brown-coloured, sticky wool. The leaves are leathery, 5–12 mm long by 1–2 mm wide, and clad below with a sticky, brown tomentum. Flowers occur as dense corymbs, 2.5–4 cm across, from December to April. Found in subalpine scrub and grasslands from about Thames southwards.

ASTERACEAE

890 Golden cottonwood, *Cassinia fulvida* var. *montana,* is a compact branching shrub, to 2 m high, with sticky branchlets clothed with a yellow-coloured tomentum, which imparts a golden colour to the shrub. The leaves, 4–8 mm long by 1 mm wide, are thick, leathery with revolute margins and each arising from an erect petiole. The flowers as in *C. vauvilliersii,* and the plant is found in similar situations from the Kaikoura Mountains southwards.

ASTERACEAE

889 Mountain cottonwood flower-head, Mt Taranaki (January)

890 Golden cottonwood spray with flower-head, Ure River bed (March)

Subalpine grass trees

Grass trees belonging to the genus *Dracophyllum* are common in subalpine situations, some 18 of the 35 known New Zealand species occurring in mountain regions. A selection of these is illustrated on these two pages.   EPACRIDACEAE

891–892 Curved-leaf grass tree, *Dracophyllum recurvum*, is found on the Central Volcanic Plateau and the Ruahine and Kaimanawa Ranges among subalpine scrub and fellfields. Leaves are 10–40 mm long (fig. 891) and flowers (fig. 892) occur as spikes, 6 mm long, from December to February.

891   Curved-leaf grass tree on Mt Ruapehu (December)

892   Flowers of curved-leaf grass tree, Mt Ruapehu (December)

893 Turpentine scrub, *Dracophyllum uniflorum*, is an erect shrub, 1 m high, with pungent, needle-like leaves 12–25 mm long, with hairy upper surfaces and ciliate margins. Flowers occur singly around the tips of lateral branches during January and February. Found from the Kaimanawa Ranges southwards in subalpine scrub, herbfields, fellfields and grasslands.

893   Turpentine scrub in flower, Mt Holdsworth (February)

894 Spreading grass tree, *Dracophyllum menziesii*, is a spreading, branching shrub, often covering large areas in damp subalpine scrub or fellfields in the South Island. Leaves are 7–20 cm long by 10–20 mm wide, and flowers occur from December to February.

894   Spreading grass tree with flowers, Mt Belle, Homer Cirque (December)

895 Needle-leaf grass tree, *Dracophyllum fili-folium*, is a shrub or small tree, to 2 m high, with thread-like leaves, 6–16 cm long, with their tips three-sided. Flowers occur in December and January, and the shrub is found in subalpine and alpine scrub and fellfields of the North Island mountains.

897–898 Fiordland grass tree, *Dracophyllum fiordense*, is a small tree, to 3 m high, usually with only one trunk and a crown of leaves (fig. 897), each 60–70 cm long by 4–5 cm wide, beneath which are drooping panicles of flowers, 10–15 cm long (fig. 898), which occur during December and January. Found only in subalpine to alpine scrub in Fiordland and western Otago.

895   Needle-leaf grass tree leaves and flowers, Mount Holdsworth (December)

896 Trailing grass tree, *Dracophyllum pronum*, is a prostrate shrub with narrow, thick, leathery leaves, 5–12 mm long and 1 mm wide, swollen at the tips and with finely serrated margins. Flowers occur singly along the branchlets during December and January. Found from subalpine scrub to fellfields on the east side of the South Island mountains.

897   Fiordland grass trees in flower, Stillwater Basin, Fiordland (December)

896   Trailing grass tree spray with flowers, Waterfall Valley, Cass River, Lake Tekapo (December)

898   Flowers of Fiordland grass tree, Stillwater Basin, Fiordland (December)

899 Torlesse grass tree, *Dracophyllum acerosum*, is a slender, erect shrub, to 2 m high, found in subalpine scrub on the eastern slopes of the mountains between the Spencer Range and Arthur's Pass. Leaves are rigid, ascending, leathery, 2–3 cm long, and pungent. Flowers occur singly near the branch tips during January and February.

900 Specimens of the large grass tree, Mt Arthur (January)

899 Torlesse grass tree spray showing the thick leathery leaves and the flowers arising singly, Lewis Pass (January)

900–901 The **large grass tree,** *Dracophyllum traversii*, is the largest New Zealand grass tree, reaching a height of 10 m, with spreading, ascending branches bearing leaves 30–60 cm long by 4–5 cm wide (fig. 900). Flowers occur terminally on the branches as panicles, 10–20 cm long, during December and January, and the seeds ripen in a large, red-brown sheath (fig. 901) by late February. Found in subalpine scrub and forest from the Mt Arthur Tableland south to about Arthur's Pass.

901 A seed-head of the large grass tree, Otari (February)

TUSSOCKLANDS AND GRASSLANDS

Among the New Zealand mountains, tussocks and carpet grasses belonging to the family Gramineae clothe large areas of the rolling slopes above the bushline, forming meadows that can extend upwards to 2,000 m. In a healthy meadow the ground among the tussocks is carpeted with grasses, sedges, creeping mat plants, herbs, bluebells, bidibids, wild Spaniards, gentians and orchids. Some occasional subshrubs also occur here, and tussock meadows often merge above into herbfields or fellfields.

902 Narrow-leaved tussock in seed in Cupola Basin (April)

902 Narrow-leaved tussock, *Chionochloa pallens*, 60 cm–2 m high, is found abundantly from Mts Hikurangi and Taranaki southwards to altitudes of 1,700 m. It is probably the most common tussock among South Island mountains. GRAMINEAE

903–904 Red tussock, *Chionochloa rubra*, to 1.6 m high, is found to 1,900 m altitude from East Cape and Mt Taranaki southwards (fig. 904) and is abundant in the South Island. Fig. 903 depicts a landscape of red tussock rolling across the Central Volcanic Plateau from Napoleon's Knob.

GRAMINEAE

903 Red tussock uplands above the Moawhango Gorge (April)

904 Red tussock plant, Lewis Pass (January)

905 Silver tussock, *Rytidosperma viride*, is up to 1 m high, with fine silvery leaves, and is found throughout New Zealand to 1,400 m altitude.

GRAMINEAE

905   Silver tussock plant with seeds, Tauhara Mountain (January)

908   Woollyhead plant in flower, Ward's Pass, Molesworth (January)

908 Woollyhead, *Craspedia uniflora*, is a soft-leaved perennial with leaves 5–12 cm long, their margins whitened by tangled, woolly hairs. Flower-stalks are woolly and bear flowers, 15–30 mm across, during December and January. Found from East Cape southwards in herbfields, fellfields and stony places.   ASTERACEAE

906 Carpet grass, *Chionochloa australis*, is found on steep slopes among the mountains of Nelson and Canterbury, up to 2,000 m, as tight, green mats. Flowers occur in December and January, followed by silvery seed-heads. Carpet grass is very slippery to walk on after snow or rain.   GRAMINEAE

906   Carpet grass in seed at Cupola Basin (April)

907 Blue bidibidi, *Acaena inermis*, forms a prostrate, spreading herb with bluish-coloured, 5–7-foliolate leaves up to 5 cm long. It is found in grasslands and fellfields and along the edges of screes among the South Island mountains up to 1,600 m. Flowers occur from December to January.

ROSACEAE

907   Blue bidibidi with flower-heads, Boulder Lake (February)

909 Red bidibidi/piripiri, *Acaena novae-zelandiae*, is a creeping and rooting, prostrate, silky-haired herb with creeping stems, up to 1 m long, forming large patches in subalpine grasslands. The globose flower-heads occur from September till February, and the seeds, which ripen from late November on, attach themselves to clothing and animal fur. Found throughout New Zealand in lowland and subalpine grasslands and open spaces.   ROSACEAE

910 Bidibidi, *Acaena viridior*, is similar to piripiri but never has red-tinged leaves. The flower-heads occur from September till February, and seed-heads from December on. The plant is found throughout New Zealand in lowland tussocklands or open spaces. ROSACEAE

910 Bidibidi plant with flower- and seed-heads, Hinakura (November)

911 Nelson mountain Spaniard, *Aciphylla ferox*, is similar in appearance to *A. horrida*, but smaller and found only among the mountains of Nelson and Marlborough, where it is common. APIACEAE

911 Plant of the Nelson mountain Spaniard in flower, Mt Travers (January)

909 Red bidibidi with seed-heads, Volcanic Plateau (November)

912 Bedstraw, *Galium perpusillum*, is a prostrate, spreading and rooting, perennial herb, forming patches to 30 cm across in damp situations between tussocks, in fellfields near streams or in partially shaded, open, rocky places. Three species of *Galium* occur in New Zealand, of which two grow in lowland areas; *G. perpusillum* is found in mountain regions. Small flowers, 2 mm across, occur in profusion from December to March. RUBIACEAE

912 Bedstraw plant in flower, Cupola Basin (December)

913 Onion-leaved orchid, *Microtis unifolia*, is 7–60 cm high, with fleshy stems and one leaf that is usually longer than the flower-spike. Flowers occur from October till February, either closely or sparingly spaced on the stem, and the plant occurs throughout New Zealand in grasslands and herbfields. ORCHIDACEAE

913 Onion-leaved orchid, close-up of spike showing detail of flowers, Volcanic Plateau (December)

914 Grassland orchid flowers, Volcanic Plateau (February)

915 Grassland orchid, red-flowered form, Ruahine Range (April)

916 Plants of odd-leaved orchid, Lewis Pass (December)

914–915 Grassland orchid, *Orthoceras strictum*, 20–60 cm high, is found in flower from December till April in herbfields and also in lowland dry, open grasslands from Northland to Nelson. Each flower-spike has 3–12 flowers, which may vary considerably in colour from spike to spike.

ORCHIDACEAE

916–917 Odd-leaved orchid, *Aporostylis bifolia*, has two leaves, with one leaf, 4–7.5 cm long, longer and broader than the other. The flower may be white or pink and is borne on a hairy stem from December to January. Found in subalpine grasslands from the Coromandel Ranges southwards to the Auckland and Campbell Islands, descending to sea-level in Stewart Island.   ORCHIDACEAE

918 Rimu-roa flower, Lewis Pass (December)

917 Odd-leaved orchid flowers, close up, Lewis Pass (December)

918 Rimu-roa, *Wahlenbergia gracilis*, is a perennial herb to 40 cm high, with upright stems and very narrow leaves, 10–40 mm long, found in both sub-alpine and lowland grasslands throughout New Zealand. Flowers and fruits occur from September till April.   CAMPANULACEAE

919–920 New Zealand bluebell, *Wahlenbergia albomarginata*, is a perennial, tufted herb with elliptic or spathulate leaves, 5–40 mm long by 1–10 mm wide, arranged as rosettes. Flowers, 10–30 mm across, may be erect, inclined or drooping and are borne on slender stems from November till February. Found throughout the South Island mountains in grasslands, herbfields and fellfields between 600 and 1,550 m altitude.

CAMPANULACEAE

920 New Zealand bluebell, close-up of plant and flowers, Tarndale, Molesworth (January)

919 New Zealand bluebell plant in flower, Mt Robert (January)

921 Marlborough bluebell, *Wahlenbergia trichogyna*, is an Australian plant that grows in New Zealand only in grasslands on the hills of Marlborough. Flowers occur in profusion from November to February. CAMPANULACEAE

922 Flowers of Maori bluebell, Volcanic Plateau (December)

921 Marlborough bluebell plant in flower, Otari (January)

922 Maori bluebell, *Wahlenbergia pygmaea*, forms a low-growing, hairless, tufted perennial herb with serrate-margined leaves, 15 mm long by 2–3 mm wide, arranged as rosettes. Flowers are held erect on slender stalks, 2–5 cm high, from November to February. Found in subalpine and alpine grasslands and herbfields from the Volcanic Plateau to Fiordland. CAMPANULACEAE

923–924 Benmore gentian, *Gentiana astonii*, is an erect, spreading plant, forming large plants up to 1 m across on limestone outcrops in montane to sub-alpine grasslands and herbfields along river valleys of the Kaikoura Coast between the Clarence and Ure Rivers. The leaves are in pairs, 2 cm long by 1–2 mm wide, and flowers, 2–3 cm across, occur singly in great profusion during March and April.

GENTIANACEAE

923 Benmore gentian flowers, close up, Upper Ure River on slopes of Mt Benmore (March)

924 Plant of Benmore gentian in flower, slopes of Mt Benmore (March)

926–927 Prostrate snowberry, *Pernettya macro-stigma*, is a much-branching, prostrate, straggly shrub, to 20 cm high, forming patches in subalpine grasslands, herbfields, fellfields, and rocky places throughout New Zealand. Flowers (fig. 926), about 3.3 cm long, appear from November to January; fleshy berries, 4–7 mm across, each within a fleshy calyx matching the berry in colour (fig. 927), are ripe from December to May. *Pernettya* and *Gaultheria* species hybridise to produce *Gaultheria*-like plants with white or red berries.   ERICACEAE

925 Grassland daisy, *Brachycome sinclairii* var. *sinclairii*, has spathulate, lobed leaves, 2.5–7.5 cm long, arranged in rosettes. Flowers 6–15 mm across on long scapes, 10–15 mm high, occur from October to December. Found in subalpine grasslands and herbfields from East Cape southwards.   ASTERACEAE

925 Grassland daisy plant in flower, Old Man Range (December)

926 Prostrate snowberry flowers, close up, Mt Ruapehu (November)

927 Prostrate snowberry hybrid with red berries, Outerere Gorge, Volcanic Plateau (May)

Wild Spaniards or speargrasses

These are the popular names given to plants of the genus *Aciphylla*, found commonly from sea-level to 1,850 m throughout New Zealand. Thirty-nine species of *Aciphylla* are known between the coast and the high mountains, all characterised by their pinnate, sword-like or spear-like, rigid, pointed leaves. *Aciphylla* species are found in grasslands, herbfields, subalpine and alpine scrub and rocky places. A selection of the alpine species is shown here.   APIACEAE

928–929 Horrid Spaniard, *Aciphylla horrida,* forms clumps to 1 m high with flowers, to 1.5 m high (fig. 929), during December and January and is found in alpine scrub and herbfields on the eastern, dry side of the Main Divide in the South Island from Arthur's Pass to Fiordland. A similar but smaller speargrass, *A. ferox*, occurs among the mountains of Nelson and Marlborough.

928   Horrid Spaniard plants in flower, Mt Belle, Homer Tunnel (December)

929   A plant of the horrid Spaniard in flower, Mt Belle, Homer Tunnel (December)

930   Feathery Spaniard, *Aciphylla squarrosa* var. *flaccida*, is a soft-leaved form of the common Spaniard, *A. squarrosa*, flowering from December to February and found only on North Island mountains in damp alpine and subalpine scrub.

930   Feathery Spaniard plants in flower, Mt Holdsworth (February)

931   Subalpine Spaniard plants in flower, Wapiti Lake, Fiordland (December)

931   Subalpine Spaniard, *Aciphylla pinnatifida*, is a small Spaniard with leaves to 20 cm long, each conspicuously marked by a deep yellow-coloured, central stripe. Flowers occur during December and January. Found in alpine herbfields of western Otago and Fiordland.

932   Plants of the giant Spaniard in flower, Tasman Glacier morain (December)

933   Pigmy speargrass plant in flower, Waterfall Valley, Cass River, Lake Tekapo (December)

934   Close-up of flowers and leaf of pigmy speargrass, Waterfall Valley, Cass River, Lake Tekapo (December)

932   The **giant Spaniard**, *Aciphylla scott-thomsonii*, forms a large clump to 3 m high, bearing flower-heads on stalks, to 4 m tall, during December and January. Found in subalpine scrub and fellfields from about Mt Cook southwards on the drier side of the mountains.

933–934   **Pigmy speargrass**, *Aciphylla monroi*, is a small, tufted speargrass (fig. 933), found in subalpine grasslands, herbfields and rocky places up to 1,700 m among the mountains of Nelson, Marlborough and North Canterbury. Flowers (fig. 934) occur during December and January.

935   **Armstrong's speargrass**, *Aciphylla montana*, is a small speargrass found only among the mountains of the South Island from the Arrowsmith Range south to the Harris Mountains. Flowers on stalks, to 60 cm high, occur during December and January. A similar speargrass, *A. lyallii*, is found among the mountains of Fiordland.

936   Tararua speargrass plant in flower, Mt Holdsworth (December)

935   Armstrong's speargrass plant in flower, bank of Godley River, Lake Tekapo (December)

936   **Tararua speargrass**, *Aciphylla dissecta*, is a small plant, up to 40 cm high, found only in subalpine grasslands on the Tararua Range. Flowers occur during November and December.

937 Wild Spaniard, *Aciphylla colensoi*, is a large plant, to 1 m high, with distinct bluish-coloured, rigid, sword-like, pungent leaves, 30–50 cm long. Sweet-scented flowers occur on long pole-like stalks, 2.5 m high, arising from the centre of the plant, from November to February. Found in subalpine and alpine grasslands and herbfields from East Cape to Canterbury.

937   Wild Spaniard in flower, Mt Holdsworth (February)

938   Little mountain heath, *Pentachondra pumila*, is a slender dwarf shrub, forming patches to 40 cm across in subalpine and alpine grasslands, herbfields, fellfields and along the edges of alpine bogs, occasionally also in lowland grasslands and sand dunes. Flowers occur from November to February, and the large berries, 6–12 mm across, occur from December to April, often with the flower corollas remaining attached.   EPACRIDACEAE

938   Little mountain heath plant with flowers and berries, Jack's Pass (January)

939–941 Prostrate coprosma, *Coprosma petrei*, *Coprosma pumila*, are creeping and rooting coprosmas, forming large, dense, flat mats among grasslands, herbfields and stony places in subalpine regions throughout New Zealand. Flowers are erect and conspicuous during November and December (fig. 939, *C. petrei*). The drupes, 6–8 mm long, are translucent, ripe by March and persist on the plants until December (fig. 940, *C. petrei* & fig. 941, *C. pumila*); they may be greenish white, green, pale blue, orange-red, purplish red or claret in colour.
RUBIACEAE

939   Prostrate coprosma, *Coprosma petrei*, plant in flower, Key Summit (January)

940   Prostrate *C. petrei* plant with drupes, Molesworth (March)

941   Prostrate coprosma, *Coprosma pumila*, plant with drupes, Cupola Basin (December)

942-943 Brown-stemmed coprosma, *Coprosma acerosa* var. *brunnea*, is a sprawling plant with interlacing branches forming flattened mats, to 2 m across, in subalpine grasslands, especially along river terraces and among rocks between 600 and 2,000 m throughout New Zealand. Flowers occur from August to October and the translucent drupes, 5-6 mm long, ripen during March and April, those of the North Island being pale blue with darker stripes (fig. 942), those of the South Island a deep rich blue (fig. 943). RUBIACEAE

Maori onion, *Bulbinella* species

944-948 Maori onion is the popular name for six species of perennial lilies, 30-60 cm high, found in subalpine grasslands and herbfields from Lake Taupo and Mt Taranaki southwards. Three species are illustrated here; all have conspicuous yellow flowers, which occur from October to January, and in the South Island these plants often grow in great numbers over wide areas, making a brilliant display (fig. 944) when in bloom. Flower stalks can reach 45 cm high, with racemes to 15 cm long and flowers, each 10-14 mm across, occupying the upper third. *B. hookeri* is found from the Volcanic Plateau south to the Waiau River. *B angustifolia* occurs from south of the Waiau River to Southland. *B. gibbsii* var. *balanifera* is a shorter plant than *B. hookeri* and is found along the west side of the Southern Alps into Fiordland. ASPHODELACEAE

942 Brown-stemmed coprosma showing interlacing branches, leaves and pale drupes, Upper Waipunga River (April)

944 Maori onion, *Bulbinella hookeri*, clothes the Acheron River Valley floor (January)

943 Brown-stemmed coprosma with deep blue drupes, Jack's Pass (March)

945 Maori onion, *B. hookeri*, plant in flower, Acheron River Valley (January)

946 *B. hookeri* flowers, close up, Acheron River Valley (January)

948 Maori onion, *Bulbinella angustifolia*, flowers, Thomas River (December)

947 Maori onion, *Bulbinella gibbsii* var. *balanifera*, plant in flower, Wapiti Lake, Fiordland (December)

New Zealand subalpine orchids
Many species of orchids occur in our subalpine and alpine regions, but only two of the more striking species are illustrated here.   ORCHIDACEAE

949 Broad-leaved thelymitra flower, Volcanic Plateau (December)

949 **Broad-leaved thelymitra**, *Thelymitra decora*, is up to 50 cm high, with a single, keeled, broad leaf, 10 mm wide; flowers occur in racemes of 1–10 flowers during November and December. Found throughout New Zealand in grasslands and herbfields to 1,250 m altitude.   ORCHIDACEAE

950   Long-leaved thelymitra
      flowers, Volcanic Plateau
      (November)

951   Long-leaved thelymitra
      flowers, close up,
      Kauaeranga Valley, Billy
      Goat Track (December)

952   Tall mountain sedge in
      flower, Boulder Lake
      (January)

950–951   Long-leaved thelymitra, *Thelymitra longi-folia,* (fig. 950), is 7–45 cm high with a single, fleshy, keeled leaf, 3–18 mm wide, found in grasslands and herbfields throughout New Zealand. Flowers, 8–18 mm across (fig. 951), occur during November and December and may be white, blue or pink in colour.   ORCHIDACEAE

952   Tall mountain sedge, *Gahnia rigida,* has erect leaves with long, drooping tips and inrolled margins. Flower-stems, 60 cm–2 m high, occur during January and February, and the plant is found in subalpine regions among the mountains of Nelson and Marlborough.   CYPERACEAE

953   Tall tufted sedge, plume of seeds, Mt Ruapehu
      (May)

953   Tall tufted sedge, *Gahnia procera,* is up to 1 m high, with flower-stems, 60 cm–1 m long, bearing panicles of flowers from October to December. Seeds, each 6 mm long (fig. 953), ripen from February to May. Found in subalpine and lowland grasslands from North Cape to Westland.

                                                CYPERACEAE

954   Black-stemmed daisy, *Lagenophora petiolata,* is a spreading plant that forms patches of rosettes of coarsely serrated, orbicular leaves 10–20 mm across, on petioles 2–3 cm long, at the nodes. Flowers, about 8 mm across, arise on long black stalks from November to February. Found throughout New Zealand and the Kermadec Islands in open places in grasslands and shrublands. ASTERACEAE

954   Black-stemmed daisy plant in flower, Ruahine
      Range (February)

HERBFIELDS AND FELLFIELDS

Herbfields occur in the upper belt of the subalpine region, tending to merge below into the tussocklands and above into the fellfields. The herbfields form extensive areas on the more gentle slopes in the mountains and are characterised by an abundance of tall and medium-sized herbs, sometimes intermingled with tussocks and grasses and an occasional shrub. Fellfields are steeper stony areas with little soil and extend above the herbfields to the permanent snow-line. Fellfield plants are mostly low-growing species, forming more open associations among the rocks and capable of surviving on poor soils in a harsh climate.

955–956 Giant buttercup/Mt Cook lily, *Ranunculus lyallii*, is the largest ranunculus plant in the world (fig. 955), attaining a height of 1–1.5 m with leaves 13–20 cm across. Found in herbfields, especially alongside creeks and streams, from Marlborough to Fiordland, it produces large panicles of white flowers (fig. 956), each 4–5 cm across, from October till January. Although common, it is local in occurrence but makes an unforgettable sight when found clothing an entire slope with these sparkling flowers.

RANUNCULACEAE

955   Giant buttercup in flower, Homer Cirque (December)

956   Close-up of flowers of giant buttercup, Homer Cirque (December)

957   Lobe-leaf buttercup flowers and leaves, Mt Holdsworth (December)

957   Lobe-leaf buttercup, *Ranunculus clivalis*, is a slender, lax but erect buttercup with a few deeply lobed leaves, 5–10 cm across, on slender petioles 7–15 cm long. Flowers, 2–3 cm across, occur sparingly on stalks up to 60 cm high during December and January. Found in wet situations in subalpine grasslands and herbfields and shrubland where the delicate stems are supported by surrounding vegetation, from the Ruahine Range south to the mountains of north-west Nelson.   RANUNCULACEAE

958   Creeping matipo, *Myrsine nummularia*, is a prostrate, rambling and trailing shrub, found among herbfields, fellfields and along scree edges throughout New Zealand. Flowers occur from October till February, and the berries mature during April and May.   MYRSINACEAE

958   Creeping matipo spray with berries, Arthur's Pass (April)

959   Mountain mikimiki, *Cyathodes empetrifolia*, is a prostrate shrub with wiry branches up to 40 cm long and hairy branchlets, to 15 cm high, bearing thick, narrow leaves, 3–5 mm long, with recurved margins. Flowers occur from November to February and the red fruits, 3–5 mm long, from January till April. Found in subalpine herbfields, fellfields, grasslands and boggy places throughout New Zealand.   EPACRIDACEAE

959   Mountain mikimiki plant with fruits, Volcanic Plateau (February)

960   Yellow rock daisy, *Brachyglottis lagopus*, is a small herb clothed with hairs and recognised by its soft reticulated leaves, 3–15 cm long by 3–10 cm wide, on stout hairy petioles, 3–10 cm long. Flowers, 2–4 cm across, on hairy stems up to 35 cm high, occur from January to March, and the plant is found in herbfields, fellfields and grasslands, mostly in the shade of other plants, from the Ruahine Range to Otago.   ASTERACEAE

960   Yellow rock daisy plant in flower, Mt Holdsworth (February)

961–962 Mountain snowberry, *Gaultheria depressa* var. *novae-zelandiae*, forms a prostrate, creeping, rooting shrub with interlacing branches and hairy branchlets bearing thick, leathery, crenulate leaves, 5–10 mm long by 4–6 mm wide, on short petioles (fig. 961). Flowers start in November and continue till February, followed by white or pink berries (fig. 962), 3–4 mm across, from January onwards. Found in subalpine and alpine herbfields and fellfields, grasslands, boggy and rocky places from the Kaimanawa Mountains to Stewart Island.

ERICACEAE

963 Alpine groundsel, *Senecio bellidioides*, is a small plant with rounded, rugose, leathery leaves, up to 5 cm long, on hairy petioles with, usually, hairy margins; the leaves are often appressed to the ground. Flowers, 2–3 cm across, arise on black hairy stalks, up to 30 cm long, from November till January and the plant is found in subalpine herbfields and grasslands throughout the South Island.

ASTERACEAE

963 Alpine groundsel plant in flower, Lewis Pass (December)

961 Mountain snowberry plant in full flower, Homer Cirque (December)

962 Berries of mountain snowberry, Lewis Pass (January)

964 Brockie's bluebell, *Wahlenbergia brockiei*, forms clumps of leafy rosettes crowded together; the leaves are 10–30 mm long by 1–2 mm wide. Flowers, 16–20 mm across, arise on slender stems, to 10 cm high, from each rosette from November to January. Found growing on limestone soils near Castle Hill.

CAMPANULACEAE

964 Brockie's bluebell plant in flower, Otari (January)

965   Alpine avens in flower, Mt Lucretia (January)

966   Mountain heath, showing flowers and leaves, Sugarloaf, Cass (November)

965   Alpine avens, *Geum uniflorum*, is a herbaceous plant, forming large patches in damp areas of subalpine and alpine herbfields, fellfields and open rocky places from the Nelson mountains to Otago. The rounded, hairy-margined leaves arise directly from the root stock, and flowers, to 2.5 cm across, occur during January and February.

ROSACEAE

966–967   Patotara/mountain heath, *Leucopogon suaveolens*, is a prostrate, branching shrub, forming large patches or dense hummocks, 8–15 cm high and up to 1 m across, in herbfields, fellfields, grasslands and exposed places from the Volcanic Plateau southwards. The leaves, 5–9 mm long, are characterised by five parallel veins on each underside, and the leaf margins are hairy. Flowers, each about 7 mm long, arise as 2–5-flowered terminal racemes from November to February. The fruit, about 3 mm across, may be white, pink or crimson and occurs from January to April.   EPACRIDACEAE

967   Mountain heath, spray showing fruit and leaves, Volcanic Plateau (February)

968   Spray of tall pinatoro, showing flowers and leaves, Mt Ruapehu (December)

968   Tall pinatoro, *Pimelea buxifolia*, is a shrub, to 1 m high, found in fellfields and alpine grasslands among the North Island mountains. Leaves, 5–10 mm long by 3–5 mm wide, are keeled, and flowers occur from September to April. A very similar plant, *P. traversii*, is found in fellfields throughout the South Island mountains.

THYMELAEACEAE

969–970 Mountain pinatoro, *Pimelea oreophila,* is a small shrub, 10–15 cm high, with ascending branches; the older branches are scarred by rings left by falling leaves, the younger branches are hairy. Leaves, 3–6 mm long by 1–3 mm wide, have hairy margins, and the sweet-scented flowers, 6 mm across, which appear from October to March, give rise to white, ovoid berries, 2 mm across, from January on. Found in subalpine herbfields, grasslands and exposed places throughout New Zealand.

THYMELAEACEAE

971–972 Common drapetes, *Drapetes dieffenbachii,* is a sprawling, prostrate plant, forming patches up to 30 cm across in subalpine herbfields, fellfields and grasslands from the Coromandel Ranges south to Stewart Island. Leaves, 2.5–3.5 mm long, are appressed to the branches and taper from base to apex. Clusters of flowers occur terminally on the branches from November to January.

THYMELAEACEAE

971   Common drapetes plant in flower, Mt Holdsworth (December)

969   Mountain pinatoro spray, close up, showing flowers, leaves and branch, Sugarloaf, Cass (November)

970   Mountain pinatoro flowers, Sugarloaf, Cass (November)

972   Flowers of common drapetes, close up, Mt Holdsworth (December)

973–974 Matted ourisia, *Ourisia vulcanica,* is a spreading and rooting herb, forming a mat 10–15 cm across. Its thick, fleshy, hairless leaves, 10–25 mm long by 6–15 mm wide, are appressed to the ground, and flowers, 15–20 mm across, on hairy stalks up to 12 cm high, occur from October to January. Found in dry, sunny places on herbfields, fellfields and exposed places on the Kaimanawa Ranges and the Volcanic Plateau.   SCROPHULARIACEAE

973 Matted ourisia flowers and flower-buds, Mt Ruapehu (December)

974 Matted ourisia plant with flowers, Mt Ruapehu (December)

Mountain daisies

Daisies belonging to the genus *Celmisia* are among the most common plants found in the New Zealand mountains. Some have already been met with among the tussocklands but the most species by far are found on the herbfields and fellfields of our mountains. A selection of these is illustrated here.

ASTERACEAE

975–976 Brown mountain daisy, *Celmisia traversii,* is the only *Celmisia* with a rich brown-coloured, velvety tomentum on its leaves and stems (fig. 976). The plant (fig. 975) occurs in herbfields from north-west Nelson south to the Lewis Pass, and flowers, 4–5 cm across, occur during December and January.

975 Brown mountain daisy plant in flower, Mt Lucretia (January)

976 Close-up of leaves of brown mountain daisy, Mt Lucretia (January)

977 Strap-leaved daisy plant in
flower, Fog Peak (January)

977 Strap-leaved daisy, *Celmisia angustifolia,* is a
small, woody plant with leathery, strap-like leaves,
3–5 cm long, flowering from December to February,
and found in subalpine fellfields and herbfields from
Arthur's Pass to the Humboldt Mountains.

**980–981 Large mountain daisy/tikumu/silvery
cotton plant,** *Celmisia coriacea,* forms a large, tufted
herb, with leaves to 60 cm long, silvery above when
young but green when old (fig. 981). The large, sym-
metrical flower (fig. 980), 5–12 cm across, occurs on
fluffy, woolly stems, to 75 cm high, from December
to February. Found in alpine herbfields, fellfields
and grasslands throughout the South Island.

978–979 White cushion daisy, *Celmisia sessiliflora,*
forms a spreading perennial herb with rigid leaves
in tight rosettes, which together can form a mat (fig.
978) 1 m across. Sessile flowers, 10–20 mm across
(fig. 979), occur during December and January.
Found throughout the South Island in subalpine and
alpine herbfields and fellfields.

978 White cushion daisy plant in flower, Wapiti
Lake, Fiordland (December)

979 Close-up of flower of white cushion daisy, Mt
Robert (January)

980 Large mountain daisy, showing the symmetry of
the flower and flower-bud, Arthur's Pass
(January)

981 Large mountain daisy plants in full flower,
Jack's Pass (January)

982   Armstrong's daisy, plants in full flower, Arthur's
      Pass (January)

982–983   **Armstrong's daisy**, *Celmisia armstrongi*
(fig. 982), and **Lance-leaved daisy**, *Celmisia lanceolata* (fig. 983), are both tufted herbs with rigid,
lance-like leaves up to 35 cm long; those of *lanceolata*
each have a distinct, stout, yellowish orange-coloured
midrib; those of *armstrongi* each have a yellowish
orange band down each side of the midrib. Both
flower during January and February and are found
in herbfields and fellfields among the mountains of
the South Island.

983   Lance-leaved daisy plants with flowers, Stillwater
      Basin, Fiordland (December)

984   **Sticky-stalked daisy**, *Celmisia hieracifolia*, is
a grassland to herbfield daisy found from the Ruahine and Tararua Ranges south to the mountains of
Nelson. Flowers, 3–4 cm across, occur on sticky,
hairy stalks during January and February.

984   Sticky-stalked daisy in flower, Brown
      Cow Pass, Boulder Lake (February)

985   **Haast's daisy**, *Celmisia haastii*, forms patches
to 75 cm across; the leaves, 4–7 cm long, are
leathery, green above with distinct longitudinal
grooves and with satiny hairs below. Flowers, up to
4 cm across, occur during December and January.
The whole plant is very sticky and is found throughout South Island mountains in herbfields and fellfields.

985   Haast's daisy plants in flower, Temple Basin
      (January)

986 Purple-stalked daisy, *Celmisia petiolata*, forms a large, tufted herb with smooth, silky leaves, 7–15 cm long, having purple lateral veins and flat, purple petioles. Flowers, 5–7 cm across, occur on purple, hairy stalks, 15–20 cm high, during December and January. Found in western herbfields and fellfields from the Spencer Mountains southwards.

986 Purple-stalked daisy in flower, Arthur's Pass (January)

987 Boulder Lake daisy, *Celmisia parva*, is found only near Boulder Lake and on the Paparoa Range in herbfields and grasslands where the climate tends to be wetter. The leaves, 5–15 cm long by 10–15 mm wide, each tapering to its apex, have recurved, faintly toothed margins and dense appressed hairs below. Flowers, 2–3 cm across, occur during January and February.

987 Boulder Lake daisy in flower, Douglas Ridge, Boulder Lake (February)

988 Pigmy daisy, *Celmisia lateralis*, is a sprawling plant with branching stems 30 cm long and thick, dense, overlapping, hairless leaves, 6–8 mm long by 1–2 mm wide. Flowers, 10–20 mm across, occur on 8 cm long, slender stalks during January and February. Found only in herbfields and fellfields of the mountains of north-west Nelson and the Paparoa Range.

988 Pigmy daisy plant with flowers, Lead Hill (February)

989 Dagger-leaf daisy, *Celmisia petriei*, is a tufted daisy with rigid, thick, sharply pointed, dagger-like leaves, 20–30 cm long by 10–20 mm wide, bright green above with two prominent ridges. Flowers, 3–4 cm across, occur on stout, woolly stems, 20–50 cm high, from December to February. Found in herb- and fellfields from the Nelson mountains to Fiordland.

989 Dagger-leaf daisy in flower, Douglas Ridge, Boulder Lake (February)

990 Cotton daisy/cotton plant, tikumu, *Celmisia spectabilis*, forms a stout, rosulate, tufted herb, sometimes forming patches. Leaves, 10–15 cm long by 10–25 mm wide, are very thick and leathery, with the upper surface smooth and shining but the lower surface densely clothed by soft, matted, buff or white hairs. Flowers, 4–5 cm across, on white woolly stalks, 8–25 cm high, occur from December to February. Found in subalpine grasslands, herbfields and fellfields from Mt Hikurangi south to North Otago.

990   Cotton daisy plant in flower, Porter's Pass (January)

991 Allan's daisy, *Celmisia allanii*, forms a loosely branched plant with leaf remains on the branches. The leaves, 3–4 cm long by 10–15 mm wide, are thin, flexible and densely clothed all over with soft white hairs. Flowers, 3–4 cm across, occur on hairy stalks during January and February. Found in herbfields and grasslands from the Nelson mountains to the Lewis Pass.

991   Allan's daisy plant in flower, Mt Lucretia (January)

992 Dainty daisy, *Celmisia gracilenta*, is a slender, tufted daisy with tough, inflexible leaves, 10–15 cm long by 2–4 mm wide, each tapering to a sharp apex and with recurved margins, rolled almost to the midrib, to cover the lower surface of appressed, satiny hairs. Flowers, 10–20 mm across, on slender stalks occur from November to February. Found all over New Zealand in herbfields and subalpine or alpine bogs and grasslands from the Coromandel Peninsula southwards. I have also found it growing splendidly beside the sea at Wharariki Beach at the base of Farewell Spit (fig. 1008).

992   Flowers and leaves of dainty daisy, Mt Tongariro (November)

993 Marlborough daisy, *Celmisia monroi*, grows as a single plant or as a mass of branching stock, with rigid leaves, 7–15 cm long by 5–20 mm wide, the upper surface grooved longitudinally and covered by a thin, silvery skin. Flowers, 4 cm across, occur during December and January. Found in subalpine herbfields and fellfields of Marlborough and at the Woodside Gorge.

994 Larch-leaf daisy, *Celmisia laricifolia*, is a mat-forming plant with branches clothed by leaf remains and bearing tufts of narrow, pointed, silvery green, larch-like leaves, 10–15 mm long by 1–1.5 mm wide, with the margins rolled almost to the midrib. Flowers, 10–20 mm across, occur during January and February. Found in fellfields and exposed places throughout the mountains of the South Island.

993   Marlborough daisy plant in
      flower, Woodside Gorge
      (December)

994   Larch-leaf daisy plant in
      flower, Temple Basin
      (January)

995   Sticky daisy plant in flower,
      Mt Torlesse Range (January)

995   Sticky daisy, *Celmisia viscosa,* forms patches up to 1 m across in herb and fellfields and grasslands of the eastern slopes of the South Island mountains. the leaves, 6–8 cm long by 6–9 mm wide, are moderately sticky, and flowers, 2–4 cm across, occur on sticky stalks, 15–30 cm high, from December to February.

996   Mt Taranaki daisy, *Celmisia major* var. *brevis,* is found in herb- and fellfields on Mt Taranaki. The thick, leathery leaves taper to an acute apex with distinct midrib, recurved margins and satiny hairs below. Flowers, 2.5–3 cm across, occur from December to February.

996   Mt Taranaki daisy plants in
      flower, Mt Taranaki (December)

997   Downy daisy plants in flower, Mt Taranaki
      (January)

997   Downy daisy, *Celmisia glandulosa,* is a creeping, rooting plant, forming rosettes of leaves 15–25 mm long and 10–15 mm wide, covered by many minute glands and having finely serrate margins. Flowers, 2 cm across, arise on slender stalks, 7–12 cm high, during December and January. Found throughout New Zealand in herbfields, fellfields, grasslands and exposed places.

998 Musk daisies, *Celmisia du-rietzii* and *Celmisia discolor*, are both strongly musk-scented, sprawling or prostrate herbs, often forming huge patches in alpine herbfields, fellfields, grasslands and rocky places throughout the South Island mountains. Older sections of branches are covered with dead leaves, the younger sections with rosulate tufts of sticky leaves, 2–4 cm long by 8–12 mm wide; in *discolor* the upper surface is lightly clothed with white hairs, in *du-rietzii* it is bare. Flowers, 2–3 cm across, on stalks 10–15 cm high, occur during December and January.

998 Musk daisies *C. discolor*, in full flower, Arthur's Pass (January)

New Zealand gentians

Nineteen species of gentians are known from the mountain regions of New Zealand. All flower in the late summer or autumn. Their flowers are mostly white and the species are difficult to separate. A selection is illustrated here.   GENTIANACEAE

999 Ridge-stemmed gentian, *Gentiana tereticaulis*, is an erect herb with ridged stems, 24–45 cm high. Flowers, to 2 cm across, are both terminal and subterminal, with the upper cauline leaves sessile. Found in herbfields and fellfields from the Nelson Mountains to Fiordland.

1000 Alpine gentian, *Gentiana patula*, is a sprawling herb with stems rising at their tips to 20–50 cm to bear cymes of flowers, each 2–2.5 cm across, during January and February. Found in subalpine and alpine herbfields and grasslands from the Tararua Ranges southwards.

1001 Townson's gentian, *Gentiana townsonii*, forms a herb up to 30 cm high, with crowded, fleshy, basal leaves. The plant is found in subalpine herbfields and grasslands from north-west Nelson to Arthur's Pass and flowers during January and February.

999 Ridge-stemmed gentian flowers, Cupola Basin (April)

1000 Alpine gentian flowers, Mt Holdsworth (February)

1001 Townson's gentian flowers, Boulder Lake (January)

1002 North-west Nelson gentian plant in flower, Brown Cow Pass, Boulder Lake (February)

1003 Pink gentian flowers, Upper Cobb Valley (February)

1004 Snow gentian, mauve flowers, McKenzie Pass (March)

1002 North-west Nelson gentian, *Gentiana spenceri,* is a herb about 15 cm high, with flowers, about 2 cm across, occurring from February to April as umbels each surrounded by a whorl of 5–7 spathulate leaves. Found in shaded situations in subalpine herbfields among the mountains of north-west Nelson and Westland.

1003 Pink gentian, *Gentiana tenuifolia,* is a perennial with finely ridged, erect stems, up to 40 cm high, bearing masses of pink flowers, 2.5 cm across, during February and March. Found in subalpine herbfields from north-west Nelson to the Lewis Pass.

1004–1005 Snow gentian, *Gentiana matthewsii,* forms a tall gentian, branching and dividing from the base (fig. 1005), and found in subalpine to alpine herbfields and fellfields from the Humboldt Mountains to Fiordland. Flowers, about 2.5 cm across, occur in profusion from January to March and can be white or pale mauve (fig. 1004).

1006 Common New Zealand gentian, *Gentiana bellidifolia,* is a herb with crowded tufts of thick, fleshy, elliptic basal leaves, 10–15 mm long by 5–7 mm wide. Flowers, each about 2 cm across, occur during March and April as terminal cymes of 2–6 flowers each, on several stems 5–15 cm high. Found from Mt Hikurangi to Fiordland in damp places in herbfields and grasslands.

1005 Large plant of snow gentian with white flowers, Gertrude Cirque, Fiordland (January)

1006 Common New Zealand gentian flowers, Arthur's Pass (April)

PLANT ASSOCIATIONS

Most New Zealand plants will grow singly, as isolated specimens, but in their natural state they normally grow in intimate and often complicated associations and communities. Forests, the most complicated and extensive communities, exert a profound effect upon climate and help to minimise land erosion. Along the New Zealand sea coasts and in the mountains delicate, often fragile plant associations also play essential roles in land stabilisation and climate regulation.

1007 Looking north-east from Mt Lucretia over the Lewis Pass showing the characteristic demarcation line between the beech forest and the subalpine and alpine regions (January)

1009 An alpine meadow on Jack's Pass, North Canterbury (January).

1008 An aberrant association of the subalpine *Celmisia gracilenta* growing in sand at Wharariki Beach, Farewell Spit. Other New Zealand *Celmisia* also occur in this aberrant fashion around the Fiordland Sounds.

1010 A rain forest interior in a Westland podocarp forest shows the nature of a primeval forest community in New Zealand, Fox Glacier (January).

GLOSSARY

acuminate: tapering to a fine point

adpressed: pressed closely together but not joined

alveolate: deeply and closely pitted

apetalous: not having any petals

appressed: closely applied to

aril: an outgrowth or appendage to a seed

attenuated: gradually tapering to a point

axillary: the upper angle between two structures

calyx: the outer whorl of the parts of a flower, usually green, and made up of the separate or fused sepals

capsule: a dry fruit that splits at maturity to release its seeds

carpidium: the scale at the base of the cone in Gymnosperms

ciliated: fringed with hairs along the margin

cladodes: flattened stems having the function of leaves

corymb: a flat topped cluster of flowers in which the longer stalked outer flowers open first

costa: a rib, especially the midvein of a leaf

crenate: having rounded shallow teeth

cupule: a cup-like structure occurring at the base of some flowers

cyme: an inflorescence in which the principal stalk ends with a flower and in which other flowers are produced at the ends of lateral stalks

decumbent: lying along the ground but with the tip ascending

dentate: having sharp teeth at right angles to the margin

dioecious: having male and female flowers on different plants

divaricating: spreading at very wide angles

drupe: a fruit with a seed enclosed in a bony cover surrounded by a fleshy layer, a "stone fruit"

emarginate: having a narrow notch at the apex

fascicles: a closely bunched cluster

glabrous: without any hairs

globose, globate: a round-shaped structure

hybrid: an organism produced from interbreeding of two different species

imbricate: overlapping, like the tiles on a roof

inflorescence: an arrangement of flowers or a collection of flowering parts

lanceolate: lance-shaped, tapering from about one third from base towards the apex

leaflet: a single element of a compound leaf

monoecious: having male and female flowers on the same plant

montane: a zone between the lowland and subalpine regions

obovate: egg-shaped

ochreous: dull yellow coloured with often a tinge of red

panicle: a branching inflorescence of flowers with each flower on a stalk

pedicel: the stalk that holds a single flower

peduncle: the stalk of a solitary flower or the stalk of a compound flower head

perianth: the outer whorl of a flower when it is not distinctly formed from
    sepals and petals; a general term for the calyx and the corolla together

petiole: the stalk of a leaf

pinna: a primary division of a divided leaf

pinnate: with the parts divided and arranged on either side of the axis:
    compound

pinnatifid, pinnatisect: compound

raceme: an unbranched elongate inflorescence with flowers having stalks; the
    flowers at the base being the oldest

rachis: the main axis of an inflorescence; plural is rachides

reticulated: forming a network

revolute: rolled outwards or rolled to the lower side

rhizome: an underground, spreading stem

rosulate: forming a small rosette

rugose: wrinkled

serrated: sharply toothed with teeth pointing forwards or outwards

sessile: without a stalk

spathulate: spoon-shaped

strobilus: a cone-shaped structure as in pines and lycopods; plural is strobili

terminal: borne at the apex of a stem

tomentum: a more or less dense covering of matted or appressed hairs,
    described as tomentose

trifoliate: having three leaves

trifoliolate: having three leaflets

umbel: a usually flat-topped inflorescence having the pedicels arising from a
    common centre point

unifoliate: having a single leaf

REFERENCE WORKS

Adams, Nancy M., *Mountain Flowers of New Zealand*. A.H. & A.W. Reed, Wellington, 1965.

Allan, H.H., *Flora of New Zealand: Volume 1*. Government Printer, Wellington, 1961.

Brooker, S.G.; Cambie, R.C.; and Cooper, R.C., *New Zealand Medicinal Plants*. Revised edition, Heinemann, Auckland, 1987.

Cartman, Joe, *Growing New Zealand Alpine Plants*. Reed Methuen, Auckland, 1985.

Cheeseman, T.F., *Manual of the New Zealand Flora*. Government Printer, Wellington, 1925.

Cockayne, L., *New Zealand Plants and Their Story*. Government Printer, Wellington, 1910.
—— *The Vegetation of New Zealand*. Third Edition, Engelmann, Leipzig, 1958.

Cockayne, L. and Turner, E.P., *The Trees of New Zealand*. Second edition, Government Printer, Wellington, 1958.

Connor, H.E., *The Poisonous Plants in New Zealand*. New Zealand DSIR Bulletin 99, Second edition, 1977.

Crowe, Andrew, *A Field Guide to the Native Edible Plants of New Zealand*. Collins, Auckland, 1981.

Eagle, A.S., *Eagle's Trees and Shrubs of New Zealand in Colour*. Collins, Auckland, 1975.
—— *Eagle's Trees and Shrubs of New Zealand in Colour*. Second series, Collins, Auckland, 1982.
—— *100 Shrubs and Climbers of New Zealand*. Collins, Auckland, 1978.

Featon, E.H., *The Art Album of New Zealand Flora*. Bock & Cousins, London, 1889.

Fisher, Muriel E.; Satchell, E.; and Watkins, Janet M., *Gardening with New Zealand Plants, Shrubs and Trees*. Collins, Auckland, 1970.

Given, David R., *Rare and Endangered Plants of New Zealand*. A.H. & A.W. Reed, Wellington, 1981.

Healy, A.J. and Edgar, E., *Flora of New Zealand: Volume 3*. Government Printer, Wellington, 1980.

Johnson, Marguerite, *New Zealand Flowering Plants*. Caxton Press, 1968.

Johnson, Peter and Brooke, Pat, *Wetland Plants in New Zealand*. DSIR Publishing, Wellington, 1989.

Laing, R.M. and Blackwell, E.W., *Plants of New Zealand*, Fourth edition, Whitcombe and Tombs Ltd., Wellington, 1940.

Malcolm, Bill and Nancy, *New Zealand's Alpine Plants Inside and Out*. Craig Potton, Nelson, 1988.

Mark, A.F. and Adams, N.M., *New Zealand Alpine Plants*. A.H. & A.W. Reed, Wellington, 1973.

Metcalf, L.J. *The Cultivation of New Zealand Trees and Shrubs*. A.H. & A.W. Reed, Wellington, 1972.

Moore, L.B., and Adams, Nancy M., *Plants of the New Zealand Coast*. Pauls Book Arcade, Auckland, 1963.

Moore, L.B. and Edgar, E., *Flora of New Zealand: Volume 2*. Government Printer, Wellington, 1970.

Moore, L.B. and Irwin, J.B., *The Oxford Book of New Zealand Plants*. Oxford University Press, Wellington, 1978.

Philipson, W.R. and Hearn, D., *Rock Garden Plants of the Southern Alps*. Caxton Press, Christchurch, 1962.

Poole, A.L. and Adams, N.M., *Trees and Shrubs of New Zealand*. Revised edition, Government Printer, Wellington, 1979.

Richards, E.C., *Our New Zealand Trees and Flowers*. Third edition, Simpson and Williams Ltd, Christchurch, 1956.

Salmon, J.T., *The Native Trees of New Zealand*. A.H. & A.W. Reed, Wellington, 1980.

—— *Collins Guide to the Alpine Plants of New Zealand*. Collins, Auckland, 1985.

—— *A Field Guide to the Native Trees of New Zealand*. Reed Methuen, Auckland, 1986.

Sampson, F. Bruce, *Early New Zealand Botanical Art*. Reed Methuen, Auckland, 1985.

Webb, Colin; Johnson, Peter; and Sykes, Bill, *Flowering Plants of New Zealand*. DSIR, Christchurch, 1990.

Wilson, Catherine M. and Given, David R., *Threatened Plants of New Zealand*. DSIR Publishing, Wellington, 1989.

Wilson, H.D., *Field Guide, Wild Plants of Mount Cook National Park*. Field Guide Publications, Christchurch, 1978.

—— *Field Guide, Stewart Island Plants*. Field Guide Publications, Christchurch, 1982.

INDEX

S0-AEN-258

The United States Government Manual 2011

Office of the Federal Register
National Archives and Records Administration

Revised March 14, 2011

Raymond A. Mosley,
Director of the Federal Register.

David S. Ferriero,
Archivist of the United States.

On the cover: This edition of *The United States Government Manual* marks the 75th anniversary of the *Federal Register* and celebrates its role as the official journal of the rules and regulations, notices, and Presidential documents of the Federal government of the United States of America. The cover features the majestic stone eagle from the National Archives Building in Washington, DC, designed by noted architect John Russell Pope. Symbolic of protection, the eagle serves as a reminder of the National Archives and Records Administration's role as guardian of this Nation's records, and thus served as the inspiration for NARA's new agency logo. On the back, the design incorporates text from the very first document published in the *Federal Register.*

Before the advent of the *Federal Register,* American citizens had no single, authoritative source of information about their legal rights and responsibilities under Federal agency programs. Since, March 14, 1936, the *Federal Register* has been published, every Federal business day, to provide public notice of the regulatory activities and official documents of the Federal executive branch agencies and the President. Final rules published in the *Federal Register* are ultimately codified in the *Code of Federal Regulations,* a related publication of the Office of the Federal Register, which is part of the National Archives.

With the aid of rapidly changing technology, on its 75th anniversary, the *Federal Register* has transformed itself from a print-only publication to include the online publication known as Federal Register 2.0 (FR 2.0). On July 26, 2010, the National Archives Office of the Federal Register and the Government Printing Office launched FR 2.0 on FederalRegister.gov. The FR 2.0 web site is similar to a daily web newspaper. The site displays individual news sections for the topics of money, environment, world, science and technology, business and industry, and health and public welfare. FR 2.0 takes advantage of social media and integrates seamlessly with Regulations.gov and the Unified Agenda to make it easy for users to submit comments directly into the official e-Rulemaking docket and view the history of rulemaking activity through a regulatory timeline. The next phase in its evolution is the development of a mobile application for displaying the *Federal Register* on smartphones and tablet devices.

We extend special thanks to the Creative and Digital Media Services at the Government Printing Office for its artistic contributions in designing the cover.

For sale by the Superintendent of Documents, U.S. Government Printing Office
Internet: bookstore.gpo.gov   Phone: toll free (866) 512–1800; DC area (202) 512–1800
Fax: (202) 512–2250   Mail: Stop SSOP, Washington, DC 20402–0001

ISBN 978–0–16–087470–3

Preface

As the official handbook of the Federal Government, *The United States Government Manual* provides comprehensive information on the agencies of the legislative, judicial, and executive branches. The *Manual* also includes information on quasi-official agencies; international organizations in which the United States participates; and boards, commissions, and committees.

A typical agency description includes a list of principal officials, a summary statement of the agency's purpose and role in the Federal Government, a brief history of the agency, including its legislative or executive authority, a description of its programs and activities, and a "Sources of Information" section. This last section provides information on consumer activities, contracts and grants, employment, publications, and many other areas of public interest.

The *Manual* is also currently available and periodically updated on its own website. The U.S. Government Manual website (usgovernmentmanual.gov) is jointly administered by the Office of the Federal Register (OFR)/Government Printing Office (GPO) partnership. The website offers three ways to find information about Government agencies and organizations by entering a term in the keyword search box, browsing categories, or using "The Government of the United States" site map for an overview of the Government.

The 2011 *Manual* was prepared by the Presidential and Legislative Publications Unit, Office of the Federal Register. Alfred W. Jones was Managing Editor; Matthew R. Regan was Chief Editor, assisted by Heather McDaniel, Lois Davis, Joseph Frankovic, Martin Franks, Joshua Liberatore, and Joseph Vetter.

THE FEDERAL REGISTER AND ITS SPECIAL EDITIONS

The *Manual* is published as a special edition of the *Federal Register* (see 1 CFR 9.1). Its focus is on programs and activities. Persons interested in detailed organizational structure, the regulatory documents of an agency, or Presidential documents should refer to the Federal Register or one of its other special editions, described below.

Issued each Federal working day, the *Federal Register* provides a uniform system for publishing Presidential documents, regulatory documents with general applicability and legal effect, proposed rules, notices, and documents required to be published by statute.

The *Code of Federal Regulations* is an annual codification of the general and permanent rules published in the *Federal Register*. The *Code* is divided into 50 titles that represent broad areas subject to Federal regulation. The *Code* is kept up to date by the individual issues of the *Federal Register*.

The *Compilation of Presidential Documents* serves as a timely, up-to-date reference source for the public policies and activities of the President. It contains remarks, news conferences, messages, statements, and other Presidential material of a public nature issued by the White House. The *Compilation of Presidential Documents* collection is composed of the *Daily Compilation of Presidential Documents* and its predecessor, the *Weekly Compilation of Presidential Documents*.

A companion publication to the *Compilation of Presidential Documents* is the *Public Papers of the Presidents*, which contains public Presidential documents and speeches in convenient book form. Volumes of the *Public Papers* have been published for every President since Herbert Hoover, with the exception of Franklin D. Roosevelt, whose papers were published privately.

OTHER OFFICE OF THE FEDERAL REGISTER PUBLICATIONS

The Office of the Federal Register publishes slip laws, which are pamphlet prints of each public and private law enacted by Congress. Slip laws are compiled annually as the *United States Statutes at Large*. The *Statutes* volumes contain all public and private laws and concurrent resolutions enacted during a session of Congress; recommendations for executive, legislative, and judicial salaries; reorganization plans; proposed and ratified amendments to the Constitution; and Presidential proclamations. Included with many of these documents are sidenotes, *U.S. Code* and statutes citations, and a summary of their legislative histories.

ELECTRONIC SERVICES

The Office of the Federal Register maintains an Internet site for public law numbers, the Federal Register's public inspection list, and information on the Office and its activities at www.archives.gov/federal-register. This site also contains links to the texts of *The United States Government Manual*, public laws, the *Compilation of Presidential Documents*, the *Federal Register*, and the *Code of Federal Regulations* (both as officially published on a quarterly basis and a new unofficial, daily updated version, the e-CFR) in electronic format through the GPO's Federal Digital System (FDsys) at www.fdsys.gov. For more information, contact the GPO Customer Contact Center, U.S. Government Printing Office. Phone, 202–512–1800, or 866–512–1800 (toll-free). Email, gpo@custhelp.com.

INQUIRIES

For inquiries concerning *The United States Government Manual* and other publications of the Office of the Federal Register, call 202–741–6000, write to the Director, Office of the Federal Register, National Archives and Records Administration, Washington, DC 20408, or email fedreg.info@nara.gov.

SALES

The publications of the Office of the Federal Register are available for sale by writing to the Superintendent of Documents, P.O. Box 371954, Pittsburgh, PA 15250–7954. Publications are also available for sale through the GPO's online bookstore at http://bookstore.gpo.gov, the GPO bookstore located in Washington, DC, and the retail sales outlet in Laurel, MD. Telephone inquiries should be directed to 202–512–1800, 866–512–1800 (toll-free), or 202–512–2104 (fax).

Contents

Declaration of Independence

Action of Second Continental Congress, July 4, 1776

IN CONGRESS, JULY 4, 1776.

THE UNANIMOUS DECLARATION of the thirteen united STATES OF AMERICA,

WHEN in the Course of human events, it becomes necessary for one people to dissolve the political bands which have connected them with another, and to assume among the powers of the earth, the separate and equal station to which the Laws of Nature and of Nature's God entitle them, a decent respect to the opinions of mankind requires that they should declare the causes which impel them to the separation.

We hold these truths to be self-evident, that all men are created equal, that they are endowed by their Creator with certain unalienable Rights, that among these are Life, Liberty and the pursuit of Happiness.—That to secure these rights, Governments are instituted among Men, deriving their just powers from the consent of the governed,— That whenever any Form of Government becomes destructive of these ends, it is the Right of the People to alter or to abolish it, and to institute new Government, laying its foundation on such principles and organizing its powers in such form, as to them shall seem most likely to effect their Safety and Happiness. Prudence, indeed, will dictate that Governments long established should not be changed for light and transient causes; and accordingly all experience hath shewn, that mankind are more disposed to suffer, while evils are sufferable, than to right themselves by abolishing the forms to which they are accustomed. But when a long train of abuses and usurpations, pursuing invariably the same Object evinces a design to reduce them under absolute Despotism, it is their right, it is their duty, to throw off such Government, and to provide new Guards for their future security.—Such has been the patient sufferance of these Colonies; and such is now the necessity which constrains them to alter their former Systems of Government. The history of the present King of Great Britain is a history of repeated injuries and usurpations, all having in direct object the establishment of an absolute Tyranny over these States. To prove this, let Facts be submitted to a candid world.—He has refused his Assent to Laws, the most wholesome and necessary for the public good.—He has forbidden his Governors to pass Laws of immediate and pressing importance, unless suspended in their operation till his Assent should be obtained; and when so suspended, he has utterly neglected to attend to them.—He has refused to pass other Laws for the accommodation of large districts of people, unless those people would relinquish the right of Representation in the Legislature, a right inestimable to them and formidable to tyrants only.—He has called together legislative bodies at places unusual, uncomfortable, and distant from the depository of their public Records, for the sole purpose of fatiguing them into compliance with his measures.—He has dissolved Representative Houses repeatedly, for opposing with manly firmness his invasions on the rights of the people.—He has refused for a long time, after such dissolutions, to cause others to be elected; whereby the Legislative powers, incapable of Annihilation, have returned to the People at large

1

for their exercise; the State remaining in the mean time exposed to all the dangers of invasion from without, and convulsions within.—He has endeavoured to prevent the population of these States; for that purpose obstructing the Laws for Naturalization of Foreigners; refusing to pass others to encourage their migrations hither, and raising the conditions of new Appropriations of Lands.—He has obstructed the Administration of Justice, by refusing his Assent to Laws for establishing Judiciary powers.—He has made Judges dependent on his Will alone, for the tenure of their offices, and the amount and payment of their salaries.—He has erected a multitude of New Offices, and sent hither swarms of Officers to harrass our people, and eat out their substance.—He has kept among us, in times of peace, Standing Armies without the Consent of our legislatures.—He has affected to render the Military independent of and superior to the Civil power.—He has combined with others to subject us to a jurisdiction foreign to our constitution, and unacknowledged by our laws; giving his Assent to their Acts of pretended Legislation:—For Quartering large bodies of armed troops among us:—For protecting them, by a mock Trial, from punishment for any Murders which they should commit on the Inhabitants of these States:—For cutting off our Trade with all parts of the world:—For imposing Taxes on us without our Consent:—For depriving us in many cases, of the benefits of Trial by Jury:—For transporting us beyond Seas to be tried for pretended offences—For abolishing the free System of English Laws in a neighbouring Province, establishing therein an Arbitrary government, and enlarging its Boundaries so as to render it at once an example and fit instrument for introducing the same absolute rule into these Colonies:—For taking away our Charters, abolishing our most valuable Laws, and altering fundamentally the Forms of our Governments:—For suspending our own Legislatures, and declaring themselves invested with power to legislate for us in all cases whatsoever.—He has abdicated Government here, by declaring us out of his Protection and waging War against us.—He has plundered our seas, ravaged our Coasts, burnt our towns, and destroyed the lives of our people.—He is at this time transporting large Armies of foreign Mercenaries to compleat the works of death, desolation and tyranny, already begun with circumstances of Cruelty & perfidy scarcely paralleled in the most barbarous ages, and totally unworthy the Head of a civilized nation.—He has constrained our fellow Citizens taken Captive on the high Seas to bear Arms against their Country, to become the executioners of their friends and Brethren, or to fall themselves by their Hands.—He has excited domestic insurrections amongst us, and has endeavoured to bring on the inhabitants of our frontiers, the merciless Indian Savages, whose known rule of warfare, is an undistinguished destruction of all ages, sexes and conditions.—In every stage of these Oppressions We have Petitioned for Redress in the most humble terms: Our repeated Petitions have been answered only by repeated injury. A Prince whose character is thus marked by every act which may define a Tyrant, is unfit to be the ruler of a free people.—Nor have We been wanting in attentions to our Brittish brethren. We have warned them from time to time of attempts by their legislature to extend an unwarrantable jurisdiction over us. We have reminded them of the circumstances of our emigration and settlement here. We have appealed to their native justice and magnanimity, and we have conjured them by the ties of our common kindred to disavow these usurpations, which, would inevitably interrupt our connections and correspondence. They too have been deaf to the voice of justice and of consanguinity. We must, therefore, acquiesce in the necessity, which denounces our Separation, and hold them, as we hold the rest of mankind, Enemies in War, in Peace Friends.

WE, THEREFORE, the Representatives of the UNITED STATES OF AMERICA, in General Congress, Assembled, appealing to the Supreme Judge of the world for the rectitude of our intentions, do, in the Name, and by Authority of the good People of these Colonies, solemnly publish and declare, That these United Colonies are, and of Right ought to be FREE AND INDEPENDENT STATES; that they are Absolved from all Allegiance to the British Crown, and that all political connection between them and the State of Great Britain, is and ought to be totally dissolved; and that as Free and Independent

States, they have full Power to levy War, conclude Peace, contract Alliances, establish Commerce, and to do all other Acts and Things which Independent States may of right do. And for the support of this Declaration, with a firm reliance on the protection of divine Providence, we mutually pledge to each other our Lives, our Fortunes and our sacred Honor.

The 56 signatures on the Declaration appear in the positions indicated:

Column 1

Georgia:
   Button Gwinnett
   Lyman Hall
   George Walton

Column 2

North Carolina:
   William Hooper
   Joseph Hewes
   John Penn

South Carolina:
   Edward Rutledge
   Thomas Heyward, Jr.
   Thomas Lynch, Jr.
   Arthur Middleton

Column 3

Massachusetts:
   John Hancock

Maryland:
   Samuel Chase
   William Paca
   Thomas Stone
   Charles Carroll of
   Carrollton

Virginia:
   George Wythe
   Richard Henry Lee
   Thomas Jefferson
   Benjamin Harrison
   Thomas Nelson, Jr.
   Francis Lightfoot Lee
   Carter Braxton

Column 4

Pennsylvania:
   Robert Morris
   Benjamin Rush
   Benjamin Franklin
   John Morton
   George Clymer
   James Smith
   George Taylor
   James Wilson
   George Ross

Delaware:
   Caesar Rodney
   George Read
   Thomas McKean

Column 5

New York:
   William Floyd
   Philip Livingston
   Francis Lewis
   Lewis Morris

New Jersey:
   Richard Stockton
   John Witherspoon
   Francis Hopkinson
   John Hart
   Abraham Clark

Column 6

New Hampshire:
   Josiah Bartlett
   William Whipple

Massachusetts:
   Samuel Adams
   John Adams
   Robert Treat Paine
   Elbridge Gerry

Rhode Island:
   Stephen Hopkins
   William Ellery

Connecticut:
   Roger Sherman
   Samuel Huntington
   William Williams
   Oliver Wolcott

New Hampshire:
   Matthew Thornton

For more information on the Declaration of Independence and the Charters of Freedom, see http://archives.gov/exhibits/charters/declaration.html

Constitution of the United States

Note: The following text is a transcription of the Constitution in its original form. Items that are underlined have since been amended or superseded.

Preamble

WE THE PEOPLE of the United States, in order to form a more perfect union, establish justice, insure domestic tranquility, provide for the common defense, promote the general welfare, and secure the blessings of liberty to ourselves and our posterity, do ordain and establish this Constitution for the United States of America.

Article I

Section 1. All legislative powers herein granted shall be vested in a Congress of the United States, which shall consist of a Senate and House of Representatives.

Section 2. The House of Representatives shall be composed of members chosen every second year by the people of the several states, and the electors in each state shall have the qualifications requisite for electors of the most numerous branch of the state legislature.

No person shall be a Representative who shall not have attained to the age of twenty five years, and been seven years a citizen of the United States, and who shall not, when elected, be an inhabitant of that state in which he shall be chosen.

Representatives and direct taxes shall be apportioned among the several states which may be included within this union, according to their respective numbers, which shall be determined by adding to the whole number of free persons, including those bound to service for a term of years, and excluding Indians not taxed, three fifths of all other Persons. The actual Enumeration shall be made within three years after the first meeting of the Congress of the United States, and within every subsequent term of ten years, in such manner as they shall by law direct. The number of Representatives shall not exceed one for every thirty thousand, but each state shall have at least one Representative; and until such enumeration shall be made, the state of New Hampshire shall be entitled to chuse three, Massachusetts eight, Rhode Island and Providence Plantations one, Connecticut five, New York six, New Jersey four, Pennsylvania eight, Delaware one, Maryland six, Virginia ten, North Carolina five, South Carolina five, and Georgia three.

When vacancies happen in the Representation from any state, the executive authority thereof shall issue writs of election to fill such vacancies.

The House of Representatives shall choose their speaker and other officers; and shall have the sole power of impeachment.

Section 3. The Senate of the United States shall be composed of two Senators from each state, chosen by the legislature thereof, for six years; and each Senator shall have one vote.

Immediately after they shall be assembled in consequence of the first election, they shall be divided as equally as may be into three classes. The seats of the Senators of the first class shall be vacated at the expiration of the second year, of the second class at the expiration of the fourth year, and the third class at the expiration of the sixth year, so that one third may be chosen every second year; and if vacancies happen by resignation, or otherwise, during the recess of the legislature of any state, the executive thereof may make temporary appointments until the next meeting of the legislature, which shall then fill such vacancies.

No person shall be a Senator who shall not have attained to the age of thirty years, and been nine years a citizen of the United States and who shall not, when elected, be an inhabitant of that state for which he shall be chosen.

The Vice President of the United States shall be President of the Senate, but shall have no vote, unless they be equally divided.

The Senate shall choose their other officers, and also a President pro tempore, in the absence of the Vice President, or when he shall exercise the office of President of the United States.

The Senate shall have the sole power to try all impeachments. When sitting for that purpose, they shall be on oath or affirmation. When the President of the United States is tried, the Chief Justice shall preside: And no person shall be convicted without the concurrence of two thirds of the members present.

Judgment in cases of impeachment shall not extend further than to removal from office, and disqualification to hold and enjoy any office of honor, trust or profit under the United States: but the party convicted shall nevertheless be liable and subject to indictment, trial, judgment and punishment, according to law.

Section 4. The times, places and manner of holding elections for Senators and Representatives, shall be prescribed in each state by the legislature thereof; but the Congress may at any time by law make or alter such regulations, except as to the places of choosing Senators.

The Congress shall assemble at least once in every year, and such meeting shall be on the first Monday in December, unless they shall by law appoint a different day.

Section 5. Each House shall be the judge of the elections, returns and qualifications of its own members, and a majority of each shall constitute a quorum to do business; but a smaller number may adjourn from day to day, and may be authorized to compel the attendance of absent members, in such manner, and under such penalties as each House may provide.

Each House may determine the rules of its proceedings, punish its members for disorderly behavior, and, with the concurrence of two thirds, expel a member.

Each House shall keep a journal of its proceedings, and from time to time publish the same, excepting such parts as may in their judgment require secrecy; and the yeas and nays of the members of either House on any question shall, at the desire of one fifth of those present, be entered on the journal.

Neither House, during the session of Congress, shall, without the consent of the other, adjourn for more than three days, nor to any other place than that in which the two Houses shall be sitting.

Section 6. The Senators and Representatives shall receive a compensation for their services, to be ascertained by law, and paid out of the treasury of the United States. They shall in all cases, except treason, felony and breach of the peace, be privileged from arrest during their attendance at the session of their respective Houses, and in going to and returning from the same; and for any speech or debate in either House, they shall not be questioned in any other place.

No Senator or Representative shall, during the time for which he was elected, be appointed to any civil office under the authority of the United States, which shall have been created, or the emoluments whereof shall have been increased during such time: and no person holding any office under the United States, shall be a member of either House during his continuance in office.

Section 7. All bills for raising revenue shall originate in the House of Representatives; but the Senate may propose or concur with amendments as on other Bills.

Every bill which shall have passed the House of Representatives and the Senate, shall, before it become a law, be presented to the President of the United States; if he approve he shall sign it, but if not he shall return it, with his objections to that House in which it shall have originated, who shall enter the objections at large on their journal, and proceed to reconsider it. If after such reconsideration two thirds of that House shall agree to pass the bill, it shall be sent, together with the objections, to the other House, by which it shall likewise be reconsidered, and if approved by two thirds of that House, it shall become a law. But in all such cases the votes of both Houses shall be determined by yeas and nays, and the names of the persons voting for and against the bill shall be entered on the journal of each House respectively. If any bill shall not be returned by the President within ten days (Sundays excepted) after it shall have been presented to him, the same shall be a law, in like manner as if he had signed it, unless the Congress by their adjournment prevent its return, in which case it shall not be a law.

Every order, resolution, or vote to which the concurrence of the Senate and House of Representatives may be necessary (except on a question of adjournment) shall be presented to the President of the United States; and before the same shall take effect, shall be approved by him, or being disapproved by him, shall be repassed by two thirds of the Senate and House of Representatives, according to the rules and limitations prescribed in the case of a bill.

Section 8. The Congress shall have power to lay and collect taxes, duties, imposts and excises, to pay the debts and provide for the common defense and general welfare of the United States; but all duties, imposts and excises shall be uniform throughout the United States;

To borrow money on the credit of the United States;

To regulate commerce with foreign nations, and among the several states, and with the Indian tribes;

To establish a uniform rule of naturalization, and uniform laws on the subject of bankruptcies throughout the United States;

To coin money, regulate the value thereof, and of foreign coin, and fix the standard of weights and measures;

To provide for the punishment of counterfeiting the securities and current coin of the United States;

To establish post offices and post roads;

To promote the progress of science and useful arts, by securing for limited times to authors and inventors the exclusive right to their respective writings and discoveries;

To constitute tribunals inferior to the Supreme Court;

To define and punish piracies and felonies committed on the high seas, and offenses against the law of nations;

To declare war, grant letters of marque and reprisal, and make rules concerning captures on land and water;

To raise and support armies, but no appropriation of money to that use shall be for a longer term than two years;

To provide and maintain a navy;

To make rules for the government and regulation of the land and naval forces;

To provide for calling forth the militia to execute the laws of the union, suppress insurrections and repel invasions;

To provide for organizing, arming, and disciplining, the militia, and for governing such part of them as may be employed in the service of the United States, reserving to the states respectively, the appointment of the officers, and the authority of training the militia according to the discipline prescribed by Congress;

To exercise exclusive legislation in all cases whatsoever, over such District (not exceeding ten miles square) as may, by cession of particular states, and the acceptance of Congress, become the seat of the government of the United States, and to exercise like authority over all places purchased by the consent of the legislature of the state in which the same shall be, for the erection of forts, magazines, arsenals, dockyards, and other needful buildings;—And

To make all laws which shall be necessary and proper for carrying into execution the foregoing powers, and all other powers vested by this Constitution in the government of the United States, or in any department or officer thereof.

Section 9. The migration or importation of such persons as any of the states now existing shall think proper to admit, shall not be prohibited by the Congress prior to the year one thousand eight hundred and eight, but a tax or duty may be imposed on such importation, not exceeding ten dollars for each person.

The privilege of the writ of habeas corpus shall not be suspended, unless when in cases of rebellion or invasion the public safety may require it.

No bill of attainder or ex post facto Law shall be passed.

No capitation, or other direct, tax shall be laid, unless in proportion to the census or enumeration herein before directed to be taken.

No tax or duty shall be laid on articles exported from any state.

No preference shall be given by any regulation of commerce or revenue to the ports of one state over those of another: nor shall vessels bound to, or from, one state, be obliged to enter, clear or pay duties in another.

No money shall be drawn from the treasury, but in consequence of appropriations made by law; and a regular statement and account of receipts and expenditures of all public money shall be published from time to time.

No title of nobility shall be granted by the United States: and no person holding any office of profit or trust under them, shall, without the consent of the Congress, accept of any present, emolument, office, or title, of any kind whatever, from any king, prince, or foreign state.

Section 10. No state shall enter into any treaty, alliance, or confederation; grant letters of marque and reprisal; coin money; emit bills of credit; make anything but gold and silver coin a tender in payment of debts; pass any bill of attainder, ex post facto law, or law impairing the obligation of contracts, or grant any title of nobility.

No state shall, without the consent of the Congress, lay any imposts or duties on imports or exports, except what may be absolutely necessary for executing it's inspection laws: and the net produce of all duties and imposts, laid by any state on imports or exports, shall be for the use of the treasury of the United States; and all such laws shall be subject to the revision and control of the Congress.

No state shall, without the consent of Congress, lay any duty of tonnage, keep troops, or ships of war in time of peace, enter into any agreement or compact with another state, or with a foreign power, or engage in war, unless actually invaded, or in such imminent danger as will not admit of delay.

Article II

Section 1. The executive power shall be vested in a President of the United States of America. He shall hold his office during the term of four years, and, together with the Vice President, chosen for the same term, be elected, as follows:

Each state shall appoint, in such manner as the Legislature thereof may direct, a number of electors, equal to the whole number of Senators and Representatives to which the State may be entitled in the Congress: but no Senator or Representative, or person holding an office of trust or profit under the United States, shall be appointed an elector.

The electors shall meet in their respective states, and vote by ballot for two persons, of whom one at least shall not be an inhabitant of the same state with themselves. And they shall make a list of all the persons voted for, and of the number of votes for each; which list they shall sign and certify, and transmit sealed to the seat of the government of the United States, directed to the President of the Senate. The President of the Senate shall, in the presence of the Senate and House of Representatives, open all the certificates, and the votes shall then be counted. The person having the greatest number of votes shall be the President, if such number be a majority of the whole number of electors appointed; and if there be more than one who have such majority, and have an equal number of votes, then the House of Representatives shall immediately choose by ballot one of them for President; and if no person have a majority, then from the five highest on the list the said House shall in like manner choose the President. But in choosing the President, the votes shall be taken by States, the representation from each state having one vote; A quorum for this purpose shall consist of a member or members from two thirds of the states, and a majority of all the states shall be necessary to a choice. In every case, after the choice of the President, the person having the greatest number of votes of the electors shall be the Vice President. But if there should remain two or more who have equal votes, the Senate shall choose from them by ballot the Vice President.

The Congress may determine the time of choosing the electors, and the day on which they shall give their votes; which day shall be the same throughout the United States.

No person except a natural born citizen, or a citizen of the United States, at the time of the adoption of this Constitution, shall be eligible to the office of President; neither shall any person be eligible to that office who shall not have attained to the age of thirty five years, and been fourteen Years a resident within the United States.

In case of the removal of the President from office, or of his death, resignation, or inability to discharge the powers and duties of the said office, the same shall devolve on the Vice President, and the Congress may by law provide for the case of removal, death, resignation or inability, both of the President and Vice President, declaring what officer shall then act as President, and such officer shall act accordingly, until the disability be removed, or a President shall be elected.

The President shall, at stated times, receive for his services, a compensation, which shall neither be increased nor diminished during the period for which he shall have been elected, and he shall not receive within that period any other emolument from the United States, or any of them.

Before he enter on the execution of his office, he shall take the following oath or affirmation:—"I do solemnly swear (or affirm) that I will faithfully execute the office of President of the United States, and will to the best of my ability, preserve, protect and defend the Constitution of the United States."

Section 2. The President shall be commander in chief of the Army and Navy of the United States, and of the militia of the several states, when called into the actual service of the United States; he may require the opinion, in writing, of the principal officer in each of the executive departments, upon any subject relating to the duties of their respective offices, and he shall have power to grant reprieves and pardons for offenses against the United States, except in cases of impeachment.

He shall have power, by and with the advice and consent of the Senate, to make treaties, provided two thirds of the Senators present concur; and he shall nominate, and by and with the advice and consent of the Senate, shall appoint ambassadors, other public ministers and consuls, judges of the Supreme Court, and all other officers of the United States, whose appointments are not herein otherwise provided for, and which shall be established by law: but the Congress may by law vest the appointment of such inferior officers, as they think proper, in the President alone, in the courts of law, or in the heads of departments.

The President shall have power to fill up all vacancies that may happen during the recess of the Senate, by granting commissions which shall expire at the end of their next session.

Section 3. He shall from time to time give to the Congress information of the state of the union, and recommend to their consideration such measures as he shall judge necessary and expedient; he may, on extraordinary occasions, convene both Houses, or either of them, and in case of disagreement between them, with respect to the time of adjournment, he may adjourn them to such time as he shall think proper; he shall receive ambassadors and other public ministers; he shall take care that the laws be faithfully executed, and shall commission all the officers of the United States.

Section 4. The President, Vice President and all civil officers of the United States, shall be removed from office on impeachment for, and conviction of, treason, bribery, or other high crimes and misdemeanors.

Article III

Section 1. The judicial power of the United States, shall be vested in one Supreme Court, and in such inferior courts as the Congress may from time to time ordain and establish. The judges, both of the supreme and inferior courts, shall hold their offices during good behaviour, and shall, at stated times, receive for their services, a compensation, which shall not be diminished during their continuance in office.

Section 2. The judicial power shall extend to all cases, in law and equity, arising under this Constitution, the laws of the United States, and treaties made, or which shall be made, under their authority;—to all cases affecting ambassadors, other public ministers and consuls;—to all cases of admiralty and maritime jurisdiction;—to controversies to which the United States shall be a party;—to controversies between two or more states;—between a state and citizens of another state;—between citizens of different states;—between citizens of the same state claiming lands under grants of different states, and between a state, or the citizens thereof, and foreign states, citizens or subjects.

In all cases affecting ambassadors, other public ministers and consuls, and those in which a state shall be party, the Supreme Court shall have original jurisdiction. In all the other cases before mentioned, the Supreme Court shall have appellate jurisdiction, both as to law and fact, with such exceptions, and under such regulations as the Congress shall make.

The trial of all crimes, except in cases of impeachment, shall be by jury; and such trial shall be held in the state where the said crimes shall have been committed; but when not committed within any state, the trial shall be at such place or places as the Congress may by law have directed.

Section 3. Treason against the United States, shall consist only in levying war against them, or in adhering to their enemies, giving them aid and comfort. No person shall be convicted of treason unless on the testimony of two witnesses to the same overt act, or on confession in open court.

The Congress shall have power to declare the punishment of treason, but no attainder of treason shall work corruption of blood, or forfeiture except during the life of the person attainted.

Article IV

Section 1. Full faith and credit shall be given in each state to the public acts, records, and judicial proceedings of every other state. And the Congress may by general laws prescribe the manner in which such acts, records, and proceedings shall be proved, and the effect thereof.

Section 2. The citizens of each state shall be entitled to all privileges and immunities of citizens in the several states.

A person charged in any state with treason, felony, or other crime, who shall flee from justice, and be found in another state, shall on demand of the executive authority of the state from which he fled, be delivered up, to be removed to the state having jurisdiction of the crime.

No person held to service or labor in one state, under the laws thereof, escaping into another, shall, in consequence of any law or regulation therein, be discharged from such service or labor, but shall be delivered up on claim of the party to whom such service or labor may be due.

Section 3. New states may be admitted by the Congress into this union; but no new states shall be formed or erected within the jurisdiction of any other state; nor any state be formed by the junction of two or more states, or parts of states, without the consent of the legislatures of the states concerned as well as of the Congress.

The Congress shall have power to dispose of and make all needful rules and regulations respecting the territory or other property belonging to the United States; and nothing in this Constitution shall be so construed as to prejudice any claims of the United States, or of any particular state.

Section 4. The United States shall guarantee to every state in this union a republican form of government, and shall protect each of them against invasion; and on application of the legislature, or of the executive (when the legislature cannot be convened) against domestic violence.

Article V

The Congress, whenever two thirds of both houses shall deem it necessary, shall propose amendments to this Constitution, or, on the application of the legislatures of two thirds of the several states, shall call a convention for proposing amendments, which, in either case, shall be valid to all intents and purposes, as part of this Constitution, when ratified by the legislatures of three fourths of the several states, or by conventions in three fourths thereof, as the one or the other mode of ratification may be proposed by the Congress; provided that no amendment which may be made prior to the year one thousand eight hundred and eight shall in any manner affect the first and fourth clauses in the ninth section of the first article; and that no state, without its consent, shall be deprived of its equal suffrage in the Senate.

Article VI

All debts contracted and engagements entered into, before the adoption of this Constitution, shall be as valid against the United States under this Constitution, as under the Confederation.

This Constitution, and the laws of the United States which shall be made in pursuance thereof; and all treaties made, or which shall be made, under the authority of the United States, shall be the supreme law of the land; and the judges in every state shall be bound thereby, anything in the Constitution or laws of any State to the contrary notwithstanding.

The Senators and Representatives before mentioned, and the members of the several state legislatures, and all executive and judicial officers, both of the United States and of the several states, shall be bound by oath or affirmation, to support this Constitution; but no religious test shall ever be required as a qualification to any office or public trust under the United States.

Article VII

The ratification of the conventions of nine states, shall be sufficient for the establishment of this Constitution between the states so ratifying the same.

Signers

Done in convention by the unanimous consent of the states present the seventeenth day of September in the year of our Lord one thousand seven hundred and eighty seven and of the independence of the United States of America the twelfth. *In witness whereof We have hereunto subscribed our Names,*

G° Washington—Presid^t
and deputy from Virginia

New Hampshire	John Langdon
	Nicholas Gilman
Massachusetts	Nathaniel Gorham
	Rufus King
Connecticut	W^m: Saml Johnson
	Roger Sherman
New York	Alexander Hamilton
New Jersey	Wil: Livingston
	David Brearly
	Wm Paterson
	Jona: Dayton
Pennsylvania	B. Franklin
	Thomas Mifflin
	Robt Morris
	Geo. Clymer

	Thos FitzSimons
	Jared Ingersoll
	James Wilson
	Gouv Morris
Delaware	Geo: Read
	Gunning Bedford jun
	John Dickinson
	Richard Bassett
	Jaco: Broom
Maryland	James McHenry
	Dan of St Thos Jenifer
	Danl Carroll
Virginia	John Blair—
	James Madison Jr.
North Carolina	Wm Blount
	Richd Dobbs Spaight
	Hu Williamson
South Carolina	J. Rutledge
	Charles Cotesworth Pinckney
	Charles Pinckney
	Pierce Butler
Georgia	William Few
	Abr Baldwin

Amendments

Note: The first ten Amendments were ratified December 15, 1791, and form what is known as the Bill of Rights.

Amendment 1

Congress shall make no law respecting an establishment of religion, or prohibiting the free exercise thereof; or abridging the freedom of speech, or of the press; or the right of the people peaceably to assemble, and to petition the government for a redress of grievances.

Amendment 2

A well regulated militia, being necessary to the security of a free state, the right of the people to keep and bear arms, shall not be infringed.

Amendment 3

No soldier shall, in time of peace be quartered in any house, without the consent of the owner, nor in time of war, but in a manner to be prescribed by law.

Amendment 4

The right of the people to be secure in their persons, houses, papers, and effects, against unreasonable searches and seizures, shall not be violated, and no warrants shall issue, but upon probable cause, supported by oath or affirmation, and particularly describing the place to be searched, and the persons or things to be seized.

Amendment 5

No person shall be held to answer for a capital, or otherwise infamous crime, unless on a presentment or indictment of a grand jury, except in cases arising in the land or naval forces, or in the militia, when in actual service in time of war or public danger; nor shall any person be subject for the same offense to be twice put in jeopardy of life or limb; nor shall be compelled in any criminal case to be a witness against himself, nor be deprived of life, liberty, or property, without due process of law; nor shall private property be taken for public use, without just compensation.

Amendment 6

In all criminal prosecutions, the accused shall enjoy the right to a speedy and public trial, by an impartial jury of the state and district wherein the crime shall have been committed, which district shall have been previously ascertained by law, and to be informed of the nature and cause of the accusation; to be confronted with the witnesses against him; to have compulsory process for obtaining witnesses in his favor, and to have the assistance of counsel for his defense.

Amendment 7

In suits at common law, where the value in controversy shall exceed twenty dollars, the right of trial by jury shall be preserved, and no fact tried by a jury, shall be otherwise reexamined in any court of the United States, than according to the rules of the common law.

Amendment 8

Excessive bail shall not be required, nor excessive fines imposed, nor cruel and unusual punishments inflicted.

Amendment 9

The enumeration in the Constitution, of certain rights, shall not be construed to deny or disparage others retained by the people.

Amendment 10

The powers not delegated to the United States by the Constitution, nor prohibited by it to the states, are reserved to the states respectively, or to the people.

Amendment 11

(Ratified February 7, 1795)

The judicial power of the United States shall not be construed to extend to any suit in law or equity, commenced or prosecuted against one of the United States by citizens of another state, or by citizens or subjects of any foreign state.

Amendment 12

(Ratified July 27, 1804)

The electors shall meet in their respective states and vote by ballot for President and Vice-President, one of whom, at least, shall not be an inhabitant of the same state with themselves; they shall name in their ballots the person voted for as President, and in distinct ballots the person voted for as Vice-President, and they shall make distinct lists of all persons voted for as President, and of all persons voted for as Vice-President, and of the number of votes for each, which lists they shall sign and certify, and transmit sealed to the seat of the government of the United States, directed to the President of the Senate;—The President of the Senate shall, in the presence of the Senate and House of Representatives, open all the certificates and the votes shall then be counted;—the person having the greatest number of votes for President, shall be the President, if such number be a majority of the whole number of electors appointed; and if no person have such majority, then from the persons having the highest numbers not exceeding three on the list of those voted for as President, the House of Representatives shall choose immediately, by ballot, the President. But in choosing the President, the votes shall be taken by states, the representation from each state having one vote; a quorum for this purpose shall consist of a member or members from two-thirds of the states, and a majority of all the states shall be necessary to a choice. And if the House of Representatives shall not choose a President whenever the right of choice shall devolve upon them, before the fourth day of March next following, then the Vice-President shall act as President, as in the case of the death or other constitutional disability of the President. The person having the greatest number of votes as Vice-President, shall be the Vice-President, if such number be a majority of the whole number of electors appointed, and if no person have a majority, then from the two highest numbers on the list, the Senate shall choose the Vice-President; a quorum for the purpose shall consist of two-thirds of the whole number of Senators, and a majority of the whole number shall be necessary to a choice. But no person constitutionally ineligible to the office of President shall be eligible to that of Vice-President of the United States.

Amendment 13

(Ratified December 6, 1865)

Section 1. Neither slavery nor involuntary servitude, except as a punishment for crime whereof the party shall have been duly convicted, shall exist within the United States, or any place subject to their jurisdiction.

Section 2. Congress shall have power to enforce this article by appropriate legislation.

Amendment 14

(Ratified July 9, 1868)

Section 1. All persons born or naturalized in the United States, and subject to the jurisdiction thereof, are citizens of the United States and of the state wherein they reside. No state shall make or enforce any law which shall abridge the privileges or immunities of citizens of the United States; nor shall any state deprive any person of life, liberty, or property, without due process of law; nor deny to any person within its jurisdiction the equal protection of the laws.

Section 2. Representatives shall be apportioned among the several states according to their respective numbers, counting the whole number of persons in each state, excluding Indians not taxed. But when the right to vote at any election for the choice of electors for President and Vice President of the United States, Representatives in Congress, the executive and judicial officers of a state, or the members of the legislature thereof, is denied to any of the male inhabitants of such state, being twenty-one years of age, and citizens of the United States, or in any way abridged, except for participation in rebellion, or other crime, the basis of representation therein shall be reduced in the proportion which the number of such male citizens shall bear to the whole number of male citizens twenty-one years of age in such state.

Section 3. No person shall be a Senator or Representative in Congress, or elector of President and Vice President, or hold any office, civil or military, under the United States, or under any state, who, having previously taken an oath, as a member of Congress, or as an officer of the United States, or as a member of any state legislature, or as an executive or judicial officer of any state, to support the Constitution of the United States, shall have engaged in insurrection or rebellion against the same, or given aid or comfort to the enemies thereof. But Congress may by a vote of two-thirds of each House, remove such disability.

Section 4. The validity of the public debt of the United States, authorized by law, including debts incurred for payment of pensions and bounties for services in suppressing insurrection or rebellion, shall not be questioned. But neither the United States nor any state shall assume or pay any debt or obligation incurred in aid of insurrection or rebellion against the United States, or any claim for the loss or emancipation of any slave; but all such debts, obligations and claims shall be held illegal and void.

Section 5. The Congress shall have power to enforce, by appropriate legislation, the provisions of this article.

Amendment 15

(Ratified February 3, 1870)

Section 1. The right of citizens of the United States to vote shall not be denied or abridged by the United States or by any state on account of race, color, or previous condition of servitude.

Section 2. The Congress shall have power to enforce this article by appropriate legislation.

Amendment 16

(Ratified February 3, 1913)

The Congress shall have power to lay and collect taxes on incomes, from whatever source derived, without apportionment among the several states, and without regard to any census or enumeration.

Amendment 17

(Ratified April 8, 1913)

The Senate of the United States shall be composed of two Senators from each state, elected by the people thereof, for six years; and each Senator shall have one vote. The electors in each state shall have the qualifications requisite for electors of the most numerous branch of the state legislatures.

When vacancies happen in the representation of any state in the Senate, the executive authority of such state shall issue writs of election to fill such vacancies: Provided, that the legislature of any state may empower the executive thereof to make temporary appointments until the people fill the vacancies by election as the legislature may direct.

This amendment shall not be so construed as to affect the election or term of any Senator chosen before it becomes valid as part of the Constitution.

Amendment 18

(Ratified January 16, 1919. Repealed December 5, 1933 by Amendment 21)

Section 1. After one year from the ratification of this article the manufacture, sale, or transportation of intoxicating liquors within, the importation thereof into, or the exportation thereof from the United States and all territory subject to the jurisdiction thereof for beverage purposes is hereby prohibited.

Section 2. The Congress and the several states shall have concurrent power to enforce this article by appropriate legislation.

Section 3. This article shall be inoperative unless it shall have been ratified as an amendment to the Constitution by the legislatures of the several states, as provided in the Constitution, within seven years from the date of the submission hereof to the states by the Congress.

Amendment 19

(Ratified August 18, 1920)

The right of citizens of the United States to vote shall not be denied or abridged by the United States or by any state on account of sex.

Congress shall have power to enforce this article by appropriate legislation.

Amendment 20

(Ratified January 23, 1933)

Section 1. The terms of the President and Vice President shall end at noon on the 20th day of January, and the terms of Senators and Representatives at noon on the 3d day of January, of the years in which such terms would have ended if this article had not been ratified; and the terms of their successors shall then begin.

Section 2. The Congress shall assemble at least once in every year, and such meeting shall begin at noon on the 3d day of January, unless they shall by law appoint a different day.

Section 3. If, at the time fixed for the beginning of the term of the President, the President elect shall have died, the Vice President elect shall become President. If a President shall not have been chosen before the time fixed for the beginning of his term, or if the President elect shall have failed to qualify, then the Vice President elect shall act as President until a President shall have qualified; and the Congress may by law provide for the case wherein neither a President elect nor a Vice President elect shall have qualified, declaring who shall then act as President, or the manner in which one who is to act shall be selected, and such person shall act accordingly until a President or Vice President shall have qualified.

Section 4. The Congress may by law provide for the case of the death of any of the persons from whom the House of Representatives may choose a President whenever the right of choice shall have devolved upon them, and for the case of the death of any of the persons from whom the Senate may choose a Vice President whenever the right of choice shall have devolved upon them.

Section 5. Sections 1 and 2 shall take effect on the 15th day of October following the ratification of this article.

Section 6. This article shall be inoperative unless it shall have been ratified as an amendment to the Constitution by the legislatures of three-fourths of the several states within seven years from the date of its submission.

Amendment 21

(Ratified December 5, 1933)

Section 1. The eighteenth article of amendment to the Constitution of the United States is hereby repealed.

Section 2. The transportation or importation into any state, territory, or possession of the United States for delivery or use therein of intoxicating liquors, in violation of the laws thereof, is hereby prohibited.

Section 3. This article shall be inoperative unless it shall have been ratified as an amendment to the Constitution by conventions in the several states, as provided in the Constitution, within seven years from the date of the submission hereof to the states by the Congress.

Amendment 22

(Ratified February 27, 1951)

Section 1. No person shall be elected to the office of the President more than twice, and no person who has held the office of President, or acted as President, for more than two years of a term to which some other person was elected President shall be elected to the office of the President more than once. But this article shall not apply to any person holding the office of President when this article was proposed by the Congress, and shall not prevent any person who may be holding the office of President, or acting as President, during the term within which this article becomes operative from holding the office of President or acting as President during the remainder of such term.

Section 2. This article shall be inoperative unless it shall have been ratified as an amendment to the Constitution by the legislatures of three-fourths of the several states within seven years from the date of its submission to the states by the Congress.

Amendment 23

(Ratified March 29, 1961)

Section 1. The District constituting the seat of government of the United States shall appoint in such manner as the Congress may direct:
    A number of electors of President and Vice President equal to the whole number of Senators and Representatives in Congress to which the District would be entitled if it were a state, but in no event more than the least populous state; they shall be in addition to those appointed by the states, but they shall be considered, for the purposes of the election of President and Vice President, to be electors appointed by a state; and they shall meet in the District and perform such duties as provided by the twelfth article of amendment.

Section 2. The Congress shall have power to enforce this article by appropriate legislation.

Amendment 24

(Ratified January 23, 1964)

Section 1. The right of citizens of the United States to vote in any primary or other election for President or Vice President, for electors for President or Vice President, or for Senator or Representative in Congress, shall not be denied or abridged by the United States or any state by reason of failure to pay any poll tax or other tax.

Section 2. The Congress shall have power to enforce this article by appropriate legislation.

Amendment 25

(Ratified February 10, 1967)

Section 1. In case of the removal of the President from office or of his death or resignation, the Vice President shall become President.

Section 2. Whenever there is a vacancy in the office of the Vice President, the President shall nominate a Vice President who shall take office upon confirmation by a majority vote of both Houses of Congress.

Section 3. Whenever the President transmits to the President pro tempore of the Senate and the Speaker of the House of Representatives his written declaration that he is unable to discharge the powers and duties of his office, and until he transmits to them a written declaration to the contrary, such powers and duties shall be discharged by the Vice President as Acting President.

Section 4. Whenever the Vice President and a majority of either the principal officers of the executive departments or of such other body as Congress may by law provide, transmit to the President pro tempore of the Senate and the Speaker of the House of Representatives their written declaration that the President is unable to discharge the powers and duties of his office, the Vice President shall immediately assume the powers and duties of the office as Acting President.

Thereafter, when the President transmits to the President pro tempore of the Senate and the Speaker of the House of Representatives his written declaration that no inability exists, he shall resume the powers and duties of his office unless the Vice President and a majority of either the principal officers of the executive department or of such other body as Congress may by law provide, transmit within four days to the President pro tempore of the Senate and the Speaker of the House of Representatives their written declaration that the President is unable to discharge the powers and duties of his office. Thereupon Congress shall decide the issue, assembling within forty-eight hours for that purpose if not in session. If the Congress, within twenty-one days after receipt of the latter written declaration, or, if Congress is not in session, within twenty-one days after Congress is required to assemble, determines by two-thirds vote of both Houses that the President is unable to discharge the powers and duties of his office, the Vice President shall continue to discharge the same as Acting President; otherwise, the President shall resume the powers and duties of his office.

Amendment 26

(Ratified July 1, 1971)

Section 1. The right of citizens of the United States, who are 18 years of age or older, to vote, shall not be denied or abridged by the United States or any state on account of age.

Section 2. The Congress shall have the power to enforce this article by appropriate legislation.

Amendment 27

(Ratified May 7, 1992)

No law, varying the compensation for the services of the Senators and Representatives, shall take effect, until an election of Representatives shall have intervened.

For more information on the Constitution of the United States and the Charters of Freedom, see http://archives.gov/exhibits/charters/constitution.html

THE GOVERNMENT OF THE UNITED STATES

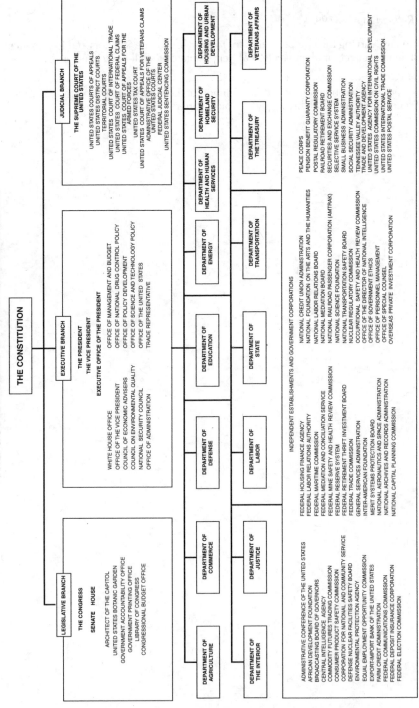

THE CONSTITUTION

LEGISLATIVE BRANCH

THE CONGRESS

SENATE   HOUSE

ARCHITECT OF THE CAPITOL
UNITED STATES BOTANIC GARDEN
GOVERNMENT ACCOUNTABILITY OFFICE
GOVERNMENT PRINTING OFFICE
LIBRARY OF CONGRESS
CONGRESSIONAL BUDGET OFFICE

EXECUTIVE BRANCH

THE PRESIDENT
THE VICE PRESIDENT

EXECUTIVE OFFICE OF THE PRESIDENT

WHITE HOUSE OFFICE
OFFICE OF THE VICE PRESIDENT
COUNCIL OF ECONOMIC ADVISERS
COUNCIL ON ENVIRONMENTAL QUALITY
NATIONAL SECURITY COUNCIL
OFFICE OF ADMINISTRATION
OFFICE OF MANAGEMENT AND BUDGET
OFFICE OF NATIONAL DRUG CONTROL POLICY
OFFICE OF POLICY DEVELOPMENT
OFFICE OF SCIENCE AND TECHNOLOGY POLICY
OFFICE OF THE UNITED STATES
TRADE REPRESENTATIVE

JUDICIAL BRANCH

THE SUPREME COURT OF THE
UNITED STATES

UNITED STATES COURTS OF APPEALS
UNITED STATES DISTRICT COURTS
TERRITORIAL COURTS
UNITED STATES COURT OF INTERNATIONAL TRADE
UNITED STATES COURT OF FEDERAL CLAIMS
UNITED STATES COURT OF APPEALS FOR THE
ARMED FORCES
UNITED STATES TAX COURT
UNITED STATES COURT OF APPEALS FOR VETERANS CLAIMS
ADMINISTRATIVE OFFICE OF THE
UNITED STATES COURTS
FEDERAL JUDICIAL CENTER
UNITED STATES SENTENCING COMMISSION

DEPARTMENT OF AGRICULTURE

DEPARTMENT OF COMMERCE

DEPARTMENT OF DEFENSE

DEPARTMENT OF EDUCATION

DEPARTMENT OF ENERGY

DEPARTMENT OF HEALTH AND HUMAN SERVICES

DEPARTMENT OF HOMELAND SECURITY

DEPARTMENT OF HOUSING AND URBAN DEVELOPMENT

DEPARTMENT OF THE INTERIOR

DEPARTMENT OF JUSTICE

DEPARTMENT OF LABOR

DEPARTMENT OF STATE

DEPARTMENT OF TRANSPORTATION

DEPARTMENT OF THE TREASURY

DEPARTMENT OF VETERANS AFFAIRS

INDEPENDENT ESTABLISHMENTS AND GOVERNMENT CORPORATIONS

ADMINISTRATIVE CONFERENCE OF THE UNITED STATES
AFRICAN DEVELOPMENT FOUNDATION
BROADCASTING BOARD OF GOVERNORS
CENTRAL INTELLIGENCE AGENCY
COMMODITY FUTURES TRADING COMMISSION
CONSUMER PRODUCT SAFETY COMMISSION
CORPORATION FOR NATIONAL AND COMMUNITY SERVICE
DEFENSE NUCLEAR FACILITIES SAFETY BOARD
ENVIRONMENTAL PROTECTION AGENCY
EQUAL EMPLOYMENT OPPORTUNITY COMMISSION
EXPORT-IMPORT BANK OF THE UNITED STATES
FARM CREDIT ADMINISTRATION
FEDERAL COMMUNICATIONS COMMISSION
FEDERAL DEPOSIT INSURANCE CORPORATION
FEDERAL ELECTION COMMISSION

FEDERAL HOUSING FINANCE AGENCY
FEDERAL LABOR RELATIONS AUTHORITY
FEDERAL MARITIME COMMISSION
FEDERAL MEDIATION AND CONCILIATION SERVICE
FEDERAL MINE SAFETY AND HEALTH REVIEW COMMISSION
FEDERAL RESERVE SYSTEM
FEDERAL RETIREMENT THRIFT INVESTMENT BOARD
FEDERAL TRADE COMMISSION
GENERAL SERVICES ADMINISTRATION
INTER-AMERICAN FOUNDATION
MERIT SYSTEMS PROTECTION BOARD
NATIONAL AERONAUTICS AND SPACE ADMINISTRATION
NATIONAL ARCHIVES AND RECORDS ADMINISTRATION
NATIONAL CAPITAL PLANNING COMMISSION

NATIONAL CREDIT UNION ADMINISTRATION
NATIONAL FOUNDATION ON THE ARTS AND THE HUMANITIES
NATIONAL LABOR RELATIONS BOARD
NATIONAL MEDIATION BOARD
NATIONAL RAILROAD PASSENGER CORPORATION (AMTRAK)
NATIONAL SCIENCE FOUNDATION
NATIONAL TRANSPORTATION SAFETY BOARD
NUCLEAR REGULATORY COMMISSION
OCCUPATIONAL SAFETY AND HEALTH REVIEW COMMISSION
OFFICE OF THE DIRECTOR OF NATIONAL INTELLIGENCE
OFFICE OF GOVERNMENT ETHICS
OFFICE OF PERSONNEL MANAGEMENT
OFFICE OF SPECIAL COUNSEL
OVERSEAS PRIVATE INVESTMENT CORPORATION

PEACE CORPS
PENSION BENEFIT GUARANTY CORPORATION
POSTAL REGULATORY COMMISSION
RAILROAD RETIREMENT BOARD
SECURITIES AND EXCHANGE COMMISSION
SELECTIVE SERVICE SYSTEM
SMALL BUSINESS ADMINISTRATION
SOCIAL SECURITY ADMINISTRATION
TENNESSEE VALLEY AUTHORITY
TRADE AND DEVELOPMENT AGENCY
UNITED STATES AGENCY FOR INTERNATIONAL DEVELOPMENT
UNITED STATES COMMISSION ON CIVIL RIGHTS
UNITED STATES INTERNATIONAL TRADE COMMISSION
UNITED STATES POSTAL SERVICE

Legislative Branch

LEGISLATIVE BRANCH

CONGRESS
One Hundred and Twelfth Congress, First Session

The Congress of the United States was created by Article I, section 1, of the Constitution, adopted by the Constitutional Convention on September 17, 1787, providing that "All legislative Powers herein granted shall be vested in a Congress of the United States, which shall consist of a Senate and House of Representatives."

The first Congress under the Constitution met on March 4, 1789, in the Federal Hall in New York City. The membership then consisted of 20 Senators and 59 Representatives.[1]

Congressional Record Proceedings of Congress are published in the Congressional Record, which is issued each day when Congress is in session. Publication of the Record began March 4, 1873. It was the first record of debate officially reported, printed, and published directly by the Federal Government. The Daily Digest of the Congressional Record, printed in the back of each issue of the Record, summarizes the proceedings of that day in each House and each of their committees and subcommittees, respectively. The Digest also presents the legislative program for each day and, at the end of the week, gives the program for the following week. Its publication was begun March 17, 1947.

Sessions Section 4 of Article I of the Constitution makes it mandatory that "The Congress shall assemble at least once in every Year. . . ." Under this provision, also, the date for convening Congress was designated originally as the first Monday in December, "unless they shall by Law appoint a different Day." Eighteen acts were passed, up to 1820, providing for the meeting of

Congress on other days of the year. From 1820 to 1934, however, Congress met regularly on the first Monday in December. In 1934 the 20th amendment changed the convening of Congress to January 3, unless Congress "shall by law appoint a different day." In addition, the President, according to Article II, section 3, of the Constitution "may, on extraordinary Occasions, convene both Houses, or either of them, and in Case of Disagreement between them, with Respect to the Time of Adjournment, he may adjourn them to such Time as he shall think proper. . . ."

Powers of Congress Article I, section 8, of the Constitution defines the powers of Congress. Included are the powers to assess and collect taxes—called the chief power; to regulate commerce, both interstate and foreign; to coin money; to establish post offices and post roads; to establish courts inferior to the Supreme Court; to declare war; and to raise and maintain an army and navy. Congress is further empowered "To provide for calling forth the Militia to execute the Laws of the Union, suppress Insurrections and repel Invasions;" and "To make all Laws which shall be necessary and proper for carrying into Execution the foregoing Powers, and all other Powers

---

[1] New York ratified the Constitution on July 26, 1788, but did not elect its Senators until July 15 and 16, 1789. North Carolina did not ratify the Constitution until November 21, 1789; Rhode Island ratified it on May 29, 1790.

vested by this Constitution in the Government of the United States, or in any Department or Officer thereof."

Amendments to the Constitution Another power vested in the Congress is the right to propose amendments to the Constitution, whenever two-thirds of both Houses shall deem it necessary. Should two-thirds of the State legislatures demand changes in the Constitution, it is the duty of Congress to call a constitutional convention. Proposed amendments shall be valid as part of the Constitution when ratified by the legislatures or by conventions of three-fourths of the States, as one or the other mode of ratification may be proposed by Congress.

Prohibitions Upon Congress Section 9 of Article I of the Constitution also imposes prohibitions upon Congress. "The Privilege of the Writ of Habeas Corpus shall not be suspended, unless when in Cases of Rebellion or Invasion the public Safety may require it." A bill of attainder or an ex post facto law cannot be passed. No export duty can be imposed. Ports of one State cannot be given preference over those of another State. "No money shall be drawn from the Treasury, but in Consequence of Appropriations made by Law. . . ." No title of nobility may be granted.

Rights of Members According to section 6 of Article I, Members of Congress are granted certain privileges. In no case, except in treason, felony, and breach of the peace, can Members be arrested while attending sessions of Congress "and in going to and returning from the same. . . ." Furthermore, the Members cannot be questioned in any other place for remarks made in Congress. Each House may expel a Member of its body by a two-thirds vote.

Enactment of Laws In order to become law, all bills and joint resolutions, except those proposing a constitutional amendment, must pass both the House of Representatives and the Senate and either be signed by the President or be passed over the President's veto by a two-thirds vote of both Houses of Congress. Section 7 of Article I states: "If any Bill shall not be returned by the President within ten Days (Sundays excepted) after it shall have been presented to him, the Same shall be a Law, in like Manner as if he had signed it, unless the Congress by their Adjournment prevent its Return, in which Case it shall not be a Law." When a bill or joint resolution is introduced in the House, the usual procedure for its enactment into law is as follows: assignment to House committee having jurisdiction; if favorably considered, it is reported to the House either in its original form or with recommended amendments; if the bill or resolution is passed by the House, it is messaged to the Senate and referred to the committee having jurisdiction; in the Senate committee the bill, if favorably considered, may be reported in the form as received from the House, or with recommended amendments; the approved bill or resolution is reported to the Senate, and if passed by that body, is returned to the House; if one body does not accept the amendments to a bill by the other body, a conference committee comprised of Members of both bodies is usually appointed to effect a compromise; when the bill or joint resolution is finally approved by both Houses, it is signed by the Speaker (or Speaker pro tempore) and the Vice President (or President pro tempore or acting President pro tempore) and is presented to the President; and once the President's signature is affixed, the measure becomes a law. If the President vetoes the bill, it cannot become a law unless it is re-passed by a two-thirds vote of both Houses.

The Senate

The Capitol, Washington, DC 20510
Phone, 202–224–3121. Internet, http://www.senate.gov.

President of the Senate (Vice President of the United States)	JOSEPH R. BIDEN, JR.
President pro tempore	DANIEL K. INOUYE
Majority Leader	HARRY REID
Minority Leader	MITCH MCCONNELL
Secretary of the Senate	NANCY ERICKSON
Sergeant at Arms	TERRANCE GAINER
Secretary for the Majority	LULA JOHNSON DAVIS
Secretary for the Minority	DAVID J. SCHIAPPA
Chaplain	BARRY BLACK

The Senate is composed of 100 Members, 2 from each State, who are elected to serve for a term of 6 years. Senators were originally chosen by the State legislatures. This procedure was changed by the 17th amendment to the Constitution, adopted in 1913, which made the election of Senators a function of the people. There are three classes of Senators, and a new class is elected every 2 years.

Senators must be residents of the State from which they are chosen. In addition, a Senator must be at least 30 years of age and must have been a citizen of the United States for at least 9 years.

Officers The Vice President of the United States is the Presiding Officer of the Senate. In the Vice President's absence, the duties are taken over by a President pro tempore, elected by that body, or someone designated by the President pro tempore.

The positions of Senate Majority and Minority Leader have been in existence only since the early years of the 20th century. Leaders are elected at the beginning of each new Congress by a majority vote of the Senators in their political party. In cooperation with their party organizations, Leaders are responsible for the design and achievement of a legislative program. This involves managing the flow of legislation, expediting noncontroversial measures, and keeping Members informed regarding proposed action on pending business. Each Leader serves as an ex officio member of his party's policymaking and organizational bodies and is aided by an assistant floor leader (whip) and a party secretary.

The Secretary of the Senate, elected by vote of the Senate, performs the duties of the Presiding Officer of the Senate in the absence of the Vice President and pending the election of a President pro tempore. The Secretary is the custodian of the seal of the Senate, draws requisitions on the Secretary of the Treasury for moneys appropriated for the compensation of Senators, officers, and employees, and for the contingent expenses of the Senate, and is empowered to administer oaths to any officer of the Senate and to any witness produced before it. The Secretary's executive duties include certification of extracts from the Journal of the Senate; the attestation of bills and joint, concurrent, and Senate resolutions; in impeachment trials, issuance, under the authority of the Presiding Officer, of all orders, mandates, writs, and precepts authorized by the Senate; and certification to the President of the United States of the advice and consent of the Senate to ratification of treaties and the names of persons confirmed or rejected upon the nomination of the President.

The Sergeant at Arms, elected by vote of the Senate, serves as the executive, chief law enforcement, and protocol officer and is the principal administrative manager for most support services in the Senate. As executive officer, the Sergeant

UNITED STATES SENATE

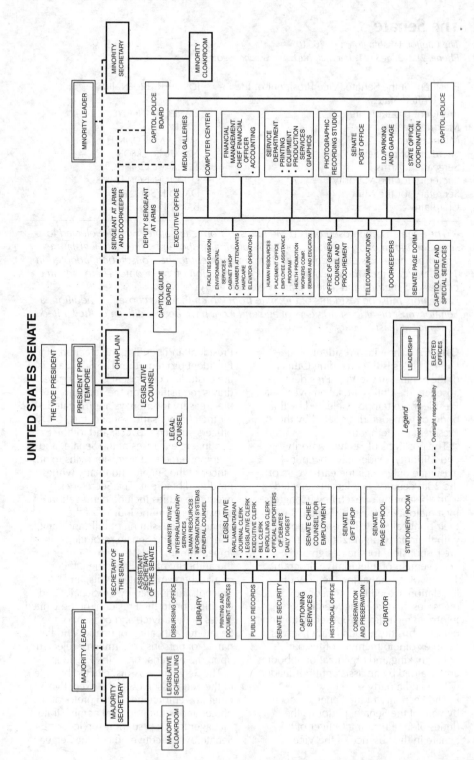

at Arms has custody of the Senate gavel; enforces Senate rules and regulations as they pertain to the Senate Chamber, the Senate wing of the Capitol, and the Senate Office Buildings; and subject to the Presiding Officer, maintains order on the Senate floor, Chamber, and galleries. As chief law enforcement officer of the Senate, the Sergeant at Arms is authorized to maintain security in the Capitol and all Senate buildings, as well as to protect Senators; to arrest and detain any person violating Senate rules; and to locate absentee Senators for a quorum. The Sergeant at Arms serves as a member of the Capitol Police Board and as its chairman each odd year. As protocol officer, the Sergeant at Arms escorts the President and other heads of state or official guests of the Senate who are attending official functions in the Capitol; makes arrangements for funerals of Senators who die in office; and assists in planning the inauguration of the President and organizing the swearing-in and orientation programs for newly elected Senators.

Committees The work of preparing and considering legislation is done largely by committees of both Houses of Congress. There are 16 standing committees in the Senate. The standing committees of the Senate are shown in the list below. In addition, there are two select committees in each House and various congressional commissions and joint committees composed of Members of both Houses. Each House may also appoint special investigating committees. The membership of the standing committees of each House is chosen by a vote of the entire body; members of other committees are appointed under the provisions of the measure establishing them.

Each bill and resolution is usually referred to the appropriate committee, which may report a bill out in its original form, favorably or unfavorably, recommend amendments, report original measures, or allow the proposed legislation to die in committee without action.

Standing Committees of the Senate

Senate Commitee	Room*
Agriculture, Nutrition, and Forestry	SR–328A
Appropriations	S–128
Armed Services	SR–228
Banking, Housing, and Urban Affairs	SD–534
Budget	SD–624
Commerce, Science, and Transportation	SD–508
Energy and Natural Resources	SD–304
Environment and Public Works	SD–410
Finance	SD–219
Foreign Relations	SD–446
Health, Education, Labor, and Pensions	SD–428
Homeland Security and Governmental Affairs	SD–340
Judiciary	SD–224
Rules and Administration	SR–305
Small Business and Entrepreneurship	SR–428A
Veterans' Affairs	SR–412

*Room numbers preceded by S are in the Senate wing of the Capitol Building; those preceded by SD are in the Dirksen Office Building; those preceded by SR are in the Russell Office Building; and those preceded by SH are in the Hart Office Building.

Special Powers of the Senate Under the Constitution, the Senate is granted certain powers not accorded to the House of Representatives. The Senate approves or disapproves certain Presidential appointments by majority vote, and treaties must be concurred in by a two-thirds vote.

Senators

[Democrats (57); Republicans (41); Independents (2); total, 100]. Members who have died or resigned appear in brackets [ ]. Room numbers preceded by SR are in the Russell Office Building (Delaware and Constitution Avenues); those preceded by SRCY are in the Russell Office Building Courtyard; those preceded by SD are in the Dirksen Office Building (First Street and Constitution Avenue); and those preceded by SH are in the Hart Office Building (Second and C Streets). The most current listing of Senators can be found on the Internet at www.senate.gov. Members' offices may be reached by phone at 202–224–3121. Internet, http://www.senate.gov.

Name	State	Room
Akaka, Daniel K. (D)	Hawaii	SH141
Alexander, Lamar (R)	Tennessee	SD455
Ayotte, Kelly (R)	New Hampshire	SR188
Barrasso, John (R)	Wyoming	SD307
Baucus, Max (D)	Montana	SH511
Begich, Mark (D)	Alaska	SR144
Bennet, Michael F. (D)	Colorado	SH702
Bingaman, Jeff (D)	New Mexico	SH703
Blumenthal, Richard (D)	Connecticut	SDG55
Blunt, Roy (R)	Missouri	SDB40C
Boozman, John (R)	Arkansas	SRCY1
Boxer, Barbara (D)	California	SH112
Brown, Scott P. (R)	Massachusetts	SR317
Brown, Sherrod (D)	Ohio	SH713
Burr, Richard (R)	North Carolina	SR217
Cantwell, Maria (D)	Washington	SD511
Cardin, Benjamin L. (D)	Maryland	SH509
Carper, Thomas R. (D)	Delaware	SH513
Casey, Robert P., Jr. (D)	Pennsylvania	SR393
Chambliss, Saxby (R)	Georgia	SR416
Coast, Daniel (R)	Indiana	SDB40E
Coburn, Tom (R)	Oklahoma	SR172
Cochran, Thad (R)	Mississippi	SD113
Collins, Susan M. (R)	Maine	SD413
Conrad, Kent (D)	North Dakota	SH530
Coons, Christopher A. (D)	Delaware	SR383
Corker, Bob (R)	Tennessee	SD185
Cornyn, John (R)	Texas	SH517
Crapo, Mike (R)	Idaho	SD239
DeMint, Jim (R)	South Carolina	SR340
Durbin, Richard J. (D)	Illinois	SH309
Ensign, John (R)	Nevada	SR119
Enzi, Mike (R)	Wyoming	SR379A
Feinstein, Dianne (D)	California	SH331
Franken, Al (D)	Minnesota	SH320
Gillibrand, Kristen E. (D)	New York	SR478
Graham, Lindsey O. (R)	South Carolina	SR290
Grassley, Charles E. (R)	Iowa	SH135
Hagan, Kay R. (D)	North Carolina	SD521
Harkin, Tom (D)	Iowa	SH731
Hatch, Orrin G. (R)	Utah	SH104
Hoeven, John (R)	North Dakota	SDG11
Hutchison, Kay Bailey (R)	Texas	SR284
Inhofe, James M. (R)	Oklahoma	SR453
Inouye, Daniel K. (D)	Hawaii	SH722
Isakson, Johnny (R)	Georgia	SR120
Johanns, Mike (R)	Nebraska	SR404
Johnson, Ron (R)	Wisconsin	SRCY2
Johnson, Tim (D)	South Dakota	SH136
Kerry, John F. (D)	Massachusetts	SR218
Kirk, Mark (R)	Illinois	SR387

Senators—Continued

[Democrats (57); Republicans (41); Independents (2); total, 100]. Members who have died or resigned appear in brackets [ ]. Room numbers preceded by SR are in the Russell Office Building (Delaware and Constitution Avenues); those preceded by SRCY are in the Russell Office Building Courtyard; those preceded by SD are in the Dirksen Office Building (First Street and Constitution Avenue); and those preceded by SH are in the Hart Office Building (Second and C Streets). The most current listing of Senators can be found on the Internet at www.senate.gov. Members' offices may be reached by phone at 202–224–3121. Internet, http://www.senate.gov.

Name	State	Room
Klobuchar, Amy (D)	Minnesota	SH302
Kohl, Herb (D)	Wisconsin	SH330
Kyl, Jon (R)	Arizona	SH730
Landrieu, Mary L. (D)	Louisiana	SH328
Lautenberg, Frank R. (D)	New Jersey	SH324
Leahy, Patrick J. (D)	Vermont	SR433
Lee, Mike (R)	Utah	SH825
Levin, Carl (D)	Michigan	SR269
Lieberman, Joseph I. (I)	Connecticut	SH706
Lugar, Richard G. (R)	Indiana	SH306
McCain, John (R)	Arizona	SR241
McCaskill, Claire (D)	Missouri	SH717
McConnell, Mitch (R)	Kentucky	SR361A
Manchin, Joe, III (D)	West Virginia	SH311
Menendez, Robert (D)	New Jersey	SH528
Merkley, Jeff (D)	Oregon	SR107
Mikulski, Barbara A. (D)	Maryland	SH503
Moran, Jerry (R)	Kansas	SRCY4
Murkowski, Lisa (R)	Alaska	SH709
Murray, Patty (D)	Washington	SR173
Nelson, Ben (D)	Nebraska	SH720
Nelson, Bill (D)	Florida	SH716
Paul, Rand (R)	Kentucky	SRCY5
Pryor, Mark L. (D)	Arkansas	SD255
Reed, Jack (D)	Rhode Island	SH728
Reid, Harry (D)	Nevada	SH522
Risch, James E. (R)	Idaho	SR483
Roberts, Pat (R)	Kansas	SH109
Rockefeller, John D., IV (D)	West Virginia	SH531
Rubio, Marco (R)	Florida	SDB40A
Sanders, Bernard (I)	Vermont	SD332
Schumer, Charles E. (D)	New York	SH313
Sessions, Jeff (R)	Alabama	SR335
Shaheen, Jeanne (D)	New Hampshire	SH520
Shelby, Richard C. (R)	Alabama	SR304
Snowe, Olympia J. (R)	Maine	SR154
Stabenow, Debbie (D)	Michigan	SH133
Tester, Jon (D)	Montana	SH724
Thune, John (R)	South Dakota	SR493
Toomey, Patrick J. (R)	Pennsylvania	SDB40B
Udall, Mark (D)	Colorado	SH317
Udall, Tom (D)	New Mexico	SH110
Vitter, David (R)	Louisiana	SH516
Warner, Mark R. (D)	Virginia	SR459A
Webb, Jim (D)	Virginia	SR248
Whitehouse, Sheldon (D)	Rhode Island	SH502
Wicker, Roger F. (R)	Mississippi	SD555
Wyden, Ron (D)	Oregon	SD223

Sources of Information

Electronic Access Specific information and legislation can be found on the Internet at http://thomas.loc.gov or www.senate.gov.

Publications The Congressional Directory, the Senate Manual, and telephone directory for the U.S. Senate may be obtained from the Superintendent of Documents, Government Printing Office, Washington, DC 20402. Internet, http://www.gpo.gov/fdsys/browse/collectiontab.action.

For further information, contact the Secretary of the Senate, The Capitol, Washington, DC 20510. Phone, 202–224–2115. Internet, http://www.senate.gov.

The House of Representatives

The Capitol, Washington, DC 20515
Phone, 202–225–3121. Internet, http://www.house.gov.

The Speaker	JOHN A. BOEHNER
Clerk	KAREN L. HAAS
Sergeant at Arms	WILSON L. LIVINGOOD
Chief Administrative Officer	DANIEL J. STRODEL
Chaplain	REV. DANIEL P. COUGHLIN

The House of Representatives comprises 435 Representatives. The number representing each State is determined by population, but every State is entitled to at least one Representative. Members are elected by the people for 2-year terms, all terms running for the same period. Representatives must be residents of the State from which they are chosen. In addition, a Representative must be at least 25 years of age and must have been a citizen for at least 7 years.

A Resident Commissioner from Puerto Rico (elected for a 4-year term) and Delegates from American Samoa, the District of Columbia, Guam, and the Virgin Islands complete the composition of the Congress of the United States. Delegates are elected for a term of 2 years. The Resident Commissioner and Delegates may take part in the floor discussions but have no vote in the full House. They do, however, vote in the committees to which they are assigned and in the Committee of the Whole House on the State of the Union.

Officers The Presiding Officer of the House of Representatives, the Speaker, is elected by the House. The Speaker may designate any Member of the House to act in the Speaker's absence.

The House leadership is structured essentially the same as the Senate, with the Members in the political parties responsible for the election of their respective leader and whips.

The elected officers of the House of Representatives include the Clerk, the Sergeant at Arms, the Chief Administrative Officer, and the Chaplain.

The Clerk is custodian of the seal of the House and administers the primary legislative activities of the House. These duties include accepting the credentials of the Members-elect and calling the Members to order at the commencement of the first session of each Congress; keeping the Journal; taking all votes and certifying the passage of bills; and processing all legislation. Through various departments, the Clerk is also responsible for floor and committee reporting services; legislative information and reference services; the administration of House reports pursuant to House rules and certain legislation including the Ethics in Government Act and the Lobbying Disclosure Act of 1995; the distribution of House documents; and administration of the House Page Program. The Clerk is also charged with supervision of the offices vacated by Members due to death, resignation, or expulsion.

HOUSE OF REPRESENTATIVES

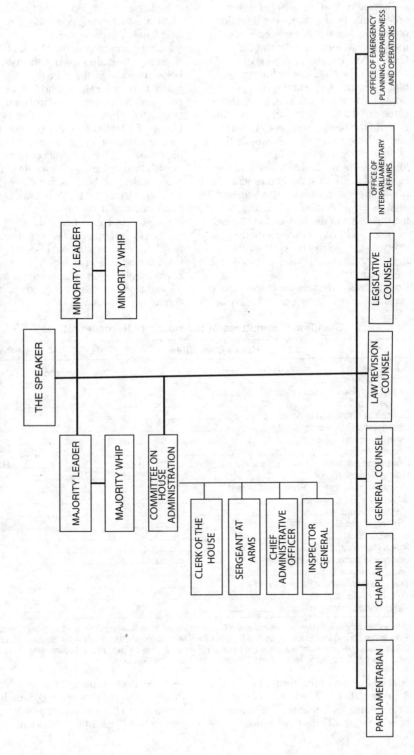

The Sergeant at Arms maintains the order of the House under the direction of the Speaker and is the keeper of the Mace. As a member of the U.S. Capitol Police Board, the Sergeant at Arms is the chief law enforcement officer for the House and serves as Board Chairman each even year. The ceremonial and protocol duties parallel those of the Senate Sergeant at Arms and include arranging the inauguration of the President of the United States, Joint Sessions of Congress, visits to the House of heads of state, and funerals of Members of Congress. The Sergeant at Arms enforces the rules relating to the privileges of the Hall of the House, including admission to the galleries, oversees garage and parking security of the House, and distributes all House staff identification cards.

Committees   The work of preparing and considering legislation is done largely by committees of both Houses of Congress. There are 19 standing committees in the House of Representatives. The standing committees of the House of Representatives are shown in the list below. In addition, there are two select committees in the House and various congressional commissions and joint committees composed of Members of both Houses. Each House may also appoint special investigating committees. The membership of the standing committees of each House is chosen by a vote of the entire body; members of other committees are appointed under the provisions of the measure establishing them.

Each bill and resolution is usually referred to the appropriate committee, which may report a bill out in its original form, favorably or unfavorably, recommend amendments, report original measures, or allow the proposed legislation to die in committee without action.

Standing Committees of the House of Representatives

House Committee	Room*
Agriculture	1301
Appropriations	H307
Armed Services	2120
Budget	207
Education and the Workforce	2181
Energy and Commerce	2125
Financial Services	2129
Foreign Affairs	2170
Homeland Security	H2–176
House Administration	1309
House Administration (Franking Office)	1313
Judiciary	2138
Natural Resources	1324
Oversight and Government Reform	2157
Rules	H312
Rules (Minority)	1627
Science, Space, and Technology	2321
Small Business	2361
Transportation and Infrastructure	2165
Veterans' Affairs	335
Ways and Means	1102

*Room numbers with three digits are in the Cannon House Office Building, four digits beginning with 1 are in the Longworth House Office Building, and four digits beginning with 2 are in the Rayburn House Office Building. Room numbers preceded by H or HT are in the House wing of the Capitol Building. Rooms preceded by LA are in the John Adams Building of the Library of Congress.

Special Powers of the House of Representatives   The House of Representatives is granted the power of originating all bills for the raising of revenue. Both Houses of Congress act in impeachment proceedings, which, according to the Constitution, may be instituted against the President,

Vice President, and all civil officers of the United States. The House of Representatives has the sole power of impeachment, and the Senate has the sole power to try impeachments.

Representatives, Delegates, and Resident Commissioners

[Republicans (241); Democrats (193); vacancies (1); total, 435 Members; 5 Delegates; 1 Resident Commissioner]. Members who have resigned appear in bold brackets [ ]. Room numbers with three digits are in the Cannon House Office Building (New Jersey and Independence Avenues), four digits beginning with 1 are in the Longworth House Office Building (between South Capitol Street and New Jersey Avenue on Independence Avenue), and four digits beginning with 2 are in the Rayburn House Office Building (between First and South Capitol Streets on Independence Avenue). Members' offices may be reached by phone at 202–225–3121. The most current listing of House Members can be found on the Internet at http://clerk.house.gov.

Name	State (District) / Territory	Room
Ackerman, Gary L. (D)	New York (5)	2111
Adams, Sandy (R)	Florida (24)	216
Aderholt, Robert B. (R)	Alabama (4)	2264
Akin, W. Todd (R)	Missouri (2)	117
Alexander, Rodney (R)	Louisiana (5)	316
Altmire, Jason (D)	Pennsylvania (4)	332
Amash, Justin (R)	Michigan (3)	114
Andrews, Robert E. (D)	New Jersey (1)	2265
Austria, Steve (R)	Ohio (7)	439
Baca, Joe (D)	California (43)	2366
Bachmann, Michele (R)	Minnesota (6)	103
Bachus, Spencer (R)	Alabama (6)	2246
Baldwin, Tammy (D)	Wisconsin (2)	2446
Barletta, Lou (R)	Pennsylvania (11)	510
Barrow, John (D)	Georgia (12)	2202
Bartlett, Roscoe G. (R)	Maryland (6)	2412
Barton, Joe (R)	Texas (6)	2109
Bass, Charles F. (R)	New Hampshire (2)	2350
Bass, Karen (D)	California (33)	408
Becerra, Xavier (D)	California (31)	1226
Benishek, Dan (R)	Michigan (1)	514
Berg, Rick (R)	North Dakota (At Large)	323
Berkley, Shelley (D)	Nevada (1)	405
Berman, Howard L. (D)	California (28)	2221
Biggert, Judy (R)	Illinois (13)	2113
Bilbray, Brian P. (R)	California (50)	2410
Bilirakis, Gus M. (R)	Florida (9)	407
Bishop, Rob (R)	Utah (1)	123
Bishop, Sanford D., Jr. (R)	Georgia (2)	2429
Bishop, Timothy H. (D)	New York (1)	306
Black, Diane (R)	Tennessee (6)	1531
Blackburn, Marsha (R)	Tennessee (7)	217
Blumenauer, Earl (D)	Oregon (3)	1502
Boehner, John A. (R)	Ohio (8)	1011
Bonner, Jo (R)	Alabama (1)	2236
Bono Mack, Mary (R)	California (45)	104
Bordallo, Madeleine Z. (D)	Guam (Delegate)	2441
Boren, Dan (D)	Oklahoma (2)	2447
Boswell, Leonard L. (D)	Iowa (3)	1026
Boustany, Charles W., Jr. (R)	Louisiana (7)	1431
Brady, Kevin (R)	Texas (8)	301
Brady, Robert A. (D)	Pennsylvania (1)	102
Braley, Bruce L. (D)	Iowa (1)	1727
Brooks, Mo (R)	Alabama (5)	1641
Broun, Paul C. (R)	Georgia (10)	325
Brown, Corrine (D)	Florida (3)	2336

Representatives, Delegates, and Resident Commissioners—Continued

[Republicans (241); Democrats (193); vacancies (1); total, 435 Members; 5 Delegates; 1 Resident Commissioner]. Members who have resigned appear in bold brackets [ ]. Room numbers with three digits are in the Cannon House Office Building (New Jersey and Independence Avenues), four digits beginning with 1 are in the Longworth House Office Building (between South Capitol Street and New Jersey Avenue on Independence Avenue), and four digits beginning with 2 are in the Rayburn House Office Building (between First and South Capitol Streets on Independence Avenue). Members' offices may be reached by phone at 202–225–3121. The most current listing of House Members can be found on the Internet at http://clerk.house.gov.

Name	State (District) / Territory	Room
Buchanan, Vern (R)	Florida (13)	221
Bucshon, Larry (R)	Indiana (8)	1123
Buerkle, Ann Marie (R)	New York (25)	1630
Burgess, Michael C. (R)	Texas (26)	2241
Burton, Dan (R)	Indiana (5)	2308
Butterfield, G. K. (D)	North Carolina (1)	2305
Calvert, Ken (R)	California (44)	2269
Camp, Dave (R)	Michigan (4)	341
Campbell, John (R)	California (48)	1507
Canseco, Francisco "Quico" (R)	Texas (23)	1339
Cantor, Eric (R)	Virginia (7)	303
Capito, Shelley Moore (R)	West Virginia (2)	2443
Capps, Lois (D)	California (23)	2231
Capuano, Michael E. (D)	Massachusetts (8)	1414
Cardoza, Dennis A. (D)	California (18)	2437
Carnahan, Russ (D)	Missouri (3)	1710
Carney, John C., Jr. (D)	Delaware (At Large)	1429
Carson, Andre (D)	Indiana (7)	425
Carter, John R. (R)	Texas (31)	409
Cassidy, William "Bill" (R)	Louisiana (6)	1535
Castor, Kathy (D)	Florida (11)	137
Chabot, Steve (R)	Ohio (1)	2351
Chaffetz, Jason (R)	Utah (3)	1032
Chandler, Ben (D)	Kentucky (6)	1504
Christensen, Donna M. (D)	Virgin Islands (Delegate)	1510
Chu, Judy (D)	California (32)	1520
Cicilline, David N. (D)	Rhode Island (1)	128
Clarke, Hansen (D)	Michigan (13)	1319
Clarke, Yvette D. (D)	New York (11)	1029
Clay, William "Lacy" (D)	Missouri (1)	2418
Cleaver, Emanuel (D)	Missouri (5)	1433
Clyburn, James E. (D)	South Carolina (6)	2135
Coble, Howard (R)	North Carolina (6)	2188
Coffman, Mike (R)	Colorado (6)	1222
Cohen, Steve (D)	Tennessee (9)	1005
Cole, Tom (R)	Oklahoma (4)	2458
Conaway, K. Michael (R)	Texas (11)	2430
Connolly, Gerald E. (D)	Virginia (11)	424
Conyers, John, Jr. (D)	Michigan (14)	2426
Cooper, Jim (D)	Tennessee (5)	1536
Costa, Jim (D)	California (20)	1314
Costello, Jerry F. (D)	Illinois (12)	2408
Courtney, Joe (D)	Connecticut (2)	215
Cravaack, Chip (R)	Minnesota (8)	508
Crawford, Eric A. "Rick" (R)	Arkansas (1)	1408
Crenshaw, Ander (R)	Florida (4)	440
Critz, Mark S. (D)	Pennsylvania (12)	1022
Crowley, Joseph (D)	New York (7)	2404
Cuellar, Henry (D)	Texas (28)	2463
Culberson, John Abney (R)	Texas (7)	2352
Cummings, Elijah E. (D)	Maryland (7)	2235

Representatives, Delegates, and Resident Commissioners—Continued

[Republicans (241); Democrats (193); vacancies (1); total, 435 Members; 5 Delegates; 1 Resident Commissioner]. Members who have resigned appear in bold brackets [ ]. Room numbers with three digits are in the Cannon House Office Building (New Jersey and Independence Avenues), four digits beginning with 1 are in the Longworth House Office Building (between South Capitol Street and New Jersey Avenue on Independence Avenue), and four digits beginning with 2 are in the Rayburn House Office Building (between First and South Capitol Streets on Independence Avenue). Members' offices may be reached by phone at 202–225–3121. The most current listing of House Members can be found on the Internet at http://clerk.house.gov.

Name	State (District) / Territory	Room
Davis, Danny K. (D)	Illinois (7)	2159
Davis, Geoff (R)	Kentucky (4)	1119
Davis, Susan A. (D)	California (53)	1526
DeFazio, Peter A. (D)	Oregon (4)	2134
DeGette, Diana (D)	Colorado (1)	2335
DeLauro, Rosa L. (D)	Connecticut (3)	2413
Denham, Jeff (R)	California (19)	1605
Dent, Charles W. (R)	Pennsylvania (15)	1009
DesJarlais, Scott (R)	Tennessee (4)	413
Deutch, Theodore E. (D)	Florida (19)	1024
Diaz-Balart, Mario (R)	Florida (21)	436
Dicks, Norman D. (D)	Washington (6)	2467
Dingell, John D. (D)	Michigan (15)	2328
Doggett, Lloyd (D)	Texas (25)	201
Dold, Robert J. (R)	Illinois (10)	212
Donnelly, Joe (D)	Indiana (2)	1530
Doyle, Michael F. (D)	Pennsylvania (14)	401
Dreier, David (R)	California (26)	233
Duffy, Sean P. (R)	Wisconsin (7)	1208
Duncan, Jeff (R)	South Carolina (3)	116
Duncan, John J., Jr. (R)	Tennessee (2)	2207
Edwards, Donna F. (D)	Maryland (4)	318
Ellison, Keith (D)	Minnesota (5)	1027
Ellmers, Renee L. (R)	North Carolina (2)	1533
Emerson, Jo Ann (R)	Missouri (8)	2230
Engel, Eliot L. (D)	New York (17)	2161
Eshoo, Anna G. (D)	California (14)	205
Faleomavaega, Eni F.H. (D)	American Samoa (Delegate)	2422
Farenthold, Blake (R)	Texas (27)	2110
Farr, Sam (D)	California (17)	1126
Fattah, Chaka (D)	Pennsylvania (2)	2301
Filner, Bob (D)	California (51)	2428
Fincher, Stephen Lee (R)	Tennessee (8)	1118
Fitzpatrick, Michael G. (R)	Pennsylvania (8)	1224
Flake, Jeff (R)	Arizona (6)	240
Fleischmann, Charles J. "Chuck" (R)	Tennessee (3)	511
Fleming, John (R)	Louisiana (4)	416
Flores, Bill (R)	Texas (17)	1505
Forbes, J. Randy (R)	Virginia (4)	2438
Fortenberry, Jeff (R)	Nebraska (1)	1514
Foxx, Virginia (R)	North Carolina (5)	1230
Frank, Barney (D)	Massachusetts (4)	2252
Franks, Trent (R)	Arizona (2)	2435
Frelinghuysen, Rodney P. (R)	New Jersey (11)	2369
Fudge, Marcia L. (D)	Ohio (11)	1019
Gallegly, Elton (R)	California (24)	2309
Garamendi, John (D)	California (10)	228
Gardner, Cory (R)	Colorado (4)	213
Garrett, Scott (R)	New Jersey (5)	2244
Gerlach, Jim (R)	Pennsylvania (6)	2442
Gibbs, Bob (R)	Ohio (18)	329

Representatives, Delegates, and Resident Commissioners—Continued

[Republicans (241); Democrats (193); vacancies (1); total, 435 Members; 5 Delegates; 1 Resident Commissioner]. Members who have resigned appear in bold brackets [ ]. Room numbers with three digits are in the Cannon House Office Building (New Jersey and Independence Avenues), four digits beginning with 1 are in the Longworth House Office Building (between South Capitol Street and New Jersey Avenue on Independence Avenue), and four digits beginning with 2 are in the Rayburn House Office Building (between First and South Capitol Streets on Independence Avenue). Members' offices may be reached by phone at 202–225–3121. The most current listing of House Members can be found on the Internet at http://clerk.house.gov.

Name	State (District) / Territory	Room
Gibson, Christopher P. (R)	New York (20)	502
Giffords, Gabrielle (D)	Arizona (8)	1030
Gingrey, Phil (R)	Georgia (11)	442
Gohmert, Louie (R)	Texas (1)	2440
Gonzalez, Charles A. (D)	Texas (20)	1436
Goodlatte, Bob (R)	Virginia (6)	2240
Gosar, Paul A. (R)	Arizona (1)	504
Gowdy, Trey (R)	South Carolina (4)	1237
Granger, Kay (R)	Texas (12)	320
Graves, Sam (R)	Missouri (6)	1415
Graves, Tom (R)	Georgia (9)	1113
Green, Al (D)	Texas (9)	2201
Green, Gene (D)	Texas (29)	2470
Griffin, Tim (R)	Arkansas (2)	1232
Griffith, H. Morgan (R)	Virginia (9)	1108
Grijalva, Raul M. (D)	Arizona (7)	1511
Grimm, Michael G. (R)	New York (13)	512
Guinta, Frank C. (R)	New Hampshire (1)	1223
Guthrie, Brett (R)	Kentucky (2)	308
Gutierrez, Luis V. (D)	Illinois (4)	2266
Hall, Ralph M. (R)	Texas (4)	2405
Hanabusa, Colleen W. (D)	Hawaii (1)	238
Hanna, Richard L. (R)	New York (24)	319
Harman, Jane (D)	California (36)	2400
Harper, Gregg (R)	Mississippi (3)	307
Harris, Andy (R)	Maryland (1)	506
Hartzler, Vicky (R)	Missouri (4)	1023
Hastings, Alcee L. (D)	Florida (23)	2353
Hastings, Doc (R)	Washington (4)	1203
Hayworth, Nan A.S. (R)	New York (19)	1440
Heck, Joseph J. (R)	Nevada (3)	132
Heinrich, Martin (D)	New Mexico (1)	336
Heller, Dean (R)	Nevada (2)	125
Hensarling, Jeb (R)	Texas (5)	129
Herger, Wally (R)	California (2)	242
Herrera Beutler, Jaime (R)	Washington (3)	1130
Higgins, Brian (D)	New York (27)	2459
Himes, James A. (D)	Connecticut (4)	119
Hinchey, Maurice D. (D)	New York (22)	2431
Hinojosa, Ruben (D)	Texas (15)	2262
Hirono, Mazie K. (D)	Hawaii (2)	1410
Holden, Tim (D)	Pennsylvania (17)	2417
Holt, Rush D. (D)	New Jersey (12)	1214
Honda, Michael M. (D)	California (15)	1713
Hoyer, Steny H. (D)	Maryland (5)	1705
Huelskamp, Tim (R)	Kansas (1)	126
Huizenga, Bill (R)	Michigan (2)	1217
Hultgren, Randy (R)	Illinois (14)	427
Hunter, Duncan (R)	California (52)	223
Hurt, Robert (R)	Virginia (5)	1516
Inslee, Jay (D)	Washington (1)	2329

Representatives, Delegates, and Resident Commissioners—Continued

[Republicans (241); Democrats (193); vacancies (1); total, 435 Members; 5 Delegates; 1 Resident Commissioner]. Members who have resigned appear in bold brackets [ ]. Room numbers with three digits are in the Cannon House Office Building (New Jersey and Independence Avenues), four digits beginning with 1 are in the Longworth House Office Building (between South Capitol Street and New Jersey Avenue on Independence Avenue), and four digits beginning with 2 are in the Rayburn House Office Building (between First and South Capitol Streets on Independence Avenue). Members' offices may be reached by phone at 202–225–3121. The most current listing of House Members can be found on the Internet at http://clerk.house.gov.

Name	State (District) / Territory	Room
Israel, Steve (D)	New York (2)	2457
Issa, Darrell E. (R)	California (49)	2347
Jackson, Jesse L., Jr. (D)	Illinois (2)	2419
Jackson Lee, Sheila (D)	Texas (18)	2160
Jenkins, Lynn (R)	Kansas (2)	1122
Johnson, Bill (R)	Ohio (6)	317
Johnson, Eddie Bernice (D)	Texas (30)	2468
Johnson, Henry C. "Hank", Jr. (D)	Georgia (4)	1427
Johnson, Sam (R)	Texas (3)	1211
Johnson, Timothy V. (R)	Illinois (15)	1207
Jones, Walter B. (R)	North Carolina (3)	2333
Jordan, Jim (R)	Ohio (4)	1524
Kaptur, Marcy (D)	Ohio (9)	2186
Keating, William R. (D)	Massachusetts (10)	315
Kelly, Mike (R)	Pennsylvania (3)	515
Kildee, Dale E. (D)	Michigan (5)	2107
Kind, Ron (D)	Wisconsin (3)	1406
King, Peter T. (R)	New York (3)	339
King, Steve (R)	Iowa (5)	1131
Kingston, Jack (R)	Georgia (1)	2372
Kinzinger, Adam (R)	Illinois (11)	1218
Kissell, Larry (D)	North Carolina (8)	1632
Kline, John (R)	Minnesota (2)	2439
Kucinich, Dennis J. (D)	Ohio (10)	2445
Labrador, Raul R. (R)	Idaho (1)	1523
Lamborn, Doug (R)	Colorado (5)	437
Lance, Leonard (R)	New Jersey (7)	426
Landry, Jeffrey M. (R)	Louisiana (3)	206
Langevin, James R. (D)	Rhode Island (2)	109
Lankford, James (R)	Oklahoma (5)	509
Larsen, Rick (D)	Washington (2)	108
Larson, John B. (D)	Connecticut (1)	1501
Latham, Tom (R)	Iowa (4)	2217
LaTourette, Steven C. (R)	Ohio (14)	2371
Latta, Robert E. (R)	Ohio (5)	1323
Lee, Barbara (D)	California (9)	2267
[Lee, Christopher J.] (R)	New York (26)	1711
Levin, Sander M. (D)	Michigan (12)	1236
Lewis, Jerry (R)	California (41)	2112
Lewis, John (D)	Georgia (5)	343
Lipinski, Daniel (D)	Illinois (3)	1717
LoBiondo, Frank A. (R)	New Jersey (2)	2427
Loebsack, David (D)	Iowa (2)	1527
Lofgren, Zoe (D)	California (16)	1401
Long, Billy (R)	Missouri (7)	1541
Lowey, Nita M. (D)	New York (18)	2365
Lucas, Frank D. (R)	Oklahoma (3)	2311
Luetkemeyer, Blaine (R)	Missouri (9)	1740
Lujan, Ben Ray (D)	New Mexico (3)	330
Lummis, Cynthia M. (R)	Wyoming (At Large)	113
Lungren, Daniel E. (R)	California (3)	2313

Representatives, Delegates, and Resident Commissioners—Continued

[Republicans (241); Democrats (193); vacancies (1); total, 435 Members; 5 Delegates; 1 Resident Commissioner]. Members who have resigned appear in bold brackets [ ]. Room numbers with three digits are in the Cannon House Office Building (New Jersey and Independence Avenues), four digits beginning with 1 are in the Longworth House Office Building (between South Capitol Street and New Jersey Avenue on Independence Avenue), and four digits beginning with 2 are in the Rayburn House Office Building (between First and South Capitol Streets on Independence Avenue). Members' offices may be reached by phone at 202–225–3121. The most current listing of House Members can be found on the Internet at http://clerk.house.gov.

Name	State (District) / Territory	Room
Lynch, Stephen F. (D)	Massachusetts (9)	2348
Mack, Connie (R)	Florida (14)	115
Maloney, Carolyn B. (D)	New York (14)	2332
Manzullo, Donald A. (R)	Illinois (16)	2228
Marchant, Kenny (R)	Texas (24)	1110
Marino, Tom (R)	Pennsylvania (10)	410
Markey, Edward J. (D)	Massachusetts (7)	2108
Matheson, Jim (D)	Utah (2)	2434
Matsui, Doris O. (D)	California (5)	222
McCarthy, Carolyn (D)	New York (4)	2346
McCarthy, Kevin (R)	California (22)	326
McCaul, Michael T. (R)	Texas (10)	131
McClintock, Tom (R)	California (4)	428
McCollum, Betty (D)	Minnesota (4)	1714
McCotter, Thaddeus G. (R)	Michigan (11)	2243
McDermott, Jim (D)	Washington (7)	1035
McGovern, James P. (D)	Massachusetts (3)	438
McHenry, Patrick T. (R)	North Carolina (10)	224
McIntyre, Mike (D)	North Carolina (7)	2133
McKeon, Howard P. "Buck" (R)	California (25)	2184
McKinley, David B. (R)	West Virginia (1)	313
McMorris Rodgers, Cathy (R)	Washington (5)	2421
McNerney, Jerry (D)	California (11)	1210
Meehan, Patrick (R)	Pennsylvania (7)	513
Meeks, Gregory W. (D)	New York (6)	2234
Mica, John L. (R)	Florida (7)	2187
Michaud, Michael H. (D)	Maine (2)	1724
Miller, Brad (D)	North Carolina (13)	1127
Miller, Candice S. (R)	Michigan (10)	1034
Miller, Gary G. (R)	California (42)	2349
Miller, George (D)	California (7)	2205
Miller, Jeff (R)	Florida (1)	2416
Moore, Gwen (D)	Wisconsin (4)	2245
Moran, James P. (D)	Virginia (8)	2239
Mulvaney, Mick (R)	South Carolina (5)	1004
Murphy, Christopher S. (D)	Connecticut (5)	412
Murphy, Tim (R)	Pennsylvania (18)	322
Myrick, Sue Wilkins (R)	North Carolina (9)	230
Nadler, Jerrold (D)	New York (8)	2334
Napolitano, Grace F. (D)	California (38)	1610
Neal, Richard E. (D)	Massachusetts (2)	2208
Neugebauer, Randy (R)	Texas (19)	1424
Noem, Kristi L. (R)	South Dakota (At Large)	226
Norton, Eleanor Holmes (D)	District of Columbia (Delegate)	2136
Nugent, Richard B. (R)	Florida (5)	1517
Nunes, Devin (R)	California (21)	1013
Nunnelee, Alan (R)	Mississippi (1)	1432
Olson, Pete (R)	Texas (22)	312
Olver, John W. (D)	Massachusetts (1)	1111
Owens, William L. (D)	New York (23)	431
Palazzo, Steven M. (D)	Mississippi (4)	331

Representatives, Delegates, and Resident Commissioners—Continued

[Republicans (241); Democrats (193); vacancies (1); total, 435 Members; 5 Delegates; 1 Resident Commissioner]. Members who have resigned appear in bold brackets [ ]. Room numbers with three digits are in the Cannon House Office Building (New Jersey and Independence Avenues), four digits beginning with 1 are in the Longworth House Office Building (between South Capitol Street and New Jersey Avenue on Independence Avenue), and four digits beginning with 2 are in the Rayburn House Office Building (between First and South Capitol Streets on Independence Avenue). Members' offices may be reached by phone at 202–225–3121. The most current listing of House Members can be found on the Internet at http://clerk.house.gov.

Name	State (District) / Territory	Room
Pallone, Frank, Jr. (D)	New Jersey (6)	237
Pascrell, Bill, Jr. (D)	New Jersey (8)	2370
Pastor, Ed (D)	Arizona (4)	2465
Paul, Ron (R)	Texas (14)	203
Paulsen, Erik (R)	Minnesota (3)	127
Payne, Donald M. (D)	New Jersey (10)	2310
Pearce, Steve (R)	New Mexico (2)	2432
Pelosi, Nancy (D)	California (8)	235
Pence, Mike (R)	Indiana (6)	100
Perlmutter, Ed (D)	Colorado (7)	1221
Peters, Gary C. (D)	Michigan (9)	1609
Peterson, Collin C. (D)	Minnesota (7)	2211
Petri, Thomas E. (R)	Wisconsin (6)	2462
Pierluisi, Pedro R. (D)	Puerto Rico (Resident Commissioner)	1213
Pingree, Chellie (D)	Maine (1)	1318
Pitts, Joseph R. (R)	Pennsylvania (16)	420
Platts, Todd Russell (R)	Pennsylvania (19)	2455
Poe, Ted (R)	Texas (2)	430
Polis, Jared (D)	Colorado (2)	501
Pompeo, Mike (R)	Kansas (4)	107
Posey, Bill (R)	Florida (15)	120
Price, David E. (D)	North Carolina (4)	2162
Price, Tom (R)	Georgia (6)	403
Quayle, Benjamin (R)	Arizona (3)	1419
Quigley, Mike (D)	Illinois (5)	1124
Rahall, Nick J., II (D)	West Virginia (3)	2307
Rangel, Charles B. (D)	New York (15)	2354
Reed, Tom (R)	New York (29)	1037
Rehberg, Denny (R)	Montana (At Large)	2448
Reichert, David G. (R)	Washington (8)	1730
Renacci, James B. (R)	Ohio (16)	130
Reyes, Silvestre (D)	Texas (16)	2210
Ribble, Reid J. (R)	Wisconsin (8)	1513
Richardson, Laura (D)	California (37)	1330
Richmond, Cedric L. (D)	Louisiana (2)	415
Rigell, E. Scott (R)	Virginia (2)	327
Rivera, David (R)	Florida (25)	417
Roby, Martha (R)	Alabama (2)	414
Roe, David P. (R)	Tennessee (1)	419
Rogers, Harold (R)	Kentucky (5)	2406
Rogers, Mike (R)	Michigan (8)	324
Rogers, Mike (R)	Alabama (3)	133
Rohrabacher, Dana (R)	California (46)	2300
Rokita, Todd (R)	Indiana (4)	236
Rooney, Thomas J. (R)	Florida (16)	1529
Roskam, Peter J. (R)	Illinois (6)	227
Ros-Lehtinen, Ileana (R)	Florida (18)	2206
Ross, Dennis A. (R)	Florida (12)	404
Ross, Mike (D)	Arkansas (4)	2436
Rothman, Steven R. (D)	New Jersey (9)	2303
Roybal-Allard, Lucille (D)	California (34)	2330

Representatives, Delegates, and Resident Commissioners—Continued

[Republicans (241); Democrats (193); vacancies (1); total, 435 Members; 5 Delegates; 1 Resident Commissioner]. Members who have resigned appear in bold brackets [ ]. Room numbers with three digits are in the Cannon House Office Building (New Jersey and Independence Avenues), four digits beginning with 1 are in the Longworth House Office Building (between South Capitol Street and New Jersey Avenue on Independence Avenue), and four digits beginning with 2 are in the Rayburn House Office Building (between First and South Capitol Streets on Independence Avenue). Members' offices may be reached by phone at 202–225–3121. The most current listing of House Members can be found on the Internet at http://clerk.house.gov.

Name	State (District) / Territory	Room
Royce, Edward R. (R)	California (40)	2185
Runyan, Jon (R)	New Jersey (3)	1239
Ruppersberger, C. A. Dutch (D)	Maryland (2)	2453
Rush, Bobby L. (D)	Illinois (1)	2268
Ryan, Paul (R)	Wisconsin (1)	1233
Ryan, Tim (D)	Ohio (17)	1421
Sablan, Gregorio Kilili Camacho (D)	Northern Mariana Islands (Delegate)	423
Sanchez, Linda T. (D)	California (39)	2423
Sanchez, Loretta (D)	California (47)	1114
Sarbanes, John P. (D)	Maryland (3)	2444
Scalise, Steve (R)	Louisiana (1)	429
Schakowsky, Janice D. (D)	Illinois (9)	2367
Schiff, Adam B. (D)	California (29)	2411
Schilling, Robert T. (R)	Illinois (17)	507
Schmidt, Jean (R)	Ohio (2)	2464
Schock, Aaron (R)	Illinois (18)	328
Schrader, Kurt (D)	Oregon (5)	• 314
Schwartz, Allyson Y. (D)	Pennsylvania (13)	1227
Schweikert, David (R)	Arizona (5)	1205
Scott, Austin (R)	Georgia (8)	516
Scott, David (D)	Georgia (13)	225
Scott, Robert C. "Bobby" (D)	Virginia (3)	1201
Scott, Tim (R)	South Carolina (1)	1117
Sensenbrenner, F. James, Jr. (R)	Wisconsin (5)	2449
Serrano, Jose E. (D)	New York (16)	2227
Sessions, Pete (R)	Texas (32)	2233
Sewell, Terri A. (D)	Alabama (7)	1133
Sherman, Brad (D)	California (27)	2242
Shimkus, John (R)	Illinois (19)	2452
Shuler, Heath (D)	North Carolina (11)	229
Shuster, Bill (R)	Pennsylvania (9)	204
Simpson, Michael K. (R)	Idaho (2)	2312
Sires, Albio (D)	New Jersey (13)	2342
Slaughter, Louise McIntosh (D)	New York (28)	2469
Smith, Adam (D)	Washington (9)	2402
Smith, Adrian (R)	Nebraska (3)	503
Smith, Christopher H. (R)	New Jersey (4)	2373
Smith, Lamar (R)	Texas (21)	2409
Southerland, Steve, II (R)	Florida (2)	1229
Speier, Jackie (D)	California (12)	211
Stark, Fortney H. "Pete" (D)	California (13)	239
Stearns, Cliff (R)	Florida (6)	2306
Stivers, Steve (R)	Ohio (15)	1007
Stutzman, Marlin A. (R)	Indiana (3)	1728
Sullivan, John (R)	Oklahoma (1)	434
Sutton, Betty (D)	Ohio (13)	1519
Terry, Lee (R)	Nebraska (2)	2331
Thompson, Bennie G. (D)	Mississippi (2)	2466
Thompson, Glenn (R)	Pennsylvania (5)	124
Thompson, Mike (D)	California (1)	231
Thornberry, Mac (R)	Texas (13)	2209

Representatives, Delegates, and Resident Commissioners—Continued

[Republicans (241); Democrats (193); vacancies (1); total, 435 Members; 5 Delegates; 1 Resident Commissioner]. Members who have resigned appear in bold brackets [ ]. Room numbers with three digits are in the Cannon House Office Building (New Jersey and Independence Avenues), four digits beginning with 1 are in the Longworth House Office Building (between South Capitol Street and New Jersey Avenue on Independence Avenue), and four digits beginning with 2 are in the Rayburn House Office Building (between First and South Capitol Streets on Independence Avenue). Members' offices may be reached by phone at 202–225–3121. The most current listing of House Members can be found on the Internet at http://clerk.house.gov.

Name	State (District) / Territory	Room
Tiberi, Patrick J. (R)	Ohio (12)	106
Tierney, John F. (D)	Massachusetts (6)	2238
Tipton, Scott R. (R)	Colorado (3)	218
Tonko, Paul (D)	New York (21)	422
Towns, Edolphus (D)	New York (10)	2232
Tsongas, Niki (D)	Massachusetts (5)	1607
Turner, Michael R. (R)	Ohio (3)	2454
Upton, Fred (R)	Michigan (6)	2183
Van Hollen, Chris (D)	Maryland (8)	1707
Velazquez, Nydia M. (D)	New York (12)	2302
Visclosky, Peter J. (D)	Indiana (1)	2256
Walberg, Tim (R)	Michigan (7)	418
Walden, Greg (R)	Oregon (2)	2182
Walsh, Joe (R)	Illinois (8)	432
Walz, Timothy J. (D)	Minnesota (1)	1722
Wasserman Schultz, Debbie (D)	Florida (20)	118
Waters, Maxine (D)	California (35)	2344
Watt, Melvin L. (D)	North Carolina (12)	2304
Waxman, Henry A. (D)	California (30)	2204
Webster, Daniel (R)	Florida (8)	1039
Weiner, Anthony D. (D)	New York (9)	2104
Welch, Peter (D)	Vermont (At Large)	1404
West, Allen B. (R)	Florida (22)	1708
Westmoreland, Lynn A. (R)	Georgia (3)	2433
Whitfield, Ed (R)	Kentucky (1)	2368
Wilson, Frederica A. (D)	Florida (17)	208
Wilson, Joe (R)	South Carolina (2)	2229
Wittman, Robert J. (R)	Virginia (1)	1317
Wolf, Frank R. (R)	Virginia (10)	241
Womack, Steve (R)	Arkansas (3)	1508
Woodall, Rob (R)	Georgia (7)	1725
Woolsey, Lynn C. (D)	California (6)	2263
Wu, David (D)	Oregon (1)	2338
Yarmuth, John A. (D)	Kentucky (3)	435
Yoder, Kevin (R)	Kansas (3)	214
Young, C. W. "Bill" (R)	Florida (10)	2407
Young, Don (R)	Alaska (At Large)	2314
Young, Todd C. (R)	Indiana (9)	1721

Sources of Information

Electronic Access Specific information and legislation can be found on the Internet at http://thomas.loc.gov or http://clerk.house.gov.

Publications The Congressional Directory, telephone directories for the House of Representatives, and the House Rules and Manual may be obtained from the Superintendent of Documents, Government Printing Office, Washington, DC 20402. Internet, http://www.gpo.gov/fdsys/browse/collectiontab.action.

For further information, contact the Clerk, The Capitol, Washington, DC 20515. Phone, 202–225–7000. Internet, http://clerk.house.gov.

ARCHITECT OF THE CAPITOL

U.S. Capitol Building, Washington, DC 20515
Phone, 202–228–1793. Internet, http://www.aoc.gov.

Architect of the Capitol	STEPHEN T. AYERS
Deputy Architect/Chief Operating Officer	(VACANCY)
Assistant Architect of the Capitol	MICHAEL G. TURNBULL
Superintendent, U.S. Capitol	CARLOS ELIAS
Superintendent, Capitol Grounds	TED BECHTOL
Chief Executive Officer, Capitol Visitor Center	TERRI ROUSE
Chief Administrative Officer	DAVID FERGUSON
Chief Financial Officer	PAULA LETTICE
Director of Congressional and External Relations	MIKE CULVER
General Counsel	PETER KUSHNER
Superintendent, House Office Buildings	WILLIAM WEIDEMEYER
Inspector General	CAROL BATES
Superintendent, Library of Congress Buildings and Grounds	GREGORY SIMMONS
Director, Planning and Project Management	ANNA FRANZ
Director, Safety, Fire, and Environmental Programs	SUSAN ADAMS
Director, Security Programs	KENNETH EADS
Superintendent, Senate Office Buildings	ROBIN MOREY
Superintendent, U.S. Supreme Court	JAMES YELLMAN
Director, U.S. Botanic Garden	HOLLY H. SHIMIZU
Director of Utilities, U.S. Capitol Power Plant	MARK WEISS

The Architect of the Capitol maintains the U.S. Capitol and the buildings and grounds of the Capitol complex.

In addition to the Capitol, the Architect is responsible for the upkeep of all of the congressional office buildings, the Library of Congress buildings, the U.S. Supreme Court building, the Thurgood Marshall Federal Judiciary Building, the Capitol Power Plant, the Capitol Police headquarters, and the Robert A. Taft Memorial. The Architect performs his duties in connection with the Senate side of the Capitol and the Senate office buildings subject to the approval of the Senate Committee on Rules and Administration. In matters of general policy in connection with the House office buildings, his activities are subject to the approval and direction of the House Office Building Commission. The Architect is under the direction of the Speaker in matters concerning the House side of the Capitol. He is subject to the oversight of the Committee on House Administration with respect to many administrative matters affecting operations on the House side of the Capitol complex. In addition, the Architect of the Capitol serves as the Acting Director of the U.S. Botanic Garden under the Joint Committee on the Library.

The position of Architect of the Capitol was historically filled by Presidential appointment for an indefinite term. Legislation enacted in 1989 provides that the Architect is to be appointed for a term of 10 years by the President, with the advice and consent of the Senate, from a list of three candidates recommended by a congressional commission. Upon confirmation by the Senate, the Architect becomes an official of the legislative branch as an officer of Congress. He is eligible for reappointment after completion of his term.

Projects carried out by the Architect of the Capitol include operating the

Capitol Visitor Center; conservation of murals and decorative paintings in the Capitol; improvement of speech-reinforcement, electrical, and fire-protection systems in the Capitol and congressional office buildings; work on security improvements within the Capitol complex; renovation, restoration, and modification of the interiors and exteriors of the Thomas Jefferson and John Adams Buildings of the Library of Congress and provision of off-site book storage facilities for the Library; and facility management of the Thurgood Marshall Federal Judiciary Building.

For further information, contact the Office of the Architect of the Capitol, U.S. Capitol Building, Washington, DC 20515. Phone, 202–228–1793. Internet, http://www.aoc.gov.

UNITED STATES BOTANIC GARDEN

Office of Executive Director, 245 First Street, SW., Washington, DC 20024
Phone, 202–225–6670. Internet, http://www.usbg.gov.

Conservatory, 100 Maryland Avenue, SW., Washington, DC 20001
Phone, 202–226–8333.

Production Facility, 4700 Shepherd Parkway, SW., Washington, DC 20032
Phone, 202–226–4780.

Director (Architect of the Capitol)	STEPHEN T. AYERS, *Acting*
Executive Director	HOLLY H. SHIMIZU

The United States Botanic Garden informs visitors about the importance and value of plants to the well-being of humankind and earth's ecosystems.

The United States Botanic Garden (USBG) is one of the oldest botanic gardens in North America. The Garden highlights the diversity of plants worldwide, as well as their aesthetic, cultural, economic, therapeutic, and ecological significance. The USBG encourages plant appreciation and the growth of botanical knowledge through artistic plant displays, exhibits, educational programs, and curation of a large collection of plants. It fosters plant conservation by serving as a repository for endangered species. Uniquely situated at the heart of the U.S. Government, the Garden seeks to promote the exchange of ideas and information relevant to its mission among national and international visitors and policymakers.

The Garden's collections include orchids, epiphytes, bromeliads, carnivorous plants, ferns, cycads, cacti, succulents, medicinal plants, rare and endangered plants, and plants valued as sources of food, beverages, fibers, cosmetics, and industrial products.

The U.S. Botanic Garden's facilities include the Conservatory, the National Garden, Bartholdi Park, an administration building, and an off-site production facility. The Conservatory, one of the largest structures of its kind in this country, reopened on December 11, 2001, after undergoing major renovation that required more than 4 years to complete. In addition to upgraded amenities for visitors, it features 12 exhibit and plant display areas.

The National Garden opened on October 1, 2006. Located on three acres adjacent to the west side of the Conservatory, the National Garden comprises a First Ladies Water Garden, a Butterfly Garden, a Rose Garden celebrating our national flower, a Lawn Terrace, a Regional Garden of native Mid-Atlantic plants, and an amphitheater where visitors may relax and enjoy the stunning views of the U.S. Capitol.

Outdoor plantings are also showcased in Bartholdi Park, a home-landscape demonstration area located across from the Conservatory. Each of the displays is sized and scaled for suitability in an urban or suburban setting. The gardens display ornamental plants that perform well in this region arrayed in a variety of styles and themes. Also located in this park is Bartholdi Fountain, created by Frederic Auguste Bartholdi (1834–1904), sculptor of the Statue of Liberty. After undergoing extensive restoration and modifications to save both energy and water, Bartholdi Fountain was re-installed in 2010.

The U.S. Botanic Garden's staff is organized into horticulture, operations, administration, and public programs divisions. Programs for the public are listed in a quarterly calendar of events and also on the Garden's Web site. A horticultural hotline and e-mail address are available to answer questions from the public.

The U.S. Botanic Garden was founded in 1820 under the auspices of the Columbian Institute for the Promotion of Arts and Sciences, an organization that was the outgrowth of an association known as the Metropolitan Society, which received its charter from Congress on April 20, 1818. The Garden continued under the direction of the Institute until 1837, when the Institute ceased to exist as an active organization.

In June 1842, the U.S. Exploring Expedition under the command of Captain Charles Wilkes returned from its 4-year voyage with a wealth of information, artifacts, pressed-plant specimens, and living plants from around the world. The living plants were temporarily placed on exhibit on a lot behind the old Patent Office under the care of William D. Brackenridge, the Expedition's botanist. By November 1842, the plants were moved into a greenhouse built there with funds appropriated by Congress. Subsequently,

the greenhouse was expanded with two additions and a small growing area to care for the burgeoning collection. In 1843, stewardship of the collection was placed under the direction and control of the Joint Committee on the Library, which had also assumed responsibility for publication of the results of the Expedition. Expansion of the Patent Office in 1849 necessitated finding a new location for the botanical collections.

The act of May 15, 1850 (9 Stat. 427), provided for the relocation of the Botanic Garden under the direction of the Joint Committee on the Library. The site selected was on the National Mall at the west end of the Capitol Grounds, practically the same site the Garden occupied during the period it functioned under the Columbian Institute. This site was later enlarged, and the main area continued to serve as the principal Garden site from 1850 to 1933, when the Garden was relocated to its present site.

Although the Government had assumed responsibility for the maintenance and stewardship of the plant collection in 1842, the two functions were divided between the Commissioner of Public Buildings and the Joint Committee on the Library, respectively. In 1856, in recognition of their increasing stature, the collections and their associated operations and facilities were officially named the United States Botanic Garden, and the Joint Committee on the Library assumed jurisdiction over both its direction and maintenance (11 Stat. 104). An annual appropriation has been provided by Congress since 1856.

Presently, the Joint Committee on the Library has supervision over the U.S. Botanic Garden through the Architect of the Capitol, who has held the title of Acting Director since 1934.

For further information concerning the United States Botanic Garden, contact the Public Program Division, 245 First Street SW., Washington, DC 20024. Phone, 202–225–8333. Plant Hotline, 202–226–4785. Email, usbg@aoc.gov. Internet, http://www.usbg.gov.

GOVERNMENT ACCOUNTABILITY OFFICE

441 G Street NW., Washington, DC 20548
Phone, 202–512–3000. Internet, http://www.gao.gov.

Comptroller General of the United States	GENE L. DODARO, *Acting*
Chief Operating Officer	GENE L. DODARO
Chief Administrative Officer	SALLYANNE HARPER
Deputy Chief Administrative Officer	CHERYL WHITAKER
General Counsel	GARY L. KEPPLINGER
Inspector General	FRANCES GARCIA
Chief Human Capital Officer	PATRINA M. CLARK
Chief Information Officer	HOWARD WILLIAMS, *Acting*
Controller/Administrative Services Office	PAMELA J. LARUE
Managing Director, Acquisition and Sourcing Management	PAUL L. FRANCIS
Managing Director, Applied Research and Methods	NANCY KINGSBURY
Managing Director, Defense Capabilities and Management	JANET ST. LAURENT
Managing Director, Congressional Relations	RALPH DAWN
Managing Director, Education, Workforce, and Income Security	BARBARA D. BOVBJERG
Managing Director, Field Offices	DENISE HUNTER
Managing Director, Financial Management and Assurance	JEANETTE M. FRANZEL
Managing Director, Forensic Audits and Special Investigations	GREGORY D. KUTZ
Managing Director, Financial Markets and Community Investments	RICHARD J. HILLMAN
Managing Director, Health Care	MARJORIE KANOFF
Managing Director, Homeland Security and Justice	CATHLEEN BERRICK
Managing Director, International Affairs and Trade	JACQUELYN WILLIAMS-BRIDGERS
Managing Director, Information Technology	JOEL WILLEMSSEN
Managing Director, Knowledge Services	CATHERINE TETI
Managing Director, Natural Resources and Environment	PATRICIA DALTON
Managing Director, Opportunity and Inclusiveness	REGINALD E. JONES
Managing Director, Physical Infrastructure	KATHERINE SIGGERUD
Managing Director, Professional Development Program	DAVID CLARK
Managing Director, Public Affairs	CHARLES YOUNG
Managing Director, Quality and Continuous Improvement	TIMOTHY BOWLING
Managing Director, Strategic Issues	J. CHRISTOPHER MIHM
Managing Director, Strategic Planning and External Liaison	HELEN HSING

The Government Accountability Office is the investigative arm of the Congress and is charged with examining all matters relating to the receipt and disbursement of public funds.

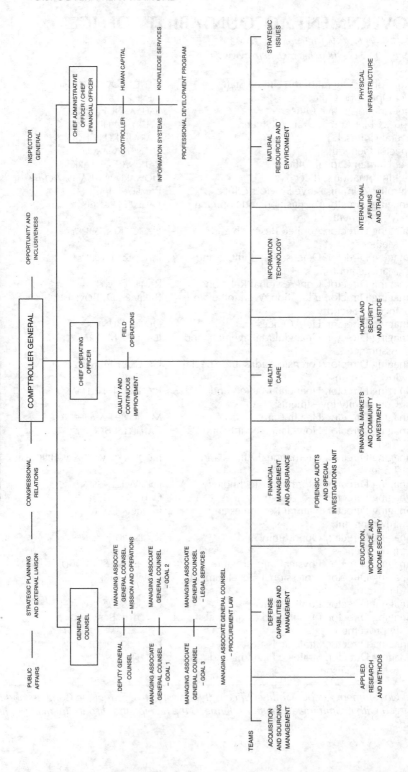

GOVERNMENT ACCOUNTABILITY OFFICE

The Government Accountability Office (GAO) is an independent, nonpartisan Agency that works for Congress. GAO is often called the "congressional watchdog" because it investigates how the Federal Government spends taxpayer dollars. The GAO was established as the General Accounting Office by the Budget Accounting Act of 1921 (31 U.S.C. 702). It was renamed the Government Accountability Office pursuant to the GAO Capital Reform Act of 2004 (31 U.S.C. 702 note).

Activities

GAO gathers information to help Congress determine how effectively executive branch agencies are doing their jobs. GAO's work routinely answers such basic questions as whether Government programs are meeting their objectives or providing good service to the public. Ultimately, GAO ensures that Government is accountable to the American people.

To help Senators and Representatives arrive at informed policy decisions, GAO provides them with information that is accurate, timely, and balanced. The Office supports congressional oversight by evaluating how well Government policies and programs are working; auditing Agency operations to determine whether Federal funds are being spent efficiently, effectively, and appropriately; investigating allegation of illegal and improper activities; and issuing legal decisions and opinions.

With virtually the entire Federal Government subject to its review, GAO issues a steady stream of products— more than 1,000 reports and hundreds of testimonies by GAO officials each year. GAO's familiar "blue book" reports meet short-term immediate needs for information on a wide range of Government operations. These reports also help Congress better understand issues that are newly emerging, long term in nature, and with more far-reaching impacts. GAO's work translates into a wide variety of legislative actions, improvements in Government operations, and billions of dollars in financial benefits for the American people.

For further information, contact the Office of Public Affairs, Government Accountability Office, 441 G Street NW., Washington, DC 20548. Phone, 202–512–4800. Internet, http://www.gao.gov.

GOVERNMENT PRINTING OFFICE

732 North Capitol Street NW., Washington, DC 20401
Phone, 202–512–1800. Internet, http://www.gpo.gov.

Public Printer of the United States	WILLIAM J. BOARMAN
Deputy Public Printer	(VACANCY)
Assistant Public Printer, Operations	JIM BRADLEY
Chief of Staff	DAVITA VANCE-COOKS
Assistant Public Printer, Superintendent of Documents	MARY ALICE BAISH
Inspector General	RODOLFO RAMIERZ, JR.
Director, Congressional, Public, and Employee Communications	ANDREW M. SHERMAN
Manager, Public Relations	GARY SOMERSET
Director, Employee Communications	JEFFREY BROOKE
General Counsel	DREW SPALDING
Director, Equal Employment Opportunity	NADINE L. ELZY
Director, Quality Assurance and Policy Planning	JOHN VAN SANTEN
Managing Director, Library Services and Content Management	TED PRIEBE, *Acting*

Managing Director, Business Products and Services	HERBERT H. JACKSON
Managing Director, Publication and Information Sales	HERBERT H. JACKSON
Managing Director, Plant Operations	OLIVIER A. GIROD
Managing Director, Security and Intelligent Documents	STEPHEN G. LEBLANC
Managing Director, Official Journals of Government	JIM BRADLEY
Managing Director, Customer Services	R.T. SULLIVAN, *Acting*
Chief Human Capital Officer	WILLIAM T. HARRIS
Chief Acquisition Officer	SHEREE YOUNG
Chief Financial Officer	STEVEN T. SHEDD
Chief Information Officer	CHUCK RIDDLE
Chief Technology Officer, Programs, Strategy and Technology	RICHARD G. DAVIS
Director, Labor Relations	T. MICHAEL FRAZIER
Director, Security Service	LAMONT VERNON
Director, Sales and Marketing	BRUCE SEGER

The Government Printing Office produces, procures, and disseminates printed and electronic publications of the Congress, executive departments, and establishments of the Federal Government.

The Government Printing Office (GPO) opened for business on March 4, 1861. GPO's duties are defined in title 44 of the U.S. Code. The Public Printer, who serves as the head of GPO, is appointed by the President and confirmed by the Senate.

Activities

Headquartered in Washington, DC, with a total employment of approximately 2,200, GPO is responsible for the production and distribution of information products and services for all three branches of the Federal Government. GPO is the Federal Government's primary centralized resource for producing, procuring, cataloging, indexing, authenticating, disseminating and preserving the official information products of the U.S. Government in digital and tangible forms.

While many of our Nation's most important products, such as the Congressional Record and Federal Register, are produced at GPO's main plant, the majority of the Government's printing needs are met through a longstanding partnership with America's printing industry. GPO procures 75 percent of all printing orders through private sector vendors across the country, competitively buying products and services from thousands of private sector companies in all 50 States. The contracts cover the entire spectrum of printing and publishing services and are available to fit almost any firm from the largest to the smallest.

GPO disseminates Federal information products through a sales program, distribution network of more than 1,200 Federal libraries nationwide, and via GPO's Federal Digital System (FDsys). More than 286,000 Federal Government document titles are available to the public at www.fdsys.gov.

Printed copies of many documents, ranging from Supreme Court opinions to reports from the Bureau of Labor Statistics, may also be purchased as follows:

To order in person, please visit the GPO Main Bookstore at 710 North Capitol Street NW., Washington, DC, (corner of North Capitol and H Streets) from 8 a.m. to 4 p.m., eastern standard time. To order online, visit the GPO Online Bookstore at bookstore.gpo.gov. To order by phone or inquire about an order, call 866–512–1800 or, in the Washington, DC, metro area, call

GOVERNMENT PRINTING OFFICE

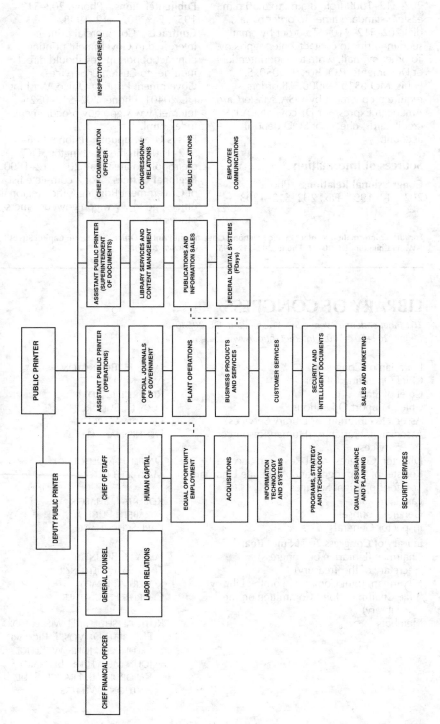

202–512–1800 from 8 a.m. to 5:30 p.m., eastern standard time. To order by fax, dial 202–512–2104. To order by email, send inquiries to contactcenter@gpo.gov. To order by mail, write to Superintendent of Documents, P.O. Box 979050, St. Louis, MO 63197–9000. All orders require prepayment by VISA, MasterCard, American Express, or Discover/NOVUS credit cards, check, or SOD deposit account.

Sources of Information

Congressional Relations Phone, 202–512–1991. Fax, 202–512–1293.

Public Relations Phone, 202–512–1957. Fax, 202–512–1998.

Contracts Commercial printers interested in Government printing contract opportunities should direct inquiries to Customer Services, Government Printing Office, Washington, DC 20401. Phone, 202–512–0526. Internet, www.gpo.gov/procurement/index.html.

FDsys User Support Phone, 866–512–1800. In the Washington, DC, metropolitan area, call 202–512–1800.

Regional Offices For a complete list of Government Printing Office regional offices, go to www.gpo.gov/customers/offices.htm.

For further information, contact Public Relations, Government Printing Office, 732 North Capitol Street NW., Washington, DC 20401. Phone, 202–512–1957. Fax, 202–512–1998.

LIBRARY OF CONGRESS

101 Independence Avenue SE., Washington, DC 20540
Phone, 202–707–5000. Internet, http://www.loc.gov.

Librarian of Congress	JAMES H. BILLINGTON
Chief Operating Officer	(VACANCY)
Chief of Staff	ROBERT DIZARD, JR.
Chief, Support Operations	LUCY D. SUDDRETH
Associate Librarian for Library Services	DEANNA MARCUM
Associate Librarian for Human Resources Services	DENNIS HANRATTY
Director, Congressional Research Service	DANIEL P. MULHOLLAN
Register of Copyrights and Associate Librarian for Copyright Services	MARYBETH PETERS
Law Librarian	ROBERTA I. SHAFFER
General Counsel	ELIZABETH PUGH
Inspector General	KARL SCHORNAGEL
Library of Congress Trust Fund Board	
Chairman (Librarian of Congress)	JAMES H. BILLINGTON
(Secretary of the Treasury)	TIMOTHY F. GEITHNER
(Chairman, Joint Committee on the Library)	ROBERT A. BRADY
(Vice Chairman, Joint Committee on the Library)	CHARLES E. SCHUMER
Members	RUTH ALTSHULER, EDWIN L. COX, ELISABETH DE VOS, J. RICHARD FREDERICKS, JOHN W. KLUGE, JOHN MEDVECKIS, BERNARD RAPOPORT, B. FRANCIS SAUL, II, ANTHONY WELTERS

The Library of Congress is the national library of the United States, offering diverse materials for research including the world's most extensive collections in many areas such as American history, music, and law.

The Library of Congress was established by Act of April 24, 1800 (2 Stat. 56), appropriating $5,000 "for the purchase of such books as may be necessary for the use of Congress . . . ." The Library's scope of responsibility has been widened by subsequent legislation (2 U.S.C. 131–168d). The Librarian, appointed by the President with the advice and consent of the Senate, directs the Library.

The Library's first responsibility is service to Congress. As the Library has developed, its range of service has expanded to include the entire governmental establishment and the public at large, making it a national library for the United States and a global resource through its Web site at www. loc.gov.

Activities

Collections   The Library's extensive collections are universal in scope. They include books, serials, and pamphlets on every subject and in a multitude of languages and research materials in many formats, including maps, photographs, manuscripts, motion pictures, and sound recordings. Among them are the most comprehensive collections of Chinese, Japanese, and Russian language books outside Asia and the former Soviet Union; volumes relating to science and legal materials outstanding for American and foreign law; the world's largest collection of published aeronautical literature; and the most extensive collection in the Western Hemisphere of books printed before 1501 A.D.

The manuscript collections relate to manifold aspects of American history and civilization and include the personal papers of most of the Presidents from George Washington through Calvin Coolidge. The music collections contain volumes and pieces—manuscript and published—from classic works to the newest popular compositions. Other materials available for research include maps and views; photographic records

from the daguerreotype to the latest news photo; recordings, including folksongs and other music, speeches, and poetry readings; prints, drawings, and posters; government documents, newspapers, and periodicals from all over the world; and motion pictures, microforms, audio and video tapes, and digital products.

Reference Resources   Admission to the various research facilities of the Library is free. No introduction or credentials are required for persons over high school age. Readers must register by presenting valid photo identification with a current address, and for certain collections there are additional requirements. As demands for service to Congress and Federal Government agencies increase, reference service available through correspondence has become limited. The Library must decline some requests and refer correspondents to a library within their area that can provide satisfactory assistance. While priority is given to inquiries pertaining to its holdings of special materials or to subjects in which its resources are unique, the Library does attempt to provide helpful responses to all inquirers. Online reference service is also available through the "Ask a Librarian" site, at www.loc.gov/rr/askalib.

Copyrights   With the enactment of the second general revision of the U.S. copyright law by Act of July 8, 1870 (16 Stat. 212–217), all activities relating to copyright, including deposit and registration, were centralized in the Library of Congress. The Copyright Act of 1976 (90 Stat. 2541) brought all forms of copyrightable authorship, both published and unpublished, under a single statutory system which gives authors protection immediately upon creation of their works. Exclusive rights granted to authors under the statute include the right to reproduce and prepare derivative works, distribute copies or phonorecords, perform and display the work publicly, and in the case of sound recordings, to perform the work publicly by means of a digital audio transmission. Works

LIBRARY OF CONGRESS

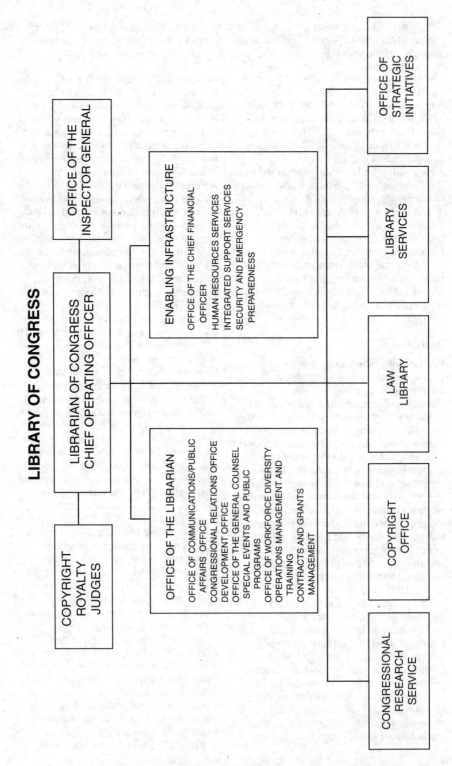

OFFICE OF THE INSPECTOR GENERAL

LIBRARIAN OF CONGRESS CHIEF OPERATING OFFICER

COPYRIGHT ROYALTY JUDGES

ENABLING INFRASTRUCTURE

OFFICE OF THE CHIEF FINANCIAL OFFICER
HUMAN RESOURCES SERVICES
INTEGRATED SUPPORT SERVICES
SECURITY AND EMERGENCY PREPAREDNESS

OFFICE OF THE LIBRARIAN

OFFICE OF COMMUNICATIONS/PUBLIC AFFAIRS OFFICE
CONGRESSIONAL RELATIONS OFFICE
DEVELOPMENT OFFICE
OFFICE OF THE GENERAL COUNSEL
SPECIAL EVENTS AND PUBLIC PROGRAMS
OFFICE OF WORKFORCE DIVERSITY
OPERATIONS MANAGEMENT AND TRAINING
CONTRACTS AND GRANTS MANAGEMENT

OFFICE OF STRATEGIC INITIATIVES

LIBRARY SERVICES

LAW LIBRARY

COPYRIGHT OFFICE

CONGRESSIONAL RESEARCH SERVICE

eligible for copyright include literary works (books and periodicals), musical works, dramatic works, pantomimes and choreographic works, pictorial, graphic, and sculptural works, motion pictures, sound recordings, vessel hull designs, mask works, and architectural works. Serving in its capacity as a national registry for creative works, the Copyright Office registers more than 500,000 claims to copyright annually (representing more than 800,000 works) and is a major source of acquisitions for the universal collections of the Library of Congress. Most of the information available on paper is also accessible online, at www.loc.gov/copyright.

Extension of Service   The Library extends its service through the following: an interlibrary loan system; the photoduplication, at reasonable cost, of books, manuscripts, maps, newspapers, and prints in its collections; the sale of sound recordings, which are released by its Recording Laboratory; the exchange of duplicates with other institutions; the sale of CD–ROM cataloging tools and magnetic tapes and the publication in book format or microform of cumulative catalogs, which make available the results of the expert bibliographical and cataloging work of its technical personnel; a centralized cataloging program whereby the Library of Congress acquires material published all over the world, catalogs it promptly, and distributes cataloging information in machine-readable form and other means to the Nation's libraries; a cooperative cataloging program whereby the cataloging of data, by name authority and bibliographic records, prepared by other libraries becomes part of the Library of Congress database and is distributed through the Cataloging Distribution Service; a cataloging-in-publication program in cooperation with American publishers for printing cataloging information in current books; the National Serials Data Program, a national center that maintains a record of serial titles to which International Standard Serial Numbers have been assigned and serves, with this file, as the United States Register; and the development of general schemes of classification (Library of Congress and Dewey Decimal), subject headings, and cataloging, embracing the entire field of printed matter.

Furthermore, the Library provides for the following: the preparation of bibliographical lists responsive to the needs of Government and research; the maintenance and the publication of cooperative publications; the publication of catalogs, bibliographical guides, and lists, and of texts of original manuscripts and rare books in the Library of Congress; the circulation in traveling exhibitions of items from the Library's collections; the provision of books in Braille, electronic access to Braille books on the Internet, "talking books," and books on tape for the blind and the physically handicapped through 134 cooperating libraries throughout the Nation; the distribution of its electronic materials via the Internet; and the provision of research and analytical services on a fee-for-service basis to agencies in the executive and judicial branches.

American Folklife Center   The Center was established in the Library of Congress by Act of January 2, 1976 (20 U.S.C. 2102 et seq.). It supports, preserves, and presents American folklife by receiving and maintaining folklife collections, scholarly research, field projects, performances, exhibitions, festivals, workshops, publications, and audiovisual presentations. The Center has conducted projects in many locations across the country, such as the ethnic communities of Chicago, IL; southern Georgia; a ranching community in northern Nevada; the Blue Ridge Parkway in southern Virginia and northern North Carolina; and the States of New Jersey, Rhode Island, and Montana. The projects have provided large collections of recordings and photographs for the Archive of Folk Culture. The Center administers the Federal Cylinder Project, which is charged with preserving and disseminating music and oral traditions recorded on wax cylinders dating from the late 1800s to the early 1940s. A cultural conservation study was developed at the Center, in cooperation with the Department of the Interior, pursuant to a congressional mandate. Various conferences, workshops, and symposia are given throughout the year.

The American Folklife Center maintains and administers the Archive of Folk Culture, an extensive collection of ethnographic materials from this country and around the world. It is the national repository for folk-related recordings, manuscripts, and other unpublished materials. The Center administers the Veterans History Project, which records and preserves the first-person accounts of war veterans. It also participates in StoryCorps, a program to record and collect oral histories from people from all walks of life. This collection also resides in the American Folklife Center. The Center's reading room contains over 3,500 books and periodicals; a sizable collection of magazines, newsletters, unpublished theses, and dissertations; field notes; and many textual and some musical transcriptions and recordings.

The Folklife Center News, a quarterly newsletter, and other informational publications are available upon request. Many Center publications and a number of collections are available online through the Internet, at www.loc.gov/folklife.

For further information, call 202–707–5510.

Center for the Book   The Center was established in the Library of Congress by an Act of October 13, 1977 (2 U.S.C. 171 et seq.), to stimulate public interest in books, reading, and libraries, and to encourage the study of books and print culture. The Center is a catalyst for promoting and exploring the vital role of books, reading, and libraries, nationally and internationally. As a partnership between the Government and the private sector, the Center for the Book depends on tax-deductible contributions from individuals and corporations to support its programs.

The Center's activities are directed toward the general public and scholars. The overall program includes reading promotion projects with television and radio networks, symposia, lectures, exhibitions, special events, and publications. More than 80 national

educational and civic organizations participate in the Center's annual reading promotion campaign.

All 50 States, the District of Columbia, and the U.S. Virgin Islands have established statewide book centers that are affiliated with the Center for the Book in the Library of Congress. State centers plan and fund their own projects, involving members of the State's "community of the book," including authors, readers, prominent citizens, and public officials who serve as honorary advisers.

The Center also administers the position of the National Ambassador for Young People's Literature in collaboration with the Children's Book Council. For more information on the Center and the Library's literacy promotion activities, go to www.Read.gov.

For further information, contact the Center for the Book. Phone, 202–707–5221. Fax, 202–707–0269. Email, cfbook@loc.gov.

National Film Preservation Board   The National Film Preservation Board, established by the National Film Preservation Act of 1992 (2 U.S.C. 179) and reauthorized by the National Film Preservation Act of 2005 (2 U.S.C. 179n), serves as a public advisory group to the Librarian of Congress. The Board works to ensure the survival, conservation, and increased public availability of America's film heritage, including advising the Librarian on the annual selection of films to the National Film Registry and counseling the Librarian on development and implementation of the national film preservation plan. Key publications are Film Preservation 1993: A Study of the Current State of American Film Preservation; Redefining Film Preservation: A National Plan; and Television and Video Preservation 1997: A Study of the Current State of American Television and Video Preservation.

For further information, call 202–707–5912.

National Sound Recording Preservation Board   The National Sound Recording Preservation Board, established by the National Recording Preservation Act of 2000 (2 U.S.C. 1701 note), includes three major components: a National

Recording Preservation Advisory Board, which brings together experts in the field, a National Recording Registry, and a fundraising foundation, all of which are conducted under the auspices of the Library of Congress. The purpose of the Board is to and implement a national plan for the long-term preservation and accessibility of the Nation's audio heritage. It also advises the Librarian on the selection of culturally, aesthetically, or historically significant sound recordings to be included on the National Recording Registry. The national recording preservation program will set standards for future private and public preservation efforts and will be conducted in conjunction with the Library's Packard Campus for Audio-Visual Conservation in Culpeper, VA.

For further information, call 202–707–5856.

Preservation   The Library provides technical information related to the preservation of library and archival material. A series of handouts on various preservation and conservation topics has been prepared by the Preservation Office. Information and publications are available from the Office of the Director for Preservation, Library of Congress, Washington, DC 20540–4500. Phone, 202–707–1840.

Sources of Information

Books for the Blind and Physically Handicapped   Talking and Braille books and magazines are distributed through 134 regional and subregional libraries to blind and physically handicapped residents of the United States and its territories. Qualified users can also register for Web-Braille, an Internet-based service. Information is available at public libraries throughout the United States and from the headquarters office, National Library Service for the Blind and Physically Handicapped, Library of Congress, 1291 Taylor Street NW., Washington, DC 20542–4960. Phone, 202–707–5100.

Cataloging Data Distribution
Cataloging and bibliographic information in the form of microfiche catalogs, book catalogs, magnetic tapes, CD–

ROM cataloging tools, bibliographies, and other technical publications is distributed to libraries and other institutions. Information about ordering materials is available from the Cataloging Distribution Service, Library of Congress, Washington, DC 20541–4910. Phone, 202–707–6100. TDD, 202–707–0012. Fax, 202–707–1334. Email, cdsinfo@ mail.loc.gov. Card numbers for new publications and Electronic Preassigned Control Numbers for publishers are available from the Cataloging in Publication Division, Library of Congress, Washington, DC 20541–4910. Phone, 202–707–6345.

Contracts   Persons seeking information about conducting business with the Library of Congress should visit the Library's Web site at www.loc.gov/about/ business.

Copyright Services   Information about the copyright law (title 17 of the U.S. Code), the method of securing copyright, and copyright registration procedures may be obtained by writing to the Copyright Office, Library of Congress, 101 Independence Avenue SE., Washington, DC 20559–6000. Phone, 202–707–3000. Copyright information is also available through the Internet at www.loc.gov/copyright. Registration application forms may be ordered by calling the forms hotline at 202– 707–9100. Copyright records may be researched and reported by the Copyright Office for a fee; for an estimate, call 202–707–6850. Members of the public may use the copyright card catalog in the Copyright Office without charge. The database of Copyright Office records cataloged from January 1, 1978, to the present is available through the Internet at www.loc.gov/copyright/rb.html. The Copyright Information Office is located in Room LM–401, James Madison Memorial Building, 101 Independence Avenue SE., Washington, DC 20559–6000. It is open to the public Monday through Friday, 8:30 a.m. to 5 p.m., except for Federal holidays.

Employment   Employment inquiries should be directed to Human Resources Services, Library of Congress, 101 Independence Avenue SE.,

Washington, DC 20540–2200. Vacancy announcements and applications are also available from the Employment Office, Room LM–107, 101 Independence Avenue SE., Washington, DC 20540. Phone, 202–707–4315. Internet, www. loc.gov/hr/employment.

Photoduplication Service   Copies of manuscripts, prints, photographs, maps, and book material not subject to copyright and other restrictions are available for a fee. Order forms for photo reproduction and price schedules are available from the Photoduplication Service, Library of Congress, 101 Independence Avenue SE., Washington, DC 20540–4570. Phone, 202–707–5640.

Exhibitions   Throughout the year, the Library offers free exhibitions featuring items from its collections. The new interactive Library of Congress Experience may be viewed Monday through Saturday, 8:30 a.m. to 4:30 p.m., in the Thomas Jefferson Building. For more information, call 202–707–4604. To view current and past exhibitions, go to www.myloc.gov.

Publications   Library of Congress publications are available through the Internet at www.loc.gov/shop. The Library of Congress Information Bulletin is published 11 times a year and may be viewed online at www.loc.gov/loc/lcib/. The calendar of public events is also available online at www.loc.gov/loc/events and is available by mail to persons within 100 miles of Washington, DC. Send request to be added to the mailing list to Office Systems Services, Mail and Distribution Management Section, Library of Congress, 101 Independence Avenue SE., Washington, DC 20540–9441 or email pao@loc.gov.

Reference and Bibliographic Services   Guidance is offered to readers in identifying and using the material in the Library's collections, and reference service is provided to those with inquiries who have exhausted local, State, and regional resources. Persons requiring

services that cannot be performed by the Library staff can be supplied with names of private researchers who work on a fee-for-service basis. Requests for information should be directed to the Reference Referral Service, Library of Congress, 101 Independence Avenue SE., Washington, DC 20540–4720. Phone, 202–707–5522. Fax, 202–707–1389. They may also be submitted online through "Ask a Librarian," www.loc.gov/rr/askalib/.

Research and Reference Services in Science and Technology   Reference specialists in the Science, Technology, and Business Division provide a free service in answering brief technical inquiries entailing a bibliographic response. Requests for reference services should be directed to the Science, Technology, and Business Division, Library of Congress, Science Reference Section, 101 Independence Avenue SE., Washington, DC 20540–4750. Phone, 202–707–5639. Internet, www.loc.gov/rr/scitech.

Research Services in General Topics   Federal Government agencies can procure directed research and analytical products on foreign and domestic topics using the collections of the Library of Congress through the Federal Research Division. Science, technology, humanities, and social science topics of research are conducted by staff specialists exclusively on behalf of Federal agencies on a fee-for-service basis. Requests for service should be directed to the Federal Research Division, Marketing Office, Library of Congress, Washington, DC 20540–4840. Phone, 202–707–9133. Fax, 202–707–3920.

Visiting the Library of Congress   Guided tours of the Library are offered to the public Monday through Friday at 10:30 and 11:30 a.m. and 1:30, 2:30, and 3:30 p.m. and on Saturday at 10:30 and 11:30 a.m. and 1:30 and 2:30 p.m. For more information about scheduling tours for groups of 10 or more, contact the Visitor Services Office at 202–707–0919.

For further information, contact the Public Affairs Office, Library of Congress, 101 Independence Avenue SE., Washington, DC 20540–8610. Phone, 202–707–2905. Fax, 202–707–9199. Email, pao@loc.gov. Internet, http://www.loc.gov.

Congressional Research Service

101 Independence Avenue SE., Washington, DC 20540
Phone, 202–707–5000. Internet, http://www.loc.gov.

Director, Congressional Research Service   DANIEL P. MULHOLLAN

The Congressional Research Service (CRS) provides confidential and nonpartisan policy analysis exclusively to the United States Congress. CRS assists Congress in the legislative process by identifying and clarifying current policy problems, exploring potential implications of proposed policies, monitoring and assessing program implementation and oversight, and helping congressional staff understand legislative procedures and processes.

For further information, call 202–707–5700.

CONGRESSIONAL BUDGET OFFICE

Second and D Streets SW., Washington, DC 20515
Phone, 202–226–2600. Internet, http://www.cbo.gov.

Director	DOUGLASS W. ELMENDORF
Deputy Director	ROBERT A. SUNSHINE
Associate Director for Communications	MELISSA MERSON
Associate Director of Economic Analysis	JEFFREY KLING
Associate Director for Legislative Affairs	EDWARD DAVIS
General Counsel	MARK P. HADLEY
Deputy General Counsel	T.J. McGRATH
Assistant Director for Budget Analysis	PETE FONTAINE
Assistant Director for Financial Analysis	DEBORAH LUCAS
Assistant Director for Health and Human Resources	BRUCE VAVRICHEK
Assistant Director for Macroeconomic Analysis	ROBERT A. DENNIS
Assistant Director for Management, Business, and Information Services	RODERICK GOODWIN
Assistant Director for Microeconomic Studies	JOSEPH KILE
Assistant Director for National Security	DAVID MOSHER
Assistant Director for Tax Analysis	FRANK SAMMARTINO

The Congressional Budget Office provides the Congress with economic and budgetary analyses and with information and estimates required for the Congressional budget process.

The Congressional Budget Office (CBO) was established by the Congressional Budget Act of 1974 (2 U.S.C. 601), which also created a procedure by which the United States Congress considers and acts upon the annual Federal budget. This process enables the Congress to have an overview of the Federal budget and to make overall decisions regarding spending and taxing levels and the deficit or surplus these levels incur.

Activities

The Congressional Budget Office's chief responsibility under the Budget Act is to assist the Congressional budget committees with drafting and enforcing the annual budget resolution, which serves as a blueprint for total levels of Government spending and revenues in a fiscal year. Once completed, the budget resolution guides the action of other Congressional committees in drafting

CONGRESSIONAL BUDGET OFFICE

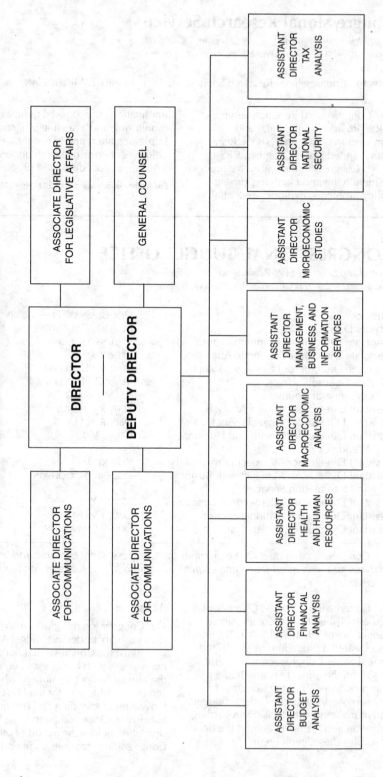

subsequent spending and revenue legislation within their jurisdiction.

To support this process, the Congressional Budget Office prepares reports that provide budgetary and economic projections, analyze the proposals set forth in the President's budget request, and detail alternative spending and revenue options for lawmakers to consider. CBO also provides estimates of the costs of bills approved by Congressional committees and tracks the progress of spending and revenue legislation in a "scorekeeping" system. CBO's cost estimates and scorekeeping system help the budget committees determine whether the budgetary effects of individual proposals are consistent with the most recent spending and revenue targets.

Upon Congressional request, CBO also produces reports analyzing specific policy and program issues that are significant for the budget. In keeping with the agency's nonpartisan role, its analyses do not present policy recommendations and routinely disclose their underlying assumptions and methods. This open and nonpartisan stance has been instrumental in preserving the credibility of the agency's analyses.

Baseline Budget Projections and Economic Forecasts   Each year, the Congressional Budget Office prepares a report on the budget and economic outlook for a specific planning horizon, typically a span of 10 years, to provide the Congress with a baseline against which to measure the effects of proposed changes in spending and tax laws. This baseline represents CBO's best judgments about how the economy and other factors will affect spending under existing laws and policies. To construct this baseline, CBO creates its own economic forecasts, which cover major economic variables such as gross domestic product, unemployment, inflation, and interest rates, as well as other important economic indicators. CBO draws this information from an ongoing analysis of daily economic events and data, comparison with major commercial forecasting services, consultation with economists both within and outside the Federal Government, and the advice of the CBO Panel of Economic Advisers.

Analysis of the President's Budget   Each year, the Congressional Budget Office provides an independent "reestimate" of the budgetary impact of the President's budget proposals using CBO's economic assumptions and estimating techniques. This allows the Congress to compare the administration's spending and revenue proposals to CBO's baseline projections and to other proposals using a consistent set of economic and technical assumptions.

Cost Estimates for Bills   The Budget Act requires CBO to produce a cost estimate for every bill "reported out" by a Congressional committee, showing how the legislation would affect spending or revenues over the next 5 years or more. CBO also frequently prepares cost estimates for use in formulating floor amendments and working out the final form of legislation in conference committees.

Scorekeeping   One of the Congressional Budget Office's most important functions is to keep track of all spending and revenue legislation considered each year so that the Congress can know whether it is acting within the levels set by the budget resolution, in a process known as "scorekeeping." CBO's scorekeeping system keeps track of all bills affecting the budget from the time they are reported out of committee to the time they are enacted into law, providing the budget and appropriations committees with frequent tabulations of Congressional action on both spending and revenue bills.

Federal Mandates   As required by the Unfunded Mandates Reform Act of 1995, the Congressional Budget Office provides committees (with the exception of the appropriations committees) with a statement regarding the costs of Federal mandates in reported legislation and how they might affect State, local, and tribal governments, as well as the private sector. If the 5-year direct costs of an intergovernmental or private-sector mandate exceed specified thresholds, CBO must provide an estimate of those costs (if feasible) and the basis of the

estimate. CBO generally includes both intergovernmental and private-sector mandate statements with its cost estimate for each committee-approved bill.
Budgetary and Economic Policy Issues
The Congressional Budget Office also analyzes specific program and policy

issues that affect the Federal budget and the economy. Generally, requests for these analyses come from the Chairman or Ranking Member of a committee or subcommittee or from the leadership of either party in the House or Senate.

For further information, contact the Management, Business, and Information Services Division, Congressional Budget Office, Second and D Streets SW., Washington, DC 20515. Phone, 202–226–2600. Fax, 202–226–2714. Internet, http://www.cbo.gov.

Judicial Branch

JUDICIAL BRANCH

THE SUPREME COURT OF THE UNITED STATES

United States Supreme Court Building, One First Street NE., Washington, DC 20543
Phone, 202–479–3000. Internet, http://www.supremecourt.gov.

Members

Chief Justice of the United States	JOHN G. ROBERTS, JR.
Associate Justices	ANTONIN SCALIA, ANTHONY M. KENNEDY, CLARENCE THOMAS, RUTH BADER GINSBURG, STEPHEN G. BREYER, SAMUEL A. ALITO, JR., SONIA SOTOMAYOR, ELENA KAGAN

Officers

Counselor to the Chief Justice	JEFFREY P. MINEAR
Clerk	WILLIAM K. SUTER
Legal Officer	SCOTT S. HARRIS
Curator	CATHERINE E. FITTS
Director of Data Systems	(VACANCY)
Librarian	JUDITH GASKELL
Marshal	PAMELA TALKIN
Public Information Officer	KATHLEEN L. ARBERG
Reporter of Decisions	FRANK D. WAGNER

Article III, section 1, of the Constitution of the United States provides that "[t]he judicial Power of the United States, shall be vested in one supreme Court, and in such inferior Courts as the Congress may from time to time ordain and establish."

The Supreme Court of the United States was created in accordance with this provision and by authority of the Judiciary Act of September 24, 1789 (1 Stat. 73). It was organized on February 2, 1790. Article III, section 2 of the Constitution defines the jurisdiction of the Supreme Court.

The Supreme Court is comprised of the Chief Justice of the United States and such number of Associate Justices as may be fixed by Congress, which is currently fixed at eight (28 U.S.C. 1). The President nominates the Justices with the advice and consent of the Senate. Article III, section 1, of the Constitution further provides that "[t]he Judges, both of the supreme and inferior Courts, shall hold their Offices during good Behaviour, and shall, at stated Times, receive for their Services, a Compensation, which shall not be diminished during their Continuance in Office."

Court officers assist the Court in the performance of its functions. They include the Counselor to the Chief Justice, the Clerk, the Reporter of Decisions, the Librarian, the Marshal, the Legal Officer, the Curator, the Director of Data Systems, and the Public Information Officer.

Appellate Jurisdiction Appellate jurisdiction has been conferred upon the Supreme Court by various statutes under the authority given Congress by the Constitution. The basic statute effective at this time in conferring and controlling jurisdiction of the Supreme Court may be found in 28 U.S.C. 1251, 1253, 1254, 1257–1259, and various special statutes. Congress has no authority to change the original jurisdiction of this Court.

Rulemaking Power   Congress has from time to time conferred upon the Supreme Court power to prescribe rules of procedure to be followed by the lower courts of the United States.

Court Term   The term of the Court begins on the first Monday in October and lasts until the first Monday in October of the next year. Approximately 8,000–10,000 cases are filed with the Court in the course of a term, and some 1,000 applications of various kinds are filed each year that can be acted upon by a single Justice.

Access to Facilities   The Supreme Court is open to the public from 9 a.m. to 4:30 p.m., Monday through Friday, except on Federal holidays. Unless the Court or Chief Justice orders otherwise, the Clerk's office is open from 9 a.m. to 5 p.m., Monday through Friday, except on Federal legal holidays. The library is open to members of the bar of the Court, attorneys for the various Federal departments and agencies, and Members of Congress.

For further information concerning the Supreme Court, contact the Public Information Office, United States Supreme Court Building, One First Street NE., Washington, DC 20543. Phone, 202–479–3211. Internet, http://www.supremecourt.gov.

LOWER COURTS

Article III of the Constitution declares, in section 1, that the judicial power of the United States shall be invested in one Supreme Court and in "such inferior Courts as the Congress may from time to time ordain and establish." The Supreme Court has held that these constitutional courts ". . . share in the exercise of the judicial power defined in that section, can be invested with no other jurisdiction, and have judges who hold office during good behavior, with no power in Congress to provide otherwise."

United States Courts of Appeals

The courts of appeals are intermediate appellate courts created by act of March 3, 1891 (28 U.S.C. ch. 3), to relieve the Supreme Court of considering all appeals in cases originally decided by the Federal trial courts. They are empowered to review all final decisions and certain interlocutory decisions (18 U.S.C. 3731; 28 U.S.C. 1291, 1292) of district courts. They also are empowered to review and enforce orders of many Federal administrative bodies. The decisions of the courts of appeals are final except as they are subject to review on writ of certiorari by the Supreme Court.

The United States is divided geographically into 12 judicial circuits, including the District of Columbia.

Each circuit has a court of appeals (28 U.S.C. 41, 1294). Each of the 50 States is assigned to one of the circuits. The territories and the Commonwealth of Puerto Rico are assigned variously to the first, third, and ninth circuits. There is also a Court of Appeals for the Federal Circuit, which has nationwide jurisdiction defined by subject matter. At present each court of appeals has from 6 to 28 permanent circuit judgeships (179 in all), depending upon the amount of judicial work in the circuit. Circuit judges hold their offices during good behavior as provided by Article III, section 1, of the Constitution. The judge senior in commission who is under 70 years of age (65 at inception of term), has been in office at least 1 year, and has not previously been chief judge, serves as the chief judge of the circuit for a 7-year term. One of the Justices of the Supreme Court is assigned as circuit justice for each of the 13 judicial circuits. Each court of appeals normally hears cases in panels consisting of three judges but may sit en banc with all judges present.

The judges of each circuit (except the Federal Circuit) by vote determine the size of the judicial council for the circuit, which consists of the chief judge and an equal number of circuit and district judges. The council considers the state of Federal judicial business in the circuit and may "make all necessary and appropriate orders for [its] effective and expeditious administration . . ." (28 U.S.C. 332).

The chief judge of each circuit may summon periodically a judicial conference of all judges of the circuit, including members of the bar, to discuss the business of the Federal courts of the circuit (28 U.S.C. 333). The chief judge of each circuit and a district judge elected from each of the 12 geographical circuits, together with the chief judge of the Court of International Trade, serve as members of the Judicial Conference of the United States, over which the Chief Justice of the United States presides. This is the governing body for the administration of the Federal judicial system as a whole (28 U.S.C. 331).

To obtain a complete list of judges, court officials, and official stations of the United States Courts of Appeals for the Federal Circuit, as well as information on opinions and cases before the court, consult the Judicial Circuit Web sites listed below.

List of Judicial Circuit Web Sites—United States Courts of Appeals

Circuit	URL
District of Columbia Circuit	http://www.cadc.uscourts.gov
First Circuit	http://www.ca1.uscourts.gov
Second Circuit	http://www.ca2.uscourts.gov
Third Circuit	http://www.ca3.uscourts.gov
Fourth Circuit	http://www.ca4.uscourts.gov
Fifth Circuit	http://www.ca5.uscourts.gov
Sixth Circuit	http://www.ca6.uscourts.gov
Seventh Circuit	http://www.ca7.uscourts.gov
Eighth Circuit	http://www.ca8.uscourts.gov
Ninth Circuit	http://www.ca9.uscourts.gov
Tenth Circuit	http://www.ca10.uscourts.gov/
Eleventh Circuit	http://www.ca11.uscourts.gov/

United States Court of Appeals for the Federal Circuit

This court was established under Article III of the Constitution pursuant to the Federal Courts Improvement Act of 1982 (28 U.S.C. 41, 44, 48), as successor to the former United States Court of Customs and Patent Appeals and the United States Court of Claims. The jurisdiction of the court is nationwide (as provided by 28 U.S.C. 1295) and includes appeals from the district courts in patent cases; appeals from the district courts in contract, and certain other civil actions in which the United States is a defendant; and appeals from final decisions of the U.S. Court of International Trade, the U.S. Court of Federal Claims, and the U.S. Court of Appeals for Veterans Claims. The jurisdiction of the court also includes the review of administrative rulings by the Patent and Trademark Office, U.S. International Trade Commission, Secretary of Commerce, agency boards of contract appeals, and the Merit Systems Protection Board, as well as rulemaking of the Department of Veterans Affairs; review of decisions of the U.S. Senate Committee on Ethics concerning discrimination claims of

Senate employees; and review of a final order of an entity to be designated by the President concerning discrimination claims of Presidential appointees.

The court consists of 12 circuit judges. It sits in panels of three or more on each case and may also hear or rehear a case en banc. The court sits principally in Washington, DC, and may hold court wherever any court of appeals sits (28 U.S.C. 48).

To obtain a complete list of judges and court officials of the United States Courts of Appeals for the Federal Circuit, as well as information on opinions and cases before the court, consult the following Web site: http://www.cafc.uscourts.gov.

United States District Courts

The district courts are the trial courts of general Federal jurisdiction. Each State has at least one district court, while the larger States have as many as four. There are 89 district courts in the 50 States, plus the one in the District of Columbia. In addition, the Commonwealth of Puerto Rico has a district court with jurisdiction corresponding to that of district courts in the various States.

At present, each district court has from 2 to 28 Federal district judgeships, depending upon the amount of judicial work within its territory. Only one judge is usually required to hear and decide a case in a district court, but in some limited cases it is required that three judges be called together to comprise the court (28 U.S.C. 2284). The judge senior in commission who is under 70 years of age (65 at inception of term), has been in office for at least 1 year, and has not previously been chief judge, serves as chief judge for a 7-year term. There are 645 permanent district judgeships in the 50 States and 15 in the District of Columbia. There are seven district judgeships in Puerto Rico. District judges hold their offices during good behavior as provided by Article III, section 1, of the Constitution. However, Congress may temporary judgeships for a court with the provision that when a future vacancy occurs in that district, such vacancy shall not be filled. Each district court has one or more United States magistrate judges and bankruptcy judges, a clerk, a United States attorney, a United States marshal, probation officers, court reporters, and their staffs. The jurisdiction of the district courts is set forth in title 28, chapter 85, of the United States Code and at 18 U.S.C. 3231.

Cases from the district courts are reviewable on appeal by the applicable court of appeals.

Territorial Courts

Pursuant to its authority to govern the Territories (Art. IV, sec. 3, clause 2, of the Constitution), Congress has established district courts in the territories of Guam and the Virgin Islands. The District Court of the Canal Zone was abolished on April 1, 1982, pursuant to the Panama Canal Act of 1979 (22 U.S.C. 3601 note). Congress has also established a district court in the Northern Mariana Islands, which presently is administered by the United States under a trusteeship agreement with the United Nations. These Territorial courts have jurisdiction not only over the subjects described in the judicial article of the Constitution but also over many local matters that, within the States, are decided in State courts. The District Court of Puerto Rico, by contrast, is established under Article III, is classified like other "district courts," and is called a "court of the United States" (28 U.S.C. 451). There is one judge each in Guam and the Northern Mariana Islands, and two in the Virgin Islands. The judges in these courts are appointed for terms of 10 years.

For further information concerning the lower courts, contact the Administrative Office of the United States Courts, Thurgood Marshall Federal Judiciary Building, One Columbus Circle NE., Washington, DC 20544. Phone, 202–502–2600.

United States Court of International Trade

This court was originally established as the Board of United States General Appraisers by act of June 10, 1890, which conferred upon it jurisdiction theretofore held by the district and circuit courts in actions arising under the tariff acts (19 U.S.C. ch. 4). The act of May 28, 1926 (19 U.S.C. 405a), created the United States Customs Court to supersede the Board; by acts of August 7, 1939, and June 25, 1948 (28 U.S.C. 1582, 1583), the court was integrated into the United States court structure, organization, and procedure. The act of July 14, 1956 (28 U.S.C. 251), established the court as a court of record of the United States under Article III of the Constitution. The Customs Court Act of 1980 (28 U.S.C. 251) constituted the court as the United States Court of International Trade.

The Court of International Trade has jurisdiction over any civil action against the United States arising from Federal laws governing import transactions. This includes classification and valuation cases, as well as authority to review certain agency determinations under the Trade Agreements Act of 1979 (19 U.S.C. 2501) involving antidumping and countervailing duty matters. In addition, it has exclusive jurisdiction of civil actions to review determinations as to the eligibility of workers, firms, and communities for adjustment assistance under the Trade Act of 1974 (19 U.S.C. 2101). Civil actions commenced by the United States to recover customs duties, to recover on a customs bond, or for certain civil penalties alleging fraud or negligence are also within the exclusive jurisdiction of the court.

The court is composed of a chief judge and eight judges, not more than five of whom may belong to any one political party. Any of its judges may be temporarily designated and assigned by the Chief Justice of the United States to sit as a court of appeals or district court judge in any circuit or district. The court has a clerk and deputy clerks, a librarian, court reporters, and other supporting personnel. Cases before the court may be tried before a jury. Under the Federal Courts Improvement Act of 1982 (28 U.S.C. 1295), appeals are taken to the U.S. Court of Appeals for the Federal Circuit, and ultimately review may be sought in appropriate cases in the Supreme Court of the United States.

The principal offices are located in New York, NY, but the court is empowered to hear and determine cases arising at any port or place within the jurisdiction of the United States.

For further information, contact the Clerk, United States Court of International Trade, One Federal Plaza, New York, NY 10278–0001. Phone, 212–264–2814.

Judicial Panel on Multidistrict Litigation

The Panel, created by act of April 29, 1968 (28 U.S.C. 1407), and consisting of seven Federal judges designated by the Chief Justice from the courts of appeals and district courts, is authorized to temporarily transfer to a single district, for coordinated or consolidated pretrial proceedings, civil actions pending in different districts that involve one or more common questions of fact.

For further information, contact the Clerk, Judicial Panel on Multidistrict Litigation, Room G–255, Thurgood Marshall Federal Judiciary Building, One Columbus Circle NE., Washington, DC 20002–8041. Phone, 202–502–2800.

SPECIAL COURTS

United States Court of Federal Claims

717 Madison Place NW., Washington, DC 20005–1086
Phone, 202–357–6400. Internet, http://www.uscfc.uscourts.gov.

The U.S. Court of Federal Claims has jurisdiction over claims seeking money judgments against the United States. A claim must be founded upon the United States Constitution; an act of Congress; the regulation of an executive department; an express or implied-in-fact contract with the United States; or damages, liquidated or unliquidated, in cases not sounding in tort. Judges in the U.S. Court of Federal Claims are appointed by the President for 15-year terms, subject to Senate confirmation. Appeals are to the U.S. Court of Appeals for the Federal Circuit.

For further information, contact the Clerk's Office, United States Court of Federal Claims, 717 Madison Place NW., Washington, DC 20005–1086. Phone, 202–357–6400. Internet, http://www.uscfc.uscourts.gov.

United States Court of Appeals for the Armed Forces

450 E Street NW., Washington, DC 20442–0001
Phone, 202–761–1448. Fax, 202–761–4672. Internet, http://www.armfor.uscourts.gov.

This court was established under Article I of the Constitution of the United States pursuant to act of May 5, 1950, as amended (10 U.S.C. 867). Subject only to certiorari review by the Supreme Court of the United States in a limited number of cases, the court serves as the final appellate tribunal to review court-martial convictions of all the Armed Forces. It is exclusively an appellate criminal court, consisting of five civilian judges who are appointed for 15-year terms by the President with the advice and consent of the Senate.

The court is called upon to exercise jurisdiction to review the record in all cases extending to death; certified to the court by a Judge Advocate General of an armed force or by the General Counsel of the Department of Transportation, acting for the Coast Guard; or petitioned by accused who have received a sentence of confinement for 1 year or more, and/or a punitive discharge.

The court also exercises authority under the All Writs Act (28 U.S.C. 1651(a)).

In addition, the judges of the court are required by law to work jointly with the senior uniformed lawyer from each armed force, the Chief Counsel of the Coast Guard, and two members of the public appointed by the Secretary of Defense, to make an annual comprehensive survey and to report annually to the Congress on the operation and progress of the military justice system under the Uniform Code of Military Justice, and to recommend improvements wherever necessary.

For further information, contact the Clerk, United States Court of Appeals for the Armed Forces, 450 E Street NW., Washington, DC 20442–0001. Phone, 202–761–1448. Fax, 202–761–4672. Internet, http://www. armfor.uscourts.gov.

United States Tax Court

400 Second Street NW., Washington, DC 20217–0002
Phone, 202–521–0700. Internet, http://www.ustaxcourt.gov.

The United States Tax Court is a court of record under Article I of the Constitution of the United States (26 U.S.C. 7441). The court was created as the United States Board of Tax Appeals by the Revenue Act of 1924 (43 Stat. 336). The name was changed to the Tax Court of the United States by the Revenue Act of 1942 (56 Stat. 957). The Tax Reform Act of 1969 (83 Stat. 730) established the court under Article I and then changed its name to the United States Tax Court.

The court comprises 19 judges who are appointed by the President to 15-year terms and subject to Senate confirmation. The court also has varying numbers of both senior judges (who may be recalled by the chief judge to perform further judicial duties) and special trial judges (who are appointed by the chief judge and may hear and decide a variety of cases). The court's jurisdiction is set forth in various sections of title 26 of the U.S. Code.

The offices of the court and its judges are in Washington, DC. However, the court has national jurisdiction and schedules trial sessions in more than 70 cities in the United States. Each trial session is conducted by one judge, senior judge, or special trial judge. Court proceedings are open to the public and are conducted in accordance with the court's Rules of Practice and Procedure and the rules of evidence applicable in trials without a jury in the U.S. District Court for the District of Columbia. A fee of $60 is charged for the filing of a petition. Practice before the court is limited to practitioners admitted under the court's Rules of Practice and Procedure.

Decisions entered by the court, other than decisions in small tax cases, may be appealed to the regional courts of appeals and, thereafter, upon the granting of a writ of certiorari, to the Supreme Court of the United States. At the option of petitioners, simplified procedures may be used in small tax cases. Small tax cases are final and not subject to review by any court.

For further information, contact the Office of the Clerk of the Court, United States Tax Court, 400 Second Street NW., Washington, DC 20217–0002. Phone, 202–521–0700. Internet, http://www.ustaxcourt.gov.

United States Court of Appeals for Veterans Claims

Suite 900, 625 Indiana Avenue NW., Washington, DC 20004–2950
Phone, 202–501–5970. Internet, http://www.uscourts.cavc.gov.

The United States Court of Veterans Appeals was established on November 18, 1988 (102 Stat. 4105, 38 U.S.C. 7251) pursuant to Article I of the Constitution, and given exclusive jurisdiction to review decisions of the Board of Veterans Appeals. The court was renamed the United States Court of Appeals for Veterans Claims by the Veterans Programs Enhancement Act of 1998 (38 U.S.C. 7251 note). The court may not review the schedule of ratings for disabilities or actions of the Secretary in adopting or revising that schedule. Decisions of the Court of Appeals for Veterans Claims may be appealed to the United States Court of Appeals for the Federal Circuit.

The court consists of seven judges appointed by the President, with the advice and consent of the Senate, for 15-year terms. One of the judges serves as chief judge.

The court's principal office is in the District of Columbia, but the court can also act at any place within the United States.

For further information, contact the Clerk, United States Court of Appeals for Veterans Claims, Suite 900, 625 Indiana Avenue NW., Washington, DC 20004–2950. Phone, 202–501–5970. Internet, http://www. uscourts.cavc.gov.

ADMINISTRATIVE OFFICE OF THE UNITED STATES COURTS

One Columbus Circle NE., Washington, DC 20544
Phone, 202–502–2600. Internet, http://www.uscourts.gov.

Director	JAMES C. DUFF
Deputy Director	JILL C. SAYENGA
Associate Director and General Counsel	WILLIAM R. BURCHILL, JR.
Assistant Director, Office of Judicial Conference Executive Secretariat	LAURA C. MINOR
Assistant Director, Office of Legislative Affairs	CORDIA A. STROM
Assistant Director, Office of Public Affairs	DAVID A. SELLERS
Assistant Director, Office of Court Administration	NOEL J. AUGUSTYN
Assistant Director, Office of Defender Services	THEODORE J. LIDZ
Assistant Director, Office of Facilities and Security	ROSS EISENMAN
Assistant Director, Office of Finance and Budget	GEORGE H. SCHAFER
Assistant Director, Office of Human Resources	PATRICIA J. FITZGIBBONS
Assistant Director, Office of Information Technology	HOWARD J. GRANDIER
Assistant Director, Office of Internal Services	DOREEN G.B. BYDUME
Assistant Director, Office of Judges Programs	PETER G. MCCABE
Assistant Director, Office of Probation and Pretrial Services	JOHN M. HUGHES

The Administrative Office of the United States Courts supports and serves the nonjudicial, administrative business of the United States Courts, including the maintenance of workload statistics and the disbursement of funds appropriated for the maintenance of the U.S. judicial system.

The Administrative Office of the United States Courts was created by act of August 7, 1939 (28 U.S.C. 601). The Office was established November 6, 1939. Its Director and Deputy Director are appointed by the Chief Justice of the United States after consultation with the Judicial Conference.

Administering the Courts The Director is the administrative officer of the courts of the United States (except the Supreme Court). Under the guidance of the Judicial Conference of the United States the Director is required, among other things, to supervise all administrative matters relating to the offices of clerks and other clerical and administrative personnel of the courts; to examine the state of the dockets of the courts, secure information as to the courts' need of assistance, and prepare and transmit quarterly to the chief judges of the circuits statistical data and reports as

to the business of the courts; to submit to the annual meeting of the Judicial Conference of the United States, at least 2 weeks prior thereto, a report of the activities of the Administrative Office and the state of the business of the courts; to fix the compensation of employees of the courts whose compensation is not otherwise fixed by law; to regulate and pay annuities to widows and surviving dependent children of judges; to disburse moneys appropriated for the maintenance and operation of the courts; to examine accounts of court officers; to regulate travel of judicial personnel; to provide accommodations and supplies for the courts and their clerical and administrative personnel; to establish and maintain programs for the certification and utilization of court interpreters and the provision of special interpretation services in the courts; and to perform such other duties as may be assigned

ADMINISTRATIVE OFFICE OF THE UNITED STATES COURTS

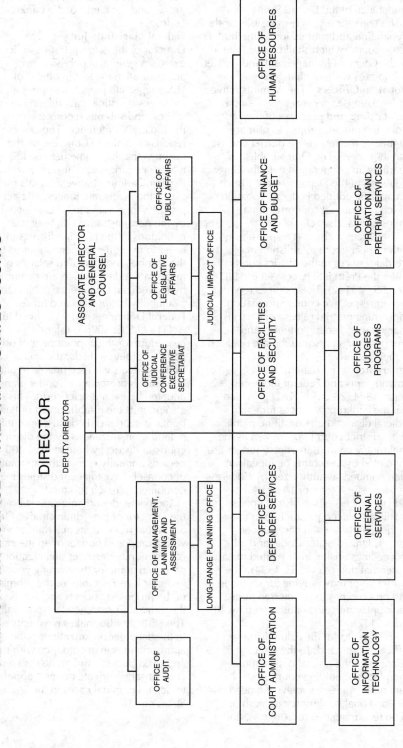

DIRECTOR
DEPUTY DIRECTOR

ASSOCIATE DIRECTOR AND GENERAL COUNSEL

OFFICE OF PUBLIC AFFAIRS

OFFICE OF LEGISLATIVE AFFAIRS

JUDICIAL IMPACT OFFICE

OFFICE OF JUDICIAL CONFERENCE EXECUTIVE SECRETARIAT

OFFICE OF AUDIT

OFFICE OF MANAGEMENT, PLANNING AND ASSESSMENT

LONG-RANGE PLANNING OFFICE

OFFICE OF COURT ADMINISTRATION

OFFICE OF DEFENDER SERVICES

OFFICE OF FACILITIES AND SECURITY

OFFICE OF FINANCE AND BUDGET

OFFICE OF HUMAN RESOURCES

OFFICE OF INFORMATION TECHNOLOGY

OFFICE OF INTERNAL SERVICES

OFFICE OF JUDGES PROGRAMS

OFFICE OF PROBATION AND PRETRIAL SERVICES

by the Supreme Court or the Judicial Conference of the United States.

The Director is also responsible for the preparation and submission of the budget of the courts, which shall be transmitted by the Office of Management and Budget to Congress without change.

Probation Officers   The Administrative Office exercises general supervision of the accounts and practices of the Federal probation offices, subject to primary control by the respective district courts that they serve. The Office publishes quarterly, in cooperation with the Bureau of Prisons of the Department of Justice, a magazine entitled Federal Probation, which is a journal "of correctional philosophy and practice."

The Director also has responsibility with respect to the establishment of pretrial services in the district courts under the Pretrial Services Act of 1982 (18 U.S.C. 3152). These offices report to their respective courts information concerning pretrial release of persons charged with Federal offenses and supervise such persons who are released to their custody.

Bankruptcy   The Bankruptcy Amendments and Federal Judgeship Act of 1984 (28 U.S.C. 151) provided that the bankruptcy judges for each judicial district shall constitute a unit of the district court to be known as the bankruptcy court. Bankruptcy judges are appointed by the courts of appeals in such numbers as authorized by Congress and serve for a term of 14 years as judicial officers of the district courts.

This act placed jurisdiction in the district courts over all cases under title 11, United States Code, and all proceedings arising in or related to cases under that title (28 U.S.C. 1334). The district court may provide for such cases and proceedings to be referred to its bankruptcy judges (as authorized by 28 U.S.C. 157).

The Director of the Administrative Office recommends to the Judicial Conference the official duty stations and places of holding court of bankruptcy judges, surveys the need for additional bankruptcy judgeships to be recommended to Congress, and

determines the staff needs of bankruptcy judges and the clerks of the bankruptcy courts.

Federal Magistrate Judges   The Director of the Administrative Office exercises general supervision over administrative matters in offices of U.S. magistrate judges, compiles and evaluates statistical data relating to such offices, and submits reports thereon to the Judicial Conference. The Director reports annually to Congress on the business that has come before U.S. magistrate judges and also prepares legal and administrative manuals for the use of the magistrate judges. The act provides for surveys to be conducted by the Administrative Office of the conditions in the judicial districts in order to make recommendations as to the number, location, and salaries of magistrate judges, which are determined by the Judicial Conference subject to the availability of appropriated funds.

Federal Defenders   The Criminal Justice Act (18 U.S.C. 3006A) establishes the procedure for the appointment of private panel attorneys in Federal criminal cases for individuals who are unable to afford adequate representation, under plans adopted by each district court. The act also permits the establishment of Federal public defender or Federal community defender organizations by the district courts in districts where at least 200 persons annually require the appointment of counsel. Two adjacent districts may be combined to reach this total.

Each defender organization submits to the Director of the Administrative Office an annual report of its activities along with a proposed budget or, in the case of community defender organizations, a proposed grant for the coming year. The Director is responsible for the submission of the proposed budgets and grants to the Judicial Conference for approval. The Director also makes payments to the defender organizations out of appropriations in accordance with the approved budgets and grants, as well as compensating private counsel appointed to defend criminal cases in the United States courts.

Sources of Information

Bankruptcy Judges Division. Phone, 202–502–1900.

Budget Division. Phone, 202–502–2100.

Defender Services Division. Phone, 202–502–3030.

General Counsel. Phone, 202–502–1100.

Human Resources Division. Phone, 202–502–3100.

Judicial Conference Executive Secretariat. Phone, 202–502–2400.

Legislative Affairs Office. Phone, 202–502–1700.

Magistrate Judges Division. Phone, 202–502–1830.

Office of Probation and Pretrial Services. Phone, 202–502–1610.

Public Affairs Office. Phone, 202–502–2600.

Statistics Division. Phone, 202–502–1440.

For further information, contact the Administrative Office of the United States Courts, Thurgood Marshall Federal Judiciary Building, One Columbus Circle NE., Washington, DC 20544. Phone, 202–502–2600. Internet, http://www.uscourts.gov.

FEDERAL JUDICIAL CENTER

Thurgood Marshall Federal Judiciary Building, One Columbus Circle NE., Washington, DC 20002–8003
Phone, 202–502–4000. Internet, http://www.fjc.gov.

Director	BARBARA J. ROTHSTEIN
Deputy Director	JOHN S. COOKE
Director, Education Division	BRUCE M. CLARKE
Director, Systems Innovation and Development Office	TED E. COLEMAN
Director, Research Division	JAMES B. EAGLIN
Director, International Judicial Relations Office	MIRA GUR-ARIE
Director, Federal Judicial History Office	BRUCE A. RAGSDALE
Director, Communications Policy and Design Office	SYLVAN A. SOBEL

The Federal Judicial Center is the judicial branch's agency for policy research and continuing education.

The Federal Judicial Center was created by act of December 20, 1967 (28 U.S.C. 620), to further the development and adoption of improved judicial administration in the courts of the United States.

The Center's basic policies and activities are determined by its Board, which is composed of the Chief Justice of the United States, who is permanent Chair of the Board by statute, and two judges of the U.S. courts of appeals, three judges of the U.S. district courts, one bankruptcy judge, and one magistrate judge, all of whom are elected for 4-year terms by the Judicial Conference of the United States. The Director of the

Administrative Office of the United States Courts is also a permanent member of the Board.

Pursuant to statute, the Center carries out the following duties: develops and administers orientation and continuing education programs for Federal judges, Federal defenders, and nonjudicial court personnel, including probation officers, pretrial services officers, and clerks' office employees; conducts empirical and exploratory research and evaluation on Federal judicial processes, court management, and sentencing and its consequences, usually for the committees of the Judicial Conference or the courts themselves; produces research reports, training manuals, satellite broadcasts, video programs, computer based training, and periodicals about the Federal courts; provides guidance and advice and maintains data and records to assist those interested in documenting and conserving the history of the Federal courts; and cooperates with and assists other agencies and organizations in providing advice to improve the administration of justice in the courts of foreign countries.

Sources of Information

For general information about the Federal Judiciary Center, including a directory of telephone and fax numbers for its component offices and divisions, visit: www.fjc.gov/public/home.nsf/pages/104.
Electronic Access   Selected Federal Judicial Center publications, Federal judicial history databases, and various educational resources are available at www.fjc.gov.
Publications   Single copies of most Federal Judicial Center publications are available free of charge. Phone, 202–502–4153. Fax, 202–502–4077.

For further information, contact the Federal Judicial Center, Thurgood Marshall Federal Judiciary Building, One Columbus Circle NE., Washington, DC 20002–8003. Phone, 202–502–4000. Internet, http://www.fjc. gov.

UNITED STATES SENTENCING COMMISSION

Suite 2–500, South Lobby, One Columbus Circle NE., Washington, DC 20002–8002
Phone, 202–502–4500. Internet, http://www.ussc.gov.

Chair	WILLIAM K. SESSIONS, III
Vice Chairs	WILLIAM B. CARR, JR., RUBEN CASTILLO, KETANJI BROWN JACKSON
Commissioners	DABNEY L. FRIEDRICH, RICARDO H. HINOJOSA, BERYL A. HOWELL
Commissioners (ex officio)s	ISAAC FULWOOD, JR., JONATHAN J. WROBLEWSKI
Staff Director	JUDITH W. SHEON
General Counsel	KENNETH P. COHEN
Public Affairs Officer	MICHAEL COURLANDER
Director of Administration and Planning	SUSAN M. BRAZEL
Director and Chief Counsel of Training	PAMELA G. MONTGOMERY
Director of Legislative and Public Affairs	LISA A. RICH
Director of Research and Data	GLENN R. SCHMITT

The United States Sentencing Commission develops sentencing guidelines and policies for the Federal court system.

The United States Sentencing Commission was established as an independent agency in the judicial branch of the Federal Government by the Sentencing Reform Act of 1984 (28 U.S.C. 991 et seq. and 18 U.S.C. 3551 et seq.). The Commission establishes sentencing guidelines and policies for the Federal courts, advising them of the appropriate form and severity of

punishment for offenders convicted of Federal crimes.

The Commission is composed of seven voting members appointed by the President with the advice and consent of the Senate for 6-year terms, and two nonvoting members. One of the voting members is appointed Chairperson.

The Commission evaluates the effects of the sentencing guidelines on the criminal justice system, advises Congress regarding the modification or enactment of statutes relating to criminal law and sentencing matters, establishes a research and development program on sentencing issues, and performs other related duties.

In executing its duties, the Commission promulgates and distributes to Federal courts and to the U.S. probation system guidelines to be consulted in determining sentences to be imposed in criminal cases, general policy statements regarding the application of guidelines, and policy statements on the appropriate use of probation and supervised release revocation provisions. These sentencing guidelines and policy statements are designed to further the purposes of just punishment, deterrence, incapacitation, and rehabilitation; provide fairness in meeting the purposes of sentencing; avoid unwarranted disparity; and reflect advancement in the knowledge of human behavior as it relates to the criminal justice process.

In addition, the Commission provides training, conducts research on sentencing-related issues, and serves as an information resource for Congress, criminal justice practitioners, and the public.

Sources of Information

Electronic Access   Commission information and materials may be obtained through the Internet at www. ussc.gov.
Guideline Application Assistance Helpline   Phone, 202–502–4545.
Public Information   Information concerning Commission activities is available from the Office of Publishing and Public Affairs. Phone, 202–502–4590.

For further information, contact the Office of Publishing and Public Affairs, United States Sentencing Commission, Suite 2–500, South Lobby, One Columbus Circle NE., Washington, DC 20002–8002. Phone, 202–502–4590. Internet, http://www.ussc.gov.

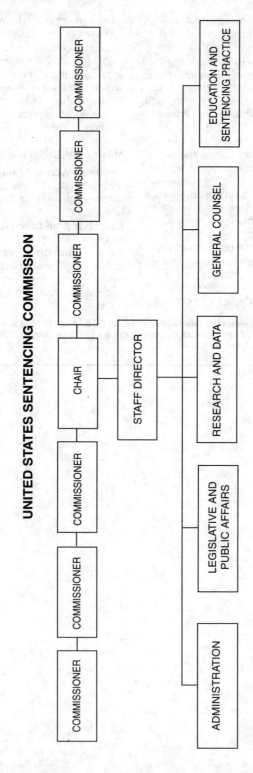

UNITED STATES SENTENCING COMMISSION

Executive Branch

EXECUTIVE BRANCH: THE PRESIDENT

THE PRESIDENT

THE PRESIDENT OF THE UNITED STATES — BARACK OBAMA

Article II, section 1, of the Constitution provides that "[t]he executive Power shall be vested in a President of the United States of America. He shall hold his Office during the Term of four Years, . . . together with the Vice President, chosen for the same Term . . . ." In addition to the powers set forth in the Constitution, the statutes have conferred upon the President specific authority and responsibility covering a wide range of matters (United States Code Index).

The President is the administrative head of the executive branch of the Government, which includes numerous agencies, both temporary and permanent, as well as the 15 executive departments. **The Cabinet** The Cabinet, a creation of custom and tradition dating back to George Washington's administration, functions at the pleasure of the President. Its purpose is to advise the President upon any subject, relating to the duties of the respective offices, on which he requests information (pursuant to Article II, section 2, of the Constitution).

The Cabinet is composed of the Vice President and the heads of the 15 executive departments—the Secretaries of Agriculture, Commerce, Defense, Education, Energy, Health and Human Services, Homeland Security, Housing and Urban Development, Interior, Labor, State, Transportation, Treasury, and Veterans Affairs, and the Attorney General. Additionally, in the Obama administration, Cabinet-level rank has been accorded to the Chief of Staff to the President; the Administrator, Environmental Protection Agency; the Chair, Council of Economic Advisers; the Director, Office of Management and Budget; the U.S. Permanent Representative to the United Nations; and the U.S. Trade Representative.

THE VICE PRESIDENT

THE VICE PRESIDENT — JOSEPH R. BIDEN, JR.

Article II, section 1, of the Constitution provides that the President "shall hold his Office during the Term of four Years, . . . together with the Vice President . . . ." In addition to his role as President of the Senate, the Vice President is empowered to succeed to the Presidency, pursuant to Article II and the 20th and 25th amendments to the Constitution.

The executive functions of the Vice President include participation in Cabinet meetings and, by statute, membership on the National Security Council and the Board of Regents of the Smithsonian Institution.

THE EXECUTIVE OFFICE OF THE PRESIDENT

Under authority of the Reorganization Act of 1939 (5 U.S.C. 133–133r, 133t note), various agencies were transferred to the Executive Office of the President by the President's Reorganization Plans I and II of 1939 (5 U.S.C. app.), effective July 1, 1939. Executive Order 8248 of September 8, 1939, established the divisions of the Executive Office and defined their functions. Subsequently, Presidents have used Executive orders, reorganization plans, and legislative initiatives to reorganize the Executive Office to make its composition compatible with the goals of their administrations.

White House Office

1600 Pennsylvania Avenue NW., Washington, DC 20500
Phone, 202–456–1414. Internet, http://www.whitehouse.gov.

Assistant to the President and Chief of Staff	WILLIAM M. DALEY
Assistant to the President and Deputy Chief of Staff for Operations	JAMES A. MESSINA
Assistant to the President and Deputy Chief of Staff for Policy	MONA K. SUTPHEN
Assistant to the President and Senior Advisor	DAVID PLOUFFE
Counselor to the President	PETER M. ROUSE
National Security Advisor	THOMAS E. DONILON
Assistant to the President and Deputy National Security Advisor	DENIS R. MCDONOUGH
Assistant to the President for Homeland Security and Counterterrorism	JOHN O. BRENNAN
Assistant to the President and Cabinet Secretary	CHRISTOPHER P. LU
Assistant to the President and Counsel to the President	ROBERT F. BAUER
Assistant to the President and Director, Office of Legislative Affairs	PHILIP M. SCHILIRO
Assistant to the President and Director of Political Affairs	PATRICK H. GASPARD
Assistant to the President and Director of Scheduling and Advance	ALYSSA M. MASTROMONACO
Assistant to the President and Director of Speechwriting	JONATHAN E. FAVREAU
Assistant to the President and Press Secretary	JAMES "JAY" CARNEY
Assistant to the President and Staff Secretary	ELIZABETH M. BROWN
Assistant to the President and Director of Communications	H. DANIEL PFEIFFER
Assistant to the President for Energy and Climate Change	CAROL M. BROWNER
Senior Advisor and Assistant to the President for Intergovernmental Affairs and Public Engagement	VALERIE B. JARRETT
Assistant to the President for Special Projects	STEPHANIE CUTTER
Assistant to the President for Management and Administration	BRADLEY J. KILEY
Assistant to the President, Chief of Staff to the First Lady and Counsel	SUSAN S. SHER

Deputy Assistant to the President and Chief of Staff for National Security Operations	MARK W. LIPPERT
Deputy Assistant to the President and Deputy Cabinet Secretary	ELIZABETH S. SMITH
Deputy Assistant to the President and Deputy Counsel to the Presidents	SUSAN M. DAVIES, MARY B. DEROSA, KATHRYN RUEMMLER
Deputy Assistant to the President and Deputy Director of Communications	JENNIFER R. PSAKI
Deputy Assistant to the President and Deputy Staff Secretary	PETER F. RUNDLET
Deputy Assistant to the President and Director of Advance and Operations	EMMETT S. BELIVEAU
Deputy Assistant to the President and Director of Appointments and Scheduling	DANIELLE M. CRUTCHFIELD
Deputy Assistant to the President and Director of Intergovernmental Affairs	CECILIA MUNOZ
Deputy Assistant to the President and Director of Policy and Projects for the First Lady	JOCELYN C. FRYE
Deputy Assistant to the President and Director of the Office of Public Engagement	CHRISTINA M. TCHEN
Deputy Assistant to the President and Director, Office of Urban Affairs	ADOLFO CARRION, JR
Deputy Assistant to the President and National Security Council Chief of Staff	DENIS R. MCDONOUGH
Deputy Assistant to the President and Principal Deputy Counsel to the President	DANIEL J. MELTZER
Deputy Assistant to the President for Energy and Climate Change	HEATHER R. ZICHAL
Deputy Assistant to the President and Director of Presidential Personnel	NANCY D. HOGAN
Deputy Assistant to the President for Legislative Affairs	LISA M. KONWINSKI
Deputy Assistant to the President for Legislative Affairs and House Liaison	DANIEL A. TURTON
Deputy Assistant to the President for Legislative Affairs and Senate Liaison	SHAWN P. MAHER
Deputy Assistant to the President for Management and Administration	HENRY F. DE SIO, JR.

The White House Office serves the President in the performance of the many detailed activities incident to his immediate office.

The President's staff facilitates and maintains communication with the Congress, the heads of executive agencies, the press and other information media, and the general public. The various Assistants to the President aid the President in such matters as he may direct.

Office of the Vice President

Eisenhower Executive Office Building, Washington, DC 20501
Phone, 202–456–7549.

Assistant to the President and Chief of Staff to the Vice President	RON KLAIN

Deputy Assistant to the President and Deputy Chief of Staff to the Vice President·	ALAN HOFFMAN
Deputy Assistant to the President and National Security Affairs Advisor to the Vice President	TONY BLINKEN
Deputy Assistant to the President and Domestic Policy Advisor to the Vice President	TERRELL MCSWEENY
Deputy Assistant to the President and Economic Advisor to the Vice President	JARED BERNSTEIN
Special Assistant to the President and Intergovernmental Affairs Advisor to the Vice President	EVAN RYAN
Assistant to the Vice President and Counsel to the Vice President	CYNTHIA HOGAN
Assistant to the Vice President for Communications	SHAILAGH MURRAY
Press Secretary for the Vice President	ELIZABETH ALEXANDER
Assistant to the Vice President for Legislative Affairs	SUDAFI HENRY
Assistant to the Vice President for Management and Administration	DENISE MAES
Assistant to the Vice President and Director of Scheduling	ELISABETH HIRE
Assistant to the Vice President and Director of Advance	PETE SELFRIDGE
Chief of Staff to Dr. Jill Biden	CATHY RUSSELL
Executive Assistant to the Vice President	MICHELE SMITH

The Office of the Vice President serves the Vice President in the performance of the many detailed activities incident to his immediate office.

Council of Economic Advisers

1800 G Street NW., Washington, DC 20502
Phone, 202–395–5084. Internet, http://www.whitehouse.gov/cea.

Chair	AUSTAN D. GOOLSBEE
Members	CECILIA E. ROUSE, (VACANCY)

The Council of Economic Advisers performs an analysis and appraisal of the national economy for the purpose of providing policy recommendations to the President.

The Council of Economic Advisers (CEA) was established in the Executive Office of the President by the Employment Act of 1946 (15 U.S.C. 1023). It now functions under that statute and Reorganization Plan No. 9 of 1953 (5 U.S.C. app.), effective August 1, 1953.

The Council consists of three members appointed by the President with the advice and consent of the Senate. One of the members is designated by the President as Chairman.

The Council analyzes the national economy and its various segments; advises the President on economic developments; appraises the economic programs and policies of the Federal Government; recommends to the President policies for economic growth and stability; assists in the preparation of the economic reports of the President to the Congress; and prepares the Annual Report of the Council of Economic Advisers.

For further information, contact the Council of Economic Advisers, 1800 G Street NW., Washington, DC 20502. Phone, 202–395–5084. Internet, http://www.whitehouse.gov/cea.

Council on Environmental Quality

722 Jackson Place NW., Washington, DC 20503
Phone, 202–395–5750 or 202–456–6224. Fax, 202–456–2710. Internet, http://www.whitehouse.
gov/administration/eop/ceq.

Chair	NANCY H. SUTLEY
Deputy Director and General Counsel	GARY GUZY
Senior Counsel	EDWARD A. BOLING
Chief of Staff	JON CARSON
Associate Director for Climate Change	JASON BORDOFF
Associate Director for Communications	CHRISTINE GLUNZ
Associate Director for Communities, Environmental Protection and Green Jobs	NICOLE BUFFA
Associate Director for Lands and Water Ecosystems	MICHAEL BOOTS
Associate Director for Legislative Affairs	JESSICA MAHER
Associate Director for NEPA Oversight	HORST GRECZMIEL
Associate Director for Policy Outreach	AMELIA SALZMAN

The Council on Environmental Quality formulates and recommends national policies and initiatives to improve the environment.

The Council on Environmental Quality (CEQ) was established within the Executive Office of the President by the National Environmental Policy Act of 1969 (NEPA) (42 U.S.C. 4321 et seq.). The Environmental Quality Improvement Act of 1970 (42 U.S.C. 4371 et seq.) established the Office of Environmental Quality (OEQ) to provide professional and administrative support for the Council. The Council and OEQ are collectively referred to as the Council on Environmental Quality, and the CEQ Chair, who is appointed by the President and confirmed by the Senate, serves as the Director of OEQ.

The Council develops policies which bring together the Nation's social, economic, and environmental priorities, with the goal of improving Federal decisionmaking. As required by NEPA, CEQ evaluates, coordinates, and mediates Federal activities. It advises and assists the President on both national and international environmental policy matters. CEQ also oversees Federal agency and department implementation of NEPA.

For further information, contact the Information Office, Council on Environmental Quality, 722 Jackson Place NW., Washington, DC 20503. Phone, 202–395–5750. Fax, 202–456–2710. Internet, http://www. whitehouse.gov/administration/eop/ceq.

National Security Council

Eisenhower Executive Office Building, Washington, DC 20504
Phone, 202–456–1414. Internet, http://www.whitehouse.gov/nsc.

Members

The President	BARACK OBAMA
The Vice President	JOSEPH R. BIDEN, JR.
The Secretary of State	HILLARY RODHAM CLINTON
The Secretary of Defense	ROBERT M. GATES

Statutory Advisers
Director of National Intelligence
Chairman, Joint Chiefs of Staff

JAMES R. CLAPPER, JR.
ADM. MICHAEL G. MULLEN, USN

Standing Participants
The Secretary of the Treasury
Chief of Staff to the President
Counsel to the President
National Security Advisor
Assistant to the President for Economic Policy

TIMOTHY F. GEITHNER
PETER M. ROUSE, *Acting*
ROBERT F. BAUER
THOMAS E. DONILON
LAWRENCE H. SUMMERS

Officials
Assistant to the President for National Security
    Affairs
Assistant to the President for National Security
    Affairs and Deputy National Security Adviser

THOMAS E. DONILON

DENIS R. MCDONOUGH

The National Security Council was established by the National Security Act of 1947, as amended (50 U.S.C. 402). The Council was placed in the Executive Office of the President by Reorganization Plan No. 4 of 1949 (5 U.S.C. app.).

The National Security Council is chaired by the President. Its statutory members, in addition to the President, are the Vice President and the Secretaries of State and Defense. The Chairman of the Joint Chiefs of Staff is the statutory military adviser to the Council, and the Director of National Intelligence is its intelligence adviser. The Secretary of the Treasury, the U.S. Representative to the United

Nations, the Assistant to the President for National Security Affairs, the Assistant to the President for Economic Policy, and the Chief of Staff to the President are invited to all meetings of the Council. The Attorney General and the Director of National Drug Control Policy are invited to attend meetings pertaining to their jurisdictions; other officials are invited, as appropriate.

The Council advises and assists the President in integrating all aspects of national security policy as it affects the United States—domestic, foreign, military, intelligence, and economic—in conjunction with the National Economic Council.

For further information, contact the National Security Council, Eisenhower Executive Office Building, Washington, DC 20504. Phone, 202–456–1414. Internet, http://www.whitehouse.gov/nsc.

Office of Administration

Eisenhower Executive Office Building, 1650 Pennsylvania Avenue, NW., Washington, DC 20503 Phone, 202–456–2861. Internet, http://www.whitehouse.gov/oa.

Director
Deputy Director
Chief Financial Officer
Chief Information Officer
Chief Procurement and Contract Management
    Officer
Director for Equal Employment Opportunity
Director for Office of Security and Emergency
    Preparedness
General Counsel

CAMERON MOODY
(VACANCY)
ALLYSON LAACKMAN
BROOK COLANGELO
ALTHEA KIREILIS

CLARA PATTERSON
JOHN GILL

DENISE MAES

The Office of Administration was formally established within the Executive Office of the President by Executive

Order 12028 of December 12, 1977. The Office provides administrative support services to all units within the Executive

Office of the President. The services provided include information, personnel, technology, and financial management; data processing; library and research services; security; legislative liaisons; and general office operations such as mail, messenger, printing, procurement, and supply services.

For further information, contact the Office of the Director, Office of Administration, Washington, DC 20503. Phone, 202–456–2861. Internet, http://www.whitehouse.gov/oa.

Office of Management and Budget

New Executive Office Building, Washington, DC 20503
Phone, 202–395–3080. Internet, http://www.whitehouse.gov/omb.

Director	JACOB J. LEW
Deputy Director	ROB NABORS, *Acting*
Deputy Director for Management	JEFFREY D. ZIENTS
Chief of Staff	MELISSA GREEN
Administrator, Office of Federal Procurement Policy	DANIEL GORDON
Administrator, Office of Information and Regulatory Affairs	CASS R. SUNSTEIN
Assistant Director for Management and Operations	LAUREN E. WRIGHT
Assistant Director for Budget	COURTNEY TIMBERLAKE
Assistant Director for Legislative Reference	JAMES J. JUKES
Associate Director for Communications	KENNETH BAER
Associate Director for Economic Policy	ALEXANDRE MAS
Associate Director for Education, Income Maintenance and Labor	ROBERT GORDON
Associate Director for General Government Programs	XAVIER BRIGGS
Associate Director for Information Technology and E–Government	VIVEK KUNDRA
Associate Director for Legislative Affairs	KATHERINE ELTRICH
Associate Director for National Security Programs	STEVE KOSIAK
Associate Director for Natural Resource Programs	SALLY ERICSSON
Controller, Office of Federal Financial Management	DANIEL I. WERFEL
General Counsel	PREETA D. BANSAL
Associate Director for Health	KEITH FONTENOT
Executive Associate Director	JEFFREY LIEBMAN

The Office of Management and Budget evaluates, formulates, and coordinates management procedures and program objectives within and among Federal departments and agencies. It also controls the administration of the Federal budget, while routinely providing the President with recommendations regarding budget proposals and relevant legislative enactments.

The Office of Management and Budget (OMB), formerly the Bureau of the Budget, was established in the Executive Office of the President pursuant to Reorganization Plan No. 1 of 1939 (5 U.S.C. app.).

The Office's primary functions are: to assist the President in developing and maintaining effective government by

OFFICE OF MANAGEMENT AND BUDGET

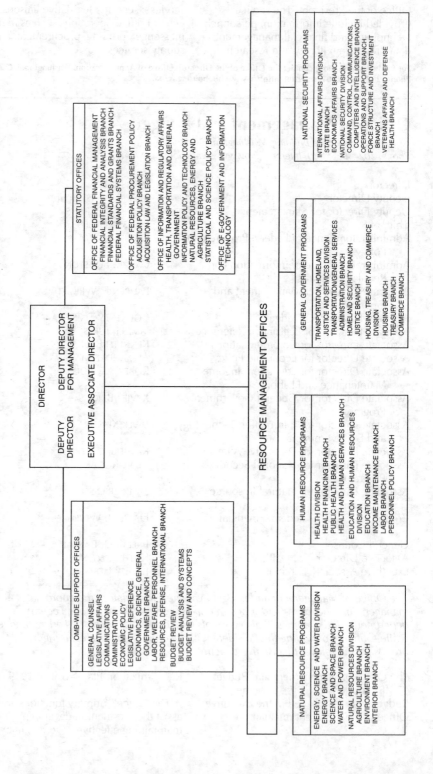

DIRECTOR

DEPUTY DIRECTOR
FOR MANAGEMENT

DEPUTY
DIRECTOR

EXECUTIVE ASSOCIATE DIRECTOR

STATUTORY OFFICES

OFFICE OF FEDERAL FINANCIAL MANAGEMENT
FINANCIAL INTEGRITY AND ANALYSIS BRANCH
FINANCIAL STANDARDS AND GRANTS BRANCH
FEDERAL FINANCIAL SYSTEMS BRANCH

OFFICE OF FEDERAL PROCUREMENT POLICY
ACQUISITION POLICY BRANCH
ACQUISITION LAW AND LEGISLATION BRANCH

OFFICE OF INFORMATION AND REGULATORY AFFAIRS
HEALTH, TRANSPORTATION AND GENERAL
GOVERNMENT
INFORMATION POLICY AND TECHNOLOGY BRANCH
NATURAL RESOURCES, ENERGY AND
AGRICULTURE BRANCH
STATISTICAL AND SCIENCE POLICY BRANCH

OFFICE OF E-GOVERNMENT AND INFORMATION
TECHNOLOGY

OMB-WIDE SUPPORT OFFICES

GENERAL COUNSEL
LEGISLATIVE AFFAIRS
COMMUNICATIONS
ADMINISTRATION
ECONOMIC POLICY
LEGISLATIVE REFERENCE
ECONOMICS, SCIENCE, GENERAL
GOVERNMENT BRANCH
LABOR, WELFARE, PERSONNEL BRANCH
RESOURCES, DEFENSE, INTERNATIONAL BRANCH
BUDGET REVIEW
BUDGET ANALYSIS AND SYSTEMS
BUDGET REVIEW AND CONCEPTS

RESOURCE MANAGEMENT OFFICES

NATIONAL SECURITY PROGRAMS

INTERNATIONAL AFFAIRS DIVISION
STATE BRANCH
ECONOMICS AFFAIRS BRANCH
NATIONAL SECURITY DIVISION
COMMAND, CONTROL, COMMUNICATIONS,
COMPUTERS AND INTELLIGENCE BRANCH
OPERATIONS AND SUPPORT BRANCH
FORCE STRUCTURE AND INVESTMENT
BRANCH
VETERANS AFFAIRS AND DEFENSE
HEALTH BRANCH

GENERAL GOVERNMENT PROGRAMS

TRANSPORTATION, HOMELAND,
JUSTICE AND SERVICES DIVISION
TRANSPORTATION/GENERAL SERVICES
ADMINISTRATION BRANCH
HOMELAND SECURITY BRANCH
JUSTICE BRANCH
HOUSING, TREASURY AND COMMERCE
DIVISION
HOUSING BRANCH
TREASURY BRANCH
COMMERCE BRANCH

HUMAN RESOURCE PROGRAMS

HEALTH DIVISION
HEALTH FINANCING BRANCH
PUBLIC HEALTH BRANCH
HEALTH AND HUMAN SERVICES BRANCH
EDUCATION AND HUMAN RESOURCES
DIVISION
EDUCATION BRANCH
INCOME MAINTENANCE BRANCH
LABOR BRANCH
PERSONNEL POLICY BRANCH

NATURAL RESOURCE PROGRAMS

ENERGY, SCIENCE AND WATER DIVISION
ENERGY BRANCH
SCIENCE AND SPACE BRANCH
WATER AND POWER BRANCH
NATURAL RESOURCES DIVISION
AGRICULTURE BRANCH
ENVIRONMENT BRANCH
INTERIOR BRANCH

reviewing the organizational structure and management procedures of the executive branch to ensure that the intended results are achieved; to assist in developing efficient coordinating mechanisms to implement Government activities and to expand interagency cooperation; to assist the President in preparing the budget and in formulating the Government's fiscal program; to supervise and control the administration of the budget; to assist the President by clearing and coordinating departmental advice on proposed legislation and by making recommendations effecting Presidential action on legislative enactments, in accordance with past practice; to assist in developing regulatory reform proposals and programs for paperwork reduction, especially reporting burdens of the public; to assist in considering, clearing, and, where necessary, preparing proposed Executive orders and proclamations; to plan and develop information systems that provide the President with program performance data; to plan, conduct, and promote evaluation efforts that assist the President in assessing program objectives, performance, and efficiency; to keep the President informed of the progress of activities by Government agencies with respect to work proposed, initiated, and completed, together with the relative timing of work between the several agencies of the Government, all

to the end that the work programs of the several agencies of the executive branch of the Government may be coordinated and that the moneys appropriated by the Congress may be expended in the most economical manner, barring overlapping and duplication of effort; and to improve the economy, efficiency, and effectiveness of the procurement processes by providing overall direction of procurement policies, regulations, procedures, and forms.

Sources of Information

Employment   Delegated examining is used for filling positions, such as economist, program examiners, and program analyst. Inquiries on employment should be directed to the Human Resources Division, Office of Administration, Washington, DC 20500. Phone, 202–395–1088.

Inquiries   Contact the Management and Operations Division, Office of Management and Budget, New Executive Office Building, Washington, DC 20503. Phone, 202–395–3080. Fax, 202–395–3504. Internet, www.whitehouse.gov/omb.

Publications   The Budget of the U.S. Government and The Budget System and Concepts are available for sale by the Superintendent of Documents, Government Printing Office, Washington, DC 20402.

For further information, contact the Office of Management and Budget, New Executive Office Building, Washington, DC 20503. Phone, 202–395–3080. Internet, http://www.whitehouse.gov/omb.

EDITORIAL NOTE: The Office of National Drug Control Policy did not meet the publication deadline for submitting updated information of its activities, functions, and sources of information as required by the automatic disclosure provisions of the Freedom of Information Act (5 U.S.C. 552(a)(1)(A)).

Office of National Drug Control Policy

Executive Office of the President, Washington, DC 20503
Phone, 202–395–6700. Fax, 202–395–6708. Internet, http://www.ondcp.gov.

Director	R. GIL KERLIKOWSKE
Deputy Director	A. THOMAS MCLELLAN
Deputy Director for Demand Reduction	DAVID K. MINETA
Deputy Director for State, Local, and Tribal Affairs	BENJAMIN B. TUCKER
Deputy Director for Supply Reduction	PATRICK M. WARD, *Acting*
Associate Director, Performance and Budget	JON E. RICE
Associate Director, Legislative Affairs	DEBORAH J. WALKER, *Acting*
Associate Director, Management and Administration	MICHELE C. MARX
Associate Director, Public Affairs	RAFAEL LEMAITRE, *Acting*
General Counsel	DANIEL R. PETERSEN, *Acting*
Chief Scientist, Counter-Drug Technology Assessment Center	DAVID W. MURRAY
Associate Director, National Youth Anti-Drug Media Campaign	ROBERT W. DENNISTON

The Office of National Drug Control Policy assists the President in establishing policies, priorities, and objectives in the National Drug Control Strategy. It also provides budget, program, and policy recommendations on the efforts of National Drug Control Program agencies.

The Office of National Drug Control Policy was established by the National Narcotics Leadership Act of 1988 (21 U.S.C. 1501 et seq.), effective January 29, 1989, reauthorized through the Office of National Drug Control Policy Reauthorization Act of 1988 (21 U.S.C. 1701 et seq.), and again reauthorized through the Office of National Drug Control Policy Reauthorization Act of 2006 (21 U.S.C. 1701 et seq.).

The Director of National Drug Control Policy is appointed by the President with the advice and consent of the Senate. The Director is assisted by a Deputy Director, a Deputy Director for Demand Reduction, a Deputy Director for Supply Reduction, and a Deputy Director for State, Local, and Tribal Affairs.

The Director is responsible for establishing policies, objectives, priorities, and performance measurements for the National Drug Control Program, as well as for annually promulgating drug control strategies and supporting reports and a program budget, which the President submits to Congress. The Director advises the President regarding necessary changes in the organization, management, budgeting, and personnel allocation of Federal agencies enforcing drug activities. The Director also notifies Federal agencies if their policies do not comply with their responsibilities under the National Drug Control Strategy. Additionally, the Office has direct programmatic responsibility for the Drug-Free Communities Program, the National Youth Anti-Drug Media Campaign, the various programs under the Counter-Drug Technology Assessment Center, and the High Intensity Drug Trafficking Areas Program.

Sources of Information

Employment   Inquiries regarding employment should be directed to the Personnel Section, Office of National Drug Control Policy. Phone, 202–395–6695.

Publications   To receive publications on drugs and crime, access specific drug-related data, obtain customized bibliographic searches, and learn more about data availability and other information resources, contact the Drugs and Crime Clearinghouse. Phone, 800–666–3332. Fax, 301–519–5212. Internet, www.ondcp.gov.

For further information, contact the Office of National Drug Control Policy, Executive Office of the President, Washington, DC 20503. Phone, 202–395–6700. Fax, 202–395–6708. Internet, http://www.ondcp.gov.

Office of Policy Development

The Office of Policy Development is comprised of the Domestic Policy Council and the National Economic Council, which are responsible for advising and assisting the President in the formulation, coordination, and implementation of domestic and economic policy. The Office of Policy Development also provides support for other policy development and implementation activities as directed by the President.

Domestic Policy Council

Room 469, Eisenhower Executive Office Building, Washington, DC 20502
Phone, 202–456–5594. Internet, http://www.whitehouse.gov/dpc.

Assistant to the President and Director of the Domestic Policy Council	MELODY C. BARNES
Deputy Assistant to the President and Deputy Director of the Domestic Policy Council	HEATHER A. HIGGINBOTTOM
Deputy Assistant to the President and Director, Office of Social Innovation and Civic Participation	SONAL R. SHAH

The Domestic Policy Council was established August 16, 1993, by Executive Order 12859. The Council oversees development and implementation of the President's domestic policy agenda and ensures coordination and communication among the heads of relevant Federal offices and agencies.

National Economic Council

Room 235, Eisenhower Executive Office Building, Washington, DC 20502
Phone, 202–456–2800. Internet, http://www.whitehouse.gov/nec.

Assistant to the President for Economic Policy and Director of the National Economic Council	LAWRENCE H. SUMMERS
Deputy Assistant to the President for Economic Policys	DIANA FARRELL, JASON L. FURMAN

The National Economic Council was created January 25, 1993, by Executive Order 12835, to coordinate the economic policymaking process and provide economic policy advice to the President. The Council also ensures that economic policy decisions and programs are consistent with the President's stated goals, and monitors the implementation of the President's economic goals.

Office of Science and Technology Policy

New Executive Office Building, 725 17th Street NW., Washington, DC 20502
Phone, 202–456–7116. Fax, 202–456–6021. Internet, http://www.ostp.gov.

Director	JOHN P. HOLDREN
Chief of Staff	JIM KOHLENBERGER
Deputy Chief of Staff and Assistant Director	TED WACKLER
Deputy Director for Policy	TOM KALIL
Assistant Director At-Large	STEVE FETTER
General Counsel	RACHAEL LEONARD
Assistant Director, Federal Research and Development	KEI KOIZUMI
Assistant Director, Strategic Communicators / Senior Policy Analyst	RICK WEISS
Assistant Director, International Relations	JOAN ROLF
Assistant Director, Legislative Affairs	DONNA PIGNATELLI
Assistant Director, Social and Behavioral Science	DIANE DIEULIIS, *Acting*
Assistant Director, Physical Sciences and Engineering	STEPHEN MERKOWITZ
Assistant Director, Life Sciences	DIANE DIEULIIS
Assistant Director, Biotechnology	MIKE STEBBINS
Associate Director and Chief Technology Officer	ANEESH CHOPRA
Deputy Chief Technology Officer, Open Government	BETH NOVECK
Deputy Chief Technology Officer, Internet Policy	ANDREW MCLAUGHLIN
Deputy Chief Technology Officer, Telecommunications	SCOTT DEUTCHMAN
Assistant Director, Nanotechnology	TRAVIS EARLES
Assistant Director, Information Technology Research and Development	CHRIS GREER
Assistant Director, Space and Aeronautics	DAMON WELLS
Assistant Director, Energy Research and Development	KEVIN HURST
Assistant Director, National Security and Emergency Preparedness	MARK LEBLANC
Assistant Director, Defense Programs	ARUN SERAPHIN
Associate Director, Environment	SHERE ABBOTT
Assistant Director, Climate Adaptation and Assessment	KATHY JACOBS
Executive Director, President's Council of Advisors on Science and Technology	DEBORAH STINE
Co-Chair, President's Council of Advisors on Science and Technology	ERIC LANDER
Executive Director, National Science and Technology Council	CHRISTYL JOHNSON
Director, National Coordination Office for the National Nanotechnology Initiative	CLAYTON TEAGUE

Director, National Coordination Office for   GEORGE STRAWN
    Networking and Information Technology
    Research and Development

The Office of Science and Technology Policy was established within the Executive Office of the President by the National Science and Technology Policy, Organization, and Priorities Act of 1976 (42 U.S.C. 6611).

The Office serves as a source of scientific, engineering, and technological analysis and judgment for the President with respect to major policies, plans, and programs of the Federal Government. In carrying out this mission, the Office advises the President of scientific and technological considerations involved in areas of national concern, including the economy, national security, health, foreign relations, and the environment; evaluates the scale, quality, and effectiveness of the Federal effort in science and technology; provides advice and assistance to the President, the Office of Management and Budget, and Federal agencies throughout the Federal budget development process; and assists the President in providing leadership and coordination for the research and development programs of the Federal Government.

For further information, contact the Office of Science and Technology Policy, New Executive Office Building, 725 17th Street NW., Washington, DC 20502. Phone, 202–456–7116. Fax, 202–456–6021. Internet, http:// www.ostp.gov.

EDITORIAL NOTE: The Office of the United States Trade Representative did not meet the publication deadline for submitting updated information of its activities, functions, and sources of information as required by the automatic disclosure provisions of the Freedom of Information Act (5 U.S.C. 552(a)(1)(A)).

Office of the United States Trade Representative

600 Seventeenth Street NW., Washington, DC 20508
Phone, 202–395–3230. Internet, http://www.ustr.gov.

United States Trade Representative	RONALD KIRK
Deputy U.S. Trade Representative (Washington)	DEMETRIOS MARANTIS
Deputy U.S. Trade Representative (Geneva)	MICHAEL PUNKE
Chief of Staff	LISA GARCIA
General Counsel	TIMOTHY REIF
Senior Counsel	PETER COWHEY
Assistant U.S. Trade Representative for Administration	FRED AMES
Assistant U.S. Trade Representative for Agricultural Affairs	ISLAM SIDDIQUI
Assistant U.S. Trade Representative for Southeast Asia and the Pacific	BARBARA WEISEL
Assistant U.S. Trade Representative for Congressional Affairs	LUIS JIMENEZ
Assistant U.S. Trade Representative for Economic Affairs	WILLIAM SHPIECE, *Acting*
Assistant U.S. Trade Representative for Environment and Natural Resources	MARK LINSCOTT
Assistant U.S. Trade Representative for Europe and the Middle East	DANIEL MULLANEY
Assistant U.S. Trade Representative for Small Business, Market Access & Industrial Competitiveness	JAMES SANFORD
Assistant U.S. Trade Representative for Intergovernmental Affairs and Public Engagement	MYESHA WARD
Assistant U.S. Trade Representative for Japan, Korea, and Asia Pacific Economic Cooperation Affairs	WENDY CUTLER
Assistant U.S. Trade Representative for China Affairs	CLAIRE READE
Assistant U.S. Trade Representative for South Asia	MICHAEL J. DELANEY
Assistant U.S. Trade Representative for Monitoring and Enforcement	STEVEN FABRY
Assistant U.S. Trade Representative for Africa	FLORIZELLE LISER
Assistant U.S. Trade Representative for Policy Coordination	CARMEN SURO-BREDIE
Assistant U.S. Trade Representative for Services and Investment	CHRISTINE BLISS
Assistant U.S. Trade Representative for Intellectual Property and Innovation	STANFORD MCCOY
Assistant U.S. Trade Representative for Labor	LEWIS KARESH

Assistant U.S. Trade Representative for World Trade Organization (WTO) and Multilateral Affairs	CHRISTOPHER S. WILSON
Assistant U.S. Trade Representative for the Americas	EVERETT EISSENSTAT
Assistant U.S. Trade Representative for Public and Media Affairs	CAROL GUTHRIE
Assistant U.S. Trade Representative for Trade & Development	MARY E. RYCKMAN
Deputy Chief of Mission (Geneva)	DAVID SHARK

The United States Trade Representative is responsible for directing all trade negotiations of and formulating trade policy for the United States.

The Office of the United States Trade Representative was created as the Office of the Special Representative for Trade Negotiations by Executive Order 11075 of January 15, 1963. The Trade Act of 1974 (19 U.S.C. 2171) established the Office as an agency of the Executive Office of the President charged with administering the trade agreements program.

The Office is responsible for setting and administering overall trade policy. It also provides that the United States Trade Representative shall be chief representative of the United States for the following: all activities concerning the General Agreement on Tariffs and Trade; discussions, meetings, and negotiations in the Organization for Economic Cooperation and Development when such activities deal primarily with trade and commodity issues; negotiations in the U.N. Conference on Trade and Development and other multilateral institutions when such negotiations deal primarily with trade and commodity issues; other bilateral and multilateral negotiations when trade, including East-West trade, or commodities is the primary issue; negotiations under sections 704 and 734 of the Tariff Act of 1930 (19 U.S.C. 1671c and 1673c); and negotiations concerning direct investment incentives and disincentives and bilateral investment issues concerning barriers to investment.

The Omnibus Trade and Competitiveness Act of 1988 codified these prior authorities and added additional authority, including the implementation of section 301 actions (regarding enforcement of U.S. rights under international trade agreements).

The Office is headed by the United States Trade Representative, a Cabinet-level official with the rank of Ambassador, who is directly responsible to the President. There are three Deputy United States Trade Representatives, who also hold the rank of Ambassador, two located in Washington and one in Geneva. The Chief Agricultural Negotiator also holds the rank of Ambassador.

The United States Trade Representative serves as an ex officio member of the Boards of Directors of the Export-Import Bank and the Overseas Private Investment Corporation and serves on the National Advisory Council for International Monetary and Financial Policy.

For further information, contact the Office of Public Affairs, Office of the United States Trade Representative, 600 Seventeenth Street NW., Washington, DC 20506. Phone, 202–395–3230. Internet, http://www.ustr.gov.

OFFICE OF THE UNITED STATES TRADE REPRESENTATIVE

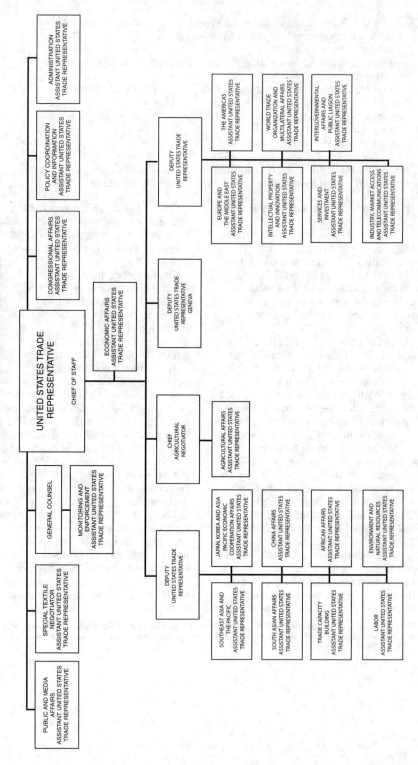

EXECUTIVE BRANCH: DEPARTMENTS

DEPARTMENT OF AGRICULTURE

1400 Independence Avenue SW., Washington, DC 20250
Phone, 202–720–4623. Internet, http://www.usda.gov.

Secretary of Agriculture	THOMAS J. VILSACK
Deputy Secretary	KATHLEEN MERRIGAN
Director, Office of Communications	CHRIS MATHER
Inspector General	PHYLLIS K. FONG
General Counsel	STEVE SILVERMAN, *Acting*
Assistant Secretary for Congressional Relations	KRYSTA HARDEN
Assistant Secretary for Administration	PEARLIE REED
Assistant Secretary for Civil Rights	JOE LEONARD
Chief Information Officer	CHRIS SMITH
Chief Financial Officer	JON HOLLADAY, *Acting*
Chief Economist	JOSEPH GLAUBER
Under Secretary for Natural Resources and Environment	HARRIS SHERMAN
Chief, Forest Service	THOMAS TIDWELL
Chief, Natural Resources Conservation Service	DAVID WHITE
Under Secretary for Farm and Foreign Agricultural Services	JAMES MILLER
Administrator, Farm Service Agency	JONATHAN COPPESS
Administrator, Foreign Agricultural Service	JOHN BREWER
Administrator, Risk Management Agency	WILLIAM MURPHY
Under Secretary for Rural Development	DALLAS TONSAGER
Administrator, Rural Business-Cooperative Service	JUDY CANALES
Administrator, Rural Housing Service	TAMMYE TREVINO
Administrator, Rural Utilities Service	JONATHAN ADELSTEIN
Under Secretary for Food, Nutrition, and Consumer Services	KEVIN CONCANNON
Administrator, Food and Nutrition Service	JULIE PARADIS
Director, Center for Nutrition Policy and Promotion	RAJ ANAND
Under Secretary for Food Safety	JERALD MANDE, *Acting*
Administrator, Food Safety and Inspection Service	ALFRED V. ALMANZA
Under Secretary for Research, Education, and Economics	MOLLY JAHN
Administrator, Agricultural Research Service	EDWARD B. KNIPLING
Director, National Institute of Food and Agriculture	ROGER N. BEACHY

Administrator, Economic Research Service	KATHERINE SMITH
Director, National Agricultural Library	SIMON Y. LIN
Administrator, National Agricultural Statistics Service	CYNTHIA CLARK
Under Secretary for Marketing and Regulatory Programs	EDWARD M. AVALOS
Administrator, Agricultural Marketing Service	RAYNE PEGG
Administrator, Animal and Plant Health Inspection Service	CINDY J. SMITH
Administrator, Grain Inspection, Packers, and Stockyards Administration	J. DUDLEY BUTLER
Chief Judge, Administrative Law Judges	PETER DAVENPORT

[For the Department of Agriculture statement of organization, see the Code of Federal Regulations, Title 7, Part 2]

The Department of Agriculture provides leadership on food, agricultural, and environmental issues by developing agricultural markets, fighting hunger and malnutrition, conserving natural resources, and ensuring standards of food quality through safeguards and inspections.

The Department of Agriculture (USDA) was created by act of May 15, 1862 (7 U.S.C. 2201).

In carrying out its work in the program mission areas, USDA relies on the support of departmental administration staff, as well as the Office of the Chief Financial Officer, Office of the Chief Information Officer, Office of Communications, Office of Congressional and Intergovernmental Relations, Office of the Inspector General, and the Office of the General Counsel.

Rural Development

USDA's rural development mission is to increase the economic opportunities of rural Americans and improve their quality of life. To accomplish this, USDA works to foster new cooperative relationships among Government, industry, and communities. As a capital investment bank, USDA provides financing for rural housing and community facilities, business and cooperative development, telephone and high-speed Internet access, electric, water, and sewer infrastructure. Approximately 800 rural development field offices, staffed by 7,000 employees, provide frontline delivery of rural development loan and grant programs at the local level.

Rural Business-Cooperative Programs To meet business credit needs in underserved areas, USDA rural development business programs are usually leveraged with commercial, cooperative, or other private sector lenders. USDA's rural development business programs are listed below.

Business and Industry Guaranteed Loans This program generates jobs and stimulates rural economies by providing financial backing for rural businesses. Loan proceeds may be used for working capital, machinery and equipment, buildings and real estate, and certain types of debt refinancing.

Business Enterprise These grants help public bodies, nonprofit corporations, and federally recognized Indian tribal groups finance and facilitate development of small and emerging private business enterprises located in rural areas. Grant funds can pay for the acquisition and development of land and the construction of buildings, plants, equipment, access streets and roads, parking areas, utility and service extensions, refinancing, and fees for professional services, as well as technical assistance and related training, startup costs and working capital, financial assistance to a third party, production of television programs targeted to rural residents, and rural distance learning networks.

Business Opportunities This program promotes sustainable economic

DEPARTMENT OF AGRICULTURE

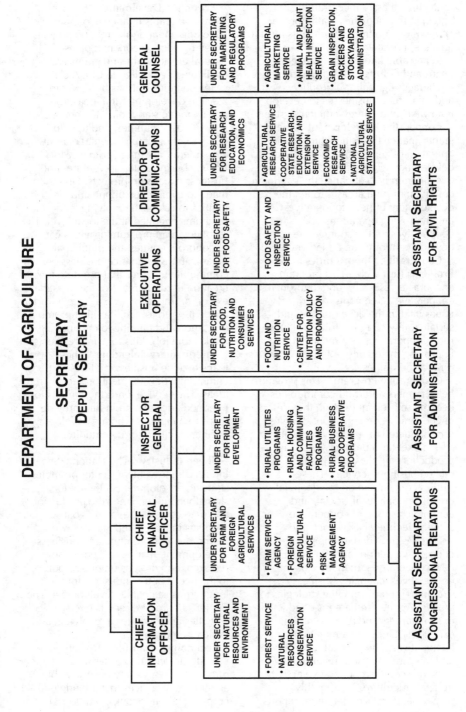

SECRETARY
DEPUTY SECRETARY

GENERAL COUNSEL

DIRECTOR OF COMMUNICATIONS

EXECUTIVE OPERATIONS

INSPECTOR GENERAL

CHIEF FINANCIAL OFFICER

CHIEF INFORMATION OFFICER

UNDER SECRETARY FOR MARKETING AND REGULATORY PROGRAMS
• AGRICULTURAL MARKETING SERVICE
• ANIMAL AND PLANT HEALTH INSPECTION SERVICE
• GRAIN INSPECTION, PACKERS AND STOCKYARDS ADMINISTRATION

UNDER SECRETARY FOR RESEARCH EDUCATION, AND ECONOMICS
• AGRICULTURAL RESEARCH SERVICE
• COOPERATIVE STATE RESEARCH, EDUCATION, AND EXTENSION SERVICE
• ECONOMIC RESEARCH SERVICE
• NATIONAL AGRICULTURAL STATISTICS SERVICE

UNDER SECRETARY FOR FOOD SAFETY
• FOOD SAFETY AND INSPECTION SERVICE

UNDER SECRETARY FOR FOOD, NUTRITION AND CONSUMER SERVICES
• FOOD AND NUTRITION SERVICE
• CENTER FOR NUTRITION POLICY AND PROMOTION

UNDER SECRETARY FOR RURAL DEVELOPMENT
• RURAL UTILITIES PROGRAMS
• RURAL HOUSING AND COMMUNITY FACILITIES PROGRAMS
• RURAL BUSINESS AND COOPERATIVE PROGRAMS

UNDER SECRETARY FOR FARM AND FOREIGN AGRICULTURAL SERVICES
• FARM SERVICE AGENCY
• FOREIGN AGRICULTURAL SERVICE
• RISK MANAGEMENT AGENCY

UNDER SECRETARY FOR NATURAL RESOURCES AND ENVIRONMENT
• FOREST SERVICE
• NATURAL RESOURCES CONSERVATION SERVICE

ASSISTANT SECRETARY FOR CIVIL RIGHTS

ASSISTANT SECRETARY FOR ADMINISTRATION

ASSISTANT SECRETARY FOR CONGRESSIONAL RELATIONS

development in rural communities with exceptional needs. Funds are provided for technical assistance, training, and planning activities that improve economic conditions. Applicants must be located in rural areas.

Renewable Energy and Efficiency Loans and Grants This program encourages agricultural producers and small rural businesses to develop renewable and energy-efficient systems.

Cooperative Development These grants finance the establishment and operation of centers for cooperative development. The primary purpose of this program is to enhance the economic condition of rural areas through the development of new cooperatives and improving operations of existing cooperatives and encourage the development of value-added ventures.

Cooperative Opportunities and Problems Research This program encourages research, funded through cooperative agreements, on critical issues vital to the development and sustainability of agricultural and other rural cooperatives as a means of improving the quality of life in America's rural communities.

Cooperative Program This program helps farmers and rural communities become self-reliant through the use of cooperative organizations. Studies are conducted to support cooperatives that market farm products, purchase production supplies, and perform related business services. These studies concentrate on the financial, organizational, legal, social, and economic aspects of cooperative activity. Technical assistance and research is provided to improve cooperative performance in organizing new cooperatives, merging existing cooperatives, changing business structures, and developing strategies for growth. Applied research is conducted to give farmers and rural communities expert assistance pertaining to their cooperatives. The program also collects and publishes statistics regarding the role and scope of cooperative activity in U.S. agriculture. A bimonthly magazine, Rural Cooperatives, reports

current developments and research for cooperative management leadership.

Economic Development These loans and grants finance economic development and job creation projects based on sound economic plans in rural areas. Loans and grants are available to any eligible USDA electric or telecommunications borrower to assist in developing rural areas from an economic standpoint, to generate new job opportunities, and to help retain existing employment. Loans at zero interest are made primarily to finance business startup ventures and business expansion projects. Grants are made to eligible telephone and electric utilities to establish revolving loan programs operated at the local level. The revolving loan program provides capital to nonprofit entities and municipal organizations to finance business or community facilities that promote job creation in rural areas, for facilities that extend or improve medical care to rural residents, and for facilities that promote education and training to enhance marketable job skills for rural residents.

Intermediary Relending These loans finance business facilities and community development projects in rural areas. The Service lends these funds to intermediaries, which in turn provide loans to recipients who are developing business facilities or community development projects

Sheep Industry The National Sheep Industry Improvement Center promotes strategic development activities to strengthen and enhance the production and marketing of sheep and goat products in the United States. It works to improve infrastructure and business development, market and environmental research, and designs unique responses to the needs of the industries for their long-term sustainable development. The Center's board of directors oversees its activities and operates a revolving fund for loans and grants.

Technology Transfer This program provides information to farmers and other rural users on a variety of sustainable agricultural practices that include both cropping and livestock operations. It offers reliable, practical information on production techniques and practices that reduce costs and that are environmentally friendly. Farmers can request such information by telephone at 800–346–9140.

Rural Housing Programs Rural Development provides affordable rental housing, homeownership opportunities, and essential community facilities to rural Americans through a broad array of direct loan, guarantee, and grant programs. Rural residents and communities may inquire about any of these programs through local and State rural development offices. It also provides financial and management assistance through the following services: guaranteed single-family housing (SFH) loans which guarantee loans made by commercial lenders to moderate-income rural residents with sufficient income and acceptable credit, who may lack the downpayment to secure a loan without assistance; direct SFH loans made available to people with incomes less than 80 percent of area median, to build, purchase, and repair rural homes; home improvement and repair loans and grants for owner-occupants to remove health and safety hazards from a home; mutual self-help housing technical assistance grants for nonprofit organizations and public bodies to help groups of six to eight lower income families to build their own homes by providing "sweat equity," which reduces the families' mortgages; rural housing site loans for private or public nonprofit organizations to purchase sites for the development of housing for lower income families; direct and guaranteed multifamily housing loans for private nonprofit corporations, consumer cooperatives, State or local public agencies, and individuals or organizations operating on a nonprofit or limited profit basis to provide rental or cooperative housing in rural areas for persons of very low, low, and moderate income; farm labor housing loans and grants enabling farmers, public or private nonprofit organizations, or units of local government to build, buy, or rehabilitate farm labor housing; housing preservation grants made to a public body or public or

private nonprofit organization to provide assistance to homeowners and landlords to repair and rehabilitate housing for lower income families in rural areas; housing for the homeless, SFH real estate-owned (REO) property to nonprofit organizations or public bodies for transitional housing for the homeless and to the Federal Emergency Management Agency to house families affected by natural disasters; and community program loans, direct and guaranteed loans, and grants for public and quasi-public bodies, nonprofit associations, and Indian tribes for essential community facilities such as health care centers, public safety buildings and vehicles, and child care centers.

Rural Utilities Programs Rural Development helps finance rural electric and telecommunications utilities in obtaining financing and administers a nationwide water and waste loan and grant program to improve the quality of life and promote economic development in rural America. A total of 890 rural electric and 800 rural telecommunications utilities in 47 States, Puerto Rico, the Virgin Islands, Guam, the Republic of the Marshall Islands, the Northern Mariana Islands, and the Federated States of Micronesia have received financial assistance. It also provides assistance through the programs that are mentioned here. The electric program provides loans for improving electric service to persons in rural areas, including construction of electric generating plants and transmission and distribution lines to provide reliable electric service. The telecommunications program provides loans and grants to improve telecommunications service and high-speed Internet access in rural areas. The water and waste direct and guaranteed loan program provides assistance to develop water and wastewater systems, including solid waste disposal and storm drainage in rural areas, cities, and towns. Water and waste disposal grants assist in reducing water and waste disposal costs to a reasonable level for users of the system. Emergency community water assistance grants provide assistance to rural communities experiencing a significant decline in quantity or quality of drinking water. Technical assistance and training grants are available to nonprofit organizations to provide rural water and waste system officials with technical assistance and training on a wide range of issues relating to the delivery of water and waste service to rural residents. Solid waste management grants are available for nonprofit organizations and public bodies to provide technical assistance and training to rural areas and towns to reduce or eliminate pollution of water resources and improve planning and management of solid waste facilities. The rural water circuit rider technical assistance program provides technical assistance to rural water systems to solve operational, financial, and management challenges. The distance learning and telemedicine program provides financing to help rural schools and health care providers purchase or improve telecommunications facilities and equipment to bring educational and medical resources to rural areas that otherwise might be unavailable. Rural Development also guarantees loans from the Department of the Treasury's Federal Financing Bank (FFB), which lends to borrowers, primarily for large-scale electric and telecommunication facilities. It may also guarantee electric and telecommunications loans from private sources.

For further information, contact the Rural Development Legislative and Public Affairs Staff, Department of Agriculture, Stop 0705, 1400 Independence Avenue SW., Washington, DC 20250–0320. Phone, 202–720–4323.

Marketing and Regulatory Programs

This mission area includes marketing and regulatory programs other than those concerned with food safety.

Agricultural Marketing Service   The Agricultural Marketing Service (AMS) was established by the Secretary of Agriculture on April 2, 1972, under the authority of Reorganization Plan No. 2 of 1953 (5 U.S.C. app.) and other authorities. The Service administers standardization, grading, certification,

market news, marketing orders, research and promotion, and regulatory programs.

Market News   The Service provides current, unbiased information to producers, processors, distributors, and others to assist them in the orderly marketing and distribution of farm commodities. Information is collected on supplies, shipments, prices, location, quality, condition, and other market data on farm products in specific markets and marketing areas. The data is disseminated nationally via the Internet and other electronic means and is shared with several countries. The Service also assists other countries in developing their own marketing information systems.

Standardization, Grading, and Classing   Nearly 600 grade standards have been established for some 230 agricultural commodities to help buyers and sellers trade on agreed-upon quality levels. Standards are developed with the benefit of views from those in the industries directly affected and others interested. The Service also participates in developing international commodity standards to facilitate trade.

Grading and classing services are provided to certify the grade and quality of products. These grading services are provided to buyers and sellers of live cattle, swine, sheep, meat, poultry, eggs, rabbits, fruits, vegetables, tree nuts, peanuts, dairy products, and tobacco. Classing services are provided to buyers and sellers of cotton and cotton products. These services are mainly voluntary and are provided upon request and for a fee. The Service is also responsible for testing seed.

Laboratory Testing   The Service provides microbiological, chemical, and other scientific laboratory support to its commodity and food procurement programs, testing peanuts for aflatoxin and seeds for germination and purity. The Agency also carries out quality assurance and safety oversight activities for its milk market laboratories, resident grading programs, and State and private laboratory programs.

The Service also administers the Pesticide Data Program, which, in cooperation with States, samples and analyzes 33 agricultural commodities for pesticides residue. It shares residue test results with the Environmental Protection Agency and other public agencies.

Food Quality Assurance Under a Governmentwide quality assurance program, AMS is responsible for the development and revision of specifications used by Federal agencies in procuring food for military and civilian uses. The Service coordinates and approves certification programs designed to ensure that purchased products conform to the specification requirements.

Regulatory Programs The Service administers several regulatory programs designed collectively to protect producers, handlers, and consumers of agricultural commodities from financial loss or personal injury resulting from careless, deceptive, or fraudulent marketing practices. Such regulatory programs encourage fair trading practices in the marketing of fruits and vegetables, require truth in seed labeling and in advertising. The Service provides voluntary laboratory analyses of egg products and monitors the disposition of restricted shell eggs, a potential health hazard.

Marketing Agreements and Orders The Service administers marketing agreements and orders to establish and maintain orderly marketing conditions for certain commodities. Milk marketing orders establish minimum prices that handlers or distributors are required to pay producers. Programs for fruits, vegetables, and related specialty crops like nuts and spearmint oil promote product quality control and help stabilize supplies and market prices. In some cases, they also authorize research and market development activities, including advertising supported by assessments that handlers pay. Through the orderly marketing of commodities facilitated by these programs, the interests of both producers and consumers are protected.

Plant Variety Protection Program The Service administers a program that provides for the issuance of certificates of plant variety protection. These certificates afford developers of novel varieties of sexually reproduced plants exclusive rights to sell, reproduce, import, or export such varieties, or use them in the production of hybrids or different varieties for a period of 20 years for nonwoody plants and 25 years for woody plants.

Research and Promotion Programs The Service monitors certain industry-sponsored research, promotion, and information programs authorized by Federal laws. These programs provide farmers and processors with a means to finance and operate various research, promotion, and information activities for agricultural products including cotton, potatoes, soybeans, sorghum, peanuts, popcorn, mushrooms, blueberries, avocados, mangoes, watermelon, honey, eggs, milk and dairy products, beef, pork, and lamb.

Transportation Programs The Service is also responsible for the promotion of an efficient transportation system for rural America that begins at the farm gate and moves agricultural and other rural products through the Nation's highways, railroads, airports, and waterways, and into the domestic and international marketplace. To accomplish this, AMS conducts economic studies and analyses of these systems and represents agricultural and rural transportation interests in policy and regulatory forums. To provide direct assistance to the transportation community, AMS supplies research and technical information to producers, producer groups, shippers, exporters, rural communities, carriers, governmental agencies, and universities.

National Organic Program Through the National Organic Program, the Service develops, implements, and administers national production, handling, and labeling standards for organic food production. Organic production integrates cultural, biological, and mechanical practices to foster cycling of resources, promote ecological balance, and conserve biodiversity.

Farmers Markets and Local Food Marketing The Service also helps to facilitate the distribution of U.S. agricultural products by designing marketing facilities, improving farmers markets and other direct-to-consumer marketing activities, researching and developing marketing channels, and providing educational information.

Microbiological Data Program and Pesticide Data Program These programs manage the collection, analysis, and reporting of foodborne pathogens and pesticides on agricultural commodities in the U.S. food supply.

Pesticide Recordkeeping Program The Agricultural Marketing Service manages the Pesticide Recordkeeping Program in coordination with State agencies, the National Agricultural Statistics Service, and the Environmental Protection Agency. The Service has developed educational programs and works with State agencies in inspecting applicator records.

For further information, contact the Public Affairs Staff, Agricultural Marketing Service, Department of Agriculture, Room 2532, South Agriculture Building, Stop 0273, Washington, DC 20250. Phone, 202–720–8998, or visit the Web site at www.ams. usda.gov.

Animal and Plant Health Inspection Service [For the Animal and Plant Health Inspection Service statement of organization, see the Code of Federal Regulations, Title 7, Part 371]

The Animal and Plant Health Inspection Service was reestablished by the Secretary of Agriculture on March 14, 1977, pursuant to authority contained in 5 U.S.C. 301 and Reorganization Plan No. 2 of 1953 (5 U.S.C. app.).

The Service was established to conduct regulatory and control programs to protect and improve animal and plant health for the benefit of man and the environment. In cooperation with State governments, the agency administers Federal laws and regulations pertaining to animal and plant health and quarantine, humane treatment of animals, and the control and eradication of pests and diseases. Regulations to prevent the introduction or interstate spread of certain animal or plant pests or diseases are also enforced by the Service. It also carries out research and operational activities to reduce crop and livestock depredation caused by birds, rodents, and predators.

Biotechnology Regulatory Services Biotechnology regulatory officials are responsible for regulating the importation, movement, and field release of genetically engineered plants, insects, microorganisms, and any other known organism or potential plant pest.

Biotechnology regulations are designed to ensure that genetically engineered organisms, such as herbicide-tolerant cotton or virus-resistant papayas, are just as safe for agriculture and the environment as traditionally bred crop varieties. In regulating biotechnology, the Service works in concert with the Environmental Protection Agency and the Food and Drug Administration, agencies that also play important roles in protecting agriculture, a safe food supply, and the environment. Its involvement begins when a person or organization wishes to import, move across a State line, or field-test a genetically engineered plant. These activities are subject to the Service's permitting and notification system.

Plant Protection and Quarantine Plant protection officials are responsible for programs to control or eradicate plant pests and diseases. These programs are carried out in cooperation with the States involved, other Federal agencies, farmers, and private organizations. Pest control programs use a single tool or a combination of pest control techniques, both chemical and nonchemical, which are both effective and safe.

Plant protection officials develop Federal regulations and policies that prohibit or restrict the entry of foreign pests and plants, plant products, animal products and byproducts, and other materials that may harbor pests or diseases. These regulations and policies help protect agricultural production and natural resources from pests and diseases. The Service continues to take regulatory action on prohibited or restricted products.

Veterinary Services Animal health officials are responsible for programs to protect and improve the health, quality, and marketability of U.S. animals and animal products. The programs are carried out through cooperative links with States, foreign governments, livestock producers, and other Federal agencies.

Service officials exclude, control, and eradicate animal pests and diseases by carrying out eradication and control programs for certain diseases, providing diagnostic services, and gathering and disseminating information regarding animal health in the United States through land, air, and ocean ports. They also certify as to the health status of animals and animal products being exported to other countries and respond to animal disease incursions or epidemics that threaten the health status of U.S. livestock and poultry.

The Service also administers a Federal law intended to ensure that all veterinary biological products, whether developed by conventional or new biotechnological procedures, used in the diagnosis, prevention, and treatment of animal disease are safe, pure, potent, and effective. The Service regulates firms that manufacture veterinary biological products subject to the act, including licensing the manufacturing establishment and its products, inspecting production facilities and production methods, and testing products under a surveillance program.

Animal Care   The Service administers Federal laws concerned with the humane care and handling of all warm-blooded animals bought, sold, and transported in commerce and used or intended for use as pets at the wholesale level or used or intended for use in exhibitions or for research purposes. The agency also enforces the Horse Protection Act of 1970, which prohibits the soring of horses at shows and sales.

International Services   Service activities in the international arena include conducting cooperative plant and animal pest and disease control, eradication, and surveillance programs in foreign countries. These programs provide a first line of defense for the United States against threats such as screwworm, medfly, foot-and-mouth disease, and other exotic diseases and pests. The Service also provides international representation concerning sanitary and phytosanitary technical trade issues and manages programs for overseas preclearance of commodities, passengers, and U.S. military activities.

Wildlife Services   Wildlife services officials cooperate with States, counties, local communities, and agricultural producer groups to reduce crop and livestock depredations caused by birds, rodents, and predators. Using methods and techniques that are biologically sound, environmentally acceptable, and economically feasible, they participate in efforts to educate and advise farmers and ranchers on proper uses of control methods and techniques; suppress serious nuisances and threats to public health and safety caused by birds, rodents, and other wildlife in urban and rural communities; and work with airport managers to reduce risks of bird strikes. In addition, they conduct research into predator-prey relationships, new control methods, and more efficient and safe uses of present methods such as toxicants, repellants and attractants, biological controls, scare devices, and habitat alteration.

For further information, contact Legislative and Public Affairs, Animal and Plant Health Inspection Service, Department of Agriculture, Washington, DC 20250. Phone, 202–720–2511.

Grain Inspection, Packers, and Stockyards Administration   The Grain Inspection, Packers, and Stockyards Administration (GIPSA) was established in 1994 to facilitate the marketing of livestock, poultry, meat, cereals, oilseeds, and related agricultural products and to promote fair and competitive trading practices for the overall benefit of consumers and American agriculture. The Agency's mission is carried out in two different segments of American agriculture. The Federal Grain Inspection Service provides the U.S. grain market with Federal quality standards and a uniform system for applying them. The Packers and Stockyards Programs (P&SP) enforces the Packers and Stockyards Act of 1921 (P&S Act), 7 U.S.C. 181 et seq., to promote fair and competitive marketing environments for the livestock, meat, and poultry industries. GIPSA also certifies State central filing systems for notification of liens against farm products. GIPSA is responsible for establishing official U.S. standards for grain and other assigned commodities and for administering a nationwide official inspection and weighing system.

Inspection The United States Grain Standards Act requires most U.S. export grain to be officially inspected. At export port locations, inspection is performed by GIPSA or by State agencies that have been delegated export inspection authority by the Administrator. For domestic grain marketed at inland locations, the Administrator designates private and State agencies to provide official inspection services upon request. Both export and domestic services are provided on a fee-for-service basis.

Weighing Official weighing of U.S. export grain is performed at port locations by GIPSA or by State agencies that have been delegated export weighing authority by the Administrator. For domestic grain marketed at inland locations, the weighing services may be provided by GIPSA or by designated private or State agencies. Upon request, weighing services are provided on a fee-for-service basis.

Standardization The Administration is responsible for establishing, maintaining, and revising official U.S. standards for corn, wheat, rye, oats, barley, flaxseed, sorghum, soybeans, triticale, sunflower seed, canola, and mixed grain. It is authorized to perform applied research to develop methods to improve accuracy and uniformity in grading grain. It is also responsible for standardization and inspection activities for rice, dry beans, peas, lentils, hay, straw, hops, and related processed grain commodities. Although standards no longer exist for hay, straw, and hops, GIPSA maintains inspection procedures for and retains authority to inspect these commodities.

Methods Development The Administration's methods development activities include applied research or tests that produce new or improved techniques for measuring grain quality. Examples include new knowledge gained through study of how to establish the framework for real-time grain inspection

and develop reference methods to maintain consistency and standardization in the grain inspection system and the comparison of different techniques for evaluation of end-use quality in wheat.

Packers and Stockyards Activities Through the administration of the Packers and Stockyards Act, GIPSA prohibits unfair, deceptive, and unjust discriminatory practices by market agencies, dealers, stockyards, packers, swine contractors, and live poultry dealers in the livestock, meat packing, and poultry industries. GIPSA fosters fair competition and ensures payment protection for growers and farmers. To this end, the agency performs various regulatory functions, including investigating alleged violations of the Act, auditing regulated entities, verifying the accuracy of scales, and monitoring industry trends to protect consumers and members of the livestock, meat, and poultry industries.

The agency also is responsible for the Truth-in-Lending Act and the Fair Credit Reporting Act as each relates to persons and firms subject to the Packers and Stockyards Act. GIPSA carries out the Secretary's responsibilities under section 1324 of the Food Security Act of 1985 pertaining to State-established central filing systems to prenotify buyers, commission merchants, and selling agents of security interests against farm products. GIPSA administers the section of the statute commonly referred to as the "Clear Title" provision and certifies qualifying State systems.

For further information, contact the Grain Inspection, Packers, and Stockyards Administration, Department of Agriculture, Washington, DC 20250. Phone, 202–720–0219.

Food Safety

Food Safety and Inspection Service
The Food Safety and Inspection Service (FSIS) was established by the Secretary of Agriculture on June 17, 1981, pursuant to authority contained in 5 U.S.C. 301 and Reorganization Plan No. 2 of 1953 (5 U.S.C. app.). FSIS is responsible for ensuring that the Nation's commercial supply of meat, poultry, and egg products

is safe, wholesome, and correctly labeled and packaged.

Meat, Poultry, and Egg Products Inspection FSIS sets public health performance standards for food safety by carrying out inspection and enforcement activities for all raw and processed meat, poultry, and egg products, including imported products; ensuring the food supply is safe for use as human food; and working to better understand, predict, and prevent contamination of meat, poultry, and egg products to improve consumer health.

FSIS conducts mandatory inspections in Federal facilities for meat, poultry, and egg production, including cattle, swine, goats, sheep, horses and other equines, chickens, turkeys, ducks, geese, and guinea fowl, and also provides voluntary inspection for animals not covered under mandatory inspection regulations such as buffalo, rabbit, and deer. The Service tests samples of meat, poultry, and egg products for microbial and chemical contaminants to monitor trends for enforcement purposes and conducts food defense activities to protect against contamination.

FSIS also monitors meat, poultry, and egg products throughout storage, distribution, and retail channels and ensures regulatory compliance to protect the public, including detention of products, voluntary product recalls, court-ordered seizures of products, administrative suspension and withdrawal of inspection, and referral of violations for criminal and civil prosecution. FSIS monitors State inspection programs that inspect meat and poultry products sold only within the State in which they were produced.

FSIS administers a program designed to provide that humane methods are employed in the slaughtering of livestock and in the handling of livestock in connection with slaughter.

FSIS maintains a toll-free meat and poultry hotline (phone, 888–674–6854; TTY, 800–256–7072) to answer questions in English and Spanish about the safe handling of meat, poultry, and egg products. The hotline's hours are Monday through Friday, from 10 a.m.

to 4 p.m., eastern standard time, year round. An extensive selection of food safety messages in English and Spanish is available at the same number 24 hours a day.

"Ask Karen," a Web-based virtual representative tool providing answers to consumer questions about food safety, may be accessed at www.askkaren.gov.

For further information, contact the Director, Food Safety Education Staff, Food Safety and Inspection Service, Department of Agriculture, Beltsville, Maryland 20705. Phone, 301–344–4755. Fax, 301–504–0203. Email, MPHotline.fsis@usda.gov. Internet, http://www.fsis.usda.gov.

Food, Nutrition, and Consumer Services

The mission of Food, Nutrition, and Consumer Services is to reduce hunger and food insecurity, in partnership with cooperating organizations, by providing access to food, a healthful diet, and nutrition education to children and needy people in a manner that supports American agriculture.

Food and Nutrition Service The Food and Nutrition Service (FNS) administers the USDA food assistance programs. These programs, which serve one in six Americans, represent our Nation's commitment to the principle that no one in this country should fear hunger or experience want. They provide a Federal safety net to people in need. The goals of the programs are to provide needy persons with access to a more nutritious diet, to improve the eating habits of the Nation's children, and to help America's farmers by providing an outlet for distributing foods purchased under farmer assistance authorities.

The Service works in partnership with the States in all its programs. State and local agencies determine most administrative details regarding distribution of food benefits and eligibility of participants, and FNS provides commodities and funding for additional food and to cover administrative costs. FNS administers the following food assistance programs:

The Food Stamp Program provides food benefits through State and local welfare agencies to needy persons to increase their food purchasing power. The benefits are used by program participants to buy food in retail stores approved by the Food and Nutrition Service to accept and redeem the benefits.

The Special Supplemental Nutrition Program for Women, Infants, and Children (WIC) improves the health of low-income pregnant, breastfeeding, and nonbreastfeeding postpartum women, and infants and children up to 5 years of age by providing them with specific nutritious food supplements, nutrition education, and health care referrals.

The WIC Farmers' Market Nutrition Program provides WIC participants with increased access to fresh produce. WIC participants receive coupons to purchase fresh fruits and vegetables from authorized farmers.

The Commodity Supplemental Food Program provides a package of foods monthly to low-income pregnant, postpartum, and breastfeeding women, their infants and children under age 6, and the elderly. Nutrition education is also provided through this program.

The National School Lunch Program supports nonprofit food services in elementary and secondary schools and in residential childcare institutions. More than half of the meals served through these institutions are free or at reduced cost.

The School Breakfast Program supplements the National School Lunch Program by supporting schools in providing needy children with free or low-cost breakfasts that meet established nutritional standards.

The Special Milk Program for Children provides milk for children in those schools, summer camps, and childcare institutions that have no federally supported meal programs.

The Child and Adult Care Food Program provides cash and commodities for meals for preschool and school-age children in childcare facilities and for functionally impaired adults in facilities that provide nonresidential care for such individuals.

The Summer Food Service Program for Children helps various organizations get nutritious meals to needy preschool and school-age children during the summer months and during school vacations.

The Emergency Food Assistance Program provides State agencies with commodities for distribution to food banks, food pantries, soup kitchens, and other charitable institutions throughout the country, with administrative funds to assist in distribution.

The Food Distribution Program on Indian Reservations and the Trust Territories provides an extensive package of commodities monthly to low-income households on or near Indian reservations in lieu of food stamps. This program is administered at the local level by Indian tribal organizations or State agencies.

The Nutrition Program for the Elderly provides cash and commodities to States for meals for senior citizens. The food is delivered through senior citizen centers or meals-on-wheels programs.

The Nutrition Assistance Programs for Puerto Rico and the Northern Marianas are block grant programs that replace the Food Stamp Programs in these two territories and provide cash and coupons to resident participants.

The Nutrition Education and Training Program grants funds to States for the development and dissemination of nutrition information and materials to children and for training of food service and teaching personnel.

For further information, contact the Public Information Officer, Food and Nutrition Service, Department of Agriculture, Alexandria, VA 22302. Phone, 703–305–2286. Internet, http://www.usda. gov.

Center for Nutrition Policy and Promotion

The Center coordinates nutrition policy in USDA and provides overall leadership in nutrition education for the American public. It also coordinates with the Department of Health and Human Services in the review, revision, and dissemination of the Dietary Guidelines for Americans, the Federal Government's statement of nutrition policy formed by a consensus of scientific and medical professionals.

For further information, contact the Office of Public Information, Center for Nutrition Policy and Promotion, Suite 200, 1120 20th Street NW., Washington, DC 20036–3406. Phone, 202–418– 2312. Internet, http://www.cnpp.usda.gov.

Farm and Foreign Agricultural Services

Farm Service Agency   The Farm Service Agency (FSA) administers farm commodity, disaster, and conservation programs for farmers and ranchers, and makes and guarantees farm emergency, ownership, and operating loans through a network of State and county offices.

Farm Commodity Programs   FSA manages commodity programs such as the direct and countercyclical program, commodity and livestock disaster programs, marketing assistance loan programs, noninsured crop disaster assistance programs, and the tobacco transition payment program. It administers commodity loan programs for wheat, rice, corn, grain sorghum, barley, oats, oilseeds, peanuts, upland and extra-long-staple cotton, and sugar. FSA provides operating personnel for the Commodity Credit Corporation (CCC), a Government-owned and -operated organization. CCC provides short-term loans using the commodity as collateral. These loans provide farmers with interim financing and facilitate orderly marketing of farm commodities throughout the year.

Farm Loan Programs   FSA makes and guarantees loans to family farmers and ranchers to purchase farmland and finance agricultural production. These programs help farmers who are temporarily unable to obtain private commercial credit. These may be beginning farmers who have insufficient net worth to qualify for commercial credit, who have suffered financial setbacks from natural disasters, or who have limited resources with which to establish and maintain profitable farming operations.

Noninsured Crop Disaster Assistance Program (NAP)   NAP provides catastrophic crop loss protection for crops not covered by Federal crop insurance. Crops that are eligible include commercial crops grown for food and fiber, floriculture, ornamental nursery products, Christmas tree crops, turfgrass sod, seed crops, aquaculture (including ornamental fish such as goldfish), and industrial crops. Losses resulting from natural disasters not covered by the crop insurance policy may also be eligible for NAP assistance. NAP does not include trees grown for wood, paper, or pulp products.

Other Emergency Assistance   There are FSA programs to assist farmers who encounter natural disasters from drought, flood, freeze, tornadoes, and other calamities. Eligible producers can be compensated for crop losses, livestock feed losses, and tree damage and for the cost of rehabilitating eligible farmlands damaged by natural disaster. Low-interest loans for eligible farmers can help cover production and physical losses in counties declared disaster areas.

The largest component of USDA disaster assistance is the Crop Disaster Program (CDP), which has provided more than $3 billion in financial relief to farmers, ranchers, foresters, and other agricultural producers who incurred losses because of recent adverse weather conditions.

Conservation Programs   FSA's conservation programs include enhancement of wildlife habitat and water and air quality. The Conservation Reserve Program is the Federal Government's single-largest environmental improvement program on private lands. It safeguards millions of acres of topsoil from erosion, improves air quality, increases wildlife habitat, and protects ground and surface water by reducing water runoff and sedimentation. In return for planting a protective cover of grass or trees on vulnerable property, the owner receives a rental payment each year of a multiyear contract. Cost-share payments are also available to help establish permanent areas of grass, legumes, trees, windbreaks, or plants that improve water quality and give shelter and food to wildlife.

Commodity Operations   FSA's commodity operations system facilitates the storage, management, and disposition of commodities used to meet humanitarian needs abroad. It administers the United States Warehouse Act (USWA), which authorizes the Secretary of Agriculture to license warehouse operators who store agricultural products. Warehouse operators that apply must meet the USDA standards established within the USWA and its regulations. Under the milk price support program, the Commodity Credit Corporation buys surplus butter, cheese, and nonfat dry milk from processors at announced prices to support the price of milk. These purchases help maintain market prices at the legislated support level, and the surplus commodities are used for hunger relief both domestically and internationally. FSA's commodity operations system also coordinates with other Government agencies to provide surplus commodities for various programs and also purchases commodities for the National School Lunch Program and other domestic feeding programs.

For further information, contact the Public Affairs Branch, Farm Service Agency, Department of Agriculture, Stop 0506, 1400 Independence Avenue SW., Washington, DC 20250. Phone, 202–720–5237. Internet, http://www.fsa.usda.gov.

Commodity Credit Corporation   The Commodity Credit Corporation (CCC) stabilizes, supports, and protects farm income and prices, assists in maintaining balanced and adequate supplies of agricultural commodities and their products, and facilitates the orderly distribution of commodities.

CCC carries out assigned foreign assistance activities, such as guaranteeing the credit sale of U.S. agricultural commodities abroad. Major emphasis is also being directed toward meeting the needs of developing nations. Agricultural commodities are supplied and exported to combat hunger and malnutrition and to encourage economic development in developing countries. In addition, under the Food for Progress Program, CCC supplies commodities to provide assistance to developing democracies.

For further information, contact the Information Division, Foreign Agricultural Service, Department of Agriculture, Stop 1004, 1400 Independence Avenue SW., Washington, DC 20250. Phone, 202–720–7115. Fax, 202–720–1727.

Risk Management Agency   The Risk Management Agency (RMA), via the Federal Crop Insurance Corporation (FCIC), oversees and administers the crop insurance program under the Federal Crop Insurance Act.

Crop insurance is offered to qualifying producers through 16 private sector crop insurance companies. Under the new Standard Reinsurance Agreement (SRA), RMA provides reinsurance, pays premium subsidies, reimburses insurers for administrative and operating costs and oversees the financial integrity and operational performance of the delivery system. RMA bears much of the noncommercial insurance risk under the SRA, allowing insurers to retain commercial insurance risks or reinsure those risks in the private market.

In 2006, the Federal crop insurance program provided producers with more than $44 billion in protection on approximately 246 million acres through about 1.2 million policies. There are 22 insurance plans available and 26 active pilot programs in various stages of development.

RMA also works closely with the private sector to find new and innovative ways to provide expanded coverage. This includes risk protection for specialty crops, livestock and forage, and rangeland and pasture. Thus, RMA is able to reduce the need for ad hoc disaster bills and available coverage caused by long-term production declines that result from extended drought in many areas.

Additional information about RMA can be found on its Web site, www.rma.usda.gov, including agency news, State profiles, publications, announcements on current issues, summaries of insurance sales, pilot programs, downloadable crop policies, and agency-sponsored events. The site also features online tools, calculators, and applications.

For further information, contact the Office of the Administrator, Risk Management Agency, Department of Agriculture, Stop 0801, 1400 Independence Avenue SW., Washington, DC 20250. Phone, 202–690–2803. Internet, http://www.rma.usda.gov.

Foreign Agriculture Service   The Foreign Agricultural Service (FAS) works to improve foreign market access for U.S. products, to build new markets, to improve the competitive position of U.S. agriculture in the global marketplace, and to provide food aid and technical assistance to foreign countries.

FAS has the primary responsibility for USDA's activities in the areas of international marketing, trade agreements and negotiations, and the collection and analysis of international statistics and market information. It also administers the USDA's export credit guarantee and food aid programs. FAS helps increase income and food availability in developing nations by mobilizing expertise for agriculturally led economic growth.

FAS also enhances U.S. agricultural competitiveness through a global network of agricultural economists, marketing experts, negotiators, and other specialists. FAS agricultural counselors, attaches, trade officers, and locally employed staff are stationed in over 90 countries to support U.S. agricultural interests and cover 140 countries.

In addition to agricultural affairs offices in U.S. embassies, agricultural trade offices also have been established in a number of key foreign markets and function as service centers for U.S. exporters and foreign buyers seeking market information.

Reports prepared by our overseas offices cover changes in policies and other developments that could affect U.S. agricultural exports. FAS staff in U.S. embassies around the world assess U.S. export marketing opportunities and respond to the daily informational needs of those who develop, initiate, monitor, and evaluate U.S. food and agricultural policies and programs.

In addition to data collection, FAS also maintains a worldwide agricultural reporting system based on information from U.S. agricultural traders, remote sensing systems, and other sources. Analysts in Washington, DC, prepare production forecasts, assess export marketing opportunities, and track changes in policies affecting U.S. agricultural exports and imports.

FAS programs help U.S. exporters develop and maintain markets for hundreds of food and agricultural products, from bulk commodities to brand name items. Formal market promotion activities are carried out chiefly in cooperation with agricultural trade associations, State-regional trade groups, small businesses, and cooperatives that plan, manage, and contribute staff resources and funds to support these efforts. FAS also provides guidance to help exporters locate buyers and provides assistance through a variety of other methods. This includes supporting U.S. participation in several major trade shows and a number of single-industry exhibitions each year.

For further information, contact the Public Affairs Division, Foreign Agricultural Service, Stop 1004, 1400 Independence Avenue, SW., Department of Agriculture, Washington, DC 20250–1004. Phone, 202–720–7115. Fax, 202–720–1727. Internet, http://www.fas.usda.gov.

Research, Education, and Economics

This mission area's main focus is to create, apply, and transfer knowledge and technology to provide affordable food and fiber, ensure food safety and nutrition, and support rural development and natural resource needs of people by conducting integrated national and international research, information, education, and statistical programs and services that are in the national interest.

Agricultural Research Service The Agricultural Research Service (ARS) conducts research to develop and transfer solutions to agricultural problems of high national priority. It provides information access and dissemination to ensure high-quality safe food and other agricultural products; assess the nutritional needs of Americans; sustain a competitive agricultural economy; enhance the natural resource base and the environment; and provide economic opportunities for rural citizens, communities, and society as a whole.

Research activities are carried out at 103 domestic locations (including Puerto Rico and the U.S. Virgin Islands) and 5 overseas locations. Much of this research is conducted in cooperation with partners in State universities and experiment stations, other Federal agencies, and private organizations. National Programs, headquartered in Beltsville, MD, is the focal point in the overall planning and coordination of ARS's research programs. Day-to-day management of the respective programs for specific field locations is assigned to eight area offices.

ARS also includes the National Agricultural Library (NAL), which is the primary resource in the United States for information about food, agriculture, and natural resources and serves as an electronic gateway to a widening array of scientific literature, printed text, and agricultural images. NAL serves USDA and a broad customer base including policymakers, agricultural specialists, research scientists, and the general public. NAL works with other agricultural libraries and institutions to advance open and democratic access to information about agriculture and the Nation's agricultural knowledge.

For further information, contact the Agricultural Research Service, Department of Agriculture, 1400 Independence Avenue SW., Washington, DC 20250. Phone, 202–720–3656. Fax, 202–720–5427. Internet, http://www.ars.usda.gov.

Cooperative State Research, Education, and Extension Service   The Cooperative State Research, Education, and Extension Service (CSREES) links the research and education resources and activities of USDA and works with academic and land-grant institutions throughout the Nation. In cooperation with its partners and customers, CSREES advances a global system of research, extension, and higher education in the food and agricultural sciences and related environmental and human sciences to benefit people, communities, and the Nation.

CSREES's programs increase and provide access to scientific knowledge; strengthen the capabilities of land-grant and other institutions in research, extension, and higher education; increase access to and use of improved communication and network systems; and promote informed decisionmaking by producers, consumers, families, and community leaders to improve social conditions in the United States and around the world. These conditions include improved agricultural and other economic enterprises; safer, cleaner water, food, and air; enhanced stewardship and management of natural resources; healthier, more responsible

and more productive individuals, families, and communities; and a stable, secure, diverse, and affordable national food supply.

CSREES provides research, extension, and education leadership through programs in plant and animal systems; natural resources and environment; economic and community systems; families, 4–H, and nutrition; competitive research and integrated research, education, and extension programs and awards management; science and education resources development; and information systems and technology management.

CSREES's partnership with the land-grant universities is critical to the effective shared planning, delivery, and accountability for research, higher education, and extension programs.

For further information, contact the Communications Staff, Cooperative State Research, Education, and Extension Service, Department of Agriculture, 1400 Independence Avenue SW., Washington, DC 20250–2207. Phone, 202–720–4651. Fax, 202–690–0289.

Economic Research Service   The mission of the Economic Research Service (ERS) is to inform and enhance public and private decisionmaking on economic and policy issues related to agriculture, food, the environment, and rural development.

Activities to support this mission and the following goals involve research and development of economic and statistical indicators on a broad range of topics including, but not limited to, global agricultural market conditions, trade restrictions, agribusiness concentration, farm and retail food prices, foodborne illnesses, food labeling, nutrition, food assistance programs, worker safety, agrichemical usage, livestock waste management, conservation, sustainability, genetic diversity, technology transfer, rural infrastructure, and rural employment.

Research results and economic indicators on such important agricultural, food, natural resource, and rural issues are fully disseminated to public and private decisionmakers through published and electronic reports and articles; special staff analyses, briefings, presentations, and papers; databases; and individual contacts. Through such activities, ERS provides public and private decisionmakers with economic and related social science information and analysis in support of the Department's goals of enhancing economic opportunities for agricultural producers; supporting economic opportunities and quality of life in rural America; enhancing the protection and safety of U.S. agriculture and food; improving U.S. nutrition and health; and enhancing the natural resource base and environment. More information on ERS's program is available at www.ers.usda.gov.

For further information, contact the Information Services Division, Economic Research Service, Department of Agriculture, Washington, DC 20036–5831. Phone, 202–694–5100. Fax, 202–694–5641.

National Agricultural Statistics Service
The National Agricultural Statistics Service (NASS) prepares estimates and reports on production, supply, price, chemical use, and other items necessary for the orderly operation of the U.S. agricultural economy.

The reports include statistics on field crops, fruits and vegetables, dairy, cattle, hogs, sheep, poultry, aquaculture, and related commodities or processed products. Other estimates concern farm numbers, farm production expenditures, agricultural chemical use, prices received by farmers for products sold, prices paid for commodities and services, indexes of prices received and paid, parity prices, farm employment, and farm wage rates.

The Service prepares these estimates through a complex system of sample surveys of producers, processors, buyers, and others associated with agriculture. Information is gathered by mail, telephone, personal interviews, and field visits.

NASS is responsible for conducting the Census of Agriculture. The Census of Agriculture is taken every 5 years and provides comprehensive data on the agricultural economy down to the county level. Periodic reports are also issued on aquacultures, irrigation, and horticultural specialties.

The Service performs reimbursable survey work and statistical consulting services for other Federal and State agencies and provides technical assistance for developing agricultural data systems in other countries.

For further information, contact the Executive Assistant to the Administrator, National Agricultural Statistics Service, Department of Agriculture, Washington, DC 20250–2000. Phone, 202–720–2707. Fax, 202–720–9013.

Natural Resources and Environment

This mission area is responsible for fostering sound stewardship of 75 percent of the Nation's total land area. Ecosystems are the underpinning for the Department's operating philosophy in this area in order to maximize stewardship of our natural resources. This approach ensures that products, values, services, and uses desired by people are produced in ways that sustain healthy, productive ecosystems.

Forest Service [For the Forest Service statement of organization, see the Code of Federal Regulations, Title 36, Part 200.1]

The Forest Service was created by the Transfer Act of February 1, 1905 (16 U.S.C. 472), which transferred the Federal forest reserves and the responsibility for their management from the Department of the Interior to the Department of Agriculture. The mission of the Forest Service is to achieve quality land management under the sustainable, multiple-use management concept to meet the diverse needs of people. Its objectives include: advocating a conservation ethic in promoting the health, productivity, diversity, and beauty of forests and associated lands; listening to people and responding to their diverse needs in making decisions; protecting and managing the national forests and grasslands to best demonstrate the sustainable, multiple-use management concept; providing technical and financial assistance to State and private forest landowners, encouraging them toward active stewardship and quality land management in meeting their specific objectives; providing technical and financial assistance to cities and communities to improve their natural environment by planting trees and caring for their forests; providing international technical assistance and scientific exchanges to sustain and enhance global resources and to encourage quality land management; assisting States and communities in using the forests wisely to promote rural economic development and a quality rural environment; developing and providing scientific and technical knowledge, improving our capability to protect, manage, and use forests and rangelands; and providing work, training, and education to the unemployed, underemployed, elderly, youth, and the disadvantaged.

National Forest System The Service manages 155 national forests, 20 national grasslands, and 8 land utilization projects on over 191 million acres in 44 States, the Virgin Islands, and Puerto Rico under the principles of multiple-use and sustained yield. The Nation's tremendous need for wood and paper products is balanced with the other vital, renewable resources or benefits that the national forests and grasslands provide: recreation and natural beauty, wildlife habitat, livestock forage, and water supplies. The guiding principle is the greatest good to the greatest number in the long run.

These lands are protected as much as possible from wildfire, epidemics of disease and insect pests, erosion, floods, and water and air pollution. Burned areas get emergency seeding treatment to prevent massive erosion and stream siltation. Roads and trails are built where needed to allow for closely regulated timber harvesting and to give the public access to outdoor recreation areas and provide scenic drives and hikes. Picnic, camping, water sports, skiing, and other areas are provided with facilities for public convenience and enjoyment. Timber harvesting methods are used that will protect the land and streams, assure rapid renewal of the forest, provide food and cover for wildlife and fish, and have minimum impact on scenic and recreation values. Local communities benefit from the logging and milling activities. These lands also provide needed oil, gas, and minerals.

Rangelands are improved for millions of livestock and game animals. The national forests provide a refuge for many species of endangered birds, animals, and fish. Some 34.6 million acres are set aside as wilderness and 175,000 acres as primitive areas where timber will not be harvested.

For information on the National Forest System Regions or State and Private Forestry Areas, visit our Web site at www.fs.fed.us.

Forest Research The Service performs basic and applied research to develop the scientific information and technology needed to protect, manage, use, and sustain the natural resources of the Nation's forests and rangelands. The Service's forest research strategy focuses on three major program components: understanding the structure and functions of forest and range ecosystems; understanding how people perceive and value the protection, management, and use of natural resources; and determining which protection, management, and utilization practices are most suitable for sustainable production and use of the world's natural resources.

For information on Forest Research Stations in your area, visit our Web site at http://www.fs.fed.us/research/climate/usfs-cc-research.shtml.

Natural Resources Conservation Service

[For the Natural Resources Conservation Service statement of organization, see the Code of Federal Regulations, Title 7, Parts 601 and 601]

The Natural Resources Conservation Service (NRCS), formerly the Soil Conservation Service, has national responsibility for helping America's farmers, ranchers, and other private landowners develop and carry out voluntary efforts to conserve and protect our natural resources.

Conservation Technical Assistance This is the foundation program of NRCS. Under this program, NRCS provides technical assistance to land users and units of government for the purpose of sustaining agricultural productivity and protecting and enhancing the natural resource base. This assistance is based on the voluntary cooperation of private landowners and involves comprehensive approaches to reduce soil erosion, improve soil and water quantity and quality, improve and conserve wetlands, enhance fish and wildlife habitat, improve air quality, improve pasture and range condition, reduce upstream flooding, and improve woodlands.

Emergency Watershed Protection Program   This program provides emergency assistance to safeguard lives and property in jeopardy due to sudden watershed impairment by natural disasters. Emergency work includes quickly establishing a protective plant cover on denuded land and stream banks; opening dangerously restricted channels; and repairing diversions and levees. An emergency area need not be declared a national disaster area to be eligible for help under this program.

Environmental Quality Incentive Program   This program assists producers with environmental and natural resource conservation improvements on their agricultural lands. One-half of the available funds are for conservation activities related to livestock production. Technical assistance, cost-share payments, incentive payments, and education focus on priority areas and natural resource concerns identified in cooperation with State technical committees, including such areas as nutrient management, pest management, and grazing land management.

Farmland Protection Program   This program protects soil by encouraging landowners to limit conversion of their farmland to nonagricultural uses. States, Indian tribes, or local governments administer all aspects of acquiring lands that are in the program, except when it is more effective and efficient for the Federal Government to do so.

Forestry Incentives Program   This program helps to increase the Nation's supply of products from nonindustrial private forest lands. This also ensures more effective use of existing forest lands and, over time, helps to prevent shortages and price increases for forest products. The program shares the cost incurred by landowners for tree planting and timberstand improvement.

National Cooperative Soil Survey
The National Cooperative Soil Survey provides the public with local information on the uses and capabilities of their soils. The published soil survey for a county or other designated area includes maps and interpretations that are the foundation for farm planning and other private land use decisions as well as for resource planning and policy by Federal, State, and local governments. The surveys are conducted cooperatively with other Federal, State, and local agencies and land-grant universities. The Service is the national and world leader in soil classification and soil mapping, and is now expanding its work in soil quality.

Plant Materials Program   At 26 plant materials centers across the country, NRCS tests, selects, and ensures the commercial availability of new and improved conservation plants for erosion reduction, wetland restoration, water quality improvement, streambank and riparian area protection, coastal dune stabilization, biomass production, carbon sequestration, and other needs. The Plant Materials Program is a cooperative effort with conservation districts, other Federal and State agencies, commercial businesses, and seed and nursery associations.

Resource Conservation and Development Program   This is a locally driven program, an opportunity for civic-oriented groups to work together sharing knowledge and resources in solving common problems facing their region. The program offers aid in balancing the environmental, economic, and social needs of an area. A USDA coordinator helps each designated RC&D council plan, develop, and carry out programs for resource conservation, water management, community development, and environmental enhancement.

Rural Abandoned Mine Program   This program helps protect people and the environment from the adverse effects of past coal-mining practices and promotes the development of soil and water resources on unreclaimed mine land. It provides technical and financial assistance to land users who voluntarily enter into 5- to 10-year contracts for the reclamation of eligible land and water.

Small Watersheds Program   The program helps local sponsoring groups to voluntarily plan and install watershed protection projects on private lands. These projects include flood prevention, water quality improvement, soil erosion and sediment reduction, rural and municipal water supply, irrigation water management, fish and wildlife habitat enhancement, and wetlands restoration. The Service helps local community groups, government entities, and private landowners working together using an integrated, comprehensive watershed approach to natural resource planning.

Snow Survey and Water Supply Forecasting Program   This program collects snowpack moisture data and forecasts seasonal water supplies for streams that derive most of their water from snowmelt. It helps farm operators, rural communities, and municipalities manage water resources through water supply forecasts. It also provides hydrometeorological data for regulating reservoir storage and managing streamflow. The Snow Supply Program is conducted in the Western States and Alaska.

Watershed Surveys and Planning   This program assists Federal, State, and local agencies and tribal governments in protecting watersheds from damage caused by erosion, floodwater, and sediment and conserves and develops water and land resources. Resource concerns addressed by the program include water quality, water conservation, wetland and water storage capacity, agricultural drought problems, rural development, municipal and industrial water needs, upstream flood damages, and water needs for fish, wildlife, and forest-based industries. Types of surveys and plans include watershed plans, river basin surveys and studies, flood hazard analysis, and flood plain management assistance. The focus of these plans is to identify solutions that use land treatment and nonstructural measures to solve resource problems.

Wetlands Reserve Program   Under this program, USDA purchases easements from agricultural landowners who voluntarily agree to restore and protect wetlands. Service employees help these owners develop plans to retire critical wetland habitat from crop production. The primary objectives are to preserve and restore wetlands, improve wildlife habitat, and protect migratory waterfowl.

Wildlife Habitat Incentives Program   This program provides financial incentives to develop habitats for fish and wildlife on private lands. Participants agree to implement a wildlife habitat development plan, and USDA agrees to provide cost-share assistance for the initial implementation of wildlife habitat development practices. USDA and program participants enter into a cost-share agreement for wildlife habitat development, which generally lasts a minimum of 10 years from the date that the contract is signed.

For further information, contact the Management Services Division, Natural Resources Conservation Service, Department of Agriculture, P.O. Box 2890, Washington, DC 20013. Phone, 202–690–4811.

Sources of Information

Consumer Activities   Educational, organizational, and financial assistance is offered to consumers and their families in such fields as rural housing and farm operating programs, improved nutrition, family living and recreation, food stamp, school lunch, donated foods, and other food programs.

Contracts and Small Business Activities   To obtain information about contracting or subcontracting opportunities, attending small business outreach activities, or how to do business with USDA, contact the Office of Small and Disadvantaged Business Utilization. Phone, 202–720–7117. Internet, http://www.usda.gov/da/smallbus.

Employment   Most jobs in the Department are in the competitive service and are filled by applicants who have established eligibility under an appropriate examination administered by the Office of Personnel Management or Department Special Examining Units. General employment information is available at www.usajobs.opm.gov.

Whistleblower Hotline   Persons wishing to register complaints of alleged improprieties concerning the Department should contact one of the regional offices or the Inspector General's whistleblower hotline. Phone, 800–424–9121 (toll free, outside Washington, DC); 202–690–1622 (within the Washington, DC, metropolitan area); or 202–690–1202 (TDD). Fax, 202–690–2474.

Reading Rooms   Reading Rooms are located at the headquarters of each USDA agency. Use the contact information provided in the "For further information" sections in the program description text above to inquire about locations, hours, and availability.

Speakers   Contact the nearest Department of Agriculture office or county Extension agent. In the District of Columbia, contact the Office of Public Liaison, Office of Communications, Department of Agriculture, Washington, DC 20250. Phone, 202–720–2798.

For further information concerning the Department of Agriculture, contact the Office of Communications, Department of Agriculture, Washington, DC 20250. Phone, 202–720–4623. Internet, http://www.usda.gov.

Graduate School

600 Maryland Avenue SW., Suite 300, Washington, DC 20024–2520
Phone, 888–744–4723.

Executive Director	JERRY ICE
Deputy Executive Director	LYNN EDWARDS

The Graduate School was established by act of May 15, 1862 (7 U.S.C. 2201). It is a continuing education school offering career-related training to adults. Courses are planned with the assistance of Government professionals and specialists. The Graduate School's objective is to improve Government services by providing needed continuing education and training opportunities for Government employees and agencies.

The faculty is mostly part-time and is drawn from throughout Government and the community at large. They are selected because of their professional and specialized knowledge and experience and thus bring a practicality and experience to their classrooms.

The School does not grant degrees, but does provide planned sequences of courses leading to certificates of accomplishment in a number of occupational and career fields important to government. Training areas include management, auditing, computer science, communications, foreign language, procurement, financial management, and others.

For further information, contact the Communications Office, Graduate School, Room 270, 600 Maryland Avenue SW., Washington, DC 20024. Phone, 888–744–4723.

DEPARTMENT OF COMMERCE

Fourteenth Street and Constitution Avenue NW., Washington, DC 20230
Phone, 202–482–2000. Internet, http://www.doc.gov.

Secretary of Commerce	GARY F. LOCKE
Deputy Secretary	DENNIS F. HIGHTOWER
Chief of Staff	ELLEN MORAN
Senior Advisor and Deputy Chief of Staff	RICK WADE
Deputy Chief of Staff	JAY REICH
Assistant Secretary for Legislative and Intergovernmental Affairs	APRIL S. BOYD
Chief Financial Officer and Assistant Secretary for Administration	SCOTT QUEHL
Chief Information Officer	SUZANNE E. HILDING
General Counsel	CAMERON F. KERRY
Inspector General	TODD J. ZINSER
Director, Office of Business Liaison	N. ANNE OLAIMEY
Director, Office of Policy and Strategic Planning	TRAVIS SULLIVAN
Director, Office of Public Affairs	KEVIN GRIFFIS
Director, Executive Secretariat	TENE DOLPHIN
Director, Office of White House Liaison	JOHN C. CONNOR

The Department of Commerce promotes the Nation's domestic and international trade, economic growth, and technological advancement by fostering a globally competitive free enterprise system, supporting fair trade practices, compiling social and economic statistics, protecting Earth's physical and oceanic resources, granting patents and registering trademarks, and providing assistance to small and minority-owned businesses.

The Department was designated as such by act of March 4, 1913 (15 U.S.C. 1501), which reorganized the Department of Commerce and Labor, created by act of February 14, 1903 (15 U.S.C. 1501), by transferring all labor activities into a new, separate Department of Labor.

Office of the Secretary

Secretary The Secretary is responsible for the administration of all functions and authorities assigned to the Department of Commerce and for advising the President on Federal policy and programs affecting the industrial and commercial segments of the national economy. The Secretary is served by the offices of Deputy Secretary, Inspector General,

General Counsel, and the Assistant Secretaries of Administration, Legislative and Intergovernmental Affairs, and Public Affairs. Other offices whose public purposes are widely administered are detailed below.

Business Liaison The Office of Business Liaison directs the business community to the offices and policy experts who can best respond to their needs by promoting proactive, responsive, and effective outreach programs and relationships with the business community. It also informs the Secretary and Department officials of the critical issues facing the business community, informs the business community of Department and administration initiatives and priorities, as well as information regarding Department resources, policies, and programs,

DEPARTMENT OF COMMERCE

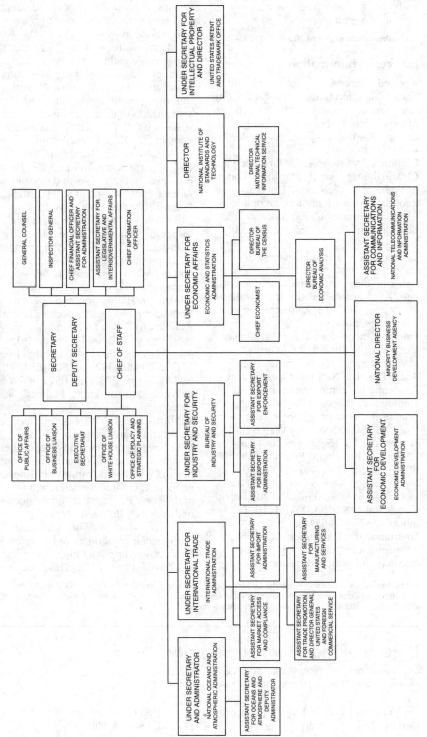

SECRETARY

DEPUTY SECRETARY

CHIEF OF STAFF

GENERAL COUNSEL

INSPECTOR GENERAL

CHIEF FINANCIAL OFFICER AND ASSISTANT SECRETARY FOR ADMINISTRATION

ASSISTANT SECRETARY FOR LEGISLATIVE AND INTERGOVERNMENTAL AFFAIRS

CHIEF INFORMATION OFFICER

OFFICE OF PUBLIC AFFAIRS

OFFICE OF BUSINESS LIAISON

EXECUTIVE SECRETARIAT

OFFICE OF WHITE HOUSE LIAISON

OFFICE OF POLICY AND STRATEGIC PLANNING

UNDER SECRETARY FOR INTELLECTUAL PROPERTY AND DIRECTOR
UNITED STATES PATENT AND TRADEMARK OFFICE

DIRECTOR
NATIONAL INSTITUTE OF STANDARDS AND TECHNOLOGY

DIRECTOR
NATIONAL TECHNICAL INFORMATION SERVICE

UNDER SECRETARY FOR ECONOMIC AFFAIRS
ECONOMIC AND STATISTICS ADMINISTRATION

DIRECTOR BUREAU OF THE CENSUS

CHIEF ECONOMIST

DIRECTOR BUREAU OF ECONOMIC ANALYSIS

ASSISTANT SECRETARY FOR COMMUNICATIONS AND INFORMATION
NATIONAL TELECOMMUNICATIONS AND INFORMATION ADMINISTRATION

NATIONAL DIRECTOR
MINORITY BUSINESS DEVELOPMENT AGENCY

ASSISTANT SECRETARY FOR ECONOMIC DEVELOPMENT
ECONOMIC DEVELOPMENT ADMINISTRATION

UNDER SECRETARY FOR INDUSTRY AND SECURITY
BUREAU OF INDUSTRY AND SECURITY

ASSISTANT SECRETARY FOR EXPORT ENFORCEMENT

ASSISTANT SECRETARY FOR EXPORT ADMINISTRATION

UNDER SECRETARY FOR INTERNATIONAL TRADE
INTERNATIONAL TRADE ADMINISTRATION

ASSISTANT SECRETARY FOR IMPORT ADMINISTRATION

ASSISTANT SECRETARY FOR MANUFACTURING AND SERVICES

ASSISTANT SECRETARY FOR MARKET ACCESS AND COMPLIANCE

ASSISTANT SECRETARY FOR TRADE PROMOTION AND DIRECTOR GENERAL UNITED STATES AND FOREIGN COMMERCIAL SERVICE

UNDER SECRETARY AND ADMINISTRATOR
NATIONAL OCEANIC AND ATMOSPHERIC ADMINISTRATION

ASSISTANT SECRETARY FOR OCEANS AND ATMOSPHERE AND DEPUTY ADMINISTRATOR

and provides general assistance to the business community.

For further information, call 202–482–1360.

Sources of Information

Age and Citizenship Age search and citizenship information is available from the Personal Census Search Unit, Bureau of the Census, National Processing Center, P.O. Box 1545, Jeffersonville, IN 47131. Phone, 812–218–3046.

Economic Development Information Clearinghouse The EDA will host on its Web site the Economic Development Information Clearinghouse, an online depository of information on economic development (Internet, www.doc.gov/eda).

Contracting and Small Business For information regarding contract opportunities, contact the Office of Small and Disadvantaged Business Utilization. Phone, 202–482–1472. Internet, www.doc.gov/osdbu.

Employment Information is available electronically through the Internet, at www.doc.gov/ohrm. Phone, 202–482–5138. The National Oceanic and Atmospheric Administration has field employment offices at the Western Administrative Support Center, Bin C15700, 7600 Sand Point Way NE., Seattle, WA 98115 (phone, 206–526–6294); 325 Broadway, Boulder, CO 80303 (phone, 303–497–6332); 601 East Twelfth Street, Kansas City, MO 64106 (phone, 816–426–2056); and 200 World Trade Center, Norfolk, VA 23510–1624 (phone, 757–441–6516).

Environment The National Oceanic and Atmospheric Administration conducts research and gathers data about the oceans, atmosphere, space, and Sun, and applies this knowledge to science and service in ways that touch the lives of all Americans, including warning of dangerous weather, charting seas and skies, guiding our use and protection of ocean and coastal resources, and improving our understanding and stewardship of the environment which sustains us all. For further information, contact the Office of Public and Constituent Affairs, National Oceanic and Atmospheric Administration, Room 6013, Fourteenth Street and Constitution Avenue NW., Washington, DC 20230. Phone, 202–482–6090. Fax, 202–482–3154. Internet, http://www.noaa.gov.

Inspector General Hotline The Office of Inspector General works to promote economy, efficiency, and effectiveness and to prevent and detect fraud, waste, abuse, and mismanagement in departmental programs and operations. Contact the Hotline, Inspector General, P.O. Box 612, Ben Franklin Station, Washington, DC 20044. Phone, 202–482–2495, or 800–424–5197 (toll free). TTD, 202–482–5923, or 800–854–8407 (toll free). Fax, 202–789–0522. Internet, www.oig.doc.gov. Email, hotline@oig.doc.gov.

Publications The titles of selected publications are printed below with the operating units responsible for their issuance. These and other publications dealing with a wide range of business, economic, environmental, scientific, and technical matters are announced in the weekly Business Service Checklist, which may be purchased from the Superintendent of Documents, Government Printing Office, Washington, DC 20402. Phone, 202–512–1800.

For further information concerning the Department of Commerce, contact the Office of Public Affairs, Department of Commerce, Fourteenth Street and Constitution Avenue NW., Room 5040, Washington, DC 20230. Phone, 202–482–3263. Internet, http://www.doc.gov.

Bureau of Industry and Security

Department of Commerce, Washington, DC 20230
Phone, 202–482–2721. Internet, http://www.bis.doc.gov.

Under Secretary for Industry and Security	ERIC HIRSCHHORN
Deputy Under Secretary	DANIEL O. HILL

Assistant Secretary for Export Administration   KEVIN WOLF
Assistant Secretary for Export Enforcement   DAVID W. MILLS
Chief Financial Officer   GAY SHRUM

[For the Bureau of Industry and Security statement of organization, see the Federal Registers of June 7, 1988, 53 FR 20881, and April 26, 2002, 67 FR 20630]

The mission of the Bureau of Industry and Security (BIS) is to advance U.S. national security, foreign policy, and economic objectives by ensuring an effective export control and treaty compliance system and promoting continued U.S. strategic technology leadership. BIS activities include regulating the export of sensitive goods and technologies in an effective and efficient manner; enforcing export control, antiboycott, and public safety laws; cooperating with and assisting other countries on export control and strategic trade issues; assisting U.S. industry to comply with international arms control agreements; and monitoring the viability of the U.S. defense industrial base and seeking to ensure that it is capable of satisfying U.S. national and homeland security needs.

Export Administration   The Office of the Assistant Secretary for Export Administration is responsible for export licenses, treaty compliance, treaty obligations relating to weapons of mass destruction, and the defense industrial and technology base. The Office regulates the export of dual-use items requiring licenses for national security, nonproliferation, foreign policy, and short supply; ensures that approval or denial of license applications is consistent with economic and security concerns; promotes an understanding of export control regulations within the business community; represents the Department in interagency and international forums relating to export controls, particularly in multilateral regimes; monitors the availability of industrial resources of national defense; analyzes the impact of export controls on strategic industries; and assesses the security consequences of certain foreign investments.

Export Enforcement   The Office of the Assistant Secretary for Export Enforcement enforces dual-use export controls. This enables exporters to take advantage of legal export opportunities while ensuring that illegal exports will be detected and either prevented or investigated and sanctioned. The Office also ensures prompt, aggressive action against restrictive trade practices; reviews visa applications of foreign nationals to prevent illegal technology transfers; and conducts cooperative enforcement activities on an international basis.

For information on the Export Enforcement field offices, visit our Web site at www.bis.doc.gov/about/ programoffices.htm.

Management and Policy Coordination
The Management and Policy Coordination (MPC) unit establishes and evaluates the Bureau's overall policy agenda, priorities, goals, unit objectives, and key metrics. MPC performs oversight of program operations and expenditures; executes or supervises the President's Management Agenda; and adjudicates appeals of licensing and enforcement decisions as part of an extended legal process involving administrative law judges and the Office of General Counsel. MPC provides guidance and coordination for the Bureau's participation in the Export Control and Related Border Security Assistance Program, which provides technical assistance to strengthen the export and transit control systems of nations that are identified as potential locations for the exporting of weapons of mass destruction, missile delivery systems, or the commodities, technologies, and equipment that can be used to design and build them.

Sources of Information

The Bureau's Web site (Internet, www. bis.doc.gov) provides information for the U.S. business community, including export news, general, subject and policy fact sheets, updates to the Export Administration regulations, Bureau program information, e-FOIA information, and export seminar event schedules. Publications available on the site include the Bureau's annual report, foreign policy controls report, and international diversification and defense market assessment guides. The Government Printing Office, in conjunction with the Bureau, has created a Web site that contains an up-to-date database of the entire export administration regulations, including the commerce control list, the commerce country chart, and the denied persons list (www.access.gpo.gov/bis/index.html). The Outreach and Educational Services Division has offices in Washington, DC (phone, 202–482–4811; fax, 202–482–2927) and on the West Coast (phone, 949–660–0144, or 408–351–3378; fax, 949–660–9347, or 408–351–3355). For enforcement-related questions, contact the partnership-in-security hotline (phone, 800–424–2980).

For further information, contact the Bureau of Industry and Security, Office of Public Affairs, Room 3897, Fourteenth Street and Constitution Avenue NW., Washington, DC 20230. Phone, 202–482–2721. Internet, http://www.bis.doc.gov.

Economic Development Administration

Department of Commerce, Washington, DC 20230
Phone, 202–482–2309. Internet, http://www.eda.gov.

Assistant Secretary for Economic Development	JOHN FERNANDEZ
Deputy Assistant Secretary	BRIAN MCGOWAN
Chief Financial Office and Director of Administration	SANDRA WALTERS

The Economic Development Administration (EDA) was created in 1965 under the Public Works and Economic Development Act (42 U.S.C. 3121) as part of an effort to target Federal resources to economically distressed areas and to help develop local economies in the United States. It was mandated to assist rural and urban communities that were outside the mainstream economy and that lagged in economic development, industrial growth, and personal income.

EDA provides grants to States, regions, and communities across the Nation to help wealth and minimize poverty by promoting a favorable business environment to attract private capital investment and higher skill, higher wage jobs through capacity building, planning, infrastructure, research grants, and strategic initiatives. Through its grant program, EDA utilizes public sector resources to facilitate an environment where the private sector risks capital and job opportunities are created.

Public works and development facilities grants support infrastructure projects that foster the establishment or expansion of industrial and commercial businesses, supporting the retention and creation of jobs.

Planning grants support the design and implementation of effective economic development policies and programs, by local development organizations, in States and communities. EDA funds a network of over 350 planning districts throughout the country.

Technical assistance provides for local feasibility and industry studies, management and operational assistance, natural resource development, and export promotion. In addition, EDA funds a network of university centers that provides technical assistance.

Research, evaluation, and demonstration funds are used to support studies about the causes of economic distress and to seek solutions to counteract and prevent such problems.

Economic adjustment grants help communities adjust to a gradual erosion or sudden dislocation of their local economic structure. This assistance provides funding for both planning and implementation to address economic change.

The Trade Adjustment Assistance program helps U.S. firms and industries injured as the result of economic globalization. A nationwide network of Trade Adjustment Assistance Centers offers low-cost, effective professional assistance to certified firms to develop and implement recovery strategies.

For information on the EDA's regional offices, visit http://www.eda.gov/AboutEDA/Regions.xml.

For further information, contact the Economic Development Administration, Department of Commerce, Washington, DC 20230. Phone, 202–482–2309. Fax, 202–273–4723. Internet, http://www.eda.gov.

Economic and Statistics Administration

Department of Commerce, Washington, DC 20230
Phone, 202–482–3727. Internet, http://www.esa.doc.gov.

Under Secretary for Economic Affairs	REBECCA M. BLANK
Deputy Under Secretary	NANCY POTOK
Associate Under Secretary for Management	KIM WHITE
Chief Economist	MARK E. DOMS
Director, Bureau of the Census	ROBERT GROVES
Director, Bureau of Economic Analysis	J. STEVEN LANDEFELD

The Economics and Statistics Administration (ESA) provides broad and targeted economic data, analyses, and forecasts for use by Government agencies, businesses, and others, as well as develops domestic and international economic policy. The Under Secretary is the chief economic adviser to the Secretary and provides leadership and executive management of the Bureau of the Census and the Bureau of Economic Analysis.

ESA provides key business, economic, and international trade information products that American business and the public can use to make informed decisions through www.stat-usa.gov.

Bureau of the Census

[For the Bureau of the Census statement of organization, see the Federal Register of Sept. 16, 1975, 40 FR 42765]

The Bureau of the Census was established as a permanent office by act of March 6, 1902 (32 Stat. 51). The major functions of the Census Bureau are authorized by the Constitution, which provides that a census of population shall be taken every 10 years, and by laws codified as title 13 of the United States Code. The law also provides that the information collected by the Census Bureau from individual persons, households, or establishments be kept strictly confidential and be used only for statistical purposes. The Census Bureau is responsible for the decennial censuses of population and housing; the quinquennial censuses of State and local governments, manufacturers, mineral industries, distributive trades, construction industries, and transportation; current surveys that provide information on many of the subjects covered in the censuses at monthly, quarterly, annual, or other intervals; compilation of current statistics on U.S. foreign trade, including data on imports, exports, and shipping; special censuses at the request and expense of State and local government units; publication of estimates and projections of the population; publication of current data on population and housing characteristics; and current reports on manufacturing, retail and wholesale trade, services, construction, imports and exports, State and local government finances and employment, and other subjects.

The Census Bureau makes available statistical results of its censuses, surveys, and other programs to the public through printed reports, CD–ROMs and DVDs, the Internet and other media, and prepares special tabulations sponsored and paid for by data users. It also produces statistical compendia, catalogs, guides, and directories that are useful in locating information on specific subjects. Upon request, the Bureau makes searches of decennial census records and furnishes certificates to individuals for use as evidence of age, relationship, or place of birth. A fee is charged for searches.

For information on the Census Bureau regional offices, visit http://www.census. gov/regions.

For further information, contact the Marketing Service Office, Bureau of the Census, Department of Commerce, Washington, DC 20233. Phone, 301–763–4636. Fax, 301–457–3842. Internet, http://www.census.gov.

Bureau of Economic Analysis

[For the Bureau of Economic Analysis statement of organization, see the Federal Register of Dec. 29, 1980, 45 FR 85496]

The Bureau of Economic Analysis (BEA) promotes a better understanding of the U.S. economy by providing the most timely, relevant, and accurate economic accounts data in an objective and cost-effective manner. BEA's economic statistics are closely watched and provide a comprehensive picture of the U.S. economy. BEA prepares national, regional, industry, and international accounts that present essential information on such issues in the world economy.

BEA's national economic statistics provide a comprehensive look at U.S. production, consumption, investment, exports and imports, and income and saving. The international transactions accounts provide information on trade in goods and services (including the balance of payments and trade), investment income, and government and private finances. In addition, the accounts measure the value of U.S. international assets and liabilities and direct investment by multinational companies.

The regional accounts provide data on total and per capita personal income by region, State, metropolitan area, and county, and on gross State product. The industry economic account provides a detailed view of the interrelationships between U.S. producers and users and the contribution to production across industries.

For further information, contact the Public Information Office, Bureau of Economic Analysis, Department of Commerce, Washington, DC 20230. Phone, 202–606–9900. Fax, 202–606–5310. Email, customerservice@ bea.gov. Internet, http://www.bea.gov.

Sources of Information

Census Publications Numerous publications presenting statistical information on a wide variety of subjects are available from the Government Printing Office, including the following: Statistical Abstract of the U.S.; Historical Statistics of the United States, Colonial Times to 1970; County and City Data Book, 1994; and State and Metropolitan Area Data Book, 1997–1998.

Census Electronic Analysis Employment opportunities, data highlights, large data files, access tools, and other material are available on the World Wide Web. Internet, www.census.gov. Email, webmaster@census.gov.

Economic Analysis Publications The Survey of Current Business (Monthly Journal) is available from the Government Printing Office. Current and historical estimates, general information, and employment opportunities are available on BEA's Web site at www.bea.gov. For more information, contact the Public Information Office. Phone, 202–606–9900. Email, webmaster@bea.gov.

For further information, contact the Economics and Statistics Administration, Department of Commerce, Washington, DC 20230. Phone, 202–482–3727. Internet, http://www.esa.doc.gov.

International Trade Administration

Department of Commerce, Washington, DC 20230
Phone, 202–482–3917. Internet, http://www.trade.gov.

Under Secretary for International Trade	FRANCISCO SANCHEZ
Deputy Under Secretary	MICHELLE O'NEILL

Assistant Secretary for Import Administration   (VACANCY)
Assistant Secretary for Market Access and   (VACANCY)
    Compliance
Assistant Secretary for Manufacturing and   NICOLE Y. LAMB-HALE
    Services
Assistant Secretary for Trade Promotion   SURESH KUMAR
    and Director of the U.S. and Foreign
    Commercial Service
Chief Financial Officer and Director of   PATTY SEFCIK
    Administration
Chief Information Officer   RENEE MACKLIN

[For the International Trade Administration statement of organization, see the Federal Register of Jan. 25, 1980, 45 FR 6148]

The International Trade Administration (ITA) was established on January 2, 1980, by the Secretary of Commerce to promote world trade and to strengthen the international trade and investment position of the United States.

ITA is headed by the Under Secretary for International Trade, who coordinates all issues concerning trade promotion, international commercial policy, market access, and trade law enforcement. The Administration is responsible for nonagricultural trade operations of the U.S. Government and supports the trade policy negotiation efforts of the U.S. Trade Representative.

Import Administration The Office of Import Administration defends American industry against injurious and unfair trade practices by administering efficiently, fairly, and in a manner consistent with U.S. international trade obligations the antidumping and countervailing duty laws of the United States. The Office ensures the proper administration of foreign trade zones and advises the Secretary on establishment of new zones; oversees the administration of the Department's textiles program; and administers programs governing watch assemblies, and other statutory import programs.

Market Access and Compliance The Office of Market Access and Compliance advises on the analysis, formulation, and implementation of U.S. international economic policies and carries out programs to promote international trade, improve access by U.S. companies to overseas markets,

and strengthen the international trade and investment position of the United States. The Office analyzes and develops recommendations for region- and country-specific international economic, trade, and investment policy strategies and objectives. In addition, the Office is responsible for implementing, monitoring, and enforcing foreign compliance with bilateral and multilateral trade agreements.

Manufacturing and Services The Manufacturing and Services unit advises on domestic and international trade and investment policies affecting the competitiveness of U.S. industry and carries on a program of research and analysis on manufacturing and services. Based on this analysis and interaction with U.S. industry, the unit Secretary develops strategies, policies, and programs to strengthen the competitive position of U.S. industries in the United States and world markets. The unit manages an integrated program that includes both industry and economic analysis, trade policy development and multilateral, regional, and bilateral trade agreements for manufactured goods and services; administers trade arrangements (other than those involving AD/CVD proceedings) with foreign governments in product and service areas; and develops and provides business information and assistance to the United States on its rights and opportunities under multilateral and other agreements.

Trade Promotion and U.S. and Foreign Commercial Service The Trade Promotion and U.S. and Foreign

Commercial Service unit directs ITA's export promotion programs, develops and implements a unified goal-setting and evaluation process to increase trade assistance to small- and medium-sized businesses, directs a program of international trade events, market research, and export-related trade information products and services; and directs programs to aid U.S. firms to compete successfully for major projects and procurements worldwide. ITA provides a comprehensive platform of export assistance services to support U.S. firms who enter or expand their presence in overseas markets, including counseling, trade events, and outreach services through 109 export assistance centers located in the United States and 158 posts located in 83 countries throughout the world. For a complete listing of ITA's export assistance centers, both in the United States and abroad, consult the Web site at www.export.gov/eac or call the Trade Information Center at 1–800–872–8723.

Sources of Information

Electronic Access   The Administration maintains a Web site, (Internet, www.trade.gov) which offers the single best place for individuals or firms seeking reports, documents, import case/regulations, texts of international agreements like NAFTA and GATT, market research, and points of contact for assistance in exporting, obtaining remedies from unfair trading practices, or receiving help with market access problems. Customers are able to review comprehensive information on how to export, search for trade information by either industry or by country, learn how to petition against unfairly priced imports, and obtain information on a number of useful international trade-related products like overseas trade leads and agent distributor reports. The Web site also features email addresses and locations for trade contacts in Washington, overseas, in major exporting centers in the United States, and in other parts of the Federal Government.

For further information, contact the International Trade Administration, Department of Commerce, Washington, DC 20230. Phone, 202–482–3917. Internet, http://www.trade.gov.

Minority Business Development Agency

Department of Commerce, Washington, DC 20230
Phone, 202–482–5061. Internet, http://www.mbda.gov.

National Director, Minority Business Development Agency	DAVID HINSON
Deputy Director	(VACANCY)
Associate Director for Management	EDITH J. MCCLOUD

[For the Minority Business Development Agency statement of organization, see the Federal Register of Mar. 17, 1972, 37 FR 5650, as amended]

The Minority Business Development Agency was established by Executive order in 1969. The Agency develops and coordinates a national program for minority business enterprise.

The Agency was created to assist minority businesses in achieving effective and equitable participation in the American free enterprise system and in overcoming social and economic disadvantages that have limited their participation in the past. The

Agency provides national policies and leadership in forming and strengthening a partnership of business, industry, and government with the Nation's minority businesses.

Business development services are provided to the minority business community through three vehicles: the minority business opportunity committees, which disseminate information on business opportunities; the minority business development centers, which provide management and technical assistance and other business development services; and electronic commerce, which includes a Web site that shows how to start a business and use the service to find contract opportunities.

The Agency promotes and coordinates the efforts of other Federal agencies in assisting or providing market opportunities for minority business. It coordinates opportunities for minority firms in the private sector. Through such public and private cooperative activities, the Agency promotes the participation of Federal, State, and local governments, and business and industry in directing resources for the development of strong minority businesses.

Sources of Information

Publications   Copies of Minority Business Today and the BDC Directory may be obtained by contacting the Office of Business Development. Phone, 202–482–6022. Comprehensive information about programs, policy, centers, and access to the job matching database is available through the Internet at www.mbda.gov.

For further information, contact the Office of the Director, Minority Business Development Agency, Department of Commerce, Washington, DC 20230. Phone, 202–482–5061. Internet, http://www.mbda.gov.

National Oceanic and Atmospheric Administration

Department of Commerce, Washington, DC 20230
Phone, 202–482–2985. Internet, http://www.noaa.gov.

Under Secretary for Oceans and Atmosphere and Administrator	JANE LUBCHENCO
Assistant Secretary for Oceans and Atmosphere and Deputy Administrator	(VACANCY)
Deputy Under Secretary for Oceans and Atmosphere	MARY M. GLACKIN
Deputy Assistant Secretary for Oceans and Atmosphere	(VACANCY)
Deputy Assistant Secretary for International Affairs	JAMES M. TURNER
General Counsel	LOIS SCHIFFER
Chief Administrative Officer	WILLIAM F. BROGLIE
Chief Financial Officer	MAUREEN E. WYLIE
Chief Information Officer	JOE KLIMAVICZ
Director, Office of Communications	JUSTIN KENNEY
Director, Office of Education	LOUISA KOCH
Director, Office of Legislative Affairs	JOHN GRAY
Director, Office of Program Analysis & Evaluation	STEVE AUSTIN
Director, Workforce Management Office	EDUARDO RIBAS

[For the National Oceanic and Atmospheric Administration statement of organization, see the Federal Register of Feb. 13, 1978, 43 FR 6128]

The National Oceanic and Atmospheric Administration (NOAA) was formed on October 3, 1970, by Reorganization Plan No. 4 of 1970 (5 U.S.C. app.).

NOAA's mission entails environmental assessment, prediction, and stewardship. It is dedicated to monitoring and assessing the state of the environment in order to make accurate and timely forecasts to protect life, property, and natural resources, as well as to promote the economic well-being of the United States and to enhance its environmental security. NOAA is committed to protecting America's ocean, coastal, and living marine resources while promoting sustainable economic development.

For a complete listing of NOAA facilities and activities in your State or Territory, visit www.legislative.noaa.gov/NIYS/index.html.

National Weather Service The National Weather Service (NWS) provides weather, water and climate warnings, forecasts and data for the United States, its territories, adjacent waters, and ocean areas. NWS data and products form a national information database and infrastructure used by Government agencies, the private sector, the public, and the global community to protect life and property and to enhance the national economy. Working with partners in Government, academic and research institutions and private industry, NWS strives to ensure their products and services are responsive to the needs of the American public. NWS data and information services support aviation and marine activities, wildfire suppression, and many other sectors of the economy. NWS supports national security efforts with long- and short-range forecasts, air quality and cloud dispersion forecasts, and broadcasts of warnings and critical information over the 800-station NOAA Weather Radio network.

For further information, contact the National Weather Service, Attention: Executive Affairs, 1325 East-West Highway, Silver Spring, MD 20910–3283. Phone, 301–713–0675. Fax, 301–713–0049. Internet, http://www.nws.noaa.gov.

National Environmental Satellite, Data, and Information Service The National Environmental Satellite, Data, and Information Service (NESDIS) operates the Nation's civilian geostationary and polar-orbiting environmental satellites. It also manages the largest collection of atmospheric, climatic, geophysical, and oceanographic data in the world. From these sources, NESDIS develops and provides, through various media, environmental data for forecasts, national security, and weather warnings to protect life and property. This data is also used to assist in energy distribution, the development of global food supplies, the management of natural resources, and

in the recovery of downed pilots and mariners in distress.

For further information, contact the National Environmental Satellite, Data, and Information Service, 1335 East-West Highway, Silver Spring, MD 20910–3283. Phone, 301–713–3578. Fax, 301–713–1249. Internet, http://www.noaa.gov/nesdis/nesdis.html.

National Marine Fisheries Service The National Marine Fisheries Service (NMFS) supports the management, conservation, and sustainable development of domestic and international living marine resources and the protection and restoration of healthy ecosystems. NMFS is involved in the stock assessment of the Nation's multi-billion-dollar marine fisheries, protecting marine mammals and threatened species, habitat conservation operations, trade and industry assistance, and fishery enforcement activities.

For further information, contact the National Marine Fisheries Service, 1315 East-West Highway, Silver Spring, MD 20910. Phone, 301–713–2239. Fax, 301–713–2258. Internet, http://www.nmfs.noaa.gov.

National Ocean Service The National Ocean Service (NOS) works to balance the Nation's use of coastal resources through research, management, and policy. NOS monitors the health of U.S. coasts by examining how human use and natural events impact coastal ecosystems. Coastal communities rely on NOS for information about natural hazards so they can more effectively reduce or eliminate the destructive effects of coastal hazards. NOS assesses the damage caused by hazardous material spills and works to restore or replace the affected coastal resources. Through varied programs, NOS protects wetlands, water quality, beaches, and wildlife. In addition, NOS provides a wide range of navigational products and data that assist vessels' safe movement through U.S. waters and provides the basic set of information that establishes the latitude, longitude, and elevation framework necessary for the Nation's surveying, navigation, positioning, and mapping activities.

For further information, contact the National Ocean Service, Room 13231, SSMC 4, 1305 East-West Highway, Silver Spring, MD 20910. Phone, 301–713–3070. Fax, 301–713–4307. Internet, http://www.nos.noaa.gov.

Office of Oceanic and Atmospheric Research The Office of Oceanic and Atmospheric Research (OAR) carries out research on weather, air quality and composition, climate variability and change, and ocean, coastal, and Great Lakes ecosystems. OAR conducts and directs its research programs in coastal, marine, atmospheric, and space sciences through its own laboratories and offices, as well as through networks of university-based programs across the country.

For further information, contact the Office of Oceanic and Atmospheric Research, Room 11458, 1315 East-West Highway, Silver Spring, MD 20910. Phone, 301–713–2458. Fax, 301–713–0163. Internet, http://www.oar.noaa.gov.

Office of Marine and Aviation Operations The Office of Marine and Aviation Operations maintains a fleet of ships and aircraft, and manages several safety programs. Ships and aircraft are used for operational data collection and research in support of NOAA's mission, the Global Earth Observation System, and the Integrated Ocean Observing System. This includes flying "hurricane hunter" aircraft into nature's most turbulent storms to collect data critical to hurricane research.

For further information, contact Office of Marine and Aviation Operations, Suite 500, 8403 Colesville Rd., Silver Spring, MD 20910. Phone, 301–713–1045.

Sources of Information

Publications and Resources The Administration provides technical memoranda, technical reports, monographs, nautical and aeronautical charts, coastal zone maps, data tapes, and a wide variety of raw and processed environmental data. Information on NOAA products is available through the Internet at www.noaa.gov. Contact the Office of Public and Constituent Affairs, Fourteenth Street and Constitution Avenue NW., Washington, DC 20230. Phone, 202–482–6090. Fax, 202–482–3154.

For further information, contact the Office of Public Affairs, National Oceanic and Atmospheric Administration, Department of Commerce, Washington, DC 20230. Phone, 202–482–4190. Internet, http://www.noaa.gov.

National Telecommunications and Information Administration

Department of Commerce, Washington, DC 20230
Phone, 202–428–1840. Internet, http://www.ntia.doc.gov.

Assistant Secretary for Communications and Information	LAWRENCE STRICKLIN
Deputy Assistant Secretary	ANNA M. GOMEZ
Chief Counsel	KATHY D. SMITH
Associate Administrator for Spectrum Management	KARL NEBBIA
Associate Administrator for Policy Analysis and Development	DANIEL J. WEITZNER
Associate Administrator for International Affairs	FIONA ALEXANDER
Associate Administrator for Telecommunications and Information Applications	BERNADETTE A. MCGUIRE-RIVERA
Director, Institute for Telecommunication Sciences	ALAN VINCENT

[For the National Telecommunications and Information Administration statement of organization, see the Federal Register of June 5, 1978, 43 FR 24348]

The National Telecommunications and Information Administration (NTIA) was established in 1978 by Reorganization Plan No. 1 of 1977 (5 U.S.C. app.)

and Executive Order 12046 of March 27, 1978 (3 CFR, 1978 Comp., p. 158), by combining the Office of Telecommunications Policy of the Executive Office of the President and the Office of Telecommunications of the Department of Commerce to form a new agency reporting to the Secretary of Commerce. NTIA operates under the authority of the National Telecommunications and Information Administration Organization Act (47 U.S.C. 901).

NTIA's principal responsibilities and functions include serving as the principal executive branch adviser to the President on telecommunications and information policy; developing and presenting U.S. plans and policies at international communications conferences and related meetings; prescribing policies for and managing Federal use of the radio frequency spectrum; serving as the principal Federal telecommunications research and engineering laboratory, through NTIA's Institute for Telecommunication Sciences, headquartered in Boulder, CO; administering Federal programs to assist telecommunication facilities, public safety organizations, and the general public with the transition to digital broadcasting; providing grants through

the Broadband Technology Opportunities Program to increase broadband accessibility in underserved areas of the United States; and providing grants through the Public Telecommunications Facilities Program to extend delivery or public telecommunications services to U.S. citizens, to increase ownership and management by women and minorities, and to strengthen the capabilities of existing public broadcasting stations to provide telecommunications services.

Sources of Information

Publications Since 1970, several hundred technical reports and memoranda, special publications, contractor reports, and other information products have been published by NTIA or its predecessor agency. The publications are available from the National Telecommunications and Information Administration, Department of Commerce, Washington, DC 20230 (phone, 202–482–1551); or the National Telecommunications and Information Administration, Institute for Telecommunication Sciences, Department of Commerce, Boulder, CO 80302 (phone, 303–497–3572). More information can be obtained by visiting the Web site at www.ntia.doc.gov.

For further information, contact the National Telecommunications and Information Administration, Department of Commerce, Washington, DC 20230. Phone, 202–482–1551. Internet, http://www.ntia.doc.gov.

National Institute of Standards and Technology

100 Bureau Drive, Gaithersburg, MD 20899
Phone, 301–975–6478. TTY, 800–877–8339. Internet, http://www.nist.gov.

Director	PATRICK D. GALLAGHER
Deputy Director	(VACANCY)
Chief Financial Officer	DAVID ROBINSON
Chief Scientist	KATHARINE GEBBIE, Acting

The National Institute of Standards and Technology (NIST) operates under the authority of the National Institute of Standards and Technology Act (15 U.S.C. 271), which amends the Organic Act of March 3, 1901 (ch. 872), that created the National Bureau of Standards (NBS) in 1901. In 1988, Congress renamed NBS

as NIST and expanded its activities and responsibilities.

NIST is a nonregulatory Federal agency within the Commerce Department. Its mission is to promote measurement science, standards, and technology to enhance productivity, facilitate trade, and improve the quality of life. NIST carries out its mission through the NIST laboratories, which conduct research to advance the U.S. technological infrastructure; the Baldrige National Quality Program, which helps U.S. businesses and other organizations improve the performance and quality of their operations; the Hollings Manufacturing Extension Partnership, which helps smaller firms adopt new manufacturing and management technologies; and the Technology Innovative Program, which provides cost-shared awards to industry and other institutions for high-risk, high-reward research in areas of critical national need.

Sources of Information

Publications Journal of Research; Publications of the Advanced Technology Program and Manufacturing Extension Partnership Program; Handbook of Mathematical Functions; Experimental Statistics; International System of Units (SI); Standard Reference Materials Catalog; Specifications, Tolerances, and Other Technical Requirements for Weighing and Measuring Devices Handbook; and Uniform Laws and Regulations Handbook are available from the Government Printing Office.

For further information, contact the National Institute of Standards and Technology, 100 Bureau Drive, Gaithersburg, MD. Phone, 301–975–6478. Fax, 301–926–1630. Email, inquiries@nist.gov. Internet, http://www.nist.gov.

National Technical Information Service Administration

5301 Shawnee Road, Alexandria, VA 22312
Phone, 703–605–6050; 888–584–8332. Internet, http://www.ntis.gov.

Director	BRUCE BORZINO
Deputy Director	(VACANCY)
Chief Financial Officer	MARY HOUFF
Chief Information Officer	KEITH SINNER

The National Technical Information Service (NTIS) operates a central clearinghouse of scientific and technical information that is useful to U.S. business and industry. NTIS collects scientific and technical information; catalogs, abstracts, indexes, and permanently archives the information; disseminates products in the forms and formats most useful to its customers; develops electronic and other media to disseminate information; and provides information processing services to other Federal agencies. NTIS receives no appropriations. Its revenue comes from two sources: the sale of technical reports to business and industry, schools and universities, State and local government offices, and the public at large and from services to Federal agencies that help them communicate more effectively with their employees and constituents.

The NTIS collection of approximately 2.5 million works covers a broad array of subjects and includes reports on the results of research and development and scientific studies on manufacturing processes, current events, and foreign and domestic trade; business and management studies; social, economic, and trade statistics; computer software and databases; health care reports, manuals, and data; environmental handbooks, regulations, economic studies, and applied technologies; directories to Federal laboratory and technical resources; and global competitive intelligence. The collection also includes audiovisual training materials in such areas as foreign languages, workplace safety and health, law enforcement, and fire services.

The NTIS Bibliographic Database is available online through commercial vendors and on CD–ROM from NTIS. Database entries since 1964 are available at the NTIS Web site. Internet, www.ntis.gov.

Sources of Information

Products and Services   For general inquiries, to place an order, or to request the Catalog of NTIS Products and

Services, contact the NTIS Sales Desk from 8 a.m. to 6 p.m. (eastern standard time). Phone, 800–553–6847. TDD, 703–487–4639. Fax, 703–605–6900. Email, info@ntis.gov. Internet, www.ntis. gov. To inquire about NTIS information services for other Federal agencies, call 703–605–6540.

For further information, contact the National Technical Information Service Administration, 5301 Shawnee Road, Arlington, VA 22312. Phone, 703–605–6050, or 888–584–8332. Internet, http://www.ntis.gov.

United States Patent and Trademark Office

600 Dulany Street, Arlington, VA 22313
Phone, 571–272–8400. Internet, http://www.uspto.gov.

Under Secretary for Intellectual Property and Director	DAVID KAPPOS
Deputy Under Secretary for Intellectual Property and Deputy Director	SHARON BARNER
Commissioner for Patents	ROBERT STOLL
Commissioner for Trademarks	LYNNE G. BERESFORD

[For the Patent and Trademark Office statement of organization, see the Federal Register of Apr. 14, 1975, 40 FR 16707]

The United States Patent and Trademark Office (USPTO) was established by the act of July 19, 1952 (35 U.S.C. 1) to promote the progress of science and the useful arts by securing for limited times to inventors the exclusive right to their respective discoveries for a certain period of time (Article I, Section 8 of the United States Constitution). The registration of trademarks is based on the commerce clause of the U.S. Constitution.

USPTO examines and issues patents. There are three major patent categories: utility patents, design patents, and plant patents. USPTO also issues statutory invention registrations and processes international patent applications.

Through the registration of trademarks, USPTO assists businesses in protecting their investments, promoting goods and services, and safeguarding consumers against confusion and deception in the marketplace. A trademark includes any distinctive word, name, symbol, device, or any combination thereof adopted and used or intended to be used by a manufacturer or merchant to identify his goods or services and distinguish them from those manufactured or sold

by others. Trademarks are examined by the Office for compliance with various statutory requirements to prevent unfair competition and consumer deception.

In addition to the examination of patent and trademark applications, issuance of patents, and registration of trademarks, USPTO advises and assists government agencies and officials in matters involving all domestic and global aspects of intellectual property. USPTO also promotes an understanding of intellectual property protection.

USPTO provides public access to patent, trademark, and related scientific and technical information. Patents and trademarks may be freely reviewed and searched online at www.uspto.gov or at designated Patent and Trademark Depository Libraries. There are 80 Patent and Trademark Depository Libraries located within the United States and the territory of Puerto Rico. Additionally, USPTO's Scientific and Technical Information Center in Alexandria, VA, houses over 120,000 volumes of scientific and technical books in various languages; 90,000 bound volumes of periodicals devoted to science and technology; the official journals of 77 foreign patent organizations; and over 40 million foreign patents on paper, microfilm, microfiche, and CD–ROM.

Sources of Information

Patents   The United States Patent and Trademark Office has priority programs for advancement of examination of certain patent applications where the invention could materially enhance the quality of the environment of mankind. For further information, contact the Commissioner for Patents, Office of Petitions, Washington, DC 20231. Phone, 703–305–9282.

General Information   General Information Concerning Patents, Basic Facts About Trademarks, Official Gazette of the United States Patent and Trademark Office, and Attorneys and Agents Registered To Practice Before the U.S. Patent and Trademark Office are available from the Government Printing Office. Publications can be accessed through the Internet at www.uspto.gov. Phone, 703–308–4357, or 800–786–9199.

For further information, contact the Office of Public Affairs, United States Patent and Trademark Office, 600 Dulany Street, Alexandria, VA 22314. Phone, 571–272–8400. Internet, http://www.uspto.gov.

DEPARTMENT OF DEFENSE

Office of the Secretary, The Pentagon, Washington, DC 20301–1155
Phone, 703–545–6700. Internet, http://www.defense.gov..

Secretary of Defense	ROBERT M. GATES
Deputy Secretary of Defense	WILLIAM LYNN, III
Under Secretary of Defense for Acquisition, Technology, and Logistics	ASHTON B. CARTER
Principal Deputy Under Secretary of Defense for Acquisition, Technology, and Logistics	FRANK KENDALL, III
Assistant Secretary of Defense for Acquisition	(VACANCY)
Assistant Secretary of Defense for Logistics and Materiel Readiness	(VACANCY)
Director, Defense Research and Engineering	ZACHARY J. LEMNIOS
Deputy Under Secretary of Defense (Business Transformation)	PAUL A. BRINKLEY
Deputy Under Secretary of Defense (Installations and Environment)	DOROTHY L. ROBYN
Deputy Under Secretary of Defense (Science and Technology)	ANDRE VAN TILBORG
Under Secretary of Defense for Policy	MICHELE A. FLOURNOY
Principal Deputy Under Secretary of Defense for Policy	JAMES N. MILLER, JR.
Assistant Secretary of Defense (International Security Affairs)	ALEXANDER R. VERSHBOW
Assistant Secretary of Defense (Special Operations and Low-Intensity Conflict)	MICHAEL G. VICKERS
Assistant Secretary of Defense (Homeland Defense and America's Security Affairs)	PAUL N. STOCKTON
Assistant Secretary of Defense (Global Security Affairs)	MICHAEL L. NACHT
Assistant Secretary of Defense (Asian and Pacific Security Affairs)	WALLACE C. GREGSON
Deputy Assistant Secretary of Defense (Plans)	JANINE A. DAVIDSON
Deputy Under Secretary of Defense (Strategy, Plans and Forces)	KATHLEEN H. HICKS
Deputy Under Secretary of Defense (Policy Integration and Chief of Staff)	PETER F. VERGA
Under Secretary of Defense for Personnel and Readiness	CLIFFORD L. STANLEY
Principal Deputy Under Secretary of Defense for Personnel and Readiness	(VACANCY)
Assistant Secretary of Defense (Reserve Affairs)	DENNIS M. MCCARTHY
Assistant Secretary of Defense (Health Affairs)	(VACANCY)
Principal Deputy Assistant Secretary of Defense (Reserve Affairs)	DAVID L. MCGINNIS
Deputy Assistant Secretary of Defense (Resources)	JOHN T. HASTINGS

Deputy Under Secretary of Defense (Program Integration)	(VACANCY)
Deputy Under Secretary of Defense (Readiness)	(VACANCY)
Deputy Under Secretary of Defense (Military Personnel Policy)	WILLIAM J. CARR
Deputy Under Secretary of Defense (Military Community and Family Policy)	ARTHUR J. MYERS, *Acting*
Deputy Under Secretary of Defense (Transition Policy Care Coordination)	NOEL C. KOCH
Deputy Under Secretary of Defense (Civilian Personnel Policy)	(VACANCY)
Deputy Under Secretary of Defense (Plans)	GAIL H. MCGINN
Under Secretary of Defense (Comptroller)/Chief Financial Officer	ROBERT F. HALE
Principal Deputy Under Secretary of Defense (Comptroller)	MICHAEL J. MCCORD
Under Secretary of Defense (Intelligence)	JAMES R. CLAPPER, JR.
Assistant Secretary of Defense (Networks and Information Integration/DOD Chief Information Officer)	(VACANCY)
Principal Assistant Secretary of Defense (Networks and Information Integration/ DOD Chief Information Officer)	CHERYL J. ROBY
Assistant Secretary of Defense (Legislative Affairs)	ELIZABETH L. KING
Assistant Secretary of Defense (Public Affairs)	DOUGLAS B. WILSON
General Counsel	JEH CHARLES JOHNSON
Director, Operational Test and Evaluation	JAMES M. GILMORE
Principal Deputy Director for Operational Test and Evaluation	DAVID W. DUMA
Director, Cost Assessment and Program Evaluation	CHRISTINE H. FOX
Inspector General	GORDON S. HEDDELL
Director of Administration and Management	MICHAEL L. RHODES

Joint Chiefs of Staff

Chairman	ADM. MICHAEL G. MULLEN, USN
Vice Chairman	GEN. JAMES E. CARTWRIGHT, USMC
Chief of Staff, Army	GEN. GEORGE W. CASEY, JR., USA
Chief of Naval Operations	ADM. GARY ROUGHEAD, USN
Chief of Staff, Air Force	GEN. NORTON A. SCHWARTZ, USAF
Commandant, Marine Corps	GEN. JAMES T. CONWAY, USMC

[For the Department of Defense statement of organization, see the Code of Federal Regulations, Title 32, Chapter I, Subchapter R]

The Department of Defense is responsible for providing the military forces needed to deter war and protect the security of our country. The major elements of these forces are the Army, Navy, Marine Corps, and Air Force, consisting of about 1.3 million men and women on active duty. They are backed, in case of emergency, by the 825,000 members of the Reserve and National Guard. In addition, there are about 600,000 civilian employees in the Defense Department. Under the President, who is also Commander in Chief, the Secretary of Defense exercises authority, direction, and control over the Department, which includes the separately organized military departments of Army, Navy, and Air Force, the Joint Chiefs of Staff providing military advice, the combatant commands, and defense agencies and field activities established for specific purposes.

The National Security Act Amendments of 1949 redesignated the National Military Establishment as the Department of Defense and established it as an executive department (10 U.S.C. 111) headed by the Secretary of Defense.

Structure The Department of Defense is composed of the Office of the Secretary of Defense; the military departments and the military services within those departments; the Chairman of the Joint Chiefs of Staff and the Joint Staff; the combatant commands; the defense agencies; DOD field activities; and such other offices, agencies, activities, and commands as may be established or designated by law or by the President or the Secretary of Defense.

Each military department is separately organized under its own Secretary and functions under the authority, direction, and control of the Secretary of Defense. The Secretary of each military department is responsible to the Secretary of Defense for the operation and efficiency of his department. Orders to the military departments are issued through the Secretaries of these departments or their designees, by the Secretary of Defense, or under authority specifically delegated in writing by the Secretary of Defense or provided by law.

The commanders of the combatant commands are responsible to the President and the Secretary of Defense for accomplishing the military missions assigned to them and exercising command authority over forces assigned to them. The operational chain of command runs from the President to the Secretary of Defense to the commanders of the combatant commands. The Chairman of the Joint Chiefs of Staff functions within the chain of command by transmitting the orders of the President or the Secretary of Defense to the commanders of the combatant commands.

Office of the Secretary of Defense

Secretary of Defense The Secretary of Defense is the principal defense policy adviser to the President and is responsible for the formulation of general defense policy and policy related to DOD and for the execution of approved policy. Under the direction of the President, the Secretary exercises authority, direction, and control over the Department of Defense.

Acquisition, Technology, and Logistics The Under Secretary of Defense for Acquisition, Technology, and Logistics is the principal staff assistant and adviser to the Secretary of Defense for all matters relating to the DOD Acquisition System; research and development; modeling and simulation; systems engineering; advanced technology; developmental test and evaluation; production; systems integration; logistics; installation management; military construction; procurement; environment, safety, and occupational health management; utilities and energy management; business management modernization; document services; and nuclear, chemical, and biological defense programs.

Intelligence The Under Secretary of Defense for Intelligence is the principal staff assistant and adviser to the Secretary and Deputy Secretary of Defense for intelligence, intelligence-related matters, counterintelligence, and security. The Under Secretary of Defense for Intelligence supervises all intelligence and intelligence-related affairs of DOD.

Networks and Information Integration The Assistant Secretary of Defense (Networks and Information Integration) is the principal staff assistant and adviser to the Secretary and Deputy Secretary of Defense for achieving and maintaining information superiority in support of DOD missions, while exploiting or denying an adversary's ability to do the same. The Assistant Secretary of Defense also serves as the Chief Information Officer.

Personnel and Readiness The Under Secretary of Defense for Personnel and Readiness is the principal staff assistant and adviser to the Secretary of Defense for policy matters relating to

DEPARTMENT OF DEFENSE

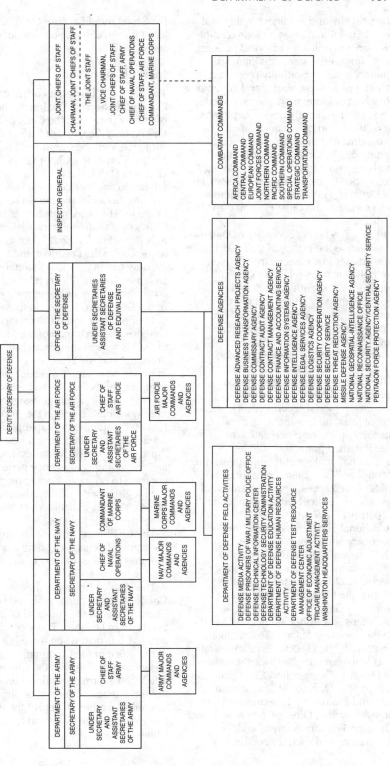

SECRETARY OF DEFENSE
DEPUTY SECRETARY OF DEFENSE

JOINT CHIEFS OF STAFF
CHAIRMAN, JOINT CHIEFS OF STAFF
THE JOINT STAFF
VICE CHAIRMAN,
JOINT CHIEFS OF STAFF
CHIEF OF STAFF, ARMY
CHIEF OF NAVAL OPERATIONS
CHIEF OF STAFF, AIR FORCE
COMMANDANT, MARINE CORPS

INSPECTOR GENERAL

COMBATANT COMMANDS
AFRICA COMMAND
CENTRAL COMMAND
EUROPEAN COMMAND
JOINT FORCES COMMAND
NORTHERN COMMAND
PACIFIC COMMAND
SOUTHERN COMMAND
SPECIAL OPERATIONS COMMAND
STRATEGIC COMMAND
TRANSPORTATION COMMAND

OFFICE OF THE SECRETARY OF DEFENSE
UNDER SECRETARIES
ASSISTANT SECRETARIES
OF DEFENSE
AND EQUIVALENTS

DEFENSE AGENCIES
DEFENSE ADVANCED RESEARCH PROJECTS AGENCY
DEFENSE BUSINESS TRANSFORMATION AGENCY
DEFENSE COMMISSARY AGENCY
DEFENSE CONTRACT AUDIT AGENCY
DEFENSE CONTRACT MANAGEMENT AGENCY
DEFENSE FINANCE AND ACCOUNTING SERVICE
DEFENSE INFORMATION SYSTEMS AGENCY
DEFENSE INTELLIGENCE AGENCY
DEFENSE LEGAL SERVICES AGENCY
DEFENSE LOGISTICS AGENCY
DEFENSE SECURITY COOPERATION AGENCY
DEFENSE SECURITY SERVICE
DEFENSE THREAT REDUCTION AGENCY
MISSILE DEFENSE AGENCY
NATIONAL GEOSPATIAL-INTELLIGENCE AGENCY
NATIONAL RECONNAISSANCE OFFICE
NATIONAL SECURITY AGENCY/CENTRAL SECURITY SERVICE
PENTAGON FORCE PROTECTION AGENCY

DEPARTMENT OF THE AIR FORCE
SECRETARY OF THE AIR FORCE
UNDER SECRETARY
AND ASSISTANT
SECRETARIES
OF THE
AIR FORCE
CHIEF OF STAFF
AIR FORCE
AIR FORCE MAJOR COMMANDS AND AGENCIES

DEPARTMENT OF THE NAVY
SECRETARY OF THE NAVY
UNDER SECRETARY
AND ASSISTANT
SECRETARIES
OF THE NAVY
CHIEF OF NAVAL OPERATIONS
COMMANDANT OF MARINE CORPS
NAVY MAJOR COMMANDS AND AGENCIES
MARINE CORPS MAJOR COMMANDS AND AGENCIES

DEPARTMENT OF THE ARMY
SECRETARY OF THE ARMY
UNDER SECRETARY
AND ASSISTANT
SECRETARIES
OF THE ARMY
CHIEF OF STAFF ARMY
ARMY MAJOR COMMANDS AND AGENCIES

DEPARTMENT OF DEFENSE FIELD ACTIVITIES
DEFENSE MEDIA ACTIVITY
DEFENSE PRISONERS OF WAR / MILITARY POLICE OFFICE
DEFENSE TECHNICAL INFORMATION CENTER
DEFENSE TECHNOLOGY SECURITY ADMINISTRATION
DEPARTMENT OF DEFENSE EDUCATION ACTIVITY
DEPARTMENT OF DEFENSE HUMAN RESOURCES ACTIVITY
DEPARTMENT OF DEFENSE TEST RESOURCE MANAGEMENT CENTER
OFFICE OF ECONOMIC ADJUSTMENT
TRICARE MANAGEMENT ACTIVITY
WASHINGTON HEADQUARTERS SERVICES

the structure and readiness of the total force. Functional areas include readiness; civilian and military personnel policies, programs, and systems; civilian and military equal opportunity programs; health policies, programs, and activities; Reserve component programs, policies, and activities; family policy, dependents' education, and personnel support programs; mobilization planning and requirements; language capabilities and programs; and the Federal Voting Assistance Program. The Under Secretary of Defense (Personnel and Readiness) also serves as the Chief Human Capital Officer.

Policy   The Under Secretary of Defense for Policy is the principal staff assistant and adviser to the Secretary of Defense for policy matters relating to overall international security policy and political-military affairs and represents the Department at the National Security Council and other external agencies regarding national security policy. Functional areas include homeland defense; NATO affairs; foreign military sales; arms limitation agreements; international trade and technology security; regional security affairs; special operations and low-intensity conflict; stability operations; integration of departmental plans and policies with overall national security objectives; drug control policy, requirements, priorities, systems, resources, and programs; and issuance of policy guidance affecting departmental programs.

Additional Staff   In addition, the Secretary and Deputy Secretary of Defense are assisted by a special staff of assistants, including the Assistant Secretary of Defense for Legislative Affairs; the General Counsel; the Inspector General; the Assistant Secretary of Defense for Public Affairs; the Assistant to the Secretary of Defense (Intelligence Oversight); the Director of Administration and Management; the Under Secretary of Defense (Comptroller)/Chief Financial Officer; the Director of Operational Test and Evaluation; Director, Business Transformation; Director, Net Assessment; Director, Program Analysis and Evaluation; and such other officers as the Secretary of Defense establishes to assist him in carrying out his duties and responsibilities.

Joint Chiefs of Staff

The Joint Chiefs of Staff consist of the Chairman; the Vice Chairman; the Chief of Staff of the Army; the Chief of Naval Operations; the Chief of Staff of the Air Force; and the Commandant of the Marine Corps. The Chairman of the Joint Chiefs of Staff is the principal military adviser to the President, the National Security Council, and the Secretary of Defense. The other members of the Joint Chiefs of Staff are military advisers who may provide additional information upon request from the President, the National Security Council, or the Secretary of Defense. They may also submit their advice when it does not agree with that of the Chairman. Subject to the authority of the President and the Secretary of Defense, the Chairman of the Joint Chiefs of Staff is responsible for assisting the President and the Secretary of Defense in providing strategic direction and planning for the Armed Forces; making recommendations for the assignment of responsibilities within the Armed Forces; comparing the capabilities of American and allied Armed Forces with those of potential adversaries; preparing and reviewing contingency plans that conform to policy guidance; preparing joint logistic and mobility plans; and recommending assignment of logistic and mobility responsibilities.

The Chairman, while so serving, holds the grade of general or admiral and outranks all other officers of the Armed Forces.

The Vice Chairman of the Joint Chiefs performs duties assigned by the Chairman, with the approval of the Secretary of Defense. The Vice Chairman acts as Chairman when there is a vacancy in the office of the Chairman or in the absence or disability of the Chairman. The Vice Chairman, while so serving, holds the grade of general or admiral and outranks all other officers of the Armed Forces except the Chairman of the Joint Chiefs of Staff.

Joint Staff The Joint Staff, under the Chairman of the Joint Chiefs of Staff, assists the Chairman and the other members of the Joint Chiefs of Staff in carrying out their responsibilities.

The Joint Staff is headed by a Director who is selected by the Chairman in consultation with the other members of the Joint Chiefs of Staff and with the approval of the Secretary of Defense. Officers assigned to serve on the Joint Staff are selected by the Chairman in approximate equal numbers from the Army, Navy, Marine Corps, and Air Force.

Combatant Commands

The combatant commands are military commands with broad continuing missions maintaining the security and defense of the United States against attack; supporting and advancing the national policies and interests of the United States and discharging U.S. military responsibilities in their assigned areas; and preparing plans, conducting operations, and coordinating activities of the forces assigned to them in accordance with the directives of higher authority. The operational chain of command runs from the President to the Secretary of Defense to the commanders of the combatant commands. The Chairman of the Joint Chiefs of Staff serves as the spokesman for the commanders of the combatant commands, especially on the administrative requirements of their commands.

For a complete listing of the combatant commands, including a map of each command's geographic area of responsibility and links to command Web sites, visit http://www.defense.gov/specials/unifiedcommand/.

Field Activities

Counterintelligence Field Activity The DOD Counterintelligence Field Activity was established in 2002 to build a Defense counterintelligence (CI) system that is informed by national goals and objectives and supports the protection of DOD personnel and critical assets from foreign intelligence services, foreign terrorists, and other clandestine or covert threats. The desired end is a transformed Defense CI system that integrates and synchronizes the counterintelligence activities of the military departments, defense agencies, Joint Staff, and combatant commands.

For further information, contact the Department of Defense Counterintelligence Field Activity, Crystal Square 5, Suite 1200, 1755 Jefferson Davis Highway, Arlington, VA 22202–3537. Phone, 703–699–7799.

Defense Technical Information Center The Defense Technical Information Center (DTIC) is a field activity in the Office of the Under Secretary of Defense (Acquisition, Technology, and Logistics). It operates under the authority, direction, and control of the Director, Defense Research and Engineering. DTIC provides defense scientific and technical information, offers controlled access to defense information, and designs and hosts more than 100 DOD Web sites. DTIC's collections include technical reports, summaries of research in progress, independent research and development material, defense technology transfer agreements, and DOD planning documents.

For further information, contact the Defense Technical Information Center, 8725 John J. Kingman Road, Fort Belvoir, VA 22060–6218. Phone, 800–225–3842. Internet, http://www.dtic.mil/dtic/index.html.

Defense Technology Security Administration The Defense Technology Security Administration (DTSA) is the central DOD point of contact for development and implementation of technology security policies governing defense articles and services and dual-use commodities. DTSA administers the development and implementation of DOD technology security policies on international transfers of defense-related goods, services, and technologies to ensure that critical U.S. military technological advantages are preserved; transfers that could prove detrimental to U.S. security interests are controlled and limited; proliferation of weapons of mass destruction and their means of delivery is prevented; diversion of defense-related goods to terrorists is prevented; legitimate defense cooperation with foreign friends and allies is supported; and the health of the defense industrial base is assured.

For further information, contact the Director, Defense Technology Security Administration, 2900 Defense Pentagon, Washington, DC 20301–2900. Phone, 703–325–3294. Fax, 703–325–6467.

Education Activity   The Department of Defense Education Activity (DODEA) was established in 1992. It consists of two subordinate organizational entities: the Department of Defense Dependents Schools (DODDS) and the Department of Defense Domestic Dependent Elementary and Secondary Schools (DDESS). DODEA formulates, develops, and implements policies, technical guidance, and standards for the effective management of Defense dependents education activities and programs. It also plans, directs, coordinates, and manages the education programs for eligible dependents of U.S. military and civilian personnel stationed overseas and stateside; evaluates the programmatic and operational policies and procedures for DODDS and DDESS; and provides education activity representation at meetings and deliberations of educational panels and advisory groups.

For further information, contact the Department of Defense Education Activity, 4040 North Fairfax Drive, Arlington, VA 22203–1635. Phone, 703–588–3200. Internet, http://www.dodea.edu.

Human Resources Field Activity   The Department of Defense Human Resources Activity (DODHRA) enhances the operational effectiveness and efficiency of a host of dynamic and diverse programs supporting the Office of the Under Secretary of Defense for Personnel and Readiness. The Field Activity supports policy development, performs cutting-edge research and expert analysis, supports readiness and reengineering efforts, manages the largest automated personnel data repositories in the world, prepares tomorrow's leaders through robust developmental programs, supports recruiting and retaining the best and brightest, and delivers both benefits and critical services to warfighters and their families.

For further information, contact the Department of Defense Human Resources Activity Headquarters, Suite 200, 4040 Fairfax Drive, Arlington, VA 22203–1613. Phone, 703–696–1036. Internet, http://www.dhra.mil.

TRICARE Management Activity   The TRICARE Management Activity (TMA) was formed in 1998 from the consolidation of the TRICARE Support Office (formerly Civilian Health and Medical Program of the Uniformed Services (CHAMPUS) headquarters), the Defense Medical Programs Activity, and the integration of health management program functions formerly located in the Office of the Assistant Secretary of Defense for Health Affairs. The mission of TMA is to manage TRICARE; manage the Defense Health Program appropriation; provide operational direction and support to the Uniformed Services in the management and administration of the TRICARE program; and administer CHAMPUS.

For further information, contact the TRICARE Management Activity, Suite 810, Skyline 5, 5111 Leesburg Pike, Falls Church, VA 22041–3206. Phone, 703–681–1730. Fax, 703–681–3665. Internet, http://www.tricare.osd.mil.

Test Resource Management   The Test Resource Management Center (TRMC) is a DOD Field Activity under the authority, direction, and control of the Under Secretary of Defense for Acquisition, Technology, and Logistics. The Center develops policy, plans for, and assesses the adequacy of the major range and test facility base to provide adequate testing in support of development, acquisition, fielding, and sustainment of defense systems. TRMC develops and maintains the test and evaluation resources strategic plan, reviews the proposed DOD test and evaluation budgets, and certifies the adequacy of the proposed budgets and whether they provide balanced support of the strategic plan. TRMC manages the Central Test and Evaluation Investment Program, the Test and Evaluation Science and Technology Program, and the Joint Mission Environment Test Capability Program.

Defense Prisoner of War/Missing Personnel Office   The Defense Prisoner of War/Missing Personnel Office (DPMO) was established in 1993 to provide centralized management of prisoner of war/missing personnel affairs within the DOD. DPMO's primary responsibilities include leadership for and policy oversight over all efforts to account for Americans still missing from past conflicts and the recovery of and accounting for those who may become isolated in hostile territory in future conflicts. The Office also provides administrative and logistical support to the U.S.-Russia Joint Commission on POW/MIAs; conducts research and analysis to help resolve cases of those unaccounted for; examines DOD documents for possible public disclosure; and, through periodic consultations and other appropriate measures, maintains viable channels of communications on POW/MIA matters between DOD and Congress, the families of the missing, and the American public.

For further information, contact the Defense Prisoner of War/Missing Personnel Office, 2400 Defense Pentagon, Washington, DC 20301–2400. Phone, 703–699–1160. Fax, 703–602–4375. Internet, http://www.dtic.mil/dpmo.

Office of Economic Adjustment   The Office of Economic Adjustment (OEA) assists communities that are adversely affected by base closures, expansions, or realignments and Defense contract or program cancellations. OEA provides technical and financial assistance to those communities and coordinates other Federal agencies' involvement through the Defense Economic Adjustment Program.

For further information, contact the Office of Economic Adjustment, Department of Defense, Suite 200, 400 Army Navy Drive, Arlington, VA 22202–4704. Phone, 703–604–6020.

Washington Headquarters Services   Washington Headquarters Services (WHS), established as a DOD Field Activity on October 1, 1977, is under the authority, direction, and control of the Director of Administration and Management. WHS provides a wide range of administrative and operational services to the Office of the Secretary of Defense, specified DOD components, Federal Government activities, and the general public. This support includes contracting and procurement; Defense facilities management; Pentagon renovation and construction; directives

and records management; financial management; library service; human resource services for executive, political, military, and civilian personnel; personnel security services; support for advisory boards and commissions; legal services and advice; information technology and data systems support; enterprise information technology infrastructure services; and planning and evaluation functions.

For further information, contact the Administration and Program Support Directorate, Washington Headquarters Services, 1155 Defense Pentagon, Washington, DC 20301–1155. Phone, 703–601–6138.

Defense Media Activity   Defense Media Activity (DMA) gathers and reports Defense news and information from all levels in the Department to the DOD family worldwide through the Armed Forces Radio and Television Network, the Internet, and printed publications. DMA reports news about individual soldiers, sailors, marines, airmen, and Defense civilian employees to the American public through the Hometown News Service. DMA provides World Wide Web infrastructure and services for DOD organizations. It collects, processes, and stores DOD imagery products created by the Department and makes them available to the American public. It trains the Department's public affairs and visual information military and civilian professionals. DMA also operates Stars and Stripes, a news and information organization, free of Government editorial control and censorship for military audiences overseas.

For further information, contact the Defense Media Activity, Department of Defense, Suite 250, 601 North Fairfax Street, Alexandria, VA 22314–2007. Phone, 703–428–1200. Internet, http://www.dma.mil.

Sources of Information

News Organizations   Newspapers and radio and television stations may subscribe to receive news releases about individual military members and Defense Department civilian employees at no cost. Phone, 210–925–6541. Email, hometown@dma.mil. Internet, www1.dmasa.dma.mil/hometown/. Official news releases and transcripts of press conferences can be found on the Internet at www.defense.gov.

Audiovisual Products   Certain Department of Defense productions on film and videotapes, CD–ROMs, and other audiovisual products such as stock footage and still photographs are available to the public. For an up-to-date, full-text, searchable listing of the Department's inventory of photographs and films of operations, exercises, and historical events or for interactive training materials, contact the Defense Imagery Management Center. Phone, 888–743–4662. Email, askdimoc@dma.mil. Internet, http://www.defenseimagery.mil.

Contracts and Small Business Activities Contact the Director, Small and Disadvantaged Business Utilization, Office of the Secretary of Defense, 3061 Defense Pentagon, Washington, DC 20301–3061. Phone, 703–588–8631.

DOD Directives and Instructions Contact the Executive Services and Communications Directorate, Washington Headquarters Services, 1155 Defense Pentagon, Washington, DC 20301–1155. Phone, 703–601–4722.

Electronic Access   Information about the following offices is available as listed below:

Office of the Secretary of Defense: Internet, http://www.defense.gov/osd.

Joint Chiefs of Staff: Internet, http://www.jcs.mil.

Central Command: Internet, http://www.centcom.mil.

Combatant Commands: Internet, http://www.defense.gov/specials/unifiedcommand.

European Command: Internet, http://www.eucom.mil.

Joint Forces Command: Internet, http://www.jfcom.mil.

Pacific Command: Internet, http://www.pacom.mil.

Northern Command: Internet, http://www.northcom.mil.

Southern Command: Internet, http://www.southcom.mil.

Strategic Command: Internet, http://www.stratcom.mil.

Transportation Command: Internet, http://www.transcom.mil.

Employment   Positions are filled by a variety of sources. Information concerning current vacancies and how to apply for positions may be found at https://storm.psd.whs.mil. Assistance in applying for positions is also available from our Human Resources Services Center Help Desk at 703–604–6219, 7:30 a.m. to 5:00 p.m. weekdays, or by writing to Washington Headquarters Services, 2521 South Clark Street, Suite 4000, Arlington, VA 22202.

Speakers   Civilian and military officials from the Department of Defense are available to speak to numerous public and private sector groups interested in a variety of defense-related topics, including the global war on terrorism. Requests for speakers should be addressed to the Director for Community Relations and Public Liaison, 1400 Defense Pentagon, Room 2C546, Washington, DC 20310–1400, or by calling 703–695–2733.

Pentagon Tours   Information on guided tours of the Pentagon may be obtained by writing to the Director, Pentagon Tours, 1400 Defense Pentagon, Room 2C546, Washington, DC 20310–1400, or calling 703–695–7778, or by sending an email to tourschd.pa@osd.mil. Internet, http://pentagon.afis.osd.mil.

For further information concerning the Department of Defense, contact the Director, Directorate for Public Inquiry and Analysis, Office of the Assistant Secretary of Defense for Public Affairs, 1400 Defense Pentagon, Washington, DC 20301–1400. Phone, 703–428–0711. Internet, http://www.defense.gov.

EDITORIAL NOTE: The Department of the Air Force did not meet the publication deadline for submitting updated information of its activities, functions, and sources of information as required by the automatic disclosure provisions of the Freedom of Information Act (5 U.S.C. 552(a)(1)(A)).

Department of the Air Force

1690 Air Force Pentagon, Washington, DC 20330–1670
Phone, 703–697–6061. Internet, http://www.af.mil.

Secretary of the Air Force	MICHAEL B. DONLEY
Chief of Staff	GEN. NORTON A. SCHWARTZ
Vice Chief of Staff	GEN. CARROL H. CHANDLER
Chief Master Sergeant	CMSGT. JAMES A. ROY
Administrative Assistant	WILLIAM A. DAVIDSON
Auditor General	THEODORE J. WILLIAMS
Acquisition	DAVID M. VAN BUREN, *Acting*
Financial Management and Comptroller	JAMIE M. MORIN
General Counsel	CHARLES A. BLANCHARD
International Affairs	MAJ. GEN. RICHARD E. PERRAUT, JR.
Installations, Environment and Logistics	TERRY A. YONKERS
Inspector General	LT. GEN. MARC E. ROGERS
Legislative Liaison	MAJ. GEN. ROBIN RAND
Manpower and Reserve Affairs	DANIEL B. GINSBERG
Public Affairs	COL. LES A. KODLICK
Small Business Programs	RONALD A. POUSSARD
Warfighting Integration and Chief Information Officer	LT. GEN. WILLIAM L. SHELTON
Manpower and Personnel	LT. GEN. RICHARD Y. NEWTON, III
Intelligence, Surveillance and Reconnaissance	LT. GEN. DAVID A. DEPTULA
Air, Space and Information Operations, Plans and Requirements	LT. GEN. DANIEL J. DARNELL
Logistics, Installation and Mission Support	LT. GEN. LOREN M. RENO
Strategic Plans and Programs	LT. GEN. CHRISTOPHER D. MILLER
Analyses, Assessments and Lessons Learned	JACQUELINE R. HENNINGSEN
Strategic Deterrence and Nuclear Integration Office	MAJ. GEN. WILLIAMS A. CHAMBERS
Chief of Chaplains	MAJ. GEN. CECIL R. RICHARDSON
Air Force Historian	CLARENCE R. ANDEREGG
Judge Advocate General	LT. GEN. RICHARD C. HARDING
Chief, Air Force Reserve	LT. GEN. CHARLES E. STENNER, JR.
Safety	MAJ. GEN. FREDERICK F. ROGGERO
Surgeon General	LT. GEN. CHARLES B. GREEN
Chief Scientist	WERNER J.A. DAHM
Test and Evaluation	JOHN T. MANCLARK
Chief, National Guard Bureau	GEN. CRAIG R. MCKINLEY
Director, Air National Guard	LT. GEN. HARRY M. WYATT, III

The Department of the Air Force is responsible for defending the United States through control and exploitation of air and space.

The Department of the Air Force (USAF) was established as part of the National Military Establishment by the National Security Act of 1947 (61 Stat. 502) and came into being on September 18, 1947. The National Security Act Amendments of 1949 redesignated the National Military Establishment as the Department of Defense, established it as an executive department, and made the Department of the Air Force a military department within the Department of Defense (63 Stat. 578). The Department of the Air Force is separately organized under the Secretary of the Air Force. It operates under the authority, direction, and control of the Secretary of Defense (10 U.S.C. 8010). The Department consists of the Office of the Secretary of the Air Force, the Air Staff, and field organizations.

Secretary   The Secretary is responsible for matters pertaining to organization, training, logistical support, maintenance, welfare of personnel, administrative, recruiting, research and development, and other activities prescribed by the President or the Secretary of Defense.

Air Staff   The mission of the Air Staff is to furnish professional assistance to the Secretary, the Under Secretary, the Assistant Secretaries, and the Chief of Staff in executing their responsibilities.

Field Organizations   The major commands, field operating agencies, and direct reporting units together represent the field organizations of the Air Force. These are organized primarily on a functional basis in the United States and on an area basis overseas. These commands are responsible for accomplishing certain phases of the worldwide activities of the Air Force. They also are responsible for organizing, administering, equipping, and training their subordinate elements for the accomplishment of assigned missions.

Major Commands: Continental U.S. Commands

Air Combat Command   This Command operates Air Force bombers and CONUS-based, combat-coded fighter and attack aircraft. It organizes, trains, equips, and maintains combat-ready forces for rapid deployment and employment while ensuring strategic air defense forces are ready to meet the challenges of peacetime air sovereignty and wartime air defense.

Air Force Materiel Command   This Command advances, integrates, and uses technology to develop, test, acquire, and sustain weapons systems. It also performs single-manager continuous product and process improvement throughout a product's life cycle.

Air Mobility Command   This Command provides airlift, air refueling, special air mission, and aeromedical evacuation for U.S. forces. It also supplies forces to theater commands to support wartime tasking.

Air Force Reserve Command   This Command supports the Air Force mission of defending the Nation through control and exploitation of air and space. It plays an integral role in the day-to-day Air Force mission and is not a force held in reserve for possible war or contingency operations.

Air Force Space Command   This Command operates space and ballistic missile systems, including ballistic missile warning, space control, spacelift, and satellite operations.

Air Force Special Operations Command   This Command provides the air component of U.S. Special Operations Command, deploying specialized air power and delivering special operations combat power.

Air Education and Training Command   This Command recruits, assesses, commissions, educates, and trains Air Force enlisted and officer personnel. It provides basic military training, initial and advanced technical training, flying training, and professional military and degree-granting professional education. The Command also conducts joint, medical service, readiness, and Air Force security assistance training.

Major Commands: Overseas Commands

Pacific Air Forces   The Command is responsible for planning, conducting, and coordinating offensive and defensive air operations in the Pacific and Asian theaters.

United States Air Forces in Europe   The Command plans, conducts, controls, coordinates, and supports air and space operations to achieve United States national and NATO objectives.

For a list of active Major Commands, go to http://www.af.mil/publicwebsites/index.asp.

Field Activities

Air National Guard   The Center performs the operational and technical tasks associated with manning, equipping, and training Air National Guard units to required readiness levels.

Base Closures   The Agency serves as the Federal real property disposal agent and provides integrated executive management for Air Force bases in the United States as they are closed under the delegated authorities of the Base Closure and Realignment Act of 1988 and the Defense Base Closure and Realignment Act of 1990.

Communications   The Agency ensures that command, control, communications, and computer systems used by USAF warfighters are integrated and interoperable. It develops and validates C4 architectures, technical standards, technical reference codes, policies, processes and procedures, and technical solutions, supporting information superiority through technical excellence.

Emergency Preparedness   The Office is responsible for Air Force-related national security emergency preparedness functions, including military support to civil authorities, civil defense, and law enforcement agencies and planning for continuity of operations during emergencies.

Engineering   The Agency maximizes Air Force civil engineers' capabilities in base and contingency operations by providing tools, practices, and professional support for readiness, training, technical support, management practices, automation support, vehicles and equipment, and research, development, and acquisition consultation.

Environmental Quality   The Center provides the Air Force with services in environmental remediation, compliance, planning, and pollution prevention, as well as construction management and facilities design.

Flight Standards   The Agency performs worldwide inspection of airfields, navigation systems, and instrument approaches. It provides flight standards to develop Air Force instrument requirements and certifies procedures and directives for cockpit display and navigation systems. It also provides air traffic control and airlift procedures and evaluates air traffic control systems and airspace management procedures.

Historic Publications   The Office researches, writes, and publishes books and other studies on Air Force history and provides historical support to Air Force headquarters.

Historical Research   The Agency serves as a repository for Air Force historical records and provides research facilities for scholars and the general public.

Intelligence   The Agency provides intelligence services to support Air Force operations through flexible collection, tailored air and space intelligence, weapons monitoring, and information warfare products and services.

Medical Operations   The Agency assists the USAF Surgeon General in developing plans, programs, and policies for the medical service, aerospace medicine, clinical investigations, quality assurance, health promotion, family advocacy, bioenvironmental engineering, military public health, and radioactive material management.

Modeling and Simulation   The Agency implements policies and standards and supports field operations in the areas of modeling and simulation.

News   The Agency gathers information and packages and disseminates electronic and printed news and information products. It manages and operationally controls Air Force Internal Information, the Army and Air Force Hometown News Service, the Air Force Broadcasting Service, and the Air Force Armed Forces Radio and Television outlets worldwide; operates the Air Force hotline; and provides electronic information through the Air Force bulletin board and the Internet.

Nuclear Weapons Monitoring   The Air Force Technical Applications Center monitors compliance with various nuclear treaties. It provides real-time reporting of nuclear weapons tests and operates a global network of sensors and analytical laboratories to monitor foreign nuclear activity. It conducts research and development of proliferation detection technologies for all weapons of mass destruction.

Real Estate   The Agency acquires, manages, and disposes of land for the Air Force worldwide and maintains a complete land and facilities inventory.

Weather Services   The Service provides centralized weather services to the Air Force, Army joint staff, designated unified commands, and other agencies, ensuring standardization of procedures and interoperability within the USAF weather system and assessing its technical performance and effectiveness.

Direct Reporting Units

Air Force Communication and Information Center   The Center applies information technology to improve operations processes and manages all Air Force information technology systems.

Air Force District of Washington   The Air Force District of Washington (AFDW) provides support for Headquarters Air Force and other Air Force units in the National Capital Region.

Curtis E. LeMay Center for Doctrine Development and Education   The LeMay Center leads in the development of Air Force operational-level doctrine and establishes the Air Force's position in Joint and multinational doctrine. The Center assists in the development, analysis, and wargaming of air, space, and cyberspace power concepts, doctrine, and strategy. Through wargames and military education, it also educates Air Force, Joint Defense, and multinational communities on warfighting doctrine.

Air Force Operational Test and Evaluation Center   The Center plans and conducts test and evaluation procedures to determine operational effectiveness and suitability of new or modified USAF systems and their capacity to meet mission needs.

Air Force Security Forces Center   The Center ensures quick and effective security responses to protect U.S. personnel around the globe.

U.S. Air Force Academy   The Academy provides academic and military instruction and experience to prepare future USAF career officers. Graduates receive Bachelor of Science degrees in 1 of 26 academic majors and commissions as second lieutenants.

For a list of active direct reporting units and field operating agencies, go to http://www.afhra.af.mil/organizationalrecords/druandfoa.asp.

For further information concerning the Department of the Air Force, contact the Office of the Director of Public Affairs, Department of the Air Force, 1690 Air Force Pentagon, Washington, DC 20330–1670. Phone, 703–697–6061. Internet, http://www.af.mil.

Department of the Army

The Pentagon, Washington, DC 20310
Phone, 703–695–6518. Internet, http://www.army.mil.

Secretary of the Army	JOHN M. McHUGH
Under Secretary of the Army	JOSEPH W. WESTPHAL
Assistant Secretary of the Army (Acquisition, Logistics, and Technology)	MALCOLM ROSS O'NEILL
Assistant Secretary of the Army (Civil Works)	JO-ELLEN DARCY
Assistant Secretary of the Army (Financial Management)	MARY SALLY MATIELLA
Assistant Secretary of the Army (Installations and Environment)	(VACANCY)
Assistant Secretary of the Army (Manpower and Reserve Affairs)	THOMAS R. LAMONT
General Counsel	(VACANCY)

Administrative Assistant to the Secretary of the Army	JOYCE E. MORROW
Chief Information Officer, G–6	LT. GEN. JEFFREY A. SORENSON
Inspector General	LT. GEN. R. STEVEN WHITCOMB
Auditor General	RANDALL EXLEY
Deputy Under Secretary of the Army	JOYCE E. MORROW, *Acting*
Deputy Under Secretary of the Army (Business Transformation)	(VACANCY)
Chief of Legislative Liaison	MAJ. GEN. JAMES C. MCCONVILLE
Chief of Public Affairs	BRIG. GEN. LEWIS M. BOONE
Director, Small Business Programs	TRACEY L. PINSON

Office of the Chief of Staff

Chief of Staff, United States Army	GEN. GEORGE W. CASEY, JR.
Vice Chief of Staff	GEN. PETER W. CHIARELLI
Director of the Army Staff	LT. GEN. DAVID H. HUNTOON
Vice Director of the Army Staff	JAMES B. GUNLICKS

Army Staff

Deputy Chief of Staff, G–1	LT. GEN. THOMAS P. BOSTICK
Deputy Chief of Staff, G–2	LT. GEN. RICHARD P. ZAHNER
Deputy Chief of Staff, G–3/5/7	LT. GEN. DANIEL P. BOLGER
Deputy Chief of Staff, G–4	LT. GEN. MITCHELL H. STEVENSON
Deputy Chief of Staff, G–8	LT. GEN. ROBERT P. LENNOX
Chief, Army Reserve	LT. GEN. JACK C. STULTZ
Chief, National Guard Bureau	GEN. CRAIG R. MCKINLEY
Chief of Engineers	LT. GEN. ROBERT L. VAN ANTWERP, JR.
Surgeon General	LT. GEN. ERIC B. SCHOOMAKER
Assistant Chief of Staff for Installation Management	LT. GEN. RICK LYNCH
Chief of Chaplains	MAJ. GEN. DOUGLAS L. CARVER
Provost Marshal General	BRIG. GEN. COLLEEN L. MCGUIRE
Judge Advocate General	LT. GEN. DANA K. CHIPMAN
Director, Army National Guard	MAJ. GEN. RAYMOND CARPENTER, *Acting*

Commands

Commanding General, U.S. Army Forces Command	GEN. JAMES D. THURMAN
Commanding General, U.S. Army Training and Doctrine Command	GEN. MARTIN E. DEMPSEY
Commanding General, U.S. Army Materiel Command	GEN. ANN E. DUNWOODY

Army Service Component Commands

Commanding General, U.S. Army Central	LT. GEN. WILLIAM G. WEBSTER
Commanding General, U.S. Army North	LT. GEN. GUY C. SWAN, III
Commanding General, U.S. Army South	MAJ. GEN. KEITH M. HUBER
Commanding General, U.S. Army Europe	GEN. CARTER F. HAM
Commanding General, U.S. Army Pacific	LT. GEN. BENJAMIN R. MIXON
Commanding General, Eighth U.S. Army	LT. GEN. JOSEPH F. FIL, JR.
Commanding General, U.S. Army Special Operations Command	LT. GEN. JOHN F. MULHOLLAND, JR.
Commanding General, U.S. Army Military Surface Deployment and Distribution Command	MAJ. GEN. JAMES L. HODGE

Army Service Component Commands

Commanding General, U.S. Army Space and Missile Defense Command/Army Strategic Command	LT. GEN. KEVIN T. CAMPBELL

Direct Reporting Units

Commanding General, U.S. Army Network Enterprise Technology Command/9th Signal Command	MAJ. GEN. SUSAN S. LAWRENCE
Commanding General, U.S. Army Medical Command	LT. GEN. ERIC B. SCHOOMAKER
Commanding General, U.S. Army Intelligence and Security Command	MAJ. GEN. MARY A. LEGERE
Commanding General, U.S. Army Criminal Investigation Command	BRIG. GEN. COLLEEN L. MCGUIRE
Commanding General, U.S. Army Corps of Engineers	LT. GEN. ROBERT L. VAN ANTWERP, JR.
Commanding General, U.S. Army Military District of Washington	MAJ. GEN. KARL R. HORST
Commanding General, U.S. Army Test and Evaluation Command	MAJ. GEN. ROGER A. NADEAU
Superintendent, U.S. Military Academy	LT. GEN. FRANKLIN L. HAGENBECK
Commanding General, U.S. Army Reserve Command	LT. GEN. JACK C. STULTZ
Director, U.S. Army Acquisition Support Center	CRAIG SPISAK
Commanding General, U.S. Army Installation Management Command	LT. GEN. RICK LYNCH

The mission of the Department of the Army is to organize, train, and equip active duty and Reserve forces for the preservation of peace, security, and the defense of our Nation. As part of our national military team, the Army focuses on land operations; its soldiers must be trained with modern arms and equipment and be ready to respond quickly. The Army also administers programs aimed at protecting the environment, improving waterway navigation, controlling floods and beach erosion, and developing water resources. It provides military assistance to Federal, State, and local government agencies, including natural disaster relief assistance.

The American Continental Army, now called the United States Army, was established by the Continental Congress on June 14, 1775, more than a year before the Declaration of Independence. The Department of War was established as an executive department at the seat of government by act approved August 7, 1789 (1 Stat. 49). The Secretary of War was established as its head. The National Security Act of 1947 (50 U.S.C. 401) created the National Military Establishment, and the Department of War was designated the Department of the Army. The title of its Secretary became Secretary of the Army (5 U.S.C. 171). The National Security Act Amendments of 1949 (63 Stat. 578) provided that the Department of the

Army be a military department within the Department of Defense.

Secretary

The Secretary of the Army is the senior official of the Department of the Army. Subject to the direction, authority, and control of the President as Commander in Chief and of the Secretary of Defense, the Secretary of the Army is responsible for and has the authority to conduct all affairs of the Department of the Army, including its organization, administration, operation, efficiency, and such other activities as may be prescribed by the President or the Secretary of Defense as authorized by law.

For further information, call 703–695–2422.

Army Staff

Presided over by the Chief of Staff, the Army Staff is the military staff of the Secretary of the Army. It is the duty of the Army Staff to perform the following functions: prepare for deployment of the Army and for such recruiting, organizing, supplying, equipping, training, mobilizing, and demobilizing of the Army as will assist the execution of any power, duty, or function of the Secretary or the Chief of Staff; investigate and report upon the efficiency of the Army and its preparation for military operations; act as the agent of the Secretary of the Army and the Chief of Staff in coordinating the action of all organizations of the Department of the Army; and perform such other duties not otherwise assigned by law as may be prescribed by the Secretary of the Army.

Program Areas

Civil Functions   Civil functions of the Department of the Army include the Civil Works Program, the Nation's major Federal water resources development activity involving engineering works such as major dams, reservoirs, levees, harbors, waterways, locks, and many other types of structures; the administration of Arlington and the U.S. Soldiers' and Airmen's Home National Cemeteries; and other related matters.

History   This area includes advisory and coordination service provided on historical matters, including historical properties; formulation and execution of the Army Historical Program; and preparation and publication of histories required by the Army.

Installations   This area consists of policies, procedures, and resources for the management of installations to ensure the availability of efficient and affordable base services and infrastructure in support of military missions. It includes the review of facilities requirements and stationing, identification and validation of resource requirements, and program and budget development and justification. Other activities include support for base operations; morale, welfare, and recreation; real property maintenance

and repair; environmental programs; military construction; housing; base realignment and closure; and competitive sourcing.

Intelligence   This area includes management of Army intelligence with responsibility for policy formulation, planning, programming, budgeting, evaluation, and oversight of intelligence activities. The Army Staff is responsible for monitoring relevant foreign intelligence developments and foreign disclosure; imagery, signals, human, open-source, measurement, and signatures intelligence; counterintelligence; threat models and simulations; and security countermeasures.

Medical   This area includes management of health services for the Army and as directed for other services, agencies, and organizations; health standards for Army personnel; health professional education and training; career management authority over commissioned and warrant officer personnel of the Army Medical Department; medical research, materiel development, testing and evaluation; policies concerning health aspects of Army environmental programs and prevention of disease; and planning, programming, and budgeting for Army-wide health services.

Military Operations and Plans   This includes Army forces strategy formation; mid-range, long-range, and regional strategy application; arms control, negotiation, and disarmament; national security affairs; joint service matters; net assessment; politico-military affairs; force mobilization and demobilization; force planning, programming structuring, development, analysis, requirements, and management; operational readiness; overall roles and missions; collective security; individual and unit training; psychological operations; information operations; unconventional warfare; counterterrorism; operations security; signal security; special plans; equipment development and approval; nuclear and chemical matters; civil affairs; military support of civil defense; civil disturbance; domestic actions; command and control;

DEPARTMENT OF THE ARMY

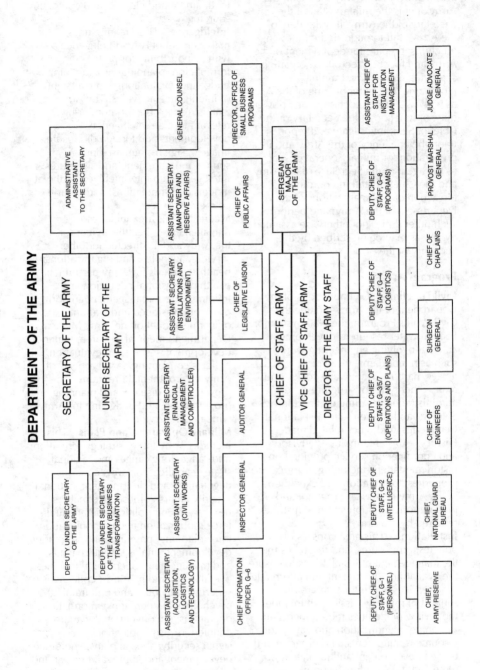

automation and communications programs and activities; management of the program for law enforcement, correction, and crime prevention for military members of the Army; special operations forces; foreign language and distance learning; and physical security.

Reserve Components This area includes management of individual and unit readiness and mobilization for Reserve Components, comprising the Army National Guard and U.S. Army Reserve.

Religious This area includes management of religious and moral leadership and chaplain support activities throughout the Department; religious ministrations, religious education, pastoral care, and counseling for Army military personnel; liaison with ecclesiastical agencies; chapel construction requirements and design approval; and career management of clergymen serving in the Chaplains Corps.

Army Commands

U.S. Army Forces Command U.S. Army Forces Command (FORSCOM) trains, mobilizes, deploys, sustains, transforms, and reconstitutes conventional forces, providing relevant and ready land power to combatant commanders worldwide in defense of the Nation at home and abroad.

For further information, contact FORSCOM. Phone, 404–464–5023. Internet, www.forscom.army.mil.

U.S. Army Training and Doctrine Command Headquartered in Fort Monroe, Virginia, U.S. Army Training and Doctrine Command (TRADOC) prepares the Army's soldiers and civilian leaders for joint warfighting through recruitment, training, and education, building a campaign-capable, expeditionary force.

For further information, contact TRADOC. Phone, 757–788–4465. Internet, www.tradoc.army.mil.

U.S. Army Materiel Command U.S. Army Materiel Command (AMC) is the Army's principal materiel developer. AMC's missions include the development of weapon systems, advanced research on future technologies, and maintenance and distribution of spare parts and equipment. AMC works closely with program executive offices, industry, academia, and other military services and Government agencies to develop, test, and acquire equipment that soldiers and units need to accomplish their missions.

For further information, contact AMC. Phone, 703–806–8010. Internet, www.amc.army.mil.

Army Service Component Commands

U.S. Army Central U.S. Army Central (USARCENT) conducts operations to attack, disrupt, and defeat terrorism; deter and defeat adversaries; deny access to weapons of mass destruction; ensure regional access; strengthen regional stability; build the self-reliance of partner nations' security forces; and protect the vital interests of the United States.

For further information, contact the USARCENT Public Affairs Office. Phone, 813–827–5895. Email, pao@centcom.mil. Internet, www.arcent.army.mil.

U.S. Army North U.S. Army North (USARNORTH) was established to support U.S. Northern Command. USARNORTH provides training to all units in the country and helps maintain readiness to support homeland defense and civil support missions.

For further information, contact USARNORTH Public Affairs Office. Phone, 210–221–0015. Email, 5aopao@arnorth.army.mil. Internet, www.arnorth.army.mil.

U.S. Army South U.S. Army South (USARSO) is a major subordinate command of U.S. Army Forces Command, Fort McPherson, Georgia, and is the Army Service Component Command of U.S. Southern Command. USARSO executes and is responsible for all Army operations within the U.S. Southern Command's area of responsibility (Central and South America and the Caribbean Islands). USARSO seeks to build regional cooperative security and increase hemispheric cooperation by planning and executing multilateral exercises and carrying out humanitarian aid, peacekeeping, engineering, and medical assistance exercises. USARSO maintains a deployable headquarters at Fort Sam

Houston, Texas, where they conduct strategic and operational planning.

For further information, contact the USARSO Public Affairs Office. Phone, 210–295–6388. Email, usarsowebmaster@samhouston.army.mil. Internet, www.usarso.army.mil.

U.S. Army Europe U.S. Army Europe provides the principal land component for U.S. European Command (EUCOM). U.S. Army Europe (USAREUR) forms, trains, and projects expeditionary forces and joint task force-capable headquarters to prosecute joint and combined operations throughout a 91-country area. As the U.S. Army's largest forward-deployed expeditionary force, USAREUR supports NATO and U.S. bilateral, multinational, and unilateral objectives. It supports U.S. Army forces in the European command area; receives and assists in the reception, staging, and onward movement and integration of U.S. forces; establishes, operates, and expands operational lines of communication; ensures regional security, access, and stability through presence and security cooperation; and supports U.S. combatant commanders and joint and combined commanders.

For further information, contact USAREUR. Phone, 011–49–6221–39–4100. Internet, www.hqusareur. army.mil.

U.S. Army Pacific U.S. Army Pacific (USARPAC) provides trained and ready forces in support of military operations and peacetime engagements in the Asia Pacific area. USARPAC carries out a cooperative engagement strategy known as the Theater Security Cooperation Program with the 43 Asian and Pacific nations within or bordering its area of responsibility. These countries include the Philippines, Thailand, Vietnam, Japan, Mongolia, Russia, China, South Korea, India, Bangladesh, Australia, New Zealand, Marshall Islands, and Papua New Guinea.

For further information, contact USARPAC. Phone, 808–438–1393. Internet, www.usarpac.army.mil.

Eighth U.S. Army Eighth U.S. Army provides forces to the commander of the U.S. Forces Korea, who in turn provides them to the commander of the

Republic of Korea/U.S. Combined Forces Command.

For further information, contact Eighth U.S. Army. Phone, 011–82–279–13–6544. Internet, http://8tharmy.korea.army.mil.

U.S. Army Special Operations Command U.S. Army Special Operations Command (USASOC) trains, equips, deploys, and sustains Army special operations forces for worldwide special operations supporting regional combatant commanders and country ambassadors. USASOC soldiers deploy to numerous countries conducting missions such as peacekeeping, humanitarian assistance, demining, and foreign internal defense. USASOC includes special forces, rangers, civil affairs, psychological operations, special operations aviation, and signal and support.

For further information, contact USASOC. Phone, 910–432–3000. Internet, www.usasoc.mil.

U.S. Army Military Surface Deployment and Distribution Command The U.S. Army Military Surface Deployment and Distribution Command (SDDC) provides global surface deployment command and control and distribution operations to meet national security objectives for the Department of Defense (DOD). This requires a presence in 24 ports worldwide as DOD's single-port manager, transportation, traffic-management services, deployment planning and engineering, and development of new technologies. SDDC is also the link between DOD shippers and the commercial surface transportation industry.

For further information, contact SDDC. Phone, 703–428–3207. Internet, www.sddc.army.mil.

U.S. Army Space and Missile Defense Command/Army Strategic Command U.S. Army Space and Missile Defense Command (SMDC/ARSTRAT) serves as the Army's specified proponent for space and national missile defense and operational integrator for theater missile defense. SMDC/ARSTRAT coordinates, integrates, and/or executes combat development, materiel development, technology, and advanced research and development for missile defense and space programs. It also serves as the Army service component command, the primary land component in support of U.S. Strategic Command missions (strategic deterrence, integrated missile defense, space operations, and cyberspace operations). In addition, SMDC/ARSTRAT conducts mission-related research and development.

For further information, contact SMDC. Phone, 703–607–1873. Internet, www.smdc.army.mil.

Sources of Information

Arlington and Soldiers' and Airmen's Home National Cemeteries For information write to the Superintendent, Arlington National Cemetery, Arlington, VA 22211–5003. Phone, 703–607–8545.
Army Historical Program For information concerning the Army Historical Program, write to the U.S. Army Center of Military History, Collins Hall, 103 Third Avenue, Fort Lesley J. McNair, Washington, DC 20319–5058. Phone, 202–685–2714. Fax, 202–685–4570. Internet, www.army.mil/cmh. Information on the preservation and utilization of historic buildings is available through the Office of Historic Properties. Phone, 703–692–9892.
Civilian Employment For information, visit the Army civilian personnel Web site (Internet, www.cpol.army.mil) or contact the civilian personnel advisory center at the desired Army installation.
Contracts Contract procurement policies and procedures are the responsibility of the Deputy for Procurement, Office of the Assistant Secretary of the Army (Acquisition, Logistics and Technology), The Pentagon, Washington, DC 20310–0103. Phone, 703–695–6154.
Environment Contact the Public Affairs Office, Office of the Chief of Public Affairs, Headquarters, Department of the Army, Washington, DC 20314–1000 (Phone, 202–761–0010); U.S. Army Environmental Command (Internet, http://aec.army.mil/usaec); or the Army Environmental Policy Institute (Internet, www.aepi.army.mil).
Films and Videos Requests for loan of Army-produced films should be addressed to the Visual Information

Support Centers of Army installations. Unclassified Army productions are available for sale from the National Audiovisual Center, National Technical Information Service, 5285 Port Royal Road, Springfield, VA 22161. Phone, 800–553–6847. Internet, www.ntis.gov/Index.aspx.
Freedom of Information and Privacy Act Requests Requests should be addressed to the Information Management Officer of the Army installation or activity responsible for the requested information.
Military Surface Deployment and Distribution Command Information concerning military transportation news and issues is available electronically through the Internet, www.sddc.army.mil/Public/Home.
Public Affairs and Community Relations For official Army information and community relations, contact the Office of the Chief of Public Affairs, Department of the Army, Washington, DC 20310–1508. Phone, 703–697–5081. During nonoffice hours, call 703–697–4200.
Publications Requests should be addressed to either the proponent listed on the title page of the document or the Information Management Officer of the Army activity that publishes the requested publication. Official texts published by Headquarters, Department of the Army, are available from the National Technical Information Service, Department of Commerce, Attn: Order Preprocessing Section, 5285 Port Royal Road, Springfield, VA 22161–2171. Phone, 703–487–4600. Internet, www.ntis.gov. (If it is uncertain which Army activity published the document, forward the request to the Publishing Division, Army Publishing Directorate, Room 1050, 2461 Eisenhower Avenue, Alexandria, VA 22331–0301. Phone, 703–325–6291. Internet, www.apd.army.mil.)
Research Information on long-range research and development plans concerning future materiel requirements and objectives may be obtained from the Commander, U.S. Army Research Development and Engineering Command, Attn: AMSRD–PA, Bldg. E5101, 5183 Blackhawk Road, Aberdeen Proving Ground, MD 21010–5424.

Small Business Activities   Assistance for small businesses and minority educational institutions to enhance their ability to participate in the Army contracting program is available through the Office of Small Business Programs, Office of the Secretary of the Army, 106 Army Pentagon, Washington, DC 20310–0106. Phone, 703–697–2868.

Speakers   Civilian organizations desiring an Army speaker may contact a nearby Army installation or write or call the Community Relations Division, Office of the Chief of Public Affairs, Department of the Army, Washington, DC 20310–1508. Phone, 703–695–6547. Requests for Army Reserve speakers may be addressed to HQDA (DAAR–PA), Washington, DC 20310–2423, or the local Army Reserve Center. Organizations in the Washington, DC, area desiring chaplain speakers may contact the Chief of Chaplains, Department of the Army, Washington, DC 20310–2700. Phone, 703–601–1140. Information on speakers may be obtained by contacting the Public Affairs Office, Office of the Chief of Engineers, Washington, DC 20314, or the nearest Corps of Engineer Division or District Office.

Military Career and Training Opportunities   Information on all phases of Army enlistments and specialized training is available by writing to the U.S. Army Recruiting Command, 1307 Third Avenue, Fort Knox, KY 40121–2725. For information about career and training opportunities, contact one of the offices that are listed below.

Army Health Professions   Headquarters U.S. Army Recruiting Command, Health Services Directorate (RCHS–OP), 1307 Third Avenue, Fort Knox, KY 40121–

2725. Phone, 502–626–0367. Email, Tanya.Beecher@usarec.army.mil. Internet, www.healthcare.goarmy.com.

Army National Guard Training Opportunities   Army National Guard, NGB–ASM, 1411 Jefferson Davis Highway, Arlington, VA 22202–3231. Phone, 703–607–5834. Internet, www.arng.army.mil.

Army Reserve Training Opportunities for Enlisted Personnel and Officers   Army Reserve Personnel Command, One Reserve Way, St. Louis, MO 63132–5200. Phone, 314–592–0000, or 800–318–5298. Internet, www.goarmyreserve.com.

Army Reserve Officers' Training Corps (ROTC)   U.S. Army Cadet Command, Recruiting, Retention and Operations Directorate, ATCC–OP, 55 Patch Road, Fort Monroe, VA 23651. Phone, 757–788–3770. Or contact a professor of military science or Army ROTC Advisor at the nearest college or university offering the program in your area. Internet, www.armyrotc.com.

Chaplain Recruiting Branch HQ   U.S. Army Recruiting Command, Attn: RCRO–SM–CH, 1307 Third Avenue, Fort Knox, KY 40121–2726. Phone, 502–626–0722, or 866–684–1571. Fax, 502–626–1213. Internet, www.chaplain.goarmy.com.

Judge Advocate General's Corps   Department of the Army, Judge Advocate Recruiting Office, 1777 North Kent Street, Suite 5200, Rosslyn, VA 20124–2194. Phone, 866–276–9524 (866–ArmyJag). Internet, www.law.goarmy.com.

U.S. Military Academy   Director of Admissions, United States Military Academy, Building 606, West Point, NY 10996. Phone, 845–938–4041. Internet, www.usma.edu.

For further information concerning the Department of the Army, contact the Office of the Chief of Public Affairs, Headquarters, Department of the Army, Washington, DC 20310–1508. Phone, 703–697–5081. Internet, http://www.army.mil.

Department of the Navy

The Pentagon, Washington, DC 20350
Phone, 703–697–7391. Internet, http://www.navy.mil.

Secretary of the Navy	RAYMOND E. MABUS
Under Secretary of the Navy	ROBERT O. WORK

Auditor General	RICHARD A. LEACH
Chief of Information	REAR ADM. DENNIS J. MOYNIHAN, USN
Chief Information Officer	ROBERT J. CAREY
Chief of Legislative Affairs	REAR ADM. MICHAEL H. MILLER, USN
General Counsel	PAUL L. OOSTBURG SANZ
Naval Inspector General	VICE ADM. ANTHONY L. WINNS, USN
Judge Advocate General	VICE ADM. JAMES W. HOUCK, JAGC, USN
Assistant Secretary (Energy, Installations and Environment)	JACKALYNE PFANNENSTIEL
Assistant Secretary (Financial Management and Comptroller)	GLADYS J. COMMONS
Assistant Secretary (Manpower and Reserve Affairs)	JUAN M. GARCIA, III
Assistant Secretary (Research, Development, and Acquisition)	SEAN J. STACKLEY

Naval Operations

Chief of Naval Operations	ADM. GARY ROUGHEAD, USN
Vice Chief of Naval Operations	ADM. JONATHAN W. GREENERT, USN
Master Chief Petty Officer of the Navy	MCPO RICK D. WEST, USN
Director, Naval Criminal Investigative Service	MARK D. CLOOKIE
Director, Naval Intelligence	VICE ADM. DAVID J. DORSETT, USN
Director, Naval Nuclear Propulsion Program	ADM. KIRKLAND H. DONALD, USN
Director, Navy Staff	VICE ADM. JOHN C. HARVEY, JR., USN
Chief of Chaplains of the Navy	REAR ADM. ROBERT BURT, CHC, USN
Chief of Naval Research/Director, Test and Evaluation and Technology Requirements	REAR ADM. NEVIN P. CARR, JR., USN
Chief of Naval Reserve	VICE ADM. DIRK DEBBINK, USN
Assistant Chief, Next Generation Enterprise Network System Program Office	REAR ADM. JOHN W. GOODWIN, USN
Deputy Chief of Naval Operations, Integration of Capabilities and Resources	VICE ADM. JOHN T. BLAKE, USN
Deputy Chief of Naval Operations, Intelligence Dominance	VICE ADM. DAVID J. DORSETT, USN
Deputy Chief of Naval Operations, Manpower, Personnel, Education, and Training	VICE ADM. MARK E. FERGUSON, III, USN
Deputy Chief of Naval Operations, Fleet Readiness and Logistics	VICE ADM. MICHAEL K. LOOSE, USN
Deputy Chief of Naval Operations, Operations, Plans and Strategy	VICE ADM. WILLIAM D. CROWDER, USN
Oceanographer and Navigator of the Navy	REAR ADM. DAVID W. TITLEY, USN
Surgeon General of the Navy	VICE ADM. ADAM M. ROBINSON, JR., MC ,USN

Shore Establishment

Chief, Naval Personnel	VICE ADM. MARK E. FERGUSON, III, USN
Chief, Bureau of Medicine and Surgery	VICE ADM. ADAM M. ROBINSON, JR., MC, USN

Shore Establishment

Commander, Naval Air Systems Command	VICE ADM. DAVID ARCHITZEL, USN
Commander, Naval Education and Training Command	REAR ADM. JOSEPH K. KILKENNY, USN
Commander, Naval Facilities Engineering Command	REAR ADM. CHRISTOPHER J. MOSSEY, USN
Commander, Naval Legal Service Command	REAR ADM. NANETTE DERENZI, JAGC, USN
Commander, Naval Meteorology and Oceanography	REAR ADM. JONATHAN W. WHITE, USN
Commander, Naval Network Warfare Command	REAR ADM. EDWARD H. DEETS, III, USN
Commander, Naval Sea Systems Command	VICE ADM. KEVIN M. MCCOY, USN
Commander, Naval Supply Systems Command	REAR ADM. MICHAEL J. LYDEN, SC, USN
Commander, Naval Warfare Development Command	REAR ADM. WENDI B. CARPENTER, USN
Commander, Office of Naval Intelligence	CAPT. J. TODD ROSS, USN
Commander, Space and Naval Warfare Systems Command	REAR ADM. MICHAEL L. BACHMANN, USN
Director, Strategic Systems Program	REAR ADM. DENNIS M. DWYER, USN
Superintendent, U.S. Naval Academy	VICE ADM. MICHAEL H. MILLER, USN

Operating Forces

Commander, Atlantic Fleet	REAR ADM. RICHARD O'HANLON, USN
Commander, Pacific Fleet	ADM. PATRICK M. WALSH, USN
Commander, Military Sealift Command	REAR ADM. MARK H. BUZBY, USN
Commander, Naval Forces Central Command	VICE ADM. WILLIAM E. GORTNEY, USN
Commander, Naval Forces Europe	ADM. MARK FITZGERALD, USN
Commander, Navy Installations Command	VICE ADM. MICHAEL C. VITALE, USN
Commander, Naval Reserve Forces Command	REAR ADM. BUZZ LITTLE, USN
Commander, Naval Special Warfare Command	REAR ADM. EDWARD G. WINTERS, USN
Commander, Operational Test and Evaluation Force	REAR ADM. DAVID A. DUNAWAY, USN

[For the Department of the Navy statement of organization, see the Code of Federal Regulations, Title 32, Part 700]

The primary mission of the Department of the Navy is to protect the United States, as directed by the President or the Secretary of Defense, by the effective prosecution of war at sea including, with its Marine Corps component, the seizure or defense of advanced naval bases; to support, as required, the forces of all military departments of the United States; and to maintain freedom of the seas.

The United States Navy was founded on October 13, 1775, when Congress enacted the first legislation creating the Continental Navy of the American Revolution. The Department of the Navy and the Office of Secretary of the Navy were established by act of April 30, 1798

(10 U.S.C. 5011, 5031). For 9 years prior to that date, by act of August 7, 1789 (1 Stat. 49), the conduct of naval affairs was under the Secretary of War.

The National Security Act Amendments of 1949 provided that the Department of

DEPARTMENT OF THE NAVY

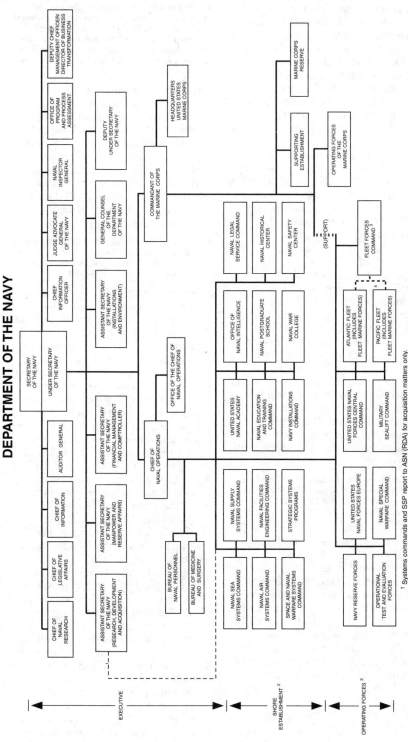

1 Systems commands and SSP report to ASN (RDA) for acquisition matters only.
2 Also includes other Echelon 2 commands and subordinate activities under the command or supervision of the designated organizations.
3 Additional duty for Commander, U.S. Atlantic Fleet.
4 For Interdeployment Training Cycle purposes, Commander, Fleet Forms Command Controls LANFLT and PACFLT assets.

the Navy be a military department within the Department of Defense (63 Stat. 578).

The Secretary of the Navy is appointed by the President as the head of the Department of the Navy and is responsible to the Secretary of Defense for the operation and efficiency of the Navy (10 U.S.C. 5031). The Department of the Navy includes the U.S. Coast Guard when it is operating as a Service in the Navy.

Secretary   The Secretary of the Navy is the head of the Department of the Navy, responsible for the policies and control of the Department of the Navy, including its organization, administration, functioning, and efficiency. The members of the Secretary's executive administration assist in the discharge of the responsibilities of the Secretary of the Navy.

Legal   The Office of the Judge Advocate General provides all legal advice and related services throughout the Department of the Navy, except for the advice and services provided by the General Counsel. It also provides legal and policy advice to the Secretary of the Navy on military justice, ethics, administrative law, claims, environmental law, operational and international law and treaty interpretation, and litigation involving these issues. The Judge Advocate General provides technical supervision for the Naval Justice School at Newport, RI.

For further information, contact the Office of the Judge Advocate General, Department of the Navy, Washington Navy Yard, Suite 3000, 1322 Patterson Avenue SE., Washington, DC 20374–5066. Phone, 202–685–5190.

Criminal Investigations   The Naval Criminal Investigative Service provides criminal investigative, counterintelligence, law enforcement and physical security, and information and personnel security support to Navy and Marine Corps organizations and personnel worldwide, both ashore and afloat. The Naval Criminal Investigative Service is comprised of law enforcement professionals who are investigators, crime laboratory technicians, technical investigative specialists, security specialists, and administrative support personnel.

For further information, contact the Director, Naval Criminal Investigative Service, Department of the Navy, 716 Sicard Street SE., Building 111, Washington Navy Yard, Washington, DC 20388–5000 (phone, 202–433–8800) or the Operations Control Center/Headquarters Duty Officer (phone, 202–433–9323).

Research   The Office of Naval Research encourages, promotes, plans, initiates, and coordinates naval research; coordinates naval research and development conducted by other agencies and offices of the Department of the Navy; and supervises, manages, and controls activities within or for the Department of the Navy relating to patents, inventions, trademarks, copyrights, and royalty payments.

For further information, contact the Public Affairs Office, Office of Naval Research, Ballston Tower One, 800 North Quincy Street, Arlington, VA 22217–5660. Phone, 703–696–5031.

Operating Forces   The operating forces of the Navy are responsible for naval operations necessary to carry out the Department of the Navy's role in upholding and advancing the national policies and interests of the United States. The operating forces of the Navy include the several fleets, seagoing forces, fleet marine forces, and other assigned Marine Corps forces, the Military Sealift Command, Naval Reserve forces, and other forces and activities as may be assigned by the President or the Secretary of the Navy. The Chief of Naval Operations is responsible for the command and administration of the operating forces of the Navy.

The Atlantic Fleet is composed of ships, submarines, and aircraft that operate throughout the Atlantic Ocean and Mediterranean Sea.

The Naval Forces Europe includes forces assigned by the Chief of Naval Operations or made available from either the Pacific or Atlantic Fleet to operate in the European theater.

The Pacific Fleet is composed of ships, submarines, and aircraft operating throughout the Pacific and Indian Oceans.

The Military Sealift Command provides ocean transportation for personnel and cargo of all components of the Department of Defense and, as authorized, for other Federal agencies; operates and maintains underway replenishment ships and other vessels providing mobile logistic support to elements of the combatant fleets; and operates ships in support of scientific projects and other programs for Federal agencies.

Other major commands of the operating forces of the Navy are the Naval Forces Central Command, Operational Test and Evaluation Force, Naval Special Warfare Command, and Naval Reserve Force.

Activities

Air Systems   The Naval Air Systems Command provides material support to the Navy and Marine Corps for aircraft, airborne weapons systems, avionics, related photographic and support equipment, ranges, and targets.

For further information, contact the Commander, Naval Air Systems Command, 47123 Buse Road, Patuxent River, MD 20670–1547. Phone, 301–757–1487.

Coast Guard   The Commandant of the Coast Guard reports to the Secretary of the Navy and the Chief of Naval Operations when the Coast Guard is operating as a service in the Navy and represents the Coast Guard before the Joint Chiefs of Staff. During such service, Coast Guard operations are integrated and uniform with Department of the Navy operations to the maximum extent possible. The Commandant of the Coast Guard organizes, trains, prepares, and maintains the readiness of the Coast Guard for the performance of national defense missions as directed. The Commandant also maintains a security capability; enforces Federal laws and regulations on and under the high seas and waters subject to the jurisdiction of the United States; and develops, establishes, maintains, and operates aids to maritime navigation and ice-breaking and rescue facilities, with due regard to the requirements of national defense.

Computers and Telecommunications   The Naval Network and Space Operations Command (NNSOC) was formed in July 2002 by the merger of

elements of Naval Space Command and Naval Network Operations Command. The command operates and maintains the Navy's space and global telecommunications systems and services, directly supports warfighting operations and command and control of naval forces, and promotes innovative technological solutions to warfighting requirements. NNSOC enables naval forces to use information and space technologies and expertise to achieve and maintain knowledge superiority essential for dominating the battle space.

For further information, contact the Commander, Naval Network and Space Operations Command, 5280 Fourth Street, Dahlgren, VA 22448–5300. Phone, 540–653–6100.

Education and Training   The Naval Education and Training Command provides shore-based education and training for Navy, certain Marine Corps, and other personnel; develops specifically designated education and training afloat programs for the fleet; provides voluntary and dependents education; and participates with research and development activities in the development and implementation of the most effective teaching and training systems and devices for optimal education and training.

For further information, contact the Commander, Naval Education and Training Command, Department of the Navy, Naval Air Station Pensacola, 250 Dallas Street, Pensacola, FL 32508–5220. Phone, 850–452–2713.

Facilities   The Naval Facilities Engineering Command provides material and technical support to the Navy and Marine Corps for shore facilities, real property and utilities, fixed ocean systems and structures, transportation and construction equipment, energy, environmental and natural resources management, and support of the naval construction forces.

For further information, contact the Commander, Naval Facilities Engineering Command and Chief of Civil Engineers, Washington Navy Yard, 1322 Patterson Avenue SE., Suite 1000, Washington, DC 20374–5065. Phone, 202–685–1423.

Intelligence   The Office of Naval Intelligence ensures the fulfillment of the intelligence requirements and responsibilities of the Department of the Navy.

For further information, contact the Commander, Office of Naval Intelligence, Department of the Navy, 4251 Suitland Road, Washington, DC 20395–5720. Phone, 301–669–3001.

Manpower   The Bureau of Naval Personnel directs the procurement, distribution, administration, and career motivation of the military personnel of the regular and reserve components of the U.S. Navy to meet the quantitative and qualitative manpower requirements determined by the Chief of Naval Operations.

For further information, contact the Bureau of Naval Personnel, Department of the Navy, Federal Office Building 2, Washington, DC 20370–5000. Phone, 703–614–2000.

Medicine   The Bureau of Medicine and Surgery directs the provision of medical and dental services for Navy and Marine Corps personnel and their dependents; administers the implementation of contingency support plans and programs to provide effective medical and dental readiness capability; provides professional and technical medical and dental service to the fleet, fleet marine force, and shore activities of the Navy; and ensures cooperation with civil authorities in matters pertaining to public health disasters and other emergencies.

For further information, contact the Bureau of Medicine and Surgery, Department of the Navy, 2300 E Street NW., Washington, DC 20373–5300. Phone, 202–762–3211.

Oceanography   The Naval Meteorology and Oceanography Command and the Naval Observatory are responsible for the science, technology, and engineering operations that are essential to explore the ocean and the atmosphere and to provide astronomical data and time for naval and related national objectives. To that end, the naval oceanographic program studies astrometry, hydrography, meteorology, oceanography, and precise time.

For further information, contact the following offices: Oceanographer of the Navy, U.S. Naval Observatory, 3450 Massachusetts Avenue NW., Washington, DC 20392–1800. Phone, 202–762–1026; Commander, Naval Meteorology and Oceanography Command, 1100 Balch Boulevard,

Stennis Space Center, MS 39529–5005. Phone, 228–688–4188; and Superintendent, U.S. Naval Observatory, 3450 Massachusetts Avenue NW., Washington, DC 20392–5100. Phone, 202–762–1438.

Sea Systems   The Naval Sea Systems Command provides material support to the Navy and Marine Corps and to the Departments of Defense and Transportation for ships, submarines, and other sea platforms, shipboard combat systems and components, other surface and undersea warfare and weapons systems, and ordnance expendables not specifically assigned to other system commands.

For further information, contact the Commander, Naval Sea Systems Command, 1333 Isaac Hull Avenue SE., Washington Navy Yard, DC 20376–1010. Phone, 202–781–1973.

Space and Naval Warfare   The Space and Naval Warfare Systems Command provides technical and material support to the Department of the Navy for space systems; command, control, communications, and intelligence systems; and electronic warfare and undersea surveillance.

For further information, contact the Commander, Space and Naval Warfare Systems Command, 4301 Pacific Highway, San Diego, CA 92110–3127. Phone, 619–524–3428.

Strategic Systems   The Office of Strategic Systems Programs provides development, production, and material support to the Navy for fleet ballistic missile and strategic weapons systems, security, training of personnel, and the installation and direction of necessary supporting facilities.

For further information, contact the Director, Strategic Systems Programs, Department of the Navy, Nebraska Avenue Complex, 287 Somers Court NW., Suite 10041, Washington, DC 20393–5446. Phone, 202–764–1608.

Supply Systems   The Naval Supply Systems Command provides supply management policies and methods and administers related support service systems for the Navy and Marine Corps.

For further information, contact the Commander, Naval Supply Systems Command, 5450 Carlisle Pike, P.O. Box 2050, Mechanicsburg, PA 17055–0791. Phone, 717–605–3565.

Warfare Development   The Navy
Warfare Development Command
plans and coordinates experiments
employing emerging operational
concepts; represents the Department
of the Navy in joint and other service
laboratories and facilities and tactical
development commands; and publishes
and disseminates naval doctrine.

For further information, contact the Commander,
Navy Warfare Development Command, 686 Cushing
Road, Sims Hall, Newport, RI 02841. Phone,
401–841–2833.

Sources of Information

Civilian Employment   Information
about civilian employment with the
Department of the Navy can be obtained
from the Office of the Deputy Assistant
Secretary of the Navy for Civilian
Human Resources Web site, www.
donhr.navy.mil. Information about
civilian employment opportunities in
the Washington, DC, metropolitan area
can be obtained from the Secretariat/
Headquarters Human Resources Office,
Navy Annex, Room 2510, Washington,
DC 20370–5240. Phone, 703–693–0888.
Consumer Activities   Research programs
of the Office of Naval Research cover
a broad spectrum of scientific fields,
primarily for the needs of the Navy, but
much information is of interest to the
public. Inquiries on specific research
programs should be directed to the Office
of Naval Research (Code 10), 800 North
Quincy Street, Arlington, VA 22217–
5660. Phone, 703–696–5031.
Contracts and Small Business Activities
Information about small businesses,
minority-owned businesses, and labor
surplus activities can be obtained from
the Office of Small and Disadvantaged
Business Utilization (SADBU),
720 Kennon Street SE., Room 207,
Washington Navy Yard, DC 20374–5015.
Phone, 202–685–6485.

Environment   For information on
environmental protection and natural
resources management programs of
the Navy and Marine Corps, contact
the Assistant Secretary of the Navy
(Installations and Environment),
Environment and Safety, 1000 Navy
Pentagon, Room 4A686, Washington, DC
20350–1000. Phone, 703–693–5080.
General Inquiries   Navy recruiting
offices and installation commanders are
available to answer general inquiries
concerning the Navy and its community
and public information programs.
The Department of the Navy Office
of Information provides accurate and
timely information about the Navy so
that the general public, the press, and
Congress may understand and assess
the Navy's programs, operations, and
needs. The Office also coordinates Navy
participation in community events and
supervises the Navy's internal information
programs. For general inquiries,
contact the Navy Office of Information,
1200 Navy Pentagon, Room 4B463,
Washington, DC 20350–1200. Phone,
703–695–0965.
Speakers and Films   Information can
be obtained on speakers, films, and
the Naval Recruiting Exhibit Center.
For information on the above Navy
items, contact the Office of Information,
Department of the Navy, 1200 Navy
Pentagon, Room 4B463, Washington, DC
20350–1200. Phone, 703–695- 0965.
Tours   To broaden the understanding
of the mission, functions, and programs
of the U.S. Naval Observatory, regular
night tours and special group day tours
are conducted. The night tours are
by reservation only and are given on
alternating Monday nights. Information
concerning activities of the observatory
and public tours may be obtained by
writing to the Superintendent, U.S. Naval
Observatory, 3450 Massachusetts Avenue
NW., Washington, DC 20392–5420.
Phone, 202–762–1438.

For further information concerning the Navy, contact the Office of Information, Department of the Navy,
1200 Navy Pentagon, Washington, DC 20350–1200. For press inquiries, phone 703–697–7391, or 703–697–
5342 Internet, http://www.navy.mil.

United States Marine Corps

Commandant of the Marine Corps, Headquarters, U.S. Marine Corps, 2 Navy Annex (Pentagon 5D773), Washington, DC 20380–1775
Phone, 703–614–1034. Internet, http://www.usmc.mil.

Commandant of the Marine Corps	GEN. JAMES T. CONWAY, USMC
Assistant Commandant of the Marine Corps	GEN. JAMES AMOS, USMC
Sergeant Major of the Marine Corps	SGT. MAJ. CARLTON KENT, USMC
Director, Marine Corps Staff	BRIG. GEN. WILLIE J. WILLIAMS, USMC
Director, Command, Control, Communications, and Computers	BRIG. GEN. GEORGE ALLEN, USMC
Deputy Commandant for Aviation	LT. GEN. GEORGE J. TRAUTMAN, USMC
Deputy Commandant for Installations and Logistics	LT. GEN. FRANK A. PANTER, JR., USMC
Deputy Commandant for Manpower and Reserve Affairs	LT. GEN. RICHARD C. ZILMER, USMC
Deputy Commandant for Plans, Policies, and Operations	LT. GEN. THOMAS D. WALDHAUSER, USMC
Deputy Commandant for Programs and Resources	LT. GEN. DUANE THISEN, USMC
Counsel for the Commandant	ROBERT D. HOGUE
Director of Administration and Resource Management	ALBERT A. WASHINGTON
Director of Marine Corps History and Museums	CHARLES P. NEIMEYER
Director of Public Affairs	COL. BRIAN F. SALAS, USMC
Director, Special Projects Directorate	TOM DAWSON
Legislative Assistant to the Commandant	BRIG. GEN. RICHARD L. SIMCOCK, II, USMC
Chaplain of the Marine Corps	REAR ADM. MARK TIDD, CHC, USN
Marine Corps Dental Officer	KENNETH R. WRIGHT
Medical Officer of the Marine Corps	REAR ADM. RICHARD R. JEFFRIES, USN
President, Permanent Marine Corps Uniform Board	BRIG. GEN. LEIF H. HENDRICKSON, USMC
Commanding General, Marine Corps Recruiting Command	MAJ. GEN. ROBERT E. MILSTEAD, JR., USMC
Commanding General, Marine Corps Combat Development Command	LT. GEN. GEORGE FLYNN, USMC
Commander, Marine Corps Systems Commands	BRIG. GEN. MICHAEL BROGAN, USMC
Commander, Marine Corps Base Quantico	COL. DANIEL J. CHOIKE, USMC

The United States Marine Corps was established on November 10, 1775, by resolution of the Continental Congress. Marine Corps composition and functions are detailed in 10 U.S.C. 5063.

The Marine Corps, which is part of the Department of the Navy, is the smallest of the Nation's combat forces and is the only service specifically tasked by Congress to be able to fight in the air, on land, and at sea. Although marines fight in each of these dimensions, they are primarily a maritime force, inextricably linked with the Navy to move from the sea to fight on land.

The Marine Corps conducts entry-level training for its enlisted marines at two bases, Marine Corps Recruit Depot, Parris Island, SC, and Marine Corps Recruit Depot, San Diego, CA. Officer candidates are evaluated at Officer Candidate School at Marine Corps Combat Development Command, Quantico, VA. Marines train to be first on the scene to respond to attacks on the United States or its interests, acts of political violence against Americans abroad, disaster relief, humanitarian assistance, or evacuation of Americans from foreign countries.

For a complete list of Marine Corps units, go to http://www.marines.mil/news/Pages/UnitDirectory.aspx.

Sources of Information

General Inquiries   Marine Corps recruiting offices, installation commanders, and Commanding Officers of Marine Corps Districts are available to answer general inquiries concerning the Marine Corps and its community and public information programs.

Speakers and Films   For information on above Marine Corps items, contact the Commandant of the Marine Corps, Headquarters, U.S. Marine Corps (PHC), Room 5E774, The Pentagon, Washington, DC 20380–1775. Phone, 703–614–4309.

Marine Corps Military Career and Training Opportunities   The Marine Corps conducts enlisted personnel and officer training programs; provides specialized skill training; participates in the Naval Reserve Officers Training Corps Program for commissioning officers in the Marine Corps; provides the Platoon Leaders Class program for commissioning officers in the Marine Corps Reserve to college freshmen, sophomores, or juniors and the Officer Candidate Class program for college graduates or seniors. Information about these programs is available at most civilian educational institutions and Navy and Marine Corps recruiting stations. Interested persons also may write directly to the Marine Corps Recruiting Command, 3280 Russell Road, Quantico, VA 22134–5103. Phone, 703–784–9454. Information about Marine Corps Reserve opportunities can be obtained from local Marine Corps recruiting stations or Marine Corps Reserve drill centers. Interested persons may also write directly to the Director, Reserve Affairs, 3280 Russell Road, Suite 507, Quantico, VA 22134–5103. Phone, 703–784–9100.

For further information regarding the Marine Corps, contact the Director of Public Affairs, Headquarters, U.S. Marine Corps, 2 Navy Annex (Pentagon 5D773), Washington, DC 20380–1775. Phone, 703–614–1492. Internet, http://www.usmc.mil.

United States Naval Academy

Annapolis, MD 21402–5018
Phone, 410–293–1500. Internet, http://www.usna.edu.

Superintendent	VICE ADM. MICHAEL H. MILLER, USN
Commandant of Midshipmen	CAPT. ROBERT E. CLARK, II, USN

The U.S. Naval Academy is the undergraduate college of the Naval Service. Through its comprehensive 4-year program, which stresses excellence in academics, physical education, professional training, conduct, and honor, the Academy prepares young men and women morally, mentally, and physically to be professional officers in the Navy and Marine Corps. All graduates receive a bachelor of science degree in 1 of 19 majors.

For further information concerning the U.S. Naval Academy, contact the Superintendent, U.S. Naval Academy, 121 Blake Road, Annapolis, MD 21402–5018. Internet, http://www.usna.edu.

Defense Agencies

Defense Advanced Research Projects Agency

3701 North Fairfax Drive, Arlington, VA 22203–1714
Phone, 703–526–6630. Internet, http://www.darpa.mil.

Director	REGINA E. DUGAN
Deputy Director	ROBERT LEHENY

The Defense Advanced Research Projects Agency is a separately organized agency within Department of Defense and is under the authority, direction, and control of the Under Secretary of Defense (Acquisition, Technology and Logistics). The Agency serves as the central research and development organization of the Department of Defense with a primary responsibility to maintain U.S. technological superiority over potential adversaries. It pursues imaginative and innovative research and development projects and conducts demonstration projects that embody technology appropriate for joint programs, programs in support of deployed forces, or selected programs of the military departments. To this end, the Agency arranges, manages, and directs the performance of work connected with assigned advanced projects by the military departments, other Government agencies, individuals, private business entities, and educational or research institutions, as appropriate.

For further information, contact the Defense Advanced Research Projects Agency, 3701 North Fairfax Drive, Arlington, VA 22203–1714. Phone, 703–526–6630. Internet, http://www.darpa.mil.

Defense Business Transformation Agency

1851 South Bell Street, Suite 1000, Arlington, VA 22240
Phone, 703–607–2110. Internet, http://www.bta.mil.

Director	DAVID M. FISHER
Deputy Director	DOULGAS W. WEBSTER

The Defense Business Transformation Agency (BTA) was established on October 7, 2005. BTA's mission is to guide the transformation of business operations throughout DOD and to deliver enterprise-level capabilities that align to warfighter needs. BTA directs improvement in business operations across DOD.

For further information, please contact the Defense Business Transformation Agency. Phone, 703–607–2110. Internet, http://www.dod.mil/dbt/contact.html.

Defense Commissary Agency

1300 E Avenue, Fort Lee, VA 23801
Phone, 804–734–8253. Internet, http://www.commissaries.com.

Director and Chief Executive Officer	PHILIP E. SAKOWITZ, JR.
Chief Operating Officer	THOMAS E. MILKS

The Defense Commissary Agency (DeCA) was established in 1990 and is under the authority, direction, and control of the Under Secretary of Defense for Personnel and Readiness and the operational supervision of the Defense Commissary Agency Board of Directors.

DeCA provides an efficient and effective worldwide system of commissaries that sell quality groceries and household supplies at low prices to members of the Armed Services community. This benefit satisfies customer demand for quality products and delivers exceptional savings while enhancing the military community's quality of life. DeCA works closely with its employees, customers, and civilian business partners to satisfy its customers and to promote the commissary benefit. The benefit fosters recruitment, retention, and readiness of skilled and trained personnel.

Sources of Information

Employment information is available at www.commissaries.com or by calling the following telephone numbers: employment (703–603–1600); small business activities (804–734–8000, extension 4–8015/4–8529); contracting for resale items (804–734–8000, extension 4–8884/4–8885); and contracting for operations support and equipment (804–734–8000, extension 4–8391/4–8830).

For further information, contact the Defense Commissary Agency, 1300 E Avenue, Fort Lee, VA 23801–1800. Phone, 800–699–5063, extension 4–8998. Internet, http://www.commissaries.com.

Defense Contract Audit Agency

8725 John J. Kingman Road, Suite 2135, Fort Belvoir, VA 22060–6219
Phone, 703–767–3200. Internet, http://www.dcaa.mil.

Director	APRIL STEPHENSON
Deputy Director	FRANCIS P. SUMMERS, JR.

The Defense Contract Audit Agency (DCAA) was established in 1965 and is under the authority, direction, and control of the Under Secretary of Defense (Comptroller)/Chief Financial Officer. DCAA performs all necessary contract audit functions for DOD and provides accounting and financial advisory services to all Defense components responsible for procurement and contract administration. These services are provided in connection with the negotiation, administration, and settlement of contracts and subcontracts to ensure taxpayer dollars are spent on fair and reasonable contract prices. They include evaluating the acceptability of costs claimed or proposed by contractors and reviewing the efficiency and economy of contractor operations. Other Government agencies may request the DCAA's services under appropriate arrangements.

DCAA manages its operations through five regional offices responsible for approximately 104 field audit offices throughout the United States and overseas. Each region is responsible for the contract auditing function in its assigned area. Point of contact information for DCAA regional offices is available at www.dcaa.mil.

For further information, contact the Executive Officer, Defense Contract Audit Agency, 8725 John J. Kingman Road, Suite 2135, Fort Belvoir, VA 22060–6219. Phone, 703–767–3265. Internet, http://www.dcaa.mil.

Defense Contract Management Agency

6350 Walker Lane, Alexandria, VA 22310–3241
Phone, 703–428–1700. Internet, http://www.dcma.mil.

Director	CHARLES E. WILLIAMS, JR., *Acting*
Deputy Director	JIM RUSSELL

The Defense Contract Management Agency (DCMA) was established by the Deputy Secretary of Defense in 2000 and is under the authority, direction, and control of the Under Secretary of Defense (Acquisition, Technology, and Logistics). DCMA is responsible for DOD contract management in support of the military departments, other DOD components, the National Aeronautics and Space Administration, other designated Federal and State agencies, foreign governments, and international organizations, as appropriate.

For further information, contact the Public Affairs Office, Defense Contract Management Agency, 6350 Walker Lane, Alexandria, VA 22310–3241. Phone, 703–428–1969. Internet, http://www.dcma.mil.

Defense Finance and Accounting Service

Crystal Mall 3, Room 920, Arlington, VA 22240–5291
Phone, 703–607–2616. Internet, http://www.dfas.mil.

Director	TERESA A. MCKAY
Principal Deputy Director	RICHARD GUSTAFSON

The Defense Finance and Accounting Service (DFAS) was established in 1991 under the authority, direction, and control of the Under Secretary of Defense (Comptroller)/Chief Financial Officer to strengthen and reduce costs of financial management and operations within DOD. DFAS is responsible for all payments to servicemembers, employees, vendors, and contractors. It provides business intelligence and finance and accounting information to DOD decisionmakers. DFAS is also responsible for preparing annual financial statements and the consolidation, standardization, and modernization of finance and accounting requirements, functions, processes, operations, and systems for DOD.

For further information, contact Defense Finance and Accounting Service Corporate Communications, Room 924, Crystal Mall 3, Arlington, VA 22240–5291. Phone, 703–607–0122. Internet, http://www.dfas.mil.

Defense Information Systems Agency

P.O. Box 4502, Arlington, VA 22204–4502
Phone, 703–607–6900. Internet, http://www.disa.mil.

Director	LT. GEN. CARROLL F. POLLETT, USA
Vice Director	REAR ADM. ELIZABETH HIGHT, USN

The Defense Information Systems Agency (DISA), established originally as the Defense Communications Agency in 1960, is under the authority, direction, and control of the Assistant Secretary of Defense (Networks and Information Integration). DISA is a combat support agency responsible for planning, engineering, acquiring, fielding, operating, and supporting global net-centric solutions to serve the needs of the President, Vice President, Secretary of Defense, and other DOD components.

For further information, contact the Public Affairs Office, Defense Information Systems Agency, P.O. Box 4502, Arlington, VA 22204–4502. Phone, 703–607–6900. Internet, http://www.disa.mil.

Defense Intelligence Agency

The Pentagon, Washington, DC 20340–5100
Phone, 703–695–0071. Internet, http://www.dia.mil.

Director	LT. GEN. RONALD L. BURGESS, JR., USA
Deputy Director	LETITIA A. LONG

The Defense Intelligence Agency (DIA) was established in 1961 and is under the authority, direction, and control of the Under Secretary of Defense for Intelligence. DIA provides timely, objective, and cogent military intelligence to warfighters, force planners, and defense and national security policymakers. DIA obtains and reports information through its field sites worldwide and the Defense Attache System; provides timely intelligence analysis; directs Defense Human Intelligence programs; operates the Joint Intelligence Task Force for Combating Terrorism and the Joint Military Intelligence College; coordinates and facilitates Measurement and Signature Intelligence activities; manages and plans collection from specialized technical sources; manages secure DOD intelligence networks; and coordinates required intelligence support for the Secretary of Defense, Joint Chiefs of Staff, Combatant Commanders, and Joint Task Forces.

For further information, contact the Public Affairs Office, Defense Intelligence Agency, Washington, DC 20340–5100. Phone, 703–695–0071. Internet, http://www.dia.mil.

Defense Legal Services Agency

The Pentagon, Washington, DC 20301–1600
Phone, 703–695–3341. Internet, http://www.dod.mil/dodgc.

Director (General Counsel)	JEH CHARLES JOHNSON
Deputy Director (Deputy General Counsel)	ROBERT S. TAYLOR

The Defense Legal Services Agency (DLSA) was established in 1981 and is under the authority, direction, and control of the General Counsel of the Department of Defense, who also serves as its Director. DLSA provides legal advice and services for specified DOD components and adjudication of personnel security cases for DOD and other assigned Federal agencies and departments. It also provides technical support and assistance for development of the Department's legislative program; coordinates positions on legislation and Presidential Executive orders; provides a centralized legislative and congressional document reference and distribution point for the Department; maintains the Department's historical legislative files; and administers programs governing standards of conduct and alternative dispute resolution.

For further information, contact the Administrative Office, Defense Legal Services Agency, Room 3A734, Washington, DC 20301–1600. Phone, 703–697–8343. Internet, http://www.dod.mil/dodgc.

Defense Logistics Agency

8725 John J. Kingman Road, Fort Belvoir, VA 22060–6221
Phone, 703–767–6200. Internet, http://www.dla.mil. •

Director	VICE ADM. ALAN S. THOMPSON, USN
Vice Director	MAJ. GEN. ARTHUR B. MORRILL, III, USAF

The Defense Logistics Agency (DLA) is under the authority, direction, and control of the Under Secretary of Defense for Acquisition, Technology, and Logistics. DLA supports both the logistics requirements of the military services and their acquisition of weapons and other materiel. It provides logistics support and technical services to all branches of the military and to a number of Federal agencies. DLA supply centers consolidate the requirements of the military services and procure the supplies in sufficient quantities to meet their projected needs. DLA manages supplies in eight commodity areas: fuel, food, clothing, construction material, electronic supplies, general supplies, industrial supplies, and medical supplies. Information on DLA's field activities and regional commands is available at www.dla.mil/ataglance.aspx.

Sources of Information

Employment   For the Washington, DC, metropolitan area, all inquiries and applications concerning job recruitment programs should be addressed to Human Resources, Customer Support Office, 3990 East Broad Street, Building 11, Section 3, Columbus, OH, 43213–0919. Phone, 877–352–4762.

Environmental Program   For information concerning the environmental program, contact the Staff Director, Environmental and Safety, Defense Logistics Agency, Attn: DSS–E, 8725 John J. Kingman Road, Fort Belvoir, VA 22060–6221. Phone, 703–767–6278.

Procurement and Small Business Activities   For information concerning procurement and small business activities, contact the Director, Small and Disadvantaged Business Utilization, Defense Logistics Agency, Attn: DB, 8725 John J. Kingman Road, Fort Belvoir, VA 22060–6221. Phone, 703–767–0192.

Surplus Sales Program   Questions concerning this program should be addressed to DOD Surplus Sales, International Sales Office, 74 Washington Avenue North, Battle Creek, MI 49017–3092. Phone, 877–352–2255.

For further information, contact the Defense Logistics Agency, 8725 John J. Kingman Road, Fort Belvoir, VA 22060–6221. Phone, 703–767–5200. Internet, http://www.dla.mil.

Defense Security Cooperation Agency

2800 Defense Pentagon, Washington, DC 20301–2800
Phone, 703–601–3700. Internet, http://www.dsca.mil.

Director	VICE ADM. JEFFREY WIERINGA, USN
Deputy Director	RICHARD MILLIES

The Defense Security Cooperation Agency (DSCA) was established in 1971 and is under the authority, direction, and control of the Under Secretary of Defense (Policy). DSCA provides traditional security assistance functions such as military assistance, international military education and training, and foreign military sales. DSCA also has program management responsibilities for humanitarian assistance, demining, and other DOD programs.

For further information, contact the Defense Security Cooperation Agency, 2800 Defense Pentagon, Washington, DC 20301–2800. Phone, 703–601–3700. Email, lpa-web@dsca.mil. Internet, http://www.dsca.mil.

Defense Security Service

1340 Braddock Place, Alexandria, VA 22314–1651
Phone, 703–325–9471. Internet, http://www.dss.mil.

Director	KATHLEEN WATSON
Deputy Director	(VACANCY)

The Defense Security Service (DSS) is under the authority, direction, and control of the Under Secretary of Defense for Intelligence. DSS ensures the safeguarding of classified information used by contractors on behalf of the DOD and 22 other executive branch agencies under the National Industrial Security Program. It oversees the protection of conventional arms, munitions, and explosives in the custody of DOD contractors; evaluates the protection of selected private sector critical assets and infrastructures (physical and cyber-based systems) and recommends measures needed to maintain operations identified as vital to DOD. DSS makes clearance determinations for industry and provides support services for DOD Central Adjudicative Facilities. It provides security education, training, and proactive awareness programs for military, civilian, and cleared industry to enhance their proficiency and awareness of DOD security policies and procedures. DSS also has a counterintelligence office to integrate counterintelligence principles into security countermeasures missions and to support the national counterintelligence strategy. Information on DSS operating locations and centers is available at http://www.dss.mil/isp/dss_oper_loc.html.

For further information, contact the Defense Security Service, Office of Congressional and Public Affairs, 1340 Braddock Place, Alexandria, VA 22314–1651. Phone, 703–325–9471. Internet, http://www.dss.mil.

Defense Threat Reduction Agency

8725 John J. Kingman Road, MS 6201, Fort Belvoir, VA 22260–5916
Phone, 703–325–2102. Internet, http://www.dtra.mil.

Director	JAMES A. TEGNELIA
Deputy Director	MAJ. GEN. RANDAL R. CASTRO, USA

The Defense Threat Reduction Agency (DTRA) was established in 1998 and is under the authority, direction, and control of the Under Secretary of Defense for Acquisition, Technology, and Logistics. DTRA's mission is to reduce the threat posed by weapons of mass destruction (WMD). DTRA covers the full range of WMD threats (chemical, biological, nuclear, radiological, and high explosive), bridges the gap between the warfighters and the technical community, sustains the nuclear deterrent, and provides both offensive and defensive technology and operational concepts to the warfighters. DTRA reduces the threat of WMD by implementing arms control treaties and executing the Cooperative Threat Reduction Program. It uses combat support, technology development, and chemical-biological defense to deter the use and reduce the impact of such weapons. DTRA also prepares for future threats by developing the technology and concepts needed to counter new WMD threats and adversaries.

For further information, contact the Public Affairs Office, Defense Threat Reduction Agency, 8725 John J. Kingman Road, MS 6201, Fort Belvoir, VA 22060–5916. Phone, 703–767–5870. Internet, http://www.dtra.mil.

Missile Defense Agency

The Pentagon, Washington, DC 20301–7100
Phone, 703–695–6420. Internet, http://www.mda.mil.

Director	LT. GEN. PATRICK J. O'REILLY, USA
Deputy Director	REAR ADM. JOSEPH A. HORN, JR., USN
Executive Director	DAVID ALTWEGG
Chief of Staff	COL. DAVID BAGNATI, USA

[For the Missile Defense Agency statement of organization, see the Code of Federal Regulations, Title 32, Part 388]

The Missile Defense Agency's (MDA) mission is to establish and deploy a layered ballistic missile defense system to intercept missiles in all phases of their flight and against all ranges of threats. This capability will provide a defense of the United States, deployed forces, allies, and friends. MDA is under the authority, direction, and control of the Under Secretary of Defense for Acquisition, Technology, and Logistics. MDA manages and directs DOD's ballistic missile defense acquisition programs and enables the Services to field elements of the overall system as soon as practicable. MDA develops and tests technologies and, if necessary, uses prototype and test assets to provide early capability. Additionally, MDA improves the effectiveness of deployed capabilities by implementing new technologies as they become available or when the threat warrants an accelerated capability.

For further information, contact the Human Resources Directorate, Missile Defense Agency, Washington, DC 20301–7100. Phone, 703–614–8740. Internet, http://www.mda.mil.

National Geospatial-Intelligence Agency

4600 Sangamore Road, Bethesda, MD 20816–5003
Phone, 301–227–7300. Internet, http://https://www1.nga.mil.

Director	VICE ADM. ROBERT B. MURRETT, USN
Deputy Director	LLOYD B. ROWLAND

The National Geospatial-Intelligence Agency (NGA), formerly the National Imagery and Mapping Agency, was established in 1996 and is under the authority, direction, and control of the Under Secretary of Defense for Intelligence. NGA is a DOD combat support agency and a member of the national intelligence community. NGA's mission is to provide timely, relevant, and accurate geospatial intelligence in support of our national security. Geospatial intelligence means the use and analysis of imagery to describe, assess, and visually depict physical features and geographically referenced activities on the Earth. Headquartered in Bethesda, MD, NGA has major facilities in the Washington, DC, Northern Virginia, and St. Louis, MO, areas with NGA support teams worldwide.

For further information, contact the Public Affairs Office, National Geospatial-Intelligence Agency, 4600 Sangamore Road, Bethesda, MD 20816–5003. Phone, 301–227–2057. Fax, 301–227–3920.

National Security Agency / Central Security Service

Fort George G. Meade, MD 20755–6248
Phone, 301–688–6524. Internet, http://www.nsa.gov.

Director	LT. GEN. KEITH B. ALEXANDER, USA
Deputy Director	JOHN C. INGLIS

The National Security Agency (NSA) was established in 1952 and the Central Security Service (CSS) was established in 1972. NSA/CSS is under the authority, direction, and control of the Under Secretary of Defense for Intelligence. As the Nation's cryptologic organization, NSA/CSS employs the Nation's premier codemakers and codebreakers. It ensures an informed, alert, and secure environment for U.S. warfighters and policymakers. The cryptologic resources of NSA/CSS unite to provide U.S. policymakers with intelligence information derived from America's adversaries while protecting U.S. Government signals and information systems from exploitation by those same adversaries.

For further information, contact the Public Affairs Office, National Security Agency/Central Security Service, Fort George G. Meade, MD 20755–6248. Phone, 301–688–6524. Internet, http://www.nsa.gov.

Pentagon Force Protection Agency
Washington, DC 20301
Phone, 703–693–3685. Internet, http://www.pfpa.mil.

Director	STEVEN E. CALVERY
Deputy Director	JONATHAN H. COFER

The Pentagon Force Protection Agency (PFPA) was established in May 2002 in response to the events of September 11, 2001, and subsequent terrorist threats facing the DOD workforce and facilities in the National Capital Region (NCR). PFPA is under the authority, direction, and control of the Director, Administration and Management, in the Office of the Secretary of Defense. PFPA provides force protection, security, and law enforcement for the people, facilities, infrastructure, and other resources at the Pentagon and for DOD activities and facilities within the NCR that are not under the jurisdiction of a military department. Consistent with the national strategy on combating terrorism, PFPA addresses threats, including chemical, biological, and radiological agents, through a strategy of prevention, preparedness, detection, and response to ensure that the DOD workforce and facilities in the NCR are secure and protected.

For further information, contact the Pentagon Force Protection Agency, Washington, DC 20301. Phone, 703–693–3685. Internet, http://www.pfpa.mil.

Joint Service Schools

Defense Acquisition University
Fort Belvoir, VA 22060–5565
Phone, 703–805–3360. Internet, http://www.dau.mil.

President	FRANK J. ANDERSON, JR.

The Defense Acquisition University (DAU), established pursuant to the Defense Acquisition Workforce Improvement Act of 1990 (10 U.S.C. 1701 note), serves as the DOD center for acquisition, technology, and logistics training; performance support; continuous learning; and knowledge sharing. DAU is a unified structure with five regional campuses and the Defense Systems Management College-School of Program Managers, which provides executive and international acquisition training. DAU's mission is to provide the training, career management, and services that enable the acquisition, technology, and logistics community to make smart business decisions and deliver timely and affordable capabilities to warfighters.

For further information, contact the Director, Operations Support Group, Defense Acquisition University, Fort Belvoir, VA 22060–5565. Phone, 800–845–7606. Internet, http://www.dau.mil.

National Defense Intelligence College
Defense Intelligence Analysis Center, Washington, DC 20340–5100
Phone, 202–231–5466. Internet, http://www.ndic.edu.

President	A. DENIS CLIFT

The National Defense Intelligence College, formerly the Joint Military Intelligence College, was established in 1962. The College is a joint service interagency

educational institution serving the
intelligence community and operates
under the authority of the Director, Defense
Intelligence Agency. Its mission is to
educate military and civilian intelligence
professionals, conduct and disseminate
relevant intelligence research, and perform
academic outreach regarding intelligence
matters. The College is authorized by

Congress to award the bachelor of science
in intelligence and master of science of
strategic intelligence. Courses are offered to
full-time students in a traditional daytime
format and for part-time students in the
evening, on Saturday, and in an executive
format (one weekend per month and a
2-week intensive summer period).

For further information, contact the Admissions Office, MCA–2, National Defense Intelligence College, Defense Intelligence Analysis Center, Washington, DC 20340–5100. Phone, 202–231–5466 or 202–231–3319. Internet, http://www.ndic.edu.

National Defense University

300 Fifth Avenue, Building 62, Fort McNair, Washington, DC 20319–5066
Phone, 202–685–3922. Internet, http://www.ndu.edu.

The National War College: 300 D Street, Building 61, Fort McNair, Washington, DC 20319–5078
Phone, 202–685–3674. Fax, 202–685–6461. Internet, http://www.ndu.edu/nwc/.

Industrial College of the Armed Forces: 408 Fourth Avenue, Building 59, Fort McNair, Washington, DC 20319–5062
Phone, 202–685–4337. Internet, http://www.ndu.edu/icaf/.

Joint Forces Staff College: Norfolk, VA 23511–1702
Phone, 757–443–6200. Internet, http://www.jfsc.ndu.edu.

Information Resources Management College: 300 Fifth Avenue, Building 62, Fort McNair, Washington, DC 20319–5066
Phone, 202–685–6300. Internet, http://www.ndu.edu/irmc.

College of International Security Affairs: 260 Fifth Avenue, Fort McNair, Washington, DC 20319–5066
Phone, 202–685–2290. Internet, http://www.ndu.edu/cisa.

President, National Defense University	VICE ADM. ANN E. RONDEAU, USN
Commandant, National War College	MAJ. GEN. ROBERT STEEL, USAF
Commandant, Industrial College of the Armed Forces	REAR ADM. GARRY HALL, USN
Commandant, Joint Forces Staff College	BRIG. GEN. KATHERINE P. KASUN, USA
Director, Information Resources Management College	ROBERT D. CHILDS
Director, College of International Security Affairs	R. JOSEPH DESUTTER

National Defense University

The mission of the National Defense University is to prepare military and civilian leaders from the United States and other countries to evaluate national and international security challenges through multidisciplinary educational and research programs, professional exchanges, and outreach.

The National Defense University was established in 1976 and incorporates the following colleges and programs: the Industrial College of the Armed Forces, the National War College, the Joint Forces Staff College, the Information Resources Management College, the College of International Security Affairs, the Institute for National Strategic Studies, the Center for the Study of Weapons of Mass Destruction, the Center for Technology and National Security Policy, the International Student Management Office, the Joint Reserve Affairs Center, CAPSTONE, the Security of Defense

Corporate Fellows Program, the NATO Education Center, the Institute for National Security Ethics and Leadership, the Center for Joint Strategic Logistics Excellence, the Center for Applied Strategic Leaders, and the Center for Complex Operations.

For further information, contact the Human Resources Directorate, National Defense University, 300 Fifth Avenue, Building 62, Fort McNair, Washington, DC 20319–5066. Phone, 202–685–2169. Internet, http://www.ndu.edu.

National War College   The National War College provides education in national security policy to selected military officers and career civil service employees of Federal departments and agencies concerned with national security. It is the only senior service college with the primary mission of offering a course of study that emphasizes national security policy formulation and the planning and implementation of national strategy. Its 10-month academic program is an issue-centered study in U.S. national security. The elective program is designed to permit each student to tailor his or her academic experience to meet individual professional development needs.

For further information, contact the Department of Administration, The National War College, 300 D Street SW., Fort McNair, Washington, DC 20319–5078. Phone, 202–685–3674. Internet, http://www.ndu.edu/nwc/.

Industrial College of the Armed Forces   The Industrial College of the Armed Forces is an educational institution that prepares selected military and civilians for strategic leadership and success in developing our national security strategy and in evaluating, marshalling, and managing resources in the execution of that strategy. The College offers an education in the understanding of the importance of industry to our national security strategy, and more importantly, the resource component of national security. The rigorous, compressed curriculum, completed in two semesters, leads to a master of science degree in national resource strategy.

For further information, contact the Director of Operations, Industrial College of the Armed Forces,

408 Fourth Avenue, Fort McNair, Washington, DC 20319–5062. Phone, 202–685–4333. Internet, http://www.ndu.edu/icaf/.

Joint Forces Staff College   The Joint Forces Staff College (JFSC) is an intermediate- and senior-level joint college in the professional military education system dedicated to the study of the principles, perspectives, and techniques of joint operational-level planning and warfare. The mission of JFSC is to educate national security professionals in the planning and execution of joint, multinational, and interagency operations in order to instill a primary commitment to joint, multinational, and interagency teamwork, attitudes, and perspectives. The College accomplishes this mission through four schools: the Joint Advanced Warfighters School, the Joint and Combined Warfighting School, the Joint Continuing and Distance Education School, and the Joint Command, Control, and Information Operations School.

For further information, contact the Public Affairs Officer, Joint Forces Staff College, 7800 Hampton Boulevard, Norfolk, VA 23511–1702. Phone, 757–443–6212. Fax, 757–443–6210. Internet, http://www.jfsc.ndu.edu.

Information Resources Management College   The Information Resources Management College provides graduate-level courses in information resources management. The College prepares leaders to direct the information component of national power by leveraging information and information technology for strategic advantage. The College's primary areas of concentration include policy, strategic planning, leadership/management, process improvement, capital planning and investment, performance- and results-based management, technology assessment, architecture, information assurance and security, acquisition, domestic preparedness, transformation, e-Government, and information operations.

For further information, contact the Registrar, Information Resources Management College, 300 Fifth Avenue, Fort McNair, Washington, DC 20319–5066. Phone, 202–685–6309. Internet, http://www.ndu.edu/irmc.

College of International Security Affairs
The College of International Security
Affairs (CISA) is one of NDU's five
colleges. CISA educates students from
across the international, interagency,
and interservice communities. CISA's
primary areas of concentration include
counterterrorism, conflict management of
stability of operations, homeland security,
and defense and international security
studies. CISA is also home to NDU's
International Counterterrorism Fellowship
Program.

For further information, contact the College of
International Security Affairs, 260 Fifth Avenue,
Fort McNair, Washington, DC 20319–5066. Phone,
202–685–2290. Internet, http://www.ndu.edu/cisa.

Uniformed Services University of the Health Sciences

4301 Jones Bridge Road, Bethesda, MD 20814–4799
Phone, 301–295–3770. Internet, http://www.usuhs.mil.

President   CHARLES L. RICE

Authorized by act of September 21,
1972 (10 U.S.C. 2112), the Uniformed
Services University of the Health
Sciences was established to educate
career-oriented medical officers for the
Military Departments and the Public
Health Service. The University currently
incorporates the F. Edward Hebert School
of Medicine (including graduate and
continuing education programs) and the
Graduate School of Nursing.

Students are selected by procedures
recommended by the Board of Regents
and prescribed by the Secretary of
Defense. The actual selection is carried
out by a faculty committee on admissions
and is based upon motivation and
dedication to a career in the uniformed
services and an overall appraisal of the
personal and intellectual characteristics
of the candidates without regard to

sex, race, religion, or national origin.
Applicants must be U.S. citizens.

Medical school matriculants will be
commissioned officers in one of the
uniformed services. They must meet
the physical and personal qualifications
for such a commission and must give
evidence of a strong commitment to
serving as a uniformed medical officer.
The graduating medical student is
required to serve a period of obligation of
not less than 7 years, excluding graduate
medical education.

Students of the Graduate School of
Nursing must be commissioned officers
of the Army, Navy, Air Force, or Public
Health Service prior to application.
Graduate nursing students must serve
a commitment determined by their
respective service.

For further information, contact the President, Uniformed Services University of the Health Sciences, 4301
Jones Bridge Road, Bethesda, MD 20814–4799. Phone, 301–295–3770. Internet, http://www.usuhs.mil.

DEPARTMENT OF EDUCATION

400 Maryland Avenue SW., Washington, DC 20202
Phone, 202–401–2000. TTY, 800–437–0833. Internet, http://www.ed.gov.

Secretary of Education	ARNE DUNCAN
Deputy Secretary	TONY MILLER
Chief of Staff	MARGOT ROGERS
Assistant Secretary for Communication and Outreach	PETER CUNNINGHAM
Assistant Secretary for Planning, Evaluation and Policy Development	CARMEL MARTIN
General Counsel	CHARLES P. ROSE
Inspector General	KATHLEEN S. TIGHE
Director, Institute of Education Sciences	JOHN EASTON
Assistant Secretary for Civil Rights	RUSSLYNN ALI
Chief Financial Officer	THOMAS SKELLY, *Acting*
Assistant Secretary for Management	WINONA H. VARNON, ACTING
Assistant Secretary for Legislation and Congressional Affairs	GABRIELLA GOMEZ
Director, Center for Faith-Based and Neighborhood Partnerships	PETER GROFF
Assistant Deputy Secretary, Office of Safe and Drug-Free Schools	KEVIN JENNINGS
Assistant Deputy Secretary, Office of Innovation and Improvement	JAMES H. SHELTON
Assistant Secretary for Special Education and Rehabilitative Services	ALEXA POSNY
Assistant Deputy Secretary and Director, Office of English Language Acquisition, Language Enhancement, and Academic Achievement for Limited English Proficient Students	RICHARD SMITH, *Acting*
Assistant Secretary for Elementary and Secondary Education	THELMA MELENDEZ
Executive Director, White House Initiative on Educational Excellence for Hispanic Americans	JUAN SEPULVEDA
Under Secretary	MARTHA KANTER
Chief Operating Officer for Federal Student Aid	WILLIAM J. TAGGART
Assistant Secretary for Postsecondary Education	DANIEL T. MADZELAN, *Acting*
Assistant Secretary for Vocational and Adult Education	BRENDA DANN-MESSIER, *Acting*
Executive Director, White House Initiative on Historically Black Colleges and Universities	JOHN WILSON
Executive Director, White House Initiative on Tribal Colleges and Universities	(VACANCY)
Executive Director, White House Initiative on Asian and Pacific Islanders	KIRAN AHUJA

The Department of Education establishes policy for, administers, and coordinates most Federal assistance to education. Its mission is to ensure equal access to education and to promote educational excellence throughout the Nation.

The Department of Education was created by the Department of Education Organization Act (20 U.S.C. 3411) and is administered under the supervision and direction of the Secretary of Education.

Secretary   The Secretary of Education advises the President on education plans, policies, and programs of the Federal Government and serves as the chief executive officer of the Department, supervising all Department activities, providing support to States and localities, and focusing resources to ensure equal access to educational excellence throughout the Nation.

Activities

Institute of Education Sciences   The Institute of Education Sciences was formally established by the Education Sciences Reform Act of 2002 (20 U.S.C. 9501 note). The Institute includes national education centers focused on research, special education, statistics, and evaluation and is the mechanism through which the Department supports the research activities needed to improve education policy and practice.

Elementary and Secondary Education   The Office of Elementary and Secondary Education directs, coordinates, and formulates policy relating to early childhood, elementary, and secondary education. Included are grants and contracts to State educational agencies and local school districts, postsecondary schools, and nonprofit organizations for disadvantaged, migrant, and Indian children; enhancement of State student achievement assessment systems; improvement of reading instruction; economic impact aid; technology; and after-school learning programs. The Office also focuses on improving K–12 education, providing children with language and cognitive development, early reading, and other readiness skills, and improving the quality of teachers and other instructional staff.

English Language Acquisition   The Office of English Language Acquisition, Language Enhancement, and Academic Achievement for Limited English Proficient Students helps children who are limited in their English, including immigrant children and youth, attain English proficiency, develop high levels of academic attainment in English, and meet the same challenging State academic content and student academic achievement standards that all children are expected to meet.

Federal Student Aid   Federal Student Aid partners with postsecondary schools and financial institutions to deliver programs and services that help students finance their education beyond high school. This includes administering postsecondary student financial assistance programs authorized under Title IV of the Higher Education Act of 1965, as amended.

Innovation and Improvement   The Office of Innovation and Improvement (OII) oversees competitive grant programs that support innovations in the educational system and disseminates the lessons learned from these innovative practices. OII administers, coordinates, and recommends programs and policy for improving the quality of activities designed to support and test innovations throughout the K–12 system in areas such as parental choice, teacher quality, use of technology in education, and arts in education. OII encourages the establishment of charter schools through planning, start-up funding, and approaches to credit enhancement for charter school facilities. OII also serves as the Department's liaison and resource to the nonpublic education community.

Postsecondary Education   The Office of Postsecondary Education (OPE) formulates Federal postsecondary education policy and administers programs that address critical national needs in support of the mission to increase access to quality postsecondary education. OPE develops policy for Federal student financial programs and

DEPARTMENT OF EDUCATION

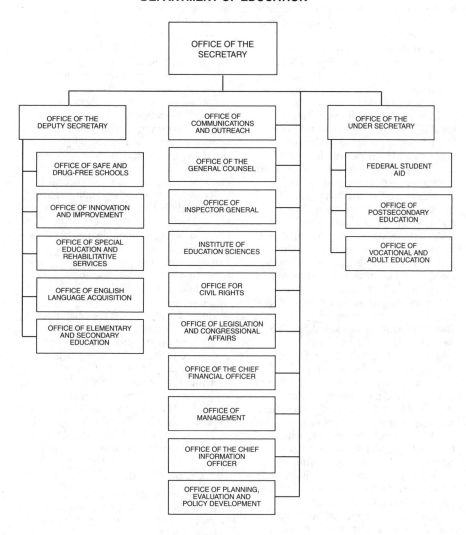

OFFICE OF THE
SECRETARY

OFFICE OF THE
DEPUTY SECRETARY

OFFICE OF
COMMUNICATIONS
AND OUTREACH

OFFICE OF THE
UNDER SECRETARY

OFFICE OF SAFE AND
DRUG-FREE SCHOOLS

OFFICE OF THE
GENERAL COUNSEL

FEDERAL STUDENT
AID

OFFICE OF INNOVATION
AND IMPROVEMENT

OFFICE OF
INSPECTOR GENERAL

OFFICE OF
POSTSECONDARY
EDUCATION

OFFICE OF SPECIAL
EDUCATION AND
REHABILITATIVE
SERVICES

INSTITUTE OF
EDUCATION SCIENCES

OFFICE OF
VOCATIONAL AND
ADULT EDUCATION

OFFICE OF ENGLISH
LANGUAGE ACQUISITION

OFFICE FOR
CIVIL RIGHTS

OFFICE OF ELEMENTARY
AND SECONDARY
EDUCATION

OFFICE OF LEGISLATION
AND CONGRESSIONAL
AFFAIRS

OFFICE OF THE CHIEF
FINANCIAL OFFICER

OFFICE OF
MANAGEMENT

OFFICE OF THE CHIEF
INFORMATION
OFFICER

OFFICE OF PLANNING,
EVALUATION AND
POLICY DEVELOPMENT

support programs that reach out to low-income, first-generation college students and communities. OPE also supports programs that strengthen the capacity of colleges and universities serving a high percentage of disadvantaged students and improve teacher quality. OPE recognizes accrediting agencies that monitor academic quality, promote innovation in higher education, and expand American educational resources for international studies and services.

Safe and Drug-Free Schools   The Office of Safe and Drug-Free Schools (OSDFS) administers, coordinates, and recommends policy for improving drug and violence prevention programs. OSDFS, in partnership with State and local educational agencies and public and private nonprofit organizations, supports and provides funding for efforts to promote safe schools, respond to crises, prevent drug and alcohol abuse, ensure the health and well-being of students, and teach students good citizenship and character. OSDFS participates in the formulation and development of program policy, legislative proposals, and administration policies related to violence and drug prevention. OSDFS also administers the Department's character, citizenship, and civic education programs.

Special Education and Rehabilitative Services   The Office of Special Education and Rehabilitative Services (OSERS) provides leadership and resources to help ensure that people with disabilities have equal opportunities to learn, work, and live as fully integrated and contributing members of society. OSERS has three components: The Office of Special Education Programs administers the Individuals with Disabilities Education Act legislation, which helps States meet the early intervention and educational needs of infants, toddlers, children, and youth with disabilities. The Rehabilitation Services Administration supports State vocational rehabilitation, independent living, and assistive technology programs that provide people with disabilities the services, technology, and job training and placement assistance they need to

gain meaningful employment and lead independent lives. The National Institute on Disability and Rehabilitation Research supports research and development programs that improve the ability of individuals with disabilities to work and live in a barrier-free, inclusive society. OSERS also supports Gallaudet University, the National Technical Institute for the Deaf, the American Printing House for the Blind, and the Helen Keller National Center.

Vocational and Adult Education   The Office of Vocational and Adult Education (OVAE) administers grant, contract, and technical assistance programs for vocational-technical education and for adult education and literacy. OVAE promotes programs that enable adults to acquire the basic literacy skills necessary to function in today's society. OVAE also helps students acquire challenging academic and technical skills and prepare for high-skill, high-wage, and high-demand occupations in the 21st-century global economy. OVAE provides national leadership and works to strengthen the role of community colleges in expanding access to postsecondary education for youth and adults in advancing workforce development.

Regional Offices   Each regional office serves as a center for the dissemination of information and provides technical assistance to State and local educational agencies and other institutions and individuals interested in Federal educational activities. Offices are located in Boston, MA; New York, NY; Philadelphia, PA; Atlanta, GA; Chicago, IL; Dallas, TX; Kansas City, MO; Denver, CO; San Francisco, CA; and Seattle, WA.

Sources of Information

Inquiries on the following categories may be directed to the specified office, Department of Education, 400 Maryland Avenue SW., Washington, DC 20202.
Contracts and Small Business Activities Call or write the Office of Small and Disadvantaged Business Utilization. Phone, 202–245–6301.
Employment   Inquiries and applications for employment and inquiries regarding

the college recruitment program should be directed to the Human Capital and Client Services. Phone, 202–401–0553.

Organization   Contact the Executive Office, Office of Management. Phone, 202–401–0690. TDD, 202–260–8956.

For further information, contact the Information Resources Center, Department of Education, Room 5E248 (FB–6), 400 Maryland Avenue SW., Washington, DC 20202. Phone, 800–USA–LEARN. Internet, http://www.ed.gov.

Federally Aided Corporations

American Printing House for the Blind

P.O. Box 6085, Louisville, KY 40206
Phone, 502–895–2405. Internet, http://www.aph.org.

President	TUCK TINSLEY, III
Chairman of the Board	W. JAMES LINTNER, JR.

Founded in 1858 as a nonprofit organization, the American Printing House for the Blind (APH) received its Federal charter in 1879 when Congress passed the Act to Promote Education of the Blind. This Act designates APH as the official supplier of educational materials adapted for students who are legally blind and who are enrolled in formal educational programs below the college level. Materials produced and distributed by APH include textbooks in Braille and large type, educational tools such as Braille typewriters and computer software and hardware, teaching aides such as tests and performance measures, and other special supplies. The materials are distributed through allotments to the States to programs serving individuals who are blind.

For further information, contact the American Printing House for the Blind, P.O. Box 6085, Louisville, KY 40206. Phone, 502–895–2405. Internet, http://www.aph.org.

Gallaudet University

800 Florida Avenue NE., Washington, DC 20002
Phone, 202–651–5000. Internet, http://www.gallaudet.edu.

President, Gallaudet University	T. ALAN HURWITZ
Chair, Board of Trustees	BENJAMIN J. SOUKUP, JR.

Gallaudet University received its Federal charter in 1864 and is currently authorized by the Education of the Deaf Act of 1986, as amended. Gallaudet is a private, nonprofit educational institution providing elementary, secondary, undergraduate, and continuing education programs for persons who are deaf. The University offers a traditional liberal arts curriculum for students who are deaf and graduate programs in fields related to deafness for students who are deaf and students who are hearing. Gallaudet also conducts a wide variety of basic and applied deafness research and provides public service programs for persons who are deaf and for professionals who work with persons who are deaf.

Gallaudet University is accredited by a number of organizations, among which are the Middle States Association of Colleges and Secondary Schools, the National Council for Accreditation of Teacher Education, and the Conference of Educational Administrators of Schools and Programs for the Deaf.

Laurent Clerc National Deaf Education Center   Gallaudet's Laurent Clerc National Deaf Education Center operates elementary and secondary education programs on the main campus of the University. These programs are authorized by the Education of the Deaf Act of 1986 (20 U.S.C. 4304, as amended) for the primary purpose of developing, evaluating, and disseminating model curricula, instructional strategies, and materials to serve individuals who are deaf or hard of hearing. The Education of the Deaf Act requires the programs to include students preparing for postsecondary opportunities other than college and students with a broad spectrum of needs, such as students who are academically challenged, come from non-English-speaking homes, have secondary disabilities, are members of minority groups, or are from rural areas.

Model Secondary School for the Deaf The school was established by act of October 15, 1966 (20 U.S.C. 693), which was superseded by the Education of the Deaf Act of 1986. The school provides day and residential facilities for secondary-age students from across the United States from grades 9 to 12, inclusively.

Kendall Demonstration Elementary School The school became the Nation's first demonstration elementary school for the deaf by act of December 24, 1970 (20 U.S.C. 695). This act was superseded by the Education of the Deaf Act of 1986. The school is a day program for students from the Washington, DC, metropolitan area from the age of onset of deafness to age 15, inclusively, but not beyond the eighth grade or its equivalent.

For further information, contact the Public Relations Office, Gallaudet University, 800 Florida Avenue NE., Washington, DC 20002. Phone, 202–651–5505. Internet, http://www.gallaudet.edu.

Howard University

2400 Sixth Street NW.,Washington, DC 20059
Phone, 202–806–6100. Internet, http://www.howard.edu.

President   SIDNEY A. RIBEAU

Howard University was established by act of March 2, 1867 (14 Stat. 438). It offers instruction in 12 schools and colleges, as follows: the colleges of arts and sciences; dentistry; engineering, architecture, and computer sciences; medicine; pharmacy, nursing, and allied health sciences; the graduate school; the schools of business; communications; divinity; education; law; and social work. In addition, Howard University has research institutes, centers, and special programs in the following areas: cancer, child development, computational science and engineering, international affairs, sickle cell disease, and the national human genome project.

For further information, contact the Office of University Communications, Howard University, 2400 Sixth Street NW., Washington, DC 20059. Phone, 202–806–0970. Internet, http://www.howard.edu.

National Institute for Literacy

1775 I Street NW., Suite 730, Washington, DC 20006
Phone, 202–233–2025. Internet, http://www.nifl.gov.

Director   DANIEL J. MILLER, *Acting*

The National Institute for Literacy provides leadership on literacy issues, including the improvement of reading instruction for children, youth, and adults. The Institute serves as a national resource on current and comprehensive literacy research, practice, and policy.

National Technical Institute for the Deaf / Rochester Institute of Technology

52 Lomb Memorial Drive, Rochester, NY 14623
Phone, 716–475–6853. Internet, http://www.ntid.edu.

President, Rochester Institute of Technology   WILLIAM W. DESTLER
Vice President, National Technical Institute for   JAMES J. DECARO
    the Deaf

The National Technical Institute for the Deaf (NTID) was established by act of June 8, 1965 (20 U.S.C. 681) to promote the employment of persons who are deaf, by providing technical and professional education. The National Technical Institute for the Deaf Act was superseded by the Education of the Deaf Act of 1986 (20 U.S.C. 4431, as amended). The U.S. Department of Education maintains a contract with the Rochester Institute of Technology (RIT) for the operation of a residential facility for postsecondary technical training and education for individuals who are deaf. The purpose of the special relationship with the host institution is to give NTID's faculty and students access to more facilities, institutional services, and career preparation options than could be otherwise provided by a national technical institute for the deaf operating independently.

NTID offers a variety of technical programs at the certificate, diploma, and associate degree levels. Degree programs include majors in business, engineering, science, and visual communications. In addition, NTID students may participate in approximately 200 educational programs available through RIT. Students are provided a wide range of support services and special programs to assist them in preparing for their careers, including tutoring, counseling, interpreting, specialized educational media, cooperative work experience, and specialized job placement. RIT and NTID are both accredited by the Middle States Association of Colleges and Secondary Schools.

NTID also conducts applied research in occupational- and employment-related aspects of deafness, communication assessment, demographics of NTID's target population, and learning processes in postsecondary education. In addition, NTID conducts training workshops and seminars related to deafness. These workshops and seminars are offered to professionals throughout the Nation who employ, work with, teach, or otherwise serve persons who are deaf.

For further information, contact the Rochester Institute of Technology, National Technical Institute for the Deaf, Department of Recruitment and Admissions, Lyndon Baines Johnson Building, 52 Lomb Memorial Drive, Rochester, NY 14623–5604. Phone, 716–475–6700. Internet, http://www.ntid.edu.

DEPARTMENT OF ENERGY

1000 Independence Avenue SW., Washington, DC 20585
Phone, 202–586–5000. Internet, http://www.energy.gov.

Secretary of Energy	STEVEN CHU
Deputy Secretary	DANIEL B. PONEMAN
Chief of Staff	ROD O'CONNER
Under Secretary for Nuclear Security and Administrator for National Nuclear Security Administration	THOMAS P. D'AGOSTINO
Deputy Administrator, Defense Programs	(VACANCY)
Deputy Administrator, Defense Nuclear Nonproliferation	KENNETH E. BAKER, *Acting*
Deputy Administrator, Naval Reactors	ADM. KIRKLAND H. DONALD, USN
Deputy Under Secretary, Counterterrorism	STEVEN AOKI
Associate Administrator, Defense Nuclear Security	BRADLEY A. PETERSON
Associate Administrator, Emergency Operations	JOSEPH J. KROL, JR.
Associate Administrator, Infrastructure and Environment	THAD T. KONOPNICKI
Associate Administrator, Management and Administration	MICHAEL C. KANE
Under Secretary of Energy	KRISTINA M. JOHNSON
Director, Civilian Radioactive Waste Management	DAVID ZABRANSKY, *Acting*
Assistant Secretary, Electricity Delivery and Energy Reliability	PATRICIA HOFFMAN, *Acting*
Assistant Secretary, Energy Efficiency and Renewable Energy	CATHY ZOI
Assistant Secretary, Environmental Management	INES R. TRIAY
Assistant Secretary, Fossil Energy	JAMES J. MARKOWSKY
Director, Legacy Management	DAVID GEISER, *Acting*
Assistant Secretary, Nuclear Energy	WARREN F. MILLER, JR.
Under Secretary for Science	STEVEN E. KOONIN
Director, Office of Science	WILLIAM BRINKMAN
Administrator, Energy Information Administration	RICHARD NEWELL
Director, Advanced Research Projects Agency-Energy	ARUN MAJUMDAR
Director, Office of the American Recovery and Reinvestment Act	MATTHEW ROGERS
Chief Financial Officer	STEVEN J. ISAKOWITZ
Chief Human Capital Officer	MICHAEL C. KANE
Chief Information Officer	WILLIAM T. TURNBULL, *Acting*
Assistant Secretary, Congressional and Intergovernmental Affairs	ELIZABETH A. NOLAN, *Acting*
Director, Economic Impact and Diversity	(VACANCY)
General Counsel	SCOTT BLAKE HARRIS
Director, Health, Safety, and Security	GLENN S. PODONSKY

Director, Hearings and Appeals	POLI A. MARMOLEJOS
Inspector General	GREGORY H. FRIEDMAN
Director, Intelligence and Counterintelligence	EDWARD B. HELD
Director, Management	INGRID KOLB
Assistant Secretary, Policy and International Affairs	DAVID SANDALOW
Director, Public Affairs	DAN LEISTIKOW

The Department of Energy's mission is to advance the national, economic, and energy security of the United States; to promote scientific and technological innovation in support of that mission; and to ensure the environmental cleanup of the national nuclear weapons complex.

The Department of Energy (DOE) was established by the Department of Energy Organization Act (42 U.S.C. 7131), effective October 1, 1977, pursuant to Executive Order 12009 of September 13, 1977. The act consolidated the major Federal energy functions into one Cabinet-level Department

Secretary The Secretary decides major energy policy and planning issues; acts as the principal spokesperson for the Department; and ensures the effective communication and working relationships with Federal, State, local, and tribal governments and the public. The Secretary is the principal adviser to the President on energy policies, plans, and programs.

Intelligence and Counterintelligence The Office of Intelligence and Counterintelligence ensures that all departmental intelligence information requirements are met and that the Department's technical, analytical, and research expertise is made available to support U.S. intelligence efforts. The Office develops and implements programs to identify, neutralize, and deter foreign government or industrial intelligence activities directed at or involving Department programs, personnel, facilities, technologies, classified information, and sensitive information. The Office ensures effective use of the U.S. Government's intelligence apparatus in support of DOE's need for information on foreign energy situations and hostile threats, information on global nuclear weapons development, nonproliferation, and foreign hydrocarbon, nuclear, and other energy production and consumption. The Office formulates all DOE intelligence and counterintelligence policy and coordinates all investigative matters with the Federal Bureau of Investigation.

For further information, contact the Office of Intelligence and Counterintelligence. Phone, 202–586–2610.

Health, Safety and Security The Office of Health, Safety, and Security develops policies to protect national security and other critical assets entrusted to the Department of Energy. It also manages security operations for departmental facilities in the national capital area.

For further information, contact the Office of Health, Safety, and Security. Phone, 301–903–3777.

Energy Programs

Renewable Energy The Office of Energy Efficiency and Renewable Energy is responsible for formulating and directing programs designed to increase the production and utilization of renewable energy (solar, biomass, wind, geothermal, alcohol fuels, etc.) and hydrogen and improving the energy efficiency of the transportation, buildings, industrial, and utility sectors through support of research and development and technology transfer activities. It also has responsibility for administering programs that provide financial assistance for State energy planning; the weatherization of housing owned by the poor and disadvantaged; implementing State and local energy conservation programs; and the promotion of energy efficient construction and renovation of Federal facilities.

DEPARTMENT OF ENERGY

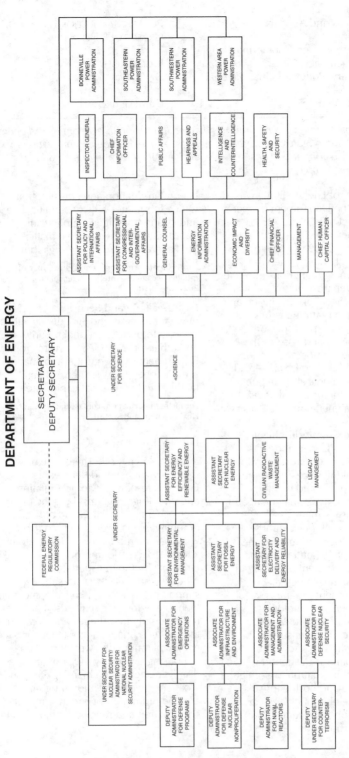

* The Deputy Secretary also serves as the Chief Operating Officer.

For further information, contact the Director of Information and Business Management Systems. Phone, 202–586–7241.

Fossil Energy   The Office of Fossil Energy is responsible for research and development of programs involving coal, petroleum, and natural gas. The fossil energy program involves applied research, exploratory development, and limited proof-of-concept testing targeted to high-risk and high-payoff endeavors. The objective of the program is to provide the general technology and knowledge base that the private sector can use to complete development and initiate commercialization of advanced processes and energy systems. The program is principally executed through the National Energy Technology Laboratory. The Office also manages the strategic petroleum reserve, the northeast home heating oil reserve, and the naval petroleum shale reserves.

For further information, contact the Office of Communications. Phone, 202–586–6803.

Nuclear Energy   The Office of Nuclear Energy manages the Department's research and development programs associated with fission and fusion energy. This includes programs relating to naval and civilian nuclear reactor development, nuclear fuel cycle, and space nuclear applications. The Office manages a program to provide radioactive and stable isotope products to various domestic and international markets for medical research, health care, and industrial research. The Office also conducts technical analyses concerning nonproliferation; assesses alternative nuclear systems and new reactor and fuel cycle concepts; manages depleted uranium hexafluoride activities, highly enriched uranium downblend, natural uranium sales, and uranium enrichment legacy activities; and evaluates proposed advanced nuclear fission energy concepts and technical improvements for possible application to nuclear powerplant systems.

For further information, contact the Director, Corporate Communications and External Affairs. Phone, 301–903–1636.

Energy Information   The Energy Information Administration is responsible for collecting, processing, publishing, and distributing data in the areas of energy resource reserves, energy production, demand, consumption, distribution, and technology. It performs analyses of energy data to assist government and nongovernment users in understanding energy trends.

For further information, contact the Director, National Energy Information Center. Phone, 202–586–6537.

Electricity Delivery and Energy Reliability   The Office of Electricity Delivery and Energy Reliability leads a national effort to modernize and expand America's electricity delivery system. The Office is responsible for the enhanced security and reliability of the energy infrastructure and facilitates the recovery from disruptions to energy supply.

For further information, contact the Office of the Director. Phone, 202–586–1411.

Advanced Research Projects Agency–Energy   The Advanced Research Projects Agency–Energy (ARPA–E) works to overcome the long-term and high-risk challenges in the development of energy technologies. ARPA–E creates and implements research and development initiatives to enhance the economic security of the United States through the development of energy technologies that reduce energy imports, improve energy efficiency, and reduce energy-related emissions. Additionally, ARPA–E ensures that the United States maintains global leadership in developing and deploying advanced energy technologies.

For further information, contact the Office of the Director. Phone, 202–287–1004.

Office of the American Recovery and Reinvestment Act   The Office of the American Recovery and Reinvestment Act promotes economic recovery by assisting individuals, organizations, and businesses most impacted by the recession. Working to foster job creation and preservation, the Office strengthens the economy by investing in technological advances in science, health, transportation, environmental protection, and other infrastructure. This assistance provides long-term economic benefits and stabilizes State and local government budgets, preventing tax increases and reductions in essential services.

For further information, contact the Office of the Director. Phone, 202–586–1989.

Nuclear Security Programs

Nuclear Security   The National Nuclear Security Administration (NNSA) was created by Congress through the National Defense Authorization Act for Fiscal Year 2000 (113 Stat. 512) to bring focus to the management of the Nation's defense nuclear security programs. Three existing organizations within the Department of Energy—Defense Programs, Defense Nuclear Nonproliferation, and Naval Reactors—were combined into a new, separately organized and managed agency within DOE, headed by an Administrator who reports to the Secretary. NNSA is responsible for strengthening United States security through military application of nuclear energy and by reducing the global threat from terrorism and weapons of mass destruction.

NNSA's service center and eight site offices (nnsa.energy.gov/about/sites.htm) provide operations oversight and contract administration for NNSA site activities, acting as the agency's risk acceptance for the site. The site offices are responsible for the following functions: the safe and secure operation of facilities under the purview of NNSA; supporting NNSA programs to ensure their success in accordance with their expectations; and ensuring the long-term viability of the site to support NNSA programs and projects.

For further information, contact the Associate Administrator for Management and Administration. Phone, 202–586–5753.

Defense Activities   The Office of the Deputy Administrator for Defense Programs directs the Nation's nuclear weapons research, development, testing, production, and surveillance program. It is also responsible for the production of the special nuclear materials used by the weapons program within the Department

and the management of defense nuclear waste and byproducts. The Office ensures the technology base for the surety, reliability, military effectiveness, and credibility of the nuclear weapon stockpile. It also manages research in inertial confinement fusion.

For further information, contact the Associate Administrator for Management and Administration. Phone, 202–586–5753.

Naval Reactors   The Office of the Deputy Administrator for Naval Reactors manages and performs research, development, design, acquisition, specification, construction, inspection, installation, certification, testing overhaul, refueling, operations procedures, maintenance, supply support, and ultimate disposition of naval nuclear propulsion plants.

For further information, contact the Deputy Administrator for Naval Reactors. Phone, 202–781–6174.

Nuclear Nonproliferation   The Office of the Deputy Administrator for Defense Nuclear Nonproliferation directs the development of the Department's policy, plans, procedures, and research and development activities relating to arms control, nonproliferation, export controls, international nuclear safety and safeguard, and surplus fissile materials inventories elimination activities.

For further information, contact the Associate Administrator for Management and Administration. Phone, 202–586–5753.

Environmental Quality Programs

Civilian Radioactive Waste Management   The Office of Civilian Radioactive Waste Management is responsible for implementation of the Nuclear Waste Policy Act of 1982, as amended (42 U.S.C. 10101 et seq.), which provides for the development of a permanent, safe geologic repository for disposal of spent nuclear fuel and high-level radioactive waste.

For further information, contact the Director for Human Capital Division Resources. Phone, 202–586–8839.

Environmental Management   The Office of the Assistant Secretary for Environmental Management manages safe cleanup and closure of sites and facilities; directs a safe and effective waste management program, including storage and disposal of transuranic and mixed low- and high-level waste; and develops and implements an applied research program to provide innovative technologies that yield permanent cleanup solutions at reduced costs.

For further information, contact the Director of Communication/External Affairs. Phone, 202–287–5591.

Legacy Management   The Office of Legacy Management manages the Department's post-closure responsibilities and ensures the future protection of human health and the environment. The Office has control and custody of legacy land, structures, and facilities and is responsible for maintaining them at levels suitable for long-term use.

For further information, contact the Director of Business Operations. Phone, 202–586–7388.

Science Program

The Office of Science supports basic research that underpins DOE missions in national security, energy, and environment; constructs and operates large scientific facilities for the U.S. scientific community; and provides the infrastructure support for 10 national laboratories and an integrated support center (www.sc.doe.gov/Field_Offices/index.htm). In terms of basic research, the Office of Science provides over 40 percent of Federal support to the physical sciences (including 90 percent of Federal support for high energy and nuclear physics), the sole support to sub-fields of national importance, such as nuclear medicine, heavy element chemistry, and magnetic fusion, and support for the research of scientists and graduate students located in universities throughout the Nation. Office of Science support for major scientific user facilities, including accelerators, synchrotron light sources, and neutron sources, means that more than 18,000 scientists per year are able to use these state-of-the-art facilities to conduct research in a wide range of fields, including biology, medicine, and materials.

For further information, contact the Director of Human Resources. Phone, 301–903–5705.

Operations and Field Offices

The vast majority of the Department's energy and physical research and development, environmental restoration, and waste management activities are carried out by contractors who operate Government-owned facilities. Management and administration of Government-owned, contractor-operated facility contracts are the major responsibility of the Department's five operations offices and three field offices.

Department operations offices provide a formal link between Department headquarters and the field laboratories and other operating facilities. They also manage programs and projects as assigned from lead headquarters program offices. Routine management guidance, coordination, oversight of the operations, field and site offices (www.energy.gov/organization/opsoffices.htm), and daily specific program direction for the operations offices is provided by the appropriate assistant secretary, office director, or program officer.

Power Administrations

The marketing and transmission of electric power produced at Federal hydroelectric projects and reservoirs is carried out by the Department's four Power Administrations. Management oversight of the Power Administrations is the responsibility of the Deputy Secretary.
Bonneville Power Administration   The Administration markets power produced by the Federal Columbia River Power System at the lowest rates, consistent with sound business practices, and gives preference to public entities.

In addition, the Administration is responsible for energy conservation, renewable resource development, and fish and wildlife enhancement under the provisions of the Pacific Northwest Electric Power Planning and Conservation Act of 1980 (16 U.S.C. 839 note).

For further information, contact the Bonneville Power Administration, 905 Eleventh Avenue NE., Portland, OR 97232–4169. Phone, 503–230–3000.

Southeastern Power Administration

The Administration is responsible for the transmission and disposition of surplus electric power and energy generated at reservoir projects in the States of West Virginia, Virginia, North Carolina, South Carolina, Georgia, Florida, Alabama, Mississippi, Tennessee, and Kentucky.

The Administration sets the lowest possible rates to consumers, consistent with sound business principles, and gives preference in the sale of such power and energy to public bodies and cooperatives.

For further information, contact the Southeastern Power Administration, 1166 Athens Tech Road, Elberton, GA 30635–4578. Phone, 706–213–3800.

Southwestern Power Administration

The Administration is responsible for the sale and disposition of electric power and energy in the States of Arkansas, Kansas, Louisiana, Missouri, Oklahoma, and Texas.

The Administration transmits and disposes of the electric power and energy generated at Federal reservoir projects, supplemented by power purchased from public and private utilities, in such a manner as to encourage the most widespread and economical use. The Administration sets the lowest possible rates to consumers, consistent with sound business principles, and gives preference in the sale of power and energy to public bodies and cooperatives.

The Administration also conducts and participates in the comprehensive planning of water resource development in the Southwest.

For further information, contact the Southwestern Power Administration, Suite 1600, Williams Center Tower One, One West Third Street, Tulsa, OK 74103–3532. Phone, 918–595–6600.

Western Area Power Administration

The Administration is responsible for the Federal electric power marketing and transmission functions in 15 Central and Western States, encompassing a geographic area of 1.3 million square miles. The Administration sells power to cooperatives, municipalities, public utility districts, private utilities, Federal and State agencies, and irrigation districts. The wholesale power customers, in turn, provide service to millions of retail consumers in the States of Arizona, California, Colorado, Iowa, Kansas, Minnesota, Montana, Nebraska, Nevada, New Mexico, North Dakota, South Dakota, Texas, Utah, and Wyoming.

The Administration is responsible for the operation and maintenance of transmission lines, substations, and various auxiliary power facilities in the aforementioned geographic area and also for planning, construction, and operation and maintenance of additional Federal transmission facilities that may be authorized in the future.

For further information, contact the Western Area Power Administration, 12155 West Alameda Parkway, Lakewood, CO 80228–1213. Phone, 720–962–7000.

Sources of Information

Consumer Information For information on the consumer impact of Department policies and operations and for other DOE consumer information, call 202–586–1908.

Contracts and Small and Disadvantaged Business Utilization Activities Information on business opportunities with the Department and its contractors is available electronically through the Internet at www.pr.doe.gov. For information on existing DOE awards, call 202–586–9051.

Electronic Access Information concerning the Department is available through the Internet at www.energy.gov.

Employment Most jobs in the Department are in the competitive service. Positions are filled through hiring individuals with Federal civil service status, but may also be filled using lists of competitive eligibles from the Office of Personnel Management or the Department's special examining units. Contact the Office of Human Capital Management. Phone, 202–586–1234.

Freedom of Information Act To obtain administrative and technical support in matters involving the Freedom of Information, Privacy, and Computer Matching Acts, call 202–586–5955 or email FOIA–Central@hq.doe.gov.

Inspector General Hotline Persons who wish to raise issues of concern regarding departmental operations, processes,

or practices or who may be aware of or suspect illegal acts or noncriminal violations should contact the hotline. Phone, 202–586–4073 or 800–541–1625. Email, ighotmail@hq.doe.gov.
Public Information Issuances, Press Releases, and Publications For media contacts, call 202–586–5575.
Public Reading Room For information materials on DOE and public access to DOE records, call 202–586–3142.
Scientific and Technical Information The Office manages a system for the centralized collection, announcement, and dissemination of and historical reference to the Department's scientific and technical information and worldwide energy information. Contact the Office of Scientific and Technical Information, 175 Oak Ridge Turnpike, Oak Ridge, TN 37830–7255. Phone, 423–576–1188.
Whistleblower Assistance Federal or DOE contractor employees wishing to make complaints of alleged wrongdoing against the Department or its contractors should call 202–586–4034.

For further information, contact the Office of Public Affairs, Department of Energy, 1000 Independence Avenue SW., Washington, DC 20585. Phone, 202–586–4940. Internet, http://www.energy.gov.

Federal Energy Regulatory Commission

888 First Street NE., Washington, DC 20426
Phone, 202–502–8055. Internet, http://www.ferc.gov.

Chairman | JOHN WELLINGHOFF
Commissioners | PHILIP D. MOELLER, JOHN R. NORRIS, MARC L. SPITZER, (VACANCY)

The Federal Energy Regulatory Commission (FERC) is an independent agency within the Department of Energy which regulates the interstate transmission of electricity, natural gas, and oil. FERC has retained many of the functions of the Federal Power Commission, such as setting rates and charges for the transportation and sale of natural gas and the transportation of oil by pipelines, as well as the valuation of such pipelines. FERC also reviews proposals to build liquefied natural gas terminals and interstate natural gas pipelines as well as licensing hydropower projects. FERC is composed of five members appointed by the President of the United States with the advice and consent of the Senate. FERC Commissioners serve 5-year terms and have an equal vote on regulatory matters. One member is designated by the President to serve as both Chairman and FERC's administrative head.

For further information, contact the Office of External Affairs. Phone, 202–502–8004 or 866–208–3372. Fax, 202–208–2106. Internet, http://www.ferc.gov.

DEPARTMENT OF HEALTH AND HUMAN SERVICES

200 Independence Avenue SW., Washington, DC 20201
Phone, 202–690–6343. Internet, http://www.hhs.gov.

Secretary of Health and Human Services	KATHLEEN SEBELIUS
Deputy Secretary	WILLIAM CORR
Chief of Staff	LAURA PETROU
Executive Secretary	DAWN SMALLS
Director, Office of Intergovernmental Affairs	PAUL DIOGUARDI
Assistant Secretary for Health, Office of Public Health and Science	HOWARD KOH
Surgeon General	REGINA M. BENJAMIN
Assistant Secretary for Administration	E.J. HOLLAND, JR.
Assistant Secretary for Financial Resources	ELLEN G. MURRAY
Assistant Secretary for Legislation	JIM ESQUEA
Assistant Secretary for Planning and Evaluation	SHERRY GLIED
Assistant Secretary for Public Affairs	JENNY BACKUS, *Acting*
Assistant Secretary for Preparedness and Response	NICOLE LURIE
Chair, Departmental Appeals Board	CONSTANCE B. TOBIAS
Chief Administrative Law Judge, Office of Medicare Hearings and Appeals	NANCY J. GRISWOLD
Director, Center for Faith-Based and Neighborhood Partnerships	ALEXIA KELLEY
Director, Office for Civil Rights	GEORGINA VERDUGO
Director, Office of Consumer Information and Insurance Oversight	JAY ANGOFF
Director, Office on Disability	HENRY CLAYPOOL
Director, Office of Global Health Affairs	NILS DAULAIRE
Director, Office of Health Reform	JEANNE LAMBREW
Director, Office of Security and Strategic Information	ARTHUR LAWRENCE
General Counsel	MARK CHILDRESS, *Acting*
Inspector General	DANIEL R. LEVINSON
National Coordinator, Office of the National Coordinator for Health Information Technology	DAVID BLUMENTHAL
Administrator, Agency for Toxic Substances and Disease Registry	THOMAS FRIEDEN
Administrator, Centers for Medicare and Medicaid Services	MARILYN TAVENNER, *Acting*
Administrator, Health Resources and Services Administration	MARY K. WAKEFIELD
Administrator, Substance Abuse and Mental Health Services Administration	PAMELA HYDE
Assistant Secretary for the Administration on Aging	KATHY J. GREENLEE
Assistant Secretary for the Administration of Children and Families	DAVID A. HANSELL, *Acting*

Commissioner, Food and Drug Administration	MARGARET HAMBURG
Director, Agency for Healthcare Research and Quality	CAROLYN CLANCY
Director, Centers for Disease Control and Prevention	THOMAS FRIEDEN
Director, Indian Health Service	YVETTE ROUBIDEAUX
Director, National Institutes of Health	FRANCIS S. COLLINS

The Department of Health and Human Services works to strengthen the public health and welfare of the American people by providing access to affordable, quality health care and childcare, ensuring the safety of food products, preparing for public health emergencies, and improving research efforts to diagnose, treat, and cure life-threatening illnesses.

The Department of Health and Human Services (HHS) was created as the Department of Health, Education, and Welfare on April 11, 1953 (5 U.S.C. app.).

Secretary The Secretary of Health and Human Services advises the President on health, welfare, and income security plans, policies, and programs of the Federal Government and directs Department staff in carrying out the approved programs and activities of the Department and promotes general public understanding of the Department's goals, programs, and objectives.

For information on the HHS regional offices, visit our Web site at http://www.hhs.gov/about.

Office of Intergovernmental Affairs The Office of Intergovernmental Affairs (IGA) serves the Secretary as the primary liaison between the Department and State, local, and tribal governments. The mission of the Office is to facilitate communication regarding HHS initiatives as they relate to State, local, and tribal governments. IGA serves the dual role of representing the State and tribal perspective in the Federal policymaking process as well as clarifying the Federal perspective to State and tribal representatives.

For further information, contact the Office of Intergovernmental Affairs, 200 Independence Avenue SW., Room 620E, Washington, DC 20201. Phone, 202–690–6060. Internet, http://www.hhs.gov/intergovernmental.

Office of Consumer Information and Insurance Oversight

The Office of Consumer Information and Insurance Oversight (OCIIO) is responsible for ensuring compliance with health insurance market rules, such as prohibitions on termination due to preexisting conditions in children. OCIIO oversees medical loss ratio rules and assists States in reviewing insurance rates. OCIIO also provides guidance and oversight for the State-based insurance exchanges, administers the preexisting conditions insurance plans and the early retiree reinsurance programs, and compiles and maintains data for a Web site providing information on insurance options.

For further information, contact the Office of Consumer Information and Insurance Oversight, Department of Health and Human Services, 200 Independence Avenue SW., Washington, DC 20201. Phone, 202–205–9924. Internet, http://www.hhs.gov/ociio/index.html.

Office of the Assistant Secretary for Preparedness and Response

The Office of the Assistant Secretary for Preparedness and Response (ASPR) was established under the Pandemic and All Hazards Preparedness Act of 2006. ASPR leads the Nation in prevention, preparation, and response efforts related to the adverse health effects of public health emergencies, natural or man-made disasters, and bioterrorism. ASPR collaborates with Federal, State, local, tribal, and international officials as well as professionals in the private health care sector to ensure a unified and strategic approach to the challenges of public health emergencies.

For further information, contact the Office of the Assistant Secretary for Preparedness and Response, Room 638–G, 200 Independence Avenue SW.,

DEPARTMENT OF HEALTH AND HUMAN SERVICES

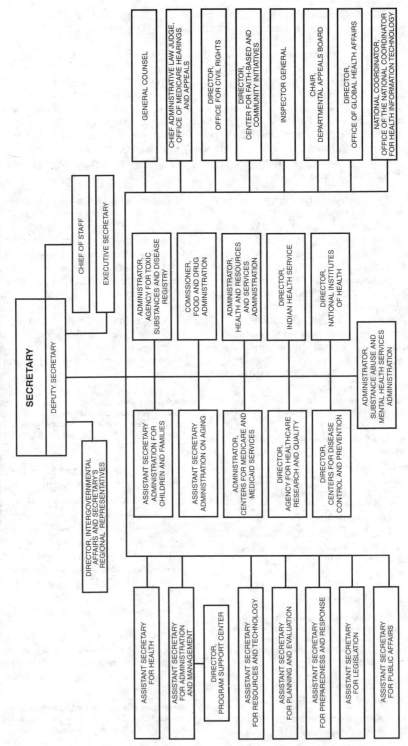

SECRETARY

DEPUTY SECRETARY

CHIEF OF STAFF

EXECUTIVE SECRETARY

DIRECTOR, INTERGOVERNMENTAL AFFAIRS AND SECRETARY'S REGIONAL REPRESENTATIVES

GENERAL COUNSEL

CHIEF ADMINISTRATIVE LAW JUDGE, OFFICE OF MEDICARE HEARINGS AND APPEALS

DIRECTOR, OFFICE FOR CIVIL RIGHTS

DIRECTOR, CENTER FOR FAITH-BASED AND COMMUNITY INITIATIVES

INSPECTOR GENERAL

CHAIR, DEPARTMENTAL APPEALS BOARD

DIRECTOR, OFFICE OF GLOBAL HEALTH AFFAIRS

NATIONAL COORDINATOR, OFFICE OF THE NATIONAL COORDINATOR FOR HEALTH INFORMATION TECHNOLOGY

ADMINISTRATOR, AGENCY FOR TOXIC SUBSTANCES AND DISEASE REGISTRY

COMISSIONER, FOOD AND DRUG ADMINISTRATION

ADMINISTRATOR, HEALTH AND RESOURCES AND SERVICES ADMINISTRATION

DIRECTOR, INDIAN HEALTH SERVICE

DIRECTOR, NATIONAL INSTITUTES OF HEALTH

ADMINISTRATOR, SUBSTANCE ABUSE AND MENTAL HEALTH SERVICES ADMINISTRATION

ASSISTANT SECRETARY ADMINISTRATION FOR CHILDREN AND FAMILIES

ASSISTANT SECRETARY ADMINISTRATION ON AGING

ADMINISTRATOR, CENTERS FOR MEDICARE AND MEDICAID SERVICES

DIRECTOR, AGENCY FOR HEALTHCARE RESEARCH AND QUALITY

DIRECTOR, CENTERS FOR DISEASE CONTROL AND PREVENTION

ASSISTANT SECRETARY FOR HEALTH

ASSISTANT SECRETARY FOR ADMINISTRATION AND MANAGEMENT

DIRECTOR, PROGRAM SUPPORT CENTER

ASSISTANT SECRETARY FOR RESOURCES AND TECHNOLOGY

ASSISTANT SECRETARY FOR PLANNING AND EVALUATION

ASSISTANT SECRETARY FOR PREPAREDNESS AND RESPONSE

ASSISTANT SECRETARY FOR LEGISLATION

ASSISTANT SECRETARY FOR PUBLIC AFFAIRS

Washington, DC 20201. Internet, http://www.phe.
gov/preparedness/pages/default.aspx.

Office of Public Health and Science

The Office of Public Health and Science
(OPHS) comprises 13 offices and 9
Presidential and secretarial advisory
committees. The Assistant Secretary
for Health heads the Office and serves
as the Secretary's senior public health
advisor. OPHS provides assistance
in implementing and coordinating
secretarial decisions for the Public
Health Service and coordination
of population-based health clinical
divisions; provides oversight of
research conducted or supported by
the Department; implements programs
that provide population-based public
health services; and provides direction
and policy oversight, through the Office
of the Surgeon General, for the Public
Health Service Commissioned Corps.
OPHS administers a wide array of
interdisciplinary programs related to
disease prevention, health promotion, the
reduction of health disparities, women's
and minority health, HIV/AIDS, vaccine
programs, physical fitness and sports,
bioethics, population affairs, blood
supply, research integrity, and human
research protections.

**For further information, contact the Office of Public
Health and Science, 200 Independence Avenue SW.,
Washington, DC 20201. Internet, http://www.hhs.
gov/ophs/index.html.**

Administration on Aging

The mission of the Administration on
Aging (AOA) is to help elderly individuals
maintain their independence and dignity
in their homes and communities through
comprehensive, coordinated, and cost-
effective systems of long-term care and
livable communities across the United
States. The agency develops policies,
plans, and programs designed to advance
the concerns and interests of older
people and their caregivers. AOA works
in partnership with the National Aging
Network to promote the development of
all-inclusive structures and home- and
community-based long-term care that is
responsive to the needs and preferences
of older people.

**For further information, contact the Administration
on Aging, 1 Massachusetts Avenue NW., Suite 4100,
Washington, DC 20201. Phone, 202–401–4634.
Internet, http://www.aoa.gov.**

Administration for Children and Families

The Administration for Children and
Families (ACF) administers programs
and provides advice to the Secretary
on issues pertaining to children, youth,
and families; child support enforcement;
community services; developmental
disabilities; family assistance;
Native American assistance; refugee
resettlement; and legalized aliens.

**For further information, contact the Administration
for Children and Families, 370 L'Enfant Promenade
SW., Washington, DC 20447. Phone, 202–401–9200.
Internet, http://www.acf.hhs.gov.**

Agency for Healthcare Research and Quality

The Agency for Healthcare Research
and Quality (AHRQ) is charged
with improving the quality, safety,
efficiency, and effectiveness of
health care for all Americans. AHRQ
supports health services research that
will improve the quality of health
care and promote evidence-based
decisionmaking.

**For further information, contact the Agency for
Healthcare Research and Quality, 540 Gaither
Road, Rockville, MD 20850. Phone, 301–427–1364.
Internet, http://www.ahrq.gov.**

Agency for Toxic Substances and Disease Registry

The Agency for Toxic Substances
and Disease Registry, as part of the
Public Health Service, is charged with
the prevention of exposure to toxic
substances and the prevention of the
adverse health effects and diminished
quality of life associated with
exposure to hazardous substances
from wastesites, unplanned releases,
and other sources of pollution present
in the environment.

**For further information, contact the Agency for
Toxic Substances and Disease Registry, MS E–61,
4770 Buford Highway NE., Atlanta, GA 30341.
Phone, 770–488–0604. Internet, http://www.atsdr.
cdc.gov.**

Centers for Disease Control and Prevention

The Centers for Disease Control and Prevention (CDC), as part of the Public Health Service, is charged with protecting the public health of the Nation by providing leadership and direction in the prevention of and control of diseases and other preventable conditions and responding to public health emergencies. Within the CDC, the following seven centers, institutes, and offices lead prevention, diagnosis, and treatment efforts for public health concerns.

For further information, contact the Centers for Disease Control and Prevention, 1600 Clifton Road, Atlanta, GA 30333. Phone, 800–232–4636. TTY, 888–232–6348. Internet, http://www.cdc.gov.

For further information, contact the Centers for Disease Control and Prevention, 1600 Clifton Road, Atlanta, GA 30333. Phone, 800–232–4636. TTY, 888–232–6348. Internet, http://www.cdc.gov.

Center for Global Health   The Center leads CDC's global health strategy, working in partnership with foreign governments and international organizations to help countries around the world to effectively plan, manage, and evaluate global health care programs. The Center works to eradicate chronic diseases and life-threatening injuries, expanding global health care programs to address the leading causes of disability, morbidity, and mortality.

National Institute for Occupational Safety and Health   The Institute plans, directs, and coordinates a national program to develop and establish recommended occupational safety and health standards and to conduct research, training, technical assistance, and related activities to assure safe and healthy working conditions for every working person.

Office of Infectious Diseases   The Office facilitates research, programs, and policies to reduce the national and international burden of infectious diseases. The Office includes the following organizational components: the National Center for HIV/AIDS, Viral Hepatitis, STD and TB Prevention; the National Center for Immunization and Respiratory Diseases; and the National Center for Emerging and Zoonotic Infectious Diseases.

Office of Noncommunicable Diseases, Injury, and Environmental Health   The Office provides strategic direction and leadership for the prevention of noncommunicable diseases, injuries, disabilities, and environmental health hazards. The Office includes the following organizational components: the National Center on Birth Defects and Developmental Disabilities; the National Center for Chronic Disease Prevention and Health Promotion; the National Center for Environmental Health; and the National Center for Injury Prevention and Control.

Office of Public Health Preparedness and Response   The Office helps the Nation prepare for and respond to urgent public health threats by providing strategic direction, coordination, and support for CDC's terrorism preparedness and emergency response activities.

Office of State, Tribal, Local, and Territorial Support   The Office provides guidance, strategic direction, oversight, and leadership in support of State, local, territorial, and tribal public health agencies, initiatives, and priorities to improve the capacity and performance of a comprehensive public health system.

Office of Surveillance, Epidemiology, and Laboratory Services   The Office provides scientific services, knowledge, and resources to promote public health, preparing for potential health threats and preventing disease, disability, and injury. The Office includes the following organizational components: the National Center for Health Statistics; the Laboratory Science, Policy and Practice Program Office; the Public Health Informatics and Technology Program Office; the Public Health Surveillance Program Office; the Epidemiology and Analysis Program Office; and the Scientific Education and Professional Development Program Office.

Centers for Medicare and Medicaid Services

The Centers for Medicare and Medicaid Services, formerly known as the Health Care Financing Administration, was created to administer the Medicare, Medicaid, and related Federal medical care programs.

For further information, contact the Centers for Medicare and Medicaid Services, Department of Health and Human Services, 7500 Security Boulevard, Baltimore, MD 21244. Phone, 410–786–3000. Internet, http://www.cms.gov.

Food and Drug Administration

The Food and Drug Administration (FDA) is responsible for protecting the public health by ensuring the safety, efficacy, and security of human and veterinary drugs, biological products, medical devices, the Nation's food supply, cosmetics, and products that emit radiation. FDA is also responsible for advancing the public health by accelerating innovations to make medicines more effective and providing the public with accurate, science-based information on medicines and food to improve their health. FDA plays a significant role in addressing the Nation's counterterrorism capability and ensuring the security of the food supply.

For further information, contact the Food and Drug Administration, 10903 New Hampshire Avenue, Silver Spring, MD 20993–0002. Phone, 888–463–6332. Internet, http://www.fda.gov.

Health Resources and Services Administration

The Health Resources and Services Administration (HRSA) improves access to health care services for people who are uninsured, isolated, or medically vulnerable. Comprising 6 bureaus and 13 offices, HRSA provides leadership and financial support to health care providers in every State and U.S. Territory. HRSA grantees provide health care to uninsured people, people living with HIV/AIDS, and pregnant women, mothers, and children. HRSA trains health professionals, improves systems of care in rural communities, and oversees organ, bone marrow, and cord blood donation. HRSA also supports programs that prepare for

bioterrorism, compensates individuals harmed by vaccination, and maintains databases that protect against health care malpractice and abuse.

For further information, contact the Office of Communications, Health Resources and Services Administration, 5600 Fishers Lane, Rockville, MD 20857. Phone, 301–443–3376. Internet, http://www.hrsa.gov.

Indian Health Service

The Indian Health Service, as part of the Public Health Service, provides a comprehensive health services delivery system for American Indians and Alaska Natives. It assists Native American tribes in developing their health programs; facilitates and assists tribes in coordinating health planning, obtaining and utilizing health resources available through Federal, State, and local programs, operating comprehensive health programs, and evaluating health programs; and provides comprehensive health care services including hospital and ambulatory medical care, preventive and rehabilitative services, and development of community sanitation facilities.

For further information, contact the Management Policy and Internal Control Staff, Indian Health Service, Suite 625A, 801 Thompson Avenue, Rockville, MD 20852. Phone, 301–443–2650. Internet, http://www.ihs.gov.

National Institutes of Health

The National Institutes of Health (NIH) supports biomedical and behavioral research domestically and abroad, conducts research in its own laboratories and clinics, trains research scientists, and develops and disseminates credible, science-based health information to the public.

For further information, contact the National Institutes of Health, 1 Center Drive, Bethesda, MD 20892. Phone, 301–496–4000. Internet, http://www.nih.gov.

Aging The Institute conducts and supports research on the aging process, age-related diseases, and other special problems and needs of older Americans. It also provides information about aging to the scientific community, health care providers, and the public.

For further information, contact the National Institute on Aging. Phone, 301–496–1752. Internet, http://www.nia.nih.gov.

Alcohol Abuse and Alcoholism   The Institute leads the national effort to reduce alcohol-related problems by conducting and supporting biomedical and behavioral research into the causes, consequences, prevention, and treatment of alcohol-use disorders.

For further information, contact the National Institute on Alcohol Abuse and Alcoholism. Phone, 301–443–3885. Internet, http://www.niaaa.nih.gov.

Allergy and Infectious Diseases   The Institute conducts and supports research to study the causes of infectious diseases and immune-mediated diseases and to develop better means of preventing, diagnosing, and treating these diseases.

For further information, contact the National Institute of Allergy and Infectious Diseases. Phone, 866–284–4107, or 301–496–5717. Internet, http://www.niaid.nih.gov.

Arthritis and Musculoskeletal and Skin Diseases   The Institute supports research into the causes, treatment, and prevention of arthritis and musculoskeletal and skin diseases; the training of basic and clinical scientists to carry out this research; and the dissemination of information on research progress in these diseases.

For further information, contact the National Institute of Arthritis and Musculoskeletal and Skin Diseases. Phone, 301–496–8190. Internet, http://www.niams.nih.gov.

Biomedical Imaging and Bioengineering   The Institute conducts, coordinates, and supports research, training, dissemination of health information, and other programs with respect to biomedical imaging, biomedical engineering, and associated technologies and modalities with biomedical applications.

For further information, contact the National Institute of Biomedical Imaging and Bioengineering. Phone, 301–451–6768. Internet, http://www.nibib.nih.gov.

Cancer   The Institute coordinates the National Cancer Program and conducts and supports research, training, and public education with regard to the cause, diagnosis, prevention, and treatment of cancer.

For further information, contact the Cancer Information Service. Phone, 800–422–6237, or 301–435–3848. Internet, http://www.cancer.gov.

Center for Information Technology
The Center provides, coordinates, and manages information technology to advance computational science.

For further information, contact the Center for Information Technology. Phone, 301–496–6203. Internet, http://www.cit.nih.gov.

Child Health and Human Development
The Institute conducts and supports basic, translational, clinical, and epidemiological research on the reproductive, rehabilitative, neurobiologic, developmental, and behavioral processes that determine, maximize, and maintain the health of children, adults, families, and populations.

For further information, contact the Eunice Kennedy Shriver National Institute of Child Health and Human Development. Phone, 800–370–2943. Internet, http://www.nichd.nih.gov.

Clinical Center   The NIH Clinical Center is the clinical research hospital for NIH. Through clinical research, physician investigators translate laboratory discoveries into better treatments, therapies, and interventions to improve the Nation's health. The Center conducts clinical and laboratory research and trains future clinical investigators. More than 400,000 patients from across the Nation have participated in clinical research studies since the Center opened in 1953. About 1,500 clinical research studies are currently in progress.

For further information, contact the Clinical Center. Phone, 301–496–2563. Internet, http://clinicalcenter.nih.gov.

Complementary and Alternative Medicine   The Center is dedicated to exploring complementary and alternative healing practices in the context of rigorous science; educating and training complementary and alternative medicine researchers; and disseminating authoritative information to the public and professionals. Through its programs, the Center seeks to facilitate the integration of safe and effective complementary and alternative practices into conventional medicine.

For further information, contact the National Center for Complementary and Alternative Medicine. Phone 888–644–6226. Internet, http://nccam.nih.gov.

Deafness and Other Communication Disorders The Institute conducts and supports biomedical and behavioral research and research training on normal and disordered processes of hearing, balance, smell, taste, voice, speech, and language, and provides health information, based on scientific discovery, to the public. The Institute conducts diverse research performed in its own laboratories and funds a program of research and research grants.

For further information, contact the National Institute on Deafness and Other Communication Disorders. Phone, 301–496–7243. Internet, http://www.nidcd.nih.gov.

Dental and Craniofacial Research The Institute conducts and supports research and research training into the causes, prevention, diagnosis, and treatment of craniofacial, oral, and dental diseases and disorders.

For further information, contact the National Institute of Dental and Craniofacial Research. Phone, 301–496–4261. Internet, http://www.nidcr.nih.gov.

Diabetes and Digestive and Kidney Diseases The Institute conducts, fosters, and supports basic and clinical research into the causes, prevention, diagnosis, and treatment of diabetes, endocrine and metabolic diseases, digestive diseases and nutrition, kidney and urologic diseases, and blood diseases.

For further information, contact the National Institute of Diabetes and Digestive and Kidney Diseases. Phone, 301–496–3583. Internet, http://www.niddk.nih.gov.

Drug Abuse The Institute's primary mission is to lead the Nation in bringing the power of science to bear on drug abuse and addiction through the strategic support and conduct of research across a broad range of disciplines and the rapid and effective dissemination and use of the results of that research to significantly improve drug abuse and addiction prevention, treatment, and policy.

For further information, contact the National Institute on Drug Abuse. Phone, 301–443–1124. Internet, http://www.nida.nih.gov.

Environmental Health Sciences The Institute reduces the burden of human illnesses and disability by understanding how the environment influences the development and progression of human disease. To have the greatest impact on preventing disease and improving human health, the Institute focuses on basic science, disease-oriented research, global environmental health, and multidisciplinary training for researchers.

For further information, contact the National Institute of Environmental Health Sciences. Phone, 919–541–3345. Internet, http://www.niehs.nih.gov.

Fogarty International Center The Center addresses global health challenges through innovative and collaborative research and training programs. It also supports and advances the NIH mission through international partnerships.

For further information, contact the Fogarty International Center. Phone, 301–496–2075. Internet, http://www.fic.nih.gov.

General Medical Sciences The Institute supports basic biomedical research and research training in areas ranging from cell biology, chemistry, and biophysics to genetics, pharmacology, and systemic response to trauma.

For further information, contact the National Institute of General Medical Sciences. Phone, 301–496–7301. Internet, http://www.nigms.nih.gov.

Heart, Lung, and Blood Diseases The Institute provides leadership for a global program in diseases of the heart, blood vessels, lung, and blood; sleep disorders; and blood resources. It conducts, fosters, and supports an integrated and coordinated program of basic research, clinical investigations and trials, observational studies, and demonstration and education projects.

For further information, contact the National Heart, Lung, and Blood Institute. Phone, 301–496–0554. Internet, http://www.nhlbi.nih.gov.

Human Genome Research The Institute, which helped lead the Human Genome Project, leads and supports a broad range of initiatives and studies aimed at understanding the structure and function of the human genome and its role in health and disease.

For further information, contact the National Human Genome Research Institute. Phone, 301–496–0844. Internet, http://www.genome.gov.

Medical Library The National Library of Medicine is the world's largest library of the health sciences. It serves as the Nation's chief medical information source, providing medical library services and extensive web-based information resources, such as PubMed, MedlinePlus, and Toxline, for both the scientific community and the general public.

For further information, contact the National Library of Medicine. Phone, 301–496–6308. Internet, http://www.nlm.nih.gov.

Mental Health The National Institute of Mental Health works to transform the understanding and treatment of mental illnesses through basic and clinical research to further the prevention, recovery, and cure of disabling mental conditions that affect millions of Americans.

For further information, contact the National Institute of Mental Health. Phone, 866–615–6464. Internet, http://www.nimh.nih.gov.

Minority Health and Health Disparities The Institute promotes minority health and works to reduce and, ultimately, eliminate health disparities. The Institute conducts and supports basic, clinical, social, and behavioral research, strengthening related infrastructure and training, fostering emerging programs and initiatives, disseminating information and resources to the public, and providing outreach to minority and other health disparate communities.

For further information, contact the National Institute on Minority Health and Health Disparities. Phone, 301–402–1366. Internet, http://www.ncmhd.nih.gov.

Neurological Disorders and Stroke The Institute's mission is to reduce the burden of neurological diseases. It conducts, fosters, coordinates, and guides research and training on the causes, prevention, diagnosis, and treatment of neurological disorders and stroke, and supports basic, translational, and clinical research in related scientific areas.

For further information, contact the Brain Resources and Information Network of the National Institute of Neurological Disorders and Stroke, P.O. Box 5801, Bethesda, MD 20824. Phone, 800–352–9424. Internet, http://www.ninds.nih.gov.

Nursing Research The Institute supports clinical and basic research to build the scientific foundation for clinical practice, prevent disease and disability, manage and eliminate symptoms caused by illness, and enhance end-of-life and palliative care. The Institute addresses current workforce challenges by training the next generation of scientists and faculty.

For further information, contact the National Institute of Nursing Research. Phone, 301–496–0207. Internet, http://www.ninr.nih.gov.

Ophthalmological Diseases The Institute conducts, fosters, and supports research on the causes, natural history, prevention, diagnosis, and treatment of disorders of the eye and visual system. It also directs the National Eye Health Education Program.

For further information, contact the National Eye Institute. Phone, 301–496–5248. Internet, http://www.nei.nih.gov.

Research Resources The Center provides laboratory scientists and clinical investigators with the tools and training to understand, detect, treat, and prevent a wide range of diseases. Through this support, scientists engage in basic laboratory research, translate these findings to animal-based studies, and apply them to patient-oriented research. Through partnerships and other collaborations, the Center supports all aspects of research, connecting researchers, patients, and communities across the Nation.

For further information, contact the National Center for Research Resources. Phone, 301–435–0888. Internet, http://www.ncrr.nih.gov.

Scientific Review The Center for Scientific Review (CSR) organizes the peer review groups that evaluate the majority of grant applications submitted to NIH. These groups include experienced and respected researchers from across the country and abroad. Since 1946, CSR has ensured that NIH grant applications receive fair, independent, expert, and timely reviews—free from inappropriate influences—so NIH can fund the most promising research. CSR also receives all incoming applications and assigns them to the NIH Institutes and Centers that fund grants.

For further information, contact the Center for Scientific Review. Phone, 301–435–1111. Internet, http://www.csr.nih.gov.

Substance Abuse and Mental Health Services Administration

The Substance Abuse and Mental Health Services Administration (SAMHSA) works to reduce the impact of substance abuse and mental illness in America's communities. In collaboration with Federal, State, local and tribal organizations, SAMHSA promotes research and practices to foster prevention, treatment, and recovery for individuals struggling with mental and substance abuse disorders.

For further information, contact the Substance Abuse and Mental Health Services Administration, 1 Choke Cherry Road, Rockville, MD 20857. Phone, 240–276–2130. Internet, http://www.samhsa.gov.

Sources of Information

Civil Rights   For information on enforcement of civil rights laws, call 800–368–1019. TDD, 800–537–7697. Internet, http://www.hhs.gov/ocr/civilrights.

Privacy Rights   For information on the HIPAA Privacy Rule or the Patient Safety Act, call 800–368–1019. TDD, 800–537–7697. Internet, http://www.hhs.gov/ocr/privacy.

Consumer Information and Insurance Oversight   Direct inquiries to Office of Consumer Information and Insurance Oversight, Department of Health and Human Services, 200 Independence Avenue SW., Washington, DC 20201. Phone, 202–205–9924. Internet, http://www.hhs.gov/ociio/index.html.

Contracts and Small Business Activities For information concerning programs, contact the Director, Office of Small and Disadvantaged Business Utilization. Phone, 202–690–7300.

Departmental Appeals Board   For operations information, call 202–565–0200, or direct inquiries to Departmental Appeals Board Immediate Office, M.S. 6127, 330 Independence Avenue

SW., Cohen Building, Room G–644, Washington, DC 20201. Internet, http://www.hhs.gov/dab.

Inspector General   General inquiries may be directed to the Office of Inspector General, Wilbur J. Cohen Building, 330 Independence Avenue SW., Washington, DC 20201. Internet, http://www.oig.hhs.gov.

Inspector General Hotline   To report fraud, waste, or abuse against Department programs, contact the Office of Inspector General, HHS–TIPS Hotline, P.O. Box 23489, L'Enfant Plaza Station, Washington, DC 20026–3489. Phone, 800–447–8477. TTY, 800–377–4950. Fax, 800–223–8164. Internet, http://www.oig.hhs.gov/fraud/hotline.

Locator   For inquiries about the location and telephone numbers of HHS offices, call 202–690–6343.

Office of Medicare Hearings and Appeals   For information concerning Medicare hearings and appeals before Administrative Law Judges, contact the Office of Medicare Hearings and Appeals at 1800 North Moore Street, Suite 1800, Arlington, Virginia 22209. Phone 703–235–0635. Internet, http://www.hhs.gov/omha.

Program Support Center   The Program Support Center provides support services to all components of the Department and Federal agencies worldwide. For information concerning fee-for-service activities in the areas of acquisitions, occupational health, information technology support and security, human resource systems, financial management, and administrative operations, contact the Program Support Center, 5600 Fishers Lane, Rockville, MD 20857. Phone, 301–443–0034. Internet, http://www.psc.gov.

Office of Public Health and Science Contact the Assistant Secretary for Health, Room 716G, 200 Independence Avenue SW., Washington, DC 20201. Phone, 202–690–7694. Internet, http://www.hhs.gov/ash.

Surgeon General   Phone, 301–443–4000. Internet, http://www.surgeongeneral.gov.

Administration on Aging   Direct inquiries to the Administration on Aging,

Washington, DC 20201. Email, aoainfo@
aoa.gov. Internet, http://www.acf.hhs.
gov/.
Aging   Contact the National Aging
Information Center. Phone, 202–619–
7501. Fax, 202–401–7620.
Elder Care Services   Contact the Elder
Care Locator. Phone, 800–677–1116.
Administration for Children and Families
Direct inquiries to the appropriate
office, Administration for Children and
Families, 370 L'Enfant Promenade SW.,
Washington, DC 20447. Phone, 202–
401–9215. Internet, www.acf.hhs.gov.
**Agency for Healthcare Research
and Quality**   Direct inquiries to the
appropriate office at the Agency for
Healthcare Research and Quality, 540
Gaither Road, Rockville, MD 20850.
Phone, 301–427–1200. Internet, http://
www.ahrq.gov.
**Agency for Toxic Substances and
Disease Registry**   Information regarding
programs and activities is available
electronically through the Internet at
www.atsdr.cdc.gov.
**Centers for Disease Control and
Prevention**   Direct inquiries to the
appropriate office at the Centers for
Disease Control and Prevention,
Department of Health and Human
Services, 1600 Clifton Road NE., Atlanta,
GA 30333.
Electronic Access   Information
regarding programs, films, publications,
employment, and activities is available
electronically through the Internet at
www.cdc.gov.
Employment   The majority of scientific
and technical positions are filled through
the Commissioned Corps of the Public
Health Service, a uniformed service of
the U.S. Government.
**Centers for Medicare and Medicaid
Services**   Direct inquiries to the
appropriate office, Centers for Medicare
and Medicaid Services, 7500 Security
Boulevard, Baltimore, MD 21244–1850.
Internet, www.cms.gov.
Electronic Access   General information
on Medicare/Medicaid is available on the
Internet at www.cms.gov. Beneficiary-
specific Medicare/Medicaid information
is available at www.medicare.gov.
General information on the Insure Kids

Now! program is available at www.
insurekidsnow.gov.
Food and Drug Administration   Direct
inquiries to the appropriate office, Food
and Drug Administration, 10903 New
Hampshire Avenue, Silver Spring, MD
20993–0002.
Electronic Access   Information on FDA
is available through the Internet at www.
fda.gov.
Employment   FDA uses various civil
service examinations and registers in
its recruitment for positions. For more
information, visit the Department's Web
site at www.hhs.gov/careers.
**Health Resources and Services
Administration**   Direct inquiries to the
appropriate office, Health Resources and
Services Administration, 5600 Fishers
Lane, Rockville, MD 20857. Internet,
http://www.hrsa.gov.
Employment   The majority of positions
are in the Federal civil service. Some
health professional positions are filled
through the Commissioned Corps of
the Public Health Service, a uniformed
service of the U.S. Government.
Indian Health Service   Direct inquiries
to the appropriate office, Indian Health
Service, 801 Thompson Avenue,
Rockville, MD 20852.
Electronic Access   Information on IHS
is available through the Internet at www.
ihs.gov.
National Institutes of Health   Direct
inquiries to the appropriate office,
National Institutes of Health, 1 Center
Drive, Bethesda, MD 20892. Phone,
301–496–4000. Internet, http://www.nih.
gov.
Employment   Information about
employment opportunities is available
at the National Institutes of Health
employment Web site at www.jobs.nih.
gov.
**Public Health Service Commissioned
Corps Officer Program**   Information
on the Commissioned Corps Officer
programs is available at NIH's Public
Health Service Commissioned Corps
Officer Web site at http://hr.od.nih.gov/
corps/default.htm.
**Substance Abuse and Mental Health
Services Administration**   Direct inquiries
to the appropriate office, Substance

Abuse and Mental Health Services Administration, 1 Choke Cherry Road, Rockville, MD 20857.

Electronic Access Information is available through the Internet at www.samhsa.gov, www.mentalhealth.org, or www.health.org.

For further information, contact the Locator, Department of Health and Human Services, 200 Independence Avenue SW., Washington, DC 20201. Phone, 202–619–0257. Internet, http://www.hhs.gov.

EDITORIAL NOTE: The Department of Homeland Security did not meet the publication deadline for submitting updated information of its activities, functions, and sources of information as required by the automatic disclosure provisions of the Freedom of Information Act (5 U.S.C. 552(a)(1)(A)).

DEPARTMENT OF HOMELAND SECURITY

Washington, DC 20528
Phone, 202–282–8000. Internet, http://www.dhs.gov.

Secretary of Homeland Security	JANET NAPOLITANO
Deputy Secretary	JANE HOLL LUTE
Chief of Staff for Policy	NOAH KROLOFF
Executive Secretary	PHILIP A. MCNAMARA
General Counsel	IVAN K. FONG
Under Secretary, Management	RAFAEL BORRAS
Under Secretary, National Protection and Programs Directorate	RAND BEERS
Assistant Secretary, Office of Cyber Security and Communications	GREG SCHAFFER
Assistant Secretary, Infrastructure Protection	TODD KEIL
Director, Federal Protective Service	PAUL DURETTE, *Acting*
Under Secretary, Science and Technology	TARA O'TOOLE
Under Secretary, Office of Intelligence and Analysis	CARYN WAGNER
Assistant Secretary, Office of Policy	DAVID HEYMAN
Assistant Secretary, Office of International Affairs	MARIKO SILVER
Assistant Secretary, Office of Policy Development	ARIF ALIKHAN
Assistant Secretary, Private Sector Office	DOUGLAS SMITH
Assistant Secretary, State and Local Law Enforcement	CHARLES F. DINSE
Assistant Secretary, Office of Intergovernmental Affairs	JULIETTE N. KAYYEM
Director, U.S. Citizenship and Immigration Services	ALEJANDRO MAYORKAS
Commandant, U.S. Coast Guard	ADM. ROBERT J. PAPP, JR., USCG
Commissioner, U.S. Customs and Border Protection	ALAN D. BERSIN
Assistant Secretary, U.S. Immigration and Customs Enforcement	JOHN T. MORTON
Administrator, Federal Emergency Management Agency	W. CRAIG FUGATE
Director, U.S. Secret Service	MARK J. SULLIVAN
Administrator ,Transportation Security Administration	JOHN S. PISTOLE

Citizenship and Immigration Services Ombudsman	JANUARY CONTRERAS
Officer for Civil Rights and Civil Liberties	MARGO SCHLANGER
Director, Office of Counternarcotics Enforcement	GRAYLING WILLIAMS
Director, Domestic Nuclear Detection Office	WILLIAM HAGAN, *Acting*
Director, Federal Law Enforcement Training Center	CONNIE L. PATRICK
Assistant Secretary, Office of Health Affairs/ Chief Medical Officer	ALEXANDER GARZA
Inspector General	RICHARD L. SKINNER
Assistant Secretary, Office of Legislative Affairs	CHANI WIGGINS
Director, Operations Coordination	RICHARD CHAVEZ, *Acting*
Chief Privacy Officer	MARY ELLEN CALLAHAN
Assistant Secretary, Office of Public Affairs	SEAN SMITH

The Department of Homeland Security leads the unified national effort to secure America. It will prevent and deter terrorist attacks and protect against and respond to threats and hazards to the Nation. The Department will ensure safe and secure borders, welcome lawful immigrants and visitors, and promote the free flow of commerce.

The Department of Homeland Security (DHS) was established by the Homeland Security Act of 2002 (6 U.S.C. 101 note). The Department came into existence on January 24, 2003, and is administered under the supervision and direction of the Secretary of Homeland Security.

Office of the Secretary

Secretary   The Secretary is charged with developing and coordinating a comprehensive national strategy to strengthen the United States against terrorist threats or attacks. In fulfilling this effort, the Secretary will advise the President on strengthening U.S. borders, providing for intelligence analysis and infrastructure protection, improving the use of science and technology to counter weapons of mass destruction, and creating a comprehensive response and recovery division.

The Office of the Secretary oversees activities with other Federal, State, local, and private entities as part of a collaborative effort to strengthen our borders, provide for intelligence analysis and infrastructure protection, improve the use of science and technology to counter weapons of mass destruction, and to a comprehensive response and recovery system. Within the Office, there

are multiple offices that contribute to the overall homeland security mission.

Directorates

Federal Emergency Management Agency
Federal Emergency Management Agency (FEMA) is responsible for leading the effort to prepare the Nation for all hazards and effectively manage Federal response and recovery efforts following any national incident. FEMA also initiates proactive mitigation activities, trains first responders, and manages the National Flood Insurance Program.

Management Directorate   The Directorate for Management is responsible for budget, appropriations, expenditure of funds, accounting and finance; procurement; human resources and personnel; information technology systems; facilities, property, equipment, and other material resources; and identification and tracking of performance measurements relating to the responsibilities of the Department.

The Directorate for Management ensures that the Department's employees have well-defined responsibilities and that managers and their employees have effective means of communicating with one another, with other governmental

DEPARTMENT OF HOMELAND SECURITY

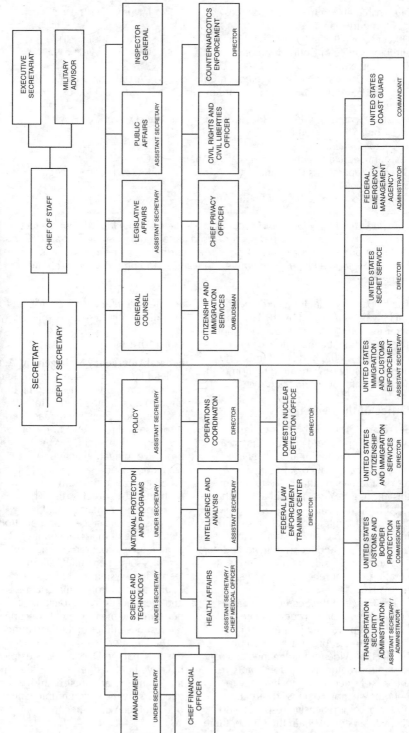

and nongovernmental bodies, and with the public they serve.

National Protection and Programs Directorate The Directorate for National Protection and Programs safeguards our critical information systems, borders, seaports, bridges, and highways by working with State, local, and private sector partners to identify threats, determine vulnerabilities, and target resources toward the greatest risks. Its functions include strengthening national risk management efforts for critical infrastructure and defining and advancing homeland security protection initiatives.

Policy Directorate The Policy Directorate develops and integrates policies, planning, and programs in order to better coordinate the Department's prevention, protection, response, and recovery missions. It is also responsible for coordinating departmentwide policies, programs, and planning; developing and communicating policies across multiple components of the homeland security network; and providing the basis and direction for departmentwide strategic planning and budget priorities.

Science and Technology Directorate The Directorate for Science and Technology is the primary research and development arm of the Department. The Directorate provides Federal, State, and local officials with the technology and capabilities to protect the homeland. Its strategic objectives are to develop and deploy systems to prevent, detect, and mitigate the consequences of chemical, biological, radiological, nuclear, and explosive attacks; develop equipment, protocols, and training procedures for response to and recovery from those attacks; enhance the Department's and other Federal, State, local, and tribal agencies' technical capabilities to fulfill their homeland security-related functions; and develop technical standards and establish certified laboratories to evaluate homeland security and emergency responder technologies for SAFETY Act certification.

Components

United States Citizenship and Immigration Services United States Citizenship and Immigration Services is responsible for the administration of immigration and naturalization adjudication functions and establishing immigration policies and priorities.

Citizenship and Immigration Services Ombudsman The Ombudsman provides recommendations for resolving individual and employer problems with United States Citizenship and Immigration Services in order to ensure national security and the integrity of the legal immigration system, increase efficiencies in administering citizenship and immigration services, and improve customer service.

Office for Civil Rights and Civil Liberties The Office provides legal and policy advice to Department leadership on civil rights and civil liberties issues, investigates and resolves complaints, and provides leadership to DHS Equal Employment Opportunity Programs.

United States Coast Guard The Coast Guard protects the public, the environment, and U.S. economic interests in the Nation's ports and waterways, along the coast, on international waters, or in any maritime region, as required, to support national security.

Office of Counternarcotics Enforcement The Office coordinates Federal policy and operations on interdicting the entry of illegal drugs into the United States and tracking and severing connections between illegal drug trafficking and terrorism.

United States Customs and Border Protection United States Customs and Border Protection is responsible for protecting our Nation's borders in order to prevent terrorists and terrorist weapons from entering the United States, while facilitating the flow of legitimate trade and travel.

Domestic Nuclear Detection Office The Office is responsible for developing a global nuclear detection architecture and acquiring and supporting the deployment of a domestic nuclear detection system to report any attempt to use nuclear or radiological material against the

United States. The office also works to enhance the nuclear detection efforts of Federal, State, territorial, tribal, and local governments and the private sector to ensure a coordinated response to such threats.

Executive Secretariat   The Executive Secretariat supports the Office of the Secretary, disseminates information and written communications within the Department, and develops, implements, and manages business processes for written communications and briefing materials. It also facilitates and manages communications with Federal agencies, the Homeland Security Council, the National Security Council, and other White House executive offices.

Federal Law Enforcement Training Center   The Center provides career-long training to law enforcement professionals from 81 Federal agencies and State, local, and international law enforcement agencies to help them fulfill their responsibilities safely and proficiently.

Chief Financial Officer   The Chief Financial Officer oversees all financial management activities relating to the programs and operations of DHS, develops and maintains an integrated accounting and financial management system, and is responsible for financial reporting and internal controls.

Office of General Counsel   The Office of General Counsel provides legal services for homeland security-related matters and ensures that DHS activities comply with all legal requirements.

Office of the Federal Coordinator for Gulf Coast Rebuilding   The Office assists the Gulf Coast region with long-term planning and coordinates the Federal Government's response to rebuild the Gulf Coast region devastated by Hurricanes Katrina and Rita.

Office of Health Affairs   The Office is responsible for providing incident management guidance, coordinates the Department's bio defense activities, and ensures a unified approach to medical preparedness. The Office also develops and maintains workforce protection and occupational health standards for DHS employees.

United States Immigration and Customs Enforcement   Immigration and Customs Enforcement is the largest investigative arm of DHS. It is responsible for identifying and shutting down vulnerabilities on the Nation's border and for economic, transportation, and infrastructure security.

Office of Inspector General   The Inspector General is responsible for conducting and supervising audits, investigations, and inspections relating to the Department's programs and operations. It also recommends ways for the Department to carry out its responsibilities in the most effective, efficient, and economical manner possible.

Office of Intelligence and Analysis   The Office is responsible for using information and intelligence from multiple sources to identify and assess current and future threats to the United States.

Office of Legislative Affairs   The Office serves as the primary liaison to Members of Congress and their staffs, the White House and executive branch, and to other Federal agencies and governmental entities that have roles in assuring national security.

Senior Military Advisor   The Senior Military Advisor is responsible for providing counsel to the Secretary and department components relating to the facilitation, coordination, and execution of policy, procedures, and preparedness activities and operations between DHS and the Department of Defense.

Office of Operations Coordination   The Office coordinates operational issues throughout the Department, integrates internal and external operations, conducts incident management, and facilitates staff planning and execution.

Chief Privacy Officer   The Privacy Officer works to minimize the impact on the individual's privacy, particularly the individual's personal information and dignity, while achieving the Department's mission.

Office of Public Affairs   The Office develops and implements a communications strategy and advises the Secretary and other DHS officials on how to effectively communicate issues and priorities of public interest.

United States Secret Service   The Secret Service protects the President and other high-level officials and investigates counterfeiting and other financial crimes, including financial institution fraud, identity theft, and computer fraud and computer-based attacks on our Nation's financial, banking, and telecommunications infrastructure.

Transportation Security Administration
The Transportation Security Administration protects the Nation's transportation systems to ensure freedom of movement for people and commerce.

Sources of Information

Electronic Access   Additional information about the Department of Homeland Security is available electronically through the Internet at www.dhs.gov.

For further information concerning the Department of Homeland Security, contact the Office of Public Affairs, Department of Homeland Security, Washington, DC 20528. Phone, 202–282–8000. Internet, http://www.dhs.gov.

DEPARTMENT OF HOUSING AND URBAN DEVELOPMENT

451 Seventh Street SW., Washington, DC 20410
Phone, 202–708–1422. Internet, http://www.hud.gov.

Secretary of Housing and Urban Development	SHAUN DONOVAN
Deputy Secretary	ROY SIMS
Chief of Staff	LAUREL A. BLATCHFORD
General Counsel	HELEN R. KANOVSKY
Inspector General	KENNETH M. DONOHUE, SR.
Assistant Secretary for Community Planning and Development	MERCEDES M. MARQUEZ
Assistant Secretary for Congressional and Intergovernmental Relations	PETER KOVAR
Assistant Secretary for Fair Housing and Equal Opportunity	JOHN TRASVINA
Assistant Secretary for Housing/Federal Housing Commissioner	DAVID STEVENS
Assistant Secretary for Policy Development and Research	RAPHAEL W. BOSTIC, *Acting*
Assistant Secretary for Public and Indian Housing	SANDRA B. HENRIQUEZ
Assistant Deputy Secretary for Field Policy and Management	PATRICIA HOBAN MOORE, *Acting*
Chief Information Officer	JERRY E. WILLIAMS
Chief Financial Officer	DOUGLAS A. CRISCITELLO, *Acting*
Chief Procurement Officer	DAVE WILLIAMSON, *Acting*
Director, Center for Faith-Based and Community Initiatives	MARK LINTON
Director, Office of Healthy Homes and Lead Hazard Control	JON GANT
Director, Office of Departmental Operations and Coordination	INEZ BANKS-DUBOSE
Director, Office of Departmental Equal Employment Opportunity	LINDA BRADFORD-WASHINGTON
General Deputy Assistant Secretary for Administration/Chief Human Capital Officer	JANIE PAYNE, *Acting*
General Deputy Assistant Secretary for Public Affairs	NEILL COLEMAN
President, Government National Mortgage Association	JOSEPH J. MURIN

The Department of Housing and Urban Development is the principal Federal agency responsible for programs concerning the Nation's housing needs, fair housing opportunities, and improvement and development of the Nation's communities.

The Department of Housing and Urban Development (HUD) was established in 1965 by the Department of Housing and Urban Development Act (42

U.S.C. 3532–3537). It was created
to administer the principal programs
which provide assistance for housing
and for the development of the Nation's
communities; to encourage the solution
of housing and community development
problems through States and localities;
and to encourage the maximum
contributions that may be made by
vigorous private homebuilding and
mortgage lending industries to housing,
community development, and the
national economy.

Although HUD administers many
programs, its six major functions are
insuring mortgages for single-family and
multifamily dwellings and extending
loans for home improvement and for the
purchasing of mobile homes; channeling
funds from investors to the mortgage
industry through the Government
National Mortgage Association;
making direct loans for construction or
rehabilitation of housing projects for the
elderly and the handicapped; providing
Federal housing subsidies for low- and
moderate-income families; providing
grants to States and communities for
community development activities; and
for promoting and enforcing fair housing
and equal housing opportunity.

Secretary   The Secretary formulates
recommendations for basic policies in
the fields of housing and community
development; encourages private
enterprise participation in housing and
community development; promotes
the growth of cities and States and
the efficient and effective use of
housing and community and economic
development resources by stimulating
private sector initiatives, public/
private sector partnerships, and public
entrepreneurship; ensures equal access
to housing and affirmatively prevents
discrimination in housing; and provides
general oversight for the Federal National
Mortgage Association.

Program Areas

Community Planning and Development
The Office administers grant programs
to help communities plan and finance
their growth and development, increase
their capacity to govern, and to provide
shelter and services for homeless
people. The Office is responsible
for the implemention of Community
Development Block Grant (CDBG)
programs for entitlement communities;
the State- and HUD-administered Small
Cities Program; community development
loan guarantees; special purpose grants
for insular areas and historically black
colleges and universities; Appalachian
Regional Commission grants; the Home
Investment in Affordable Housing
Program, which provides Federal
assistance for housing rehabilitation,
tenant-based assistance, first-time
homebuyers, and new construction for
when a jurisdiction is determined to need
new rental housing; the Department's
programs to address homelessness; the
John Heinz Neighborhood Development
Program; community outreach
partnerships; the joint community
development plan to assists institutions
of higher education working in concert
with State and local governments to
undertake activities under the CDBG
program; community adjustment and
economic diversification planning grants;
the YouthBuild Program, which provides
opportunities and assistance to very low
income high school dropouts, ages 16 to
24; empowerment zones and enterprise
communities; efforts to improve the
environment; community planning and
development efforts of other departments
and agencies, public and private
organizations, private industry, financial
markets, and international organizations.

**For further information, contact the Office of
Community Planning and Development. Phone,
202–708–2690.**

Fair Housing and Equal Opportunity
The Office administers fair housing
laws and regulations prohibiting
discrimination in public and private
housing; equal opportunity laws and
regulations prohibiting discrimination in
HUD-assisted housing and community
development programs; the fair housing
assistance grants program to provide
financial and technical assistance to
State and local government agencies to
implement local fair housing laws and
ordinances; and the Community Housing
Resources Boards program to provide

DEPARTMENT OF HOUSING AND URBAN DEVELOPMENT

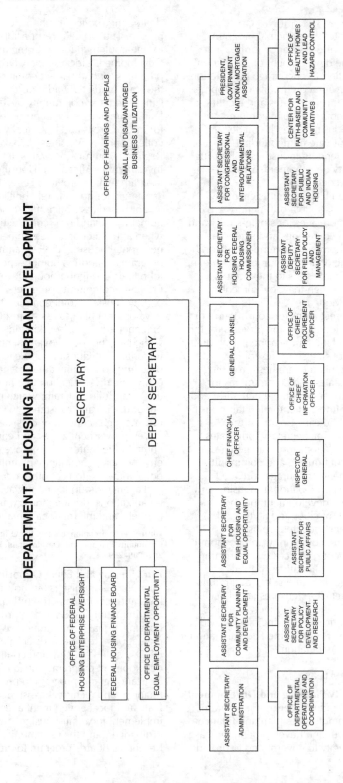

grants for fair housing activities such as outreach and education, identification of institutional barriers to fair housing, and complaint telephone hotlines.

For further information, contact the Office of Fair Housing and Equal Opportunity. Phone, 202–708–4252.

Federal Housing Enterprise Oversight The Office oversees the financial safety and soundness of the Federal National Mortgage Association (Fannie Mae) and the Federal Home Loan Mortgage Corporation (Freddie Mac) to ensure that they are adequately capitalized and operating safely.

For further information, contact the Office of Federal Housing Enterprise Oversight. Phone, 202–414–3800.

Government National Mortgage Association (GNMA) The mission of this Government corporation, also known as Ginnie Mae, is to support expanded affordable housing by providing an efficient Government-guaranteed secondary market vehicle to link the capital markets with Federal housing markets. Ginnie Mae guarantees mortgage-backed securities composed of FHA-insured or VA-guaranteed mortgage loans that are issued by private lenders and guaranteed by GNMA with the full faith and credit of the United States. Through these programs, Ginnie Mae increases the overall supply of credit available for housing by providing a vehicle for channeling funds from the securities market into the mortgage market.

For further information, contact the Government National Mortgage Association. Phone, 202–708–0926.

Housing The Office of Housing is responsible for the Department's housing functions and oversees aid for construction and financing of new and rehabilitated housing and for preservation of existing housing. The Office underwrites single-family, multifamily, property improvement, and manufactured home loans; administers special purpose programs designed specifically for the elderly, the handicapped, and the chronically mentally ill; administers assisted housing programs for low-income families who are experiencing difficulties affording standard housing; administers grants to fund resident ownership of multifamily house properties; and protects consumers against fraudulent practices of land developers and promoters.

For further information, contact the Office of Housing. Phone, 202–708–3600.

Healthy Homes and Lead Hazard Control This Office is responsible for lead hazard control policy development, abatement, training, regulations, and research. Activities of the Office include increasing public and building-industry awareness of the dangers of lead-based paint poisoning and the options for detection, risk reduction, and abatement; encouraging the development of safer, more effective, and less costly methods for detection, risk reduction, and abatement; and encouraging State and local governments to develop lead-based paint programs covering contractor certification, hazard reduction, financing, enforcement, and primary prevention, including public education.

For further information, contact the Office of Healthy Homes and Lead Hazard Control. Phone, 202–755–1785.

Public and Indian Housing The Office administers public and Indian housing programs; provides technical and financial assistance in planning, developing, and managing low-income projects; provides operating subsidies for public housing agencies (PHAs) and Indian housing authorities (IHAs), including procedures for reviewing the management of public housing agencies; administers the comprehensive improvement assistance and comprehensive grant programs for modernization of low-income housing projects to upgrade living conditions, correct physical deficiencies, and achieve operating efficiency and economy; administers programs for resident participation, resident management, home ownership, economic development and supportive services, and drug-free neighborhood programs; protects tenants from the hazards of lead-based paint poisoning by requiring PHAs and IHAs to comply with HUD regulations for the testing and removal of lead-based paint from low-income housing units; implements and monitors program requirements related to program eligibility and admission of families to public and assisted housing, and tenant income and rent requirements pertaining to continued occupancy; administers the HOPE VI and vacancy reduction programs; administers voucher and certificate programs and the Moderate Rehabilitation Program; coordinates all departmental housing and community development programs for Indian and Alaskan Natives; and awards grants to PHAs and IHAs for the construction, acquisition, and operation of public and Indian housing projects, giving priority to projects for larger families and acquisition of existing units.

For further information, contact the Office of Public and Indian Housing. Phone, 202–708–0950.

For a complete list of Department of Housing and Urban Development regional offices, go to http://portal.hud.gov/portal/page/portal/HUD/localoffices.

Sources of Information

Inquiries on the following subjects should be directed to the nearest regional office or to the specified headquarters office, Department of Housing and Urban Development, 451 Seventh Street SW., Washington, DC 20410. Phone, 202–708–0614. TDD, 202–708–1455.

Contracts   Contact the Contracting Division. Phone, 202–708–1290. Directory Locator Phone, 202–708–0614. TDD, 202–708–1455.

Directory Locator   Phone, 202–708–0614. TDD, 202–708–1455.

Employment   Inquiries and applications should be directed to the headquarters' Office of Human Resources (phone, 202–708–0408) or to the Personnel Division at the nearest regional office.

Freedom of Information Act (FOIA) Requests   Persons interested in inspecting documents or records under the Freedom of Information Act should contact the Freedom of Information Officer. Phone, 202–708–3054. Written requests should be directed to the Director, Executive Secretariat, Department of Housing and Urban Development, Room 10139, 451 Seventh Street SW., Washington, DC 20410.

HUD Hotline   The Hotline is maintained by the Office of the Inspector General as a means for individuals to report activities involving fraud, waste, or mismanagement. Phone, 202–708–4200 or 800–347–3735. TDD, 202–708–2451.

Program Information Center   The Center provides viewing facilities for information regarding departmental activities, functions, and publications and other literature to headquarters visitors. Phone, 202–708–1420.

Property Disposition   For single-family properties, contact the Property Disposition Division (phone, 202–708–0614) or the Chief Property Officer at the nearest HUD regional office. For multifamily properties, contact the Property Disposition Division (phone, 202–708–0614) or the Regional Housing Director at the nearest HUD regional office.

For further information, contact the Office of Public Affairs, Department of Housing and Urban Development, 451 Seventh Street SW., Washington, DC 20410. Phone, 202–708–0980. Internet, http://www.hud.gov.

DEPARTMENT OF THE INTERIOR

1849 C Street NW., Washington, DC 20240
Phone, 202–208–3100. Internet, http://www.doi.gov.

Secretary of the Interior	KENNETH L. SALAZAR
Deputy Secretary	DAVID HAYES
Chief Information Officer	(VACANCY)
Inspector General	EARL E. DEVANEY
Solicitor	HILARY C. TOMPKINS
Special Trustee for American Indians	(VACANCY)
Assistant Secretary for Fish and Wildlife and Parks	THOMAS STRICKLAND
Assistant Secretary for Indian Affairs	LARRY ECHO HAWK
Assistant Secretary for Insular Areas	TONY BABAUTA
Assistant Secretary for Land and Minerals Management	WILMA A. LEWIS
Assistant Secretary for Policy, Management, and Budget	RHEA S. SUH
Assistant Secretary for Water and Science	ANNE J. CASTLE
Director, Office of Insular Affairs	NIKOLAO PULA

The Department of the Interior protects America's natural resources and heritage, honors our cultures and tribal communities, and supplies the energy to power our future.

The Department of the Interior was created by act of March 3, 1849 (43 U.S.C. 1451), which transferred to it the General Land Office, the Office of Indian Affairs, the Pension Office, and the Patent Office. It was reorganized by Reorganization Plan No. 3 of 1950, as amended (5 U.S.C. app.).

The Department manages the Nation's public lands and minerals, national parks, national wildlife refuges, and western water resources and upholds Federal trust responsibilities to Indian tribes and Alaska Natives. It is also responsible for migratory wildlife conservation; historic preservation; endangered species conservation; surface-mined lands protection and restoration; mapping geological, hydrological, and biological science for the Nation; and for financial and technical assistance for the insular areas.

Secretary The Secretary of the Interior reports directly to the President and is responsible for the direction and supervision of all operations and activities of the Department. Some areas in which public purposes are broadly applied are detailed below.

Fish, Wildlife, and Parks The Office of the Assistant Secretary for Fish and Wildlife and Parks has responsibility for programs associated with the use, management, and conservation of natural resources; lands and cultural facilities associated with the National Park and National Refuge Systems; and the conservation and enhancement of fish, wildlife, vegetation, and habitat. The Office represents the Department in the coordination of marine ecosystems and biological resources programs with other Federal agencies. It also exercises secretarial direction and supervision over the U.S. Fish and Wildlife Service and the National Park Service.

Indian Affairs The Office of the Assistant Secretary for Indian Affairs

DEPARTMENT OF THE INTERIOR

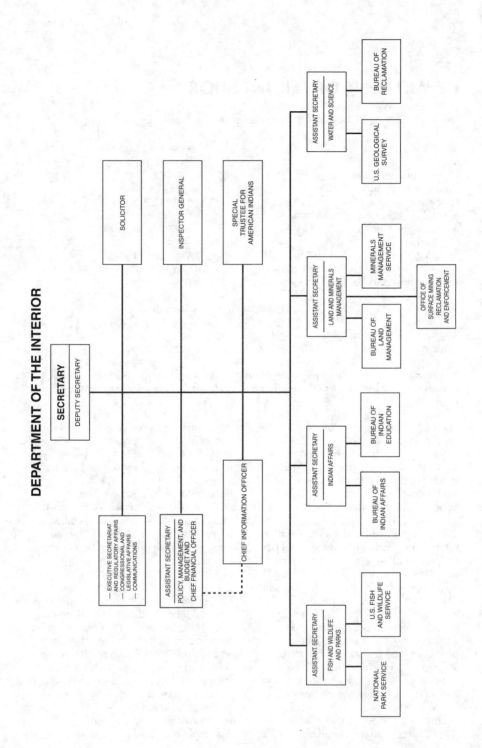

is responsible for establishing and implementing Indian policy and programs; maintaining the Federal-tribal Government-to-government relationship; assisting the Secretary of the Interior with carrying out the Department's Federal trust responsibilities; exercising direction and supervision of the Bureau of Indian Affairs and the Bureau of Indian Education; directly supervising the Federal acknowledgment of tribes, tribal self-determination and self-governance, Indian gaming, economic development, and all administrative, financial, and information resource management activities; and maintaining liaison coordination between the Department and other Federal agencies that provide services or funding to the federally recognized tribes and to the eligible American Indians and Alaska Natives.

The Office of the Special Trustee for American Indians (OST) oversees Department-wide Indian trust reform efforts to provide more effective management of and accountability for the Secretary of the Interior's trust responsibilities to Indians. OST also has programmatic responsibility for the management of financial trust assets, appraisals, and fiduciary trust beneficiary services.

Insular Areas   The Office of the Assistant Secretary for Insular Areas (IN) assists the territories of American Samoa, Guam, the U.S. Virgin Islands, and the Commonwealth of the Northern Mariana Islands in developing more efficient and effective government by providing financial and technical assistance and serves as a focal point for the management of relations between the United States and the insular areas by developing and promoting appropriate Federal policies. IN also carries out the Secretary's responsibilities that are related to the three freely associated states (the Federated States of Micronesia, the Republic of the Marshall Islands, and the Republic of Palau), the Palmyra Atoll excluded areas, and Wake Atoll's residual administration.

Land and Minerals Management   The Office of the Assistant Secretary for Land and Minerals Management

maintains administrative oversight for the Bureau of Land Management, the Minerals Management Service, and the Office of Surface Mining Reclamation and Enforcement. These bureaus are responsible for programs associated with public land management; operations management and leasing for minerals on public lands, including the Outer Continental Shelf to the outer limits of U.S. economic jurisdiction; mineral operations management on Indian lands; surface mining reclamation and enforcement functions; and management of revenues from Federal and Indian mineral leases.

Water and Science   The Office of the Assistant Secretary for Water and Science provides oversight to the U.S. Geological Survey, the Bureau of Reclamation and the Central Utah Project Completion Act Office. It provides policy direction and oversight in program areas related to water project operations, facility security and natural resource management as well as for geologic, hydrologic, cartographic, biologic, and technological research. It provides guidance in developing national water and science policies and environmental improvement.

For further information, contact the Department of the Interior, Washington, DC 20240. Phone, 202–208–3100. Internet, http://www.doi.gov.

Sources of Information

Inquiries on the following subjects should be directed to the specified office, Department of the Interior, Washington, DC 20240.

Contracts   Contact the Office of Acquisition and Property Management, Room 2607. Phone, 202–208–6352.

Electronic Access   Information is available electronically from the Department of the Interior. Internet, www.doi.gov (or see listings for specific Department components).

Employment   Direct general inquiries to the Personnel Liaison Staff, 202–208–6702, the personnel office of a specific bureau or office, or visit any of the field personnel offices.

Museum   The Interior Museum presents exhibits on the history and mission of the Department. Programs highlight bureau management of cultural and natural resources. The museum staff coordinates tours of the art and architecture of the Interior building. For more information, contact the museum staff or visit our Web site. Phone, 202–208–4743. Internet, www.doi.gov/interiormuseum.

Library   The Interior Library is a research-level collection that covers the broad range of matters related to the Department's mission. Specific collections include departmental publications, a selective depository of Federal documents, a wide array of electronic information sources available through the library Web site,

a comprehensive law collection, an extensive periodical collection, and a rare book collection consisting of 19th-century monographs on Native Americans, American history, and zoology. For more information, contact the library staff or visit our Web site. Phone, 202–208–5815. Internet, http://library.doi.gov.

Reading Room   Visit the Department of the Interior Library, Main Interior Building. Phone, 202–208–5815.

Employee Locator   To locate an employee of the Department of the Interior, call 202–208–3100.

For further information, contact the Department of the Interior, 1849 C Street NW., Washington, DC 20240. Phone, 202–208–3100. Internet, http://www.doi.gov.

United States Fish and Wildlife Service

1849 C Street NW., Washington, DC 20240
Phone, 703–358–2220. Internet, http://www.fws.gov.

Director   (VACANCY)

[For the United States Fish and Wildlife Service statement of organization, see the Code of Federal Regulations, Title 50, Subchapter A, Part 2]

The U.S. Fish and Wildlife Service is the principal Federal agency dedicated to fish and wildlife conservation. The Service's history spans nearly 140 years, dating from the establishment of its predecessor agency, the Bureau of Fisheries, in 1871. First created as an independent agency, the Bureau of Fisheries was later placed in the Department of Commerce. A second predecessor agency, the Bureau of Biological Survey, was established in 1885 in the Department of Agriculture. In 1939, the two Bureaus and their functions were transferred to the Department of the Interior. In 1940, they were consolidated into one agency and redesignated the Fish and Wildlife Service by Reorganization Plan No. 3 (5 U.S.C. app.).

The U.S. Fish and Wildlife Service is responsible for conserving, protecting, and enhancing fish, wildlife, and plants and their habitats for the continuing benefit of the American people. The Service manages the 150-million-acre National Wildlife Refuge System,

which encompasses 550 units, thousands of small wetlands, and other special management areas. It also operates 70 national fish hatcheries, 64 fishery resource offices, and 81 ecological service field stations. The Service enforces Federal wildlife laws, administers the Endangered Species Act, manages migratory bird populations, restores nationally significant fisheries, conserves and restores wildlife habitat such as wetlands, and assists foreign governments with their conservation efforts. It also oversees the Sport Fish and Wildlife Restoration Programs, which collect and distribute revenues from excise taxes on fishing and hunting equipment to State fish and wildlife agencies.

The Service is responsible for improving and maintaining fish and wildlife resources by proper management of wildlife and habitat. It also helps fulfill the public demand for recreational fishing while maintaining the Nation's fisheries at a level and in a condition that will ensure their continued survival.

Specific wildlife and fishery resource programs provide wildlife refuge management for public lands, including population control, migration and harvest surveys, and law and gaming enforcement for migratory and nonmigratory birds and mammals. Various programs also monitor hatchery production, stocking, and fishery management and provide technical assistance for coastal anadromous, Great Lakes (in cooperation with Canada), and other inland fisheries.

The Service provides leadership in identifying, protecting, and restoring endangered species of fish, wildlife, and plants. This program develops the Federal Endangered and Threatened Species List, conducts status surveys, prepares recovery plans, and coordinates national and international efforts to operate wildlife refuges.

In the area of resource management, the Service provides leadership for the protection and improvement of land and water environments (habitat preservation) that directly benefit the living natural resources and add quality to human life. The Service administers grant programs benefiting imperiled species, provides technical and financial assistance to private landowners for habitat restoration, completes environmental impact assessments and reviews of potential threats, manages Coastal Barrier Resource System mapping, monitors potential contaminants in wildlife, and studies fish and wildlife populations.

Public use and information programs include preparing informational brochures and Web sites; coordinating environmental studies on Service lands; operating visitor centers, self-guided nature trails, observation towers, and display ponds; and providing recreational activities such as hunting, fishing, and wildlife photography.

The Service's Office of Federal Assistance apportions funds for projects designed to conserve and enhance the Nation's fish and wildlife resources. The funds for the projects are generated from excise taxes on sporting arms and fishing equipment.

For a complete list of Fish and Wildlife Service regional offices, go to www.fws. gov/offices.

Sources of Information

Inquiries on the following subjects should be directed to the specified office, U.S. Fish and Wildlife Service, Department of the Interior, 1849 C Street NW., Washington, DC 20240.

Congressional/Legislative Services Congressional staffers and persons seeking information about specific legislation should call the Congressional/ Legislative Services Office. Phone, 703–358–2240.

Contracts Contact the Washington, DC, headquarters Division of Contracting and General Services (phone, 703–358–1728) or any of the regional offices.

Electronic Access The Fish and Wildlife Service offers a range of information through the Internet at www.fws.gov.

Employment For information regarding employment opportunities with the U.S. Fish and Wildlife Service, contact the Headquarters Human Capital Office (phone, 703–358–1743) or the regional office within the area you are seeking employment.

Import/Export Permits To obtain CITES permits for importing and exporting wildlife, contact the Office of Management Authority. Phone, 800–358–2104, or 703–358–2104.

Law Enforcement To obtain information about the enforcement of wildlife laws or to report an infraction of those laws, contact the Division of Law Enforcement (phone, 703–358–1949) or the nearest regional law enforcement office.

National Wildlife Refuges For general information about the National Wildlife Refuge System, as well as information about specific refuges, contact the Division of Refuges (phone, 703–358–2029) or the nearest national wildlife refuge or regional refuge office.

News Media Inquiries Specific information about the U.S. Fish and Wildlife Service and its activities is available from the Office of Public Affairs (phone, 703–358–2220) or the public affairs officer in each of the Service's regional offices.

Publications The U.S. Fish and Wildlife Service has publications available on subjects ranging from the National Wildlife Refuge System to endangered species. Some publications are only available as sales items from the Superintendent of Documents, Government Printing Office, Washington, DC 20402. Further information is available from the Publications Unit, U.S. Fish and Wildlife Service, Mail Stop NCTC, Washington, DC 20240. Phone, 800–344–9453.

For further information, contact the Office of Public Affairs, Fish and Wildlife Service, Department of the Interior, 1849 C Street NW., Washington, DC 20240. Phone, 703–358–2220. Internet, http://www.fws.gov.

National Park Service

1849 C Street NW., Washington, DC 20240
Phone, 202–208–6843. Internet, http://www.nps.gov.

Director	JONATHAN JARVIS

The National Park Service was established in the Department of the Interior on August 25, 1916 (16 U.S.C. 1).

The National Park Service is dedicated to conserving unimpaired the natural and cultural resources and values of the National Park System for the enjoyment, education, and inspiration of present and future generations. There are 392 units in the National Park System, including national parks, monuments and memorials, scenic parkways, preserves, reserves, trails, riverways, wild and scenic rivers, seashores, lakeshores, recreation areas, battlefields and battlefield parks and sites, national military parks, international historic sites, and historic sites associated with important movements, events, and personalities of the American past.

The National Park Service has a Service Center in Denver that provides planning, architectural, engineering, and other professional services. The Service is also responsible for managing a great variety of national and international programs designed to help extend the benefits of natural and cultural resource conservation and outdoor recreation throughout this country and the world.

The National Park Service develops and implements park management plans and staffs the areas under its administration. It relates the natural values and historical significance of these areas to the public through talks, tours, films, exhibits, publications, and other interpretive media. It operates campgrounds and other visitor facilities and provides lodging, food, and transportation services in many areas.

The National Park Service also administers the following programs: the State portion of the Land and Water Conservation Fund, nationwide outdoor recreation coordination and information, State comprehensive outdoor recreation planning, planning and technical assistance for the national wild and scenic rivers system, the national trails system, natural area programs, the National Register of Historic Places, national historic landmarks, historic preservation, technical preservation services, the historic American buildings survey, the historic American engineering record, and interagency archeological services.

For a complete list of National Park Service regional offices, go to http://www. nps.gov/aboutus.

Sources of Information

Contracts Contact the nearest regional office; Administrative Services Division, National Park Service, 1849 C Street NW., Washington, DC 20240. Phone, 202–354–1950; or the Denver Service Center, P.O. Box 25287, 12795 West Alameda Parkway, Denver, CO 80225. Phone, 303–969–2100.

Employment   Permanent and seasonal job opportunities with the National Park Service are posted on the USAJobs Web site: http://www.usajobs.gov. For more information about permanent careers, seasonal opportunities, and internship programs, please visit http://www.nps.gov/aboutus/workwithus.htm.

Grants   For information on grants authorized under the Land and Water Conservation Fund, contact the National Park Service, 1849 C Street NW., Washington, DC 20240. Phone, 202–354–6900. For information on grants authorized under the Historic Preservation Fund, contact the National Park Service, 1849 C Street NW., Washington, DC 20240. Phone, 202–354–2067.

Publications   Items related to the National Park Service are available from the Superintendent of Documents, Government Printing Office, Washington, DC 20401. Phone, 202–512–1800. Items available for sale include the National Park System Map and Guide (Stock No. 024–005–01261–3), and The National Parks: Index 2009–2011 (Stock No. 024–005–01269–9). Contact the Consumer Information Center, Pueblo, CO 81009, for other publications about the National Park Service available for sale. For general park and camping information, visit http://www.nps.gov or write to the National Park Service, Office of Public Inquiries, 1849 C Street NW., Washington, DC 20240.

For further information, contact the Chief, Office of Communications and Public Affairs, National Park Service, Department of the Interior, 1849 C Street NW., Washington, DC 20240. Phone, 202–208–6843. Internet, http://www.nps.gov.

United States Geological Survey

U.S. Geological Survey, Department of the Interior, 12201 Sunrise Valley Drive, Reston, VA 20192
Phone, 703–648–4000. Internet, http://www.usgs.gov. Email, ASK@usgs.gov.

Director	MARCIA MCNUTT

The U.S. Geological Survey (USGS) was established by the Organic Act of March 3, 1879 (43 U.S.C. 31).

The USGS monitors and conducts scientific research on the Earth's natural resources and environment. With specialists working in biology, geology, geography, hydrology, geospatial information, and remote sensing, the USGS studies the complex interdependent relationships between people, plants, animals, and the Earth's natural elements. The USGS holds no regulatory or management responsibilities and contributes politically neutral scientific findings to the creation of public policy. The USGS compiles statistics, reports, analyses, maps, models, and tools that forecast the consequences of various environmental strategies, and these products, created in partnership with other governmental, academic, and private organizations, provide the basis for evaluating the effectiveness of specific policies and management actions at the Federal, State, local, and tribal levels of government.

The USGS maintains a broad scope of research activities and long-term data sets, such as information relating to natural hazards, including earthquakes, floods, tsunamis, volcanoes, landslides, and coastal erosion; energy and mineral resources; and geological processes that affect the Nation's land and coasts. The USGS also collects information on the quality of surface and ground water resources; animal health, identifying and dealing with invasive species, biological species management, and ecosystems; and geospatial data, topographic maps, and satellite images critical to emergency response, homeland security, land-use planning, and resource management.

Every day, the USGS helps decisionmakers to minimize loss of life and property, manage our natural resources, and protect and enhance our quality of life.

With more than 130 years of data and experience, USGS employs 9,000 science and science-support staff, in more than 400 science centers across the United States, who work on locally, regionally, and nationally scaled studies, research projects, and sampling and monitoring sites.

Sources of Information

Contracts, Grants, and Cooperative Agreements   Write to the Office of Acquisition and Grants, 12201 Sunrise Valley Drive, National Center, Mail Stop 205G, Reston, VA 20192. Phone, 703–648–7485.

Employment   For career information at USGS, visit http://www.usgs.gov/ohr/ or contact one of the following Personnel Offices: USGS Headquarters Human Resources Office, 12201 Sunrise Valley Drive, Mail Stop 601, Reston, VA 20192. Phone, 703–648–7405; USGS Eastern Region Human Resources Office, 12201 Sunrise Valley Drive, Mail Stop 157, Reston, VA 20192. Phone, 703–648–7470; or 3850 Holcomb Bridge Road, Suite 160, Norcross, GA 30092. Phone, 770–409–7750; USGS Central Region Human Resources Office, Mail Stop 603, Box 25046, Denver, CO 80225. Phone, 303–236–9565; USGS Western Region Human Resources Office, 3020 East State University Drive, Suite 2001, Sacramento, CA 95819. Phone, 916–278–9400.

Communications   For news media and congressional inquiries, arranging interviews, and obtaining news releases and other informational products pertaining to USGS programs and activities, contact the Office of Communications at: USGS Headquarters, Office of Communications, National Center, Mail Stop 119, Reston, VA 20192. Phone, 703–648–4460; USGS Eastern Region, Office of Communications, National Center, Mail Stop 150, Reston, VA 20192. Phone, 703–648–4356;

USGS Central Region, Office of Communications, Box 25046, Denver Federal Center, Denver, CO 80225. Phone, 303–202–4744; USGS Western Region, Office of Communications, 909 First Avenue, Suite 704, Seattle, WA 98104. Phone, 206–220–4573.

General Inquiries, Maps, Publications, Scientific Reports, and Water Data   Contact USGS at 888–ASK–USGS, or email ASK@usgs.gov.

Publications and Thematic Maps   USGS scientific publications and thematic maps are available to the public through the USGS Publications Warehouse (pubs.usgs.gov), with more than 61,000 bibliographic citations for USGS reports and thematic maps. USGS technical and scientific reports and maps and nontechnical general interest publications are described in the quarterly online periodical New Publications of the U.S. Geological Survey at pubs.usgs.gov/publications.

Maps and Reports   Customers can now browse online and purchase thousands of USGS maps and reports. The USGS Store (store.usgs.gov) is an online catalog that presents thumbnail images of more than 58,000 topographic maps along with larger images of other selected maps.

Water Data   Information on the availability of and access to water data acquired by the USGS and other local, State, and Federal agencies can be obtained by calling USGS. Phone, 888–ASK–USGS. Internet, http://water.usgs.gov.

Maps, Imagery, and Publications   Maps, aerial photographs, and other USGS data sets and publications can be accessed and purchased via www.usgs.gov/pubprod.

Libraries and Reading Rooms   Reports, maps, publications, and a variety of Earth and biological information resources and historical documents are available through the USGS library system. Locations, directions, and resources are available at http://library.usgs.gov. Online reference assistance is available through Ask-A–Librarian at http://library.usgs.gov/ext_request.html. Call 1–888–ASK–USGS (888–275–8747) to speak to a librarian.

For further information, contact the U.S. Geological Survey, Department of the Interior, 12201 Sunrise Valley Drive, Reston, VA 20192. Phone, 703–648–4000. Email, ASK@usgs.gov. Internet, http://www.usgs.gov.

Office of Surface Reclamation and Enforcement

Department of the Interior, Washington, DC 20240
Phone, 202–208–2565. TDD, 202–208–2694. Internet, http://www.osmre.gov.

Director   JOSEPH PIZARCHIK

The Office of Surface Mining Reclamation and Enforcement (OSM) was established in the Department of the Interior by the Surface Mining Control and Reclamation Act of 1977 (30 U.S.C. 1211).

The Office's primary goal is to assist States in operating a nationwide program that protects society and the environment from the adverse effects of coal mining, while ensuring that surface coal mining can be done without permanent damage to land and water resources. With most coal mining States responsible for regulating coal mining and reclamation activities within their borders, OSM's main objectives are to oversee State mining regulatory and abandoned-mine reclamation programs, assist States in meeting the objectives of surface mining law, and regulate mining and reclamation activities on Federal and Indian lands, and in those States choosing not to assume primary responsibility.

The Office establishes national policy for the surface mining control and reclamation program provided for in surface mining law, reviews and approves amendments to previously approved State programs, and reviews and recommends approval of new State program submissions. Other activities include: managing the collection, disbursement, and accounting for abandoned-mine land reclamation fees; administering civil penalties programs; establishing technical standards and regulatory policy for reclamation and enforcement efforts; providing guidance for environmental considerations, research, training, and technology transfer for State, tribal, and Federal regulatory and abandoned-mine land reclamation programs; and monitoring and evaluating State and tribal regulatory programs, cooperative agreements, and abandoned-mine land reclamation programs.

Sources of Information

Contracts Contact the Procurement Branch, Office of Surface Mining, Department of the Interior, 1951 Constitution Avenue NW., Washington, DC 20240. Phone, 202–208–2839. TDD, 202–208–2737.

Employment For information on OSM employment opportunities throughout the United States, go to the jobs Web site, at https://jobs.quickhire.com/scripts/smart.exe.

For further information, contact the Office of Communications, Office of Surface Mining Reclamation and Enforcement, Department of the Interior, Washington, DC 20240. Phone, 202–208–2565. TDD, 202–208–2694. Internet, http://www.osmre.gov.

Bureau of Indian Affairs

Department of the Interior, 1849 C Street NW., Washington, DC 20240
Phone, 202–208–3710. Internet, http://www.bia.gov.

Director   MICHAEL S. BLACK

The Bureau of Indian Affairs (BIA) was created as part of the War Department in 1824 and transferred to the Department of the Interior when the latter was established in 1849. The mission of BIA is to fulfill its trust responsibilities and promote self-determination on behalf of federally recognized tribal governments, American Indians, and Alaska Natives. BIA provides services directly or through contracts, grants, or compacts to approximately 1.9 million American Indians and Alaska

Natives, members of 564 federally recognized Indian tribes in the 48 contiguous United States and Alaska.

The scope of BIA's programs is extensive, covering virtually the entire range of State and local governmental services. The programs administered by either tribes or BIA include: management of natural resources on 55 million acres of trust land, fire protection, emergency natural disaster relief, economic development programs in some of the most isolated and economically depressed areas of the United States, law enforcement, administration of tribal courts and detention centers, implementation of legislated land and water claim settlements, building, repair, and maintenance of roads and bridges, repair and maintenance of high-hazard dams, and operation of irrigation systems and agricultural programs on Federal Indian lands.

BIA works with American Indian and Alaska Native tribal governments and organizations, other Federal agencies, State and local governments, and other groups interested in the development and implementation of effective programs.

For a complete listing of Bureau of Indian Affairs Regional Offices, go to www.bia. gov/whoweare/regionaloffices/index.htm.

Sources of Information

Inquiries regarding Indian Affairs programs, including those of the Bureau of Indian Affairs, may be obtained from the Office of the Assistant Secretary for Indian Affairs, Office of Public Affairs, Department of the Interior, 1849 C Street NW., MS–3658–MIB, Washington, DC 20240. Phone, 202–208–3710.

For further information, contact the Office of the Assistant Secretary for Indian Affairs, Office of Public Affairs, Department of the Interior, 1849 C Street NW., MS–3658–MIB, Washington, DC 20240. Phone, 202–208–3710. Internet, http://www.bia.gov.

Bureau of Indian Education

Department of the Interior, 1849 C Street NW., Washington, DC 20240
Phone, 202–208–3710. Internet, http://www.bie.edu.

Director	(VACANCY)

The Bureau of Indian Education (BIE), formerly known as the Office of Indian Education Programs, provides quality educational opportunities for eligible American Indian and Alaska Native elementary, secondary, and postsecondary students from the federally recognized tribes. BIE is responsible for the direction and management of all education functions, including the formation of policies and procedures, the supervision of all program activities, and the approval of expenditure of funds appropriated for education functions.

There are 183 schools and dormitories within the BIE system serving approximately 42,000 students. Of these schools, 59 are BIE-operated and 124 are tribally controlled through BIE contracts or grants. BIE also supports 30 tribal colleges, universities, tribal technical colleges, and postsecondary schools, which includes the direct operation of Haskell Indian Nations University in Lawrence, KS, and the Southwest Indian Polytechnic Institute in Albuquerque, NM.

Sources of Information

Inquiries regarding Indian Affairs programs, including those of the Bureau of Indian Education, may be obtained from the Office of the Assistant Secretary for Indian Affairs, Office of Public Affairs, Department of the Interior, 1849 C Street NW., MS–3658–MIB, Washington, DC 20240. Phone, 202–208–3710.

For further information, contact the Office of the Assistant Secretary for Indian Affairs, Office of Public Affairs, Department of the Interior, 1849 C Street NW., MS–3658–MIB, Washington, DC 20240. Phone, 202–208–3710. Internet, http://www.bie.edu.

Bureau of Ocean Energy Management, Regulation and Enforcement

1849 C Street NW., Washington, DC 20240
Phone, 202–208–3985. Internet, http://www.boemre.gov.

Director   MICHAEL R. BROMWICH

The Minerals Management Service, the predecessor of the Bureau of Ocean Energy Management, Regulation and Enforcement (BOEMRE), was established on January 19, 1982, by Secretarial Order, to assess the nature, extent, recoverability, and value of leasable minerals on the Outer Continental Shelf. By Secretarial Order of May 19, 2010, the Secretary of the Interior established the Bureau of Ocean Energy Management, Bureau of Safety and Environmental Enforcement, and the Office of Natural Resources Revenue to assume functions formerly performed by various offices of the Minerals Management Service.

For a complete list of BOEMRE regional offices, go to www.boemre.gov/ooc/newweb/contactus.htm.

Sources of Information

Information about the Bureau of Ocean Energy Management, Regulation and Enforcement and its activities is available from the Public Affairs Office, Department of the Interior, 1849 C Street NW., Washington, DC 20240. Phone, 202–208–3985. Email, BOEMPublicAffairs@BOEMRE.GOV. Internet, http://www.boemre.gov.

For further information, contact the Office of Public Affairs, Bureau of Ocean Energy Management, 1849 C Street NW., Washington, DC 20240–7000. Phone, 202–208–3985. Email, BOEMPublicAffairs@BOEMRE. GOV. Internet, http://www.boemre.gov.

Bureau of Land Management

Department of the Interior, 1849 C Street NW., Washington, DC 20240
Phone, 202–912–7400. Internet, http://www.blm.gov.

Director   ROBERT V. ABBEY

The Bureau of Land Management (BLM) was established July 16, 1946, by the consolidation of the General Land Office (created in 1812) and the Grazing Service (formed in 1934).

The BLM manages more land (253 million acres) than any other Federal agency. This land, known as the National System of Public Lands, is primarily located in 12 Western States, including Alaska. The BLM, with a budget of about $1 billion, also administers 700 million acres of subsurface mineral estate throughout the Nation. The BLM's multiple-use mission is to sustain the health and productivity of the public lands for the use and enjoyment of present and future generations. The BLM accomplishes this by managing such

activities as outdoor recreation, livestock grazing, mineral development, and energy production, and by conserving natural, historical, cultural, and other resources on public lands.

Resources managed by the BLM include timber, solid minerals, oil and gas, geothermal energy, wildlife habitat, endangered plant and animal species, rangeland vegetation, recreation and cultural values, wild and scenic rivers, designated conservation and wilderness areas, and open space. BLM programs provide for the protection (including fire suppression when appropriate), orderly development, and use of the public lands and resources under principles of multiple use and sustained yield. Land-use plans are developed with public involvement to provide orderly use and development while maintaining and enhancing the quality of the environment. The BLM also manages watersheds to protect soil and enhance water quality; develops recreational opportunities on public lands; administers programs to protect and manage wild horses and burros; and under certain conditions, makes land available for sale to individuals, organizations, local governments, and other Federal agencies when such transfer is in the public interest. Lands may be leased to State and local government agencies and to nonprofit organizations for certain purposes.

The BLM oversees and manages the development of energy and mineral leases and ensures compliance with applicable regulations governing the extraction of these resources. It is responsible for issuing rights-of-way, leases, and permits.

The BLM is also responsible for the survey of Federal lands and establishes and maintains public land records and mining claims records.

For a complete list of Bureau of Land Management field offices, go to http://www.blm.gov.

Sources of Information

Contracts   The Bureau of Land Management (BLM) and the Department are now acquiring goods and services through the Internet Web site at http://ideasec.nbc.gov. To take advantage of future business opportunities with the BLM, you must: (1) obtain a valid Dun & Bradstreet number from Dun & Bradstreet at www.dnb.com, or by calling them at 800–333–0505, or (2) register your firm on the Central Contractor Registration System at www.ccr.gov. Also, for information about BLM's purchases, how to do business with the BLM, and BLM acquisition offices and contacts, visit the BLM National Acquisition Web site at www.blm.gov/natacq. You may also view BLM's projected purchases of goods and services, known as the Advanced Procurement Plan.

Employment   Inquiries should be directed to the National Operations Center Division of Human Resources Services, any Bureau of Land Management State Office, or the Human Capital Management Directorate, Department of the Interior, Washington, DC. Phone, 202–501–6723. For additional information on employment with the BLM, go to http://www.blm.gov/wo/st/en/res/blm_jobs.html.

General Inquiries   For general inquiries, contact any of the State offices or the Bureau of Land Management, Office of Public Affairs, Department of the Interior, Washington, DC 20240. Phone, 202–912–7400. Fax, 202–912–7181.

Publications   The annual publication, Public Land Statistics, which relates to public lands, is available from the Superintendent of Documents, Government Printing Office, Washington, DC 20402.

Reading Rooms   All State offices provide facilities for individuals who wish to examine status records, tract books, or other records relating to the public lands and their resources.

Small Business Activities   The BLM has three major buying offices that provide contacts for small business activities: the Headquarters Office in Washington, DC (phone, 202–912–7073); the National Operations Center in Lakewood, CO (phone, 303–236–9436); and the Oregon State office (phone, 503–808–6228). The acquisition plan and procurement office contacts are available through the Internet at www.blm.gov/natacq.

Speakers   Local BLM offices will arrange for speakers to explain BLM programs upon request from organizations within their areas of jurisdiction.

For further information, contact the Office of Public Affairs, Bureau of Land Management, Department of the Interior, LS–406, 1849 C Street NW., Washington, DC 20240. Phone, 202–912–7400. Internet, http://www.blm.gov.

Bureau of Reclamation

Department of the Interior, Washington, DC 20240–0001
Phone, 202–513–0575. Internet, http://www.usbr.gov.

Comissioner   MICHAEL L. CONNOR

The Bureau of Reclamation was established pursuant to the Reclamation Act of 1902 (43 U.S.C. 371 et seq.). The Bureau is the largest wholesale water supplier and the second largest producer of hydroelectric power in the United States, with operations and facilities in the 17 Western States. Its facilities also provide substantial flood control, recreation, and fish and wildlife benefits.

For a complete list of Bureau of Reclamation offices, go to http://www. usbr.gov/main/about/addresses.html.

Sources of Information

Contracts Information is available to contractors, manufacturers, and suppliers from Acquisition and Assistance Management Division, Building 67, Denver Federal Center, Denver, CO 80225. Phone, 303–445–2431. Internet, http://www.usbr.gov/mso/aamd/doing-business.html.

Employment Information on engineering and other positions is available from the Diversity and Human Resources Office, Denver, CO (phone, 303–445–2684) or from the nearest regional office. Internet, http://www. usajobs.opm.gov.

Publications Publications for sale are available through the National Technical Information Service. Phone, 703–605–6585.

For further information, contact the Public Affairs Office, Bureau of Reclamation, Department of the Interior, Washington, DC 20240–0001. Phone, 202–513–0575. Internet, http://www.usbr.gov.

DEPARTMENT OF JUSTICE

950 Pennsylvania Avenue NW., Washington, DC 20530
Phone, 202–514–2000. Internet, http://www.usdoj.gov.

Attorney General	ERIC H. HOLDER, JR.
Deputy Attorney General	GARY G. GRINDLER, *Acting*
Chief of Staff	KEVIN A. OHLSON
Associate Attorney General	THOMAS J. PERRELLI
Senior Counsel, Office of Dispute Resolution	JOANNA M. JACOBS, *Acting*
Solicitor General	NEAL KATYAL, *Acting*
Inspector General	GLENN A. FINE
Assistant Attorney General, Office of Legal Counsel	DAVID BARRON, *Acting*
Assistant Attorney General, Office of Legislative Affairs	RONALD WEICH
Assistant Attorney General, Office of Legal Policy	CHRISTOPHER SCHROEDER
Assistant Attorney General for Administration, Justice Management Division	LEE LOFTUS
Assistant Attorney General, Antitrust Division	CHRISTINE A. VARNEY
Assistant Attorney General, Civil Division	D. ANTHONY WEST
Assistant Attorney General, Civil Rights Division	THOMAS E. PEREZ
Assistant Attorney General, Criminal Division	LANNY A. BREUER
Assistant Attorney General, National Security Division	DAVID KRIS
Assistant Attorney General, Environment and Natural Resources Division	IGNACIA S. MORENO
Assistant Attorney General, Tax Division	JOHN DICICCO, *Acting*
Director, Office of Public Affairs	MATTHEW MILLER
Director, Office of Information Policy	MELANIE ANN PUSTAY
Director, Office of Intergovernmental and Public Liaison	PORTIA L. ROBERSON
Director, Executive Office for U.S. Attorneys	H. MARSHALL JARRETT
Director, Executive Office for United States Trustees	CLIFFORD J. WHITE, III
Director, Community Relations Service	ONDRAY T. HARRIS
Counsel, Office of Professional Responsibility	MARY PATRICE BROWN, *Acting*
Director, Professional Responsibility Advisory Office	JERRI U. DUNSTON
Pardon Attorney	RONALD L. RODGERS
Director, National Drug Intelligence Center	MICHAEL F. WALTHER
Trustee, Office of the Federal Detention Trustee	MICHAEL A. PEARSON
Director, Executive Office for Organized Crime Drug Enforcement Task Forces	THOMAS PADDEN, *Acting*

[For the Department of Justice statement of organization, see the Code of Federal Regulations, Title 28, Chapter I, Part 0]

The Department of Justice serves as counsel for the citizens of the United States. It represents them in enforcing the law in the public interest. Through its thousands of lawyers, investigators, and agents, the Department plays the key role in protection against criminals and subversion, ensuring healthy business competition, safeguarding the consumer, and enforcing drug, immigration, and naturalization laws.

The Department of Justice was established by act of June 22, 1870 (28 U.S.C. 501, 503, 509 note), with the Attorney General as its head. The affairs and activities of the Department of Justice are generally directed by the Attorney General.

Attorney General The Attorney General represents the United States in legal matters generally and gives advice and opinions to the President and to the heads of the executive departments of the Government when so requested. The Attorney General appears in person to represent the Government before the U.S. Supreme Court in cases of exceptional gravity or importance.

Community Relations Service The Service offers assistance to communities in resolving disputes relating to race, color, or national origin and facilitates the development of viable agreements as alternatives to coercion, violence, or litigation. It also assists and supports communities in developing local mechanisms as proactive measures to prevent or reduce racial/ethnic tensions.

For a complete list of Community Relations Service Regional Offices, visit www.justice.gov/crs/map.htm.

For further information, contact any regional office or the Director, Community Relations Service, Department of Justice, Suite 2000, 600 E Street NW., Washington, DC 20530. Phone, 202–305–2935.

Pardon Attorney The Office of the Pardon Attorney assists the President in the exercise of his pardon power under the Constitution. Generally, all requests for pardon or other forms of executive clemency, including commutation of sentences, are directed to the Pardon Attorney for investigation and review. The Pardon Attorney prepares the Department's recommendation to the President for final disposition of each application.

For further information, contact the Office of the Pardon Attorney, Department of Justice, Suite 1100, 1425 New York Avenue NW., Washington, DC 20530. Phone, 202–616–6070. Internet, http://www.usdoj.gov/pardon.

Solicitor General The Office of the Solicitor General represents the U.S. Government in cases before the Supreme Court. It decides what cases the Government should ask the Supreme Court to review and what position the Government should take in cases before the Court. It also supervises the preparation of the Government's Supreme Court briefs and other legal documents and the conduct of the oral arguments in the Court. The Solicitor General also decides whether the United States should appeal in all cases it loses before the lower courts.

For further information, contact the Executive Officer, Office of the Solicitor General, Room 5142, 950 Pennsylvania Avenue NW., RFK Justice Building (Main), Washington, DC 20530–0001.

U.S. Attorneys The Executive Office for U.S. Attorneys was created on April 6, 1953, to provide liaison between the Department of Justice in Washington, DC, and the U.S. attorneys. Its mission is to provide general executive assistance to the 94 offices of the U.S. attorneys and to coordinate the relationship between the U.S. attorneys and the organization components of the Department of Justice and other Federal agencies.

For further information, contact the Executive Office for U.S. Attorneys, Department of Justice, Room 2261, 950 Pennsylvania Avenue NW., Washington, DC 20530. Phone, 202–514–1020. Internet, http://www.usdoj.gov/usao/eousa.

U.S. Trustee Program The Program was established by the Bankruptcy Reform Act of 1978 (11 U.S.C. 101 et seq.) as a pilot effort in 10 regions encompassing 18 Federal judicial districts to promote the efficiency and protect the integrity of the bankruptcy system by identifying and helping to investigate bankruptcy fraud and abuse. It now operates nationwide except in Alabama and North Carolina.

DEPARTMENT OF JUSTICE

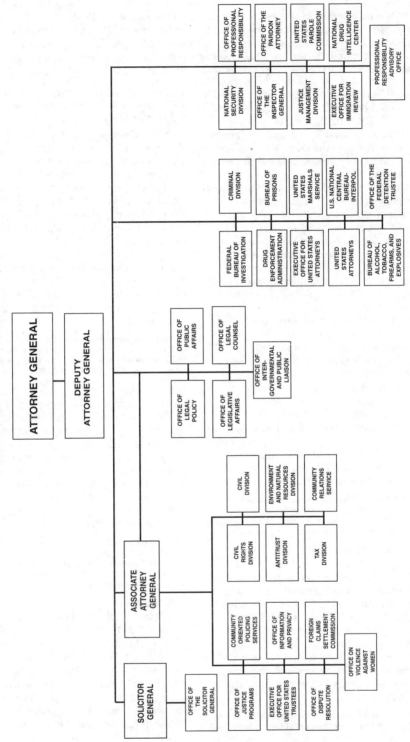

ATTORNEY GENERAL

DEPUTY ATTORNEY GENERAL

OFFICE OF PUBLIC AFFAIRS

OFFICE OF LEGAL COUNSEL

OFFICE OF LEGAL POLICY

OFFICE OF LEGISLATIVE AFFAIRS

OFFICE OF INTER-GOVERNMENTAL AND PUBLIC LIAISON

OFFICE OF PROFESSIONAL RESPONSIBILITY

OFFICE OF THE PARDON ATTORNEY

UNITED STATES PAROLE COMMISSION

NATIONAL DRUG INTELLIGENCE CENTER

NATIONAL SECURITY DIVISION

OFFICE OF THE INSPECTOR GENERAL

JUSTICE MANAGEMENT DIVISION

EXECUTIVE OFFICE FOR IMMIGRATION REVIEW

PROFESSIONAL RESPONSIBILITY ADVISORY OFFICE

CRIMINAL DIVISION

BUREAU OF PRISONS

UNITED STATES MARSHALS SERVICE

U.S. NATIONAL CENTRAL BUREAU-INTERPOL

OFFICE OF THE FEDERAL DETENTION TRUSTEE

FEDERAL BUREAU OF INVESTIGATION

DRUG ENFORCEMENT ADMINISTRATION

EXECUTIVE OFFICE FOR UNITED STATES ATTORNEYS

UNITED STATES ATTORNEYS

BUREAU OF ALCOHOL, TOBACCO, FIREARMS, AND EXPLOSIVES

ASSOCIATE ATTORNEY GENERAL

CIVIL DIVISION

ENVIRONMENT AND NATURAL RESOURCES DIVISION

COMMUNITY RELATIONS SERVICE

CIVIL RIGHTS DIVISION

ANTITRUST DIVISION

TAX DIVISION

COMMUNITY ORIENTED POLICING SERVICES

OFFICE OF INFORMATION AND PRIVACY

FOREIGN CLAIMS SETTLEMENT COMMISSION

OFFICE ON VIOLENCE AGAINST WOMEN

SOLICITOR GENERAL

OFFICE OF THE SOLICITOR GENERAL

OFFICE OF JUSTICE PROGRAMS

EXECUTIVE OFFICE FOR UNITED STATES TRUSTEES

OFFICE OF DISPUTE RESOLUTION

The Bankruptcy Abuse Prevention and Consumer Protection Act of 2005 (11 U.S.C. 101 note) significantly expanded the Program's responsibilities and provided additional tools to combat bankruptcy fraud and abuse. The Executive Office for U.S. Trustees provides day-to-day policy and legal direction, coordination, and control.

For further information, contact the Executive Office for U.S. Trustees, Department of Justice, Suite 8000, 20 Massachusetts Avenue NW., Washington, DC 20530. Phone, 202–307–1391. Internet, http://www.usdoj.gov/ust.

Divisions

Antitrust Division   The Assistant Attorney General in charge of the Antitrust Division is responsible for promoting and maintaining competitive markets by enforcing the Federal antitrust laws. This involves investigating possible antitrust violations, conducting grand jury proceedings, reviewing proposed mergers and acquisitions, preparing and trying antitrust cases, prosecuting appeals, and negotiating and enforcing final judgments. The Division prosecutes serious and willful violations of antitrust laws by filing criminal suits that can lead to large fines and jail sentences. Where criminal prosecution is not appropriate, the Division seeks a court order forbidding future violations of the law and requiring steps by the defendant to remedy the anticompetitive effects of past violations.

The Division also is responsible for acting as an advocate of competition within the Federal Government as well as internationally. This involves formal appearances in Federal administrative agency proceedings, development of legislative initiatives to promote deregulation and eliminate unjustifiable exemptions from the antitrust laws, and participation on executive branch policy task forces and in multilateral international organizations. The Division provides formal advice to other agencies on the competitive implications of proposed transactions requiring Federal approval, such as mergers of financial institutions.

For further information, contact the FOIA Unit, Antitrust Division, Department of Justice, 325 Seventh Street NW., Washington, DC 20530. Phone, 202–514–2692.

Civil Division   The Civil Division represents the United States, its departments and agencies, Members of Congress, Cabinet officers, and other Federal employees. Its litigation reflects the diversity of Government activities involving, for example, the defense of challenges to Presidential actions; national security issues; benefit programs; energy policies; commercial issues such as contract disputes, banking, insurance, fraud, and debt collection; all manner of accident and liability claims; and violations of the immigration and consumer protection laws. The Division confronts significant policy issues, which often rise to constitutional dimensions, in defending and enforcing various Federal programs and actions. Each year, Division attorneys handle thousands of cases that collectively involve billions of dollars in claims and recoveries.

The Division litigates cases in the following areas:

Commercial litigation, litigation associated with the Government's diverse financial involvements including all monetary suits involving contracts, express or implied; actions to foreclose on Government mortgages and liens; bankruptcy and insolvency proceedings; suits against guarantors and sureties; actions involving fraud against the Government, including false or fraudulent claims for Federal insurance, loans, subsidies, and other benefits such as Medicare, false or fraudulent claims for payment under Federal contracts, whistleblower suits, and Government corruption; patent, copyright, and trademark cases and suits arising out of construction, procurement, service contracts, and claims associated with contract terminations; claims for just compensation under the Fifth Amendment; claims for salary or retirement by civilian and military personnel; cases assigned by congressional reference or special legislation; and litigation involving interests of the United States in any foreign court, whether civil or criminal in nature.

Consumer litigation, including civil and criminal litigation and related matters arising under various consumer protection and public health statutes.

Federal programs, including constitutional challenges to statutes, suits to overturn Government policies and programs, challenges to the legality of Government decisions, allegations that the President has violated the Constitution or Federal law, suits to enforce regulatory statutes and to remedy or prevent statutory or regulatory violations.

The areas of litigation include:

Suits against the heads of Federal departments and agencies and other Government officials to enjoin official actions, as well as suits for judicial review of administrative decisions, orders, and regulations; suits involving national security, including suits to protect sensitive intelligence sources and materials; suits to prevent interference with Government operations; litigation concerning the constitutionality of Federal laws; and suits raising employment discrimination claims and Government personnel issues.

Immigration litigation, involving civil litigation under the Immigration and Nationality Act and related laws; district court litigation, habeas corpus review and general advice; petitions for removal order review and immigration-related appellate matters; cases pertaining to the issuance of visas and passports; and litigation arising under the legalization and employer sanction provisions of the immigration laws.

Torts, including the broad range of tort litigation arising from the operation of the Federal Government, constitutional tort claims against Federal Government officials throughout the Government, aviation disasters, environmental and occupational disease, and radiation and toxic substance exposure. It defends petitions filed pursuant to the Vaccine Injury Compensation Program and is responsible for administering the Radiation Exposure Compensation Program. It also handles maritime litigation and suits that seek personal monetary judgments against individual officers or employees.

Appellate, having primary responsibility for the litigation of Civil Division cases in the courts of appeal, and on occasion, State appeal courts. The Appellate Staff prepares Government briefs and presents oral arguments for these cases. Additionally, the Appellate Staff works with the Solicitor General's office to prepare documents filed for these cases in the Supreme Court, including briefs on the merits, petitions for certiorari, and jurisdictional statements. The Appellate Staff also works with the Solicitor General's office to obtain authorization for appellate review.

For further information, contact the Office of the Assistant Attorney General, Civil Division, Department of Justice, Tenth Street and Pennsylvania Avenue NW., Washington, DC 20530. Phone, 202–514–3301.

Civil Rights Division   The Civil Rights Division, headed by an Assistant Attorney General, was established in 1957 to secure effective Federal enforcement of civil rights. The Division is the primary institution within the Federal Government responsible for enforcing Federal statutes prohibiting discrimination on the basis of race, sex, disability, religion, citizenship, and national origin. The Division has responsibilities in the following areas:

Coordination and review of various civil rights statutes that prohibit discrimination on the basis of race, color, national origin, sex, and religion in programs and activities that receive Federal financial assistance by Federal agencies.

Criminal cases involving conspiracies to interfere with federally protected rights; deprivation of rights under color of law; the use of force or threat of force to injure or intimidate someone in their enjoyment of specific rights (such as voting, housing, employment, education, public facilities, and accommodations); interference with the free exercise of religious beliefs or damage to religious property; the holding of a worker in a condition of slavery or involuntary servitude; and interference with persons seeking to obtain or provide reproductive services.

Disability rights cases, achieving equal opportunity for people with disabilities in the United States by implementing the Americans with Disabilities Act (ADA). The section's enforcement, certification, regulatory, coordination, and technical assistance activities, combined with an innovative mediation program and a technical assistance grant program, provide an approach for carrying out the ADA's mandates. The section also carries out responsibilities under sections 504 and 508 of the Rehabilitation Act, the Help America Vote Act of 2002, the Small Business Regulatory Enforcement Fairness Act, and Executive Order 12250.

Educational opportunities litigation, involving title IV of the Civil Rights Act of 1964, the Equal Educational Opportunities Act of 1974, and title III of the Americans with Disabilities Act. In addition, the section is responsible for enforcing other statutes such as title VI of the Civil Rights Act of 1964, title IX of the Education Amendments of 1972, section 504 of the Rehabilitation Act of 1973, title II of the Americans with Disabilities Act, and the Individuals with Disabilities Education Act upon referral from other governmental agencies.

Employment litigation enforcing against State and local government employers the provisions of title VII of the Civil Rights Act of 1964, as amended, and other Federal laws prohibiting employment practices that discriminate on grounds of race, sex, religion, and national origin. The section also enforces against State and local government and private employers the provisions of the Uniformed Services Employment and Reemployment Rights Act of 1994, which prohibits employers from discriminating or retaliating against an employee or applicant for employment because of such person's past, current, or future military obligation.

Housing and Civil Enforcement statutes enforcing the Fair Housing Act, which prohibits discrimination in housing; the Equal Credit Opportunity Act, which prohibits discrimination in credit; title II of the Civil Rights Act of 1964, which prohibits discrimination in certain places of public accommodation, such as hotels, restaurants, nightclubs and theaters; title III of the Civil Rights Act of 1964, which prohibits discrimination in public facilities; and the Religious Land Use and Institutionalized Persons Act, which prohibits local governments from adopting or enforcing land use regulations that discriminate against religious assemblies and institutions or which unjustifiably burden religious exercise.

Immigration-related unfair employment practices enforcing the antidiscrimination provisions of the Immigration and Nationality Act, which protect U.S. citizens and legal immigrants from employment discrimination based upon citizenship or immigration status and national origin, unfair documentary practices relating to the employment eligibility verification process, and retaliation.

Special litigation protecting the constitutional and statutory rights of persons confined in certain institutions owned or operated by State or local governments, including facilities for individuals with mental and developmental disabilities, nursing homes, prisons, jails, and juvenile detention facilities where a pattern or practice of violations exist; civil enforcement of statutes prohibiting a pattern or practice of conduct by law enforcement agencies that violates Federal law; and protection against a threat of force and physical obstruction that injures, intimidates, or interferes with a person seeking to obtain or provide reproductive health services, or to exercise the first amendment right of religious freedom at a place of worship.

Voting cases enforcing the Voting Rights Act, the Help America Vote Act of 2002, the National Voter Registration Act, the Voting Accessibility for the Elderly and Handicapped Act, the Uniformed and Overseas Citizens Absentee Voting Act, and other Federal statutes designed to safeguard citizens' rights to vote. This includes racial and language minorities, illiterate persons, individuals with disabilities, overseas citizens, persons who change their residence shortly before a Presidential election, and persons 18 to 20 years of age.

For further information, contact the Executive Officer, Civil Rights Division, Department of Justice, 950 Pennsylvania Avenue NW., Washington, DC 20035. Phone, 202–514–4224. Internet, http://www. usdoj.gov/crt.

Criminal Division   The Criminal Division develops, enforces, and supervises the application of all Federal criminal laws, except those specifically assigned to other divisions. In addition to its direct litigation responsibilities, the Division formulates and implements criminal enforcement policy and provides advice and assistance, including representing the United States before the United States Courts of Appeal. The Division engages in and coordinates a wide range of criminal investigations and prosecutions, such as those targeting individuals and organizations that engage in international and national drug trafficking and money laundering systems or organizations and organized crime groups. The Division also approves or monitors sensitive areas of law enforcement such as participation in the Witness Security Program and the use of electronic surveillance; advises the Attorney General, Congress, the Office of Management and Budget, and the White House on matters of criminal law; provides legal advice, assistance, and training to Federal, State, and local prosecutors and investigative agencies; provides leadership for coordinating international and national law enforcement matters; and provides training and development assistance to foreign criminal justice systems. Areas of responsibility include:

Asset forfeiture and money laundering, including the prosecution of complex, sensitive, multidistrict, and international cases; formulating policy and conducting training in the money laundering and forfeiture areas; developing legislation and regulations; ensuring the uniform application of forfeiture and money laundering statutes; participating in bilateral and multilateral initiatives to develop international forfeiture and money laundering policy and promote international cooperation; adjudicating petitions for remission or mitigation of forfeited assets; distributing forfeited funds and properties to appropriate domestic and foreign law enforcement agencies and community groups within the United States; and ensuring that such agencies comply with proper usage of received funds.

Child exploitation and obscenity, including providing prosecutorial and forensic assistance to Federal prosecutors and law enforcement agents in investigating and prosecuting violators of Federal criminal statutes relating to the manufacture, distribution, receipt, or possession, of child pornography; selling, buying, or transporting women and children to engage in sexually explicit conduct; interstate or international travel to sexually abuse children; abusing children on Federal and Indian lands; transporting obscene materials in interstate or foreign commerce; international parental abduction; nonpayment of certain court-ordered child support; and contributing to the development of policy and legislative efforts related to these areas.

Computer crime and intellectual property, including cyber attacks on critical information systems, improving domestic and international infrastructure to pursue network criminals most effectively; and initiating and participating in international efforts to combat computer crime.

Enforcement, overseeing the use of the most sophisticated investigative tools at the Department's disposal; reviewing all Federal electronic surveillance requests and requests to apply for court orders permitting the use of video surveillance; authorizing or denying the entry of applicants into the Federal Witness Security Program (WSP) and coordinating and administering matters relating to all aspects of the WSP among all program components; reviewing requests for witness immunity; transfer of prisoners to and from foreign countries to serve the remainder of their prison sentences; attorney and press subpoenas; applications for S-visa status; and disclosure of grand jury information.

Fraud, including cases that focus on corporate and securities fraud schemes, financial institution fraud, insurance fraud, fraud involving Government programs such as Medicare, and international criminal activities including the bribery of foreign government officials in violation of the Foreign Corrupt Practices Act.

International affairs, including requests for international extradition and foreign evidence on behalf of Federal, State, and local prosecutors and investigators, fulfilling foreign requests for fugitives and evidence, and negotiating and implementing law enforcement treaties.

Narcotics and dangerous drugs, including statutes pertaining to controlled substances; developing and implementing domestic and international narcotics law enforcement policies and programs; developing and administering other cooperative drug enforcement strategies, such as the Bilateral Case Initiative, and projects conducted by the law enforcement and intelligence communities.

Organized crime and racketeering efforts against traditional groups and emerging groups from Asia and Europe.

Organized Crime Drug Enforcement Task Forces, combining the resources and expertise of several Federal agencies in cooperation with the Tax Division, U.S. attorneys offices, and State and local law enforcement to identify, disrupt, and dismantle major drug supply and money laundering organizations through coordinated, nationwide investigations targeting the entire infrastructure of these enterprises.

Overseas prosecutorial development, assistance, and training for prosecutors and judicial personnel in other countries to develop and sustain democratic criminal justice institutions.

Policy and legislation, developing legislative proposals and reviewing pending legislation affecting the Federal criminal justice system; reviewing and developing proposed changes to the Federal sentencing guidelines and rules; and analyzing crime policy and program issues.

Public integrity efforts to combat corruption of elected and appointed public officials at all levels of government.

Special investigations of individuals who took part in Nazi-sponsored acts of persecution abroad before and during World War II and who subsequently entered or seek to enter the United States illegally and/or fraudulently, and interagency investigation into assets looted from victims of Nazi persecution.

Terrorism, involving design, implementation, and support of law enforcement efforts, legislative initiatives, policies, and strategies relating to international and domestic terrorism.

Domestic security, enforcing Federal criminal laws relating to violent crimes, the illegal use of firearms and explosives, and alien smuggling and other immigration-related offenses.

For further information, contact the Office of the Assistant Attorney General, Criminal Division, Department of Justice, Tenth Street and Pennsylvania Avenue NW., Washington, DC 20530. Phone, 202–514–2601.

Environment and Natural Resources Division The Environment and Natural Resources Division is the Nation's environmental lawyer. The Division's responsibilities include enforcing civil and criminal environmental laws that protect America's health and environment. It also defends environmental challenges to Government activities and programs and ensures that environmental laws are implemented in a fair and consistent manner nationwide. It also represents the United States in all matters concerning the protection, use, and development of the Nation's natural resources and public lands, wildlife protection, Indian rights and claims, and the acquisition of Federal property. To carry out this broad mission, the Division litigates in the following areas:

Environmental crimes, prosecuting individuals and corporate entities violating laws designed to protect the environment.

Civil environmental enforcement, on behalf of EPA; claims for damages to natural resources filed on behalf of the Departments of the Interior, Commerce, and Agriculture; claims for contribution against private parties for contamination of public land; and recoupment of money spent to clean up certain oil spills on behalf of the U.S. Coast Guard.

Environmental defense, representing the United States in suits challenging the Government's administration of Federal environmental laws including claims that regulations are too strict or lenient and claims alleging that Federal agencies are not complying with environmental standards.

Wildlife and marine resources protection, including prosecution of smugglers and black-market dealers in protected wildlife.

Use and protection of federally owned public lands and natural resources across a broad spectrum of laws.

Indian resources protection, including establishing water rights, establishing and protecting hunting and fishing rights, collecting damages for trespass on Indian lands, and establishing reservation boundaries and rights to land.

Land acquisition for use by the Federal Government for purposes ranging from establishing public parks to building Federal courthouses.

For further information, contact the Office of the Assistant Attorney General, Environment and Natural Resources Division, Department of Justice, Tenth Street and Pennsylvania Avenue NW., Washington, DC 20530. Phone, 202–514–2701.

National Security Division The National Security Division (NSD) develops, enforces, and supervises the application of all Federal criminal laws related to the national counterterrorism and counterespionage enforcement programs, except those specifically assigned to other divisions. NSD litigates and coordinates a wide range of prosecutions and criminal investigations involving terrorism and violations of the espionage, export control, and foreign agents registration laws. It administers the Foreign Intelligence Surveillance Act and other legal authorities for national security activities; approves and monitors the use of electronic surveillance; provides legal and policy advice regarding the classification of and access to national security information; performs

prepublication review of materials written by present and former DOJ employees; trains the law enforcement and intelligence communities; and advises the Department and legislative and executive branches on all areas of national security law. NSD also serves as the Department's representative on interdepartmental boards, committees, and entities dealing with issues related to national security.

NSD also has some additional counterterrorism, counterespionage, and intelligence oversight responsibilities as follows: to promote and oversee national counterterrorism enforcement programs; develop and implement counterterrorism strategies, legislation, and initiatives; facilitate information sharing between and among the Department and other Federal agencies on terrorism threats; share information with international law enforcement officials to assist with international threat information and litigation initiatives; liaison with the intelligence, defense, and immigration communities and foreign governments on counterterrorism issues and cases; supervise the investigation and prosecution of cases involving national security, foreign relations, the export of military and strategic commodities and technology, espionage, sabotage, neutrality, and atomic energy; coordinate cases involving the application for the Classified Information Procedures Act; enforce the Foreign Agents Registration Act of 1938 and related disclosure laws; supervise the preparation of certifications and applications for orders under the Foreign Intelligence Surveillance Act (FISA); represent the United States before the Foreign Intelligence Surveillance Court; participate in the development, implementation, and review of United States intelligence policies; evaluate existing and proposed national security-related activities to determine their consistency with relevant policies and law; monitor intelligence and counterintelligence activities of other agencies to ensure conformity with Department objectives; prepare reports evaluating domestic and foreign intelligence and counterintelligence activities; and process requests to use

FISA-derived information in criminal, civil, and immigration proceedings and to disseminate that information to foreign governments.

For further information, contact the Office of the Assistant Attorney General, National Security Division, Department of Justice, Tenth Street and Pennsylvania Avenue NW., Washington, DC 20530. Phone, 202–514–5600. Internet, http://www.usdoj. gov/nsd.

Tax Division Tax Division ensures the uniform and fair enforcement of Federal tax laws in Federal and State courts. The Division conducts enforcement activities to deter specific taxpayers, as well as the taxpaying public at large, from conduct that deprives the Federal Government of its tax-related revenue. It represents the United States and its officers in all civil and criminal litigation arising under the internal revenue laws, other than proceedings in the U.S. Tax Court. Tax Division attorneys frequently join with assistant U.S. attorneys in prosecuting tax cases. Some criminal tax grand jury investigations and prosecutions are handled solely by Tax Division prosecutors, while others are delegated to assistant U.S. attorneys. Division attorneys evaluate requests by the Internal Revenue Service or U.S. attorneys to initiate grand jury investigations or prosecutions of tax crimes.

The Division handles a wide array of civil tax litigation, including the following: suits to enjoin the promotion of abusive tax shelters and to enjoin activities relating to aiding and abetting the understatement of tax liabilities of others; suits to enforce Internal Revenue Service administrative summonses that seek information essential to determine and collect taxpayers' liabilities, including summonses for records of corporate tax shelters and offshore transactions; suits brought by the United States to set aside fraudulent conveyances and to collect assets held by nominees and egos; tax refund suits challenging the Internal Revenue Service's determination of taxpayers' Federal income, employment, excise, and estate liabilities; bankruptcy litigation raising issues of the

validity, dischargeability, and priority of Federal tax claims, and the feasibility of reorganization plans; suits brought by taxpayers challenging determinations made in the collection due process proceedings before the Internal Revenue Service's Office of Appeals; and suits against the United States for damages for the unauthorized disclosure of tax return information or for damages claimed because of alleged injuries caused by Internal Revenue Service employees in the performance of their official duties.

The Division also collects judgments in tax cases. To this end, the Division directs collection efforts and coordinates with, monitors the efforts of, and provides assistance to the various U.S. attorneys' offices in collecting outstanding judgments in tax cases. The Division also works with the Internal Revenue Service, U.S. attorneys, and other Government agencies on policy and legislative proposals to enhance tax administration and handling tax cases assigned to those offices.

For further information, contact the Office of the Assistant Attorney General, Tax Division, Department of Justice, Tenth Street and Pennsylvania Avenue NW., Washington, DC 20530. Phone, 202–514–2901. Internet, http://www.usdoj. gov/tax.

Sources of Information

Disability-Related Matters Contact the Civil Rights Division's ADA Hotline. Phone, 800–514–0301. TDD, 800–514–0383. Internet, http://www.usdoj.gov/crt/ ada/adahom1.htm.
Drugs and Crime Clearinghouse Phone, 800–666–3332 (toll free).
Electronic Access Information concerning Department of Justice programs and activities is available electronically through the Internet at http://www.usdoj.gov.
Employment The Department maintains an agency-wide job line. Phone, 202–514–3397.

Attorneys' applications: Director, Office of Attorney Personnel Management, Department of Justice, Room 6150, Tenth Street and Constitution Avenue NW., Washington, DC 20530. Phone, 202–514–1432. Assistant U.S. attorney applicants should apply to individual U.S. attorneys.

United States Trustee Program: Room 770, 901 E Street NW., Washington, DC 20530. Phone, 202–616–1000.

Housing Discrimination Matters Contact the Civil Rights Division's Housing and Civil Enforcement Section. Phone, 800–896–7743.

Immigration-Related Employment Matters   The Civil Rights Division maintains a worker hotline. Phone, 800–255–7688. TDD, 800–237–2515. It also offers information for employers. Phone, 800–255–8155. TDD, 800–362–2735.

Publications and Films   The Annual Report of the Attorney General of the United States is published each year by the Department of Justice, Washington, DC 20530.

Textbooks on citizenship consisting of teacher manuals and student textbooks at various reading levels are distributed free to public schools for applicants for citizenship and are on sale to all others from the Superintendent of Documents, Government Printing Office, Washington, DC 20402. Public schools or organizations under the supervision of public schools that are entitled to free textbooks should make their requests to the appropriate Immigration and Naturalization Service Regional Office. For general information, call 202–514–3946.

The Freedom of Information Act Guide and Privacy Act Overview and the Freedom of Information Case List, both published annually, are available from the Superintendent of Documents, Government Printing Office, Washington, DC 20530.

FOIA (Stock No. 727–002–00000–6), published quarterly, is available free of charge to FOIA offices and other interested offices Governmentwide. This publication is also available from the Superintendent of Documents, Government Printing Office, Washington, DC 20402.

Guidelines for Effective Human Relations Commissions, Annual Report of the Community Relations Service, Community Relations Service Brochure, CRS Hotline Brochure, Police Use of Deadly Force: A Conciliation Handbook for Citizens and Police, Principles of Good Policing: Avoiding Violence Between Police and Citizens, Resolving Racial Conflict: A Guide for Municipalities, and Viewpoints and Guidelines on Court-Appointed Citizens Monitoring Commissions in School Desegregation are available upon request from the Public Information Office, Community Relations Service, Department of Justice, Washington, DC 20530.

A limited number of drug educational films are available, free of charge, to civic, educational, private, and religious groups.

Reading Rooms   Reading rooms are located in Washington, DC, at the following locations:

Department of Justice, Room 6505, Tenth Street and Constitution Avenue NW., Washington, DC 20530. Phone, 202–514–3775.

Board of Immigration Appeals, Suite 2400, 5107 Leesburg Pike, Falls Church, VA 22041. Phone, 703–305–0168.

National Institute of Justice, 9th Floor, 633 Indiana Avenue NW., Washington, DC 20531. Phone, 202–307–5883.

Redress for Wartime Relocation/ Internment   Contact the Civil Rights Division's Office of Redress Administration. Helpline phone, 202–219–6900. TDD, 202–219–4710. Internet, http://www.usdoj.gov.

Small Business Activities   Contract information for small businesses can be obtained from the Office of Small and Disadvantaged Business Utilization, Department of Justice, Tenth Street and Pennsylvania Avenue NW., Washington, DC 20530. Phone, 202–616–0521.

For further information concerning the Department of Justice, contact the Office of Public Affairs, Department of Justice, Tenth Street and Constitution Avenue NW., Washington, DC 20530. Phone, 202–514–2007. TDD, 202–786–5731. Internet, http://www.usdoj.gov.

Bureaus

Federal Bureau of Investigation

935 Pennsylvania Avenue NW., Washington, DC 20535
Phone, 202–324–3000. Internet, http://www.fbi.gov.

Director   ROBERT S. MUELLER, III

The Federal Bureau of Investigation (FBI) is the principal investigative arm of the United States Department of Justice. It is primarily charged with gathering and reporting facts, locating witnesses, and compiling evidence in cases involving Federal jurisdiction. It also provides law enforcement leadership and assistance to State and international law enforcement agencies.

The FBI was established in 1908 by the Attorney General, who directed that Department of Justice investigations be handled by its own staff. The Bureau is charged with investigating all violations of Federal law except those that have been assigned by legislative enactment or otherwise to another Federal agency. Its jurisdiction includes a wide range of responsibilities in the national security, criminal, and civil fields. Priority has been assigned to areas such as counterterrorism, counterintelligence, cyber crimes, internationally and nationally organized crime/drug matters, and financial crimes.

The FBI also offers cooperative services to local, State, and international law enforcement agencies. These services include fingerprint identification, laboratory examination, police training, the Law Enforcement Online communication and information service for use by the law enforcement community, the National Crime Information Center, and the National Center for the Analysis of Violent Crime.

Sources of Information

Employment For employment information, contact the Director, Washington, DC 20535, or any of the field offices or resident agencies whose addresses are listed in the front of most local telephone directories.
Publications The FBI Law Enforcement Bulletin and Uniform Crime Reports— Crime in the United States are available from the Superintendent of Documents, Government Printing Office, Washington, DC 20402.

For further information, contact the Office of Public and Congressional Affairs, Federal Bureau of Investigation, J. Edgar Hoover FBI Building, 935 Pennsylvania Avenue NW., Washington, DC 20535. Phone, 202–317–2727.

Bureau of Prisons

320 First Street NW., Washington, DC 20534
Phone, 202–307–3198. Internet, http://www.bop.gov.

Director   HARLEY G. LAPPIN

The mission of the Bureau of Prisons is to protect society by confining offenders in the controlled environments of prisons and community-based facilities that are safe, humane, cost-efficient, and appropriately secure, and that provide work and other self-improvement opportunities to assist offenders in

becoming law-abiding citizens. The Bureau has its headquarters, also known as Central Office, in Washington, DC. The Central Office is divided into nine divisions, including the National Institute of Corrections.

The Correctional Programs Division (CPD) is responsible for inmate classification and programming, including psychology and religious services, substance abuse treatment, case management, and programs for special needs offenders. CPD provides policy direction and daily operational oversight of institution security, emergency preparedness, intelligence gathering, inmate discipline, inmate sentence computations, receiving and discharge, and inmate transportation, as well as coordinating international treaty transfers and overseeing the special security needs of inmates placed in the Federal Witness Protection Program. CPD administers contracts and intergovernmental agreements for the confinement of offenders in community-based programs, community corrections centers, and other facilities, including privately managed facilities. CPD staff is also involved in the Bureau's privatization efforts.

The Industries, Education, and Vocational Training Division oversees Federal Prison Industries, or UNICOR, which is a wholly owned Government corporation that provides employment and training opportunities for inmates confined in Federal correctional facilities. Additionally, it is responsible for oversight of educational, occupational, and vocational training and leisure-time programs, as well as those related to inmate release preparation.

The National Institute of Corrections (NIC) provides technical assistance, training, and information to State and local corrections agencies throughout the country, as well as the Bureau. It also provides research assistance and documents through the NIC Information Center.

Sources of Information

Employment   For employment information, contact the Central Office, 320 First Street NW., Washington, DC 20534 (phone, 202–307–3082) or any regional or field office.

Reading Room   The reading room is located at the Bureau of Prisons, 320 First Street NW., Washington, DC 20534. Phone, 202–307–3029.

For further information, contact the Public Information Office, Bureau of Prisons, 320 First Street NW., Washington, DC 20534. Phone, 202–514–6551.

United States Marshals Service

Department of Justice, Washington, DC 20530
Phone, 202–307–9000. Internet, http://www.usmarshals.gov.

Director   JOHN F. CLARK

The United States Marshals Service is the Nation's oldest Federal law enforcement agency, having served as a vital link between the executive and judicial branches of the Government since 1789. The Marshals Service performs tasks that are essential to the operation of virtually every aspect of the Federal justice system.

The Service has these responsibilities: providing support and protection for the Federal courts, including security for 800 judicial facilities and nearly 2,000 judges and magistrates, as well as countless other trial participants such as jurors and attorneys; apprehending the majority of Federal fugitives; operating the Federal Witness Security Program and ensuring the safety of endangered Government witnesses; maintaining custody of and transporting thousands of Federal prisoners annually; executing court orders and arrest warrants; managing and selling seized property forfeited to the Government by drug traffickers and other criminals and assisting the Justice Department's asset forfeiture program; responding to emergency circumstances, including civil disturbances, terrorist incidents, and other crisis situations through its Special Operations Group; restoring order in riot and mob-violence situations; and operating the U.S. Marshals Service Training Academy.

Sources of Information

Employment For employment information, contact the Field Staffing Branch, United States Marshals Service, Department of Justice, 600 Army Navy Drive, Arlington, VA 22202–4210.

For further information, contact the Office of Public Affairs, U.S. Marshals Service, Department of Justice, Washington, DC 20530. Phone, 202–307–9065. Internet, http://www.usmarshals.gov.

International Criminal Police Organization—United States National Central Bureau

Department of Justice, Washington, DC 20530
Phone, 202–616–9000. Fax, 202–616–8400.

Director	TIMOTHY A. WILLIAMS

The U.S. National Central Bureau (USNCB) is the United States representative to INTERPOL, the International Criminal Police Organization. Also known as INTERPOL—Washington, the USNCB provides an essential communications link between the U.S. police community and their counterparts in the foreign member countries. The USNCB also serves as the United States point of contact for the European Police Office (EUROPOL), the European Union's law enforcement organization.

INTERPOL is an association of 182 countries dedicated to promoting mutual assistance among law enforcement authorities in the prevention and suppression of international crime. With no police force of its own, INTERPOL has no powers of arrest or search and seizure and therefore relies on the law enforcement authorities of its member countries. Each member country is required to have a national central bureau, such as the USNCB, to act as the primary point of contact for police matters. INTERPOL serves as a channel of communication for its member countries to cooperate in the investigation and prosecution of crime, provides a forum for discussions, working group meetings, and symposia to enable police to focus on specific areas of criminal activity affecting their countries, and issues and maintains information and databases on crime, fugitives, stolen passports and vehicles, missing persons, and humanitarian concerns, which are supplied by and can be used as a source by its member countries.

The USNCB is staffed by a permanent staff and detailed special agents from numerous Federal law enforcement agencies. The USNCB is organized into the Terrorism and Violent Crimes Division, the Economic Crimes Division, the Drug Division, the Fugitive Division, the Investigative Support Division, the Administrative Services Division, the Office of the General Counsel, and the State and Local Liaison Division (SLLD).

SLLD coordinates INTERPOL requests with 62 INTERPOL State liaison offices established in each State and the cities of New York, Boston, Chicago, Washington, Miami, San Diego, Los Angeles, San Francisco, and Seattle. The USNCB has three sub-bureaus which serve to more effectively address the law enforcement needs of U.S. territories. The sub-bureaus are located in Puerto Rico, American Samoa, and the U.S. Virgin Islands. SLLD provides the primary means of communication between foreign law enforcement authorities and domestic State and local police for the purpose of pursuing international investigations. International leads developed in criminal investigations being conducted by a State or local police entity can be pursued through their liaison office.

For further information, contact the INTERPOL–U.S. National Central Bureau, Department of Justice, Washington, DC 20530. Phone, 202–616–9000.

Drug Enforcement Administration

600–700 Army Navy Drive, Arlington, VA 22202
Phone, 202–307–1000.

Administrator	MICHELE LEONHART, *Acting*

The Drug Enforcement Administration (DEA) is the lead Federal agency in enforcing narcotics and controlled substances laws and regulations. DEA also enforces the Federal money laundering and bulk currency smuggling statutes when the funds involved in the transactions or smuggling are derived from the sale of narcotics. It was created in July 1973 by Reorganization Plan No. 2 of 1973 (5 U.S.C. app.).

DEA enforces the provisions of the controlled substances and chemical diversion and trafficking laws and regulations of the United States, and operates on a worldwide basis. It presents cases to the criminal and civil justice systems of the United States—or any other competent jurisdiction—on those significant organizations and their members involved in cultivation, production, smuggling, distribution, laundering of proceeds, or diversion of controlled substances appearing in or destined for illegal traffic in the United States. DEA disrupts and dismantles these organizations by arresting their members, confiscating their drugs, and seizing their assets; and creates, manages, and supports enforcement-related programs—domestically and internationally—aimed at reducing the availability of and demand for illicit controlled substances.

DEA's responsibilities include: investigation of major narcotic, chemical, drug-money laundering, and bulk currency smuggling violators who operate at interstate and international levels; seizure and forfeiture of assets derived from, traceable to, or intended to be used for illicit drug trafficking; seizure and forfeiture of assets derived from or traceable to drug-money laundering or the smuggling of bulk currency derived from illegal drugs; enforcement of regulations governing the legal manufacture, distribution, and dispensing of controlled substances; management of an intelligence program that supports drug investigations, initiatives, and operations worldwide; coordination with Federal, State, and local law enforcement authorities and cooperation with counterpart agencies abroad; assistance to State and local law enforcement agencies in addressing their most significant drug and drug-related violence problems; leadership and influence over international counterdrug and chemical policy and support for institution building in host nations; training, scientific research, and information exchange in support of drug traffic prevention and control; and education and assistance to the public community on the prevention, treatment, and dangers of drugs.

DEA maintains liaison with the United Nations, INTERPOL, and other organizations on matters relating to international narcotics control programs. It has offices throughout the United States and in 62 foreign countries.

Sources of Information

Controlled Substances Act Registration Information about registration under the Controlled Substances Act may be obtained from the Registration Section of the Drug Enforcement Administration, P.O. Box 28083, Central Station, Washington, DC 20038. Phone, 202–307–7255.

Employment For employment information, contact the regional offices, laboratories, or Washington Headquarters Office of Personnel.

Publications A limited selection of pamphlets and brochures is available. The most widely requested publication is Drugs of Abuse, an identification manual intended for professional use. Single copies are free.

For further information, contact the Public Affairs Section, Drug Enforcement Administration, Department of Justice, Washington, DC 20537. Phone, 202–307–7977.

Office of Justice Programs

810 Seventh Street NW., Washington, DC 20531
Phone, 202–307–0703. Internet, http://www.ojp.usdoj.gov. Email, askojp@ojp.usdoj.gov.

Assistant Attorney General	LAURIE ROBINSON

The Office of Justice Programs (OJP) was established by the Justice Assistance Act of 1984 and reauthorized in 1994 to provide Federal leadership, coordination, and assistance needed to make the Nation's justice system more efficient and effective in preventing and controlling crime. OJP is responsible for collecting statistical data and conducting analyses; identifying emerging criminal justice issues; developing and testing promising approaches to address these issues; evaluating program results; and disseminating these findings and other information to State and local governments

The Office is comprised of the following bureaus and offices: the Bureau of Justice Assistance provides funding, training, and technical assistance to State and local governments to combat violent and drug-related crime and help improve the criminal justice system; the Bureau of Justice Statistics is responsible for collecting and analyzing data on crime, criminal offenders, crime victims, and the operations of justice systems at all levels of government; the National Institute of Justice sponsors research and development programs, conducts demonstrations of innovative approaches to improve criminal justice, and develops new criminal justice technologies; the Office of Juvenile Justice and Delinquency Prevention provides grants and contracts to States to help them improve their juvenile justice systems and sponsors innovative research, demonstration, evaluation, statistics, replication, technical assistance, and training programs to help improve the Nation's understanding of and response to juvenile violence and delinquency; the Office for Victims of Crime administers

victim compensation and assistance grant programs and provides funding, training, and technical assistance to victim service organizations, criminal justice agencies, and other professionals to improve the Nation's response to crime victims; the Drug Courts Program Office supports the development, implementation, and improvement of drug courts through technical assistance and training and grants to State, local, or tribal governments and courts; the Corrections Program Office provides financial and technical assistance to State and local governments to implement corrections-related programs including correctional facility construction and corrections-based drug treatment programs; the Executive Office for Weed and Seed helps communities build stronger, safer neighborhoods by implementing the weed and seed strategy, a community-based, multidisciplinary approach to combating crime; the Office for State and Local Domestic Preparedness Support is responsible for enhancing the capacity of State and local jurisdictions to prepare for and respond to incidents of domestic terrorism involving chemical and biological agents, radiological and explosive devices, and other weapons of mass destruction; and the Office of the Police Corps and Law Enforcement Education provides college educational assistance to students who commit to public service in law enforcement, and scholarships with no service commitment to dependents of law enforcement officers who died in the line of duty.

Sources of Information

Employment For employment information, contact 633 Indiana Avenue NW., Washington, DC 20531. Phone, 202–307–0730.

For further information, contact the Department of Justice Response Center. Phone, 800–421–6770. Email, askojp@ojp.usdoj.gov. Internet, http://www.ojp.usdoj.gov.

Office on Violence Against Women

800 K Street NW., Washington, DC 20530
Phone, 202–307–6026.

Director	SUSAN B. CARBON

The Office on Violence Against Women (OVW) was established in 2005 to reduce violence against women through the implementation of the Violence Against Women Act. OVW is responsible for administering financial and technical assistance to communities that are developing programs, policies, and practices aimed at ending domestic and dating violence, sexual assault, and stalking.

For further information, contact the Office on Violence Against Women, Department of Justice, Washington, DC, 20530. Phone, 202–307–6026. Internet, http://www.ovw.usdoj.gov.

Bureau of Alcohol, Tobacco, Firearms and Explosives

650 Massachusetts Avenue NW., Washington, DC 20226
Phone, 202–927–8500. Internet, http://www.atf.gov.

Director	KENNETH E. MELSON, *Acting*

The Bureau of Alcohol, Tobacco, Firearms and Explosives (ATF) is responsible for enforcing Federal criminal laws and regulating the firearms and explosives industries. ATF, formerly known as the Bureau of Alcohol, Tobacco, and Firearms, was initially established by Department of Treasury Order No. 221, effective July 1, 1972, which transferred the functions, powers, and duties arising under laws relating to alcohol, tobacco, firearms, and explosives from the Internal Revenue Service to ATF. The Homeland Security Act of 2002 (6 U.S.C. 531) transferred certain functions and authorities of ATF to the Department of Justice and established it under its current name. ATF works, directly and through partnerships, to investigate and reduce violent crime involving firearms and explosives, acts of arson, and illegal trafficking of alcohol and tobacco products. The Bureau provides training and support to its Federal, State, local, and international law enforcement partners and works primarily in 23 field divisions across the 50 States, Puerto Rico, the U.S. Virgin Islands, and Guam. It also has foreign offices in Mexico, Canada, Colombia, and France.

For further information, contact the Office of Public Affairs, Bureau of Alcohol, Tobacco, Firearms and Explosives. Phone, 202–927–8500. Internet, http://www.atf.gov.

Boards

Executive Office for Immigration Review

Falls Church, VA 22041
Phone, 703–305–0289. Internet, http://www.usdoj.gov/eoir.

Director	THOMAS G. SNOW, *Acting*

The Executive Office for Immigration Review, under a delegation of authority from the Attorney General, is charged with adjudicating matters brought under various immigration statutes to its three administrative tribunals: the Board of Immigration Appeals, the Office of the Chief Immigration Judge, and the Office of the Chief Administrative Hearing Officer.

The Board of Immigration Appeals has nationwide jurisdiction to hear appeals from certain decisions made by immigration judges and by district directors of the Department of Homeland Security (DHS). In addition, the Board is responsible for hearing appeals involving disciplinary actions against attorneys and representatives before DHS and the Board.

Decisions of the Board are binding on all DHS officers and immigration judges unless modified or overruled by the Attorney General or a Federal court. All Board decisions are subject to judicial review in Federal court. The majority of appeals reaching the Board involve orders of removal and applications for relief from removal. Other cases before the Board include the removal of aliens applying for admission to the United States, petitions to classify the status of alien relatives for the issuance of preference immigrant visas, fines imposed upon carriers for the violation of the immigration laws, and motions for reopening and reconsideration of decisions previously rendered.

The Office of the Chief Immigration Judge provides overall direction for more than 200 immigration judges located in 53 immigration courts throughout the Nation. Immigration judges are responsible for conducting formal administrative proceedings and act independently in their decisionmaking capacity. Their decisions are administratively final, unless appealed or certified to the Board.

In removal proceedings, an immigration judge determines whether an individual from a foreign country should be admitted or allowed to stay in the United States or be removed. Judges are located throughout the United States, and each judge has jurisdiction to consider various forms of relief available under the law, including applications for asylum.

The Office of the Chief Administrative Hearing Officer is responsible for the general supervision and management of administrative law judges who preside at hearings that are mandated by provisions of immigration law concerning allegations of unlawful employment of aliens, unfair immigration-related employment practices, and immigration document fraud.

For further information, contact the Office of Legislative and Public Affairs, Executive Office for Immigration Review, Department of Justice, Falls Church, VA 22041. Phone, 703–305–0289. Internet, http://www.usdoj. gov/eoir.

United States Parole Commission

5550 Friendship Boulevard, Chevy Chase, MD 20815
Phone, 301–492–5990. Internet, http://www.usdoj.gov/uspc.

Chairman	ISAAC FULWOOD, JR.

The United States Parole Commission (USPC) makes parole release decisions for eligible Federal and District of Columbia prisoners; authorizes methods of release and conditions under which release occurs; prescribes, modifies, and monitors compliance with the terms and conditions governing offenders' behavior while on parole or mandatory or supervised release; issues warrants for violation of supervision; determines probable cause for the revocation process; revokes parole, mandatory, or supervised release; releases from supervision those offenders who are no longer a risk to public safety; and promulgates the rules, regulations, and guidelines for the exercise of USPC's authority and the implementation of a national parole policy.

USPC has sole jurisdiction over the following: Federal offenders who committed offenses before November

1, 1987; D.C. Code offenders who committed offenses before August 5, 2000; D.C. Code offenders sentenced to a term of supervised release; Uniform Code of Military Justice offenders who are in Bureau of Prison's custody; transfer treaty cases; and State probationers

and parolees in the Federal Witness Protection Program.

Sources of Information

Reading Rooms The reading room is located at 5550 Friendship Boulevard, Chevy Chase, MD 20815. Phone, 301–492–5959.

For further information, contact the U.S. Parole Commission, Department of Justice, 5550 Friendship Boulevard, Chevy Chase, MD 20815. Phone, 301–492–5990. Internet, http://www.usdoj.gov/uspc.

Office of Community Oriented Policing Services

1100 Vermont Avenue NW., Washington, DC 20530
Phone, 202–514–2058. Internet, http://www.cops.usdoj.gov.

Director   BERNARD K. MELEKIAN

The Office of Community Oriented Policing Services (COPS) was established to assist law enforcement agencies in enhancing public safety through the implementation of community policing strategies. COPS does so by providing training to enhance law enforcement officers' problem-solving and community interaction skills; encouraging law enforcement and community members to develop initiatives to prevent crime; substantially increasing the number of law enforcement officers directly interacting with the community; and supporting the development of new technologies to shift law enforcement's focus to preventing crime and disorder within their communities.

The COPS Office includes the following program divisions:

The grants administration division is responsible for developing and designing new programs to provide resources for the hiring of new officers and to further the adoption and implementation of community policing, reviewing grant applications, and assisting grantees in the implementation of their grants.

The grants monitoring division is responsible for tracking grantees' compliance with the conditions of their grants. The Division conducts site visits and reviews grantee files to ensure that COPS funds are properly used to hire officers and implement community policing. The Division also provides onsite technical assistance to grantees, office-based grant reviews, alleged noncompliance reviews, audit resolution, and collects and disseminates examples of successful community policing strategies.

The training and technical assistance division is responsible for coordinating the provision of training and technical assistance to advance the adoption, implementation, and sustaining of community policing in the thousands of communities served by the COPS Office.

The compliance division is responsible for the monitoring and coordination of the Office of Inspector General (OIG) audits and independent audits required by the Single Audit Act and serves as the liaison between grantees and auditors in the conduct and resolution of OIG audits.

For further information, contact the Office of Community Oriented Policing Services (COPS), Department of Justice, 1100 Vermont Avenue NW., Washington, DC 20530. Phone, 202–514–2058. Internet, http://www.cops.usdoj.gov.

Foreign Claims Settlement Commission of the United States

Suite 6002, 600 E Street NW., Washington, DC 20579
Phone, 202–616–6975. Fax, 202–616–6993.

Chairman	MAURICIO J. TAMARGO

The Foreign Claims Settlement Commission of the United States is a quasi-judicial, independent agency within the Department of Justice, which adjudicates claims of U.S. nationals against foreign governments, either under specific jurisdiction conferred by Congress or pursuant to international claims settlement agreements. The decisions of the Commission are final and are not reviewable under any standard by any court or other authority. Funds for payment of the Commission's awards are derived from congressional appropriations, international claims settlements, or the liquidation of foreign assets in the United States by the Departments of Justice and the Treasury.

The Commission also has authority to receive, determine the validity and amount, and provide for the payment of claims by members of the U.S. Armed Services and civilians held as prisoners of war or interned by a hostile force in Southeast Asia during the Vietnam conflict or by the survivors of such service members and civilians.

The Commission is also responsible for maintaining records and responding to inquiries related to the various claims programs it has conducted against the Governments of Albania, Bulgaria, China, Cuba, Czechoslovakia, Egypt, Ethiopia, the Federal Republic of Germany, the German Democratic Republic, Hungary, Iran, Italy, Panama, Poland, Romania, the Soviet Union, Vietnam, and Yugoslavia, as well as those authorized under the War Claims Act of 1948 and other statutes.

Sources of Information

Employment   For information of attorney positions, contact the Office of the Chief Counsel, Suite 6002, 600 E Street NW., Washington, DC 20579 (phone, 202–616–6975). For all other positions, contact the Administrative Officer, same address and phone.

Reading Room   The reading room is located at 600 E Street NW., Washington, DC 20579. Phone, 202–616–6975.

For further information, contact the Office of the Chairman, Foreign Claims Settlement Commission of the United States, Department of Justice, Suite 6002, 600 E Street NW., Washington, DC 20579. Phone, 202–616–6975. Fax, 202–616–6993.

DEPARTMENT OF LABOR

200 Constitution Avenue NW., Washington, DC 20210
Phone, 202–693–5000. Internet, http://www.dol.gov.

Secretary of Labor	HILDA L. SOLIS
Deputy Secretary	SETH D. HARRIS
Chief of Staff	KATHERINE ARCHULETA
Chief Administrative Law Judge	STEPHEN L. PURCELL, *Acting*
Chief Administrative Appeals Judge, Administrative Review Board	PAUL IGASAKI
Chief Administrative Appeals Judge, Benefits Review Board	NANCY S. DOLDER
Chairman, Employees Compensation Appeals Board	ALEC J. KOROMILAS
Director, Center for Faith-Based and Community Initiatives	PHILIP TOM
Director, Office of Job Corps	EDNA PRIMROSE
Executive Secretary	NANCY ROONEY, *Acting*
Ombudsman, Energy Employee Occupational Illness Compensation Program	MALCOLM NELSON
Administrator, Wage and Hour Division	NANCY LEPPINK, *Acting*
Assistant Secretary for Administration and Management	T. MICHAEL KERR
Assistant Secretary for Congressional and Intergovernmental Affairs	BRIAN KENNEDY
Assistant Secretary of Disability Employment Policy	KATHLEEN MARTINEZ
Assistant Secretary for Employee Benefit Security Administration	PHYLLIS C. BORZI
Assistant Secretary for Employment and Training Administration	JANE OATES
Assistant Secretary for Mine Safety and Health Administration	JOSEPH MAIN
Assistant Secretary for Occupational Safety and Health Administration	DAVID MICHAELS
Assistant Secretary for Policy	WILLIAM SPRIGGS
Assistant Secretary for Veterans' Employment and Training	RAYMOND M. JEFFERSON
Chief Economist	JESSE ROTHSTEIN
Chief Financial Officer	DAN LACEY, *Acting*
Commissioner of Labor Statistics	KEITH HALL
Deputy Under Secretary for International Affairs	SANDRA POLASKI
Director, Office of Federal Contract Compliance Programs	PATRICIA A. SHIU
Director, Office of Labor-Management Standards	JOHN LUND
Director, Office of Workers Compensation	SHELBY HALLMARK
Director, Pension Benefit Guaranty Corporation	JOSHUA GOTBAUM
Director, Women's Bureau	SARA MANZANO-DIAZ

Inspector General
Senior Advisor for Communications and Public
  Affairs
Solicitor of Labor

DANIEL PETROLE, *Acting*
CARL FILLICHIO

M. PATRICIA SMITH

The Department of Labor fosters and promotes the welfare of the job seekers, wage earners, and retirees, by improving working conditions, advancing opportunities for profitable employment, protecting retirement and health care benefits, matching workers to employers, strengthening free collective bargaining, and tracking changes in employment, prices, and other national economic measurements. The Department administers a variety of Federal labor laws to guarantee workers' rights to fair, safe, and healthy working conditions, including minimum hourly wage and overtime pay, freedom from employment discrimination, unemployment insurance, and other income support.

The Department of Labor (DOL) was created by act of March 4, 1913 (29 U.S.C. 551). A Bureau of Labor was first created by Congress by act of June 24, 1884, in the Interior Department. The Bureau of Labor later became independent as a Department of Labor without executive rank by act of June 13, 1888. It again returned to bureau status in the Department of Commerce and Labor, which was created by act of February 14, 1903 (15 U.S.C. 1501; 29 U.S.C. 1 note).

Secretary The Secretary is the principal adviser to the President on the development and execution of policies and the administration and enforcement of laws relating to wage earners, their working conditions, and their employment opportunities.

Employees' Compensation Appeals Board The Board is a three-member quasi-judicial body appointed by the Secretary which has been delegated exclusive jurisdiction by Congress to hear and make final decisions on Federal workers' compensation appeals arising under the Federal Employees' Compensation Act. The Board was created by Reorganization Plan No. 2 of 1946 (60 Stat. 1095). The Board's decisions are not reviewable and are binding upon the Office.

For further information, call the Administrative Officer. Phone, 202–693–6234. Internet, http://www.dol.gov.

Administrative Review Board The Board consists of five members appointed by the Secretary. It issues final agency decisions for appeals cases under a wide range of worker protection laws, including the McNamara O'Hara Service Contract Act and the Davis Bacon Act. The appeals cases primarily address environmental, transportation, and securities whistleblower protection; H–1B immigration provisions; child labor violations; employment discrimination; job training; seasonal and migrant workers; and Federal construction and service contracts. The Board's cases generally arise upon appeal from decisions of Department of Labor Administrative Law Judges or the Administrator of the Department's Wage and Hour Division. Depending upon the statute at issue, the parties may appeal the Board's decisions to Federal district or appellate courts and, ultimately, to the United State Supreme Court.

For further information, call 202–693–6234.

Benefits Review Board The Board consists of five members appointed by the Secretary. It adjudicates appeals cases under the Longshore and Harbor Workers' Compensation Act and the Black Lung Benefits Act. Board decisions may be appealed to the U.S. Courts of Appeals and to the U.S. Supreme Court.

For further information, call 202–693–6234.

Office of the Ombudsman for the Energy Employees Occupational Illness Compensation Program Act The Office was established in October 2004 under Part E of the Energy Employees Occupational Illness Compensation Program Act to administer a system of Federal payments to compensate certain nuclear workers for occupational

DEPARTMENT OF LABOR

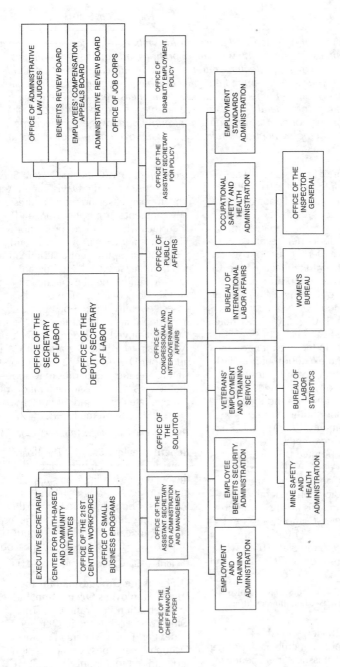

illnesses caused by exposure to toxic substances. It is a small, independent office, headed by the Ombudsman, who is appointed by the Secretary of Labor. The Office provides information to claimants on the benefits available under Parts E and B of the EEOICPA and issues annual reports detailing the complaints, grievances, and requests for assistance received by the Office.

For further information, call 202–693–5890.

The Solicitor of Labor The Office of the Solicitor provides necessary legal services to accomplish the Department's mission and goals. The Solicitor directs a broad-scale litigation effort in the Federal courts pertaining to various labor statutes administered by the Department, ranging from workers' compensation to employment discrimination.

For a complete listing of regional offices of the Office of the Solicitor, including addresses, telephone numbers, and key officials, visit www.dol.gov/sol/organizations/regions/main.htm.

For a reference to the national office divisions, visit www.dol.gov/sol/organizations/divisions/main.htm.

For further information, contact the Office of the Solicitor, Department of Labor, 200 Constitution Avenue NW., Washington, DC 20210. Phone, 202–693–5260.

Women's Bureau The Women's Bureau is responsible for promoting the status of wage-earning women, improving their working conditions, increasing their efficiency, and advancing their opportunities for profitable employment.

For a complete listing of Regional Offices of the Women's Bureau, including addresses, telephone numbers, and key officials, visit www.dol.gov/wb.

For further information, call 202–693–6710.

International Affairs The Bureau of International Labor Affairs is mandated to carry out the Secretary's international responsibilities, develop departmental policy and programs relating to international activities, and coordinate departmental international activities involving other U.S. Government agencies, intergovernmental and nongovernmental organizations.

For further information, call 202–693–4770.

Office of Inspector General The Office of Inspector General conducts audits and investigations to review the effectiveness, efficiency, economy, and integrity of all DOL programs and operations, including those performed by its contractors and grantees. The Office is unique among Inspectors General because it conducts labor racketeering investigations in employee benefit plans, labor-management relations, and internal labor union affairs.

For further information, call 202–693–5100.

Employment and Training Administration

The Employment and Training Administration (ETA) provides quality job training, employment, labor market information, and income maintenance services primarily through State and local workforce development systems. ETA also administers programs to enhance employment opportunities and business prosperity.

For a complete listing of Regional and State Offices of the Employment and Training Administration, including addresses, telephone numbers, areas served, and key officials, visit www.doleta.gov/Regions.

Office of Workforce Investment The Office of Workforce Investment (OWI) provides leadership, oversight, policy guidance, and technical assistance to the Nation's workforce investment system including the One-Stop Career Center systems, the youth and adult employment and training programs, and national programs for targeted populations. OWI oversees investments in innovative workforce solutions in high-growth sectors of the economy, including providing training through community colleges. OWI also oversees the development and dissemination of tools and information related to workforce and economic data, career guidance, and workforce skills and competencies.

For further information, call 202–693–3980.

Office of Unemployment Insurance
The Office of Unemployment Insurance

(OUI) provides national leadership, oversight, policy guidance, and technical assistance to the Federal-State unemployment compensation system. OUI also interprets Federal legislative requirements.

For more information, call 202–693–3029. Internet, http://www.unemploymentinsurance.doleta.gov.

Office of National Response The Office of National Response (ONR) is responsible for national leadership, oversight, policy guidance, funding allocations, and technical assistance for the National Emergency Grant programs for dislocated workers.

For more information, call 202–693–3500.

Office of Trade Adjustment Assistance The Office of Trade Adjustment Assistance (OTAA) is responsible for national leadership, oversight, policy guidance funding allocations, and technical assistance for dislocated workers seeking to participate in structured training programs.

For further information, call 202–693–3560.

Office of Apprenticeship The Office of Apprenticeship (OA) administers the National Registered Apprenticeship System. OA promotes the adoption by employers, labor, and other organizers of formalized, structured training programs. OA also enhances opportunities for women and minorities to participate in such programs.

For more information, call 202–693–2796. Internet, http://www.doleta.gov/oa.

Office of Foreign Labor Certification The Office of Foreign Labor Certification (OFLC) carries out the delegated responsibility of the Secretary of Labor under the Immigration and Nationality Act (INA), as amended, concerning the admission of foreign workers into the United States for employment.

In carrying out this responsibility, OFLC administers temporary nonimmigrant labor certification programs and the permanent labor certification program through ETA's National Processing Centers located, respectively, in Chicago and Atlanta.

OFLC also administers nationally the issuance of employer-requested prevailing wage determinations through ETA's National Prevailing Wage and Helpdesk Center located in Washington, DC. Prevailing wage determinations are issued for use in all nonagricultural temporary labor certification programs and the permanent labor certification program.

For more information, call 202–693–3010. Internet, http://www.foreignlaborcert.doleta.gov.

Office of Policy Development and Research The Office of Policy Development and Research (OPDR) supports ETA policies and investments to improve the public workforce system by analyzing, formulating, and recommending legislative changes and options for policy initiatives. OPDR coordinates ETA's legislative and regulatory activities, maintains the ETA portion of the Department's regulatory agenda, and provides ETA with relevant research, demonstrations, and program evaluations. OPDR also coordinates interactions with international organizations and foreign countries and disseminates advisories and publications to the Employment and Training system.

Office of Financial and Administrative Management The Office of Financial and Administrative Management (OFAM) provides leadership and direction to ensure sound management of financial resources throughout the ETA and also budget, accounting, data analysis, and technology services for the ETA. OFAM is responsible for planning, developing, promulgating, and executing policies, standards, and guidelines governing ETA management of information, budget, accounting, financial and procurement systems, personnel management, organizational analysis, technical training and life-long learning, administrative and property management services and technology. OFAM provides centralized services to ETA National Office components for procurement and for job training assistance management. The Office also plans and administers the ETA personnel and organizational management programs, which include staffing, position classification and management, employee development and utilization, and labor management and employee relations.

For further information, call 202–693–2800.

Office of Regional Innovation and Transformation The Office of Regional Management (ORM) works to improve the administration and outcomes of ETA-funded grants and programs through the work of six regional operations of grant compliance and technical assistance. The regional offices, in Boston, Philadelphia, Atlanta, Dallas, Chicago, and San Francisco, provide grant management and technical assistance to States and other grantees of the workforce investment system. ORM is responsible for facilitating two-way communications between the national office operations and the regional offices and providing information and support directly to the regional administrators and the Office of the Assistant Secretary for Employment and Training.

For further information, call 202–693–3690.

Office of Job Corps The Office of Job Corps (OJC) works to attract young adults, teach them relevant skills they need to become employable and independent, and help them secure meaningful jobs or opportunities for further education. OJC has six regional offices responsible for monitoring and oversight of Job Corps centers, outreach and admissions, and career transition services.

For a complete listing of regional offices of the Job Corps, including addresses, telephone numbers, and areas served, visit www.jobcorps.gov/contact. aspx#regional.

For a complete listing of Job Corps centers across the country, including addresses, telephone numbers, and center Web sites, visit www.jobcorps.gov/centers.aspx.

For further information, contact the Office of Job Corps, Department of Labor, 200 Constitution Avenue NW., Room N–4463, Washington, DC 20210. Phone, 202–693–3000. Internet, http://jobcorps.dol.gov.

Employee Benefits Security Administration

The Employee Benefits Security Administration (EBSA) promotes and protects the pension, health, and other benefits of the over 150 million participants and beneficiaries in over 6 million private sector employee benefit plans. In administering its responsibilities, EBSA assists workers in understanding their rights and protecting their benefits; facilitates compliance by plan sponsors, plan officials, service providers, and other members of the regulated community; encourages the growth of employment-based benefits; and deters and corrects violations of the relevant statutes. The Employee Retirement Income Security Act (ERISA) is enforced through 15 EBSA field offices nationwide and the national office in Washington, DC.

For a complete listing of Regional and District Offices of the Employee Benefits Security Administration, including addresses, telephone numbers, areas served, and key officials, visit www.dol. gov/ebsa/aboutebsa.

For further information, contact the Employee Benefits Security Administration. Phone, 866–444–3272. Internet, http://www.dol.gov/ebsa.

Office of the Assistant Secretary for Administration and Management

The Office of Small and Disadvantaged Business Utilization (OSDBU) ensures procurement opportunities for small businesses, small disadvantaged businesses, women-owned small businesses, HUBZone businesses, and businesses owned by service-disabled veterans. OSDBU serves as the Department's Ombudsman for small businesses under the Small Business Regulatory Enforcement Fairness Act (SBREFA).

For more information, please call 202–693–7299. Internet, http://www.dol.gov/oasam/programs/osdbu.

Office of Federal Contract Compliance Programs

The Office of Federal Contract Compliance Programs (OFCCP) administers and enforces three equal opportunity mandates: Executive Order 11246, as amended; section 503 of the Rehabilitation Act of 1973, as amended; and the Vietnam Era Veterans' Readjustment Assistance Act of 1974, as amended, 38 U.S.C. 4212. These mandates prohibit Federal contractors and subcontractors from discriminating on the basis of race, color, religion, sex, national origin, disability, or veteran status. They also require Federal contractors and subcontractors to take affirmative steps to ensure equal opportunity in their employment processes. OFCCP also shares responsibility with the U.S. Equal Opportunity Employment Commission in enforcing Title I of the Americans with Disabilities Act.

For a complete listing of OFCCP offices across the country, including addresses, telephone numbers, and key officials, visit www.dol.gov/ofccp/contacts/ofnation2.htm.

For further information, contact the Office of Federal Contract Compliance Programs help desk. Phone, 800–397–6251. Internet, http://www.dol.gov/ofccp/index.htm.

Wage and Hour Division

The Wage and Hour Division is responsible for planning, directing, and administering programs dealing with a variety of Federal labor legislation. These programs are designed to protect low-wage incomes; safeguard the health and welfare of workers by discouraging excessively long work hours; safeguard the health and well-being of minors; prevent curtailment of employment and earnings for students, trainees, and handicapped workers; minimize losses of income and job rights caused by indebtedness; and direct a program of farm labor contractor registration designed to protect the health, safety, and welfare of migrant and seasonal agricultural workers.

For a complete listing of Wage and Hour Division offices across the country, including addresses, telephone numbers, and key officials, visit www.dol.gov/whd/america2.htm.

For further information, contact the Office of the Administrator, Wage and Hour Division, Department of Labor, Room S–3502, 200 Constitution Avenue NW., Washington, DC 20210. Phone, 202–693–0051.

Office of Labor-Management Standards

The Office of Labor-Management Standards conducts criminal and civil investigations to safeguard the financial integrity of unions and to ensure union democracy and conducts investigative audits of labor unions to uncover and remedy criminal and civil violations of the Labor-Management Reporting and Disclosure Act and related statutes.

For a complete listing of Office of Labor-Management Standards Regional and District Offices, including addresses, telephone numbers, and key officials, visit www.dol.gov/olms/contacts/lmskeyp.htm.

For Labor-Management Reporting and Disclosure Act assistance, call 202–693–0123. For electronic forms software technical support, call 866–401–1109. For transit employee protections assistance, call 202–693–0126. Internet, http://www.dol.gov/olms.

Office of Workers' Compensation Programs

The Office of Workers' Compensation Programs is responsible for programs providing workers' compensation for Federal employees, benefits to employees in private enterprise while engaged in maritime employment on navigable waters in the United States, and benefits to coal miners who are totally disabled due to pneumoconiosis, a respiratory disease contracted after prolonged inhalation of coal mine dust, and to their survivors when the miner's death is due to pneumoconiosis.

For a complete listing of Office of Workers' Compensation Programs District Offices and contacts for Longshore and Harbor Workers', Federal Employees', Coal Mine Workers' and Energy Employees Occupation Illness Compensation, including addresses, telephone numbers, and key officials, visit www.dol.gov/owcp/owcpkeyp.htm.

For further information, contact the Office of the Director, Office of Workers' Compensation Programs, Department of Labor, Room S–3524, 200 Constitution Avenue NW., Washington, DC 20210.

Phone, 202–693–0031. Internet, http://www.dol.
gov/owcp.

Occupational Safety and Health Administration

The Occupational Safety and Health
Administration (OSHA) sets and enforces
workplace safety and health standards
and assists employers in complying
with those standards, assisting workers
in exercising their rights under Federal
labor laws. OSHA, created pursuant to
the Occupational Safety and Health Act
of 1970 (29 U.S.C. 651 et seq.) enforces
workplace safety and health rules;
promulgates strong, protective health
and safety standards; increases outreach
and help for workers and their employers
in their efforts to eliminate and control
workplace hazards; and partners with the
States that are running their own OSHA-
approved programs.

For a complete listing of OSHA
Regional and Area Offices, including
addresses, telephone numbers, and
key officials, visit www.osha.gov/html/
RAmap.html.

For further information, contact the Occupational
Safety and Health Administration, Department of
Labor, 200 Constitution Avenue NW., Washington,
DC 20210. Phone, 202–693–1999.

Mine Safety and Health Administration

The Mine Safety and Health
Administration (MSHA) is the worker
protection agency focused on prevention
of death, disease, and injury from
mining and promotion of safe and
healthful workplaces for the Nation's
miners. MSHA's approach includes
promulgating and enforcing mandatory
health and safety standards through
complete annual inspections of each
mine; targeting the most common causes
of fatal mine accidents and disasters;
reducing exposure to health risks from
mine dusts and other contaminants;
improving training of miners including
new and inexperienced miners and
contractors; improving mine emergency
response preparedness by MSHA
and the mining industry; enhancing
enforcement of miners' rights to report
hazardous conditions with protection

against retaliation; and encouraging
and enforcing a focus on prevention.
MSHA also cooperates with and
provides assistance to the States in
the development of effective State
mine safety and health programs and
contributes to the improvement and
expansion of mine safety and health
research and development.

For a complete listing of MSHA District
and Field Offices, including addresses,
telephone numbers, and key officials,
visit www.msha.gov/district/disthome.
htm.

For further information, contact the Office of
Programs, Education and Outreach Services, Mine
Safety and Health Administration, Department
of Labor, Room 2317, 1100 Wilson Boulevard,
Arlington, VA 22209–3939. Phone, 202–693–9400.

Office of Disability Employment Policy

The Office of Disability Employment
Policy (ODEP) provides national
leadership by developing and influencing
the implementation of disability
employment policies and practices
affecting the employment of people with
disabilities. ODEP's response to low labor
force participation rates among such
a large group of Americans is unique,
comprehensive, and aggressive and
includes the active involvement and
cooperation of Federal, State, and local
public and private entities, including
employers.

ODEP addresses the significant barriers
to employment faced by individuals
with disabilities by developing and
disseminating national, State, and
local disability employment policy; by
fostering implementation of innovative
strategies and practices among employers
and throughout the various systems
serving people with disabilities; by
conducting disability-related research
to build knowledge to inform policy
development; and by providing technical
assistance to service delivery systems
and employers to increase employment
opportunities and the recruitment,
retention, and promotion of people with
disabilities.

For further information, call 202–693–7880. TTY,
202–693–7881. Internet, http://www.dol.gov/odep.

Bureau of Labor Statistics

The Bureau of Labor Statistics (BLS) is the principal fact-finding agency of the Federal Government in the broad field of labor economics and statistics. The Bureau is an independent national statistical agency that collects, processes, analyzes, and disseminates essential statistical data to the American public, Congress, other Federal agencies, State and local governments, businesses, and labor. BLS also serves as a statistical resource to the Department of Labor. Data are available relating to employment, unemployment, and other characteristics of the labor force; consumer and producer prices, consumer expenditures, and import and export prices; wages and employee benefits; productivity and technological change; employment projections; occupational illness and injuries; and international comparisons of labor statistics. Most of the data are collected in surveys conducted by the Bureau, the Bureau of the Census (on a contract basis), or on a cooperative basis with State agencies.

The Bureau strives to have its data satisfy a number of criteria, including: relevance to current social and economic issues, timeliness in reflecting today's rapidly changing economic conditions, accuracy and consistently high statistical quality, and impartiality in both subject matter and presentation.

The basic data are issued in monthly, quarterly, and annual news releases; bulletins, reports, and special publications; and periodicals. Data are also made available through an electronic news service, magnetic tape, diskettes, and microfiche, as well as on the Internet at stats.bls.gov. Regional offices issue additional reports and releases, usually presenting locality or regional detail.

For a complete listing of Bureau of Labor Statistics regional offices, including addresses, telephone numbers, and key officials, visit www.bls.gov/bls/regnhome.htm.

For further information, contact the Bureau of Labor Statistics, Room 4040, 2 Massachusetts Avenue NE., Washington, DC 20212. Phone, 202–691–7800.

Veterans' Employment and Training Service

The Veterans' Employment and Training Service (VETS) is responsible for administering veterans' employment and training programs and compliance activities that help veterans and servicemembers succeed in their civilian careers. VETS administers the Jobs for Veterans State Grant program, which provides grants to States to fund personnel dedicated to serving the employment needs of veterans. VETS field staff works closely with and provides technical assistance to State employment workforce agencies to ensure that veterans receive priority of service and gain meaningful employment. VETS also administers three competitive grants programs: the Veterans Workforce Investment Program, the Homeless Veterans Reintegration Program, and the Incarcerated Veterans Transition Program. In addition, VETS prepares separating servicemembers for the civilian labor market through its Transition Assistance Program Employment Workshops.

VETS has three distinct compliance programs: the Federal Contractor Program, Veterans' Preference in Federal hiring and the Uniformed Services Employment and Reemployment Rights Act of 1994 (USERRA). With respect to Federal contractors, VETS promulgates regulations and maintains oversight of the program by assisting contractors to comply with their affirmative action and reporting obligations. Although the Office of Personnel Management is responsible for administering and interpreting statutes and regulations governing veterans' preference in Federal hiring, VETS investigates allegations that veterans' preference rights have been violated. In addition, VETS preserves servicemembers' employment and reemployment rights through its administration and enforcement of the USERRA statute. VETS conducts thorough investigations of alleged violations and conducts an extensive USERRA outreach program.

For a complete listing of Veterans' Employment and Training Service regional and State offices, including addresses, telephone numbers, and key officials, visit www.dol.gov/vets/aboutvets/contacts/main.htm#regionalstatedirectory.

For further information, contact the Assistant Secretary for Veterans' Employment and Training, Department of Labor, 200 Constitution Avenue NW., Washington, DC 20210. Phone, 202–693–4700.

Sources of Information

Contracts General inquiries may be directed to the Procurement Services Center, Room S–4307, 200 Constitution Avenue NW., Washington, DC 20210. Phone, 202–693–4570. Inquiries on doing business with the Job Corps should be directed to the Job Corps Regional Director in the appropriate Department of Labor regional office.

Electronic Access Information concerning Department of Labor agencies, programs, and activities is available at www.dol.gov.

Employment The Department of Labor's Web site (www.dol.gov) provides detailed information about job opportunities with the Department, including the address and telephone numbers of the Department's personnel offices in the regions and in Washington, DC.

Publications The Office of Public Affairs distributes fact sheets that describe the activities of the major agencies within the Department.

Employment and Training Administration The Employment and Training Administration issues periodicals such as Area Trends in Employment and Unemployment available by subscription through the Superintendent of Documents, Government Printing Office, Washington, DC 20402. Information about publications may be obtained from the Administration's Information Office. Phone, 202–219–6871.

Office of Labor-Management Standards The Office of Labor-Management Standards (OLMS) publishes the text of the Labor-Management Reporting and Disclosure Act and pamphlets that explain the reporting, election, bonding, and trusteeship provisions of the act. The pamphlets and reporting forms used by persons covered by the act are available free in limited quantities from the OLMS National Office at Room N–5616, 200 Constitution Avenue NW., Washington, DC 20210, and from OLMS field offices.

Employee Benefits Security Administration The Employee Benefits Security Administration distributes fact sheets, pamphlets, and booklets on employer obligations and employee rights under ERISA. A list of publications is available by writing to the Office of Participant Assistance, Employee Benefits Security Administration, Room N–5623, 200 Constitution Avenue NW., Washington, DC 20210. Phone, 866–444–3272. Internet, http://www.dol.gov/ebsa.

Bureau of Labor Statistics The Bureau of Labor Statistics has an information office at 2 Massachusetts Avenue NE., Room 2850, Washington, DC 20212. Phone, 202–606–5886. Periodicals include the Monthly Labor Review, Consumer Price Index, Producer Prices and Price Indexes, Employment and Earnings, Current Wage Developments, Occupational Outlook Handbook, and Occupational Outlook Quarterly. Publications are both free and for sale, but for-sale items must be obtained from the Superintendent of Documents, Government Printing Office. Inquiries may be directed to the Washington Information Office or to Bureau's regional offices.

Wage and Hour Division The Wage and Hour Division distributes numerous outreach and educational materials, including posters, fact sheets, guides, booklets, and bookmarks. Phone, 866–487–9243. Internet, http://www.wagehour.dol.gov.

Reading Rooms Department of Labor Library, Room N–2439, 200 Constitution Avenue NW., Washington, DC 20210. Phone, 202–219–6992. The Office of Labor-Management Standards maintains a Public Disclosure Room at Room N–5616, 200 Constitution Avenue NW., Washington, DC 20210. Reports filed under the Labor-Management Reporting and Disclosure Act may be examined there and purchased for 15 cents per page. Reports also may be obtained by calling the Public Disclosure Room at 202–219–7393, or by contacting an

Office field office. The Employee Benefits Security Administration maintains a Public Disclosure Room at Room N–1513, 200 Constitution Avenue NW., Washington, DC 20210. Reports filed under the Employee Retirement Income Security Act may be examined there and purchased for 15 cents per page or by calling the Public Disclosure Room at 202–693–8673.

For further information concerning the Department of Labor, contact the Office of Public Affairs, Department of Labor, Room S–1032, 200 Constitution Avenue NW., Washington, DC 20210. Phone, 202–693–4650. Internet, http://www.dol.gov.

DEPARTMENT OF STATE

2201 C Street NW., Washington, DC 20520
Phone, 202–647–4000. Internet, http://www.state.gov.

Secretary of State	HILLARY RODHAM CLINTON
Deputy Secretary of State	JAMES B. STEINBERG
Deputy Secretary of State for Management and Resources	JACOB J. LEW
Counselor and Chief of Staff	CHERYL MILLS
Executive Secretary	DANIEL B. SMITH
Under Secretary for Arms Control and International Security Affairs	ELLEN TAUSCHER
Assistant Secretary for International Security and Nonproliferation	VANN H. VAN DIEPEN, *Acting*
Assistant Secretary for Political-Military Affairs	ANDREW J. SHAPIRO
Assistant Secretary for Verification, Compliance, and Implementation	ROSE E. GOTTEMOELLER
Under Secretary for Democracy and Global Affairs	MARIA OTERO
Ambassador-at-Large for the Office to Monitor and Combat Trafficking in Persons	LUIS CDEBACA
Assistant Secretary for Democracy, Human Rights, and Labor	MICHAEL H. POSNER
Assistant Secretary for Oceans and International Environmental and Scientific Affairs	KERRI-ANN JONES
Assistant Secretary for Population, Refugees, and Migration	ERIC P. SCHWARTZ
Under Secretary for Economic, Energy, and Agricultural Affairs	ROBERT HORMATS
Assistant Secretary for Economic, Energy, and Business Affairs	JOSE W. FERNANDEZ
Under Secretary for Management	PATRICK F. KENNEDY
Assistant Secretary for Administration	STEVEN J. RODRIQUEZ, *Acting*
Assistant Secretary for Consular Affairs	JANICE L. JACOBS
Assistant Secretary for Diplomatic Security and Director of the Office of Foreign Missions	ERIC J. BOSWELL
Assistant Secretary for Information Resource Management and Chief Information Officer	SUSAN SWART
Chief Financial Officer	JAMES MILLETTE, *Acting*
Director of the Foreign Service Institute	RUTH A. WHITESIDE
Director General of the Foreign Service and Director of Human Resources	NANCY J. POWELL
Director, Office of Medical Services	THOMAS W. YUN
Director and Chief Operating Officer of Overseas Buildings Operations	ADAM NAMM, *Acting*
Under Secretary for Political Affairs	WILLIAM J. BURNS
Assistant Secretary for African Affairs	JOHNNIE CARSON
Assistant Secretary for East Asian and Pacific Affairs	KURT M. CAMPBELL

Assistant Secretary for European and Eurasian Affairs	PHILIP GORDON
Assistant Secretary for International Narcotics and Law Enforcement Affairs	DAVID JOHNSON
Assistant Secretary for International Organization Affairs	ESTHER BRIMMER
Assistant Secretary for Near Eastern Affairs	JEFFREY D. FELTMAN
Assistant Secretary for South and Central Asian Affairs	ROBERT O. BLAKE, JR.
Assistant Secretary for Western Hemisphere Affairs	ARTURO A. VALENZUELA
Under Secretary for Public Diplomacy and Public Affairs	JUDITH MCHALE
Assistant Secretary for Educational and Cultural Affairs	MAURA PALLY, *Acting*
Assistant Secretary for Public Affairs and Spokesman for the Department of State	PHILIP J. CROWLEY
Coordinator, International Information Programs	DAN SREEBNY, *Acting*
Assistant Secretary for Intelligence and Research	PHILIP S. GOLDBERG
Assistant Secretary for Legislative Affairs	RICHARD R. VERMA
Ambassador-at-Large and Coordinator for Counterterrorism	DANIEL BENJAMIN
Ambassador-at-Large of the Office of Global Women's Issues	MELANNE VERVEER
Ambassador-at-Large of the Office of War Crimes Issues	STEPHEN J. RAPP
Chief of Protocol	CAPRICIA PENAVIC MARSHALL
Director of the Office of Civil Rights	JOHN M. ROBINSON
Director, Policy Planning Staff	ANNE-MARIE SLAUGHTER
Director, U.S. Foreign Assistance	(VACANCY)
Coordinator, Office of International Energy Affairs	DAVID L. GOLDWYN
Coordinator, Office of Reconstruction and Stabilization	JOHN E. HERBST
Coordinator, Office of U.S. Global AIDS	ERIC GOOSBY
Inspector General	HAROLD W. GEISEL, *Acting*
Legal Adviser	HAROLD KOH
United States Mission to the United Nations	
United States Permanent Representative to the United Nations and Representative in the Security Council	SUSAN E. RICE
Deputy United States Representative to the United Nations	ALEJANDRO DANIEL WOLFF
United States Representative for Special Political Affairs in the United Nations	ROSEMARY DICARLO
United States Representative on the Economic and Social Council	JOHN SAMMIS
United States Representative for United Nations Management and Reform	(VACANCY)

[For the Department of State statement of organization, see the U.S. Code of Federal Regulations, Title 22, Part 5.]

The Department of State advises the President and leads the Nation in foreign policy issues to advance freedom and democracy for the American people and the international community. To this end, the Department compiles research on American overseas interests, disseminates information on foreign policy to the public, negotiates treaties and agreements with foreign nations, and represents the United States in the United Nations and other international organizations and conferences.

The Department of State was established by act of July 27, 1789, as the Department of Foreign Affairs and was renamed Department of State by act of September 15, 1789 (22 U.S.C. 2651 note).

Secretary of State The Secretary of State is responsible for the overall direction, coordination, and supervision of U.S. foreign relations and for the interdepartmental activities of the U.S. Government abroad. The Secretary is the first-ranking member of the Cabinet, is a member of the National Security Council, and is in charge of the operations of the Department, including the Foreign Service.

Regional Bureaus Foreign affairs activities worldwide are handled by the geographic bureaus, which include the Bureaus of African Affairs, European and Eurasian Affairs, East Asian and Pacific Affairs, Near Eastern Affairs, South and Central Asian Affairs, and Western Hemisphere Affairs.

Administration The Bureau of Administration provides support programs and services to Department of State and U.S. embassies and consulates. These functions include administrative policy, domestic emergency management, and management of owned or leased facilities in the United States; procurement, supply, travel, and transportation support; diplomatic pouch, domestic mail, official records, publishing, library, and language services; support to the schools abroad that educate dependents of U.S. Government employees assigned to diplomatic and consular missions; and small and disadvantaged business utilization. Direct services to the public and other Government agencies include: authenticating documents used abroad for legal and business purposes; responding to requests under the Freedom of Information and Privacy Acts and providing the electronic reading room for public reference to State Department records; and determining use

of the diplomatic reception rooms of the Harry S Truman headquarters building in Washington, DC.

For further information, contact the Bureau of Administration at 703–875–7000.

Consular Affairs The Bureau of Consular Affairs is responsible for the protection and welfare of American citizens and interests abroad; the administration and enforcement of the provisions of the immigration and nationality laws insofar as they concern the Department of State and Foreign Service; and the issuance of passports and visas and related services. Approximately 18 million passports a year are issued by the Bureau's Office of Passport Services at the processing centers in Portsmouth, NH, and Charleston, SC, and the regional agencies in Boston, MA; Chicago, IL; Aurora, CO; Honolulu, HI; Houston, TX; Los Angeles, CA; Miami, FL; New Orleans, LA; New York, NY; Philadelphia, PA; San Francisco, CA; Seattle, WA; Norwalk, CT; Detroit, MI; Minneapolis, MN; and Washington, DC. In addition, the Bureau helps secure America's borders against entry by terrorists or narcotraffickers, facilitates international adoptions, and supports parents whose children have been abducted abroad.

For further information, visit the Bureau of Consular Affairs Web site at www.travel.state.gov.

Democracy, Human Rights, and Labor The Bureau of Democracy, Human Rights, and Labor (DRL) is responsible for developing and implementing U.S. policy on democracy, human rights, labor, and religious freedom. DRL practices diplomatic engagement and advocacy to protect human rights and strengthen democratic institutions. DRL engages with governments, civil society, and in multilateral organizations to support democratic governance and human rights. DRL participates in multi-stakeholder initiatives to encourage multinational corporations

DEPARTMENT OF STATE

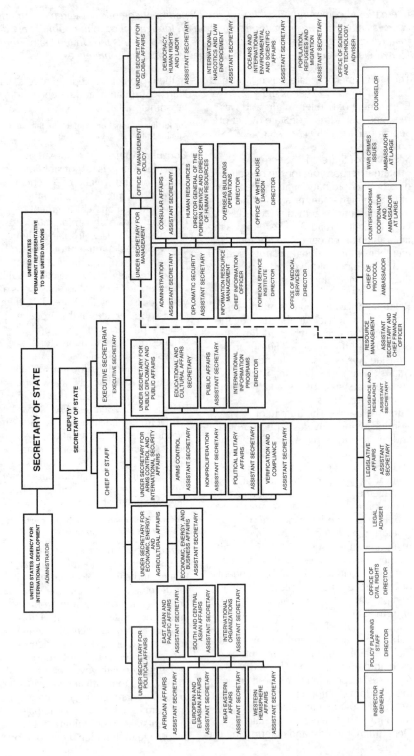

to adhere to human rights standards of conduct, including the elimination of child labor. DRL fulfills the USG reporting responsibilities on human rights and democracy, producing the annual Country Reports on Human Rights Practices, the annual International Religious Freedom report, and the Advancing Freedom and Democracy report. DRL provides targeted program assistance, through the Human Rights and Democracy Fund and other funding streams, to protect human rights and strengthen democratic institutions. DRL programs help prosecute war criminals, promote religious freedom, support workers' rights, encourage accountability in governance, facilitate freedom of expression, and freedom to access information on the Internet. DRL carries out the Congressionally-mandated responsibility to ensure that foreign military assistance and training is not provided to gross violators of human rights. DRL leads the Secretary of State's Task Force on Global Internet Freedom.

For further information, contact the Bureau of Democracy, Human Rights, and Labor at 202–647–2126.

Diplomatic Security   The Bureau of Diplomatic Security provides a secure environment to promote U.S. interests at home and abroad. The Bureau's mission includes protecting the Secretary of State and other senior Government officials, resident and visiting foreign dignitaries, and foreign missions in the United States; conducting criminal, counterintelligence, and personnel security investigations; ensuring the integrity of international travel documents, sensitive information, classified processing equipment, and management information systems; the physical and technical protection of domestic and overseas facilities of the Department of State; providing professional law enforcement and security training to U.S. and foreign personnel; and a comprehensive, multifaceted overseas security program serving the needs of U.S. missions and the resident U.S. citizens and business communities. Through the Office of Foreign Missions, the Bureau regulates the domestic activities of the foreign

diplomatic community in the areas of taxation, real property acquisitions, motor vehicle operation, domestic travel, and customs processing.

For further information, contact the Bureau of Diplomatic Security. Phone, 571–345–2507. Fax, 571–345–2527. Internet, http://www. diplomaticsecurity.state.gov.

Economic, Energy, and Business Affairs   The Bureau of Economic, Energy, and Business Affairs (EEB) promotes international trade, investment, economic development, and financial stability on behalf of the American people. EEB works to build prosperity and economic security at home and abroad by implementing policy related to international development and reconstruction, intellectual property enforcement, international energy issues, terrorism financing and economic sanctions, international communications and information policy, and aviation and maritime affairs. EEB formulates and carries out U.S. foreign economic policy and works to sustain a more democratic, secure, and prosperous world.

For further information, contact the Bureau of Economic, Energy, and Business Affairs. Phone, 202–647–7971. Fax, 202–647–5713. Internet, http://www.state.gov/e/eeb.

Educational and Cultural Affairs   The Bureau of Educational and Cultural Affairs administers the principal provisions of the Mutual Educational and Cultural Exchange Act (the Fulbright-Hays Act), including U.S. international educational and cultural exchange programs. These programs include the prestigious Fulbright Program for students, scholars, and teachers; the International Visitor Leadership Program, which brings leaders and future leaders from other countries to the United States for consultation with their professional colleagues; and professional, youth, sports, and cultural exchanges. Programs are implemented through cooperative relationships with U.S. nongovernmental organizations that support the Bureau's mission.

For further information, contact the Bureau of Educational and Cultural Affairs. Phone, 202–632–6445. Fax, 202–632–2701. Internet, http:// exchanges.state.gov/.

Foreign Missions   The Office of Foreign Missions (OFM) operates the motor vehicles, tax, customs, real property, and travel programs to regulate and serve the 175 foreign missions in the United States and approximately 55,000 foreign mission members and dependents. The Office is also an advocate for improved treatment of U.S. missions and personnel abroad. It guards the U.S. public against abuses of diplomatic privilege and preserves U.S. security interests. OFM maintains regional offices in New York, Chicago, San Francisco, Los Angeles, Miami, Houston, and Honolulu.

For further information, contact the Office of Foreign Missions. Phone, 202–895–3500. Fax, 202–736–4145.

Foreign Service Institute   The Foreign Service Institute of the Department of State is the Federal Government's primary foreign affairs-related training institution. In addition to the Department of State, the Institute provides training for more than 47 other government agencies. The Institute's more than 700 courses, including some 70 foreign language courses, range in length from 1 day to 2 years. The courses are designed to promote successful performance in each professional assignment, to ease the adjustment to other countries and cultures, and to enhance the leadership and management capabilities of the foreign affairs community.

For further information, contact the Foreign Service Institute. Phone, 703–302–6729. Fax, 703–302–7227.

Information Resource Management
The Bureau of Information Resource Management (IRM) provides the Department with the information technology it needs to carry out U.S. diplomacy in the information age. The IRM Bureau is led by the Department's Chief Information Officer. IRM establishes effective information resource management planning and policies; ensures availability of information technology systems and operations, including information technology contingency planning, to support the Department's diplomatic, consular, and management operations; exercises management responsibility for ensuring the Department's information resources meet the business requirements of the Department and provide an effective basis for knowledge sharing and collaboration within the Department and with other foreign affairs agencies and partners; exercises delegated approving authority for the Secretary of State for development and administration of the Department's computer and information security programs and policies.

For further information, contact the Bureau of Information Resource Management. Phone, 202–634–3678. Internet, http://www.state.gov/m/irm/.

Inspector General   The Office of Inspector General (OIG) conducts independent audits, inspections, and investigations to promote effective management, accountability, and positive change in the Department of State, the Broadcasting Board of Governors (BBG), and the foreign affairs community. OIG provides leadership to promote integrity, efficiency, effectiveness, and economy; prevent and detect waste, fraud, abuse, and mismanagement; identify vulnerabilities and recommend constructive solutions; offer expert assistance to improve Department and BBG operations; communicate timely, useful information that facilitates decisionmaking and achieves measurable gains; and keep the Department, BBG, and Congress informed.

For further information, contact the Office of Inspector General. Phone, 202–663–0340. Internet, http://www.oig.state.gov.

Intelligence and Research   The Bureau coordinates the activities of U.S. intelligence agencies to ensure that their overseas activities are consistent with U.S. foreign policy objectives and interests. It also provides all-source analysis which gives the Department insights and information to foreign policy questions. It organizes seminars on topics of high interest to policymakers and the intelligence community and monitors and analyzes foreign public and media opinion on key issues.

For further information, call 202–647–1080.

International Information Programs

The Bureau of International Information Programs (IIP) informs, engages, and influences international audiences about U.S. policy and society to advance America's interests. IIP is a leader in developing and implementing public diplomacy strategies that measurably influence international audiences through quality programs and cutting-edge technologies. IIP provides localized contact for U.S. policies and messages, reaching millions worldwide in English, Arabic, Chinese, French, Persian, Russian, and Spanish. IIP delivers America's message to the world through a number of key products and services. These programs reach, and are created strictly for, key international audiences, such as U.S. diplomatic missions abroad, the media, government officials, opinion leaders, and the general public in more than 140 countries around the world. They include Web and print publications, speaker programs–both traveling (live) and electronic–and information resource services. IIP orchestrates the State Department's efforts to counter anti-American disinformation/propaganda and serves as the Department's chief link with other agencies in coordinating international public diplomacy programs.

For further information, contact the Bureau of International Information Programs. Phone, 202–632–9942. Fax, 202–632–9901. Internet, http://www.state.gov/r/iip/.

International Narcotics and Law Enforcement

The Bureau of International Narcotics and Law Enforcement Affairs (INL) is responsible for developing policies and managing programs to combat and counter international narcotics production and trafficking, and to strengthen law enforcement and other rule of law institutional capabilities outside the United States. The Bureau also directs narcotics control coordinators at posts abroad and provides guidance on narcotics control, justice sector reform, and anticrime matters to the chiefs of missions. It supports the development of strong, sustainable criminal justice systems as well as training for police force and judicial officials. INL works closely with a broad range of other U.S. Government agencies.

For further information, contact the Bureau of International Narcotics and Law Enforcement Affairs. Phone, 202–647–2842. Fax, 202–736–4045.

International Organizations

The Bureau of International Organization Affairs provides guidance and support for U.S. participation in international organizations and conferences and formulates and implements U.S. policy toward international organizations, with particular emphasis on those organizations which make up the United Nations system. It provides direction in the development, coordination, and implementation of U.S. multilateral policy.

For further information, call 202–647–9326. Fax, 202–647–2175.

International Security and Nonproliferation

The Bureau of International Security and Nonproliferation (ISN), is responsible for managing a broad range of nonproliferation, counterproliferation, and arms control functions. ISN leads U.S. efforts to prevent the spread of weapons of mass destruction (nuclear, radiological, chemical, and biological weapons) related materials, and their delivery systems. It is responsible for spearheading efforts to promote international consensus on weapons of mass destruction proliferation through bilateral and multilateral diplomacy; addressing weapons of mass destruction proliferation threats posed by nonstate actors and terrorist groups by improving physical security, using interdiction and sanctions, and actively participating in the Proliferation Security Initiative; coordinating the implementation of key international treaties and arrangements, working to make them relevant to today's security challenges; working closely with the U.N., the G–8, NATO, the Organization for the Prohibition of Chemical Weapons, the International Atomic Energy Agency, and other international institutions and organizations to reduce and eliminate the threat posed by weapons of mass destruction; and supporting efforts of foreign partners to prevent, protect against, and respond to the threat or use of weapons of mass destruction by terrorists.

For further information, contact the Bureau of International Security and Nonproliferation. Phone, 202–647–9868. Fax, 202–736–4863. Internet, http://www.state.gov/t/isn.

Legal Adviser   The Office of the Legal Adviser advises the Secretary of State and other Department officials on all domestic and international legal matters relating to the Department of State, Foreign Service, and diplomatic and consular posts abroad. The Office's lawyers draft, negotiate, and interpret treaties, international agreements, domestic statutes, departmental regulations, Executive orders, and other legal documents; provide guidance on international and domestic law; represent the United States in international organization, negotiation, and treaty commission meetings; work on domestic and foreign litigation affecting the Department's interests; and represent the United States before international tribunals, including the International Court of Justice.

For further information, contact the Office of the Legal Adviser. Phone, 202–647–9598. Fax, 202–647–7096. Internet, http://www.state.gov/s/l/.

Legislative Affairs   The Bureau of Legislative Affairs coordinates legislative activity for the Department of State and advises the Secretary, the Deputy, as well as the Under Secretaries and Assistant Secretaries on legislative strategy. The Bureau facilitates effective communication between State Department officials and the Members of Congress and their staffs. Legislative Affairs works closely with the authorizing, appropriations, and oversight committees of the House and Senate, as well as with individual Members that have an interest in State Department or foreign policy issues. The Bureau also manages Department testimony before House and Senate hearings, organizes Member and staff briefings, facilitates congressional travel to overseas posts for Members and staff throughout the year, and reviews proposed legislation and

coordinates Statements of Administration Policy on legislation affecting the conduct of U.S. foreign policy. Legislative Affairs staff advises individual Bureaus of the Department on legislative and outreach strategies and coordinates those strategies with the Secretary's priorities.

For further information, contact the Bureau of Legislative Affairs. Phone, 202–647–1714.

Medical Services   The Office of Medical Services (MED) develops, manages, and staffs a worldwide primary health care system for U.S. Government employees and their eligible dependents residing overseas. In support of its overseas operations, MED approves and monitors the medical evacuation of patients, conducts pre-employment and in-service physical clearance examinations, and provides clinical referral and advisory services. MED also provides for emergency medical response in the event of a crisis at an overseas post.

For further information, contact the Office of Medical Services. Phone, 202–663–1649. Fax, 202–663–1613.

Oceans and International Environmental and Scientific Affairs   The Bureau of Oceans and International Environmental and Scientific Affairs (OES) serves as the foreign policy focal point for international oceans, environmental, and scientific efforts. OES projects, protects, and promotes U.S. global interests in these areas by articulating U.S. foreign policy, encouraging international cooperation, and negotiating treaties and other instruments of international law. The Bureau serves as the principal adviser to the Secretary of State on international environment, science, and technology matters and takes the lead in coordinating and brokering diverse interests in the interagency process, where the development of international policies or the negotiation and implementation of relevant international agreements are concerned. The Bureau seeks to promote the peaceful exploitation of outer space, develop and coordinate policy on international health issues, encourage government-to-government scientific cooperation, and prevent the destruction and degradation of the planet's natural resources and the global environment.

For further information, contact the Bureau of Oceans and International Environmental and Scientific Affairs. Phone, 202–647–6961. Fax, 202–647–0217.

Overseas Buildings Operations The Bureau of Overseas Buildings Operations (OBO) directs the worldwide overseas buildings program for the Department of State and the U.S. Government community serving abroad under the authority of the chiefs of mission. Along with the input and support of other State Department bureaus, foreign affairs agencies, and Congress, OBO sets worldwide priorities for the design, construction, acquisition, maintenance, use, and sale of real properties and the use of sales proceeds. OBO also serves as the Single Real Property Manager of all overseas facilities under the authority of the chiefs of mission.

For further information, contact the Bureau of Overseas Buildings Operations. Phone, 703–875–4131. Fax, 703–875–5043. Internet, http://www.state.gov/obo.

Political-Military Affairs The Bureau of Political-Military Affairs is the principal link between the Departments of State and Defense and is the Department of State's lead on operational military matters. The Bureau provides policy direction in the areas of international security, security assistance, military operations, defense strategy and policy, counterpiracy measures, and defense trade. Its responsibilities include coordinating the U.S. Government's response to piracy in the waters off the Horn of Africa, securing base access to support the deployment of U.S. military forces overseas, negotiating status of forces agreements, coordinating participation in coalition combat and stabilization forces, regulating arms transfers, directing military assistance to U.S. allies, combating illegal trafficking in small arms and light weapons, facilitating the education and training of international peacekeepers and foreign military personnel, managing humanitarian mine action programs, and assisting other countries in reducing the availability of man-portable air defense systems.

For further information, contact the Bureau of Political-Military Affairs. Phone, 202–647–5104. Fax, 202–736–4413. Internet, http://www.state.gov/t/pm.

Population, Refugees, and Migration The Bureau of Population, Refugees, and Migration directs the Department's population, refugee, and migration policy development. It administers U.S. contributions to international organizations for humanitarian assistance- and protection-related programs on behalf of refugees, conflict victims, and internally displaced persons, and provides U.S. contributions to nongovernmental organizations which provide assistance and protection to refugees abroad. The Bureau oversees the annual admissions of refugees to the United States for permanent resettlement, working closely with the Department of Homeland Security, the Department of Health and Human Services, and various State and private voluntary agencies. It coordinates U.S. international population policy and promotes its goals through bilateral and multilateral cooperation. It works closely with the U.S. Agency for International Development, which administers U.S. international population programs. The Bureau also coordinates the Department's international migration policy through bilateral and multilateral diplomacy. The Bureau oversees efforts to encourage greater participation in humanitarian assistance and refugee resettlement on the part of foreign governments and uses humanitarian diplomacy to increase access and assistance to those in need in the absence of political solutions.

For further information, contact the Bureau of Population, Refugees, and Migration. Phone, 202–453–9271. Fax, 202–453–9294. Internet, http://www.state.gov/g/prm.

Public Affairs The Bureau of Public Affairs (PA) carries out the Secretary's foreign policy objectives and helps American and foreign audiences understand the importance of foreign affairs. Led by the Assistant Secretary, who also serves as Department spokesman, the Bureau pursues the

State Department's mission to inform the American people and foreign audiences and to relay concerns and comments back to policymakers.

For further information, contact the Bureau of Public Affairs. Phone, 202–647–6575.

Protocol   The Chief of Protocol is the principal adviser to the U.S. Government, the President, the Vice President, and the Secretary of State on matters of diplomatic procedure governed by law or international custom and practice. The Office is responsible for arranging visits of foreign chiefs of state, heads of government, and other high officials to the United States; organizing credential presentations of newly arrived Ambassadors, as presented to the President and to the Secretary of State; operating the President's guest house, Blair House; organizing delegations representing the President at official ceremonies abroad; conducting official ceremonial functions and public events; interpreting the official order of precedence; conducting outreach programs of cultural enrichment and substantive briefings of the Diplomatic Corps; accrediting of over 118,000 embassy, consular, international organization, and other foreign government personnel, members of their families, and domestics throughout the United States; determining entitlement to diplomatic or consular immunity; publishing of diplomatic and consular lists; resolving problems arising out of diplomatic or consular immunity, such as legal and police matters; and approving the opening of embassy and consular offices in conjunction with the Office of Foreign Missions.

For further information, contact the Office of the Chief of Protocol. Phone, 202–647–2663. Fax, 202–647–1560.

Resource Management   The Bureau of Resource Management, led by the Chief Financial Officer, integrates strategic planning, budgeting, and performance to secure departmental resources. The Bureau manages all departmental strategic and performance planning; budgeting and resource management for operation accounts; global financial services, including accounting, disbursing, and payroll; issuance of financial statements and oversight of the Department's management control program; coordination of national security resources and remediation of vulnerabilities within the Department's global critical infrastructure; and management of the International Cooperative Administrative Support Services Program.

For further information, contact the Bureau of Resource Management. Phone, 202–647–7490. Internet, http://www.state.gov/s/d/rm/.

Verification, Compliance, and Implementation   The Bureau of Verification, Compliance, and Implementation is responsible for ensuring and verifying compliance with international arms control, nonproliferation, and disarmament agreements and commitments. The Bureau also leads negotiation and implementation efforts with respect to strategic arms control, most recently the new START Treaty, and conventional forces in Europe. The Bureau is the principal policy representative to the intelligence community with regard to verification and compliance matters, and uses this role to promote, preserve, and enhance key collection and analytic capabilities and to ensure that intelligence verification, compliance, and implementation requirements are met. The Bureau staffs and manages treaty implementation commissions, creating negotiation and implementation policy for agreements and commitments, and developing policy for future arms control, nonproliferation, and disarmament arrangements. It also provides support to arms control, nonproliferation, and disarmament policymaking, including information technology support and secure government-to-government communication linkages with foreign treaty partners. The Bureau is also responsible for preparing verifiability assessments on proposals and agreements, and reporting these to Congress as required. The Bureau also prepares the President's Annual Report to Congress on Adherence to and Compliance With Arms Control,

Nonproliferation, and Disarmament Agreements and Commitments, as well as the reports required by the Iran, North Korea, and Syria Nonproliferation Act.

For further information, contact the Bureau of Verification, Compliance, and Implementation. Phone, 202–647–5315. Fax, 202–647–1321.

Foreign Service   To a great extent the future of our country depends on the relations we have with other countries, and those relations are conducted principally by the U.S. Foreign Service. Trained representatives stationed worldwide provide the President and the Secretary of State with much of the raw material from which foreign policy is made and with the recommendations that help shape it.

Ambassadors are the personal representatives of the President and report to the President through the Secretary of State. Ambassadors have full responsibility for implementation of U.S. foreign policy by any and all U.S. Government personnel within their country of assignment, except those under military commands. Their responsibilities include negotiating agreements between the United States and the host country, explaining and disseminating official U.S. policy, and maintaining cordial relations with that country's government and people.

For a complete listing of Foreign Service posts, including addresses, telephone numbers, and key officials, visit www.usembassy.gov.

Sources of Information

Contracts   General inquiries may be directed to the Office of Acquisitions Management (A/LM/AQM), Department of State, P.O. Box 9115, Arlington, VA 22219. Phone, 703–516–1706. Fax, 703–875–6085.

Diplomatic and Official Passports Inquirers for these types of passports should contact their respective travel offices. The U.S. Government only issues these types of passports to individuals traveling abroad in connection with official employment. For additional information, please refer to the Consular Affairs Web site. Internet, http://travel.state.gov..

Electronic Access   The Department's Bureau of Public Affairs, Office of Public Communication, coordinates the dissemination of public electronic information for the Department. The main Web site (www.state.gov) and the Secretary's Web site (www.state.gov/secretary/) provide comprehensive, up-to-date information on foreign policy, support for U.S. businesses and careers, and the counterterrorism rewards program and much more. The Bureau of Consular Affairs Web site (www.travel.state.gov) provides travel warnings and other information designed to help Americans travel safely abroad, as well as information on U.S. passports and visas and downloadable applications. The State Department Electronic Reading Room (www.state.gov/m/a/ips/) uses new information technologies to enable access to unique historical records of international significance which have been made available to the public under the Freedom of Information Act or as a special collection.

Employment   Inquiries about employment in the Foreign Service should be directed to HR/REE, Room H–518, 2401 E Street NW., Washington, DC 20522. Phone, 202–261–8888. Internet, www.careers.state.gov. Information about civil service positions in the Department of State and copies of civil service job announcements can be accessed at www.careers.state.gov. Individual questions may be directed to cspapps@state.gov. Job information staff is also available to answer questions from 8:30 a.m. to 4:30 p.m. eastern time on Federal workdays. Phone, 202–663–2176.

Freedom of Information Act and Privacy Act Requests   Requests from the public for Department of State records should be addressed to the Director, Office of Information Programs and Services, Department of State, SA–2, 515 Twenty-second Street NW., Washington, DC 20522–6001. Phone, 202–261–8300. Individuals are requested to indicate on the outside of the envelope the statute under which they are requesting access:

FOIA REQUEST or PRIVACY REQUEST. A public reading room, where unclassified and declassified documents may be inspected, is located in the Department of State, SA–2, 515 Twenty-second Street NW., Washington, DC 20522–6001. Phone, 202–261–8484. Additional information about the Department's FOIA program can be found on the FOIA electronic reading room (http://www.state.gov/m/a/ips/).

International Adoptions   Inquiries regarding adoption of foreign children by private U.S. citizens should be directed to the Office of Children's Issues, CA/OCS/CI, Department of State, SA–29, 2201 C Street NW., Washington, DC 20520–4818. Phone, 888–407–4747 or 202–501–4444 (international). Internet, http://adoption.state.gov..

Missing Persons, Emergencies, and Deaths of Americans Abroad   For information concerning missing persons, emergencies, travel warnings, overseas voting, judicial assistance, and arrests or deaths of Americans abroad, contact the Office of American Citizens Services and Crisis Management, Department of State. Phone, 888–407–4747 or 202–501–4444 (international). Correspondence should be directed to this address: Overseas Citizens Services, Bureau of Consular Affairs, Department of State, Washington, DC 20520. Inquiries regarding international parental child abduction should be directed to the Office of Children's Issues, CA/OCS/CI, Department of State, SA–29, 2201 C Street NW., Washington, DC 20520–4818. Phone, 888–407–4747 or 202–501–4444 (international). Internet, http://www.travel.state.gov.

Passports   Passport information is available through the Internet at http://travel.state.gov. For information on where to apply for a passport nationwide go to http://iafdb.travel.state.gov. For passport questions, travel emergencies, or to make an appointment at any Regional Passport Agency, call the National Passport Information Center at 887–4-

USA–PPT (887–487–2778) (TDD/TTY: 888–874–7793). Passport information is available 24 hours, 7 days a week; customer service representatives are available Monday-Friday 8 a.m. to 10 p.m., eastern standard time, excluding Federal holidays. Correspondence can be submitted via Internet at http://travel.state.gov/passport/about or can be directed to the appropriate regional agency (Internet, http://travel.state.gov/passport/) or the Correspondence Branch, Passport Services, Room 510, 1111 Nineteenth Street NW., Washington, DC 20524.

Publications   Publications that are produced on a regular basis include Background Notes and the Foreign Relations series. The Bureau of Public Affairs also occasionally publishes brochures and other publications to inform the public of U.S. diplomatic efforts. All publications are available at www.state.gov.

Small Business Information   Information about doing business with the Department of State is available from the Office of Small and Disadvantaged Business Utilization. The publication, A Guide to Doing Business With the Department of State, the current Forecast of Contracting Opportunities, and small business links are available online. Phone, 703–875–6822. Internet, http://www.state.gov/m/a/sdbu/.

Telephone Directory   The Department's telephone directory can be accessed online. Internet, http://www.state.gov/m/a/gps/directory/.

Tips for U.S. Travelers Abroad   Tips for Americans Traveling Abroad contains extensive information about traveling and living in foreign countries is available at http://www.travel.state.gov/travel/living. Additional information for travelers is available at http://www.travel.state.gov/travel/tips/brochures.

Visas   To obtain information on visas for foreigners wishing to enter the United States, call 202–663–1225. Internet, http://www.travel.state.gov/visa.

For further information, contact the Office of Public Communication, Public Information Service, Bureau of Public Affairs, Department of State, Washington, DC 20520. Phone, 202–647–6575. Internet, http://www.state.gov.

DEPARTMENT OF TRANSPORTATION

1200 New Jersey Avenue SE., Washington, DC 20590
Phone, 202–366–4000. Internet, http://www.dot.gov.

Secretary of Transportation	RAY LAHOOD
Deputy Secretary	JOHN PORCARI
Under Secretary for Policy	ROY KIENTIZ
General Counsel	ROBERT RIVKIN
Assistant Secretary for Administration	LINDA J. WASHINGTON
Assistant Secretary for Aviation and International Affairs	SUSAN KURLAND
Assistant Secretary for Budget and Programs and Chief Financial Officer	CHRIS BERTRAM
Assistant Secretary for Governmental Affairs	DANA GRESHAM
Assistant Secretary for Transportation Policy	POLLY TROTTENBERG
Chief of Staff	JOAN DEBOER
Chief Information Officer	NITIN PRADHAN
White House Liaison	NATE TURNBULL
Director, Civil Rights	CAMILLE HAZEUR
Director, Drug and Alcohol Policy and Compliance	JIM L. SWART
Director, Executive Secretariat	CAROL DARR
Director, Public Affairs	JILL ZUCKMAN
Director, Small and Disadvantaged Business Utilization	BRANDON NEAL
Director, Intelligence, Security, and Emergency Response	MICHAEL LOWDER
Inspector General	CAL SCOVEL

[For the Department of Transportation statement of organization, see the Code of Federal Regulations, Title 49, Part 1, Subpart A]

The Department of Transportation establishes national transportation policy for highway planning, and construction, motor carrier safety, urban mass transit, railroads, aviation, and the safety of waterways, ports, highways, and pipelines.

The Department of Transportation (DOT) was established by act of October 15, 1966, as amended (49 U.S.C. 102 and 102 note), "to assure the coordinated, effective administration of the transportation programs of the Federal Government" and to develop "national transportation policies and programs conducive to the provision of fast, safe, efficient, and convenient transportation at the lowest cost consistent therewith." It became operational in April 1967 and was comprised of elements transferred from eight other major departments and agencies.

Secretary The Department of Transportation is administered by the Secretary of Transportation, who is the principal adviser to the President in all matters relating to Federal transportation programs.

Under Secretary The Under Secretary for Policy serves as a principal policy adviser to the Secretary and provides leadership in policy development for the Department.

DEPARTMENT OF TRANSPORTATION

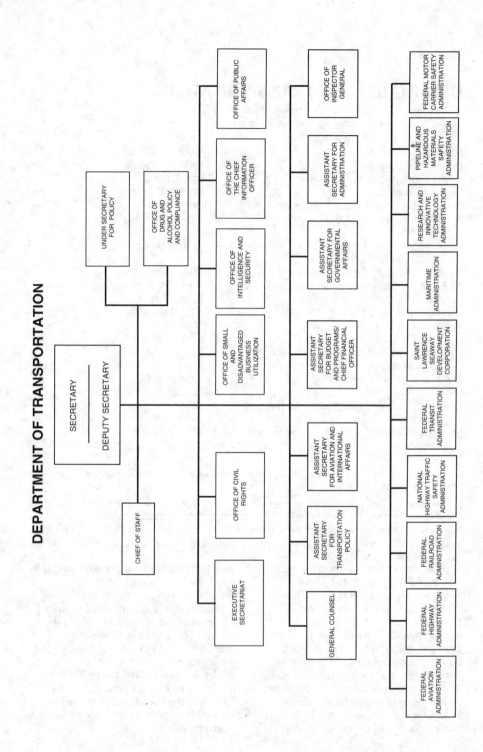

Aviation and International Affairs

The Office of the Assistant Secretary for Aviation and International Affairs has principal responsibility for the development, review, and coordination of policy for international transportation, and for development, coordination, and implementation of policy relating to economic regulation of the airline industry. The Office licenses U.S. and foreign carriers to serve in international air transportation and conducts carrier fitness determinations for carriers serving the United States. The Office also participates in negotiations with foreign governments to develop multilateral and bilateral aviation and maritime policies on a wide range of international transportation and trade matters and to coordinate cooperative agreements for the exchange of scientific and technical information between nations. In addition to these responsibilities, the Office resolves complaints concerning unfair competitive practices in domestic and international air transportation, establishes international and intra-Alaska mail rates, determines the disposition of requests for approval and immunization from the antitrust laws of international aviation agreements, and administers the essential air service program.

For further information, call 202–366–8822.

Drug and Alcohol Policy and Compliance

The Office ensures that the national and international drug and alcohol policies and goals of the Secretary are developed and carried out in a consistent, efficient, and effective manner within the transportation industries. The Office provides expert advice, counsel, and recommendations to the Secretary regarding drugs and alcohol as they pertain to the Department of Transportation and testing within the transportation industry.

For further information, contact the Office of Drug and Alcohol Policy and Compliance. Phone, 202–366–3784.

Intelligence, Security, and Emergency Response

The Office ensures development, coordination, and execution of plans and procedures for the Department of Transportation to balance transportation security requirements with safety, mobility, and economic needs of the Nation through effective intelligence, security, preparedness, and emergency response programs. The Office monitors the Nation's transportation network on a continuous basis; advises the Secretary on incidents affecting transportation systems; provides leadership on national preparedness, response, and transportation security matters; briefs the Secretary on intelligence relevant to the transportation sector; performs DOT's National Response Framework Emergency Support Function responsibilities; coordinates DOT participation in emergency preparedness and response exercises under the National Training and Exercise Program; administers DOT's Continuity of Government and Continuity of Operations programs; and serves as the DOT representative for emergency planning for civil aviation support to NATO and other allies.

For further information, contact the Office of Intelligence, Security, and Emergency Response. Phone, 202–366–6525.

Transportation Policy

The Office of the Assistant Secretary for Transportation Policy has principal responsibility for analysis, development, articulation, and review of policies and plans for all modes of transportation. The Office develops, coordinates, and evaluates public policy on safety, energy, and environmental initiatives which affect air, surface, marine, and pipeline transportation and maintains policy and economic oversight of regulatory programs and legislative initiatives of the Department. The Office also analyzes the economic and institutional implications of current and emerging transportation policy issues, transportation infrastructure finances, and new transportation technologies.

For further information, contact the Office of Transportation Policy. Phone, 202–366–4416.

Sources of Information

Inquiries for information on the following subjects should be directed to the specified office, Department of Transportation, Washington, DC 20590, or to the address indicated.

Civil Rights For information on equal employment opportunity, nondiscrimination in DOT employment and transportation services, or DOT's Disadvantaged Business Enterprise certification appeals program, contact the Director, Departmental Office of Civil Rights. Phone, 202–366–4648. Internet, http://www.dot.gov/ost/docr.

Consumer Activities For information about air travelers' rights or for assistance in resolving consumer problems with providers of commercial air transportation services, contact the Consumer Affairs Division (phone, 202–366–2220). To report vehicle safety problems, obtain information on motor vehicle and highway safety, or to request consumer information publications, call the National Highway Traffic Safety Administration's 24-hour auto safety hotline. Phone, 202–366–0123 or 800–424–9393.

Contracts Contact the Office of the Senior Procurement Executive. Phone, 202–366–4263.

Employment The principal occupations in the Department are air traffic controller, aviation safety specialist, electronics maintenance technician, engineer (civil, aeronautical, automotive, electrical, highway, and general), administrative/management, and clerical. For further information, contact the Transportation Administrative Service Center DOT Connection, Room PL–402, 1200 New Jersey Avenue SE., Washington, DC 20590. Phone, 202–366–9391 or 800–525–2878.

Environment Inquiries on environmental activities and programs should be directed to the Assistant Secretary for Transportation Policy, Office of Transportation Policy Development, Washington, DC 20590. Phone, 202–366–4416.

Films Many films on transportation subjects are available for use by educational institutions, community groups, private organizations, etc. Requests for specific films relating to a particular mode of transportation may be directed to the appropriate operating administration.

Fraud, Waste, and Abuse To report, contact the Office of Inspector General hotline, P.O. Box 23178, Washington, DC 20026–0178. Phone, 202–366–1461 or 800–424–9071.

Publications The Department and its operating agencies issue publications on a wide variety of subjects. Many of these publications are available from the issuing agency or for sale from the Government Printing Office and the National Technical Information Service, 5285 Port Royal Road, Springfield, VA 22151. Contact the Department or the specific agency at the addresses indicated in the text.

Reading Rooms Contact the Department of Transportation Dockets, PL–401, 400 Seventh Street SW., Washington, DC 20590. Phone, 800–647–5527. Administrations and their regional offices maintain reading rooms for public use. Contact the specific administration at the address indicated in the text. Other reading rooms include: Department of Transportation Library, Room 2200, 1200 New Jersey Avenue SE., Washington, DC 20590 (phone, 202–366–0745); Department of Transportation/TASC Law Library, Room 2215, 1200 New Jersey Avenue SE., Washington, DC 20590 (phone, 202–366–0749); Department of Transportation Library, FB–10A Branch, Room 930, 800 Independence Avenue SW., Washington, DC 20591 (phone, 202–267–3115); and Department of Transportation Library, Transpoint Branch, B–726, 2100 Second Street SW., Washington, DC 20593 (phone, 202–267–2536).

Speakers The Department of Transportation and its operating administrations and regional offices make speakers available for civic, labor, and community groups. Contact the specific agency or the nearest regional office at the address indicated in the text.

Surface Transportation Board Proceedings and Public Records
Requests for public assistance with pending or potential proceedings of the Board should be addressed to the Office of Public Assistance, Governmental Affairs, and Compliance, Surface Transportation Board, 395 E Street SW., Washington, DC 20423–0001. Phone, 202–245–0238. Requests for access to the Board's public records should be made to the Office of the Secretary, Surface Transportation Board, 395 E Street SW., Washington, DC 20423–0001. Phone, 202–245–0232.

Telephone Directory The Department of Transportation telephone directory is available for sale by the Superintendent of Documents, Government Printing Office, Washington, DC 20402.

For further information concerning the Department of Transportation, contact the Office of Public Affairs, Department of Transportation, 1200 New Jersey Avenue SE., Washington, DC 20590. Phone, 202–366–5580. Internet, http://www.dot.gov.

Federal Aviation Administration

800 Independence Avenue SW., Washington, DC 20591
Phone, 202–366–4000; 866–835–5322. Internet, http://www.faa.gov.

Administrator	RANDOLPH BABBITT
Deputy Administrator	J. DAVID GRIZZLE, *Acting*
Chief Counsel	J. DAVID GRIZZLE
Chief Operating Officer, Air Traffic Organization	HENRY P. KRAKOWSKI
Assistant Administrator for Civil Rights	FANNY RIVERA
Assistant Administrator for Financial Services/ Chief Financial Officer	RAMESH K. PUNWANI
Assistant Administrator for Government and Industry	RODERICK D. HALL
Assistant Administrator for Human Resource Management	VENTRIS C. GIBSON
Assistant Administrator for Information/Chief Information Officer	DAVID M. BOWEN
Assistant Administrator for International Aviation	DI REIMOLD, *Acting*
Assistant Administrator for Policy, Planning, and Environment	NANCY LOBUE, *Acting*
Assistant Administrator for Public Affairs	SASHA J. JOHNSON
Assistant Administrator for Region and Center Operations	PAULA LEWIS
Assistant Administrator for Security and Hazardous Materials	CLAUDIA MANNO
Associate Administrator for Airports	CATHERINE M. LANG, *Acting*
Associate Administrator for Aviation Safety	MARGARET GILLIGAN
Associate Administrator for Commercial Space Transportation	GEORGE NIELD

The Federal Aviation Administration (FAA), formerly the Federal Aviation Agency, was established by the Federal Aviation Act of 1958 (72 Stat. 731). The agency became a component of the Department of Transportation in 1967 pursuant to the Department of Transportation Act (49 U.S.C. 106). The mission of the FAA is to regulate civil aviation and U.S. commercial space transportation, maintain and operate air traffic control and navigation systems for both civil and military aircraft, and develop and administer programs relating to aviation safety and the National Airspace System.

Activities

Air Navigation Facilities The agency is responsible for the location, construction or installation, maintenance, operation, and quality assurance of Federal visual and electronic aids to air navigation. The agency operates and maintains voice/data communications equipment, radar facilities, computer systems, and visual display equipment at flight service stations, airport traffic control towers, and air route traffic control centers.

Airport Programs The agency maintains a national plan of airport requirements, administers a grant program for development of public use airports to assure and improve safety and to meet current and future airport capacity needs, evaluates the environmental impacts of airport development, and administers an airport noise compatibility program with the goal of reducing noncompatible uses around airports. It also develops standards and technical guidance on airport planning, design, safety, and operations and provides grants to assist public agencies in airport system and master planning and airport development and improvement.

Airspace and Air Traffic Management The safe and efficient utilization of the navigable airspace is a primary objective of the agency. To meet this objective, it operates a network of airport traffic control towers, air route traffic control centers, and flight service stations. It develops air traffic rules and regulations and allocates the use of the airspace. It also provides for the security control of air traffic to meet national defense requirements.

Civil Aviation Abroad Under the Federal Aviation Act of 1958 and the International Aviation Facilities Act (49 U.S.C. app. 1151), the agency encourages aviation safety and civil aviation abroad by exchanging aeronautical information with foreign aviation authorities; certifying foreign repair stations, airmen, and mechanics; negotiating bilateral airworthiness agreements to facilitate the import and export of aircraft and components; and providing technical assistance and training in all areas of the agency's expertise. It provides technical representation at international conferences, including participation in the International Civil Aviation Organization and other international organizations.

Commercial Space Transportation The agency regulates and encourages the U.S. commercial space transportation industry. It licenses the private sector launching of space payloads on expendable launch vehicles and commercial space launch facilities. It also sets insurance requirements for the protection of persons and property and ensures that space transportation activities comply with U.S. domestic and foreign policy.

Registration The agency provides a system for registering aircraft and recording documents affecting title or interest in the aircraft, aircraft engines, propellers, appliances, and spare parts.

Research, Engineering, and Development The research, engineering, and development activities of the agency are directed toward providing the systems, procedures, facilities, and devices needed for a safe and efficient system of air navigation and air traffic control to meet the needs of civil aviation and the air defense system. The agency also performs an aeromedical research function to apply knowledge gained from its research program and the work of others to the safety and promotion of civil aviation and the health, safety, and efficiency of agency employees. The agency also supports development and testing of improved aircraft, engines, propellers, and appliances.

Safety Regulation The Administration issues and enforces rules, regulations, and minimum standards relating to the manufacture, operation, and maintenance of aircraft, as well as the rating and certification (including medical) of airmen and the certification of airports serving air carriers. It performs flight inspection of air navigation facilities in the United States and, as required, abroad.

Test and Evaluation The agency conducts tests and evaluations of specified items such as aviation systems, subsystems, equipment, devices, materials, concepts, or procedures at any phase in the cycle of their development from conception to acceptance and implementation, as well as assigned independent testing at key decision points.

Other Programs   The agency administers the aviation insurance program under the defense materials system with respect to priorities and allocation for civil aircraft and civil aviation operations. The agency develops specifications for the preparation of aeronautical charts. It publishes current information on airways and airport service and issues technical publications for the improvement of safety in flight, airport planning and design, and other aeronautical activities. It serves as the executive administration for the operation and maintenance of the Department of Transportation automated payroll and personnel systems.

For a complete list of Federal Aviation Administration field offices, go to www.faa.gov.

For further information, contact the Office of Communications, Federal Aviation Administration, Department of Transportation, 800 Independence Avenue SW., Washington, DC 20591. Phone, 202–267–3883. Fax, 202–267–5039.

Federal Highway Administration

1200 New Jersey Avenue SE., Washington, DC 20590
Phone, 202–366–0650. Internet, http://www.fhwa.dot.gov.

Administrator	VICTOR M. MENDEZ
Deputy Administrator	GREGORY NADEAU
Executive Director	JEFFREY F. PANIATI
Chief Counsel	KAREN J. HEDLUND
Chief Financial Officer	ELISSA K. KONOVE
Associate Administrator for Administration	PATRICIA A. PROSPERI
Associate Administrator for Civil Rights	ALAN MASUDA
Associate Administrator for Federal Lands Highway	JOHN R. BAXTER
Associate Administrator for Infrastructure	KING W. GEE
Associate Administrator for Operations	JEFFREY A. LINDLEY
Associate Administrator for Planning, Environment, and Realty	GLORIA M. SHEPHERD
Associate Administrator for Policy and Governmental Affairs	SHAILEN P. BHATT
Associate Administrator for Public Affairs	CATHY ST. DENIS
Associate Administrator for Research, Development, and Technology	MICHAEL F. TRENTACOSTE
Associate Administrator for Safety	JOSEPH S. TOOLE

The Federal Highway Administration (FHWA) was established as an agency of the Department of Transportation by the Department of Transportation Act (49 U.S.C. 104). Title 23 of the United States Code and other supporting legislation authorize the Administration's various activities.

FHWA's mission is to improve mobility on our Nation's highways through national leadership, innovation, and program delivery. The Administration works with Federal, State, and local agencies as well as other stakeholders and partners to preserve and improve the National Highway System, which includes the Interstate System and other roads of importance for national defense and mobility. The FHWA works to improve highway safety and minimize traffic congestion on these and other key facilities. The FHWA bears the responsibility of ensuring that America's roads and highways remain safe, technologically up-to-date, and environmentally-friendly.

Through surface transportation programs, innovative and traditional financing mechanisms, and new types of pavement and operational technology, FHWA increases the efficiency by which people and goods move throughout the Nation. The Administration also works to improve the efficiency of highway and road connections to other modes of transportation. The Federal-aid Highway Program's budget is primarily divided between Federal-aid funding and the Federal Lands Highway Program.

Programs

Federal-aid Highway Program   FHWA manages the Federal-aid Highway Program, which provides financial and technical assistance to States for constructing and improving the Nation's transportation infrastructure. The program includes the provision of engineering standards and policies, technical expertise, and other assistance related to the maintenance of highways, rural and urban roads, bridges, tunnels, hydraulic/geotechnical structures, and other engineering activities. Projects associated with the Federal-aid highway program include the National Highway System, Surface Transportation Program, Highway Bridge Program, Congestion Mitigation and Air Quality Improvement Program,

Intelligent Transportation Systems Program, Transportation Infrastructure Finance and Innovation Act Program, the Emergency Relief Program, and the Federal Lands Highway Program.

Federal Lands Highway Program   The Federal Lands Highway Program (FLHP) funds and gives technical assistance to a coordinated program of public roads servicing the transportation needs of Federal and Indian lands. The Program provides funding for public roads and highways on Federal and tribal lands that are not a State or local government responsibility. The planning, construction, and improvement of highways and bridges in national forests and parks, other federally owned land, and tribal lands benefit from FLHP funding.

Field and Division Offices

The FHWA consists of a Headquarters office in Washington, DC; a Federal-aid division office in each State, the District of Columbia, and Puerto Rico; four metropolitan offices in New York City, Philadelphia, Chicago, and Los Angeles serving as extensions of the corresponding Federal-aid division offices; and three Federal Lands Highway division offices.

For a complete list of Federal Highway Administration field and division offices, go to www.fhwa.dot.gov/field.html.

For further information, contact the Department of Transportation, Federal Highway Administration, Office of Information and Management Services, 1200 New Jersey Avenue, SE., Washington, DC 20590. Phone, 202–366–0534. Internet, http://www.fhwa.dot.gov.

Federal Railroad Administration

1120 Vermont Avenue NW., Washington, DC 20590
Phone, 202–493–6000. Internet, http://www.fra.dot.gov.

Administrator	JOSEPH C. SZABO
Deputy Administrator	KAREN RAE
Chief Counsel	KAREN J. HEDLUND
Associate Administrator for Financial Management and Administration/Chief Financial Officer	MARGARET B. REID
Associate Administrator for Railroad Policy and Development	MARK YACHMETZ
Associate Administrator for Railroad Safety/ Chief Safety Officer	JO STRANG

Director, Public Affairs	WARREN FLATAU, *Acting*
Director, Public Engagement	TIMOTHY BARKLEY
Director, Office of Civil Rights	CALVIN GIBSON

The Federal Railroad Administration was created pursuant to section 3(e)(1) of the Department of Transportation Act of 1966 (49 U.S.C. 103). The purpose of the Administration is to promulgate and enforce rail safety regulations, administer railroad financial assistance programs, conduct research and development in support of improved railroad safety and national rail transportation policy, provide for the rehabilitation of Northeast Corridor rail passenger service, and consolidate government support of rail transportation activities.

Activities

Passenger and Freight Services The Administration oversees and provides financial assistance to Amtrak and administers financial assistance programs to demonstrate high-speed rail technology, to reduce grade crossing hazards in high-speed rail corridors, to provide for investments in small freight railroads and other rail projects, to plan for high-speed rail projects, and to plan and deploy magnetic levitation technology.

Railroad Safety The Administration administers and enforces the Federal laws and related regulations designed to promote safety on railroads; exercises jurisdiction over all areas of rail safety under the Rail Safety Act of 1970, such as track maintenance, inspection standards, equipment standards, and operating practices. Railroad and related industry equipment, facilities, and records are inspected and required reports reviewed. In addition, the Administration educates the public about safety at highway-rail grade crossings and the danger of trespassing on rail property.

Research and Development The Administration's ground transportation research and development program seeks to advance all aspects of intercity ground transportation and railroad safety pertaining to the physical sciences and engineering, in order to improve railroad safety and ensure that railroads continue to be a viable national transportation resource.

Transportation Test Center The Administration tests and evaluates conventional and advanced railroad systems and components at the Transportation Test Center near Pueblo, CO. Private sector companies and the Governments of the United States, Canada, and Japan use the facility to explore the operation of conventional and advanced systems under controlled conditions. It is used by Amtrak for the testing of new high-speed locomotives and trains and by the Federal Transit Administration for testing urban rapid transit vehicles.

For further information, contact the Transportation Technology Center, Pueblo, CO 81001. Phone, 719–584–0507.

For a complete list of Federal Railroad Administration regional offices, go to http://www.fra.dot.gov.

For further information, contact the Office of Congressional and Public Affairs, Federal Railroad Administration, Department of Transportation, 1200 New Jersey Avenue SE., Washington, DC 20590. Phone, 202–493–6024. Internet, http://www.fra.dot.gov.

National Highway Traffic Safety Administration

1200 New Jersey Avenue SE., Washington, DC 20590
Phone, 202–366–9550; 888–327–4236 (toll free). Internet, http://www.nhtsa.gov.

Administrator	DAVID L. STRICKLAND
Deputy Administrator	RONALD L. MEDFORD
Director of Communications	JULIA PISCITELLI

Chief Counsel	O. KEVIN VINCENT
Director, Office of Civil Rights	(VACANCY)
Director, Office of Intergovernmental Affairs	(VACANCY)
Senior Associate Administrator for Policy and Operations	GREGORY WALTER
Supervisor, Executive Secretariat	BERNADETTE W. MILLINGS
Director, Office of Human Resources	DARLENE PEOPLES
Associate Administrator for Planning, Administrative, and Financial Management	REBECCA PENNINGTON
Associate Administrator, Communications and Consumer Information	SUSAN GORCOWSKI
Chief Information Officer	COLLEEN COGGINS
Senior Associate Administrator for Traffic Injury Control	BRIAN MCLAUGHLIN
Associate Administrator for Research and Program Development	JEFFREY MICHAEL
Associate Administrator for Regional Operations and Program Delivery	MARLENE K. MARKISON
Senior Associate Administrator for Vehicle Safety	(VACANCY)
Director, Strategic Planning and Domestic and Global Integration for Vehicle Safety	JOSEPH CARRA
Associate Administrator for Rulemaking	STEPHEN R. KRATZKE
Associate Administrator for Enforcement	DANIEL C. SMITH
Associate Administrator for Applied Vehicle Safety Research	JOHN MADDOX
Associate Administrator, National Center for Statistics and Analysis	MARILENA AMONI

[For the National Highway Traffic Safety Administration statement of organization, see the Code of Federal Regulations, Title 49, Part 501]

The National Highway Traffic Safety Administration (NHTSA) was established by the Highway Safety Act of 1970 (23 U.S.C. 401 note) to help reduce the number of deaths, injuries, and economic losses resulting from motor vehicle crashes on the Nation's highways.

The Administration carries out programs relating to the safety performance of motor vehicles and related equipment; administers the State and community highway safety program with the FHWA; regulates the Corporate Average Fuel Economy program; investigates and prosecutes odometer fraud; carries out the National Driver Register Program to facilitate the exchange of State records on problem drivers; conducts studies and operates programs aimed at reducing economic losses in motor vehicle crashes and repairs; performs studies, conducts demonstration projects, and promotes programs to reduce impaired driving, increase seat belt use, and reduce risky driver behaviors; and issues theft prevention standards for passenger and nonpassenger motor vehicles.

Activities

Research and Development The Administration provides a foundation for the development of motor vehicle and highway safety program standards by analyzing data and researching, developing, testing, and evaluating motor vehicles, motor vehicle equipment, and advanced technologies, and collecting and analyzing crash data. The research program covers numerous areas affecting safety and includes laboratory-testing facilities to obtain necessary basic data. NHTSA strives to encourage industry to adopt advanced motor vehicle safety designs, elevate public awareness of safety potentials, and provide a base for vehicle safety information.

Regional Operations and Program Delivery   The Administration administers State highway safety grant programs, authorized by the Safe, Accountable, Flexible, Efficient Transportation Equity Act: A Legacy for Users. The Highway Safety formula grant program provides funds to the States, Indian nations, and the territories each year to support safety programs, particularly in the following national priority areas: occupant protection, impaired driving, police traffic services, emergency medical services, data/traffic records, motorcycle safety, pedestrian and bicycle safety, speed control, and roadway safety. Incentive grants are also used to encourage States to implement effective impaired driving, occupant protection, motorcycle safety, and data improvement programs.

Rulemaking   The Administration issues Federal motor vehicle safety standards that prescribe safety features and levels of safety-related performance for vehicles and vehicular equipment. It conducts the New Car Assessment Program, under which passenger cars, light trucks, and vans are subjected to high-speed crashes in order to test their frontal and side impact safety performance. Separate tests are conducted to assess the vehicles' resistance to rollovers. Results from these tests are made public in order to help consumers choose the safest motor vehicles. The Administration educates consumers on using vehicle safety features. To promote maximum feasible fuel economy, it manages a program establishing and revising fleet average fuel economy standards for passenger car and light truck manufacturers. The Administration also carries out an anti-theft program, which includes issuing rules requiring the designation of likely high-theft vehicles that must meet parts-marking requirements and calculating and publishing annual motor vehicle theft rates.

Enforcement   The Office of Enforcement identifies and investigates problems with motor vehicles and vehicular equipment. If the Office determines that a vehicle or equipment suffers from a safety-related defect or that it does not meet all applicable Federal motor vehicle safety standards, the Office will seek a recall, which requires manufacturers to notify owners and remedy the defect free of charge. The Office monitors recalls to ensure that owners are being notified in a timely manner and the scope of the recall and the remedy are adequate to correct the problem.

National Center for Statistics and Analysis   The Administration maintains a collection of scientific and technical information related to motor vehicle safety and operates the National Center for Statistics and Analysis, whose activities include the development and maintenance of national highway-crash data collection systems and related analysis efforts. These comprehensive motor vehicle safety information resources serve as documentary reference points for Federal, State, and local agencies, as well as industry, universities, and the public.

Communications and Consumer Information   The Administration develops, directs, and implements communications based on NHTSA policy and programs, including public awareness campaigns such as "Click It or Ticket" and "Over the Limit. Under Arrest." It also manages the toll-free Motor Vehicle Auto Safety Hotline to identify safety problems in motor vehicles and equipment. Consumers can call the hotline (phone, 888–327–4236; TDD, 800–424–9153 or 202–366–7800 in the Washington, DC area) 24 hours a day, 7 days a week, to report safety-related problems. English and Spanish speaking representatives are available between 8 a.m. and 10 p.m. eastern standard time, Monday through Friday, except Federal holidays. Consumers can also reach the hotline via the Internet at www.nhtsa.dot.gov/hotline. These calls form the basis for investigations and, ultimately, recalls if safety-related defects are identified. The hotline also provides information and literature to consumers about vehicle and child-seat recalls, New Car Assessment Program test results, and a variety of other highway safety information.

For a complete list of National Highway Traffic Safety Administration Regional Offices, go to www.nhtsa.dot.gov/nhtsa/whatis/regions.

For further information, contact the Office of Communications and Consumer Information, National Highway Traffic Safety Administration, Department of Transportation, 1200 New Jersey Avenue SE., Washington, DC 20590. Phone, 202–366–9550. Internet, http://www.nhtsa.dot.gov..

Federal Transit Administration

1200 New Jersey Avenue SE., Washington, DC 20590
Phone, 202–366–4043. Internet, http://www.fta.dot.gov.

Administrator	PETER ROGOFF
Deputy Administrator	THERESE M. MCMILLAN
Executive Director	MATTHEW J. WELBES
Associate Administrator for Administration	ANN LINNERTZ
Associate Administrator for Budget and Policy	ROBERT J. TUCCILLO
Associate Administrator for Communications and Congressional Affairs	BRIAN D. FARBER
Associate Administrator for Planning	SUSAN BORINSKY
Associate Administrator for Program Management	(VACANCY)
Associate Administrator for Research, Demonstration, and Innovation	VINCENT VALDES
Chief Counsel	DORVAL CARTER
Director, Office of Civil Rights	CHERYL HERSHEY

[For the Federal Transit Administration statement of organization, see the Code of Federal Regulations, Title 49, Part 601]

The Federal Transit Administration (FTA) (formerly the Urban Mass Transportation Administration) was established as an operating administration of the Department of Transportation by section 1 of Reorganization Plan No. 2 of 1968 (5 U.S.C. app. 1), effective July 1, 1968. FTA's mission is to assist in developing improved mass transportation, encourage the planning and establishment of areawide mass transportation systems, and provide financial assistance to State and local governments to finance mass transportation systems and carry out national transit goals and policy.

Programs

Capital Investment Grants are authorized to assist in financing the acquisition, construction, reconstruction, and improvement of facilities and equipment for use in mass transportation service in urban areas. There are three categories of funds available under the capital investment program: fixed guideway modernization, rolling stock renewal, safety-related improvements, and signal and power modernization;

new starts funds for construction of new fixed guideway service; and bus funds for acquiring buses and rolling stock, ancillary equipment, and the construction of bus facilities.

For further information, call 202–366–2053.

Elderly and Persons With Disabilities

The program provides financial assistance to private nonprofit agencies to meet the transportation needs of elderly persons and persons with disabilities where services provided by public operators are unavailable, insufficient, or inappropriate; to public bodies approved by the State to coordinate services for elderly persons or persons with disabilities; or to public bodies which certify to the Governor that no nonprofit corporation or association is readily available in an area to provide the service. Funds are allocated by formula to the States. Local organizations apply for funding through a designated State agency.

For further information, call 202–366–2053.

Job Access and Reverse Commute

Grants The program makes funding available to public agencies and nonprofit organizations to pay the capital and operating costs of delivering new or expanded job access or reverse commute services, and to promote the use of transit during nontraditional work hours, as well as encourage employer-based transportation strategies and use of transit pass programs. The program provides competitive grants for job access projects implementing new or expanded transportation services for transporting welfare recipients and low-income persons to and from jobs and needed employment support services such as child care and reverse commute projects implementing new or expanded general purpose public transportation services to transport residents of urban, rural, and suburban areas to suburban employment centers.

For further information, call 202–366–0176. Internet, www.fta.dot.gov/funding/grants.

Non-Urbanized Area Assistance The Administration provides capital and operating assistance for public transportation in non-urbanized areas. Funds are allocated to the Governor, and the program is administered at the State level by the designated transportation agency. Assistance is provided for planning, administrative, and program development activities; coordination of public transportation programs; vehicle acquisition; and other capital investments in support of transit services tailored to the needs of elderly individuals and individuals with disabilities and other individuals who depend upon transit for their basic mobility.

Planning The program provides financial assistance in meeting the transportation planning needs of metropolitan planning organizations by allocating funds to States which, in turn, they allocate to the metropolitan planning organizations. Assistance is available for transportation planning, technical assistance studies, demonstrations, management training, and cooperative research.

For further information, call 202–366–6385.

Research and Technology The Administration seeks to improve public transportation for America's communities

by delivering products and services that are valued by its customers and by assisting transit agencies in better meeting the needs of their customers. To accomplish these goals, it partners with the transportation industry to undertake research, development, and education that will improve the quality, reliability, and cost-effectiveness of transit in America and that leads to increases in transit ridership.

Transit research and technology efforts include joint partnership agreements with both public and private research organizations, transit providers, and industry to promote innovation in public transportation services, management, and operational practices; advanced technologies that assist the study, design, and demonstration of fixed-guideways, bus and rapid transit, fuel-cell-powered transit buses, and advanced propulsion control for rail transit; and international mass transportation programs that promote American transit products and services overseas.

For further information, call 202–366–4052. Internet, www.fta.dot.gov/research.

Rural Transportation Assistance The Rural Transportation Assistance Program allocates funds annually to the States to provide assistance for transit research, technical assistance, training, and related support activities for transit providers serving non-urbanized areas. Additional funds are used at the national level for developing training materials, developing and maintaining a national clearinghouse on rural transit activities and information, and providing technical assistance through peer practitioners to promote exemplary techniques and practices.

For further information, call 202–366–2053.

Safety The Administration's safety program supports State and local agencies in fulfilling their responsibility for the safety and security of urban mass transportation facilities and services, through the encouragement and sponsorship of safety and security planning, training, information collection and analysis, drug control programs, system/safety assurance reviews, generic research, and other cooperative government/industry activities.

For further information, call 202–366–4020.

Training and Technical Assistance
Through the National Transit Institute (NTI), the Administration develops and offers training courses for improving transit planning, operations, workforce performance, and productivity. NTI courses are conducted at sites across the United States on a wide variety of subjects, ranging from multimodal planning to management development, third-party contracting, safety, and security. Current NTI course offerings are available online at www.ntionline.com.

For further information, call 202–366–6635.

For a complete list of Federal Transit Administration offices, go to www.fta.dot.gov.

For further information, contact the Office of Communications and Congressional Affairs, Federal Transit Administration, Department of Transportation, 1200 New Jersey Avenue SE., Washington, DC 20590. Phone, 202–366–4043. Internet, http://www.fta.dot.gov.

Maritime Administration

1200 New Jersey Avenue SE., Washington, DC 20590
Phone, 202–366–5807; 800–996–2723. Internet, http://www.marad.dot.gov.

Administrator	DAVID T. MATSUDA, *Acting*
Deputy Administrator	DAVID T. MATSUDA
Assistant Administrator	JAMES E. CAPONITI
Associate Administrator for Administration/ Chief Information Officer	PAULA EWEN
Associate Administrator for National Security	KEVIN M. TOKARSKI
Associate Administrator for Intermodal, System Development	H. KEITH LESNICK
Associate Administrator for Environment and Compliance	JOSEPH A. BYRNE
Associate Administrator for Business and Workforce Development	(VACANCY)
Associate Administrator for Budget and Programs/ Chief Financial Officer	DAVID J. RIVAIT
Chief Counsel	DENISE KREPP
Director of Congressional and Public Affairs	CHERON WICKER
Director of Civil Rights	DAVID J. ADAMS
Secretary, Maritime Subsidy Board	CHRISTINE GURLAND
Superintendent, United States Merchant Marine Academy	SASHI N. KUMAR, *Acting*

The Maritime Administration was established by Reorganization Plan No. 21 of 1950 (5 U.S.C. app.). The Maritime Act of 1981 (46 U.S.C. 1601) transferred the Maritime Administration to the Department of Transportation. The Administration manages programs to aid in the development, promotion, and operation of the U.S. merchant marine. It is also charged with organizing and directing emergency merchant ship operations.

The Maritime Administration administers subsidy programs to pay the difference between certain costs of operating ships under the U.S. flag and foreign competitive flags on essential services, and the difference between the costs of constructing ships in U.S. and foreign shipyards. It provides financing guarantees for the construction, reconstruction, and reconditioning of ships; and enters into capital construction fund agreements that grant tax deferrals on moneys to be used for the acquisition, construction, or reconstruction of ships.

The Administration constructs or supervises the construction of merchant-type ships for the Federal Government. It helps industry generate increased business for U.S. ships and conducts programs to develop ports, facilities, and intermodal transport, and to promote domestic shipping.

It conducts program and technical studies and administers a war risk insurance program that insures operators and seamen against losses caused by hostile action if domestic commercial insurance is not available.

Under emergency conditions the Maritime Administration charters Government-owned ships to U.S. operators, requisitions or procures ships owned by U.S. citizens, and allocates them to meet defense needs.

It maintains a national defense reserve fleet of Government-owned ships that it operates through ship managers and general agents when required in national defense interests. An element of this activity is the Ready Reserve force consisting of a number of ships available for quick-response activation.

The Administration regulates sales to aliens and transfers to foreign registry of ships that are fully or partially owned by U.S. citizens. It also disposes of Government-owned ships found nonessential for national defense.

The Administration operates the U.S. Merchant Marine Academy, Kings Point, NY, where young people are trained to become merchant marine officers, and conducts training in shipboard firefighting at Toledo, OH. It also administers a Federal assistance program for the maritime academies operated by the States of CA, ME, MA, MI, NY, and TX.

For a complete list of Maritime Administration offices, go to www.marad. dot.gov/about_us_landing_page/gateway_ offices/Gateway_Presence.htm.

For further information, contact the Office of Congressional and Public Affairs, Maritime Administration, Department of Transportation, 1200 New Jersey Avenue SE., Washington, DC 20590. Phone, 202–366–5807 or 800–996–2723. Internet, http://www.marad.dot.gov.

Saint Lawrence Seaway Development Corporation

1200 New Jersey Avenue SE., Washington, DC 20590
Phone, 202–366–0091; 800–785–2779. Fax, 202–366–7147. Internet, http://www.seaway.dot.gov.
180 Andrews Street, Massena, NY 13662
Phone, 315–764–3200.

Washington, DC

Administrator	COLLISTER JOHNSON, JR.
Deputy Administrator	CRAIG H. MIDDLEBROOK
Chief of Staff	ANITA K. BLACKMAN
Chief Counsel	CARRIE LAVIGNE
Director of Budget and Programs	KEVIN P. O'MALLEY
Director of Congressional Affairs and Public Relations	NANCY ALCALDE
Director of Trade Development	REBECCA A. MCGILL

Massena, NY

Associate Administrator	SALVATORE L. PISANI
Deputy Associate Administrator	CAROL A. FENTON
Chief Financial Officer	MARSHA SIENKIEWICZ
Human Resources Officer	JULIE A. KUENZLER
Director of Engineering	THOMAS A. LAVIGNE
Director of Lock Operations and Marine Services	LORI K. CURRAN
Director of Maintenance	KARL J. LIVINGSTON

The Saint Lawrence Seaway Development Corporation was established by the Saint Lawrence Seaway Act of May 13, 1954 (33 U.S.C. 981–990) and became an operating administration of the Department of Transportation in 1966.

The Corporation, working cooperatively with the Saint Lawrence Seaway Management Corporation (SLSMC) of Canada, is dedicated to operating and maintaining a safe, reliable, and efficient deep draft waterway between the Great Lakes and the Atlantic Ocean. It ensures the safe transit of commercial and noncommercial vessels through the two U.S. locks and the navigation channels of the Saint Lawrence Seaway System. The Corporation works jointly with SLSMC on all matters related to rules and regulations, overall operations, vessel inspections, traffic control, navigation aids, safety, operating dates, and trade development programs.

The Great Lakes/Saint Lawrence Seaway System extends from the Atlantic Ocean to the Lake Superior ports of Duluth/Superior, a distance of 2,342 miles. The Corporation's main customers are vessel owners and operators, Midwest States and Canadian provinces, Great Lakes port communities, shippers and receivers of domestic and international cargo, and the Great Lakes/Saint Lawrence Seaway Systems maritime and related service industries. International and domestic commerce through the Seaway contributes to the economic prosperity of the entire Great Lakes region.

For further information, contact the Director of Trade Development and Public Affairs, Saint Lawrence Seaway Development Corporation, Department of Transportation, 1200 New Jersey Avenue SE., Washington, DC 20590. Phone, 202–366–0091. Fax, 202–366–7147. Internet, http://www.seaway.dot.gov.

Pipeline and Hazardous Materials Safety Administration

1200 New Jersey Avenue SE., Washington, DC 20590
Phone, 202–366–4433. Internet, http://www.phmsa.dot.gov.

Administrator	CYNTHIA QUARTERMAN
Deputy Administrator	(VACANCY)
Assistant Administrator/Chief Safety Officer	CYNTHIA DOUGLAS
Chief Counsel	BIZUNESH SCOTT
Associate Administrator for Management and Administration	(VACANCY)
Associate Administrator for Pipeline Safety	JEFFREY WIESE
Associate Administrator for Hazardous Materials Safety	MAGDY EL-SIBAIE
Director, Office of Civil Rights	HELEN HAGIN
Director, Office of Governmental, International, and Public Affairs	(VACANCY)

The Pipeline and Hazardous Materials Safety Administration was established on February 20, 2005. It is responsible for hazardous materials transportation and pipeline safety.

Hazardous Materials

The Office of Hazardous Materials Safety develops and issues regulations for the safe and secure transportation of hazardous materials by all modes, excluding bulk transportation by water. The regulations cover shipper and carrier operations, packaging and container specifications, and hazardous materials definitions. The Office provides training and outreach to help shippers and carriers meet the requirements of the hazardous material regulations.

The Office is also responsible for the enforcement of regulations other than those applicable to a single mode of transportation. The Office manages a fee-funded grant program to assist States in planning for hazardous materials emergencies and to assist States and Indian tribes with training for hazardous materials emergencies. Additionally, the Office maintains a national safety program to safeguard food and certain other products from contamination during motor or rail transportation.

For further information, call 202–366–0656. Internet, hazmat.dot.gov.

For a complete listing of Office of Hazardous Materials Safety offices, go to www.phmsa.dot.gov/hazmat/about/org.

Pipelines

The Office of Pipeline Safety's (OPS) mission is to ensure the safety, security, and environmental protection of the Nation's pipeline transportation system. The Office establishes and enforces safety and environmental standards for transportation of gas and hazardous liquids by pipeline. OPS also analyzes data, conducts education and training, promotes damage prevention, and conducts research and development for pipeline safety. Through OPS-administered grants-in-aid, States that voluntarily assume regulatory jurisdiction of pipelines can receive funding for up to 50 percent of the costs for their intrastate pipeline safety programs. OPS engineers inspect most interstate pipelines and other facilities not covered by the State programs. The Office also implements the Oil Pollution Act of 1990 by providing approval for and testing of oil pipeline spill response plans.

For further information, call 202–366–4595.

For a complete list of Office of Pipeline Safety's regional offices, go to www. phmsa.dot.gov/public/contact.

For further information, contact the Office of Governmental, International and Public Affairs, Pipeline and Hazardous Materials Safety Administration, Department of Transportation, Suite 8406, 1200 New Jersey Avenue SE., Washington, DC 20590. Phone, 202–366–4831. Internet, http://www.phmsa.dot.gov.

Research and Innovative Technology Administration

1200 New Jersey Avenue SE., Washington, DC 20590
Phone, 202–366–7582. Internet, http://www.rita.dot.gov. Email, info.rita@dot.gov.

Administrator	PETER H. APPEL
Deputy Administrator	ROBERT L. BERTINI
Chief Counsel	GREGORY D. WINFREE
Chief Financial Officer	KATHY MONTGOMERY
Associate Administrator for Administration	MARIA LEFEVRE
Associate Administrator, Intelligent Transportation Systems Joint Program Office	ROBERT L. BERTINI, *Acting*
Associate Administrator for Research, Development and Technology	JAN BRECHT-CLARK
Director, Bureau of Transportation Statistics	STEVEN DILLINGHAM
Director, Office of Civil Rights	LINDA BOWMAN
Director, Office of Government, International, and Public Affairs	JANE MELLOW
Director, Volpe National Transportation Systems Center	ROBERT C. JOHNS
Director, Transportation Safety Institute	CURTIS J. TOMPKINS, *Acting*
Public Affairs Contact	KIM RIDDLE

The Research and Innovative Technology Administration (RITA) was created under the Norman Y. Mineta Research and Special Programs Improvement Act (49 U.S.C. 101 note). RITA coordinates, facilitates, and reviews the Department's research and development programs and activities; performs comprehensive transportation statistics research, analysis, and reporting; and promotes the use of innovative technologies to improve our Nation's transportation system. RITA brings together important DOT data, research, and technology transfer assets and provides strategic direction and oversight of DOT's Intelligent Transportation Systems Program.

RITA is composed of the staff from the Office of Research, Development, and Technology; the Volpe National Transportation Systems Center; the Transportation Safety Institute; and the Bureau of Transportation Statistics.

For further information, contact the Research and Innovative Technology Administration, Department of Transportation, 1200 New Jersey Avenue SE., Washington, DC 20590. Phone, 202–366–4180. Email, info. rita@dot.gov. Internet, http://www.rita.dot.gov.

Federal Motor Carrier Safety Administration

1200 New Jersey Avenue SE., Washington, DC 20590
Phone, 202–366–2519. Internet, http://www.fmcsa.dot.gov.

Administrator	ANNE S. FERRO
Deputy Administrator	WILLIAM A. BRONROTT
Assistant Administrator/Chief Safety Officer	ROSE A. MCMURRAY
Regulatory Ombudsman	STEVEN LAFRENIERE
Director for Administration/Chief Financial Officer	KATHLEEN O'SULLIVAN
Associate Administrator for Enforcement and Program Delivery	WILLIAM QUADE
Associate Administrator for Research and Information Technology/Chief Information Officer	TERRY SHELTON
Associate Administrator of Field Operations	(VACANCY)
Associate Administrator for Policy and Program Development	LARRY MINOR
Chief Counsel	ALAIS GRIFFIN
Director, Office of Civil Rights	KENNIE MAY
Director, Office of Communications	CANDICE TOLLIVER
Associate Director for Governmental Affairs	CURTIS L. JOHNSON

The Federal Motor Carrier Safety Administration was established within the Department of Transportation on January 1, 2000, pursuant to the Motor Carrier Safety Improvement Act of 1999 (49 U.S.C. 113).

Formerly a part of the Federal Highway Administration, the Federal Motor Carrier Safety Administration's primary mission is to prevent commercial motor vehicle-related fatalities and injuries. Activities of the Administration contribute to ensuring safety in motor carrier operations through strong enforcement of safety regulations, targeting high-risk carriers and commercial motor vehicle drivers; improving safety information systems and commercial motor vehicle technologies; strengthening commercial motor vehicle equipment and operating standards; and increasing safety awareness. To accomplish these activities, the Administration works with Federal, State, and local enforcement agencies, the motor carrier industry, labor safety interest groups, and others.

Activities

Commercial Drivers' Licenses The Administration develops standards to test and license commercial motor vehicle drivers.

Data and Analysis The Administration collects and disseminates data on motor carrier safety and directs resources to improve motor carrier safety.

Regulatory Compliance and Enforcement
The Administration operates a program to improve safety performance and remove high-risk carriers from the Nation's highways.

Research and Technology The Administration coordinates research and development to improve the safety of motor carrier operations and commercial motor vehicles and drivers.

Safety Assistance The Administration provides States with financial assistance for roadside inspections and other commercial motor vehicle safety programs. It promotes motor vehicle and motor carrier safety.

Other Activities The Administration supports the development of unified motor carrier safety requirements and procedures throughout North America. It participates in international technical organizations and committees to help share the best practices in motor carrier safety throughout North America and the rest of the world. It enforces regulations ensuring safe highway transportation of hazardous materials and has established a task force to identify and investigate those carriers of household goods that have exhibited a substantial pattern of consumer abuse.

For a complete list of Federal Motor Carrier Safety Administration field offices, go to www.fmcsa.dot.gov/about/aboutus.htm

For further information, contact the Federal Motor Carrier Safety Administration, 1200 New Jersey Avenue SE., Washington, DC 20590. Phone, 202–366–2519. Internet, http://www.fmcsa.dot.gov.

Surface Transportation Board

395 E Street SW., Washington, DC 20423–0001
Phone, 202–245–0245. Internet, http://www.stb.dot.gov.

Chairman	DANIEL R. ELLIOTT, III
Vice Chairman	FRANCIS P. MULVEY
Commissioner	CHARLES D. NOTTINGHAM
Director, Office of Public Assistance, Governmental Affairs and Compliance	MATTHEW T. WALLEN
Director, Office of Economics, Environmental Analysis, and Administration	LELAND L. GARDNER
Director, Office of Proceedings	RACHEL D. CAMPBELL
General Counsel	ELLEN D. HANSON

The Surface Transportation Board was established in 1996 by the Interstate Commerce Commission (ICC) Termination Act of 1995 (49 U.S.C. 10101 et seq.) as an independent adjudicatory body organizationally housed within the Department of Transportation with jurisdiction over certain surface transportation economic regulatory matters formerly under ICC jurisdiction. The Board consists of three members, appointed by the President with the advice and consent of the Senate for 5-year terms.

The Board adjudicates disputes and regulates interstate surface transportation through various laws pertaining to the different modes of surface transportation. The Board's general responsibilities include the oversight of firms engaged in transportation in interstate and foreign commerce to the extent that it takes place within the United States, or between or among points in the contiguous United States and points in Alaska, Hawaii, or U.S. Territories or possessions. Surface transportation matters under the Board's jurisdiction in general include railroad rate and service issues, rail restructuring transactions (mergers, line sales, line construction, and line abandonments), and labor matters related thereto; certain trucking company, moving van, and noncontiguous ocean shipping company rate matters; certain intercity passenger bus company structure, financial, and operational matters; and certain pipeline matters not regulated by the Federal Energy Regulatory Commission.

In performing its functions, the Board is charged with promoting, where appropriate, substantive and procedural regulatory reform and providing an efficient and effective forum for the resolution of disputes. Through the granting of exemptions from regulations where warranted, the streamlining of its decisionmaking process and the regulations applicable thereto, and the consistent and fair application of legal and equitable principles, the Board seeks to provide an effective forum for efficient dispute resolution and facilitation of appropriate market-based business transactions. Through rulemakings and case disposition, it strives to develop new and better ways to analyze unique and complex problems, to reach fully justified decisions more quickly, to reduce the costs associated with regulatory oversight, and to encourage private sector negotiations and resolutions to problems, where appropriate.

For further information, contact the Office of Public Assistance, Governmental Affairs, and Compliance, Surface Transportation Board, 395 E Street SW., Washington, DC, 20423–0001. Phone, 202–245–0230. Internet, http://www.stb.dot.gov.

DEPARTMENT OF THE TREASURY

1500 Pennsylvania Avenue NW., Washington, DC 20220
Phone, 202–622–2000. Internet, http://www.treas.gov.

Secretary of the Treasury	TIMOTHY F. GEITHNER
Deputy Secretary of the Treasury	NEAL S. WOLIN
Treasurer of the United States	ROSIE RIOS
Chief of Staff	MARK PATTERSON
Inspector General	ERIC THORSON
Treasury Inspector General for Tax Administration	J. RUSSELL GEORGE
Under Secretary (Domestic Finance)	(VACANCY)
Assistant Secretary (Financial Institutions)	MICHAEL S. BARR
Assistant Secretary (Financial Markets)	MARY J. MILLER
Assistant Secretary (Financial Stability)	HERBERT M. ALLISON
Fiscal Assistant Secretary	RICK GREGG
Assistant Secretary (Economic Policy)	ALAN B. KRUEGER
General Counsel	GEORGE W. MADISON
Under Secretary (International Affairs)	LAEL BRAINARD
Assistant Secretary (International Finance)	CHARLES COLLYNS
Assistant Secretary (International Markets and Development)	MARISA LAGO
Assistant Secretary (Legislative Affairs)	KIM N. WALLACE
Assistant Secretary for Management/Chief Financial Officer/Chief Performance Officer	DANIEL TANGHERLINI
Assistant Secretary (Public Affairs)	(VACANCY)
Assistant Secretary (Tax Policy)	(VACANCY)
Under Secretary (Terrorism and Financial Intelligence)	STUART LEVEY
Assistant Secretary (Intelligence and Analysis)	HOWARD MENDELSOHN, *Acting*
Assistant Secretary (Terrorist Financing)	DAVID S. COHEN

The Department of the Treasury serves as financial agent for the U.S. Government, manufacturing coins and currency, enforcing financial laws, and recommending economic, tax, and fiscal policies.

The Treasury Department was created by act of September 2, 1789 (31 U.S.C. 301 and 301 note). Many subsequent acts have figured in the development of the Department, delegating new duties to its charge and establishing the numerous bureaus and divisions that now comprise the Treasury.

Secretary As a major policy adviser to the President, the Secretary has primary responsibility for recommending domestic and international financial, economic, and tax policy; formulating broad fiscal policies that have general significance for the economy; and managing the public debt. The Secretary also oversees the activities of the Department in carrying out its major law enforcement responsibility; in serving as the financial agent for the U.S. Government; and in manufacturing coins, currency, and other products for customer agencies. The Secretary also serves as the Government's chief financial officer.

289

DEPARTMENT OF THE TREASURY

SECRETARY

DEPUTY SECRETARY

CHIEF OF STAFF

DEPUTY CHIEF OF STAFF

COUNSELOR

UNDER SECRETARY FOR INTERNATIONAL AFFAIRS

- ASSISTANT SECRETARY (INTERNATIONAL AFFAIRS)
  - DEPUTY ASSISTANT SECRETARY (TECHNICAL ASSISTANCE POLICY)
  - DEPUTY ASSISTANT SECRETARY (INTERNATIONAL MONETARY AND FINANCIAL POLICY)
  - DEPUTY ASSISTANT SECRETARY (TRADE AND INVESTMENT POLICY)
  - DEPUTY ASSISTANT SECRETARY (INTERNATIONAL DEVELOPMENT, DEBT AND ENVIRONMENTAL POLICY)
  - DEPUTY ASSISTANT SECRETARY (AFRICA, THE MIDDLE EAST AND SOUTH ASIA)
  - DEPUTY ASSISTANT SECRETARY (EURASIA)
  - DEPUTY ASSISTANT SECRETARY (INVESTMENT AND DEVELOPMENT POLICY)

ASSISTANT SECRETARY (LEGISLATIVE AFFAIRS)
- DEPUTY ASSISTANT SECRETARY (TAX AND BUDGET)
- DEPUTY ASSISTANT SECRETARY (BANKING AND FINANCE)
- DEPUTY ASSISTANT SECRETARY (INTERNATIONAL)
- DEPUTY ASSISTANT SECRETARY (APPROPRIATIONS AND MANAGEMENT)

ASSISTANT SECRETARY (PUBLIC AFFAIRS)
- DEPUTY ASSISTANT SECRETARY (PUBLIC AFFAIRS)
- DEPUTY ASSISTANT SECRETARY (PUBLIC LIAISON)

ASSISTANT SECRETARY (ECONOMIC POLICY)
- DEPUTY ASSISTANT SECRETARY (MACROECONOMIC ANALYSIS)
- DEPUTY ASSISTANT SECRETARY (POLICY COORDINATION)

GENERAL COUNSEL
- DEPUTY GENERAL COUNSEL
- LEGAL DIVISION

UNDER SECRETARY FOR DOMESTIC FINANCE

- FISCAL ASSISTANT SECRETARY
  - DEPUTY ASSISTANT SECRETARY (FISCAL OPERATIONS AND POLICY)
  - DEPUTY ASSISTANT SECRETARY (ACCOUNTING POLICY)
  - DEPUTY ASSISTANT SECRETARY (CRITICAL INFRASTRUCTURE PROTECTION AND COMPLIANCE POLICY)
- ASSISTANT SECRETARY (FINANCIAL INSTITUTIONS)
  - DEPUTY ASSISTANT SECRETARY (FINANCIAL INSTITUTIONS POLICY)
  - DEPUTY ASSISTANT SECRETARY (FINANCIAL EDUCATION)
  - DIRECTOR, COMMUNITY DEVELOPMENT FINANCIAL INSTITUTIONS FUND
- ASSISTANT SECRETARY (FINANCIAL MARKETS)
  - DEPUTY ASSISTANT SECRETARY (GOVERNMENT FINANCIAL POLICY)
  - DEPUTY ASSISTANT SECRETARY (FEDERAL FINANCE)
  - ASSISTANT SECRETARY (TERRORISM FINANCING AND FINANCIAL CRIME)
  - DIRECTOR, OFFICE OF FOREIGN ASSETS CONTROL

ASSISTANT SECRETARY (TAX POLICY)
- DEPUTY ASSISTANT SECRETARY (TAX POLICY)
- DEPUTY ASSISTANT SECRETARY (REGULATORY AFFAIRS)
- DEPUTY ASSISTANT SECRETARY (TAX ANALYSIS)
- DEPUTY ASSISTANT SECRETARY (REGULATORY, TARIFFS AND INTERNATIONAL ENFORCEMENT)

TREASURER OF THE UNITED STATES

ASSISTANT SECRETARY MANAGEMENT AND CHIEF FINANCIAL OFFICER
- DEPUTY CHIEF FINANCIAL OFFICER
- CHIEF INFORMATION OFFICER
- DEPUTY ASSISTANT SECRETARY (WORKFORCE MANAGEMENT)
- DIRECTOR INTELLIGENCE AND SECURITY OPERATIONS
- DEPUTY ASSISTANT SECRETARY (MANAGEMENT AND BUDGET)
- DEPUTY ASSISTANT SECRETARY (OPERATIONS)
- DEPUTY ASSISTANT SECRETARY AND CHIEF HUMAN CAPITAL OFFICER
- DIRECTOR, OFFICE OF DC PENSIONS

TREASURY BUREAUS

- INTERNAL REVENUE SERVICE
- OFFICE OF THE COMPTROLLER OF THE CURRENCY
- OFFICE OF THRIFT SUPERVISION
- UNITED STATES MINT
- BUREAU OF ENGRAVING AND PRINTING
- OFFICE OF INSPECTOR GENERAL
- INSPECTOR GENERAL TAX ADMINISTRATION
- FINANCIAL MANAGEMENT SERVICE
- BUREAU OF THE PUBLIC DEBT
- FINANCIAL CRIMES ENFORCEMENT NETWORK
- ALCOHOL TAX AND TRADE ADMINISTRATION BUREAU

[1] Assistant Secretary (Management) and Chief Financial Officer is Treasury's Chief Operating Officer.

Activities

Economic Policy The Office of the Assistant Secretary for Economic Policy assists policymakers in the determination of economic policies. The Office analyzes domestic and international economic issues and developments in the financial markets, assists in the development of official economic projections, and works closely with Federal Government agencies to develop economic forecasts underlying the yearly budget process.

Enforcement The Office of the Assistant Secretary for Enforcement coordinates Treasury law enforcement matters, including the formulation of policies for Treasury enforcement activities, and cooperates on law enforcement matters with other Federal agencies. It oversees the Alcohol and Tobacco Tax and Trade Bureau, charged with collecting excise taxes on alcoholic beverages and tobacco products; the Office of Financial Enforcement, assisting in implementing the Bank Secrecy Act and administering related Treasury regulations; and the Office of Foreign Assets Control, controlling assets in the United States of "blocked" countries and the flow of funds and trade to them.

Financial Institutions The Office of the Assistant Secretary for Financial Institutions exercises policy direction and control over Department activities relating to the substance of proposed legislation pertaining to the general activities and regulation of private financial intermediaries and relating to other Federal regulatory agencies.

Fiscal Affairs The Office of the Fiscal Assistant Secretary supervises the administration of the Government's fiscal affairs. It manages the cash position of the Treasury and projects and monitors debt subject to limit; directs the performance of the fiscal agency functions of the Federal Reserve Banks; conducts Governmentwide accounting and cash management activities; exercises supervision over depositories of the United States; and provides management overview of investment practices for Government trusts and other accounts.

International Affairs The Office of the Assistant Secretary for International Affairs advises and assists policymakers in the formulation and execution of policies dealing with international financial, economic, monetary, trade, investment, environmental, and energy policies and programs. The work of the Office is organized into groups responsible for monetary and financial policy; international development, debt, and environmental policy; trade and investment policy; economic and financial technical assistance; and geographical areas (Asia, the Americas, Africa, Eurasia, and Latin America). The staff offices performing these functions conduct financial diplomacy with industrial and developing nations and regions; work toward improving the structure and operations of the international monetary system; monitor developments in foreign exchange and other markets and official operations affecting those markets; facilitate structural monetary cooperation through the International Monetary Fund and other channels; oversee U.S. participation in the multilateral development banks and coordinate U.S. policies and operations relating to bilateral and multilateral development lending programs and institutions; formulate policy concerning financing of trade; coordinate policies toward foreign investments in the United States and U.S. investments abroad; and analyze balance of payments and other basic financial and economic data, including energy data, affecting world payment patterns and the world economic outlook.

Tax Policy The Office of the Assistant Secretary for Tax Policy advises and assists the Secretary and the Deputy Secretary in the formulation and execution of domestic and international tax policies and programs. These functions include analysis of proposed tax legislation and tax programs; projections of economic trends affecting tax bases; studies of effects of alternative tax measures; preparation of official estimates of Government receipts for the President's annual budget messages; legal advice and analysis on domestic and

international tax matters; assistance in the
development and review of tax legislation
and domestic and international tax
regulations and rulings; and participation
in international tax treaty negotiations
and in maintenance of relations with
international organizations on tax
matters.

Treasurer of the United States The
Office of the Treasurer of the United
States was established on September
6, 1777. The Treasurer was originally
charged with the receipt and custody
of Government funds, but many of
these functions have been assumed by
different bureaus of the Department of
the Treasury. In 1981, the Treasurer was
assigned responsibility for oversight of
the Bureau of Engraving and Printing and
the United States Mint. The Treasurer
reports to the Secretary through the
Assistant Secretary for Management/Chief
Financial Officer.

Treasury Inspector General The
Treasury Inspector General for Tax
Administration (TIGTA) was established
in January 1999, in accordance with the
Internal Revenue Service Restructuring
and Reform Act of 1998, to provide
independent oversight of the Internal
Revenue Service programs and activities.
TIGTA is charged with monitoring the
Nation's tax laws to ensure that the
IRS acts with efficiency, economy,
and effectiveness toward program
accomplishment; ensuring compliance
with applicable laws and regulations,
preventing, detecting, and deterring
fraud, waste, and abuse; investigating
activities or allegations related to fraud,
waste, and abuse by IRS personnel; and
protecting the IRS against attempts to
corrupt or threaten its employees.

For further information concerning the departmental
offices, contact the Public Affairs Office,
Department of the Treasury, 1500 Pennsylvania
Avenue NW., Washington, DC 20220. Phone,
202–622–2960.

Sources of Information

Contracts Write to the Director, Office
of Procurement, Suite 400–W, 1310 G
Street NW., Washington, DC 20220.
Phone, 202–622–0203.

Environment Environmental statements
prepared by the Department are available
for review in the Departmental Library.
Information on Treasury environmental
matters may be obtained from the
Office of the Assistant Secretary of the
Treasury for Management and Chief
Financial Officer, Treasury Department,
Washington, DC 20220. Phone, 202–
622–0043.

General Inquiries For general
information about the Treasury
Department, including copies of news
releases and texts of speeches by high-
level Treasury officials, write to the
Office of the Assistant Secretary (Public
Affairs and Public Liaison), Room
3430, Departmental Offices, Treasury
Department, Washington, DC 20220.
Phone, 202–622–2920.

Inspector General For general
information, contact the Assistant
Inspector General for Management
at 202–927–5200, or visit the Office
of Inspector General (OIG) Web site
at http://www.treas.gov/inspector-
general. To report the possible existence
of a Treasury activity constituting
mismanagement, gross waste of funds,
abuse of authority, a substantial and
specific danger to the public health and
safety, or a violation of law, rules, or
regulations (not including the Internal
Revenue Service, which reports to the
Treasury Inspector General for Tax
Administration), contact the OIG by
phone at 800–359–3898; by fax at
202–927–5799; or by email at Hotline@
oig.treas.gov; or write to Treasury OIG
Hotline, Office of Inspector General,
1500 Pennsylvania Avenue NW.,
Washington, DC 20220. For Freedom
of Information Act/Privacy Act requests,
write to Freedom of Information Act
Request, Treasury OIG, Office of
Counsel, Suite 510, 740 15th Street NW.,
Washington, DC 20220.

Reading Room The Reading Room is
located in the Treasury Library, Room
1428, Main Treasury Building, 1500
Pennsylvania Avenue NW., Washington,
DC 20220. Phone, 202–622–0990.

**Small and Disadvantaged Business
Activities** Write to the Director, Office
of Small and Disadvantaged Business

Utilization, Suite 400–W, 1310 G Street NW., Washington, DC 20220. Phone, 202–622–0530.

Tax Legislation   Information on tax legislation may be obtained from the Assistant Secretary (Tax Policy), Departmental Offices, Treasury Department, Washington, DC 20220. Phone, 202–622–0050.

Telephone Directory   The Treasury Department telephone directory is available for sale by the Superintendent of Documents, Government Printing Office, Washington, DC 20402.

Treasury Inspector General for Tax Administration   Individuals wishing to report fraud, waste, or abuse against or by IRS employees should write to the Treasury Inspector General for Tax Administration, P.O. Box 589, Ben Franklin Station, Washington, DC 20044–0589. Phone, 800–366–4484. Email, complaints@tigta.treas.gov.

For further information, contact the Public Affairs Office, Department of the Treasury, 1500 Pennsylvania Avenue NW., Washington, DC 20220. Phone, 202–622–2960. Internet, http://www.treas.gov.

Alcohol and Tobacco Tax and Trade Bureau

1310 G Street NW., Washington, DC 20220
Phone, 202–453–2000. Internet, http://www.ttb.gov.

Administrator	JOHN MANFREDA
Deputy Administrator	MARY RYAN
Assistant Administrator (Field Operations)	(VACANCY)
Assistant Administrator (Headquarters)	BILL FOSTER
Assistant Administrator (Information Resources/ Chief Information Officer)	ROBERT HUGHES
Assistant Administrator (Management/Chief Financial Officer)	CHERI MITCHELL
Chief Counsel	ROBERT TOBIASSEN

The Alcohol and Tobacco Tax and Trade Bureau (TTB) administers and enforces the existing Federal laws and Tax Code provisions related to the production and taxation of alcohol and tobacco products. TTB also collects all excise taxes on the manufacture of firearms and ammunition.

For further information, contact the Administrator's Office, Alcohol and Tobacco Tax and Trade Bureau. Phone, 202–453–2000. Internet, http://www.ttb.gov.

Office of the Comptroller of the Currency

250 E Street SW., Washington, DC 20219
Phone, 202–874–5000. Internet, http://www.occ.treas.gov.

Comptroller	JOHN C. DUGAN
Chief of Staff and Public Affairs	JOHN G. WALSH
Chief Information Officer	BAJINDER N. PAUL
Deputy to the Chief of Staff and Liaison to the Federal Deposit Insurance Corporation	WILLIAM A. ROWE, III
Ombudsman	LARRY L. HATTIX
First Senior Deputy Comptroller	JULIE L. WILLIAMS
Senior Deputy Comptroller for the Office of Management and Chief Financial Officer	THOMAS R. BLOOM
Senior Deputy Comptroller for Midsize/ Community Bank Supervision	JENNIFER C. KELLY

Senior Deputy Comptroller and Chief National Bank Examiner	TIMOTHY W. LONG
Senior Deputy Comptroller for Economics	MARK LEVONIAN
Senior Deputy Comptroller for Large Bank Supervision	DOUGLAS W. ROEDER

[For the Office of the Comptroller of the Currency statement of organization, see the Code of Federal Regulations, Title 12, Part 4]

The Office of the Comptroller of the Currency (OCC) was created February 25, 1863 (12 Stat. 665), as a bureau of the Department of the Treasury. Its primary mission is to regulate national banks. The Office is headed by the Comptroller, who is appointed for a 5-year term by the President with the advice and consent of the Senate.

The Office regulates national banks by its power to examine banks; approves or denies applications for new bank charters, branches, or mergers; takes enforcement actions—such as bank closures—against banks that are not in compliance with laws and regulations; and issues rules, regulations, and interpretations on banking practices.

The Office supervises approximately 2,100 national banks, including their trust activities and overseas operations. Each bank is examined annually through a nationwide staff of approximately 1,900 bank examiners supervised in 4 district offices. The Office is independently funded through assessments of the assets of national banks.

Sources of Information

For Freedom of Information Act requests, contact the Manager, Disclosure Services and Administrative Operations, Communications Division, 250 E Street SW., Mail Stop 3–2, Washington, DC 20219 (phone, 202–874–4700; fax, 202–874–5274). For information about contracts, contact the Acquisition Management Division at 250 E Street SW., Washington, DC 20219 (phone, 202–874–5040; fax, 202–874–5625). For information regarding national bank examiner employment opportunities (generally hired at the entry level through a college recruitment program), contact the Director for Human Resources Operations, 250 E Street SW., Washington, DC 20219 (phone, 202–874–4500; fax, 202–874–4655). Publications are available from the Communications Division, 250 E Street SW., Washington, DC 20219 (phone, 202–874–4700; fax, 202–874–5263).

For further information, contact the Communications Division, Office of the Comptroller of the Currency, 250 E Street SW., Mail Stop 3–2, Washington, DC 20219. Phone, 202–874–4700.

Bureau of Engraving and Printing

Fourteenth and C Streets SW., Washington, DC 20228
Phone, 202–874–3019. Internet, http://www.moneyfactory.com.

Director	LARRY R. FELIX
Deputy Director	PAMELA J. GARDINER
Associate Director (Chief Financial Officer)	LEONARD R. OLIJAR
Associate Director (Chief Information Officer)	PETER O. JOHNSON
Associate Director (Eastern Currency Facility)	JON J. CAMERON
Associate Director (Western Currency Facility)	CHARLENE WILLIAMS
Associate Director (Management)	SCOTT WILSON
Associate Director (Product and Technology Development)	JUDITH DIAZ MYERS
Chief Counsel	KEVIN RICE

The Bureau of Engraving and Printing operates on basic authorities conferred by act of July 11, 1862 (31 U.S.C. 303), and additional authorities contained in past appropriations made to the Bureau that are still in force. Operations are financed by a revolving fund established in 1950 in accordance with Public Law 81–656. The Bureau is headed by a Director who is selected by the Secretary of the Treasury.

The Bureau designs, prints, and finishes all of the Nation's paper currency and many other security documents, including White House invitations and military identification cards. It also is responsible for advising and assisting Federal agencies in the design and production of other Government documents that, because of their innate value or for other reasons, require security or counterfeit-deterrence characteristics.

The Bureau also operates a second currency manufacturing plant in Fort Worth, TX (9000 Blue Mound Road, 76131). Phone, 817–231–4000.

Sources of Information

Address inquiries on the following subjects to the specified office, Bureau of Engraving and Printing.

Contracts and Small Business Activities Information relating to contracts and small business activities may be obtained by contacting the Office of Acquisition. Phone, 202–874–2065.

Employment Information regarding employment opportunities and required qualifications is available from the Office of Human Resources. Phone, 202–874–2633.

Freedom of Information Act Requests Inquiries should be directed to 202–874–3733.

General Inquiries Requests for information about the Bureau, its products, or numismatic and philatelic interests should be directed to 202–874–3019.

Mail Order Sales Uncut sheets of currency, engraved Presidential portraits, historical engravings of national landmarks, and other souvenirs and mementos are available for purchase by phone at 800–456–3408 and the Internet at www.moneyfactory.com.

Tours Tours of the Bureau's facilities are provided throughout the year according to the schedules listed below. Up-to-the-minute tour information is available on the Bureau's Web site at www.moneyfactory.com.

Washington, DC Peak season, March through August, 9 a.m. until 10:45 a.m. and 12:30 p.m. until 2 p.m. Tickets are required for all tours. Tours begin every 15 minutes, with the last tour beginning at 2 p.m. The times between 11 a.m. and 12:15 p.m. are reserved for school and other groups. The ticket booth is located on Raoul Wallenberg Place (formerly Fifteenth Street) and opens at 8 a.m. Tour tickets are free. The ticket booth remains open for the morning and evening tours until all tickets have been distributed. Lines form early and tickets go quickly, typically by 9 a.m. during peak season. Tickets are distributed on a first-come, first-served basis. Lines organize on Raoul Wallenberg Place. Evening tours, April through August, 5 p.m. until 7 p.m. Tours are offered every 15 minutes. Non-peak season, September through February, 9 a.m. until 2 p.m. No tickets are necessary for tours during this time. Lines organize on Fourteenth Street. No tours are given on weekends, Federal holidays, or between Christmas and New Year's Day. Information about the Washington, DC, Tour and Visitor Center can be obtained by calling 202–874–2330 or 866–874–2330.

Fort Worth, TX Peak season, June and July, 11 a.m. until 5 p.m. Tour hours are every 30 minutes, from 11 a.m. until 5 p.m. The Tour and Visitor Center is open from 10:30 a.m. until 6:30 p.m. Non-peak season, August through May, 9 a.m. until 2 p.m. The Visitor Center is open from 8:30 a.m. until 3:30 p.m. No tours are given on weekends, Federal holidays, or between Christmas and New Year's Day. Information about the Fort Worth Tour and Visitor Center can be obtained by calling 817–231–4000 or 866–865–1194.

For further information, contact the Office of External Relations, Bureau of Engraving and Printing, Department of the Treasury, Room 533–M, Fourteenth and C Streets SW., Washington, DC 20228. Phone, 202–874–3019. Fax, 202–874–3177. Internet, http://www.moneyfactory.com.

Financial Management Service

401 Fourteenth Street SW., Washington, DC 20227
Phone, 202–874–6740. Internet, http://www.fms.treas.gov.

Commissioner	DAVID A. LEBRYK
Deputy Commissioner	WANDA J. ROGERS
Director, Legislative and Public Affairs	MELODY BARRETT
Chief Counsel	MARGARET MARQUETTE
Assistant Commissioner, Business Architecture Office (Chief Business Architect)	JOHN KOPEC
Assistant Commissioner, Debt Management Services	SCOTT H. JOHNSON
Assistant Commissioner, Federal Finance	KRISTINE CONRATH
Assistant Commissioner, Governmentwide Accounting and Agency Services	DAVID REBICH
Assistant Commissioner, Information Resources (Chief Information Officer)	CHUCK SIMPSON
Assistant Commissioner, Management (Chief Financial Officer)	KENT KUYUMJIAN, *Acting*
Assistant Commissioner, Payment Management	SHERYL MORROW

The Financial Management Service (FMS) provides central payment services to Federal program agencies, operates the Federal Government's collections and deposit systems, provides Governmentwide accounting and reporting services, and manages the collection of delinquent debt owed to the Federal Government. FMS has four regional financial centers located in Austin, TX; Kansas City, MO; Philadelphia, PA; and San Francisco, CA; and one debt collection center in Alabama.

Accounting   FMS gathers and publishes Governmentwide financial information that is used by the public and private sectors to monitor the Government's financial status and establish fiscal and monetary policies. These publications include the Daily Treasury Statement, the Monthly Treasury Statement, the Treasury Bulletin, the U.S. Government Annual Report, and the Financial Report of the U.S. Government.

Collections   FMS administers the world's largest collection system, gathering nearly $2.86 trillion annually through a network of more than 9,000 financial institutions. It also manages the collection of Federal revenues such as individual and corporate income tax deposits, customs duties, loan repayments, fines, and proceeds from leases.

FMS and IRS manage the Electronic Federal Tax Payment System (www.eftps.gov), which allows individuals and businesses to pay Federal taxes through the Internet. EFTPS–OnLine also provides such features as an instant, printable acknowledgment for documenting each transaction, the ability to schedule advance payments, and access to payment history.

The Treasury Offset Program is one of the methods used to collect delinquent debt. FMS uses the program to withhold Federal payments, such as Federal income tax refunds, Federal salary payments, and Social Security benefits, to recipients with delinquent debts, including past-due child support obligations and State and Federal income tax debt.

Electronic Commerce   Through its electronic money programs, FMS offers new payment and collection technologies using the Internet and card technologies to help Federal agencies modernize their cash management activities. Examples include stored-value cards used on military bases, point-of-sale check conversion, and Internet credit card collection programs.

Payments   Each year, FMS disburses more than 1 billion non-Defense payments, totaling nearly $2.3 trillion, to a wide variety of recipients, such as those individuals who receive Social Security, IRS tax refunds, and veterans' benefits. For fiscal year 2009, nearly 81 percent of these payments were issued electronically. FMS issues the remainder of its payments, nearly 197 million annually, by check.

Sources of Information

Inquiries on the following subjects should be directed to the specified office, Financial Management Service, 401 Fourteenth Street SW., Washington, DC 20227. Fax, 202–874–7016.

Contracts   Director, Acquisition Management Division, Room 428. Phone, 202–874–6910.

Employment   Human Resources Division, Room 170A, 3700 East-West Highway, Hyattsville, MD 20782. Phone, 202–874–8090. TDD, 202–874–8825.

For further information, contact the Office of Legislative and Public Affairs, Financial Management Service, Department of the Treasury, Room 555, 401 Fourteenth Street SW., Washington, DC 20227. Phone, 202–874–6750. Internet, http://www.fms.treas.gov.

Internal Revenue Service

1111 Constitution Avenue NW., Washington, DC 20224
Phone, 202–622–5000. Internet, http://www.irs.gov.

Commissioner of Internal Revenue	DOUGLAS H. SHULMAN
Deputy Commissioner for Operations Support	MARK A. ERNST
Deputy Commissioner for Services and Enforcement	STEVEN T. MILLER
Chief Counsel	WILLIAM WILKINS
Commissioner, Large and Midsize Business Division	HEATHER C. MALOY
Commissioner, Small Business/Self-Employed Division	CHRISTOPHER WAGNER
Commissioner, Tax Exempt and Government Entities Division	SARAH H. INGRAM
Commissioner, Wage and Investment Division	RICHARD E. BYRD
Chief Financial Officer	ALISON DOONE
Chief, Agency-Wide Shared Services	DAVID GRANT
Chief, Appeals	DIANE RYAN
Chief, Communications and Liaison	FRANK M. KEITH
Chief, Criminal Investigation	VICTOR SONG
Chief Technology Officer	TERRY V. MILHOLLAND
National Taxpayer Advocate	NINA E. OLSON
Director, Office of Research, Analysis and Statistics	ROSEMARY MARCUSS
Chief Human Capital Officer	JAMES FALCONE
Director, Privacy, Information Protection and Data Security	DEBORAH G. WOLF
Director, Office of Professional Responsibility	KAREN L. HAWKINS
Director, Whistleblower Office	STEVE A. WHITLOCK

The Office of the Commissioner of Internal Revenue was established by act of July 1, 1862 (26 U.S.C. 7802).

The Internal Revenue Service (IRS) is responsible for administering and enforcing the internal revenue laws and related statutes, except those relating to alcohol, tobacco, firearms, and explosives. Its mission is to collect the proper amount of tax revenue, at the least cost to the public, by efficiently applying the tax law with integrity and fairness. To achieve that purpose, the IRS strives to achieve the highest possible degree of voluntary compliance in accordance with the tax laws and regulations; advises the public of their rights and responsibilities; determines the extent of compliance and the causes of noncompliance; properly administers and enforces the tax laws; and continually searches for and implements new, more efficient ways of accomplishing its mission. IRS ensures satisfactory resolution of taxpayer complaints; provides taxpayer service and education; determines, assesses, and collects internal revenue taxes; determines pension plan qualifications and exempt organization status; and prepares and issues rulings and regulations to supplement the provisions of the Internal Revenue Code.

The source of most revenues collected is the individual income tax and the social insurance and retirement taxes. Other major sources are corporate income, excise, estate, and gift taxes. Congress first received authority to levy taxes on the income of individuals and corporations in 1913, pursuant to the 16th Amendment of the Constitution.

Sources of Information

Audiovisual Materials   Films providing information on the American tax system, examination and appeal rights, and the tax responsibilities of running a small business are available. Some of the films are also available in Spanish. The films can be obtained by contacting any territory office. Also available are audio and video cassette tapes that provide step-by-step instructions for preparing basic individual income tax forms. These tapes are available in many local libraries.

Contracts   Write to the Internal Revenue Service (OS:A:P), 1111 Constitution Avenue NW., Washington, DC 20224. Phone, 202–283–1710.

Customer Service   The Internal Revenue Service provides year-round tax information and assistance to taxpayers, primarily through its Web site and toll-free telephone system, which also includes telephone assistance to deaf and hearing-impaired taxpayers who have access to a teletypewriter or television/phone. The toll-free numbers are listed in local telephone directories and in the annual tax form packages. Taxpayers may also visit agency offices for help with their tax problems. Individual preparation is available for handicapped or other individuals unable to use the group preparation method. Foreign language tax assistance is also available at many locations. The IRS encourages taxpayers to use the resources available at www.irs.gov for assistance with their tax questions or to locate electronic filing sources.

Educational Programs   The Service provides, free of charge, general tax information publications and booklets on specific tax topics. Taxpayer information materials also are distributed to major television networks and many radio and television stations, daily and weekly newspapers, magazines, and specialized publications. Special educational materials and films are provided for use in high schools and colleges. Individuals starting a new business are given specialized materials and information at small business workshops, and community colleges provide classes based on material provided by the Service. The community outreach tax assistance program provides assistance to community groups. Through the volunteer income tax assistance program and the tax counseling for the elderly program, the Service recruits, trains, and supports volunteers who offer free tax assistance to low-income, elderly, military, and non-English-speaking taxpayers. Materials, films, and information on the educational programs can be obtained by contacting any territory office.

Employment   For information, write to the recruitment coordinator at any of the territory offices.

Publications The Annual Report—Commissioner of Internal Revenue, the Internal Revenue Service Data Book, and periodic reports of statistics of income are available from the Superintendent of Documents, Government Printing Office, Washington, DC 20402. Examination of Returns, Appeal Rights, and Claims for Refund; Your Federal Income Tax; Farmer's Tax Guide; Tax Guide for Small Business; and other publications are available at Internal Revenue Service offices free of charge.

Reading Rooms Public reading rooms are located in the national office and in each territory office.

Speakers Arrangements for speakers on provisions of the tax law and operations of the Internal Revenue Service for professional and community groups may be made by writing to the Senior Commissioner's Representative or, for national organizations only, to the Communications Division at the IRS National Headquarters in Washington, DC.

Taxpayer Advocate Each district has a problem resolution staff which attempts to resolve taxpayer complaints not satisfied through regular channels.

For further information, contact the Internal Revenue Service, Department of the Treasury, 1111 Constitution Avenue NW., Washington, DC 20224. Phone, 202–622–5000.

United States Mint

801 Ninth Street NW., Washington, DC 20220
Phone, 202–354–7200. Internet, http://www.usmint.gov.

Director	EDMUND C. MOY
Deputy Director	ANDREW BRUNHART
Chief Counsel	DAN SHAVER
Associate Director, Finance	PATRICIA M. GREINER
Associate Director, Information Technology	ANDREW BRUNHART, *Acting*
Associate Director, Manufacturing	RICHARD PETERSON
Associate Director, Sales and Marketing	B.B. CRAIG
Associate Director, Protection	DENNIS O'CONNOR

The establishment of a mint was authorized by act of April 2, 1792 (1 Stat. 246). The Bureau of the Mint was established by act of February 12, 1873 (17 Stat. 424), and recodified on September 13, 1982 (31 U.S.C. 304, 5131). The name was changed to United States Mint by Secretarial order dated January 9, 1984.

The primary mission of the Mint is to produce an adequate volume of circulating coinage for the Nation to conduct its trade and commerce. The Mint also produces and sells numismatic coins, American Eagle gold and silver bullion coins, and national medals. In addition, the Fort Knox Bullion Depository is the primary storage facility for the Nation's gold bullion.

The U.S. Mint maintains sales centers at the Philadelphia and Denver Mints

and at Union Station in Washington, DC. Public tours are conducted, with free admission, at the Philadelphia and Denver Mints.

For a complete list of U.S. Mint field facilities, go to www.usmint.gov/about_the_mint.

Sources of Information

Contracts and Employment Inquiries should be directed to the facility head of the appropriate field office or to the Director of the Mint.

Numismatic Services The United States Mint maintains public exhibit and sales areas at the Philadelphia and Denver Mints and at Union Station in Washington, DC. Brochures and order forms for official coins, medals, and other numismatic items are available through the Internet at www.usmint.gov.

Publications   The CFO Annual Financial Report is available from the United States Mint, Department of the Treasury, 801 Ninth Street NW., Washington, DC 20220. Phone, 202–354–7800.

For further information, contact the United States Mint, Department of the Treasury, 801 Ninth Street NW., Washington, DC 20220. Phone, 202–354–7200.

Bureau of the Public Debt

799 Ninth Street NW., Washington, DC 20239–0001
Phone, 202–504–3500. Internet, http://www.publicdebt.treas.gov.

Commissioner	VAN ZECK
Deputy Commissioner	ANITA SHANDOR
Chief Counsel	PAUL WOLFTEICH
Director (Public and Legislative Affairs Staff)	KIM TREAT
Executive Director (Government Securities Regulation Staff)	LORI SANTAMORENA
Assistant Commissioner (Financing)	DARA SEAMAN
Assistant Commissioner (Information Technology)	KIMBERLY A. MCCOY
Assistant Commissioner (Public Debt Accounting)	DEBRA HINES
Assistant Commissioner (Retail Securities)	JOHN R. SWALES, III
Assistant Commissioner (Management Services)	FRED PYATT
Executive Director (Administration Resource Center)	CYNTHIA Z. SPRINGER

The Bureau of the Public Debt was established on June 30, 1940, pursuant to the Reorganization Act of 1939 (31 U.S.C. 306).

The Bureau's mission is to borrow the money needed to operate the Federal Government, account for the resulting public debt, and provide reimbursable support to Federal agencies. The Bureau fulfills its mission through five programs: wholesale securities services, Government agency investment services, retail securities services, summary debt accounting, and franchise services.

The Bureau auctions and issues Treasury bills, notes, and bonds and manages the U.S. Savings Bond Program. It issues, services, and redeems bonds through a nationwide network of issuing and paying agents. It provides daily and other periodic reports to account for the composition and size of the debt. In addition, the Bureau implements the regulations for the Government securities market. These regulations provide for investor protection while maintaining a fair and liquid market for Government securities.

Sources of Information

Electronic Access   Information about the public debt, U.S. Savings Bonds, Treasury bills, notes, and bonds, and other Treasury securities is available through the Internet at www.treasurydirect.gov. Forms and publications may be ordered electronically at the same address.

Employment   General employment inquiries should be addressed to the Bureau of the Public Debt, Division of Human Resources, Recruitment and Classification Branch, Parkersburg, WV 26106–1328. Phone, 304–480–6144.

Savings Bonds   Savings bonds are continuously on sale at more than 40,000 financial institutions and their branches in virtually every locality in the United States. Information about bonds is provided by such issuing agents. Savings bonds may also be purchased and held in an online account. Current rate information is available at www. treasurydirect.gov or by calling 800–

4US–BOND (800–487–2663). Requests for information about all series of savings bonds, savings notes, and retirement plans or individual retirement bonds should be addressed to the Bureau of the Public Debt, Department of the Treasury, 200 Third Street, Parkersburg, WV 26106–1328. Phone, 304–480–7711.

Treasury Securities Information inquiries regarding the purchase of Treasury bills, bonds, and notes should be addressed to a Treasury Direct contact center or to the Bureau of the Public Debt, 200 Third Street, Parkersburg, WV 26106–1328. Phone, 800–722–2678.

For more information, contact the Director, Public and Legislative Affairs, Office of the Commissioner, Bureau of the Public Debt, Washington, DC 20239–0001. Phone, 202–504–3502. Internet, http://www.publicdebt.treas.gov.

Office of Thrift Supervision

1700 G Street NW., Washington, DC 20552
Phone, 202–906–6000. Internet, http://www.ots.treas.gov.

Director	JOHN E. BOWMAN, *Acting*
Deputy Director, Examinations, Supervision and Consumer Protection	THOMAS A. BARNES
Chief Counsel	DEBORAH DAKIN, *Acting*
Chief Financial Officer	HANS E. HEINDENREICH
Managing Director, External Affairs	BARBARA L. SHYCOFF
Chief Information Officer	WAYNE G. LEISS
Managing Director, Human Capital	AVELINO L. RODRIGUEZ

The Office of Thrift Supervision (OTS) regulates Federal- and State-chartered savings institutions. Created by the Financial Institutions Reform, Recovery, and Enforcement Act of 1989, its mission is to effectively and efficiently supervise thrift institutions in a manner that encourages a competitive industry to meet housing and other credit and financial services needs and ensure access to financial services for all Americans.

OTS is headed by a Director appointed by the President, with the advice and consent of the Senate, for a 5-year term. The Director is responsible for the overall direction and policy of the agency. OTS is responsible for examining and supervising thrift institutions in the four OTS regions to ensure the safety and soundness of the industry; ensuring that thrifts comply with consumer protection laws and regulations; conducting a regional quality assurance program to ensure consistent applications of policies and procedures; developing national policy guidelines to enhance statutes and regulations and to establish programs to implement new policy and

law; issuing various financial reports, including the quarterly report on the financial condition of the thrift industry; preparing regulations, bulletins, other policy documents, congressional testimony, and official correspondence on matters relating to the condition of the thrift industry, interest rate risk, financial derivatives, and economic issues; and prosecuting enforcement actions relating to thrift institutions.

Sources of Information

Electronic Access Information about OTS and institutions regulated by OTS is available through the Internet at www.ots.treas.gov.
Employment Inquiries about employment opportunities with the Office of Thrift Supervision should be directed to the Human Resources Office. Phone, 202–906–6171.
Freedom of Information Act/Privacy Act Requests For information not readily available from the Web site, please submit requests to the Office of Thrift Supervision, Dissemination Branch, 1700 G Street NW., Washington, DC 20552.

Email, publicinfo@ots.treas.gov. Fax, 202–906–7755.

General Information General information about OTS may be obtained by calling 202–906–6000.

Publications Publications that provide information and guidance regarding the thrift industry are available at the "Public Information" link on the Web site.

For further information, contact External Affairs, Office of Thrift Supervision, 1700 G Street NW., Washington, DC 20552. Phone, 202–906–6677. Fax, 202–906–7477. Internet, http://www.ots.treas.gov.

DEPARTMENT OF VETERANS AFFAIRS

810 Vermont Avenue NW., Washington, DC 20420
Phone, 202–461–0000. Internet, http://www.va.gov.

Secretary of Veterans Affairs	ERIC K. SHINSEKI
Deputy Secretary	W. SCOTT GOULD
Chief of Staff	JOHN R. GINGRICH
General Counsel	WILL A. GUNN
Chairman, Board of Veterans' Appeals	JAMES P. TERRY
Inspector General	GEORGE J. OPFER
Executive Director for Office of Acquisition, Logistics, and Construction	GLENN D. HAGGSTROM
Director, Office of Small and Disadvantaged Business Utilization	TIM FOREMAN
Director, Office of Employment Discrimination Complaint Adjudication	MAXANNE WITKIN
Director, Center for Women Veterans	IRENE TROWELL-HARRIS
Director, Center for Minority Veterans	LUCRETIA MCCLENNEY
Advisory Committee Management Officer	VIVIAN DRAKE, *Acting*
Director, Center for Faith-Based and Neighborhood Partnerships	E. TERRI LAVELLE
Nongovernmental Organization Gateway Initiative Office	DOUG CARMON
Director, Federal Recovery Coordination Office	KAREN GUICE
Veterans' Service Organizations Liaison	KEVIN SECOR
Director, Office of Survivors Assistance	(VACANCY)
Assistant Secretary for Management	W. TODD GRAMS, *Acting*
Assistant Secretary for Information and Technology	ROGER W. BAKER
Assistant Secretary for Policy and Planning	RAUL PEREA-HENZE
Assistant Secretary for Operations, Security, and Preparedness	JOSE D. RIOJAS
Assistant Secretary for Human Resources and Administration	JOHN U. SEPULVEDA
Assistant Secretary for Public and Intergovernmental Affairs	L. TAMMY DUCKWORTH
Assistant Secretary for Congressional and Legislative Affairs	JOAN M. EVANS
Under Secretary for Benefits, Veterans Benefits Administration	MICHAEL WALCOFF, *Acting*
Under Secretary for Health, Veterans Health Administration	ROBERT A. PETZEL
Under Secretary for Memorial Affairs, National Cemetery Administration	STEVE L. MURO, *Acting*

The Department of Veterans Affairs operates programs to benefit veterans and members of their families. Benefits include compensation payments for disabilities or death related to military service; pensions; education and rehabilitation; home loan guaranty; burial; and a medical care program incorporating nursing homes, clinics, and medical centers.

303

The Department of Veterans Affairs (VA) was established as an executive department by the Department of Veterans Affairs Act (38 U.S.C. 201 note). It is comprised of three organizations that administer veterans programs: the Veterans Health Administration, the Veterans Benefits Administration, and the National Cemetery Administration. Each organization has field facilities and a central office component. Staff offices support the overall function of the Department and its Administrations.

Activities

Advisory Committee Management Office The Advisory Committee Management Office serves as the coordinating office for the Department's 25 Federal advisory committees. It is responsible for establishing clear and uniform goals, standards, and procedures for advisory committee activities. It is also responsible for ensuring that VA advisory committee operations are in compliance with the provisions of the Federal Advisory Committee Act.

Office of Acquisition, Logistics, and Construction The Office of Acquisition, Logistics, and Construction (OALC) is a multifunctional organization responsible for directing the acquisition, logistics, construction, and leasing functions within the Department of Veterans Affairs. The Executive Director, OALC, is also the Chief Acquisition Officer for the Department of Veterans Affairs.

Cemeteries The National Cemetery Administration (NCA) is responsible for the management and oversight of more than 128 national cemeteries in the United States and Puerto Rico, as well as 33 soldiers' lots, Confederate cemeteries, and other monument sites. Burial in a national cemetery is available to eligible veterans and their spouses and dependent children. At no cost to the family, a national cemetery burial includes the gravesite, graveliner, opening and closing of the grave, headstone or marker, and perpetual care as part of a national shrine. If a veteran is buried in a private cemetery, anywhere in the world, NCA will provide a headstone or marker. NCA's State Cemetery Grants Program provides funds to State and tribal governments to establish, expand, or improve State-operated veterans' cemeteries. NCA issues Presidential Memorial Certificates to recognize the service of honorably discharged servicemembers or veterans.

Center for Minority Veterans The Center for Minority Veterans (CMV), established by Public Law 103–446 (108 Stat. 4645), promotes the use of VA benefits, programs, and services by minority veterans. The CMV focuses on the unique and special needs of African-Americans, Hispanics, Asian-Americans, Pacific Islanders, and Native Americans, which include American Indians, Native Hawaiians, and Alaska Natives.

Center for Women Veterans The Center for Women Veterans (CWV), established by Public Law 103–446 (108 Stat. 4645), reports to the Secretary's Office and oversees the Department's programs for women veterans. The CWV Director serves as the primary advisor to the Secretary on all matters related to policies, legislation, programs, issues, and initiatives affecting women veterans. The CWV ensures that women veterans receive benefits and services on par with male veterans; VA programs are responsive to gender-specific needs of women veterans; outreach is performed to improve women veterans' awareness of services, benefits, and eligibility criteria; and women veterans are treated with dignity and respect.

Health Services The Veterans Health Administration provides hospital, nursing home, and domiciliary care, and outpatient medical and dental care to eligible veterans of military service in the Armed Forces. It conducts both individual medical and health-care delivery research projects and multi-hospital research programs, and it assists in the education of physicians and dentists and with training of many other health care professionals through affiliations with educational institutions and organizations.

Veterans Benefits The Veterans Benefits Administration provides information, advice, and assistance to veterans, their dependents, beneficiaries,

DEPARTMENT OF VETERANS AFFAIRS

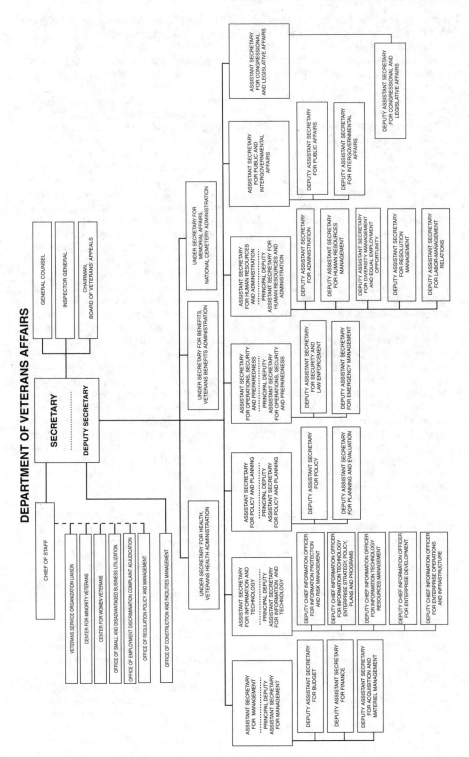

- SECRETARY
- DEPUTY SECRETARY

- GENERAL COUNSEL
- INSPECTOR GENERAL
- CHAIRMAN, BOARD OF VETERANS APPEALS

- CHIEF OF STAFF

- VETERANS SERVICE ORGANIZATION LIAISON
- CENTER FOR MINORITY VETERANS
- CENTER FOR WOMEN VETERANS
- OFFICE OF SMALL AND DISADVANTAGED BUSINESS UTILIZATION
- OFFICE OF EMPLOYMENT DISCRIMINATION COMPLAINT ADJUDICATION
- OFFICE OF REGULATION POLICY AND MANAGEMENT

- OFFICE OF CONSTRUCTION AND FACILITIES MANAGEMENT

- UNDER SECRETARY FOR MEMORIAL AFFAIRS, NATIONAL CEMETERY ADMINISTRATION
- UNDER SECRETARY FOR BENEFITS, VETERANS BENEFITS ADMINISTRATION
- UNDER SECRETARY FOR HEALTH, VETERANS HEALTH ADMINISTRATION

- ASSISTANT SECRETARY FOR CONGRESSIONAL AND LEGISLATIVE AFFAIRS
- ASSISTANT SECRETARY FOR PUBLIC AND INTERGOVERNMENTAL AFFAIRS
  - DEPUTY ASSISTANT SECRETARY FOR PUBLIC AFFAIRS
  - DEPUTY ASSISTANT SECRETARY FOR INTERGOVERNMENTAL AFFAIRS
- DEPUTY ASSISTANT SECRETARY FOR CONGRESSIONAL AND LEGISLATIVE AFFAIRS

- ASSISTANT SECRETARY FOR HUMAN RESOURCES AND ADMINISTRATION
  - PRINCIPAL DEPUTY ASSISTANT SECRETARY FOR HUMAN RESOURCES AND ADMINISTRATION
  - DEPUTY ASSISTANT SECRETARY FOR ADMINISTRATION
  - DEPUTY ASSISTANT SECRETARY FOR HUMAN RESOURCES MANAGEMENT
  - DEPUTY ASSISTANT SECRETARY FOR DIVERSITY MANAGEMENT AND EQUAL EMPLOYMENT OPPORTUNITY
  - DEPUTY ASSISTANT SECRETARY FOR RESOLUTION MANAGEMENT
  - DEPUTY ASSISTANT SECRETARY FOR LABOR-MANAGEMENT RELATIONS

- ASSISTANT SECRETARY FOR OPERATIONS, SECURITY AND PREPAREDNESS
  - PRINCIPAL DEPUTY ASSISTANT SECRETARY FOR OPERATIONS, SECURITY AND PREPAREDNESS
  - DEPUTY ASSISTANT SECRETARY FOR SECURITY AND LAW ENFORCEMENT
  - DEPUTY ASSISTANT SECRETARY FOR EMERGENCY MANAGEMENT

- ASSISTANT SECRETARY FOR POLICY AND PLANNING
  - PRINCIPAL DEPUTY ASSISTANT SECRETARY FOR POLICY AND PLANNING
  - DEPUTY ASSISTANT SECRETARY FOR POLICY
  - DEPUTY ASSISTANT SECRETARY FOR PLANNING AND EVALUATION

- ASSISTANT SECRETARY FOR INFORMATION AND TECHNOLOGY
  - PRINCIPAL DEPUTY ASSISTANT SECRETARY FOR INFORMATION AND TECHNOLOGY
  - DEPUTY CHIEF INFORMATION OFFICER FOR INFORMATION PROTECTION AND RISK MANAGEMENT
  - DEPUTY CHIEF INFORMATION OFFICER FOR INFORMATION TECHNOLOGY ENTERPRISE STRATEGY, POLICY, PLANS AND PROGRAMS
  - DEPUTY CHIEF INFORMATION OFFICER FOR INFORMATION TECHNOLOGY RESOURCES MANAGEMENT
  - DEPUTY CHIEF INFORMATION OFFICER FOR ENTERPRISE DEVELOPMENT
  - DEPUTY CHIEF INFORMATION OFFICER FOR ENTERPRISE OPERATIONS AND INFRASTRUCTURE

- ASSISTANT SECRETARY FOR MANAGEMENT
  - PRINCIPAL DEPUTY ASSISTANT SECRETARY FOR MANAGEMENT
  - DEPUTY ASSISTANT SECRETARY FOR BUDGET
  - DEPUTY ASSISTANT SECRETARY FOR FINANCE
  - DEPUTY ASSISTANT SECRETARY FOR ACQUISITION AND MATERIEL MANAGEMENT

representatives, and others applying for VA benefits. It also cooperates with the Department of Labor and other Federal, State, and local agencies in developing employment opportunities for veterans and referral for assistance in resolving socioeconomic, housing, and other related problems.

The Compensation and Pension Service is responsible for claims for disability compensation and pension, specially adapted housing, accrued benefits, adjusted compensation in death cases, and reimbursement for headstone or marker; allowances for automobiles and special adaptive equipment; special clothing allowances; emergency officers' retirement pay; survivors' claims for death compensation, dependency and indemnity compensation, death pension, and burial and plot allowance claims; forfeiture determinations; and a benefits protection program for minors and incompetent adult beneficiaries.

The Education Service administers the Montgomery GI Bill program and other programs which provide education benefits to qualified active-duty members, veterans, certain dependents of veterans, and members of the Selected and Ready Reserve. The Service also checks school records to ensure that they comply with the pertinent law, approves courses for the payment of educational benefits, and administers a work-study program. Additional details are available at www.gibill.va.gov.

The Insurance Service's operations for the benefit of servicemembers, veterans, and their beneficiaries are available through the regional office and insurance center (phone, 800–669–8477) in Philadelphia, PA, which provides the full range of activities necessary for a national life insurance program. Activities include the complete maintenance of individual accounts, underwriting functions, life and death insurance claims awards, and any other insurance-related transactions. The Service also administers the Veterans Mortgage Life Insurance Program for those disabled veterans who receive a VA grant for specially adapted housing and supervises the Servicemembers' Group

Life Insurance Program and the Veterans' Group Life Insurance Program.

The Loan Guaranty Service is responsible for operations that include appraising properties to establish their values; approving grants for specially adapted housing; supervising the construction of new residential properties; establishing the eligibility of veterans for the program; evaluating the ability of a veteran to repay a loan and the credit risk; making direct loans to Native American veterans to acquire a home on trust land; servicing and liquidating defaulted loans; and disposing of real estate acquired as the consequence of defaulted loans.

The Vocational Rehabilitation and Employment Service provides outreach, motivation, evaluation, counseling, training, employment, and other rehabilitation services to service-connected disabled veterans. Vocational and educational counseling, as well as the evaluation of abilities, aptitudes, and interests are provided to veterans and servicepersons. Counseling, assessment, education programs, and, in some cases, rehabilitation services are available to spouses and children of totally and permanently disabled veterans as well as surviving orphans, widows, and widowers of certain deceased veterans.

Vocational training and rehabilitation services are available to children with spina bifida having one or both parents who served in the Republic of Vietnam during the Vietnam era, or served in certain military units in or near the demilitarized zone in Korea, between September 1, 1967 and August 31, 1971.

Veterans' Appeals   The Board of Veterans' Appeals (BVA) is responsible for entering the final appellate decisions in claims of entitlement to veterans' benefits and for deciding certain matters concerning fees charged by attorneys and agents for representation of veterans before the VA and requests for revision of prior BVA decisions on the basis of clear and unmistakable error. Final BVA decisions are appealable to the U.S. Court of Appeals for Veterans Claims.

Field Facilities   The Department's operations are handled through the following field facilities: cemeteries, domiciliaries, medical centers, outpatient clinics, and regional offices. Cemeteries provide burial services to veterans, their spouses, and dependent children. Domiciliaries provide the least intensive level of inpatient medical care, including necessary ambulatory medical treatment, rehabilitation, and support services, in a structured environment to veterans who are unable because of their disabilities to provide adequately for themselves in the community. Medical centers provide eligible beneficiaries with medical and other health care services equivalent to those provided by private-sector institutions, augmented in many instances by services to meet the special requirements of veterans. Outpatient clinics provide eligible beneficiaries with ambulatory care. Regional offices grant benefits and services provided by law for veterans, their dependents, and beneficiaries within an assigned territory; furnish information regarding VA benefits and services; adjudicate claims and make awards for disability compensation and pension; conduct outreach and information dissemination; provide support and assistance to various segments of the veteran population to include former prisoners of war, minority veterans, homeless veterans, women veterans and elderly veterans; supervise payment of VA benefits to incompetent beneficiaries; provide vocational rehabilitation and employment training; administer educational benefits; guarantee loans for purchase, construction, or alteration of homes; process grants for specially adapted housing; process death claims; and assist veterans in exercising rights to benefits and services.

A complete listing of the Department's field facilities is available at www2. va.gov/directory/guide/home.asp.

Sources of Information

Audiovisuals   Persons interested in the availability of VA video productions or exhibits for showing outside of VA may write to the Chief, Media Services Division (032B), Department of Veterans Affairs, 810 Vermont Avenue NW., Washington, DC 20420. Phone, 202–461–5282. Email, vacomedia-photoservices@va.gov.

Contracts   Information on business opportunities with the VA can be found at www1.va.gov/oamm/oa/dbwva/index. cfm. Additional information is available at the Office of Acquisition and Material Management Web site at www1.va.gov/oamm. Information on solicitations issued by VA is available at www.va.gov/oamm/busopp/index.htm.

Small Business Programs   Persons seeking information on VA's small business programs may call 800–949–8387 or 202–565–8124. The Office of Small and Disadvantaged Business Utilization Web site contains a considerable amount of information about these programs. Internet, www. va.gov/osdbu.

Veterans Business Ownership Services   The Center for Veterans Enterprise assists veterans who want to open or expand a business. This Center is a component of the Office of Small and Disadvantaged Business Utilization. Phone, 866–584–2344. Internet, www.vetbiz.gov. Email, vacve@mail.va.gov.

Employment   The Department of Veterans Affairs employs physicians, dentists, podiatrists, optometrists, nurses, nurse anesthetists, physician assistants, expanded-function dental auxiliaries, registered respiratory therapists, certified respiratory technicians, licensed physical therapists, occupational therapists, pharmacists, and licensed practical or vocational nurses under VA's excepted merit system. This system does not require civil service eligibility. Other professional, technical, administrative, and clerical occupations, such as veterans claims examiners, secretaries, and management analysts, exist in VA that do require civil service eligibility. Persons interested in employment should contact the human resources services office at their nearest VA facility or search the VA Web site at www.va.gov/jobs. All qualified applicants will receive consideration for appointments without regard to race, religion, color, national origin, sex, political affiliation, or any nonmerit factor.

Freedom of Information Act Requests
Inquiries should be directed to the
Assistant Secretary for Information and
Technology, Freedom of Information
Service (005R1C), 810 Vermont Avenue
NW., Washington, DC 20420. Phone,
202–461–7450.

Inspector General Inquiries and Hotline
Publicly available documents and
information on the VA Office of Inspector
General are available at www.va.gov/
oig. Complaints may be sent by mail to
the VA Inspector General (53E), P.O. Box
50410, Washington, DC 20091–0410.
Hotline phone, 800–488–8244. Email,
vaoighotline@va.gov.

**Medical Center (Hospital) Design,
Construction, and Related Services**
Construction projects for VA medical
centers and other facilities in excess of
$4 million are managed and controlled
at the VA central office, located in
Washington, DC. Projects requiring
design, construction, and other related
services are advertised on the FirstGov
Web site at www.firstgov.gov. Submit
project-specific qualifications (SF
254 and SF 255) to the Director, A/E
Evaluation and Program Support Team
(181A), 810 Vermont Avenue NW.,
Washington, DC 20420. Additional
information regarding the selection
process can be found on the VA Office of
Facilities Management Web site at www.
cfm.va.gov. Construction projects for VA
medical centers and other facilities which
are less than $4 million are managed
and controlled at the individual medical
centers. For information regarding these
specific projects, contact the Acquisition
and Materiel Management Office at each
individual VA medical center. Addresses
and additional information on VA
medical centers can be found on the VA
Web site at www.va.gov/facilities.

News Media Representatives may
contact VA through the nearest regional
Office of Public Affairs: Atlanta (404–
929–5880); Chicago (312–980–4235);
Dallas (817–385–3720); Denver (303–
914–5855); Los Angeles (310–268–4207);
New York (212–807–3429); Washington,
DC (202–530–9360). National media
may contact the Office of Public Affairs
in the VA Central Office, 810 Vermont

Avenue NW., Washington, DC 20420.
Phone, 202–461–6000.

Publications The Annual Performance
and Accountability Report may be
obtained (in single copies), without
charge, from the Office of Budget (041),
810 Vermont Avenue NW., Washington,
DC 20420.

The 2010 VA pamphlet Federal
Benefits for Veterans, Dependents
and Survivors (80–98–1) is available
for sale by the Superintendent of
Documents, Government Printing Office,
Washington, DC 20402. This publication
is also available at www1.va.gov/opa/
publication.

The Board of Veterans Appeals Index
(I–01–1), an index to appellate decisions,
is available on microfiche in annual
accumulation from July 1977 through
December 1994. The quarterly indexes
may be purchased for $7 and annual
cumulative indexes for $22.50. The
VADEX/CITATOR of Appellate Research
Materials is a complete printed quarterly
looseleaf accumulation of research
material which may be purchased for
$175 with binder and for $160 without
binder. The Vadex Infobase, a computer-
searchable version of the VADEX, is also
available on diskettes for $100 per copy.
These publications may be obtained
by contacting Promisel and Korn, Inc.
Phone, 301–986–0650. Archived BVA
decisions are available at www.va.gov.

An April 2002 pamphlet entitled How
Do I Appeal (01–02–02A) is available
at www.bva.va.gov. Printed copies can
be obtained at Mail Processing Section
(014), Board of Veterans' Appeals, 810
Vermont Avenue, NW., Washington, DC
20420. There is no charge for individual
copies. A large quantity of pamphlets may
be purchased from the Superintendent of
Documents, Government Printing Office,
Washington, DC 20402. Call 202–512–
1800 or visit www.gpoacess.gov/index.
html for more information.

The VA pamphlet, A Summary of
Department of Veteran Affairs Benefits
(27–82–2), may be obtained without
charge from any VA regional office.

Interments in VA National Cemeteries,
VA NCA–IS–1, details eligibility
information and contains a list of both

national and State veterans cemeteries. Copies may be obtained without charge from the National Cemetery Administration (41C1), 810 Vermont Avenue NW., Washington, DC 20420. Call 800–827–1000 or visit www.cem. va.gov for more information.

For further information, contact the Office of Public and Intergovernmental Affairs, Department of Veterans Affairs, 810 Vermont Avenue NW., Washington, DC 20420. Phone, 202–273–6000. Internet, http://www. va.gov/opa.

EXECUTIVE BRANCH: INDEPENDENT AGENCIES AND GOVERNMENT CORPORATIONS

ADMINISTRATIVE CONFERENCE OF THE UNITED STATES

1120 Twentieth Street NW., 7th Floor South, Washington, DC 20036
Internet, http://www.acus.gov..

Chair	PAUL R. VERKUIL
Executive Director	(VACANCY)
Director of Research and Policy	JONATHAN R. SIEGEL
General Counsel	(VACANCY)
Deputy General Counsel	(VACANCY)
Council	
Vice Chair	PREETA D. BANSAL
Members	RONALD A. CASS, MARIANO-FLORENTINO CUELLAR, MICHAEL FITZPATRICK, JULIUS GENACHOWSKI, THEODORE OLSON, THOMAS E. PEREZ, THOMASINA V. ROGERS, JANE C. SHERBURNE, PATRICIA MCGOWAN WALD

The Conference develops recommendations for improving the fairness and effectiveness of procedures by which Federal agencies administer regulatory, benefit, and other Government programs.

The Administrative Conference of the United States was established as a permanent independent agency by the Administrative Conference Act (5 U.S.C. 591–596) enacted in 1964. The Conference was the successor to two temporary Administrative Conferences during the Eisenhower and Kennedy administrations.

The Conference ceased operations on October 31, 1995, due to termination of funding by Congress. Congress reauthorized the Conference in 2004 and again in 2008. The 2004 legislation expanded its responsibilities to include specific attention to achieving more effective public participation and efficiency, reducing unnecessary litigation, and improving the use of science in the rulemaking process. Funding was approved in 2009, and the Conference was officially re-established in March 2010.

ADMINISTRATIVE CONFERENCE OF THE UNITED STATES

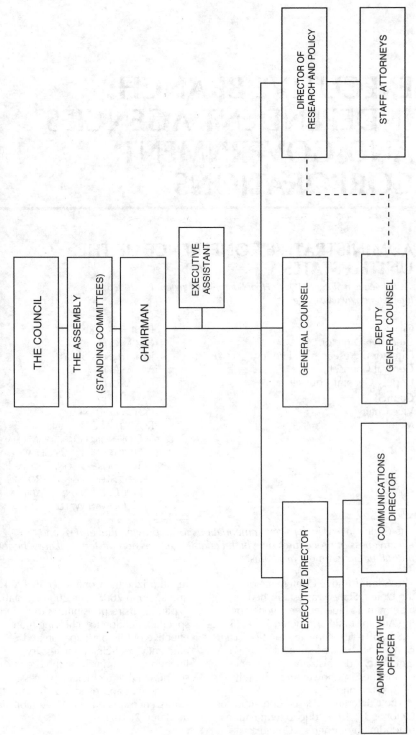

By statute the Administrative Conference has no fewer than 75 and no more than 101 members, a majority of whom are Government officials. The Chairman of the Conference is appointed by the President with the advice and consent of the Senate for a 5-year term. The Council, which acts as the executive board, consists of the Chairman and 10 other members appointed by the President for 3-year terms. Federal officials named to the Council may constitute no more than half of the total Council membership. Members representing the private sector are appointed by the Chairman, with the approval of the Council, for 2-year terms. The Chairman is the only full-time, compensated member.

The entire membership is divided into committees, each assigned a broad area of interest such as adjudication, administration, governmental processes, judicial review, regulation, or rulemaking. The membership meeting in plenary session constitutes the Assembly of the Conference, which by statute must meet at least once, and customarily meets twice, each year.

Activities

Subjects for inquiry are developed by the Chairman and approved by the Council. The committees conduct thorough studies of these subjects and propose recommendations, based on supporting reports, ordinarily prepared for the Conference by expert consultants. Recommendations are evaluated by the Council and, if ready for Assembly consideration, are distributed to the membership with the supporting reports and placed on the agenda of the next plenary session. The deliberations of the committees and Assembly are open to the public.

The Chairman is authorized to encourage the departments and agencies to adopt the recommendations of the Conference and is required to transmit to the President and to Congress an annual report and interim reports concerning the activities of the Conference, including reports on the implementation of its recommendations.

Recommendations adopted by the Conference may call for new legislation or for action on the part of affected agencies. A substantial number of recommendations were implemented prior to the termination of Conference activities in 1995.

The Chairman may make independent inquiries into procedural matters, including matters proposed by individuals inside or outside the Government. The purpose of such inquiries is to determine whether the problems should be made the subject of Conference study in the interest of developing fair and effective procedures.

Upon the request of the head of a department or agency, the Chairman is authorized to furnish advice and assistance on matters of administrative procedure. The Conference may collect information and statistics from departments and agencies and publish such reports as it considers useful for evaluating and improving administrative processes. The Conference also serves as a forum for the interchange among departments and agencies of information that may be useful in improving administrative practices and procedures.

Sources of Information

The Conference will make available, principally through the Internet, copies of its recommendations and reports, as well as information about work currently in progress.

For further information, contact the Office of the Chairman, Administrative Conference of the United States, 1120 Twentieth Street NW., 7th Floor South, Washington, DC 20036. Internet, http://www.acus.gov.

AFRICAN DEVELOPMENT FOUNDATION

1400 I Street NW., Suite 1000, Washington, DC 20005
Phone, 202–673–3916. Fax, 202–673–3810. Internet, http://www.usadf.gov.

Board of Directors

Chairman	JACK LESLIE
Vice Chairman	JOHN AGWUNOBI
Board Members	JULIUS E. COLES, MORGAN M. DAVIS, (3 VACANCIES)

Staff

President	LLOYD O. PIERSON
General Counsel	DORIS MASON MARTIN
Chief Financial Officer and Director for Strategic Planning	WILLIAM E. SCHUERCH
Chief Information Officer	(VACANCY)
Director of Management and Administration and Chief Human Capital Officer	M. CATHERINE GATES
Director of Legislative and Public Affairs	(VACANCY)

[For the African Development Foundation statement of organization, see the Code of Federal Regulations, Title 22, Part 1501]

The African Development Foundation promotes development and empowerment in Africa and enhances and strengthens U.S. relations with Africa through effective development assistance.

The African Development Foundation was established by the African Development Foundation Act (22 U.S.C. 290h) as a Government corporation to support the self-help efforts of the poor in Africa.

The Foundation invests in private and nongovernmental organizations in Africa to promote and support innovative enterprise development, generate jobs, and increase incomes of the poor. It seeks to expand local institutional and financial capacities to foster entrepreneurship, ownership, and community-based economic development.

The Foundation also works within the United States, in African countries, and with other nation states to gather and expand resources for grassroots development. It achieves this through strategic partnerships with U.S. and international private sector corporations, African host governments, U.S. and other government agencies, and philanthropic organizations.

For further information, contact the Director of Legislative and Public Affairs, African Development Foundation, 1400 I Street NW., Suite 1000, Washington, DC 20005–2248. Phone, 202–673–3916. Fax, 202–673–3810. Email, info@usadf.gov. Internet, http://www.usadf.gov.

BROADCASTING BOARD OF GOVERNORS

330 Independence Avenue SW., Washington, DC 20237
Phone, 202–203–4545. Internet, http://www.bbg.gov.

Chairman	(VACANCY)
Board Members	JOAQUIN F. BLAYA, BLANQUITA WALSH CULLUM, D. JEFFREY HIRSCHBERG, STEVEN J. SIMMONS, (4 VACANCIES)

(Secretary of State, ex officio)	HILLARY RODHAM CLINTON
Executive Director	JEFFREY N. TRIMBLE
General Counsel	(VACANCY)
Chief Financial Officer	MARYJEAN BUHLER
Director, International Broadcasting Bureau	(VACANCY)
Deputy Director, International Broadcasting Bureau	DANFORTH W. AUSTIN, *Acting*
Director, Office of Public Affairs	LETITIA KING
Director, Office of Strategic Planning and Performance Measurement	BRUCE SHERMAN
Director, Office of New Media	REBECCA MCMENAMIN
Director, Office of Marketing and Program Placement	DOUG BOYNTON
Director, Office of Performance Review	KELU CHAO
Director, Office of Civil Rights	DELIA L. JOHNSON
Director, Office of Contracts	VIVIAN B. GALLUPS
Director, Office of Human Resources	DONNA GRACE
Director, Office of Security	MICHAEL LAWRENCE
Director, Office of Policy	CHARLES GOOLSBY
Director, Office of Engineering and Technical Services and Chief Information Officer	ANDRE MENDES
Director, Voice of America	DANFORTH W. AUSTIN
Director, Office of Cuba Broadcasting	PEDRO V. ROIG
President, Radio Free Europe/Radio Liberty	JEFFREY GEDMIN
President, Radio Free Asia	LIBBY LIU
President, Middle East Broadcasting Networks	BRIAN T. CONNIFF

The Broadcasting Board of Governors promotes freedom and democracy by broadcasting accurate, objective, and balanced news and information about the United States and the world to audiences abroad.

The Broadcasting Board of Governors (BBG) became an independent agency on October 1, 1999, by authority of the Foreign Affairs Reform and Restructuring Act of 1998 (22 U.S.C. 6501 note). It is composed of nine members. Eight members are appointed by the President and confirmed by the Senate; the ninth, an ex officio member, is the Secretary of State.

The BBG serves as the governing body for all nonmilitary U.S. broadcasting and provides programming in 56 languages via radio, television, and the Internet. The BBG broadcast services include the Voice of America, the Office of Cuba Broadcasting, Radio Free Europe/Radio Liberty, Radio Free Asia, and the Middle East Broadcasting Networks.

All BBG broadcast services adhere to the broadcasting standards and principles of the International Broadcasting Act of 1994, which include reliable, accurate, and comprehensive news; balanced and comprehensive presentations of U.S. thought, institutions, and policies, as well as discussions about those policies; information about developments throughout the world; and a variety of opinions from nations around the world.

Activities

International Broadcasting Bureau

The International Broadcasting Bureau (IBB) provides human resource, EEO, procurement, security, information technology, public affairs, administrative, research, and program evaluation services to the Voice of America and Radio/TV Marti. It also provides marketing, program placement, and transmission services for all the BBG broadcast organizations, managing a global network of transmitting sites and an extensive system of leased satellite and fiber optic circuits, along with a rapidly growing Internet delivery system.

BROADCASTING BOARD OF GOVERNORS

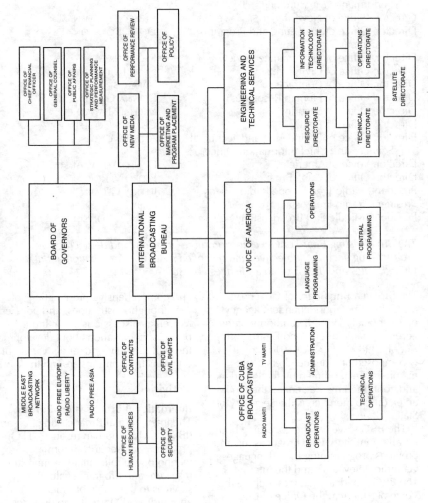

The IBB Office of Policy produces U.S. Government editorials broadcast daily on VOA. Internet, www.ibb.gov.

Voice of America   Voice of America (VOA) is an independent international multimedia broadcasting service of the U.S. Government. VOA broadcasts more than 1,000 hours a week, which include U.S. and international news, features, and educational and cultural programs about the United States and the world, to more than 115 million people worldwide. VOA programs are produced and broadcast in 44 languages. More than 1,200 VOA affiliate radio and television stations around the world download programs through satellite and digital audio (MP3) technology. Internet, www.voanews.com.

Radio Free Europe/Radio Liberty   Radio Free Europe/Radio Liberty (RFE/RL) is an international communications service to Eastern Europe, the Caucasus, and Central and Southwestern Asia. Concentrating on domestic events and international news, RFE/RL has for more than 50 years provided its 35 million listeners with balanced and reliable information to bolster democratic development and civil society in countries struggling to overcome autocratic institutions, violations of human rights, centralized economies, ethnic and religious hostilities, regional conflicts, and controlled media. A not-for-profit corporation, RFE/RL is funded by a Federal grant from the Broadcasting Board of Governors. Internet, www.rferl.org.

Radio Free Asia   Radio Free Asia (RFA) is an independent, nonprofit organization broadcasting and publishing information online in nine East Asian languages, including Burmese, Cantonese, Khmer, Korean, Lao, Mandarin, Tibetan, Uighur, and Vietnamese. RFA provides news and information related specifically to the people in East Asia who are censored from officially sanctioned domestic media. All language services are staffed and directed by native speakers and maintain Web sites which offer podcasting, syndication, and feedback options, while several services also offer regular toll-free hotlines to callers in Asia. RFA is funded by an annual grant from the Broadcasting Board of Governors. Internet, www.rfa.org.

Middle East Broadcasting Networks   The Middle East Broadcasting Networks, Inc., is a nonprofit corporation that operates the Arabic language Alhurra TV and Radio Sawa networks. Alhurra TV broadcasts to 22 countries in the Middle East via satellite. Its schedule includes up-to-the-minute newscasts, documentaries, discussion programs, and other programs on a variety of subjects. Radio Sawa broadcasts on FM and AM to major Middle Eastern countries. It also seeks to reach a significant portion of the under-30 population with a combination of news, opinion features, and a blend of mainstream Western and Arabic music. Internet, www.radiosawa.com and www.alhurra.com.

Office of Cuba Broadcasting   The Office of Cuba Broadcasting oversees Radio Marti and TV Marti. These two Spanish language services provide news about Cuba and the world, features, and entertainment programs aimed at Cuba. Based in Miami, these comprehensive and timely broadcasts offer Cubans the opportunity to receive unfiltered and accurate information. Radio and TV Marti are disseminated through medium wave (AM), shortwave, Internet, satellite, and special transmissions. Internet, www.martinoticias.org.

For further information, contact the Office of Public Affairs, Broadcasting Board of Governors, 330 Independence Avenue SW., Washington, DC 20237. Phone, 202–203–4959. Fax, 202–203–4960. Email, publicaffairs@ibb.gov. Internet, http://www.bbg.gov.

CENTRAL INTELLIGENCE AGENCY

Washington, DC 20505
Phone, 703–482–0623. Internet, http://www.cia.gov.

Director	LEON E. PANETTA
Deputy Director	MICHAEL J. MORELL
Associate Deputy Director	STEPHANIE O'SULLIVAN
General Counsel	STEPHEN W. PRESTON
Director of Public Affairs	PAUL J. GIMIGLIANO, *Acting*
Director of Intelligence	FRAN P. MOORE
Director, the National Clandestine Service	MICHAEL SULICK
Director of Science and Technology	GLENN A. GAFFNEY
Director, Center for the Study of Intelligence	PETER S. USOWSKI

[For the Central Intelligence Agency statement of organization, see the Code of Federal Regulations, Title 32, Part 1900]

The Central Intelligence Agency collects, evaluates, and disseminates vital information on political, military, economic, scientific, and other developments abroad needed to safeguard national security.

The Central Intelligence Agency was established by the National Security Act of 1947, as amended (50 U.S.C. 401 et seq.). It now functions under that statute, Executive Order 12333 of December 4, 1981, the Intelligence Reform and Terrorism Prevention Act of 2004 (50 U.S.C. 401 note), and other laws, Executive orders, regulations, and directives.

The Central Intelligence Agency is headed by a Director, who is appointed by the President with the advice and consent of the Senate.

The Central Intelligence Agency does the following: collects intelligence from human sources and other appropriate means, but it does not carry out internal security functions nor exercise police, subpoena, or law enforcement powers; correlates, evaluates, and disseminates intelligence related to national security; provides overall direction for and coordination of intelligence collecting outside the United States by U.S.

Intelligence Community elements authorized to engage in human source collection. In coordination with other departments, agencies, or authorized elements of the United States Government, it ensures that resources are used effectively and that adequate consideration is given to the risks to those involved in such collection and to the United States; carries out other intelligence-related functions and duties necessary for safeguarding national security as the President or the Director of National Intelligence (DNI) may direct; and coordinates, under the direction of the DNI and consistent with section 207 of the Foreign Service Act of 1980, relationships between elements of the U.S. Intelligence Community and the intelligence or security services of foreign governments or international organizations in matters of national security or intelligence that is acquired clandestinely.

For further information, contact the Central Intelligence Agency, Office of Public Affairs, Washington, DC 20505. Phone, 703–482–0623. Fax, 703–482–1739. Internet, http://www.cia.gov.

COMMODITY FUTURES TRADING COMMISSION

1155 Twenty-first Street NW., Washington, DC 20581
Phone, 202–418–5000. Fax, 202–418–5521. Internet, http://www.cftc.gov..

Chairman	GARY GENSLER
Commissioners	BARTHOLOMEW H. CHILTON,
	MICHAEL V. DUNN, SCOTT D.
	O'MALIA, JILL E. SOMMERS
General Counsel	TERRY S. ARBIT
Executive Director	MADGE BOLINGER
Director, Division of Market Oversight	RICHARD SHILTS
Director, Division of Clearing and Intermediary Oversight	ANANDA RADHAKRISHNAN
Director, Division of Enforcement	STEPHEN OBIE
Chief Economist	GEOFFREY PRICE

[For the Commodity Futures Trading Commission statement of organization, see the Code of Federal Regulations, Title 17, Part 140]

The Commodity Futures Trading Commission protects market users and the public from fraud, manipulation, and abusive practices related to the sale of commodity futures and options, and to foster open, competitive, and financially sound commodity futures and option markets.

The Commodity Futures Trading Commission, the Federal regulatory agency for futures trading, was established by the Commodity Futures Trading Commission Act of 1974 (7 U.S.C. 4a). The Commission began operation in April 1975, and its authority to regulate futures trading was renewed by Congress in 1978, 1982, 1986, 1992, 1995, and 2000.

The Commission consists of five Commissioners who are appointed by the President, with the advice and consent of the Senate. One Commissioner is designated by the President to serve as Chairman. The Commissioners serve staggered 5-year terms, and by law no more than three Commissioners can belong to the same political party.

The Commission has six major operating components: the Divisions of Market Oversight, Clearing and Intermediary Oversight, and Enforcement and the Offices of the Executive Director, General Counsel, and Chief Economist.

Activities

The Commission regulates trading on the U.S. futures markets, which offer commodity futures and options contracts. It regulates these markets in order to ensure the operational integrity of the futures markets. The Commission regulates two tiers of markets: designated contract markets and registered derivatives transaction execution facilities. It also exercises more limited regulatory or enforcement authority over other types of markets. Additionally, the Commission regulates derivatives clearing organizations. Each board of trade that operates a designated contract market must own or have a relationship with a derivatives clearing organization that provides clearing services for each futures contract executed.

The Commission also regulates the activities of numerous commodity trading professionals, including brokerage houses (futures commission merchants), futures industry salespersons (associated persons), commodity trading advisers, commodity pool operators, and floor brokers and traders.

The Commission's regulatory and enforcement efforts are designed to foster transparent and financially sound markets, encourage market competition and efficiency, ensure market integrity, and protect market participants and the public from fraud, manipulation, and abusive practices. It oversees the rules

COMMODITY FUTURES TRADING COMMISSION

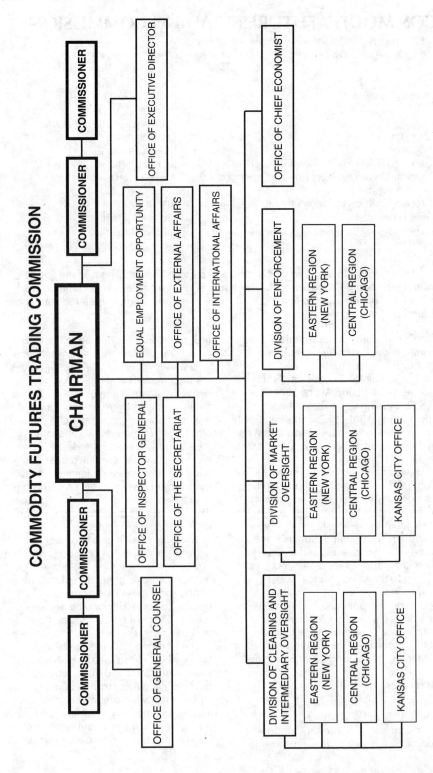

under which designated contract markets and derivatives clearing organizations operate and monitors enforcement of those rules. The Commission reviews the terms of futures contracts and registers firms and individuals who handle customer funds or give trading advice. It also protects the public by enforcing rules that require customer funds be kept in separate accounts, away from accounts maintained by firms for their own use or maintained on behalf of the firm, and that such customer accounts be marked to present market value at the close of trading each day.

The Commission maintains regional offices in Chicago, IL, and New York, NY, where many of the Nation's designated contract markets are located. An additional regional office is located in Kansas City, MO.

For further information, contact the Office of External Affairs, Commodity Futures Trading Commission, 1155 Twenty-first Street NW., Washington, DC 20581. Phone, 202–418–5080. Internet, http://www.cftc.gov.

CONSUMER PRODUCT SAFETY COMMISSION

4330 East-West Highway, Bethesda, MD 20814
Phone, 301–504–7923. Internet, http://www.cpsc.gov.

Chair	INEZ TENEBAUM
Commissioners	ROBERT ADLER, THOMAS H. MOORE, NANCY NORD, ANNE NORTHUP
General Counsel	CHERYL A. FALVEY
Director, Office of Congressional Relations	JOHN HORNER
Director, Office of the Secretary	TODD A. STEVENSON
Freedom of Information Officer	ALBERTA MILLS
Director, Office of Equal Employment Opportunity and Minority Enterprise	KATHLEEN V. BUTTREY
Executive Director	MARUTA BUDETTI
Deputy Executive Director	JACQUELINE ELDER
Inspector General	CHRISTOPHER W. DENTEL
Director, Office of Human Resources Management	DONNA M. SIMPSON
Director, Office of International Programs and Intergovernmental Affairs	RICHARD O'BRIEN
Assistant Executive Director, Office of Information and Technology Services	PATRICK D. WEDDLE
Director, Office of Information and Public Affairs	SCOTT J. WOLFSON
Director, Office of Financial Management, Planning and Evaluation	EDWARD E. QUIST

[For the Consumer Product Safety Commission statement of organization, see the Code of Federal Regulations, Title 16, Part 1000]

The Consumer Product Safety Commission protects the public by reducing the risk of injuries and deaths from consumer products.

The Consumer Product Safety Commission was established as an independent regulatory agency by the Consumer Product Safety Act (15 U.S.C. 2051 et seq.) in 1973 and reauthorized by the Consumer Product Safety Improvement Act of 2008. The Commission consists of up to five members, who are appointed by the President with the advice and consent of the Senate, for 7-year terms.

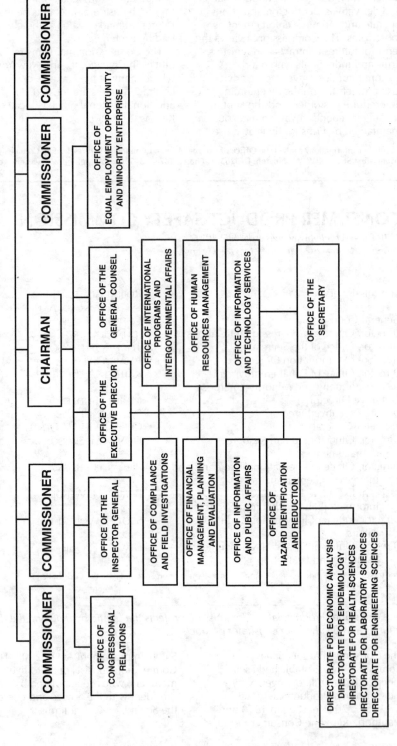

CONSUMER PRODUCT SAFETY COMMISSION

The Commission implements provisions of the Flammable Fabrics Act (15 U.S.C. 1191); Poison Prevention Packaging Act of 1970 (15 U.S.C. 1471); Federal Hazardous Substances Act (15 U.S.C. 1261); act of August 2, 1956 (15 U.S.C. 1211), prohibiting the transportation of refrigerators without door safety devices; Children's Gasoline Burn Prevention Act (15 U.S.C. 2056 note); and Virginia Graeme Baker Pool and Spa Safety Act (15 U.S.C. 8001 et seq.).

Activities

To help protect the public from unreasonable risks of injury associated with consumer products, the Commission requires manufacturers to report defects in products that could present substantial hazards; requires, where appropriate, corrective action with respect to specific substantially hazardous consumer products already in commerce; collects information on consumer product-related injuries and maintains a comprehensive Injury Information Clearinghouse; conducts research on consumer product hazards; encourages and assists in the development of voluntary standards related to the safety of consumer products; establishes, where appropriate, mandatory consumer product standards; bans, where appropriate, hazardous consumer products; and conducts outreach programs for consumers, industry, and local governments.

The Commission also has a special project to reach as many Americans as possible with important, lifesaving safety information. The Neighborhood Safety Network is an effort to disseminate safety information to hard-to-reach populations by partnering with other organizations within these populations. Organizations may register for this program at www.cpsc.gov/nsn/nsn.html.

Sources of Information

Consumer Information The Commission operates a toll-free Consumer Product Safety Hotline, 800–638–2772 (English and Spanish); and a teletypewriter for the hearing-impaired, 800–638–8270 (or in Maryland only, 800–492–8140). The Commission has several Web sites where additional safety information may be obtained. Information specific to recalls can be obtained at ccess.gpo.gov/cgi-ed. All-Terrain Vehicle safety information can be obtained from East-West Highway, Bethesda, MD 20814. Phone, 301–504–7923. Internet, www.cpsc.gov.

Reading Room A public information room is maintained at the Commission's headquarters.

For further information, contact the Office of Information and Public Affairs, Consumer Product Safety Commission, 4330 East-West Highway, Bethesda, MD 20814. Phone, 301–504–7908. Email, info@cpsc.gov. Internet, http://www.cpsc.gov.

EDITORIAL NOTE: The Corporation for National and Community Service did not meet the publication deadline for submitting updated information of its activities, functions, and sources of information as required by the automatic disclosure provisions of the Freedom of Information Act (5 U.S.C. 552(a)(1)(A)).

CORPORATION FOR NATIONAL AND COMMUNITY SERVICE

1201 New York Avenue NW., Washington, DC 20525
Phone, 202–606–5000. Internet, http://www.nationalservice.gov.

Chair	MARK D. GEARAN
Vice Chair	ERIC J. TANENBLATT
Members	JULIE FISHER CUMMINGS, MARK D. GEARAN, HYEPIN IM, JAMES PALMER, STAN SOLOWAY, ERIC J. TANENBLATT, LAYSHA WARD, (5 VACANCIES)
Chief Executive Officer	PATRICK CORVINGTON
Inspector General	KENNETH BACH, *Acting*
Chief Financial Officer	WILLIAM ANDERSON
Chief Human Capital Officer	RAYMOND LIMON
Chief of Program Operations	KRISTIN MCSWAIN
Director, AmeriCorps*NCCC	MIKEL HERRINGTON
Director, AmeriCorps*State and National	JOHN GOMPERT
Director, AmeriCorps*VISTA	PAUL DAVIS, *Acting*
Director, Civil Rights and Inclusiveness	LIZ A. HONNOLL
Director, Corporate Relations	RHONDA TAYLOR, *Acting*
Director, Emergency Management	COLLEEN CLAY

The Corporation for National and Community Service engages in community-based service and volunteering.

The Corporation was established on October 1, 1993, by the National and Community Service Trust Act of 1993 (42 U.S.C. 12651 et seq.). In addition to creating several new service programs, the Act consolidated the functions and activities of the former Commission on National and Community Service and the Federal agency ACTION.

For more than a decade, the Corporation for National and Community Service, through its Senior Corps, AmeriCorps, and Learn and Serve America programs, has mobilized a new generation of engaged citizens. This year, more than 2 million individuals of all ages and backgrounds will serve through those programs to help thousands of national and community nonprofit organizations, faith-based groups, schools, and local agencies meet local needs in education, the environment, public safety, homeland security, and other critical areas. National and community service programs work closely with traditional volunteer organizations to broaden, deepen, and strengthen the ability of America's volunteers to contribute not only to their community, but also to our Nation.

The Corporation is a Federal corporation governed by a 15-member bipartisan Board of Directors, appointed by the President with the advice and consent of the Senate. The Board has responsibility for overall policy direction of the Corporation's activities and has the power to make all final grant decisions, approve the strategic plan and annual budget, and advise and make

CORPORATION FOR NATIONAL AND COMMUNITY SERVICE

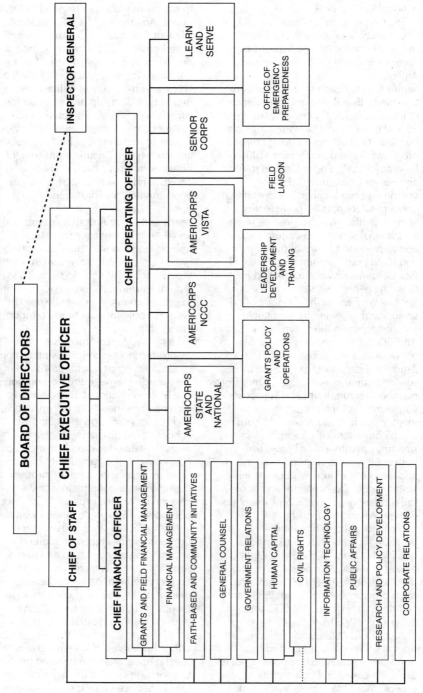

recommendations to the President and the Congress regarding changes in the national service laws.

The Corporation for National and Community Service oversees three major service initiatives: Senior Corps, AmeriCorps, and Learn and Serve America.

Senior Corps   Each year Senior Corps taps the skills, talents, and experience of more than 500,000 Americans age 55 and older to meet a wide range of community challenges through three programs: Retired and Senior Volunteers Program (RSVP), Foster Grandparents, and Senior Companions. RSVP volunteers help local police departments conduct safety patrols, participate in environmental projects, provide intensive educational services to children and adults, respond to natural disasters, and recruit other volunteers. Foster Grandparents serve as tutors and mentors to young people with special needs. Senior Companions help homebound seniors and other adults maintain independence in their own homes.

AmeriCorps   AmeriCorps provides opportunities for more than 75,000 Americans each year to give intensive service to their communities. AmeriCorps members recruit, train, and supervise community volunteers, tutor and mentor youth, build affordable housing, teach computer skills, clean parks and streams, run after-school programs, and help communities respond to disasters and nonprofit groups to become self-sustaining. In exchange for a year of full-time service, AmeriCorps members earn an education award of $4,725 that can be used to pay for college or graduate school, or to pay back qualified student loans. Since 1994 almost 500,000 Americans have served in AmeriCorps. AmeriCorps has three main programs: AmeriCorps*State and National, AmeriCorps*NCCC, and AmeriCorps*VISTA.

AmeriCorps*State and National operates through national and local nonprofit organizations, public agencies, and faith-based and community groups. More than three-quarters of AmeriCorps grant funding goes to Governor-appointed State service commissions, which in turn award grants to nonprofit groups to respond to local needs. AmeriCorps*NCCC (National Civilian Community Corps) is a team-based, residential program for men and women from ages 18 to 24 that combines the best practices of civilian service with aspects of military service, including leadership and team building. AmeriCorps*VISTA (Volunteers in Service to America) members serve full-time, for 1 year, in nonprofits, public agencies, and faith-based organizations to fight illiteracy, improve health services, build and expand business, increase housing opportunities, and bridge the digital divide.

Learn and Serve America Learn and Serve America engages more than 1 million students in community service linked to academic achievement and the development of civic skills. This type of learning, called service learning, improves communities while preparing young people for a lifetime of responsible citizenship.

Learn and Serve America provides grants to schools, colleges, and nonprofit groups to support its efforts. Grants are awarded through the Corporation and State educational agencies and commissions, nonprofit organizations, and higher education associations. School-based programs receive grants through State educational agencies or nonprofits, while community-based programs apply for funding through the same State commissions that coordinate AmeriCorps grants or through nonprofits. Higher education institutions and associations apply directly to the Corporation for grants. In addition to providing grants, Learn and Serve America serves as a resource on service and service-learning to teachers, faculty members, schools, and community groups and promotes student service through the Presidential Freedom Scholarships.

Other Initiatives   Other programs and special initiatives administered by the Corporation's umbrella include: King Day of Service Grants, which support community organizations in their efforts to engage local citizens in service on the Martin Luther King, Jr., Federal holiday; and the President's Council on Service and Civic Participation, which presents the President's Volunteer Service Award to citizens of all ages and backgrounds who have demonstrated a sustained commitment to service. The Corporation also provides extensive training and technical assistance to support and assist State service commissions and local service programs.

The Corporation and its programs work with the USA Freedom Corps, established on January 29, 2002, by Executive Order 13254. USA Freedom Corps is a White House initiative to foster a culture of citizenship, service, and responsibility, and help all Americans answer the President's call to service.

Sources of Information

Electronic Access Information regarding the Corporation's programs and activities is available on the Internet at www.nationalservice.gov. Information for persons interested in joining AmeriCorps is available at www.americorps.gov. Information for persons interested in joining Senior Corps is at www.seniorcorps.gov. Information on the USA Freedom Corps is available at www.usafreedomcorps.gov.

General Information To obtain additional information regarding AmeriCorps, call 800–942–2677. For Senior Corps programs, call 800–424–8867. TDD, 202–565–2799. For USA Freedom Corps, call 877–872–2677.

Grants All notices of available funds are made through the grants.gov Web site. State program offices and commissions on national and community service are located in most States and are the best source of information on programs in specific States or communities. To contact State offices or State commissions, visit www.nationalservice.gov/contactus.html.

For further information, contact the Corporation for National and Community Service, 1201 New York Avenue NW., Washington, DC 20525. Phone, 202–606–5000. Internet, http://www.nationalservice.gov.

DEFENSE NUCLEAR FACILITIES SAFETY BOARD

625 Indiana Avenue NW., Suite 700, Washington, DC 20004
Phone, 202–694–7000. Fax, 202–208–6518. Internet, http://www.dnfsb.gov.

Chairman	PETER S. WINOKUR
Vice Chairman	JOHN E. MANSFIELD
Members	JOSEPH F. BADER, LARRY W. BROWN, JESSIE H. ROBERSON
General Counsel	RICHARD A. AZZARO
General Manager	BRIAN GROSNER
Technical Director	TIMOTHY DWYER

The Defense Nuclear Facilities Safety Board reviews and evaluates the content and implementation of standards relating to the design, construction, operation, and decommissioning of defense nuclear facilities of the Department of Energy.

The Defense Nuclear Facilities Safety Board was established as an independent agency on September 29, 1988, by the Atomic Energy Act of 1954, as amended (42 U.S.C. 2286–2286i).

The Board is composed of five members appointed by the President with the advice and consent of the Senate. Members of the Board are appointed from among United States citizens who are respected experts in the field of nuclear safety.

Activities

The Defense Nuclear Facilities Safety Board reviews and evaluates the content and implementation of standards

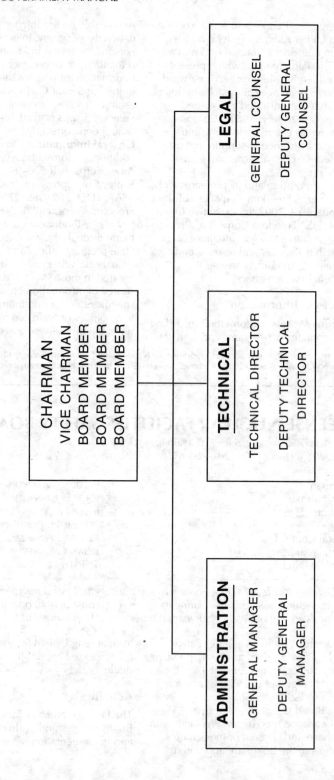

DEFENSE NUCLEAR FACILITIES SAFETY BOARD

CHAIRMAN
VICE CHAIRMAN
BOARD MEMBER
BOARD MEMBER
BOARD MEMBER

LEGAL
GENERAL COUNSEL
DEPUTY GENERAL COUNSEL

TECHNICAL
TECHNICAL DIRECTOR
DEPUTY TECHNICAL DIRECTOR

ADMINISTRATION
GENERAL MANAGER
DEPUTY GENERAL MANAGER

for defense nuclear facilities of the Department of Energy (DOE); investigates any event or practice at these facilities which may adversely affect public health and safety; and reviews and monitors the design, construction, and operation of facilities. The Board makes recommendations to the Secretary of Energy concerning DOE defense nuclear facilities to ensure adequate protection of public health and safety. In the event that any aspect of operations, practices, or occurrences reviewed by the Board is determined to present an imminent or severe threat to public health and safety, the Board transmits its recommendations directly to the President.

For further information, contact the Defense Nuclear Facilities Safety Board, 625 Indiana Avenue NW., Suite 700, Washington, DC 20004. Phone, 202–694–7000. Internet, http://www.dnfsb.gov.

ENVIRONMENTAL PROTECTION AGENCY

1200 Pennsylvania Avenue NW., Washington, DC 20460–0001
Phone, 202–272–0167. Internet, http://www.epa.gov.

Administrator	LISA P. JACKSON
Deputy Administrator	BOB PERCIASEPE
Associate Administrator for External Affairs and Environmental Education	SETH OSTER
Associate Administrator for Congressional and Intergovernmental Relations	ARVIN GANESAN, *Acting*
Associate Administrator for Policy, Economics, and Innovation	LISA HEINZERLING
Associate Administrator for Homeland Security	JUAN REYES, *Acting*
Chief Judge, Office of Administrative Law Judges	SUSAN L. BIRO
Director, Executive Secretariat	ERIC WACHTER
Director, Office of Children's Health	PETER GREVATT
Director, Office of Civil Rights	KAREN D. HIGGINBOTHAM
Director, Office of Cooperative Environmental Management	RAFAEL DELEON
Director, Office of Executive Services	DIANE N. BAZZLE
Director, Office of Small Business Programs	JEANETTE L. BROWN
Director, Science Advisory Board	VANESSA T. VU
Lead Environmental Appeals Judge, Environmental Appeals Board	CHARLES SHEEHAN
Assistant Administrator for Administration and Resources Management	CRAIG E. HOOKS
Assistant Administrator for Air and Radiation	GINA MCCARTHY
Assistant Administrator for Enforcement and Compliance Assurance	CYNTHIA GILES
Assistant Administrator for Environmental Information and Chief Information Officer	LINDA TRAVERS, *Acting*
Assistant Administrator for Tribal and International Affairs	MICHELLE DEPASS
Assistant Administrator for Chemical Safety and Pollution Prevention	STEVE OWENS
Assistant Administrator for Research and Development	PAUL ANASTAS
Assistant Administrator for Solid Waste and Emergency Response	MATHY STANISLAUS

Assistant Administrator for Water	PETER L. SILVA
Chief Financial Officer	BARBARA J. BENNETT
General Counsel	SCOTT FULTON
Inspector General	BILL RODERICK, *Acting*

The Environmental Protection Agency protects human health and safeguards the environment.

The Environmental Protection Agency (EPA) was established in the executive branch as an independent agency pursuant to Reorganization Plan No. 3 of 1970 (5 U.S.C. app.), effective December 2, 1970. It was created to facilitate coordinated and effective governmental action on behalf of the environment. The Agency is designed to serve as the public's advocate for a livable environment.

Core Functions

Air and Radiation The Office of Air and Radiation develops national programs, policies, regulations, and standards for air quality, emission standards for stationary and mobile sources, and emission standards for hazardous air pollutants; conducts research and disseminates information on indoor air pollutants; provides technical direction, support, and evaluation of regional air activities; offers training in the field of air pollution control; gives technical assistance to States and agencies having radiation protection programs, including radon mitigation programs and a national surveillance and inspection program for measuring radiation levels in the environment; and provides technical support and policy direction to international efforts to reduce global and transboundary air pollution and its effects.

For further information, call 202–564–7400.

Water The Office of Water develops national programs, technical policies, and regulations for water pollution control and water supplies; protects ground water, drinking water, and marine and estuarine habitats; controls pollution runoff; develops water quality standards and effluent guidelines; supports regional water activities; develops programs for

technical assistance and technology transfer; and offers water quality training.

For further information, call 202–564–5700.

Solid Waste and Emergency Response The Office of Solid Waste and Emergency Response provides policy, guidance, and direction for EPA's hazardous waste and emergency response programs. It develops policies, standards, and regulations for hazardous waste treatment, storage, and disposal; develops and implements programs to prevent and detect leakage from underground storage tanks and to clean up ensuing contamination; provides technical assistance in safe waste management; administers the Brownfields program which advocates for redevelopment and reuse of contaminated land; and manages the Superfund toxic waste cleanup program to respond to hazardous waste sites and chemical and oil spill accidents.

For further information, call 202–566–0200.

Chemical Safety and Pollution Prevention The Office of Chemical Safety and Pollution Prevention supports the public's right to know about industrial chemicals; prevents pollution through innovative strategies; evaluates and regulates pesticides and industrial chemicals to safeguard all Americans; establishes safe levels for pesticide residues on food; formulates national strategies for control of bioaccumulative, and toxic substances; and develops scientific criteria for assessing chemical substances, standards for test protocols for chemicals, rules and procedures for industry reporting, and scientific information for the regulation of pesticides and toxic chemicals.

For further information, call 202–564–2902.

Research and Development The Office of Research and Development (ORD)

ENVIRONMENTAL PROTECTION AGENCY

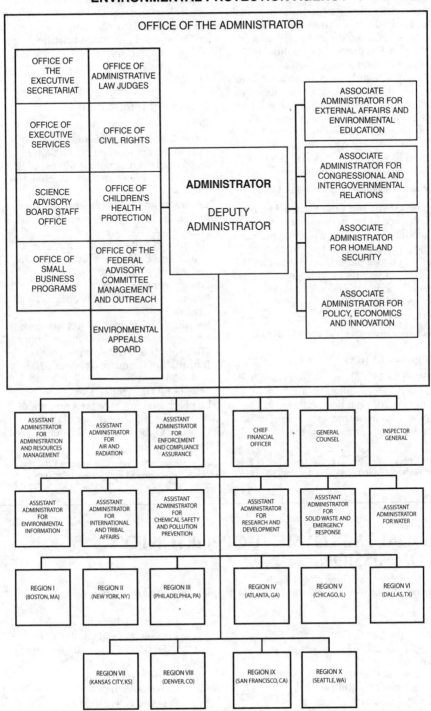

OFFICE OF THE ADMINISTRATOR

OFFICE OF THE EXECUTIVE SECRETARIAT

OFFICE OF ADMINISTRATIVE LAW JUDGES

OFFICE OF EXECUTIVE SERVICES

OFFICE OF CIVIL RIGHTS

SCIENCE ADVISORY BOARD STAFF OFFICE

OFFICE OF CHILDREN'S HEALTH PROTECTION

OFFICE OF SMALL BUSINESS PROGRAMS

OFFICE OF THE FEDERAL ADVISORY COMMITTEE MANAGEMENT AND OUTREACH

ENVIRONMENTAL APPEALS BOARD

ADMINISTRATOR

DEPUTY ADMINISTRATOR

ASSOCIATE ADMINISTRATOR FOR EXTERNAL AFFAIRS AND ENVIRONMENTAL EDUCATION

ASSOCIATE ADMINISTRATOR FOR CONGRESSIONAL AND INTERGOVERNMENTAL RELATIONS

ASSOCIATE ADMINISTRATOR FOR HOMELAND SECURITY

ASSOCIATE ADMINISTRATOR FOR POLICY, ECONOMICS AND INNOVATION

ASSISTANT ADMINISTRATOR FOR ADMINISTRATION AND RESOURCES MANAGEMENT

ASSISTANT ADMINISTRATOR FOR AIR AND RADIATION

ASSISTANT ADMINISTRATOR FOR ENFORCEMENT AND COMPLIANCE ASSURANCE

CHIEF FINANCIAL OFFICER

GENERAL COUNSEL

INSPECTOR GENERAL

ASSISTANT ADMINISTRATOR FOR ENVIRONMENTAL INFORMATION

ASSISTANT ADMINISTRATOR FOR INTERNATIONAL AND TRIBAL AFFAIRS

ASSISTANT ADMINISTRATOR FOR CHEMICAL SAFETY AND POLLUTION PREVENTION

ASSISTANT ADMINISTRATOR FOR RESEARCH AND DEVELOPMENT

ASSISTANT ADMINISTRATOR FOR SOLID WASTE AND EMERGENCY RESPONSE

ASSISTANT ADMINISTRATOR FOR WATER

REGION I (BOSTON, MA)

REGION II (NEW YORK, NY)

REGION III (PHILADELPHIA, PA)

REGION IV (ATLANTA, GA)

REGION V (CHICAGO, IL)

REGION VI (DALLAS, TX)

REGION VII (KANSAS CITY, KS)

REGION VIII (DENVER, CO)

REGION IX (SAN FRANCISCO, CA)

REGION X (SEATTLE, WA)

conducts and supports high-quality research for understanding and resolving the Nation's most serious environmental threats. ORD develops methods and technologies to reduce exposures to pollution and prevent its creation. It prepares health and ecological risk assessments and makes recommendations for sound risk management strategies in order to assure that highest risk pollution problems receive optimum remediation. ORD manages the Science To Achieve Results Program, which awards research grants to scientists in universities and environmental science students.

For further information, call 202–564–6620.

Enforcement and Compliance Assurance

The Office of Enforcement and Compliance Assurance (OECA) manages a national criminal enforcement, forensics, and training program. OECA also manages EPA's regulatory, site remediation, and Federal facilities enforcement and compliance assurance programs.

For further information, call 202–564–2440.

Regional Offices

EPA's 10 regional offices are committed to the development of strong local programs for pollution abatement. The Regional Administrators are responsible for accomplishing, within their

regions, the Agency's national program objectives. They develop, propose, and implement an approved regional program for comprehensive and integrated environmental protection activities.

For more information, visit our Web site at www.epa.gov/epahome/where. htm.

Sources of Information

Requests for information on the following subjects should be directed or sent by mail to the appropriate organization listed below. Our mailing address is 1200 Pennsylvania Avenue NW., Washington, DC 20460.

Contracts and Procurement   Office of Acquisition Management. Phone, 202–564–4310.

Grants and Fellowships   Office of Grants and Debarment. Internet, www. epa.gov/ogd.

Employment   Office of Human Resources. Internet, www.epa.gov/ careers.

Freedom of Information Act Requests Freedom of Information Officer. Phone, 202–566–1667. Email, hq.foia@epa.gov.

Information Resources   Phone, 202– 564–6665.

Telephone Directory   Directories may be purchased from the Superintendent of Documents, Government Printing Office, P.O. Box 37194, Pittsburgh, PA 15250–7954.

For further information, contact the Office of Public Affairs, Environmental Protection Agency, 1200 Pennsylvania Avenue NW., Washington, DC 20460–0001. Phone, 202–564–4355. Internet, http://www.epa. gov.

EQUAL EMPLOYMENT OPPORTUNITY COMMISSION

131 M Street NE., Washington, DC 20507
Phone, 202–663–4900. TTY, 202–663–4444. Internet, http://www.eeoc.gov.

Chair	JACQUELINE BERRIEN
Commissioners	CONSTANCE S. BARKER, CHAI
	FELDBLUM, VICTORIA A. LIPNIC
Commissoner	STUART J. ISHIMARU
Executive Officer	STEPHEN LLEWELLYN
Chief Operating Officer	CLAUDIA WITHERS
General Counsel	P. DAVID LOPEZ

Inspector General	MILTON MAYO, *Acting*
Director, Office of Communications and Legislative Affairs	BRETT BRENNER, *Acting*
Director, Office of Equal Opportunity	VERONICA VILLALOBOS
Director, Office of Federal Operations	CARLTON M. HADDEN
Legal Counsel	PEGGY R. MASTROIANNI, *Acting*
Director, Office of Field Programs	NICHOLAS INZEO
Chief Financial Officer	JEFFREY SMITH
Director, Office of Human Resources	LISA WILLIAMS
Director, Office of Information Technology	KIMBERLY HANCHER
Director, Office of Research, Information, and Planning	DEIDRE FLIPPEN

The Equal Employment Opportunity Commission enforces laws prohibiting employment discrimination based on race, color, gender, religion, national origin, age, disability, or genetic information.

The Equal Employment Opportunity Commission (EEOC) was created by Title VII of the Civil Rights Act of 1964 (42 U.S.C. 2000e-4), and became operational July 2, 1965. Laws under the EEOC's enforcement mission include Title VII of the Civil Rights Act of 1964 (42 U.S.C. 2000e et seq.), the Age Discrimination in Employment Act of 1967 (29 U.S.C. 621 et seq.), sections of the Rehabilitation Act of 1973 (29 U.S.C. 791 et seq.), the Equal Pay Act of 1963 (29 U.S.C. 206), Title I of the Americans with Disabilities Act of 1990 (42 U.S.C. 12101 et seq.), and sections of the Civil Rights Act of 1991 (105 Stat. 1071).

The EEOC is a bipartisan commission comprising five members appointed by the President, with the advice and consent of the Senate, for staggered 5-year terms. The President designates a Chairman and Vice Chairman. In addition to the members of the Commission, the President appoints a General Counsel, with the advice and consent of the Senate, to support the Commission and provide direction, coordination, and supervision of the EEOC's litigation program. The General Counsel serves for a term of 4 years.

Activities

Enforcement   The EEOC enforces its statutory, regulatory, policy, and program responsibilities through its headquarters-based Office of Field Programs, Office of General Counsel, and 53 field offices.

The field offices receive charges of discrimination from the public and use a variety of resolution methods, tailored to each charge, from voluntary mediation to full-scale investigation and conciliation. The field staff is responsible for achieving a wide range of objectives that focus on the quality, timeliness, and appropriateness of individual, class, and systemic charges; for securing relief for victims of discrimination in accordance with Commission policies; for counseling individuals about their rights under the laws enforced by the EEOC; and for conducting outreach and technical assistance programs. The Office of General Counsel conducts the Commission's litigation in U.S. District Courts and Courts of Appeal.

For information on the nearest field office, visit our Web site at www.eeoc.gov/field/index.cfm.

Complaints Against the Federal Government   The EEOC establishes the procedures for Federal employees and job applicants to file complaints of employment discrimination or retaliation. The agency charged with discrimination is responsible for informal counseling and, if a complaint is filed and accepted, investigating the claims raised therein. At the conclusion of the investigation, complainants may request a hearing before an EEOC administrative judge or that the agency issue a final decision on the matter. The agency's final decision or final action after a hearing may be appealed to the Commission.

EQUAL EMPLOYMENT OPPORTUNITY COMMISSION

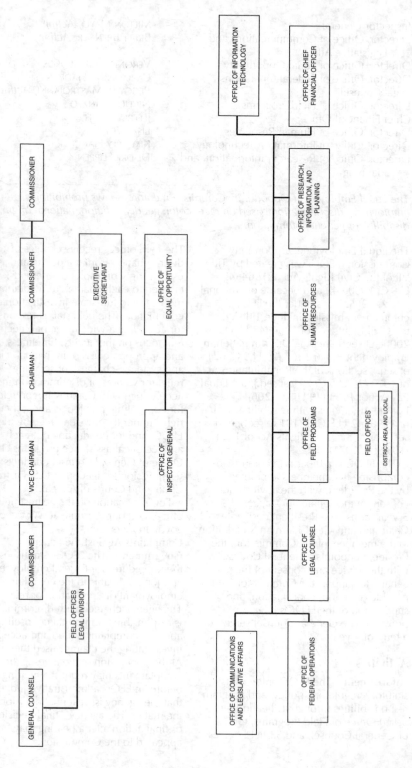

In addition, the Office of Federal Operations provides oversight and technical assistance concerning equal employment opportunity (EEO) complaint adjudication and the maintenance of affirmative employment programs throughout the Federal Government. Using the guidance and principles contained in the EEOC's EEO Management Directive 715, the Commission monitors and evaluates Federal agencies' affirmative employment programs under Title VII and section 501 of the Rehabilitation Act and ensures that all Federal employees compete on a fair and level playing field.

Other Activities   The Commission promotes voluntary compliance with EEO statutes through a variety of educational and technical assistance activities. The Commission's outreach and education programs provide general information about the EEOC, its mission, rights and responsibilities under the statutes enforced by the Commission, and the charge/complaint process. EEOC representatives are available, on a limited basis, at no cost, to make presentations and participate in meetings, conferences, and seminars with employee and employer groups, professional associations, students, nonprofit entities, community organizations, and other members of the general public.

The Commission also offers more in-depth training tailored to employers for a fee. This training is available to private employers and State, local, and Federal government personnel through the EEOC Training Institute. The EEOC Training Institute provides a wide variety of training to assist in educating managers and employees on the laws enforced by the EEOC and how to prevent and eliminate discrimination in the workplace. The EEOC develops policy guidance and provides technical assistance to employers and employees and coordinates with other agencies and stakeholders regarding the statutes and regulations it enforces.

The Commission also publishes data on the employment status of minorities and women through six employment surveys covering private employers, apprenticeship programs, labor unions, State and local governments, elementary and secondary schools, and colleges and universities. This collection of data is shared with selected Federal agencies and is made available, in appropriate form, for public use.

Sources of Information

Electronic Access   Information regarding the programs and activities of the Commission is available through the Internet at www.eeoc.gov. EEOC's most popular publications may be downloaded from that site in PDF format for easy reproduction.

Employment   The Commission hires in many different job categories, including information intake representatives, office automation assistants, investigators, paralegals, program analysts, attorneys, mediators, and social scientists. Employment opportunities are posted on www.usajobs.gov. Employment inquiries may also be directed to the appropriate district office or the Office of Human Resources, Equal Employment Opportunity Commission, 131 M Street NE., Washington, DC 20507. Phone, 202–663–4306.

General Inquiries   A nationwide toll-free telephone number links callers with the appropriate field office where charges may be filed. Phone, 800–669–4000. TTY, 800–669–6820.

Media Inquiries   Inquiries from representatives of the media should be made to the Office of Communications, Office of Communications and Legislative Affairs, 131 M Street NE., Washington, DC 20507. Phone, 202–663–4191. TTY, 202–663–4494. Email, newsroom@eeoc.gov.

Information About Survey Forms Information about EEO survey forms, no. 1–6, can be obtained from the Office of Research Information Planning, 131 M Street NE., Washington, DC 20507. Phone, 202–663–3362.

Publications   Publications not available on www.eeoc.gov may be obtained by phone or fax. Phone, 800–669–3362. TTY, 800–800–3302. Fax, 513–489–8692.

Reading Room EEOC Library, 131 M Street NE., Washington, DC 20507. Phone, 202–663–4630.

Speakers Office of the Executive Secretariat, 131 M Street NE., Washington, DC 20507. Phone, 202–663–4070. TTY, 202–663–4494.

For further information, contact the Equal Employment Opportunity Commission, 131 M Street NE., Washington, DC 20507. Phone, 202–663–4191. Internet, http://www.eeoc.gov.

EXPORT-IMPORT BANK OF THE UNITED STATES

811 Vermont Avenue NW., Washington, DC 20571.
Phone, 202–565–3946; 800–565–3946. Internet, http://www.exim.gov.

President/Chairman	FRED P. HOCHBERG
Vice Chairman	(VACANCY)
Directors	DIANE FARRELL, BIJAN R. KIAN, (VACANCY)
Executive Vice President/Chief Operating Officer	ALICE ALBRIGHT
Senior Vice President, Export Finance	JOHN A. MCADAMS
Senior Vice President/Chief Financial Officer	JOHN F. SIMONSON
Vice President, Treasurer	DAVID SENA
Vice President, Controller	(VACANCY)
Vice President, Asset Management	FRANCES I. NWACHUKU
Senior Vice President, Small Business	CHARLES TANSEY
Vice President, Transportation Portfolio Management	B. MICHELE DIXEY
Senior Vice President /General Counsel	JONATHAN CORDONE
Vice President, Trade Finance and Insurance	JEFFREY A. ABRAMSON
Vice President, Business Credit	PAMELA S. BOWERS
Vice President, Strategic Initiatives	RAYMOND J. ELLIS
Vice President, Structured Finance	BARBARA A. O'BOYLE
Vice President, Transportation	ROBERT A. MORIN
Vice President, Short-Term Trade Finance	WALTER KOSCIOW
Senior Vice President, Credit and Risk Management	KENNETH M. TINSLEY
Vice President, Credit Review and Compliance	WALTER HILL, JR.
Vice President, Engineering and Environment	JAMES A. MAHONEY, JR.
Vice President, Credit Underwriting	DAVID W. CARTER
Vice President, Country Risk and Economic Analysis	WILLIAM A. MARSTELLER
Vice President, Domestic Business Development	WAYNE L. GARDELLA
Senior Vice President, Resource Management	MICHAEL CUSHING
Chief Information Officer	FERNANDA F. YOUNG
Senior Vice President, Policy and Planning	JAMES C. CRUSE
Vice President, Policy Analysis	HELENE WALSH
Vice President, Operations and Data Quality	MICHELE A. KUESTER
Senior Vice President, Communications	STEPHANIE O'KEEFE
Vice President, Public Affairs	PHILLIP S. COGAN
Senior Vice President, Congressional Affairs	J. GRAY SASSER
Vice President, Congressional Affairs	WILLIAM E. HELLERT
Inspector General, Acting	OSVALDO GRATACOS

The Export-Import Bank assists in financing the export of U.S. goods and services to international markets.

The Export-Import Bank of the United States (Ex-Im Bank), established in 1934, operates as an independent agency of the U.S. Government under the authority of the Export-Import Bank Act of 1945, as amended (12 U.S.C. 635 et seq.). Its Board of Directors consists of a President and Chairman, a First Vice President and Vice Chair, and three other Directors. All are appointed by the President with the advice and consent of the Senate.

Ex-Im Bank's mission is to help American exporters meet government-supported financing competition from other countries, so that U.S. exports can compete for overseas business on the basis of price, performance, and service, and in doing so help and sustain U.S. jobs. The Bank also fills gaps in the availability of commercial financing for creditworthy export transactions.

Ex-Im Bank is required to find a reasonable assurance of repayment for each transaction it supports. Its legislation requires it to meet the financing terms of competitor export credit agencies, but not to compete with commercial lenders. Legislation restricts the Bank's operation in some countries and its support for military goods and services.

Activities

Ex-Im Bank is authorized to have loans, guarantees, and insurance outstanding at any one time in aggregate amount not in excess of $100 billion. It supports U.S. exporters through a range of diverse programs. These programs are offered under four broad categories of export financing: working capital guarantees, export credit insurance, loan guarantees, and direct loans.

Ex-Im Bank initiated several changes to enhance its support to small business. It created a new position, Senior Vice President for Small Business, who reports directly to the Bank's President and Chairman. In addition, the Bank's regional offices are now dedicated exclusively to small business outreach and support. The Bank also established a Small Business Committee to coordinate, evaluate, and make recommendations on Bank functions necessary for a successful small business strategy.

Regional Offices

The Export-Import Bank operates five regional offices. Internet, www.exim.gov/contact/contactus.cfm.

For further information, contact the Export-Import Bank, Business Development Office, 811 Vermont Avenue NW., Washington, DC 20571. Phone, 202–565–3946 or 800–565–3946. Internet, http://www.exim.gov.

FARM CREDIT ADMINISTRATION

1501 Farm Credit Drive, McLean, VA 22102–5090
Phone, 703–883–4000. Fax, 703–734–5784. Internet, http://www.fca.gov.

Chairman/Chief Executive Officer	LELAND A. STROM
Member of the Boards	KENNETH A. SPEARMAN, JILL LONG THOMPSON
Secretary to the Board	ROLAND E. SMITH
Chief Operating Officer	WILLIAM J. HOFFMAN
Director, Office of Congressional and Public Affairs	MICHAEL A. STOKKE
General Counsel	CHARLES R. RAWLS
Inspector General	CARL A. CLINEFELTER
Director, Office of Examination and Chief Examiner	THOMAS G. MCKENZIE

Director, Office of Regulatory Policy   ANDREW D. JACOB
Director, Office of Secondary Market Oversight   S. ROBERT COLEMAN
Director, Office of Management Services   STEPHEN G. SMITH

[For the Farm Credit Administration statement of organization, see the Code of Federal Regulations, Title 12, Parts 600 and 611]

The Farm Credit Administration ensures the safe and sound operation of the banks, associations, affiliated service organizations, and other entities of the Farm Credit System, and protects the interests of the public and those who borrow from Farm Credit institutions or invest in Farm Credit securities.

The Farm Credit Administration (FCA) was established as an independent financial regulatory agency in the executive branch of the Federal Government by Executive Order 6084 on March 27, 1933. FCA carries out its responsibilities by conducting examinations of the various Farm Credit lending institutions, which are Farm Credit Banks, the Agricultural Credit Bank, Agricultural Credit Associations, and Federal Land Credit Associations.

FCA also examines the service organizations owned by the Farm Credit lending institutions, as well as the National Consumer Cooperative Bank.

FCA policymaking is vested in the Farm Credit Administration Board, whose three full-time members are appointed to 6-year terms by the President, with the advice and consent of the Senate. One member of the Board is designated by the President as Chairman and serves as the Administration's chief executive officer. The Board is responsible for approving rules and regulations, providing for the examination and regulation of and reporting by Farm Credit institutions, and establishing the policies under which the Administration operates. Board meetings are regularly held on the second Thursday of the month and are subject to the Government in the Sunshine Act. Public announcements of these meetings are published in the Federal Register.

The lending institutions of the Farm Credit System were established to provide adequate and dependable credit and closely related services to farmers, ranchers, and producers or harvesters of aquatic products; persons engaged in providing on-the-farm services; rural homeowners; and associations of farmers, ranchers, and producers or harvesters of aquatic products, or federations of such associations that operate on a cooperative basis and are engaged in marketing, processing, supply, or business service functions for the benefit of their members. Initially capitalized by the United States Government, the Farm Credit lending institutions are organized as cooperatives and are completely owned by their borrowers. The loan funds provided to borrowers by these institutions are obtained primarily through the sale of securities to investors in the Nation's capital markets.

The Agricultural Credit Act of 1987, as amended (12 U.S.C. 2279aa-1), established the Federal Agricultural Mortgage Corporation (commonly known as "Farmer Mac"). The Corporation, designated as part of the Farm Credit System, is a federally chartered instrumentality of the United States and promotes the development of a secondary market for agricultural real estate and rural housing loans. Farmer Mac also provides guarantees for the timely payment of principal and interest on securities representing interests in or obligations backed by pools of agricultural real estate loans. The Administration is responsible for the examination and regulation of Farmer Mac to ensure the safety and soundness of its operations.

The Administration manages regulations under which Farm Credit institutions operate. These regulations implement the Farm Credit Act of 1971, as amended (12 U.S.C. 2001), and have the force and effect of law. Similar to the authorities of other Federal regulators of financial institutions, the Administration's authorities include the power to issue cease-and-desist orders, to levy civil

FARM CREDIT ADMINISTRATION

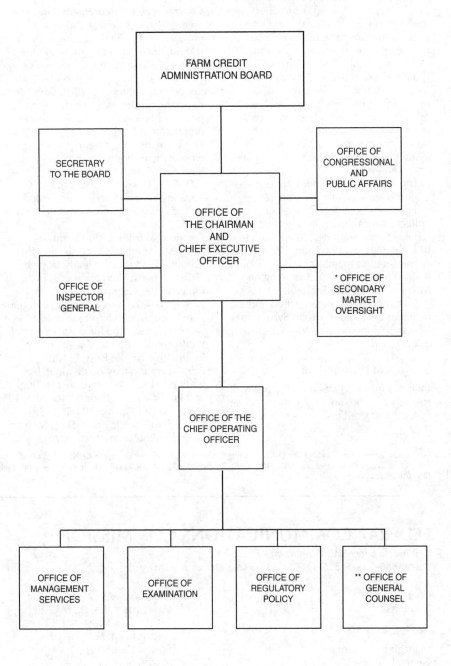

* Reports to the Board for policy and to the Chief Executive Officer for administration.

** Maintains a confidential advisory relationship with each of the Board members.

monetary penalties, to remove officers and directors of Farm Credit institutions, and to establish financial and operating reporting requirements. Although it is prohibited from participation in routine management or operations of Farm Credit institutions, the Administration is authorized to become involved in these institutions' management and operations when the Farm Credit Act or its regulations have been violated, when taking an action to correct an unsafe or unsound practice, or when assuming a formal conservatorship over an institution.

The Administration does not operate on funds appropriated by Congress; it derives income from assessments collected from the institutions that it regulates and examines. In addition to its headquarters in McLean, VA, the Administration maintains four field offices located in Aurora, CO; Bloomington, MN; Irving, TX; and Sacramento, CA.

Authority for the organization and activities of the Farm Credit System may be found in the Farm Credit Act of 1971, as amended.

Sources of Information

Inquiries for information on the following subjects may be directed to the specified office, Farm Credit Administration, 1501 Farm Credit Drive, McLean, VA 22102–5090.

Contracts and Procurement   Inquiries regarding the Administration's procurement and contracting activities should be directed in writing to the Office of Management Services. Phone, 703–883–4378. TTY, 703–883–4200. Requests for proposals, invitations for bids, and requests for quotations are posted online at www.fca.gov/about/procurement.html.

Employment   Inquiries regarding employment should be directed to the Office of Management Services. Phone, 703–883–4135. TTY, 703–883–4200. Vacancy announcements are posted online at www.fca.gov/about/careers. html.

Freedom of Information Requests Requests for agency records must be submitted in writing, clearly labeled "FOIA Request" and addressed to the Office of the General Counsel. Phone, 703–883–4020 (voice and TTY). Requests may be submitted through the Internet at www.fca.gov/home/freedom_info.html.

Publications   Publications and information on the Farm Credit Administration may be obtained by writing to the Office of Congressional and Public Affairs. Phone, 703–883–4056 (voice and TTY). Fax, 703–790–3260. Email, info-line@fca.gov. They are also posted online at www.fca.gov.

For further information, contact the Office of Congressional and Public Affairs, Farm Credit Administration, 1501 Farm Credit Drive, McLean, VA 22102–5090. Phone, 703–883–4056. Email, info-line@fca.gov. Internet, http://www.fca.gov.

FEDERAL COMMUNICATIONS COMMISSION

445 Twelfth Street SW., Washington, DC 20554
Phone, 888–225–5322. TTY, 888–835–5322. Internet, http://www.fcc.gov.

Chairman	JULIUS GENACHOWSKI
Commissioners	MEREDITH A. BAKER, MIGNON CLYBURN, MICHAEL J. COPPS, ROBERT M. McDOWELL
Managing Director	STEVEN VANROEKEL
General Counsel	AUSTIN SCHLICK
Inspector General	DAVID L. HUNT, *Acting*
Chief, Consumer and Governmental Affairs Bureau	JOEL GURIN

Chief, Enforcement Bureau	MICHELE ELLISON
Chief, International Bureau	MINDEL DE LA TORRE
Chief, Media Bureau	WILLIAM T. LAKE
Chief, Office of Administrative Law Judges	RICHARD L. SIPPEL
Chief, Office of Engineering and Technology	JULIUS KNAPP
Chief, Office of Strategic Planning and Policy Analysis	PAUL DE SA
Chief, Wireless Telecommunications Bureau	RUTH MILKMAN
Chief, Wireline Competition Bureau	SHARON GILLETT
Chief, Public Safety and Homeland Security Bureau	JAMES A. BARNETT, JR.
Director, Office of Communications Business Opportunities	THOMAS REED
Director, Office of Legislative Affairs	TERRI GLAZE
Director, Office of Media Relations	DAVID FISKE
Director, Office of Workplace Diversity	THOMAS WYATT

[For the Federal Communications Commission statement of organization, see the Code of Federal Regulations, Title 47, Part 0]

The Federal Communications Commission regulates interstate and foreign communications by radio, television, wire, satellite, and cable.

The Federal Communications Commission (FCC) was created by the Communications Act of 1934 (47 U.S.C. 151 et seq.) to regulate interstate and foreign communications by wire and radio in the public interest. The scope of FCC regulation includes radio and television broadcasting; telephone, telegraph, and cable television operation; two-way radio and radio operators; and satellite communication.

The Commission comprises five members, who are appointed by the President with the advice and consent of the Senate. One of the members is designated by the President as Chairman.

Activities

Media Bureau The Media Bureau develops, recommends, and administers policy and licensing programs for the regulation of electronic media, including cable television, multichannel video programming distribution, broadcast television and radio, and satellite services in the United States and its territories. The Bureau also conducts rulemaking proceedings, studies and analyzes electronic media services; resolves waiver petitions, declaratory rulings, and adjudications related to electronic media services; and processes applications for authorization, assignment, transfer, and renewal of media services, including AM, FM, TV, the cable TV relay service, and related matters.

For further information, contact the Media Bureau. Phone, 202–418–7200, or 888–225–5322.

Wireline Competition Bureau The Wireline Competition Bureau advises and makes recommendations to the FCC. The Bureau ensures choice, opportunity, and fairness in the development of wireline communications; assesses the present and future wireline communication needs of the Nation; encourages the development and widespread availability of wireline communication services; promotes investment in wireline communication infrastructure; and reviews and coordinates orders, programs, and actions initiated by other bureaus and offices in matters affecting wireline communications to ensure consistency with overall FCC policy.

For further information, contact the Wireline Competition Bureau. Phone, 202–418–1500, or 888–225–5322.

Consumer and Governmental Affairs Bureau The Consumer and Governmental Affairs Bureau develops and administers the FCC's consumer and governmental affairs policies and initiatives. The Bureau facilitates public

FEDERAL COMMUNICATIONS COMMISSION

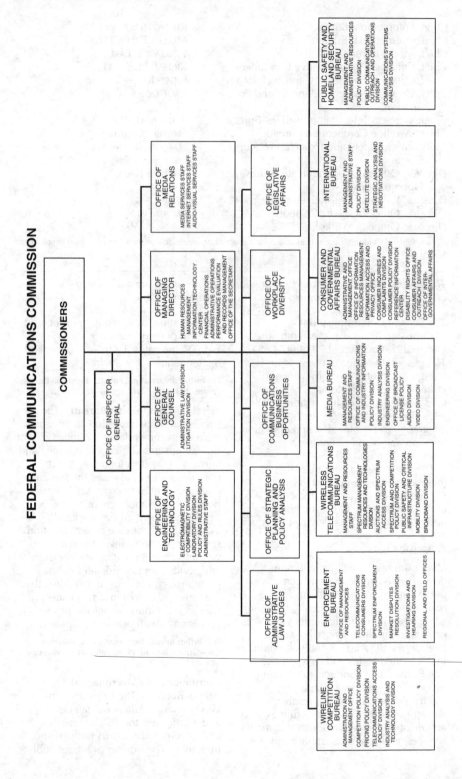

COMMISSIONERS

OFFICE OF INSPECTOR GENERAL

OFFICE OF ENGINEERING AND TECHNOLOGY
ELECTROMAGNETIC COMPATIBILITY DIVISION
LABORATORY DIVISION
POLICY AND RULES DIVISION
ADMINISTRATIVE STAFF

OFFICE OF GENERAL COUNSEL
ADMINISTRATIVE LAW DIVISION
LITIGATION DIVISION

OFFICE OF MANAGING DIRECTOR
HUMAN RESOURCES MANAGEMENT
INFORMATION TECHNOLOGY CENTER
FINANCIAL OPERATIONS
ADMINISTRATIVE OPERATIONS
PERFORMANCE EVALUATION AND RECORDS MANAGEMENT
OFFICE OF THE SECRETARY

OFFICE OF MEDIA RELATIONS
MEDIA SERVICES STAFF
INTERNET SERVICES STAFF
AUDIO-VISUAL SERVICES STAFF

OFFICE OF ADMINISTRATIVE LAW JUDGES

OFFICE OF STRATEGIC PLANNING AND POLICY ANALYSIS

OFFICE OF COMMUNICATIONS BUSINESS OPPORTUNITIES

OFFICE OF WORKPLACE DIVERSITY

OFFICE OF LEGISLATIVE AFFAIRS

ENFORCEMENT BUREAU
OFFICE OF MANAGEMENT AND RESOURCES
TELECOMMUNICATIONS CONSUMERS DIVISION
SPECTRUM ENFORCEMENT DIVISION
MARKET DISPUTES RESOLUTION DIVISION
INVESTIGATIONS AND HEARING DIVISION
REGIONAL AND FIELD OFFICES

WIRELESS TELECOMMUNICATIONS BUREAU
MANAGEMENT AND RESOURCES STAFF
SPECTRUM MANAGEMENT RESOURCES AND TECHNOLOGIES DIVISION
AUCTIONS AND SPECTRUM ACCESS DIVISION
SPECTRUM COMPETITION POLICY DIVISION
PUBLIC SAFETY AND CRITICAL INFRASTRUCTURE DIVISION
MOBILITY DIVISION
BROADBAND DIVISION

MEDIA BUREAU
MANAGEMENT AND RESOURCES STAFF
OFFICE OF COMMUNICATIONS AND INDUSTRY INFORMATION
POLICY DIVISION
INDUSTRY ANALYSIS DIVISION
ENGINEERING DIVISION
OFFICE OF BROADCAST LICENSE POLICY
AUDIO DIVISION
VIDEO DIVISION

CONSUMER AND GOVERNMENTAL AFFAIRS BUREAU
ADMINISTRATIVE AND MANAGEMENT OFFICE
OFFICE OF INFORMATION RESOURCES MANAGEMENT
INFORMATION ACCESS AND PRIVACY OFFICE
CONSUMER INQUIRIES AND COMPLAINTS DIVISION
CONSUMER POLICY DIVISION
REFERENCE INFORMATION CENTER
DISABILITY RIGHTS OFFICE
CONSUMER AFFAIRS AND OUTREACH DIVISION
OFFICE OF INTER-GOVERNMENTAL AFFAIRS

INTERNATIONAL BUREAU
MANAGEMENT AND ADMINISTRATIVE STAFF
POLICY DIVISION
SATELLITE DIVISION
STRATEGIC ANALYSIS AND NEGOTIATIONS DIVISION

PUBLIC SAFETY AND HOMELAND SECURITY BUREAU
MANAGEMENT AND ADMINISTRATIVE RESOURCES
POLICY DIVISION
PUBLIC COMMUNICATIONS OUTREACH AND OPERATIONS DIVISION
COMMUNICATIONS SYSTEMS ANALYSIS DIVISION

WIRELINE COMPETITION BUREAU
ADMINISTRATION AND MANAGEMENT OFFICE
COMPETITION POLICY DIVISION
PRICING POLICY DIVISION
TELECOMMUNICATIONS ACCESS POLICY DIVISION
INDUSTRY ANALYSIS AND TECHNOLOGY DIVISION

participation in the Commission's decisionmaking process; represents the Commission on consumer and Government committees, working groups, task forces, and conferences; works with public, Federal, State, local, and tribal agencies to develop and coordinate policies; oversees the Consumer Advisory Committee and the Intergovernmental Advisory Committee; provides expert advice and assistance regarding compliance with applicable disability and accessibility requirements, rules, and regulations; resolves informal complaints through mediation; and conducts consumer outreach and education programs.

For further information, contact the Consumer and Governmental Affairs Bureau. Phone, 202–418–1400, or 888–225–5322.

Enforcement Bureau   The Enforcement Bureau serves as the FCC's primary agency for enforcing the Communications Act, other communications statutes, and the Commission's rules and orders. The Bureau investigates and resolves complaints regarding common carriers (wireline, wireless, and international) and noncommon carriers subject to the Commission's jurisdiction under Title II of the Communications Act; radio frequency interference, equipment, and devices; accessibility to communications services and equipment for persons with disabilities; noncompliance with the lighting and marking of radio transmitting towers and pole attachment regulations; noncompliance with children's television programming commercial limits; and unauthorized construction and operation of communication facilities and false distress signals.

For further information, contact the Enforcement Bureau. Phone, 202–418–7450, or 888–225–5322. Internet, http://www.fcc.gov/eb.

International Bureau   The International Bureau serves as the FCC's principal representative in international organizations. It monitors compliance with the terms and conditions of authorizations and licenses granted by the Bureau and enforces them in conjunction with appropriate bureaus and offices; provides advice and

technical assistance to U.S. trade officials in the negotiation and implementation of telecommunications trade agreements; and promotes the international coordination of spectrum allocation and frequency and orbital assignments in order to minimize cases of international radio interference involving U.S. licenses.

For further information, contact the International Bureau. Phone, 202–418–0437, or 888–225–5322.

Wireless Telecommunications Bureau
The Wireless Telecommunications Bureau administers all domestic commercial and private wireless communication programs and rules. It addressess present and future wireless communication and spectrum needs; promotes access, efficiency, and innovation in the allocation, licensing, and use of electromagnetic spectrum; ensures choice, opportunity, and fairness in the development of wireless communication services and markets; and promotes the development and widespread availability of wireless broadband, mobile, and other wireless communication services, devices, and facilities, including through open networks. The Bureau also develops, recommends, administers, and coordinates policy for wireless communication services, including rulemaking, interpretations, and equipment standards; explains rules to and advises the public on them and provides rule-interpretation material for the Enforcement Bureau; serves as the FCC's principal policy and administrative resource for all spectrum auctions; and processes wireless service and facility authorization applications.

For further information, contact the Wireless Telecommunications Bureau. Phone, 202–418–0600, or 888–225–5322.

Public Safety and Homeland Security Bureau   The Public Safety and Homeland Security Bureau develops, recommends, and administers FCC's policies pertaining to public safety communication. This includes 911 and E911; operability and interoperability of public safety communications; communications infrastructure protection and disaster response; and network security and reliability. The Bureau also serves as a clearinghouse for public safety communication information, which encompasses priority emergency communication programs; alert and warning of U.S. citizens; continuity of government operations and operational planning; public safety outreach (e.g. first-responder organizations and hospitals); disaster management coordination and outreach; FCC 24/7 Communication Center; and studies and reports of public safety, homeland security, and disaster management issues.

For further information, contact the Public Safety and Homeland Security Bureau. Phone, 202–418–1300, or 888–225–5322. Email, pshsbinfo@fcc.gov.

Sources of Information

Consumer Assistance For general information on FCC operations, contact the Reference Center, Room CY–B523, 445 Twelfth Street SW., Washington, DC 20554. Phone, 888–225–5322. TTY, 888–835–5322.

Contracts and Procurement Direct inquiries to the Chief, Contracts and Purchasing Center. Phone, 202–418–1865.

Electronic Access Information about the FCC is also available electronically through the Internet at www.fcc.gov.

Employment and Recruitment Requests for employment information may be directed to the Recruitment and Staffing Service Center. Phone, 202–418–0130. To view or apply for job vacancies, visit www.fcc.gov/jobs/fccjobs.html.

Equal Employment Practices by Industry Direct inquiries to the FCC Consumer Center. Phone, 888–225–5322.

Ex-Parte Presentations Information concerning ex-parte presentations may be directed to the Commission's Office of General Counsel. Phone, 202–418–1720.

Federal Advisory Committee Management Direct inquiries to the Office of the Managing Director. Phone, 202–418–2178.

Fees Information concerning the FCC's fee programs is available online at www.fcc.gov/fees or by contacting the Registration System/Fee Filer Help Desk at 1–877–480–3201 (option 4).

Freedom of Information Act Requests Contact the FOIA Requester Service Center. Phone, 202–418–0212. Email, foia@fcc.gov.

Internal Equal Employment Practices Direct inquiries to the Office of Workplace Diversity. Phone, 202–418–1799.

Licensing Information concerning the FCC's licensing system is available online at www.fcc.gov/licensing.html.

Public Inspection Dockets concerning rulemaking and adjudicatory matters, copies of applications for licenses and grants, and reports required to be filed by licensees and cable system operators are maintained in the FCC's public reference rooms (some reports are held confidentially by law). The library has FCC rules and regulations on file. Phone, 202–418–0450. General information is also available from the FCC's fax-on-demand system. Phone, 202–418–2805. Additionally, each broadcasting station makes publicly available certain information about the station's operation, a current copy of the application filed for license, and nonconfidential FCC reports.

Publications The Office of Media Relations distributes publications, public notices, and press releases. Phone, 202–418–0503.

For further information, contact the Consumer Center, Federal Communications Commission, 445 Twelfth Street SW., Washington, DC 20554. Phone, 888–225–5322. TTY, 888–835–5322. Internet, http://www.fcc.gov.

FEDERAL DEPOSIT INSURANCE CORPORATION

550 Seventeenth Street NW., Washington, DC 20429
Phone, 703–562–2222. Internet, http://www.fdic.gov.

Chairman SHEILA C. BAIR
Vice Chairman MARTIN J. GRUENBERG

Director	THOMAS J. CURRY
Director, Comptroller of the Currency	JOHN C. DUGAN
Director, Office of Thrift Supervision	JOHN E. BOWMAN
Deputy to the Chairman	JASON C. CAVE
Deputy to the Chairman for Policy	MICHAEL H. KRIMMINGER
Deputy to the Chairman/Chief Financial Officer	STEVEN O. APP
Deputy to the Chairman for External Affairs	PAUL NASH
Chief of Staff	JESSE O. VILLARREAL, JR.
Deputy to the Vice Chairman	BARBARA A. RYAN
General Counsel	MICHAEL BRADFIELD
Director, Division of Administration	ARLEAS U. KEA
Director, Division of Finance	BRET D. EDWARDS
Director, Division of Information Technology/ Chief Information Officer	RUSSELL G. PITTMAN, *Acting*
Director, Division of Insurance and Research	ARTHUR J. MURTON
Director, Division of Resolutions and Receiverships	MITCHELL L. GLASSMAN
Director, Division of Supervision and Consumer Protection	SANDRA L. THOMPSON
Director, Office of Diversity and Economic Opportunity	D. MICHAEL COLLINS
Director, Office of Enterprise Risk Management	JAMES H. ANGEL, JR.
Director, Office of International Affairs	FRED S. CARNS
Director, Office of Legislative Affairs	PAUL NASH, *Acting*
Ombudsman	COTTRELL L. WEBSTER
Director, Office of Public Affairs	ANDREW S. GRAY
Chief Learning Officer	THOM H. TERWILLIGER
Inspector General	JOHN T. RYMER

The Federal Deposit Insurance Corporation preserves and promotes public confidence in U.S. financial institutions by insuring bank and thrift deposits, periodically examining State-chartered banks, and liquidating assets of failed institutions.

The Federal Deposit Insurance Corporation (FDIC) was established under the Banking Act of 1933 after numerous banks failed during the Great Depression. FDIC began insuring banks on January 1, 1934. The deposit insurance coverage on certain retirement accounts at banks or savings institutions is $250,000. The basic insurance coverage for other deposit accounts is $100,000.

The FDIC is managed by a five-person Board of Directors, all of whom are appointed by the President and confirmed by the Senate, with no more than three being from the same political party.

FDIC receives no Congressional appropriations. It is funded by insurance premiums on deposits held by insured banks and savings associations and from interest on the investment of those premiums in U.S. Government securities.

FDIC has authority to borrow up to $30 billion from the Treasury for insurance purposes.

Activities

The FDIC insures about $7 trillion of U.S. bank and thrift deposits. The insurance fund is composed of insurance premiums paid by banks and savings associations and the interest on the investment of those premiums in U.S. Government securities, as required by law. Premiums are determined by an institution's level of capitalization and potential risk to the insurance fund.

The FDIC examines about 5,250 State-chartered commercial and savings banks that are not members of the Federal Reserve System, called State nonmember banks. The FDIC also has authority to examine other types of

FEDERAL DEPOSIT INSURANCE CORPORATION

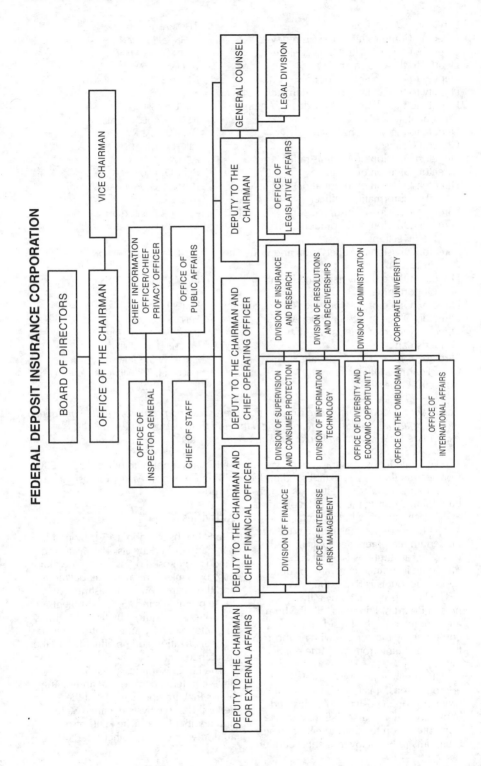

FDIC-insured institutions for deposit insurance purposes. The two types of examinations conducted are for safety and soundness and for compliance with applicable consumer laws such as the Truth in Lending Act, the Home Mortgage Disclosure Act, the Equal Credit Opportunity Act, the Fair Housing Act, and the Community Reinvestment Act. Examinations are performed on the institution's premises and off-site through computer data analysis.

A failed bank or savings association is generally closed by its chartering authority, and the FDIC is named receiver. The FDIC is required to resolve the closed institution in a manner that is least costly to the FDIC. Ordinarily, the FDIC attempts to locate a healthy institution to acquire the failed entity. If such an entity cannot be found, the FDIC pays depositors the amount of their insured funds, usually by the next business day following the closing. Depositors with funds that exceed the insurance limit often receive an advance dividend, which is a portion of their uninsured funds that is determined by an estimate of the future proceeds from liquidating the failed institution's remaining assets. Depositors with funds in a failed institution that exceed the insurance limit receive a receivership certificate for those funds and partial payments of their uninsured funds as asset disposition permits.

As part of its insurance, supervisory, and receivership responsibilities, the FDIC approves or disapproves of mergers, consolidations, and acquisitions where the resulting bank is an insured State nonmember; approves or disapproves of proposals by banks to establish and operate a new branch, close an existing branch, or move its main office from one location to another; and approves or disapproves of requests to engage as principal in activities and investments that are not permissible for a national bank. It also issues enforcement actions, including cease-and-desist orders, for specific violations or practices requiring corrective action and reviews changes in ownership or control of a bank.

Sources of Information

Consumer Information Telephone inquiries about deposit insurance and other consumer information may be directed to the Consumer Response Center at 877–275–3342, from 8 a.m. to 8 p.m., eastern standard time, Monday through Friday. For credit card complaints, call 800–378–9581, from 8:30 a.m. to 4:30 p.m., central standard time, Monday through Friday. Written inquiries may be sent to the Division of Supervision and Consumer Protection at our regional offices or to FDIC headquarters. Internet, www.fdic.gov/ regulations/community/cra_contacts. html. Email inquiries may be sent from the FDIC Web site. Internet, www.fdic. gov/deposit/deposits/index.html. FDIC's customer assistance form, for submitting an inquiry or a complaint, is available at www.fdic.gov/starsmail/index.asp. A bank's quarterly Report of Condition may be obtained at reproduction cost from the call center or free of charge from the FDIC Web site at www.fdic.gov/ Call_TFR_Rpts/.

General Inquiries Written requests for general information may be directed to the FDIC's Public Information Center, 3501 Fairfax Drive, Room E–1002, Arlington, VA 22226. Phone, 703–562– 2200 or 877–275–3342.

Public Records FDIC records are available on the FDIC Web site. Inquiries about other types of records available to the public, including records available under the Freedom of Information Act, should be directed to the Chief, FOIA/PA Group 550 17th Street NW., Washington, DC 20429, or any regional office.

Publications Publications, press releases, congressional testimony, directives to financial institutions, and other documents are available through the Public Information Center. Phone, 877–275–3342 (press 1; then press 5). Email, publicinfo@fdic.gov. Internet, www.fdic.gov/news/publications/index. html.

For further information, contact the Office of Public Affairs, Federal Deposit Insurance Corporation, 550 Seventeenth Street NW., Washington, DC 20429. Phone, 202–898–6993. Email, communications@fdic.gov. Internet, http://www.fdic.gov.

FEDERAL ELECTION COMMISSION

999 E Street NW., Washington, DC 20463
Phone, 202–694–1100; 800–424–9530. Internet, http://www.fec.gov..

Chairman	MATTHEW S. PETERSEN
Vice Chairman	CYNTHIA L. BAUERLY
Commissioners	CAROLINE C. HUNTER, DONALD F. MCGAHN, II, STEVEN T. WALTHER, ELLEN L. WEINTRAUB
Staff Director	ALEC PALMER, *Acting*
General Counsel	THOMASENIA P. DUNCAN
Inspector General	LYNNE A. MCFARLAND
Chief Financial Officer	MARY G. SPRAGUE

The Federal Election Commission provides public disclosure of campaign finance activities and ensures compliance with campaign finance laws and regulations.

The Federal Election Commission is an independent agency established by section 309 of the Federal Election Campaign Act of 1971, as amended (2 U.S.C. 437c). It comprises six Commissioners appointed by the President with the advice and consent of the Senate. The act also provides for three statutory officers—the Staff Director, the General Counsel, and the Inspector General—who are appointed by the Commission.

Activities

The Commission administers and enforces the Federal Election Campaign Act of 1971, as amended (2 U.S.C. 431 et seq.), and the Revenue Act, as amended (26 U.S.C. 1 et seq.). These laws provide for the public funding of Presidential elections, public disclosure of the financial activities of political committees involved in Federal elections, and limitations and prohibitions on contributions and expenditures made to influence Federal elections (Presidency, Senate, and House of Representatives).

Public Funding of Presidential Elections The Commission oversees the public financing of Presidential elections by certifying Federal payments to primary candidates, general election nominees, and national nominating conventions. It also audits recipients of Federal funds and may require repayments to the U.S. Treasury, if a committee makes nonqualified campaign expenditures.

Disclosure The Commission ensures public disclosure of the campaign finance activities reported by political committees supporting Federal candidates. Committee reports, filed regularly, disclose where campaign money comes from and how it is spent. The Commission places reports on the public record within 48 hours after they are received and computerizes the data contained in the reports.

Sources of Information

Congressional Affairs Office This Office serves as the primary liaison with Congress and executive branch agencies. The Office is responsible for keeping Members of Congress informed about Commission decisions and, in turn, for informing the Commission on legislative developments. For further information, call 202–694–1006 or 800–424–9530.

Employment Inquiries regarding employment opportunities should be directed to the Director, Human

Resources and Labor Relations. Phone, 202–694–1080 or 800–424–9530.

General Inquiries The Information Services Division provides information and assistance to Federal candidates, political committees, and the general public. This division answers questions on campaign finance laws, conducts workshops and seminars on the law, and provides publications and forms. For information or materials, call 202–694–1100 or 800–424–9530.

Media Inquiries The Press Office answers inquiries from print and broadcast media sources around the country, issues press releases on Commission actions and statistical data, responds to requests for information, and distributes other materials. Media representatives should direct their inquiries to the Press Office. Phone, 202–694–1220 or 800–424–9530.

Public Records The Office of Public Records, located at 999 E Street NW., Washington, DC, provides space for public inspection of all reports and statements relating to campaign finance since 1972. It is open weekdays from 9 a.m. to 5 p.m. and has extended hours during peak election periods. The public is invited to visit the Office or obtain information by calling 202–694–1120 or 800–424–9530.

Reading Room The library contains a collection of basic legal research resources on political campaign financing, corporate and labor political activity, and campaign finance reform. It is open to the public on weekdays between 9 a.m. and 5 p.m. For further information, call 202–694–1600 or 800–424–9530.

For further information, contact Information Services, Federal Election Commission, 999 E Street NW., Washington, DC 20463. Phone, 202–694–1100 or 800–424–9530. Internet, http://www.fec.gov.

FEDERAL HOUSING FINANCE AGENCY

1700 G Street NW., Washington, DC 20552
Phone, 202–414–6923. Internet, http://www.fhfa.gov.

Director	EDWARD J. DeMARCO, *Acting*
Senior Deputy Director and Chief Operating Officer	STEPHEN M. CROSS, *Acting*
Deputy Director for Enterprise Regulation	CHRIS DICKERSON
Deputy Director for Federal Home Loan Bank Regulation	STEPHEN M. CROSS
Inspector General	(VACANCY)
General Counsel	ALFRED M. POLLARD

The Federal Housing and Finance Agency promotes a stable and liquid mortgage market, affordable housing, and community investment through safety and soundness oversight of Fannie Mae, Freddie Mac, and the Federal Home Loan Banks.

The Federal Housing and Finance Agency (FHFA) was established by the Federal Housing Finance Regulatory Reform Act of 2008 (12 U.S.C. 4501 et seq.) as an independent agency in the executive branch. The FHFA is the result of the Federal Housing Finance Board and the Office of Federal Housing Enterprise Oversight merging and the Department of Housing and Urban Development's Government-sponsored enterprise mission team being transferred into the Agency.

FHFA is managed by a Director who is appointed by the President and confirmed by the Senate. FHFA's Director also serves as the Chairman of the Federal Housing Oversight Board. The Secretary of the Treasury, the Secretary of Housing and Urban Development, and

FEDERAL FINANCE HOUSING AGENCY

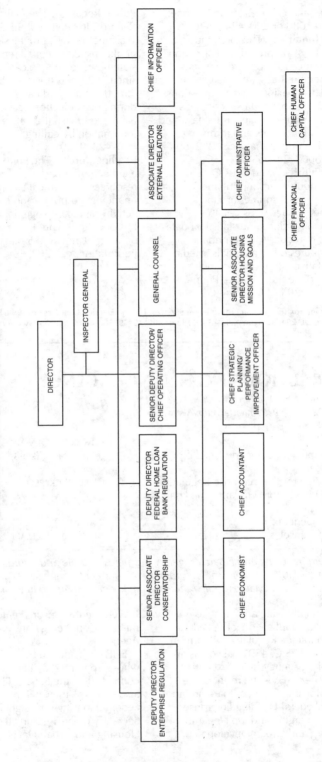

the Securities and Exchange Commission Chairman are also members of the Board.

FHFA was created to ensure the security and supervision of Fannie Mae, Freddie Mac, the 12 Federal Home Loan Banks, and the Office of Finance. The Agency's increased regulatory powers and affordable housing goals were designed to restore confidence in the 14 Government-sponsored loan enterprises, enabling these entities to provide more effective assistance to troubled mortgage markets.

Activities

FHFA practices oversight to strengthen vital components of the Nation's secondary mortgage markets. It oversees maintenance of adequate capital and internal controls; operations that foster efficient, competitive, and resilient national housing finance markets, including activities related to home mortgages for low- and moderate-income families; compliance with the rules, regulations, guidelines, and orders issued by FHFA and the authorizing statutes; and lending practices consistent with the public interest.

Sources of Information

Employment Inquiries and employment applications should be directed to the Office of Human Resources Management, FHFA, 1625 I Street NW., Washington, DC 20006 or visit our Web site at www.fhfa.gov.

Publications Written requests for copies of publications may be addressed to the Office of External Relations, FHFA, 1700 G Street NW., Washington, DC 20552 or emailed to fhfainfo@fhfa.gov.

Public Records Many FHFA records are available on online at www.fhfa.gov. Inquiries about other records available to the public, including those available under the Freedom of Information Act, should be directed by email to the Chief FOIA Officer at foia@fhfa.gov.

For further information, contact the Office of External Relations, Federal Housing Finance Agency, 1700 G Street NW., Washington, DC 20552. Phone, 866–796–5595. Email, fhfainfo@fhfa.gov. Internet, http://www.fhfa.gov.

FEDERAL LABOR RELATIONS AUTHORITY

1400 K Street NW., Washington, DC 20005
Phone, 202–218–7770. Internet, http://www.flra.gov.

Chair	CAROL W. POPE
Members	THOMAS M. BECK, ERNEST DUBESTER
Executive Director	SONNA STAMPONE
Chief Counsel	SUE MCCLUSKEY
Director, Case Intake Office/Legal Publications	GINA GRIPPANDO
Solicitor	ROSA M. KOPPEL
Inspector General	CHARLES CENTER, *Acting*
Chief Administrative Law Judge	CHARLES CENTER
General Counsel	JULIA CLARKE
Federal Service Impasses Panel	
Chairman	MARY E. JACKSTEIT
Members	THOMAS E. ANGELO, BARBARA B. FRANKLIN, EDWARD F. HARTFIELD, MARVIN E. JOHNSON, MARTIN H. MALIN, DONALD S. WASSERMAN
Executive Director	H. JOSEPH SCHIMANSKY
Foreign Service Labor Relations Board	
Chair	CAROL W. POPE

Foreign Service Labor Relations Board

Members	EARL W. HOCKENBERRY, JR., STEPHEN R. LEDFORD, (VACANCY)
General Counsel	(VACANCY)

The Federal Labor Relations Authority oversees labor-management relations between the Federal Government and its employees.

The Federal Labor Relations Authority was created as an independent establishment by Reorganization Plan No. 2 of 1978 (5 U.S.C. app.), effective January 1, 1979, pursuant to Executive Order 12107 of December 28, 1978, to consolidate the central policymaking functions in Federal labor-management relations. Its duties and authority are specified in Title VII (Federal Service Labor-Management Relations) of the Civil Service Reform Act of 1978 (5 U.S.C. 7101–7135).

The Authority comprises three members who are nominated by the President and confirmed by the Senate to a 5-year term. The Chairman of the Authority serves as the chief executive and administrative officer. The Chairman also chairs the Foreign Service Labor Relations Board. The General Counsel of the Authority investigates alleged unfair labor practices, files and prosecutes unfair labor practice complaints before the Authority, and exercises such other powers as the Authority may prescribe.

Activities

The Authority adjudicates disputes arising under the Federal Labor-Management Relations Program, deciding cases concerning the negotiability of collective bargaining agreement proposals, appeals concerning unfair labor practices and representation petitions, and exceptions to grievance arbitration awards. It also assists Federal agencies and unions in understanding their rights and responsibilities under the program through training.

The Federal Service Impasses Panel, an entity within the Authority, is assigned the function of providing assistance in resolving negotiation impasses between agencies and unions. After investigating an impasse, the Panel can either recommend procedures to the parties for the resolution of the impasse or assist the parties in resolving the impasse through whatever methods and procedures it considers appropriate, including fact finding and recommendations. If the parties do not arrive at a settlement after assistance by the Panel, the Panel may hold hearings and take whatever action is necessary to resolve the impasse.

The Foreign Service Labor Relations Board and the Foreign Service Impasse Disputes Panel administer provisions of chapter 2 of the Foreign Service Act of 1980 (22 U.S.C. 3921) concerning labor-management relations. This chapter establishes a statutory labor-management relations program for Foreign Service employees of the U.S. Government. Administrative and staff support is provided by the Federal Labor Relations Authority and the Federal Service Impasses Panel.

Sources of Information

Employment Employment inquiries and applications may be sent to the Human Resources Division. Phone, 202–218–7963. Internet, www.flra.gov/29-jobs.html.

Public Information and Publications The Authority will assist in arranging reproduction of documents and ordering transcripts of hearings. Requests for publications should be submitted to the Director, Case Intake and Publication. Phone, 202–218–7780. Internet, www.flra.gov.

For further information, contact the Office of the Executive Director, Federal Labor Relations Authority, 1400 K Street NW., Washington, DC 20005. Phone, 202–218–7949. Email, flraexecutivedirector@flra.gov. Internet, http://www.flra.gov.

FEDERAL LABOR RELATIONS AUTHORITY

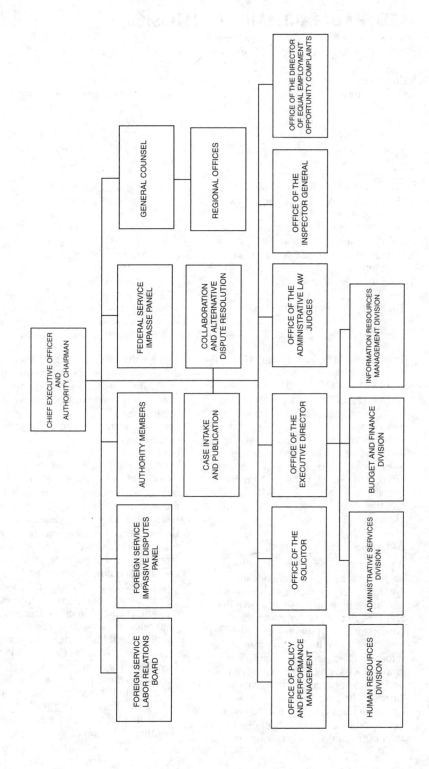

FEDERAL MARITIME COMMISSION

800 North Capitol Street NW., Washington, DC 20573–0001
Phone, 202–523–5707. Internet, http://www.fmc.gov.

Chairman	RICHARD A. LIDINSKY, JR.
Commissioners	JOSEPH E. BRENNAN, REBECCA F. DYE, MICHAEL A. KHOURI, (VACANCY)
General Counsel	(VACANCY)
Secretary	KAREN V. GREGORY
Director, Consumer Affairs and Dispute Resolution Services	VERN W. HILL
Chief, Administrative Law Judge	CLAY G. GUTHRIDGE
Director, Office of Equal Employment Opportunity	KEITH I. GILMORE
Inspector General	ADAM R. TRZECIAK
Director, Office of Managing Director	RONALD D. MURPHY
Director, Bureau of Certification and Licensing	SANDRA L. KUSUMOTO
Director, Bureau of Enforcement	PETER J. KING
Director, Bureau of Trade Analysis	AUSTIN L. SCHMITT

The Federal Maritime Commission regulates the waterborne foreign commerce of the United States. It ensures that U.S. oceanborne trades are open to all on fair and equitable terms and protects against concerted activities and unlawful practices.

The Federal Maritime Commission was established by Reorganization Plan No. 7 of 1961 (46 U.S.C. 301–307), effective August 12, 1961. It is an independent agency that regulates shipping under the following statutes: the Shipping Act of 1984, as amended (46 U.S.C. 40101–41309); Section 19 of the Merchant Marine Act, 1920 (46 U.S.C. 42101–42109); the Foreign Shipping Practices Act of 1988 (46 U.S.C. 42301–42307); and the act of November 6, 1966 (46 U.S.C. 44101–44106).

Activities

Agreements   The Commission reviews agreements by and among ocean common carriers and/or marine terminal operators, filed under section 5 of the Shipping Act of 1984, for statutory compliance as well as for likely impact on competition. It also monitors activities under all effective agreements for compliance with the provisions of law and its rules, orders, and regulations.

Tariffs   The Commission monitors and prescribes requirements to ensure accessibility and accuracy of electronic tariff publications of common carriers engaged in the foreign commerce of the United States. Special permission applications may be submitted for relief from statutory and/or Commission tariff requirements.

Service Contracts   The Commission receives and reviews filings of confidential service contracts between shippers and ocean common carriers. The Commission also monitors publication of certain essential terms of those service contracts.

Non-Vessel-Operating Common Carrier Service Arrangements   The Commission receives and reviews service arrangements entered into by non-vessel-operating common carriers and their customers. Cargo moving under these service arrangements are exempt from the tariff publication and adherence requirements of the Shipping Act, on the condition that the service arrangements must be filed with the Commission.

Licenses   The Commission issues licenses to those persons and entities in the United States who wish to carry out the business of providing freight forwarding services and non-vessel-operating common carrier services.

FEDERAL MARITIME COMMISSION

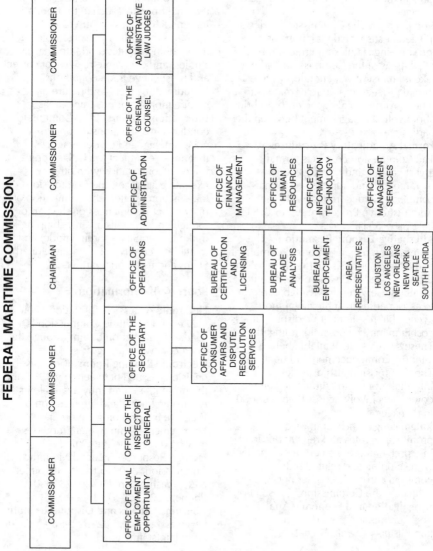

Passenger Indemnity The Commission administers the passenger indemnity provisions of the act of November 6, 1966, which require shipowners and operators to obtain certificates of financial responsibility to pay judgments for personal injury or death or to refund fares in the event of nonperformance of voyages.

Complaints The Commission reviews alleged or suspected violations of the shipping statutes and rules and regulations of the Commission and may take administrative action to institute formal proceedings, to refer matters to other governmental agencies, or to bring about voluntary agreement between the parties.

Formal Adjudicatory Procedures The Commission conducts formal investigations and hearings on its own motion and adjudicates formal complaints in accordance with the Administrative Procedure Act (5 U.S.C. note prec. 551).

Alternative Dispute Resolution The Commission reviews informal complaints and attempts to assist parties in resolving disputes. Mediation and other dispute resolution services are available in order to assist parties in achieving a more acceptable resolution to a dispute at less cost than may be possible in litigation. These services are available before and after the commencement of litigation. The Commission also provides an informal process to adjudicate certain complaints involving less than $50,000 in damages.

Rulemaking The Commission promulgates rules and regulations to interpret, enforce, and ensure compliance with shipping and related statutes by common carriers and other persons subject to the Commission's jurisdiction.

Investigation and Economic Analyses The Commission prescribes and administers programs to ensure compliance with the provisions of the shipping statutes. These programs include: education and outreach activities; the collection of information relating to field investigation of activities and practices of ocean common carriers, terminal operators, agreements among ocean common carriers and/or marine terminal operators, ocean transportation intermediaries, passenger vessel operators, and other persons subject to the shipping statutes; and rate analyses, studies, and economic reviews of current and prospective trade conditions, including the extent and nature of competition in various trade areas.

International Affairs The Commission conducts investigations of foreign governmental and carrier practices that adversely affect the U.S. shipping trade. In consultation with other executive agencies, the Commission takes action to effect the elimination of discriminatory practices on the part of foreign governments against shipping in the United States foreign trade, and to achieve comity between the United States and its trading partners.

Sources of Information

Electronic Access Information about the Federal Maritime Commission is available in electronic form through the Internet at www.fmc.gov.

Electronic Reading Room Commission decisions, issued between July 1987 and the present, are currently available in electronic format and are listed on our Web site in chronological order.

Employment Employment inquiries may be directed to the Office of Human Resources, Federal Maritime Commission, 800 North Capitol Street NW., Washington, DC 20573–0001. Phone, 202–523–5773.

Consumer Affairs and Dispute Resolution Services Phone, 202–523–5807. Email, complaints@fmc.gov.

For further information, contact the Office of the Secretary, Federal Maritime Commission, 800 North Capitol Street NW., Washington, DC 20573–0001. Phone, 202–523–5725. Fax, 202–523–0014. Email, secretary@fmc.gov. Internet, http://www.fmc.gov.

FEDERAL MEDIATION AND CONCILIATION SERVICE

2100 K Street NW., Washington, DC 20427
Phone, 202–606–8100. Internet, http://www.fmcs.gov.

Director	GEORGE H. COHEN

The Federal Mediation and Conciliation Service assists labor and management in resolving disputes in collective bargaining contract negotiation through voluntary mediation and arbitration services.

The Federal Mediation and Conciliation Service (FMCS) was created by the Labor Management Relations Act, 1947 (29 U.S.C. 172). The Director is appointed by the President with the advice and consent of the Senate.

Activities

FMCS helps prevent disruptions in the flow of interstate commerce caused by labor-management disputes by providing mediators to assist disputing parties in the resolution of their differences. Mediators have no law enforcement authority and rely wholly on persuasive techniques.

FMCS offers its services in labor-management disputes to any industry, with employees represented by a union, which affects interstate commerce. FMCS becomes involved in disputes on its own initiative or at the request of one or more of the disputants, whenever it deems that a dispute threatens to cause a substantial interruption of commerce. The Labor Management Relations Act requires that parties to a labor contract must file a dispute notice, if agreement is not reached 30 days in advance of a contract termination or reopening date. The notice must be filed with FMCS and the appropriate State or local mediation agency. FMCS is required to avoid the mediation of disputes that would have only a minor effect on interstate commerce, if State or other conciliation services are available to the parties.

Mediation   Efforts of FMCS mediators are directed toward the establishment of sound and stable labor-management relations on a continuing basis, thereby helping to reduce the incidence of work stoppages. The mediator's basic function is to encourage and promote better day-to-day relations between labor and management, so that issues arising in negotiations may be faced as problems to be settled through mutual effort rather than issues in dispute.

Arbitration   FMCS, on the joint request of employers and unions, will also assist in the selection of arbitrators from a roster of private citizens who are qualified as neutrals to adjudicate matters in dispute. For further information, contact the Office of Arbitration Services. Phone, 202–606–5111.

For further information, contact the Public Affairs Office, Federal Mediation and Conciliation Service, 2100 K Street NW., Washington, DC 20427. Phone, 202–606–8100. Internet, http://www.fmcs.gov.

FEDERAL MINE SAFETY & HEALTH REVIEW COMMISSION

601 New Jersey Avenue NW., Suite 9500, Washington, DC 20001–2021
Phone, 202–434–9900. Internet, http://www.fmshrc.gov. Email, fmshrc@fmshrc.gov.

Chairman	MARY LUCILLE JORDAN
Commissioners	ROBERT F. COHEN, JR., MICHAEL F. DUFFY, MICHAEL G. YOUNG, (VACANCY)

Chief Administrative Law Judge ROBERT J. LESNICK
General Counsel MICHAEL A. McCORD
Executive Director LISA M. BOYD

The Federal Mine Safety and Health Review Commission ensures compliance with occupational safety and health standards in the Nation's surface and underground coal, metal, and nonmetal mines.

The Federal Mine Safety and Health Review Commission is an independent, adjudicative agency established by the Federal Mine Safety and Health Act of 1977 (30 U.S.C. 801 et seq.), as amended. It provides administrative trial and appellate review of legal disputes arising from enforcement actions taken by the Department of Labor.

The Commission consists of five members who are appointed by the President with the advice and consent of the Senate and who serve staggered 6-year terms. The Chairman is appointed from among the Commissioners by the President.

The Commission and its Office of Administrative Law Judges are charged with deciding cases brought before it by the Mine Safety and Health Administration, mine operators, and miners or their representatives. These cases generally involve review of the Administration's enforcement actions, including citations, mine-closure orders, and proposals for civil penalties issued for violations of the act or the mandatory safety and health standards promulgated by the Secretary of Labor. The Commission also has jurisdiction over discrimination complaints filed by miners or their representatives in connection with their safety and health, complaints for compensation filed on behalf of miners idled as a result of mine closure orders issued by the Administration, and disputes over mine emergency response plans.

Cases brought before the Commission are assigned to the Office of Administrative Law Judges, and hearings are conducted pursuant to the requirements of the Administrative Procedure Act (5 U.S.C. 554, 556) and the Commission's procedural rules (29 CFR 2700).

A judge's decision becomes a final but nonprecedential order of the Commission 40 days after issuance unless the Commission has directed the case for review in response to a petition or on its own motion. If a review is conducted, a decision of the Commission becomes final 30 days after issuance unless a party adversely affected seeks review in the U.S. Circuit Court of Appeals for the District of Columbia or the Circuit within which the mine subject to the litigation is located.

As far as practicable, hearings are held at locations convenient to the affected mines. In addition to its Washington, DC, offices, the Office of Administrative Law Judges maintains an office in the Colonnade Center, Room 280, 1244 Speer Boulevard, Denver, CO 80204.

Sources of Information

Commission decisions are published bimonthly and are available through the Superintendent of Documents, U.S. Government Printing Office, Washington, DC 20402. The Commission's Web site includes recent decisions, a searchable database of previous decisions, procedural rules, audio recordings of recent public meetings, and other pertinent information.

Requests for Commission records should be submitted in accordance with the Commission's Freedom of Information Act regulations. Other information, including Commission rules of procedure and brochures explaining the Commission's functions, is available from the Executive Director, Federal Mine Safety and Health Review Commission, 601 New Jersey Avenue NW., Suite 9500, Washington, DC 20001–2021. Internet, www.fmshrc.gov. Email, fmshrc@fmshrc.gov.

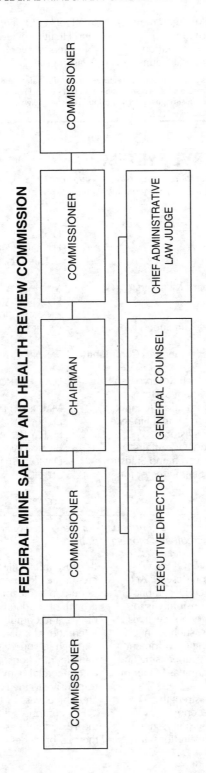

FEDERAL MINE SAFETY AND HEALTH REVIEW COMMISSION

For information on filing requirements, the status of cases before the Commission, or docket information, contact the Office of General Counsel or the Docket Office, Federal Mine Safety and Health Review Commission, 601 New Jersey Avenue NW., Suite 9500, Washington, DC 20001.

For further information, contact the Executive Director, Federal Mine Safety and Health Review Commission, 601 New Jersey Avenue NW., Suite 9500, Washington DC 20001–2021. Phone, 202–434–9905. Fax, 202–434–9906. Email, fmshrc@fmshrc.gov. Internet, http://www.fmshrc.gov.

FEDERAL RESERVE SYSTEM

Twentieth Street and Constitution Avenue NW., Washington, DC 20551
Phone, 202–452–3000. Internet, http://www.federalreserve.gov.

Chairman	BEN S. BERNANKE
Vice Chairman	DONALD L. KOHN
Members	ELIZABETH DUKE, DANIEL K. TARULLO, KEVIN M. WARSH, (2 VACANCIES)
Director, Division of Board Members	MICHELLE A. SMITH
General Counsel	SCOTT G. ALVAREZ
Secretary	JENNIFER J. JOHNSON
Director, Division of Banking Supervision and Regulation	PATRICK M. PARKINSON
Director, Division of Consumer and Community Affairs	SANDRA F. BRAUNSTEIN
Director, Division of Federal Reserve Bank Operations and Payment Systems	LOUISE L. ROSEMAN
Director, Division of Information Technology	MAUREEN HANNAN
Director, Division of International Finance	D. NATHAN SHEETS
Director, Management Division	H. FAY PETERS
Director, Division of Monetary Affairs	BRIAN MADIGAN
Director, Division of Research and Statistics	DAVID J. STOCKTON
Staff Director, Office of Staff Director for Management	STEPHEN R. MALPHRUS
Inspector General	ELIZABETH A. COLEMAN

The Federal Reserve System, the central bank of the United States, administers and formulates the Nation's credit and monetary policy.

The Federal Reserve System (FRS) was established by the Federal Reserve Act (12 U.S.C. 221), approved December 23, 1913. Its major responsibility is in the execution of monetary policy. It also performs other functions, such as the transfer of funds, handling Government deposits and debt issues, supervising and regulating banks, and acting as lender of last resort.

FRS contributes to the strength and vitality of the U.S. economy. By influencing the lending and investing activities of depository institutions and the cost and availability of money and credit, the FRS promotes the full use of human and capital resources, the growth of productivity, relatively stable prices, and equilibrium in the Nation's international balance of payments. Through its supervisory and regulatory banking functions, FRS helps maintain a commercial banking system that is responsive to the Nation's financial needs and objectives.

FRS comprises the Board of Governors; the 12 Federal Reserve Banks and their 25 branches and other facilities;

the Federal Open Market Committee; the Federal Advisory Council; the Consumer Advisory Council; the Thrift Institutions Advisory Council; and the Nation's financial institutions, including commercial banks, savings and loan associations, mutual savings banks, and credit unions.

Board of Governors   The Board comprises seven members appointed by the President with the advice and consent of the Senate. The Chairman of the Board of Governors is a member of the National Advisory Council on International Monetary and Financial Policies. The Board determines general monetary, credit, and operating policies for the System as a whole and formulates the rules and regulations necessary to carry out the purposes of the Federal Reserve Act. The Board's principal duties consist of monitoring credit conditions; supervising the Federal Reserve Banks, member banks, and bank holding companies; and regulating the implementation of certain consumer credit protection laws.

The Board has the power, within statutory limitations, to fix the requirements for reserves to be maintained by depository institutions on transaction accounts or nonpersonal time deposits. The Board reviews and determines the discount rate charged by the Federal Reserve Banks. For the purpose of preventing excessive use of credit for the purchase or carrying of securities, the Board is authorized to regulate the amount of credit that may be initially extended and subsequently maintained on any security (with certain exceptions).

Supervision of Federal Reserve Banks
The Board is authorized to make examinations of the Federal Reserve Banks, to require statements and reports from such Banks, to supervise the issue and retirement of Federal Reserve notes, to require the establishment or discontinuance of branches of Reserve Banks, and to exercise supervision over all relationships and transactions of those Banks with foreign branches.

Supervision of Bank Holding Companies
The Federal Reserve supervises and regulates bank holding companies. Its objective is to maintain the separation between banking and commerce by controlling the expansion of bank holding companies, preventing the formation of banking monopolies, restraining certain trade practices in banking, and limiting the nonbanking activities of bank holding companies. A company that seeks to become a bank holding company must obtain the prior approval of the Federal Reserve. Any company that qualifies as a bank holding company must register and file reports with the FRS.

Supervision of Banking Organizations
The Federal Reserve supervises and regulates domestic and international activities of U.S. banking organizations. It supervises State-chartered banks that are members of the System, all bank holding companies, and Edge Act and agreement corporations (corporations chartered to engage in international banking).

The Board has jurisdiction over the admission of State banks and trust companies to membership in the FRS, membership termination for these banks, the establishment of branches by these banks, and the approval of bank mergers and consolidations where the resulting institution will be a State member bank. It receives copies of condition reports submitted to the Federal Reserve Banks. It has power to examine all member banks and the affiliates of member banks and to require condition reports from them. It has authority to require periodic and other public disclosure of information with respect to an equity security of a State member bank that is held by 500 or more persons. It establishes minimum standards with respect to installation, maintenance, and operation of security devices and procedures by State member banks. It can issue cease-and-desist orders in connection with violations of law or unsafe or unsound banking practices by State member banks and to remove directors or officers of such banks in certain circumstances. It also can suspend member banks from use of the Federal Reserve System's credit facilities for using bank credit for speculation or other purposes inconsistent with the maintenance of sound credit conditions.

The Board may grant authority to member banks to establish branches in foreign countries or dependencies or insular possessions of the United States, to invest in the stocks of banks or corporations engaged in international or foreign banking, or to invest in foreign banks. It also charters, regulates, and supervises certain corporations that engage in foreign or international banking and financial activities.

The Board is authorized to issue general regulations permitting interlocking relationships in certain circumstances between member banks and organizations dealing in securities or between member banks and other banks.

The Board prescribes regulations to ensure a meaningful disclosure by lenders of credit terms so that consumers will be able to compare more readily the various credit terms available and will be informed about rules governing credit cards, including their potential liability for unauthorized use.

The Board has authority to impose reserve requirements and interest rate ceilings on branches and agencies of foreign banks in the United States, to grant loans to them, to provide them access to Federal Reserve services, and to limit their interstate banking activities.

Federal Open Market Committee   The Federal Open Market Committee comprises the Board of Governors and five of the presidents of the Reserve Banks. The Chairman of the Board of Governors is traditionally the Chairman of the Committee. The president of the Federal Reserve Bank of New York serves as a permanent member of the Committee. Four of the twelve Reserve Bank presidents rotate annually as members of the Committee.

Open market operations of the Reserve Banks are conducted under regulations adopted by the Committee and pursuant to specific policy directives issued by the Committee, which meets in Washington, DC, at frequent intervals. Purchases and sales of securities in the open market are undertaken to supply bank reserves to support the credit and money needed for long-term economic growth, to offset cyclical economic

swings, and to accommodate seasonal demands of businesses and consumers for money and credit. These operations are carried out principally in U.S. Government obligations, but they also include purchases and sales of Federal agency obligations. All operations are conducted in New York, where the primary markets for these securities are located; the Federal Reserve Bank of New York executes transactions for the Federal Reserve System Open Market Account in carrying out these operations.

Under the Committee's direction, the Federal Reserve Bank of New York also undertakes transactions in foreign currencies for the Federal Reserve System Open Market Account. The purposes of these operations include helping to safeguard the value of the dollar in international exchange markets and facilitating growth in international liquidity in accordance with the needs of an expanding world economy.

Federal Reserve Banks   The 12 Federal Reserve Banks are located in Atlanta, GA; Boston, MA; Chicago, IL; Cleveland, OH; Dallas, TX; Kansas City, MO; Minneapolis, MN; New York, NY; Philadelphia, PA; Richmond, VA; San Francisco, CA; and St. Louis, MO. Branch banks are located in Baltimore, MD; Birmingham, AL; Buffalo, NY; Charlotte, NC; Cincinnati, OH; Denver, CO; Detroit, MI; El Paso, TX; Helena, MT; Houston, TX; Jacksonville, FL; Little Rock, AR; Los Angeles, CA; Louisville, KY; Memphis, TN; Miami, FL; Nashville, TN; New Orleans, LA; Oklahoma City, OK; Omaha, NE; Pittsburgh, PA; Portland, OR; Salt Lake City, UT; San Antonio, TX; and Seattle, WA.

Reserves on Deposit   The Reserve Banks receive and hold on deposit the reserve or clearing account deposits of depository institutions. These banks are permitted to count their vault cash as part of their required reserve.

Extensions of Credit   The Federal Reserve is required to open its discount window to any depository institution that is subject to its reserve requirements on transaction accounts or nonpersonal time deposits. Discount window credit provides for Federal Reserve lending to eligible depository institutions under two basic programs. One is the adjustment credit program; the other supplies more extended credit for certain limited purposes.

Short-term adjustment credit is the primary type of Federal Reserve credit. It is available to help borrowers meet temporary requirements for funds. Borrowers are not permitted to use adjustment credit to take advantage of any spread between the discount rate and market rates.

Extended credit is provided through three programs designed to assist depository institutions in meeting longer term needs for funds. One provides seasonal credit—for periods running up to 9 months—to smaller depository institutions that lack access to market funds. A second program assists institutions that experience special difficulties arising from exceptional circumstances or practices involving only that institution. Finally, in cases where more general liquidity strains are affecting a broad range of depository institutions— such as those whose portfolios consist primarily of longer term assets—credit may be provided to address the problems of particular institutions being affected by the general situation.

Currency Issue   The Reserve Banks issue Federal Reserve notes, which constitute the bulk of money in circulation. These notes are obligations of the United States and are a prior lien upon the assets of the issuing Federal Reserve Bank. They are issued against a pledge by the Reserve Bank with the Federal Reserve agent of collateral security including gold certificates, paper discounted or purchased by the Bank, and direct obligations of the United States.

Other Powers   The Reserve Banks are empowered to act as clearinghouses and as collecting agents for depository institutions in the collection of checks and other instruments. They are also authorized to act as depositories and fiscal agents of the United States and to exercise other banking functions specified in the Federal Reserve Act. They perform a number of important functions in connection with the issue and redemption of United States Government securities.

Sources of Information

Employment   Written inquiries regarding employment should be addressed to the Director, Division of Personnel, Board of Governors of the Federal Reserve System, Washington, DC 20551.

Procurement   Firms seeking business with the Board should address their inquiries to the Director, Division of Support Services, Board of Governors of the Federal Reserve System, Washington, DC 20551.

Publications   Among the publications issued by the Board are The Federal Reserve System—Purposes and Functions, and a series of pamphlets including Guide to Business Credit and the Equal Credit Opportunity Act; Consumer Handbook; Making Deposits: When Will Your Money Be Available; and When Your Home Is On the Line: What You Should Know About Home Equity Lines of Credit. Copies of these pamphlets are available free of charge. Information regarding publications may be obtained in Room MP–510 (Martin Building) of the Board's headquarters. Phone, 202–452–3244.

Reading Room   A reading room where persons may inspect public records is located in Room B–1122 at the Board's headquarters, Twentieth Street and Constitution Avenue NW., Washington, DC 20551. Information regarding the availability of records may be obtained by calling 202–452–3684.

For further information, contact the Office of Public Affairs, Board of Governors, Federal Reserve System, Washington, DC 20551. Phone, 202–452–3204 or 202–452–3215. Internet, http://www.federalreserve.gov.

FEDERAL RETIREMENT THRIFT INVESTMENT BOARD

1250 H Street NW., Washington, DC 20005
Phone, 202–942–1600. Fax, 202–942–1676. Internet, http://www.tsp.gov.

Chairman	ANDREW M. SAUL
Members	TERRENCE A. DUFFY, DANA K. BILYEU, ALEJANDRO M. SANCHEZ, MICHAEL D. KENNEDY
Executive Director	GREGORY T. LONG
Chief Investment Officer	TRACEY A. RAY
Participant Services Director	PAMELA-JEANNE MORAN
Director of External Affairs	THOMAS J. TRABUCCO
General Counsel	THOMAS K. EMSWILER
Chief Information Officer	MARK A. HAGERTY
Chief Financial Officer	JAMES B. PETRICK
Director of Research and Strategic Planning	RENEE WILDER

The Federal Retirement Thrift Investment Board administers the Thrift Savings Plan, which provides Federal employees the opportunity to save for additional retirement security.

The Federal Retirement Thrift Investment Board was established as an independent agency by the Federal Employees' Retirement System Act of 1986 (5 U.S.C. 8351 and 8401–79). The act vests responsibility for the agency in six named fiduciaries: the five Board members and the Executive Director. The five members of the Board, one of whom is designated as Chairman, are appointed by the President with the advice and consent of the Senate and serve on the Board on a part-time basis. The members appoint the Executive Director, who is responsible for the management of the agency and the Plan.

Activities

The Thrift Savings Plan is a tax-deferred, defined contribution plan that was established as one of the three parts of the Federal Employees' Retirement System. For employees covered under the System, savings accumulated through the Plan make an important addition to the retirement benefits provided by Social Security and the System's Basic Annuity. Civil Service Retirement System employees and members of the Uniformed Services may also take advantage of the Plan to supplement their annuities.

The Board operates the Thrift Savings Plan and manages the investments of the Thrift Savings Fund solely for the benefit of participants and their beneficiaries. As part of these responsibilities, the Board maintains an account for each Plan participant, makes loans, purchases annuity contracts, and provides for the payment of benefits.

For further information, contact the Director of External Affairs, Federal Retirement Thrift Investment Board, 1250 H Street NW., Washington, DC 20005. Phone, 202–942–1640. Internet, http://www.tsp.gov.

FEDERAL TRADE COMMISSION

600 Pennsylvania Avenue NW., Washington, DC 20580
Phone, 202–326–2222. Internet, http://www.ftc.gov.

Chairman	JON LEIBOWITZ

Commissioners	WILLIAM E. KOVACIC, J. THOMAS ROSCH, EDITH RAMIREZ, JULIE BRILL
Chief of Staff	JONI LUPOVITZ
Executive Director	CHARLES SCHNEIDER
Chief Administrative Law Judge	D. MICHAEL CHAPPELL
Director, Bureau of Competition	RICHARD A. FEINSTEIN
Director, Bureau of Consumer Protection	DAVID VLADECK
Director, Bureau of Economics	JOSEPH FARRELL
Director, Office of Congressional Relations	JEANNE BUMPUS
Director, Office of International Affairs	RANDOLPH W. TRITELL
Director, Office of Policy Planning	SUSAN S. DESANTI
Director, Office of Public Affairs	CECELIA PREWETT
General Counsel	WILLARD K. TOM
Inspector General	JOHN SEEBA
Secretary of the Commission	DONALD S. CLARK

[For the Federal Trade Commission statement of organization, see the Code of Federal Regulations, Title 16, Part 0]

The Federal Trade Commission promotes consumer protections and enforces the laws that prohibit anticompetitive, deceptive, or unfair business practices.

The Federal Trade Commission (FTC) was established in 1914 by the Federal Trade Commission Act (15 U.S.C. 41–58). The Commission comprises five members appointed by the President, with the advice and consent of the Senate, for a term of 7 years. No more than three of the Commissioners may be members of the same political party. One Commissioner is designated by the President as Chairman of the Commission and is responsible for its administrative management.

Activities

FTC promotes competition and prevents general trade restraints such as price-fixing agreements, boycotts, illegal combinations of competitors, and other unfair methods of competition; prevents corporate mergers, acquisitions, or joint ventures that lessen competition or build a monopoly; and prevents pricing discrimination, exclusive dealing, tying arrangements, discrimination among competing customers by sellers, and interlocking directorates or officers' positions that may restrain competition.

FTC acts on behalf of consumers to protect them. It prevents and combats false or deceptive advertising of consumer products and services as well as other unfair or deceptive practices; promotes electronic commerce by fighting fraud on the Internet and working with other domestic and foreign agencies to develop and promote policies to safeguard online privacy of personal information; protects the privacy of consumers' personal information to prevent illegal or unwanted use of financial or other data; exposes fraudulent telemarketing schemes and protects consumers from abusive, deceptive, or unwanted telephone tactics; and enforces the National Do Not Call Registry; ensures truthful labeling of textile, wool, and fur products; requires creditors to disclose in writing certain cost information, such as the annual percentage rate, before consumers enter into credit transactions, as required by the Truth in Lending Act; protects consumers against circulation of inaccurate or obsolete credit reports and ensures that credit bureaus, consumer reporting agencies, credit grantors, and bill collectors exercise their responsibilities in a manner that is fair and equitable; educates consumers and businesses about their rights and responsibilities under Commission rules and regulations; and gathers factual data concerning economic and business

FEDERAL TRADE COMMISSION

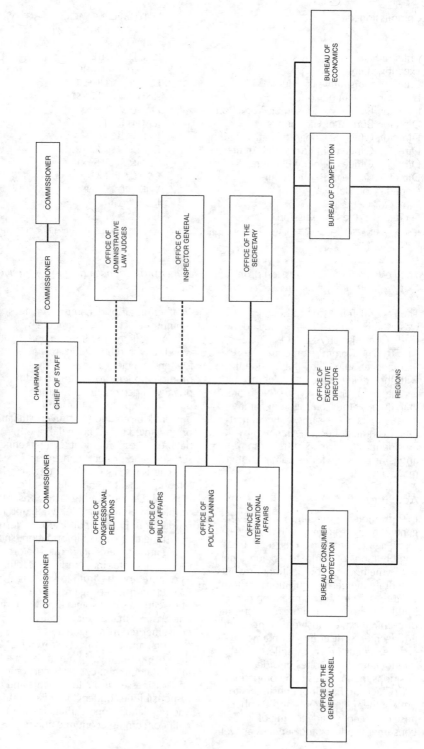

conditions and makes it available to the Congress, the President, and the public.

Competition FTC encourages competition in the American economy. It seeks to prevent unfair practices that undermine competition and attempts to prevent mergers or acquisitions of companies if the result may be to lessen competition. Under some circumstances, companies planning to merge must first give notice to the FTC and the Department of Justice's Antitrust Division and provide certain information concerning the operations of the companies involved.

Consumer Protection FTC promotes consumer protection across broad sectors of the economy. It works to increase the usefulness of advertising by ensuring that it is truthful and not misleading; reduce instances of fraudulent, deceptive, or unfair marketing practices; prevent creditors from using unlawful practices when granting credit, maintaining credit information, collecting debts, and operating credit systems; obtain redress for consumers harmed by deceptive or unfair practices; and educate businesses and the public about Commission activities.

FTC has issued rules and guidance in areas important to consumers, including health and nutrition claims in advertising, environmental advertising and labeling, children's online privacy, funeral services, telemarketing and electronic commerce, business opportunity, franchise and investment fraud, and mortgage lending and discrimination. It also has authority to bring actions to enforce its rules and certain Federal statutes.

Competition and Consumer Advocacy FTC advocates consumer interest in a competitive marketplace by encouraging courts, legislatures, and Government administrative bodies to consider efficiency and consumer welfare as important elements in their deliberations. It uses these opportunities to support procompetitive means of regulating the Nation's economy, including the elimination of anticompetitive restrictions that reduce consumer choice and the implementation of regulatory programs that protect the public and preserve competitive markets.

Compliance Activities FTC works to ensure compliance with its administrative and Federal court orders. Defendants against whom such orders have been issued are required to file reports with the FTC to substantiate their compliance. In the event compliance is not obtained, or if an order is subsequently violated, the FTC may institute contempt or civil penalty proceedings.

Cooperative Procedures FTC makes extensive use of voluntary and cooperative procedures. Through these procedures, business and industry may obtain authoritative guidance and a substantial measure of certainty as to what they may do under the laws administered by the FTC.

FTC issues, in plain language, administrative interpretations of laws enforced by the FTC. Guides provide the basis for compliance by members of a particular industry or by an industry in general. Failure to comply with the guides may result in corrective action by the FTC under applicable statutory provisions.

Enforcement FTC's law enforcement work covers actions to foster voluntary compliance with the law and formal administrative or Federal court litigation leading to mandatory orders against offenders.

Compliance may be obtained through voluntary and cooperative action by private companies in response to nonbinding staff advice, formal advisory opinions by the FTC, and guides and policy statements delineating legal requirements as to particular business practices.

Formal litigation is instituted either by issuing an administrative complaint or by filing a Federal district court complaint charging a person, partnership, or corporation with violating one or more of the statutes enforced by the FTC. If the charges are not contested or are found to be true after an administrative hearing or Federal court trial in a contested case, an administrative law judge or Federal court judge may issue an order requiring discontinuance of the unlawful practices.

Also, the FTC may request that a U.S. district court issue a preliminary or permanent injunction to halt the use of allegedly unfair or deceptive practices, to prevent an anticompetitive merger or unfair methods of competition from taking place or to prevent violations of any statute enforced by the FTC. In Federal court, the FTC may obtain other relief, including monetary redress.

Investigations   FTC investigations may originate through complaint by a consumer or competitor, the Congress, or from Federal, State, or municipal agencies. The FTC may also initiate an investigation into possible violations. Complaints may be submitted by letter or phone and should give the facts in detail accompanied by all supporting evidence. FTC also maintains electronic complaint systems that are accessible through its Web site. It is FTC policy not to disclose the identity of any complainant, except as required by law or FTC rules.

An order issued after an administrative or Federal court proceeding that requires the respondent to cease and desist or take other corrective action may be appealed. The appeals process may go as far as the Supreme Court.

Reports   The Commission prepares marketplace studies. Such reports have provided the basis for significant legislation and have also led to voluntary changes in the conduct of business, with resulting benefits to the public.

For a complete list of Federal Trade Commission regional offices, go to http://www.ftc.gov/ro/index.shtml.

Sources of Information

Contracts and Procurement   For information on contracts and procurement, contact the Assistant Chief Financial Officer for Acquisitions, Federal Trade Commission, Washington, DC 20580. Phone, 202–326–2339. Fax, 202–326–3529. Internet, www.ftc.gov.

Employment   For most positions, individuals must apply online through the Office of Personnel Management's application system, USAJOBS.

General Inquiries   To obtain general information or reach a variety of offices, contact the Federal Trade Commission at 202–326–2222. To submit a consumer complaint or to obtain free copies of all FTC consumer education materials, visit www.ftc.gov.

Publications   FTC consumer and business education publications are available through the Consumer Response Center, Federal Trade Commission, Washington, DC 20580. Phone, 877–382–4357. TTY, 866–653–4261. Internet, www.ftc.gov.

For further information, contact the Office of Public Affairs, Federal Trade Commission, 600 Pennsylvania Avenue NW., Washington, DC 20580. Phone, 202–326–2180. Fax, 202–326–3366. Internet, http://www.ftc. gov.

GENERAL SERVICES ADMINISTRATION

1800 F Street NW., Washington, DC 20405
Internet, http://www.gsa.gov.

Administrator	MARTHA N. JOHNSON
Deputy Administrator	SUSAN F. BRITA
Senior Counselor to the Administrator	STEPHEN R. LEEDS
Chief of Staff	MICHAEL J. ROBERTSON
Assistant Administrator	CATHLEEN C. KRONOPOLUS
Associate Administrator	ANTHONY E. COSTA
White House Liaison	MICHAEL J. ROBERTSON
Chairman, Civilian Board of Contract Appeals	STEPHEN M. DANIELS
Inspector General	BRIAN D. MILLER
General Counsel	KRIS E. DURMER

Associate Administrator, Office of Governmentwide Policy	KATHLEEN M. TURCO
Associate Administrator, Office of Citizen Services and Innovative Technologies	DAVID L. MCCLURE
Assiociate Administrator, Office of Communications and Marketing	SAHAR WALI
Associate Administrator, Office of Civil Rights	MADELINE C. CALIENDO
Associate Administrator, Office of Congressional and Intergovernmental Affairs	RALPH CONNER, *Acting*
Associate Administrator, Office of Small Business Utilization	JIYOUNG PARK
Associate Administrator, Office of Performance Improvement	STEVEN D. MCPEEK
Associate Administrator, Office of Emergency Response and Recovery	DARREN J. BLUE
Chief Financial Officer	MICAH M. CHEATHAM, *Acting*
Chief People Officer	GAIL T. LOVELACE
Chief Information Officer	CASEY COLEMAN
Regional Administrator, New England (Region 1)	GLENN C. ROTONDO, *Acting*
Regional Administrator, Northeast and Caribbean (Region 2)	JOHN SCORCIA, *Acting*
Regional Administrator, Mid-Atlantic (Region 3)	LINDA C. CHERO, *Acting*
Regional Administrator, Southeast Sunbelt (Region 4)	SHYAM K. REDDY
Regional Administrator, Great Lakes (Region 5)	J. DAVID HOOD, *Acting*
Regional Administrator, Heartland (Region 6)	JASON O. KLUMB
Regional Administrator, Greater Southwest (Region 7)	JUAN SALINAS
Regional Administrator, Rocky Mountain (Region 8)	SUSAN D. DAMOUR
Regional Administrator, Pacific Rim (Region 9)	JEFFREY E. NEELY, *Acting*
Regional Administrator, Northwest/Arctic (Region 10)	GEORGE E. NORTHCROFT
Regional Administrator, National Capital (Region 11)	SHARON J. BANKS, *Acting*
Commissioner, Federal Acquisition Service	STEVEN J. KEMPF, *Acting*
Commissioner, Public Buildings Service	ROBERT A. PECK

[For the General Services Administration statement of organization, see the Code of Federal Regulations, Title 41, Part 105–53]

The General Services Administration establishes policy for and provides management of Government property and records, including construction and operation of buildings; procurement and distribution of supplies; utilization and disposal of real and personal property; transportation, travel, fleet, and communications management; and management of the Governmentwide automatic data processing resources program.

The General Services Administration (GSA) was established by section 101 of the Federal Property and Administrative Services Act of 1949 (40 U.S.C. 751).
Civilian Board of Contract Appeals The Civilian Board of Contract Appeals is responsible for resolving disputes arising out of contracts between contractors and executive agencies, excluding the Defense Department, Postal Service, Postal Rate Commission, National Aeronautics and Space Administration,

and Tennessee Valley Authority. The Board also hears and decides requests for review of transportation audit rate determinations; claims by Federal civilian employees regarding travel and relocation expenses; claims for the proceeds of the sale of property of certain Federal civilian employees; cases involving the Indian Self-Determination and Education Assistance Act and the Federal Crop Insurance Corporation; and requests for arbitration to resolve disputes between applicants and the Federal Emergency Management Agency over funding for public assistance applications arising from the damage caused by Hurricanes Katrina and Rita.

In addition, the Board provides alternative dispute resolution services to executive agencies in both contract disputes which are the subject of a contracting officer's decision and other contract-related disputes. Although the Board is located within the agency, it functions as an independent tribunal.

For further information, contact the Civilian Board of Contract Appeals, General Services Administration, Washington, DC 20405. Phone, 202–606–8800. Internet, http://www.cbca.gsa.gov/.

Governmentwide Policy   The Office of Governmentwide Policy (OGP) collaborates with the Federal community to develop policies and guidelines for the management of Government property, technology, and administrative services. OGP's policymaking authority and policy support activities encompass the areas covering electronic government and information technology, acquisition, real property and the workplace, travel, transportation, personal property, aircraft, Federal motor vehicle fleet, mail, regulatory information and use of Federal Advisory Committees. OGP also provides leadership to interagency groups and facilitates Governmentwide management reform through the effective use of performance measures and best practices.

The Office of Technology Strategy provides policy guidance, architectures, uniform standards, pilots and solutions on electronic business and information technology to improve Government effectiveness and efficiency. It assists agencies on IT policy matters such as

usability, accommodation, innovative technologies, security, identity management and authentication. For further information, call 202–501–0202.

The Office of Real Property Management leads in Governmentwide real property policy development and the adoption of evolving technologies and innovative practices for managing real property resources. In addition to collaborating with Federal agencies to develop regulations and bulletins for the effective and efficient stewardship of Federal real property assets, and issuing guidance to facilitate the adoption of sustainable practices and performance measures for real property management, the Office also promotes the Governmentwide use of alternative workplace arrangements to increase space utilization efficiency and reduce carbon footprint. The Office also manages the Federal Real Property Database of all Federal Government real property assets and supports the Federal Real Property Management Council in its efforts to promote effective asset management Governmentwide. For further information, call 202–501–0856.

The Office of Federal High-Performance Green Buildings is a center of leadership and expertise in interagency greening initiatives, the deployment of advanced technologies and concepts, and large-scale sustainable real property portfolio management and operations. The Office combines authoritative knowledge of Federal processes with multidisciplinary expertise in high-performance green buildings to provide leadership within GSA, the Federal Government, and the broader commercial property market on buildings-related sustainability topics. For more information, call 202–219–1522.

The Office of Travel, Transportation, and Asset Management develops Governmentwide policies to ensure the economical and efficient management of Government assets, mail, travel, transportation, and relocation allowances. It develops regulations, collects and analyzes Governmentwide data, manages interagency policy committees, and collaborates with

GENERAL SERVICES ADMINISTRATION

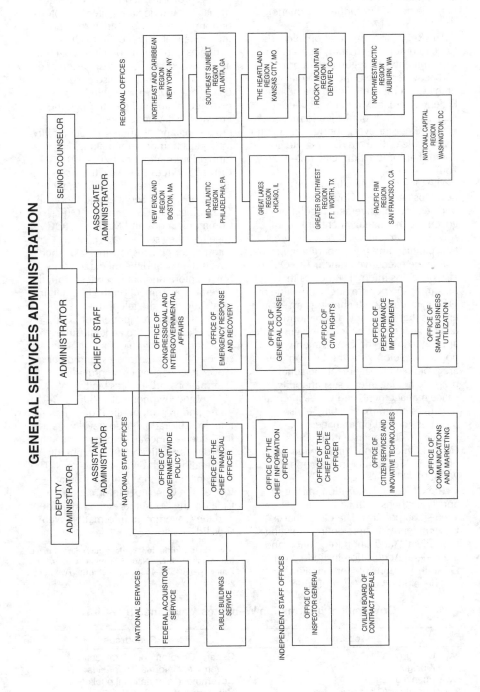

customers and stakeholders to address and facilitate Governmentwide improvements. For further information, call 202–501–1777.

The Regulatory Information Services Center compiles and analyzes data on Governmentwide regulatory information and activities. The principal publication of the Center is the Unified Agenda of Federal Regulatory and Deregulatory Actions, which is published in the Federal Register every spring and fall and is available online at www.reginfo.gov. For further information, call 202–482–7340.

The Office of Policy Initiatives develops and issues Governmentwide policies and management regulations governing the management of Federal Advisory Committees, promoting greater transparency and public participation within the executive branch. The Office also develops internal strategies that maximize the performance and value of OGP's information technology assets and ensures that GSA's internal nationwide asset management policies and guidelines are aligned with the Governmentwide policies developed by the OGP in the areas of fleet management, personal property management, sustainability, space management, and GSA's internal directives program. The Office also provides acquisition management services to customers within GSA's Central Office, supporting a wide range of actions. For further information, call 202–501–8880.

The Office of Acquisition manages and strengthens both Federal and GSA acquisition policies. The Office has a Governmentwide mission in managing the Federal Acquisition Regulation and the Federal Acquisition Institute. Within GSA, it is responsible for managing and developing GSA's acquisition

workforce, establishing GSA acquisition policy, providing acquisition oversight, managing and deciding agency protests, performing the ombudsman role for Indefinite Delivery Indefinite Quantity contracts and the metrication program, and operating the GSA suspension and debarment program. It also ensures compliance with applicable laws, regulations, and policies through the Procurement Management review processing and promotes accountability for acquisition decisionmaking. For further information, call 202–501–1043.

Citizen Services   GSA's citizen services are divided between two offices: the Office of Citizen Services and Innovative Technologies (OCSIT) and the Office of Communications and Marketing (OCM). Together, they serve as the central gateway for citizens, businesses, other governments, and the media to easily obtain information and services from the Federal Government on the Internet, in print, and over the telephone.

An important component of GSA's citizen services is the Federal Citizen Information Center (FCIC). FCIC serves citizens, businesses, and other Government agencies by providing information and services via the Web at USA.gov, by phone at 1–800–333–4636, and in print publications through Pueblo, Colorado. It also develops and implements innovative technologies that improve the delivery of Government information and services to citizens. The FCIC maintains the National Contact Center, the Consumer Information Catalog, the Consumer Action Handbook, and the Government information Web sites listed below.

For further information, contact the Federal Citizen Information Center's National Contact Center. Phone, 1–800–333–4636. Internet, http://www.info.gov.

Government Information Web Sites Maintained by FCIC

Web Site	Resources
www.USA.gov	General Government information
www.Kids.gov	Government information for kids
www.Pueblo.gsa.gov	Government informational publications
www.ConsumerAction.gov	Consumer information
www.info.gov	FCIC National Contact Center

Government Information Web Sites Maintained by FCIC—Continued

Web Site	Resources
www.GobiernoUSA.gov	General Government information in Spanish
www.Consumidor.gov	Consumer information in Spanish

Small Business Utilization   The Office of Small Business Utilization focuses on programs, policy, and outreach to assist the small business community nationwide in doing business with GSA.

For further information, contact the Office of Small Business Utilization. Phone, 202–501–1021.

Federal Acquisition Service

The Federal Acquisition Service (FAS) provides acquisition services for Federal agencies to increase overall Government efficiency. This includes the acquisition of products, services, and full-service programs in information technology, telecommunications, professional services, supplies, motor vehicles, travel and transportation, charge cards, and personal property utilization and disposal.

FAS provides multiple channels for customers to acquire the products, services, and solutions they need. Key acquisition programs include multiple awards schedules and Governmentwide acquisition contracts that provide customers easy access to a wide range of information technology, telecommunications, and professional products and services. It also provides its customers with access to the products and services they need through online Web sites such as GSA Advantage!, e-Buy, Schedules e-Library, GSA Auctions, Transportation Management Services Solutions, and a myriad of other electronic tools.

For further information, contact the Office of the Commissioner, Federal Acquisition Service. Phone, 703–605–5400.

Public Buildings Service

The Public Buildings Service (PBS) is the landlord for the civilian Federal Government. Its portfolio consists of 362 million square feet in over 9,600 assets across all 50 States, 6 U.S. Territories,

and the District of Columbia. PBS collects rent from Federal tenants, which is deposited into the Federal Buildings Fund, the principal funding mechanism for PBS.

PBS designs, builds, leases, manages, and maintains space in office buildings, courthouses, laboratories, border stations, data processing centers, warehouses, and childcare centers. It also repairs, alters, and renovates existing facilities and disposes of surplus Government properties. PBS is a leader in energy conservation and sustainable design. It preserves and maintains 480 historic properties in the Federal Government's inventory. PBS also commissions the country's most talented artists to artwork for new Federal buildings and conserves a substantial inventory of artwork from the past.

For further information, contact the Office of the Commissioner, Public Buildings Service. Phone, 202–501–1100.

Regional Offices

GSA operates 11 regional offices. For a complete list of these offices, visit www.gsa.gov/regions.

Sources of Information

Contracts   Individuals seeking to do business with the General Services Administration may obtain information through the Internet at www.gsa.gov.
Electronic Access   Information about GSA is available electronically through the Internet at www.gsa.gov.

Information about the Civilian Board of Contract Appeals is available through the Internet at www.cbca.gsa.gov.
Employment   Information regarding employment with GSA may be found at www.gsa.gov. Agency job listings are posted on www.usajobs.gov.
Fraud and Waste   Contact the Inspector General's hotline. Phone, 202–501–1780, or 800–424–5210.

Freedom of Information Act Requests
Inquiries concerning policies pertaining
to Freedom of Information Act matters
should be addressed to the GSA FOIA
Office, General Services Administration,
Room 6001, Washington, DC 20405.
Phone, 202–501–2262. Fax, 202–501–
2727.

Privacy Act Requests   Inquiries
concerning policies pertaining to Privacy
Act matters should be addressed to
GSA Privacy Act Officer, Information
Resources and Privacy Management
Division (CIB), General Services
Administration, Room 6224, Washington,
DC 20405. Phone, 202–501–0290.
Email, GSA.privacyact@gsa.gov.

Property Disposal   Inquiries about the
redistribution or competitive sale of
surplus real property should be directed
to the Office of Real Property Disposal,
Public Buildings Service, 1800 F Street
NW., Washington, DC 20405. Phone,
202–501–0084.

Public and News Media Inquiries
Inquiries from both the general public
and news media should be directed to
the Office of Communications, General
Services Administration, 1800 F Street
NW., Washington, DC 20405. Phone,
202–501–1231.

Publications   Many GSA publications
are available from the Government
Printing Office bookstore at http://
bookstore.gpo.gov. Orders and questions
about publications and paid subscriptions
should be directed to the Superintendent
of Documents, Government Printing
Office, Washington, DC 20401. Some
subscriptions may be obtained free of
charge or at cost from a Small Business
Center or GSA's Centralized Mailing List

Service (phone, 817–334–5215). If a
publication is not available through any
of these sources, contact a specific GSA
staff office, regional office, or service.
Addresses and phone numbers may be
found at www.gsa.gov.

For a free copy of the U.S.
Government TTY Directory, contact
the Federal Citizen Information Center,
Department TTY, Pueblo, CO 81009.
Phone, 888–878–3256. Internet,
www.gsa.gov/frs. For a free copy of
the quarterly Consumer Information
Catalog, including information on food,
nutrition, employment, Federal benefits,
the environment, fraud, privacy and
Internet issues, investing and credit, and
education, write to the Federal Citizen
Information Center, Pueblo, CO 81009.
Phone, 888–878–3256. Internet, www.
pueblo.gsa.gov.

For information about Federal
programs and services, call the Federal
Citizen Information Center's National
Contact Centers at 800–333–4636,
Monday through Friday from 8 a.m. to 8
p.m. eastern standard time.

For a free copy of the Federal Relay
Service Brochure, call 877–387–2001.
TTY, 202–585–1840.

Small Business Activities   Inquiries
concerning programs to assist small
businesses should be directed to the
Office of Small Business Utilization.
Phone, 202–501–1021.

Speakers   Inquiries and requests for
speakers should be directed to the
nearest regional office or the Office of
Communications and Marketing, General
Services Administration, Washington, DC
20405. Phone, 202–501–0705.

For further information concerning the General Services Administration, contact the Office of Citizen
Services, General Services Administration, Washington, DC 20405. Phone, 202–501–0705.

INTER-AMERICAN FOUNDATION

901 North Stuart Street, Tenth Floor, Arlington, VA 22203
Phone, 703–306–4301. Internet, http://www.iaf.gov.

Chair   JOHN P. SALAZAR
Vice Chair   THOMAS J. DODD

Directors	KAY K. ARNOLD, JACK C. VAUGHN, JR., ROGER W. WALLACE, (4 VACANCIES)
President	LARRY L. PALMER
General Counsel	JENNIFER HODGES
Vice President for Operations	LINDA B. KOLKO
Regional Director for South America and the Caribbean	WILBUR WRIGHT
Regional Director for Central America and Mexico	JILL WHEELER

The Inter-American Foundation supports social and economic development in Latin America and the Caribbean.

The Inter-American Foundation (IAF) was created in 1969 (22 U.S.C. 290f) as an experimental U.S. foreign assistance program. IAF is governed by a nine-person Board of Directors appointed by the President with the advice and consent of the Senate. Six members are drawn from the private sector and three from the Federal Government. The Board of Directors appoints the President of IAF.

IAF works in Latin America and the Caribbean to promote equitable, participatory, and sustainable self-help development by awarding grants directly to local organizations throughout the region. It also partners with the public and private sectors to build support and mobilize local, national, and international resources for grassroots development.

For further information, contact the Office of the President, Inter-American Foundation, 901 North Stuart Street, Tenth Floor, Arlington, VA 22203. Phone, 703–306–4301. Internet, http://www.iaf.gov.

MERIT SYSTEMS PROTECTION BOARD

1615 M Street NW., Fifth Floor, Washington, DC 20419
Phone, 202–653–7200; 800–209–8960. Fax, 202–653–7130. Internet, http://www.mspb.gov.

Chairman	SUSAN TSUI GRUNDMANN
Vice Chairman	ANNE WAGNER
Member	MARY M. ROSE
Executive Director	STEVEN V. LENKART
Clerk of the Board	WILLIAM D. SPENCER
Director, Financial and Administrative Management	ERNEST A. CAMERON, *Acting*
Director, Information Resources Management	TOMMY HWANG
Director, Office of Appeals Counsel	JAMES READ
Director, Office of Equal Employment Opportunity	CYNTHIA FERENTINOS, *Acting*
Director, Office of Policy and Evaluation	JOHN CRUM
Director, Office of Regional Operations	DEBORAH MIRON
General Counsel	JAMES EISENMANN

[For the Merit Systems Protection Board statement of organization, see the Code of Federal Regulations, Title 5, Part 1200]

The Merit Systems Protection Board protects the integrity of the Federal personnel merit systems and the rights of Federal employees.

MERIT SYSTEMS PROTECTION BOARD

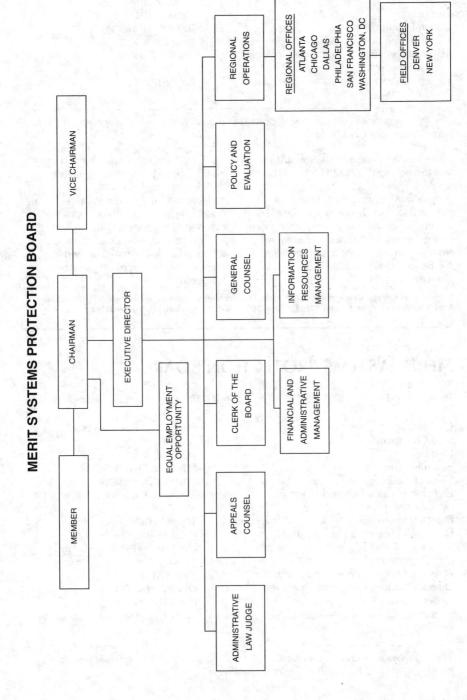

The Merit Systems Protection Board is a successor agency to the United States Civil Service Commission, established by act of January 16, 1883 (22 Stat. 403). Reorganization Plan No. 2 of 1978 (5 U.S.C. app.) redesignated part of the Commission as the Merit Systems Protection Board. The Board is comprised of three members appointed by the President with the advice and consent of the Senate.

Activities

The Board has responsibility for hearing and adjudicating appeals by Federal employees of adverse personnel actions, such as removals, suspensions, and demotions. It also resolves cases involving reemployment rights, denial of periodic step increases in pay, actions against administrative law judges, and charges of prohibited personnel practices, including charges in connection with whistleblowing.

The Board has the authority to enforce its decisions and to order corrective and disciplinary actions. An employee or applicant for employment involved in an appealable action that also involves an allegation of discrimination may ask the Equal Employment Opportunity Commission to review a Board decision. Final decisions and orders of the Board can be appealed to the U.S. Court of Appeals for the Federal Circuit.

The Board reviews regulations issued by the Office of Personnel Management and has the authority to require agencies to cease compliance with any regulation that could constitute a prohibited personnel practice. It also conducts special studies of the civil service and other executive branch merit systems and reports to the President and the Congress on whether the Federal work force is being adequately protected against political abuses and prohibited personnel practices.

For a complete list of Merit Systems Protection Board offices, visit www.mspb. gov/contact/contact.htm.

For further information, contact the Merit Systems Protection Board, 1615 M Street NW., Washington, DC 20419. Phone, 202–653–7200 or 800–209–8960. TDD, 800–877–8339. Fax, 202–653–7130. Email, mspb@ mspb.gov. Internet, http://www.mspb.gov.

NATIONAL AERONAUTICS AND SPACE ADMINISTRATION

300 E Street SW., Washington, DC 20546
Phone, 202–358–0000. Internet, http://www.nasa.gov.

Administrator	CHARLES F. BOLDEN, JR.
Deputy Administrator	LORI B. GARVER
Associate Administrator	CHRISTOPHER SCOLESE
Associate Deputy Administrator	CHARLES SCALES
Chief of Staff	DAVID P. RADZANOWSKI
Deputy Chief of Staff/White House Liaison	DAVID L. NOBLE
Assistant Associate Administrator	CHRISTYL JOHNSON
Chief Financial Officer	ELIZABETH M. ROBINSON
Chief Information Officer	LINDA Y. CURETON
Chief Scientist	(VACANCY)
Chief Technologist	ROBERT D. BRAUN
Inspector General	PAUL K. MARTIN
Chief Engineer	MICHAEL RYSCHKEWITSCH
Chief Health and Medical Officer	RICHARD S. WILLIAMS
Chief Safety and Mission Assurance Officer	BRYAN O'CONNOR

Associate Administrator, Independent Program and Cost Evaluation	W. MICHAEL HAWES
Associate Administrator, Diversity and Equal Opportunity	BRENDA R. MANUEL
Associate Administrator, Education	JAMES L. STOFAN, *Acting*
Associate Administrator, International and Interagency Relations	MICHAEL F. O'BRIEN
General Counsel	MICHAEL C. WHOLLEY
Associate Administrator, Legislative and Intergovernmental Affairs	SETH L. STATLER
Associate Administrator, Communications	BOB N. JACOBS, *Acting*
Associate Administrator, Small Business Programs	GLENN A. DELGADO
Associate Administrator, Aeronautics Research Mission Directorate	JAIWON SHIN
Associate Administrator, Exploration Systems Mission Directorate	DOUGLAS COOKE
Associate Administrator, Science Mission Directorate	ED WEILER
Associate Administrator, Space Operations Mission Directorate	WILLIAM GERSTENMAIER
Associate Administrator, Mission Support	WOODROW WHITLOW
Assistant Administrator, Human Capital Management	TONI DAWSEY
Assistant Administrator, Strategic Infrastructure	OLGA DOMINGUEZ
Executive Director, Headquarters Operations	CHRIS JEDREY
Executive Director, NASA Shared Services Center	RICHARD E. ARBUTHNOT
Assistant Administrator for Agency Operations	THOMAS LUEDTKE
Assistant Administrator, Internal Controls and Management Systems	LOUIS BECKER
Assistant Administrator, Procurement	BILL MCNALLY
Assistant Administrator, Protective Services	JACK FORSYTHE
Director, NASA Management Office	EUGENE H. TRINH
Director, Ames Research Center	S. PETE WORDEN
Director, Dryden Flight Research Center	DAVID D. MCBRIDE
Director, Glenn Research Center	RAMON LUGO, *Acting*
Director, Goddard Space Flight Center	ROBERT D. STRAIN
Director, Johnson Space Center	MICHAEL COATS
Director, Kennedy Space Center	ROBERT D. CABANA
Director, Langley Research Center	LESA ROE
Director, Marshall Space Flight Center	ROBERT M. LIGHTFOOT
Director, Stennis Space Center	PATRICK E. SCHEUERMANN
Director, Jet Propulsion Laboratory	CHARLES ELACHI

[For the National Aeronautics and Space Administration statement of organization, see the Code of Federal Regulations, Title 14, Part 1201]

The mission of the National Aeronautics and Space Administration is to pioneer the future in space exploration, scientific discovery, and aeronautics research.

The National Aeronautics and Space Administration (NASA) was established by the National Aeronautics and Space Act of 1958, as amended (42 U.S.C. 2451 et seq.).

Activities

Aeronautics Research Directorate
The Aeronautics Research Mission Directorate conducts research and

technology activities to develop the knowledge, tools, and technologies to support the development of future air and space vehicles and to support the transformation of the Nation's air transportation system. The Directorate's programs focus on cutting-edge, fundamental research in traditional aeronautical disciplines, as well as emerging fields with promising applications to aeronautics, and are conducted in conjunction with industry, academia, and other U.S. Government departments and agencies, including the Federal Aviation Administration and the Department of Defense.

For further information, call 202–358–2047.

Space Operations Mission Directorate
The Space Operations Mission Directorate (SOMD) provides the foundation for NASA's space program: space travel for human and robotic missions, in-space laboratories, and the means to return data to Earth. SOMD is responsible for many critical enabling capabilities that make possible much of the science, research, and exploration achievements of the rest of NASA. This is done through three themes: the International Space Station, Space Shuttle, and Space and Flight Support.

The International Space Station is a complex of laboratories maintained to support scientific research, technology development, and the exploration of a permanent human presence in Earth's orbit.

The Space Shuttle, first launched in 1981, provides the only current capability in the United States for human access to space. The Shuttle's focus over the next several years will be the assembly of the International Space Station, after which it will be phased out of service.

The Space and Flight Support theme encompasses space communications, launch services, and rocket propulsion testing. Space communications consists of five major elements: the Space Network or Tracking and Data Relay Satellite System, the Deep Space Network, the Near Earth Network, the NASA Integrated Services Network,

and NASA Spectrum Management. The launch services program focuses on acquisition of commercial launch services for NASA's space and Earth science missions. The rocket propulsion testing program supports the flight readiness of various liquid propulsion engines and acts as a test bed for rocket engines of the future.

For further information, call 202–358–2015.

Science Mission Directorate   The Science Mission Directorate carries out the scientific exploration of the Earth, Moon, Mars, and beyond, charting the best route of discovery. The Directorate manages and sponsors research, flight missions, advanced technology development, and related activities. It works to expand our understanding of the Earth and the Sun and the Sun's effect on the solar system environments; explore the solar system with robots to study its origins and evolution, including the origins of life within it; and explore the universe beyond, from the search for planets and life in other solar systems to the origin, evolution, and destiny of the universe itself.

For further information, call 202–358–3889, or visit www.nasascience.nasa.gov.

Exploration Systems Mission Directorate
The Exploration Systems Mission Directorate (ESMD) is responsible for technology development that enables sustained and affordable human and robotic space exploration. ESMD's activities include developing robotic missions to multiple destinations to scout human exploration targets, increasing investments in human research to prepare for long journeys beyond Earth, and developing U.S. commercial human spaceflight capabilities.

For further information, call 202–358–7246.

NASA Centers

Ames Research Center   The Ames Research Center, located in California's Silicon Valley, provides solutions to NASA's exploration questions through interdisciplinary scientific discovery and innovative technology systems. The Center provides leadership in

NATIONAL AERONAUTICS AND SPACE ADMINISTRATION

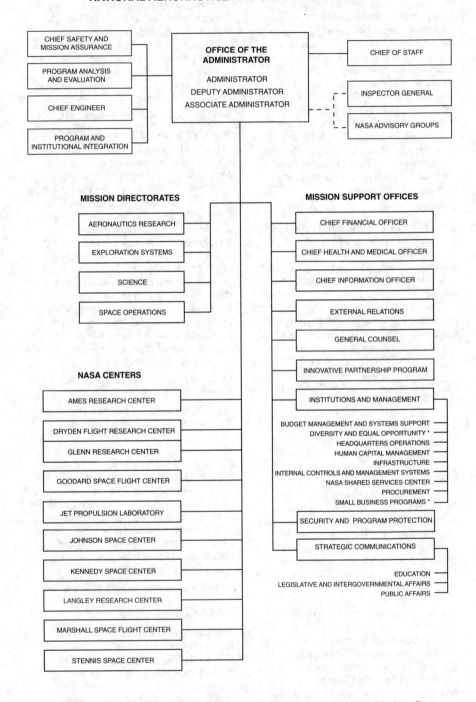

* In accordance with law or regulation, the offices of Diversity and Equal Opportunity and Small Business Programs maintain reporting relationships to the Administrator and Deputy Administrator.

astrobiology, information science, small spacecraft, advanced thermal protection systems, human factors, and the development of new tools for a safer and more efficient national airspace. It also develops unique partnerships and collaborations, exemplified by NASA's Astrobiology Institute, the NASA Research Park, and the University Affiliated Research Center.

Dryden Flight Research Center   The Dryden Flight Research Center, located in Edwards, CA, is NASA's primary installation for flight research. Since 1946, Dryden's researchers have led the way to major advancements in the design and capabilities of many civilian and military aircraft. Its workforce expertise in aeronautics and developing flight research tools and techniques, coupled with its suite of specialized laboratories and facilities, are key to the development and maturation of new vehicles.

Glenn Research Center   The Glenn Research Center, with two locations in Cleveland and Sandusky, Ohio, works with other NASA Centers to develop spaceflight systems and technologies to make a new, safer, and less expensive rocket system to return Americans to the Moon and help humanity explore the universe. The Center provides expertise in propulsion, power, communications, and testing for spacecraft and lunar systems. In aeronautics, its scientists and engineers develop new technologies to make airplanes safer, quieter, and more environmentally friendly.

Goddard Space Flight Center   The Goddard Space Flight Center, located in Greenbelt, MD, expands the knowledge of Earth and its environment, the solar system, and the universe through observations from space. The Center also conducts scientific investigations, develops and operates space systems, and advances essential technologies.

Johnson Space Center   The Lyndon B. Johnson Space Center, located in Houston, TX, leads the United States in the human exploration of space. The Center has made major advances in science, technology, engineering,

and medicine and has led the Nation's human spaceflight programs and projects. It strives to advance the Nation's exploration of the universe with its expertise in medical, biomedical, and life sciences; lunar and planetary geosciences; crew and mission operations; crew health and safety; project management; and space systems engineering. The Center also leads worldwide research in extraterrestrial materials curation and the interaction between humans and robotics, as well as the biology and physiology of humans in space.

Kennedy Space Center   The John F. Kennedy Space Center, located in Florida, is responsible for NASA's space launch operations and spaceport and range technologies. Home to NASA's three space shuttles and Launch Services Program, it manages the processing and launching of astronaut crews and associated payloads. The Center's management activities include the International Space Station segments, research experiments and supplies, and NASA's scientific and research spacecraft. These scientific and research spacecraft range from robotic landers to Earth observation satellites and space-based telescopes on a variety of launch vehicles.

Innovative technology experts at the Center support NASA's current programs and future exploration missions by developing new products and processes that benefit the space agency and consumers. The Center remains a leader in cutting-edge research and development in the areas of physics, chemistry, technology, prototype designing, engineering, environmental conservation, and renewable energy.

Langley Research Center   The Langley Research Center, located in Hampton, VA, is renowned for its scientific and technological expertise in aerospace research, systems integration, and atmospheric science. Established in 1917 as an aeronautics lab, the Center also has a rich heritage in space and science technologies. It conducts critical research in materials and structures;

aerodynamics; and hypersonic, supersonic, and subsonic flight. The Center has also developed and validated technologies to improve the effectiveness, capability, comfort, and efficiency of the Nation's air transportation system. It supports the space exploration program and space operations with systems analysis and engineering, aerosciences, materials and structures, and technology and systems development and testing. The Center continues to have a principal role in understanding and protecting our planet through atmospheric measurement, instruments, missions, and prediction algorithms. Its Engineering and Safety Center has improved mission safety by performing independent engineering assessments, testing, analysis, and evaluations to determine appropriate preventative and corrective action for problems, trends, or issues across NASA programs and projects.

Marshall Space Flight Center The George C. Marshall Space Flight Center, located in Huntsville, AL, develops and integrates the transportation and space systems required for NASA's exploration, operations, and scientific missions. The Center provides the engineering and scientific capabilities to deliver space transportation and propulsion systems, space systems development and integration, scientific and exploration instruments, and basic and applied research. It manages the Space Shuttle propulsion elements, life support systems and operations for scientific experiments aboard the International Space Station, the Chandra X-ray Observatory, the Lunar Quest Program, and Michoud Assembly Facility. Other programs and projects include the International Lunar Network and Discovery and New Frontiers Programs.

Stennis Space Center The John C. Stennis Center, located near Bay St. Louis, MS, serves as NASA's rocket propulsion testing ground. The Center provides test services not only for America's space program, but also for the Department of Defense and private sector. Its unique rocket propulsion test capabilities will be used extensively as part of the heavy lift

and propulsion technology program. The Center's Applied Science and Technology Project Office provides project management to support NASA's science and technology goals. It also supports NASA's Applied Sciences Program.

Jet Propulsion Laboratory The Laboratory is managed under contract by the California Institute of Technology in Pasadena, CA. It develops spacecraft and space sensors and conducts mission operations and ground-based research in support of solar system exploration, Earth science and applications, Earth and ocean dynamics, space physics and astronomy, and information systems technology. It is also responsible for the management of the Deep Space Network in support of NASA projects.

Sources of Information

Contracts and Small Business Activities Inquiries regarding contracting for small business opportunities with NASA should be directed to the Associate Administrator for Small Business Programs, Room 2K39, NASA Headquarters, 300 E Street SW., Washington, DC 20546. Phone, 202–358–2088.

Employment Direct all general inquiries to the NASA Shared Services Center, Stennis, MS 39529. Phone, 877–677–2123. Email, nssc-contactcenter@nasa.gov.

Library NASA Headquarters Library, Room 1J22, 300 E Street SW., Washington, DC 20546. Phone, 202–358–0168. Internet, www.hq.nasa.gov/office/hqlibrary/index.html. Email, Library@hq.nasa.gov.

OIG Hotline An individual may report crimes, fraud, waste, and abuse in NASA programs and operations by calling the OIG Hotline (800–424–9183); by writing to the NASA Inspector General (P.O. Box 23089, L'Enfant Plaza Station, Washington, DC 20026); or by sending an electronic message from the OIG's Web site (www.hq.nasa.gov/office/oig/hq/cyberhotline.html).

Publications, Speakers, Films, and Exhibit Services Several publications concerning these services can be obtained by contacting the Public Affairs

Officer of the nearest NASA Center. Publications include NASA Directory of Services for the Public, NASA Film List, and NASA Educational Publications List. The headquarters telephone directory and certain publications and picture sets are available for sale from the Superintendent of Documents, Government Printing Office, Washington, DC 20402. Telephone directories for NASA Centers are available only from the Centers. Publications and documents not available for sale from the Superintendent of Documents or the National Technical Information Service (Springfield, VA 22151) may be obtained from NASA Center's Information Center in accordance with the NASA regulation concerning freedom of information.

For further information, contact the Headquarters Information Center, National Aeronautics and Space Administration, Washington, DC 20546. Phone, 202–358–0000. Internet, http://www.nasa.gov.

NATIONAL ARCHIVES AND RECORDS ADMINISTRATION

8601 Adelphi Road, College Park, MD 20740
Phone, 866–272–6272. Internet, http://www.archives.gov.

Archivist of the United States	DAVID S. FERRIERO
Deputy Archivist of the United States	ADRIENNE C. THOMAS
Assistant Archivist for Administration	RICK JUDSON, *Acting*
Assistant Archivist for Information Services	MIKE WASH
Assistant Archivist for Presidential Libraries	SHARON FAWCETT
Assistant Archivist for Records Services	SHARON THIBODEAU, *Acting*
Assistant Archivist for Regional Records Services	DAVID WEINBERG
Chief Human Capital Officer	ANALISA ARCHER
Chief of Staff	DEBRA S. WALL
Chief Operating Officer	THOMAS E. MILLS
Deputy Chief Information Officer	CHARLES PIERCY
Director, Center for the National Archives Experience	MARVIN PINKERT
Director, Congressional Affairs	JOHN HAMILTON
Director, Information Security Oversight Office	WILLLIAM CIRA, *Acting*
Director, Office of Government Information Services	MIRIAM NISBET
Director, Office of the Federal Register	RAYMOND A. MOSLEY
Director, Policy and Planning Staff	MARY ANN HADYKA, *Acting*
Executive Director, National Historical Publications and Records Commission	KATHLEEN M. WILLIAMS
Executive for Agency Services	WILLIAM J. BOSANKO
External Affairs Liaison	DAVID MCMILLEN
General Counsel	GARY M. STERN
Public Affairs Officer	SUSAN COOPER

[For the National Archives and Records Administration statement of organization, see the Federal Register of June 25, 1985, 50 FR 26278]

The National Archives and Records Administration safeguards and preserves the records of our Government, ensuring that the people can discover, use, and learn from this documentary heritage; establishes policies and procedures for managing U.S. Government records; manages the Presidential Libraries system; and publishes the laws, regulations, and Presidential and other public documents.

The National Archives and Records Administration (NARA) is the successor agency to the National Archives Establishment, which was created in 1934 and later, in 1949, incorporated into the General Services Administration as the National Archives and Records Service in 1949. NARA was established as an independent agency in the Federal Government's executive branch of the Government by act of October 19, 1984 (44 U.S.C. 2101 et seq.), effective April 1, 1985.

Activities

Archival Program   NARA maintains the U.S. Government's most historically valuable records, ranging from the Revolutionary War era to the recent past; arranges and preserves records and prepares finding aids to facilitate their use; makes records publicly accessible online and in its research rooms; answers requests for information contained in its holdings; and provides, for a fee, copies of records. NARA holdings include the records of the U.S. House of Representatives and Senate, which are preserved and administered by the Center for Legislative Archives. Many important records are available in microfilm and on online at www.archives.gov. Archival records are maintained in NARA facilities in the Washington, DC, area. Records of exceptional local or regional interest are maintained in NARA archives located in other parts of the country. There are also nine NARA-affiliated archives holding NARA-owned records and making them available to the public.

Records Management   To ensure proper documentation of the organization, policies, and activities of the Government, NARA develops standards and guidelines for nationwide management and disposition of recorded information. It appraises Federal records and approves disposition schedules. NARA also inspects agency records and management practices, develops records management training programs, provides guidance and assistance on records management, and stores inactive records.

Office of Records Services— Washington, DC

Modern Records Program   The Modern Records Program improves the life cycle management of Federal records in all media for Government agencies. It provides formal training in Federal records management; preserves and makes available permanent electronic records; and coordinates technical assistance to Federal agencies on records creation, management, and disposition to agencies in the Washington, DC, area.

For a complete listing of Washington area records facilities, visit www.archives. gov/dc-metro.

For further information on Records Center services in the Washington, DC, area (West Virginia, Virginia, and Maryland), contact the Washington National Records Center. Phone, 301–778–1650 (Records Management Division) or 301–778–1510 (Center Operations Division).

Office of Regional Records Services

The Office of Regional Records Services assists agencies outside the Nation's capital. NARA serves nine regions and runs the National Personnel Records Center. Each region operates a full life-cycle records program, including records management operations, records centers, and archives. NARA maintains information about the records management programs of Federal agencies in each region; conducts inspections, evaluations, or surveys of agency records and records management programs; reports on its findings; and recommends improvements or necessary corrective actions. NARA also furnishes guidance and technical assistance to Federal agencies on records creation, management, and disposition; develops and conducts training aimed at improving agency records management and disposition practices; and appraises Federal records to determine whether the U.S. Government should preserve them.

Federal Records Centers   These Centers store and service noncurrent and some active Federal agency records. Services include storage of textual and special media records; storage for nonclassified and classified records;

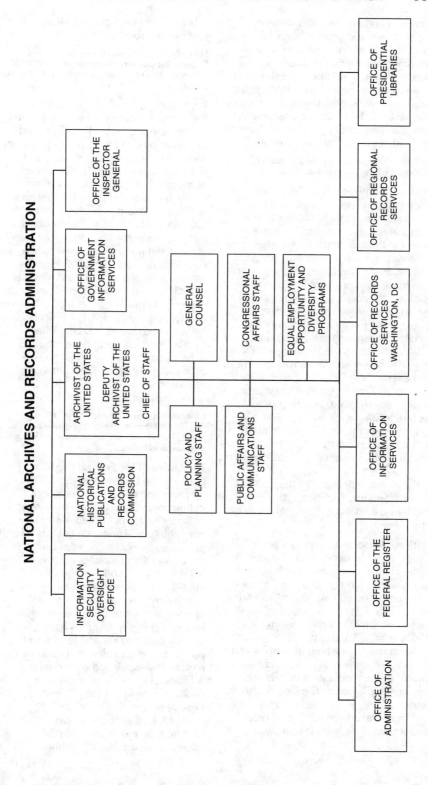

NATIONAL ARCHIVES AND RECORDS ADMINISTRATION

retrieval of records to fulfill statutory requirements and conduct daily business; special projects in response to shifts in customer demands; expedited response to congressional inquiries, litigation, and urgent business needs; and disposition services, providing for disposal of records that have reached their required retention period or transfer of permanent records to the holdings of the National Archives.

For further information on Federal Records Centers, visit www.archives.gov/frc/about.html.

Regional Archives   Archival records of exceptional regional significance are maintained in these facilities. At a Regional Archives, visitors have free access to these records, Internet resources for genealogical and historical research, and expert staff available to guide visitors. Onsite workshops teach the general public, educators, and students how to use archives for personal, professional, or academic research. Many important records, whose originals are held in NARA facilities in the Washington, DC, area, can be viewed in microform at most of these regional facilities. Archival staff arrange and preserve records and prepare finding aids to facilitate their use; make records available for use in research rooms; answer written or oral requests for information contained in NARA's holdings; and, for a fee, provide copies of records.

For a complete listing of Regional Records Services facilities, visit www.archives.gov/locations/regional-archives.html.

National Personnel Records Center   The Center stores, services, and disposes of the civilian personnel, medical, and pay records of former U.S. Civil Service employees and military personnel and medical records of discharged military personnel. Including the National Archives at St. Louis, the facility provides research rooms in which the public and Federal agency personnel can reference official military personnel folders and other related records. The Center also provides Federal agencies with services and technical advice relating to records disposition, filing and classification

schemes, and protection of vital civilian and military records.

For further information, contact the Office of Regional Records Services. Phone, 301–837–2950. Fax, 301–837–1617. Internet, http://www.archives. gov/st-louis.

Presidential Libraries   Through the Presidential libraries, which are located at sites selected by the Presidents and built with private funds, NARA preserves and makes available the records and personal papers of a particular President's administration. Each library operates research rooms and provides reference services on Presidential documents; displays the holdings of the Presidential administration; and provides public and educational programs. The records of each President since Herbert Hoover are administered by NARA. While such records were once considered personal papers, all Presidential records created on or after January 20, 1981, are declared by law to be owned and controlled by the United States and are required to be transferred to NARA at the end of the administration, pursuant to the Presidential Records Act of 1978 (44 U.S.C. 2201 et seq.).

For further information, contact the Office of Presidential Libraries. Phone, 301–837–3250. Fax, 301–837–3199. Internet, http://www.archives.gov/ presidential-libraries.

The Office of the Federal Register   The Office of the Federal Register (OFR) prepares and publishes a variety of public documents. Upon issuance, acts of Congress are published in slip law (pamphlet) form and then cumulated and published for each session of Congress in the United States Statutes at Large.

Each Federal workday, the OFR publishes the Federal Register, which contains current Presidential proclamations and Executive orders, Federal agency regulations having general applicability and legal effect, proposed agency rules, and documents required by statute to be published. All Federal regulations in force are codified annually in the Code of Federal Regulations.

Presidential speeches, news conferences, messages, and other materials released by the White

House Office of the Press Secretary are published online in the Daily Compilation of Presidential Documents and annually in the Public Papers of the Presidents. The Daily Compilation of Presidential Documents, as well as electronic versions of the previous Weekly Compilation of Presidential Documents from 1993 onward, can be accessed at www.presidentialdocuments. gov.

The United States Government Manual, published annually, serves as the official handbook of the Federal Government, providing extensive information on the legislative, judicial, and executive branches.

All of the above publications are available in both paper editions and online in electronic formats at www.ofr. gov.

For further information, contact Information Services and Technology, Office of the Federal Register. Phone, 202–741–6000. TTY, 202–741–6086. Fax, 202–741–6012. Email, fedreg.info@nara. gov. Internet, http://www.ofr.gov.

National Historical Publications and Records Commission
The National Historical Publications and Records Commission (NHPRC) is the grantmaking affiliate of the National Archives and Records Administration. Its mission is to promote the preservation and use of America's documentary heritage essential to understanding our democracy, history, and culture. NHPRC grants help State and local archives, universities, historical societies, and other nonprofit organizations solve preservation problems dealing with electronic records, improve training and techniques, strengthen archival programs, preserve and process records collections, and provide access to them through the publication of finding aids and documentary editions of the papers of the Founding Era and other themes and historical figures in American history. The NHPRC works in partnership with a national network of State historical records advisory boards. It also provides Federal leadership in public policy for the preservation of, and access to, America's documentary heritage.

For further information, contact the National Historical Publications and Records Commission. Phone, 202–357–5010. Email, nhprc@archives.gov. Internet, http://www.archives.gov/nhprc.

Information Security Oversight Office
The Information Security Oversight Office (ISOO) oversees the security classification programs in both Government and industry and reports to the President annually on their status. Executive orders 12958 and 13526 serve as the authority for ISOO. The Office receives its policy and program guidance from the National Security Council. An organizational component of the National Archives and Records Administration, ISOO's goals are to hold classification activity to the minimum necessary to protect the national security; to ensure the safeguarding of classified national security information in both Government and industry in a cost-effective and efficient manner; and to promote declassification and public access to information as soon as national security considerations permit.

For further information, contact the Information Security Oversight Office. Phone, 202–357–5250. Email, isoo@nara.gov. Internet, http://www.archives. gov/isoo.

Office of Government Information Services
Established under the OPEN Government Act of 2007 (5 U.S.C. 101 note), the Office of Government Information Services (OGIS) reviews Freedom of Information Act (FOIA) activities throughout the Government. OGIS serves as liaison between individuals making FOIA requests and administrative agencies, providing mediation services and resolving disputes as necessary. OGIS reviews policies and procedures of administrative agencies under FOIA, reviews agency compliance with FOIA, and recommends policy changes to Congress and the President to improve administration of FOIA.

For further information, contact the Office of Government Information Services. Phone, 301–837–1996. Email, ogis@nara.gov. Internet, http://www. archives.gov/ogis.

National Archives Trust Fund Board
The National Archives Trust Fund Board receives funds from the sale of reproductions of historic documents and

publications about the records, as well as from gifts and bequests. The Board invests these funds and uses income to support archival functions such as the preparation of publications that make information about historic records more widely available. Members of the Board are the Archivist of the United States, the Secretary of the Treasury, and the Chairman of the National Endowment for the Humanities.

For further information, contact the Secretary, National Archives Trust Fund Board. Phone, 301–837–3165.

Sources of Information

Calendar of Events   To be added to the mailing list for the monthly National Archives Calendar of Events or for a recorded announcement of events at the National Archives locations in Washington, DC, and College Park, MD, call 202–357–5000. TDD, 301–837–0482. Internet, www.archives.gov/calendar.

Congressional Affairs   The Congressional Affairs staff maintains contact with and responds to inquiries from congressional offices. Phone, 202–357–5100. Fax, 202–357–5959.

Contracts   Information on business opportunities with NARA is available electronically at www.fedbizopps.gov.

Management Principles and Research-Teaching Resources Workshops   NARA offers several courses on archival and records management principles and on using NARA resources for research and in the classroom. For information on public programs and workshops, contact the Center for the National Archives Experience Education Office at 202–357–5210. Fax, 202–357–5925.

Modern Archives Institute   For information, contact the Modern Archives Institute, Room 301, National Archives Building, 700 Pennsylvania Avenue NW., Washington, DC 20408–0001. Phone, 202–357–5259.

Know Your Records Program   For information about the Know Your Records program, contact the Customer Services Division, Room G–13, National Archives Building, 700 Pennsylvania Avenue, NW., Washington, DC 20408–

0001. Phone, 202–357–5333. Email, KYR@nara.gov. Internet, www.archives.gov/dc-metro/know-your-records.

Records Management Workshops   For information about records management workshops, contact the National Records Management Training Program (phone, 301–837–0660), any Regional Records Services facility, or the Office of Regional Records Services (phone, 301–837–2950). Internet, www.archives.gov/records-mgmt/training/index.html.

Federal Register Workshop   For information about the workshop, "The Federal Register: What It Is and How To Use It," call 202–741–6000.

Institute for the Editing of Historical Documents/Archives Leadership Institute   For information about the Institute for the Editing of Historical Documents or the Archives Leadership Institute at the University of Wisconsin, Madison, or fellowships in documentary editing, contact NHPRC, National Archives and Records Administration, 700 Pennsylvania Avenue NW., Washington, DC 20408–0001. Phone, 202–357–5010. Email, nhprc@archives.gov. Internet, www.archives.gov/grants.

Electronic Access   Information about NARA, its holdings and publications, and links to NARA social media sites are available electronically. Internet, www.archives.gov. Email, inquire@nara.gov.

Employment   For job opportunities, contact the nearest NARA facility or the Human Resources Operations Branch, Room 339, 9700 Page Avenue, St. Louis, MO 63132. Phone, 800–827–4898. TDD, 314–801–0886. Internet, www.archives.gov/careers/jobs.

Freedom of Information Act/Privacy Act Requests   For operational records of the National Archives and Records Administration, contact the NARA Freedom of Information Act/Privacy Act Officer, General Counsel Staff, National Archives and Records Administration, 8601 Adelphi Road, College Park, MD 20740–6001. Phone, 301–837–3642. Fax, 301–837–0293. For archival records in the custody of the Office of Records Services-Washington, DC, contact the Special Access/FOIA Staff, National Archives and Records Administration,

8601 Adelphi Road, College Park, MD 20740–6001. Phone, 301–837–3190. Fax, 301–837–1864. For archival records located at a NARA archives location outside the Washington, DC, metropolitan area, contact the facility holding the records or the Office of Regional Records Services. Phone, 301–837–2950. Fax, 301–837–1617. To determine the location of records, search NARA's online Archival Research Catalog (ARC) at www.archives.gov/research/ arc. For archival records in the custody of a Presidential library, contact the library that has custody of the records. For records in the physical custody of the Washington National Records Center or the regional Records Centers, contact the Federal agency that transferred the records to the facility. Records stored in the Records Centers remain in the legal custody of the agency that created them.

Grants For NHPRC grants, contact NHPRC, National Archives and Records Administration, 700 Pennsylvania Avenue NW., Washington, DC 20408–0001. Phone, 202–357–5010. Email, nhprc@ archives.gov. Internet, www.archives.gov/ grants.

Museum Shops Publications, document facsimiles, and document-related souvenirs are available for sale at the National Archives Shop in Washington, DC, each Presidential Library, and at the archives locations in Atlanta, GA, and Kansas City, MO. Phone, 202–357–5271. Internet, www.estore.archives.gov.

Museum Programs Contact the Center for the National Archives Experience, National Archives and Records Administration, Washington, DC 20408. Phone, 202–357–5210. Fax, 202–357– 5926.

Public Affairs The Public Affairs staff maintains contact with and responds to inquiries from the media, issues press releases and other literature, and maintains contact with organizations representing the archival profession, scholarly organizations, and other groups served by NARA. Phone, 202–357–5300.

Agency Publications Agency publications, including facsimiles of certain documents, finding aids to records, and Prologue, a scholarly journal published quarterly, are available from the Customer Service Center (NWCC2), Room 1000, National Archives at College Park, 8601 Adelphi Road, College Park, MD 20740–6001. Phone, 800–234– 8861. Fax, 301–837–0483. Internet, www.archives.gov/publications.

Records Management Publications Most records management publications are available electronically on the NARA Web site. Limited quantities of some records management publications and posters are available in hard copy format from the Life Cycle Management Division, National Archives and Records Administration, 8601 Adelphi Road, College Park, MD 20740–6001. Phone, 301–837–3560. Fax, 301–837–3699. Email, nara.recordsmgttraining@nara.gov.

Laws, Regulations, and Presidential Documents Information about laws, regulations, and Presidential documents is available from the Office of the Federal Register, National Archives and Records Administration, Washington, DC 20408. Phone, 202–741–6000. Email, fedreg. info@nara.gov. Internet, www.archives. gov/federal-register or www.ofr.gov. To subscribe to the Federal Register table of contents electronic mailing list, go to listserv.access.gpo.gov and select online mailing list archives, FEDREGTOC–L. To receive email notification of new public laws, subscribe to PENS (Public Laws Electronic Notification Service) at www.archives.gov/federal-register, "New Public Laws."

NHPRC Guidelines NHPRC guidelines are available from the NHPRC, National Archives and Records Administration, 700 Pennsylvania Avenue NW., Washington, DC 20408–0001. Phone, 202–357–5010. Email, nhprc@archives. gov. Internet, www.archives.gov/grants.

Newsletter for Research at the National Archives The Researcher News newsletter provides information needed to conduct research at the National Archives Building in Washington, DC, and the National Archives at College Park, MD. To subscribe to the electronic mailing list, send email to KYR@nara.gov.

Research Facilities Records are available for research purposes in reading rooms at the National Archives

Building, 700 Pennsylvania Avenue NW., Washington, DC (phone, 202–357–5400); at the National Archives at College Park, 8601 Adelphi Road, College Park, MD (phone, 866–272–6272); and at each Presidential Library, the National Personnel Records Center, and at NARA's 13 archives locations throughout the country. Written requests for information may be sent to any of these units, or they may be addressed to the Customer Services Division, National Archives at College Park, Room 1000, 8601 Adelphi Road, College Park, MD 20740–6001. Phone, 866–272–6272. Email, inquire@nara.gov.

Federal Register Public Inspection Desk   The Public Inspection Desk of the Office of the Federal Register is open every Federal business day for public inspection of documents scheduled for publication in the next day's Federal Register, at Suite 700, 800 North Capitol Street NW., Washington, DC. Phone, 202–741–6000. In addition, the documents currently on public inspection may be viewed at www.ofr.gov.

Speakers and Presentations   The National Archives conducts regular public programs featuring authors, films, symposia, and an annual genealogy fair related to National Archives holdings. For programs in the Washington, DC, area, more information is available by calling 202–357–5000. The calendar of events is also available by sending an email to reservations.nwe@nara.gov.

Teacher Workshops and Teaching Materials   The National Archives education specialists have developed programs to train teachers in the use of primary source material in the classroom and can provide information on how to obtain documentary teaching materials for classroom use. For further information, contact the education staff

of the Center for the National Archives Experience. Phone, 202–357–5210.

Tours   Individuals or groups may request guided tours of the exhibitions at the National Archives Building. Tours are given by reservation only and subject to availability. Individuals are requested to make arrangements at least 6 weeks in advance. Tours are scheduled for 9:45 a.m., Monday through Friday. Self-guided tours are available, through reservations, Monday through Friday from 10:15 a.m. until 90 minutes before closing. Groups are limited to 100 people. For more information and reservations, contact the Call Center. Phone, 202–357–5450 or 877–444–6777. Email, visitorservices@nara.gov. Internet, www.archives.gov/nae/visit/reserved-visits.html. Tours of the National Archives at College Park, MD, are available, through reservation, Monday through Friday from 8:00 a.m. to 4:00 p.m. Guided tours are available Monday through Thursday between 10:30 a.m. and 2:30 p.m. Group size is limited to 20 people. For more information and reservations, contact the Volunteer Program staff. Phone, 301–837–3002. Email, volunteercp@nara.gov.

Volunteer Service   Volunteer service opportunities are available at the National Archives Building and the National Archives at College Park, MD. Volunteers conduct tours, provide information in the exhibit halls, work with staff archivists in processing historic documents, and serve as research aides in the genealogical orientation room. For more information, contact the Volunteer Program staff. Phone, 202–357–5272. Fax, 202–357–5925. Email, volunteer@nara.gov. Similar opportunities exist in the Presidential Libraries and at NARA's 13 archives locations nationwide. Contact the facility closest to you for information about volunteer opportunities.

For further information, write or visit the National Archives and Records Administration, 700 Pennsylvania Avenue NW., Washington, DC 20408–0001. Phone, 202–357–5400. Email, inquire@nara.gov. Internet, http://www.archives.gov.

NATIONAL CAPITAL PLANNING COMMISSION

401 Ninth Street NW., Suite 500, Washington, DC 20004
Phone, 202–482–7200. Internet, http://www.ncpc.gov.

Chairman	L. PRESTON BRYANT, JR.
Vice Chairman	ROBERT E. MILLER
Members	HERBERT F. AMES, ARRINGTON DIXON, JOHN M. HART, GEORGE T. SIMPSON

Ex Officio

(Secretary of the Interior)	KENNETH L. SALAZAR
(Secretary of Defense)	ROBERT M. GATES
(Administrator of General Services)	MARTHA N. JOHNSON
(Chairman, Senate Committee on Homeland Security and Governmental Affairs)	JOSEPH I. LIEBERMAN
(Chairman, House Committee on Oversight and Government Reform)	EDOLPHUS TOWNS
(Mayor of the District of Columbia)	ADRIAN M. FENTY
(Chairman, Council of the District of Columbia)	VINCENT C. GRAY

Staff

Executive Director	MARCEL ACOSTA
Chief Operating Officer	BARRY S. SOCKS
Chief Urban Designer	CHRISTINE SAUM
General Counsel	ANNE R. SCHULYER
Director, Intergovernmental Affairs	JULIA A. KOSTER
Director, Physical Planning	WILLIAM G. DOWD
Director, Policy and Research	MICHAEL A. SHERMAN
Director, Urban Design and Plan Review Division	DAVID W. LEVY
Director, Administration	CHARLES J. RIEDER
Director, Office of Public Affairs	LISA N. MACSPADDEN
Secretariat	DEBORAH B. YOUNG

[For the National Capital Planning Commission statement of organization, see the Code of Federal Regulations, Title 1, Part 456.2]

The National Capital Planning Commission is the central agency for conducting planning and development activities for Federal lands and facilities in the National Capital Region. The region includes the District of Columbia and all land areas within the boundaries of Montgomery and Prince George's Counties in Maryland and Fairfax, Loudoun, Prince William, and Arlington Counties and the City of Alexandria in Virginia.

The National Capital Planning Commission was established as a park planning agency by act of June 6, 1924, as amended (40 U.S.C. 71 et seq.). Two years later its role was expanded to include comprehensive planning. In 1952, under the National Capital Planning Act, the Commission was designated the central planning agency for the Federal and District of Columbia governments.

In 1973, the National Capital Planning Act was amended by the District of Columbia Home Rule Act, which made the mayor of the District of Columbia the chief planner for the District and gave the Commission specific authority for reviewing certain District decisions. The Commission continues to serve as the central planning agency for the Federal Government in the National Capital Region.

NATIONAL CAPITAL PLANNING COMMISSION

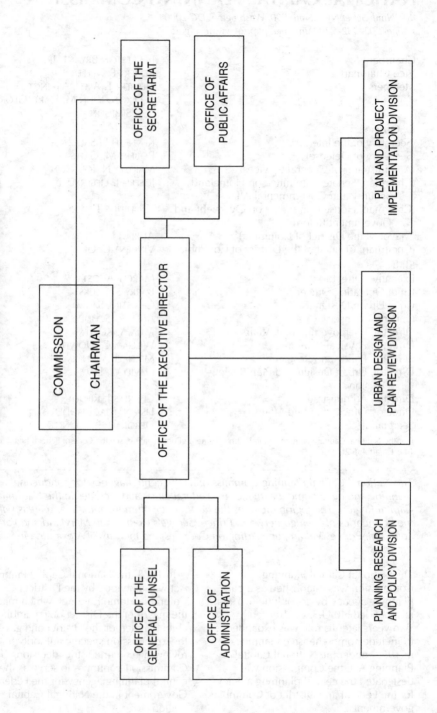

The Commission is composed of five appointed and seven ex officio members. Three citizen members, including the Chairman, are appointed by the President and two by the mayor of the District of Columbia. Presidential appointees include one resident each from Maryland and Virginia and one from anywhere in the United States. The two mayoral appointees must be District of Columbia residents.

For further information, contact the National Capital Planning Commission, 401 Ninth Street NW., Suite 500, Washington, DC 20004. Phone, 202–482–7200. Fax, 202–482–7272. Email, info@ncpc.gov. Internet, http://www.ncpc.gov.

EDITORIAL NOTE: The National Credit Union Administration did not meet the publication deadline for submitting updated information of its activities, functions, and sources of information as required by the automatic disclosure provisions of the Freedom of Information Act (5 U.S.C. 552(a)(1)(A)).

NATIONAL CREDIT UNION ADMINISTRATION

1775 Duke Street, Alexandria, VA 22314
Phone, 703–518–6300. Internet, http://www.ncua.gov.

Chairman	DEBBIE MATZ
Vice Chairman	RODNEY E. HOOD
Member	GIGI HYLAND
Executive Director	DAVID M. MARQUIS
General Counsel	ROBERT M. FENNER
Inspector General	WILLIAM DESARNO
Director, Office of Examination and Insurance	MELINDA LOVE
Director, Office of Corporate Credit Unions	SCOTT HUNT
Chief Financial Officer	MARY ANN WOODSON
Chief Information Officer	DOUG VERNER
Director, Office of Small Credit Union Initiatives	TAWANA Y. JAMES
Director, Office of Human Resources	LORRAINE PHILLIPS
Director, Public and Congressional Affairs	JOHN J. MCKECHNIE, III
Director, Office of Capital Markets and Planning	OWEN COLE
Director, Office of Consumer Protection	KENT D. BUCKHAM

[For the National Credit Union Administration statement of organization, see the Code of Federal Regulations, Title 12, Part 720]

The National Credit Union Administration is responsible for chartering, insuring, supervising, and examining Federal credit unions and administering the National Credit Union Share Insurance Fund.

The National Credit Union Administration (NCUA) was established by act of March 10, 1970 (12 U.S.C. 1752), and reorganized by act of November 10, 1978 (12 U.S.C. 226), as an independent agency in the executive branch of the Federal Government. It regulates and insures all Federal credit unions and insures State-chartered credit unions that apply and qualify for share insurance.

Activities

Chartering The Administration grants Federal credit union charters to groups sharing a common bond of occupation or association or groups within a well-defined neighborhood, community, or rural district. A preliminary investigation is made to determine if certain standards are met before granting a Federal charter.

Examinations The Administration regularly examines Federal credit unions to determine their solvency and compliance with laws and regulations and to assist credit union management and operations.

For further information, contact the Director, Office of Examination and Insurance. Phone, 703–518–6360.

Share Insurance The act of October 19, 1970 (12 U.S.C. 1781 et seq.), provides for a program of share insurance. The insurance is mandatory for Federal credit unions. State-chartered credit unions in many States are required to have Federal share insurance, and it is optional for other State-chartered credit unions. Credit union members' accounts are insured up to $100,000. The National Credit Union Share Insurance Fund requires each insured credit union to place and maintain a 1-percent deposit of its insured savings with the Fund.

For further information, contact the Director, Office of Examination and Insurance. Phone, 703–518–6360.

Supervision Supervisory activities are carried out through regular examiner contacts and through periodic policy and regulatory releases from the Administration. The Administration also identifies emerging problems and monitors operations between examinations.

For a complete list of National Credit Union Administration regional offices, visit www.ncua.gov/Contact.aspx.

Sources of Information

Consumer Assistance Questions about credit union insurance and other consumer matters can be directed to NCUA's Consumer Assistance Center at 800–755–1030, from 9 a.m. to 4 p.m., eastern standard time, Monday through Friday. After business hours, consumers may leave a recorded message.

Consumer Complaints The Administration investigates the complaints of members unable to resolve problems with their Federal credit unions. Complaints should be sent to the regional office in the State where the credit union is located.

Employment Inquiries and applications for employment should be directed to the Office of Human Resources, National Credit Union Administration, 1775 Duke Street, Alexandria, VA 22314–3428.

Federally Insured Credit Unions A list of federally insured credit union names, addresses, asset levels, and number of members is available for review at NCUA's central and regional offices. Copies of the listing are available at a nominal fee from NCUA, Publications, 1775 Duke Street, Alexandria, VA 22314–3428. Phone, 703–518–6340. A listing is also available electronically through the Internet at www.ncua.gov/indexdata.html.

Publications A listing and copies of NCUA publications are available from NCUA, Publications, 1775 Duke Street, Alexandria, VA 22314–3428. Phone, 703–518–6340. Publications are also available electronically through the Internet at www.ncua.gov.

Starting a Federal Credit Union Groups interested in forming a Federal credit union may obtain free information by writing to the appropriate regional office.

For further information, contact the Office of Public and Congressional Affairs, National Credit Union Administration, 1775 Duke Street, Alexandria, VA 22314–3428. Phone, 703–518–6330. Internet, http://www.ncua.gov.

NATIONAL FOUNDATION ON THE ARTS AND THE HUMANITIES

National Endowment for the Arts

1100 Pennsylvania Avenue NW., Washington, DC 20506
Phone, 202–682–5400. TDD, 202–682–5496. Internet, http://www.arts.gov.

Chairman	ROCCO LANDESMAN
Senior Deputy Chairman	JOAN SHIGEKAWA
Deputy Chairman, Management and Budget	LAURENCE M. BADEN
Deputy Chairman, Programs and Partnerships	PATRICE WALKER POWELL

Director of Government Affairs	ANITA DECKER
AccessAbility Director	PAULA TERRY
Administrative Services Director	KATHLEEN M. EDWARDS
Arts Education Director	SARAH B. CUNNINGHAM
Budget Officer	JOHN SOTELO
Chief Information Officer	MICHAEL BURKE
Civil Rights, Equal Employment Opportunity	ANGELIA RICHARDSON
Dance Director	DOUGLAS SONNTAG
Design Director	JASON SCHUPBACH
Federal Partnerships Coordinator	TONY TIGHE
Finance Officer	SANDRA STUECKLER
Folk and Traditional Arts Director	BARRY BERGEY
General Counsel	KAREN ELIAS, *Acting*
Grants and Contracts Officer	NICKI JACOBS
Guidelines and Panel Operations Director	JILLIAN L. MILLER
Human Resources Director	CRAIG M. MCCORD, SR.
Inspector General	AUVONETT JONES, *Acting*
International Activities Director	PENNIE OJEDA
Literature Director	JON P. PEEDE
Media Arts Director	TED LIBBEY
Museums and Visual Arts Director	ROBERT FRANKEL
Music and Opera Director	WAYNE BROWN
Presenting and Artists Communities Director	MARIO GARCIA DURHAM
Public Affairs Director	JAMIE BENNETT
Research and Analysis Director	SUNIL IYENGAR
State and Regional Partnerships Director	LAURA SCANLAN
Theater and Musical Theater Director	RALPH REMINGTON

The National Endowment for the Arts is dedicated to supporting excellence in the arts, both new and established, bringing the arts to all Americans, and providing leadership in arts education.

Through its grants and programs, the Arts Endowment brings great art to all 50 States and 6 U.S. jurisdictions, including rural areas, inner cities, and military bases. The Arts Endowment awards competitive matching grants to nonprofit organizations, units of State or local government, and federally recognized tribal communities or tribes, for projects, programs, or activities in the fields of artist communities, arts education, dance, design, folk and traditional arts, literature, local arts agencies, media arts, museums, music, musical theater, opera, presenting, theater, and visual arts. In addition, it awards competitive nonmatching individual fellowships in literature and honorary fellowships in jazz, folk and traditional arts, and opera. Forty percent of the Arts Endowment's grant funds go to the 56 State and jurisdictional arts agencies and their 6 regional arts organizations in support of arts projects in thousands of communities across the country.

Sources of Information

Grants   For information about Arts Endowment funding opportunities, contact the Public Affairs Office. Phone, 202–682–5400. TDD, 202–682–5496. Internet, www.arts.gov/grants.
Publications   To obtain a copy of the Arts Endowment's Annual Report, funding guidelines, or other publications, contact the Public Affairs Office. Phone, 202–682–5400. TDD, 202–682–5496. Internet, www.arts.gov.

For further information, contact the Public Affairs Office, National Endowment for the Arts, 1100 Pennsylvania Avenue NW., Washington, DC 20506–0001. Phone, 202–682–5400. TDD, 202–682–5496. Internet, http://www.arts.gov.

National Endowment for the Humanities

1100 Pennsylvania Avenue NW., Washington, DC 20506
Phone, 202–606–8400; 800–634–1121. Internet, http://www.neh.gov. Email, info@neh.gov.

Chairman	JIM LEACH
Deputy Chairman	CAROLE WATSON
Director, White House and Congressional Affairs	JEREMY BERNARD
Senior Adviser to the Chairman	EVA CALDERA
Inspector General	SHELDON BERNSTEIN
Chief Information Officer	BRETT BOBLEY
General Counsel	MICHAEL MCDONALD
Director, Communications	JUDY HAVEMANN
Director, We the People Program	CAROLE WATSON
Director, Office of Publications	DAVID SKINNER
Director, Federal/State Partnership	EDYTHE MANZA
Assistant Chairman for Planning and Operations	JEFFREY THOMAS
Director, Accounting	JOHN GLEASON
Director, Administrative Services	BARRY MAYNES
Director, Office of Grant Management	SUSAN DAISEY
Director, Office of Human Resources	ANTHONY MITCHELL
Director, Information Resource Management	TANYA PELTZ
Director, Office of Planning and Budget	LARRY MYERS
Assistant Chairman for Programs	ADAM WOLFSON
Director, Office of Challenge Grants	STEPHEN M. ROSS
Director, Office of Digital Humanities	BRETT BOBLEY
Director, Division of Education Programs	WILLIAM CRAIG RICE
Director, Division of Preservation and Access	NADINA GARDNER
Director, Division of Public Programs	THOMAS PHELPS
Director, Division of Research Programs	JANE AIKIN
Director, EDSITEment Partnership	CAROL PETERS

The National Endowment for the Humanities supports research, education, preservation, and public programs in the humanities.

According to the agency's authorizing legislation, the term "humanities" includes, but is not limited to, the study of the following: language, both modern and classical; linguistics; literature; history; jurisprudence; philosophy; archeology; comparative religion; ethics; the history, criticism, and theory of the arts; and those aspects of the social sciences that employ historical or philosophical approaches.

To increase understanding and appreciation of the humanities, the Endowment makes grants to individuals, groups, or institutions: schools, colleges, universities, museums, public television stations, libraries, public agencies, and nonprofit private groups.

Bridging Cultures Initiative   This initiative encourages projects that explore the ways in which cultures around the globe, as well as the many subcultures within America's borders, have influenced American society.

For further information, call 202–606–8337.

Challenge Grants   Nonprofit institutions interested in developing new sources of long-term support for educational, scholarly, preservation, and public programs in the humanities may be assisted in these efforts by a challenge grant.

For further information, call 202–606–8309.

Digital Humanities   The Office of Digital Humanities encourages and supports projects that use or study the impact

NATIONAL ENDOWMENT FOR THE HUMANITIES

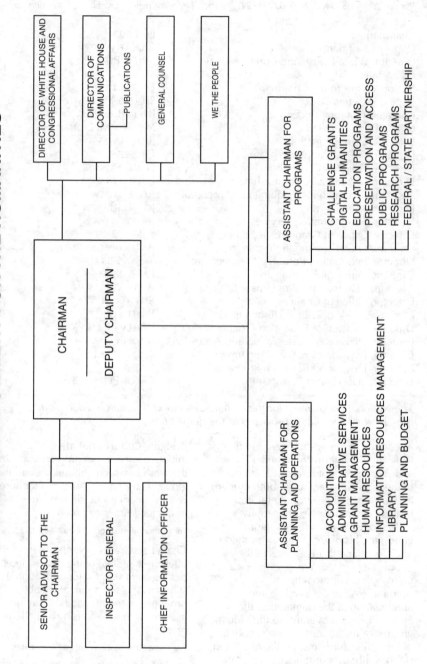

CHAIRMAN

DEPUTY CHAIRMAN

DIRECTOR OF WHITE HOUSE AND CONGRESSIONAL AFFAIRS

DIRECTOR OF COMMUNICATIONS
— PUBLICATIONS

GENERAL COUNSEL

WE THE PEOPLE

SENIOR ADVISOR TO THE CHAIRMAN

INSPECTOR GENERAL

CHIEF INFORMATION OFFICER

ASSISTANT CHAIRMAN FOR PROGRAMS
- CHALLENGE GRANTS
- DIGITAL HUMANITIES
- EDUCATION PROGRAMS
- PRESERVATION AND ACCESS
- PUBLIC PROGRAMS
- RESEARCH PROGRAMS
- FEDERAL / STATE PARTNERSHIP

ASSISTANT CHAIRMAN FOR PLANNING AND OPERATIONS
- ACCOUNTING
- ADMINISTRATIVE SERVICES
- GRANT MANAGEMENT
- HUMAN RESOURCES
- INFORMATION RESOURCES MANAGEMENT
- LIBRARY
- PLANNING AND BUDGET

of digital technology on education, preservation, public programming, and research in the humanities.

For further information, call 202–606–8401. Email, odh@neh.gov.

Education Through grants to educational institutions and fellowships to scholars and teachers, this division strengthens sustained thoughtful study of the humanities at all levels of education.

For further information, call 202–606–8500.

Federal/State Partnership Humanities committees in each of the 50 States, the Virgin Islands, Puerto Rico, the District of Columbia, the Northern Mariana Islands, American Samoa, and Guam receive grants from the Endowment, which they in turn grant to support humanities programs at the local level.

For further information, call 202–606–8254.

Preservation and Access This division supports projects that will create, preserve, and increase the availability of resources important for research, education, and public programming in the humanities.

For further information, call 202–606–8570.

Public Programs This division strives to fulfill the Endowment's mandate "to increase public understanding of the humanities" by supporting those institutions and organizations that develop and present humanities programming for general audiences.

For further information, call 202–606–8269.

Research This division promotes original research in the humanities by providing grants for significant research projects.

For further information, call 202–606–8200.

We the People Program This program is designed to encourage and enhance the teaching, study, and understanding of American history, culture, and democratic principles.

For further information, call 202–606–8337.

Sources of Information

Employment National Endowment for the Humanities job vacancies are posted on USA Jobs. Internet, www.usajobs.gov.
Grants Those interested in applying for a grant in the humanities should visit www.neh.gov for information and guidelines related to grant programs offered by the National Endowment for the Humanities. For further information, call 202–606–8400. Applications for grants must be submitted through www.grants.gov.
Publications Humanities, the Endowment's bimonthly review, is available by subscription ($24 domestic, $33.60 foreign) through the U.S. Government Printing Office, P.O. Box 979050, St. Louis, MO 63197–9000 or by phone at 202–512–1800.

For further information, contact the Office of Communications, National Endowment for the Humanities, Room 510, 1100 Pennsylvania Avenue NW., Washington, DC 20506. Phone, 202–606–8400 or 800–634–1121. TDD, 202–606–8282 or 866–372–2930. Email, info@neh.gov. Internet, http://www.neh.gov.

Institute for Museum and Library Sciences

1800 M Street NW., Ninth Floor, Washington, DC 20036
Phone, 202–653–4657. Internet, http://www.imls.gov. Email, imlsinfo@imls.gov.

Director	MARSHA L. SEMMEL, *Acting*
Deputy Director for Library Services	MARY L. CHUTE
Deputy Director for Museum Services	MARSHA L. SEMMEL
Deputy Director for Policy, Planning, Research, and Communications	MAMIE BITTNER
Chief of Staff	KATE FERNSTROM
General Counsel	NANCY E. WEISS
Human Resources Director	ALICE Y. MACKLIN

Counselor to the Director	SCHROEDER CHERRY
Chief Information Officer	DEREK SCARBROUGH
Grants Management Officer	MARY ESTELLE KENNELLY
Director, Strategic Partnerships	MARSHA L. SEMMEL
Associate Deputy Director for Library Services	JOYCE RAY
Associate Deputy Director for State Programs	LAURIE BROOKS
Associate Deputy Director for Research and Statistics	CARLOS A. MANJARREZ
Associate Deputy Director for Museum Services	CHRISTOPHER J. REICH
Special Events and Board Liaison	ELIZABETH LYONS

The Institute of Museum and Library Services creates strong libraries and museums that connect people to information and ideas.

The Institute of Museum and Library Services (IMLS) was established within the National Foundation on the Arts and the Humanities by the Museum and Library Services Act of September 30, 1996 (110 Stat. 3009), which amended the Museum Services Act (20 U.S.C. 961 et seq.). The Institute combines administration of Federal museum programs formerly carried out by the Institute of Museum Services and Federal library programs formerly carried out by the Department of Education. The Institute's Director is appointed by the President with the advice and consent of the Senate and is authorized to make grants to museums and libraries. The Director receives policy advice on museum and library programs from the National Museum and Library Services Board, which is comprised of 20 members appointed by the President, the Director, the Deputy Director for the Office of Museum Services, and the Deputy Director for the Office of Library Services.

In addition to providing distinct programs of support for museums and libraries, IMLS encourages collaboration between these community resources. It is the primary source of Federal support for the Nation's 123,000 libraries and 17,000 museums. The Institute's library programs help libraries use new technologies to identify, preserve, and share library and information resources across institutional, local, and State boundaries and to reach those for whom library use requires extra effort or special materials. Museum programs strengthen museum operations, improve care of collections, increase

professional development opportunities, and enhance the community service role of museums.

IMLS awards grants to all types of museums and libraries. Eligible museums include art, history, general, children's, natural history, science and technology, as well as historic houses, zoos, and aquariums, botanical gardens and arboretums, nature centers, and planetariums. Eligible libraries include public, school, academic, research, and special libraries. The Institute makes grants that improve electronic sharing of information and expand public access to an increasing wealth of information and services.

Native American Library Services This program provides small grants to tribes and Alaska Native villages for core library operations, technical assistance, and enhancement grants to promote innovative practices.

Native Hawaiian Library Services This program provides grants to nonprofit organizations that primarily serve and represent Native Hawaiians.

National Leadership Grants This program provides grants to enhance the quality of library and museum services nationwide. Awarded projects demonstrate national impact and generate results—whether new tools, research, models, services, practices, or alliances—that can be widely adapted or replicated to extend the benefit of Federal support.

Museums for America This program provides funds to aid museums in advancing their capacity to serve a wider, more diverse public through education, partnerships, and technology.

Laura Bush 21st-Century Librarians Program   This program supports efforts to recruit and educate the next generation of librarians and the faculty who will prepare them for careers in library science.

21st-Century Museum Professionals   This program supports the preparation of museum professionals for the future by updating and expanding their knowledge and skills.

Native American/Native Hawaiian Museum Services   This program enables Native American tribes and organizations that primarily serve Native Hawaiians to benefit their communities and audiences through strengthened museum services in the areas of programming, professional development, and enhancement of museum services.

Conservation Project Support   This program awards matching grants to help museums identify conservation needs and priorities and perform activities to ensure the safekeeping of their collections.

Museum Grants for African American History and Culture   This program enables African American museums to gain knowledge and abilities in the areas of management, operations, programming, collections care, and other museum skills.

Museum Assessment Program   IMLS helps support the cost of the Museum Assessment Program through a cooperative agreement with the American Association of Museums. The program is designed to help museums assess their strengths and weaknesses and plan for the future.

Conservation Assessment Program   IMLS helps support the cost of the Conservation Assessment Program through a cooperative agreement with Heritage Preservation. The program is designed to support a 2-day site visit by a conservation professional to perform the assessment and up to 3 days to write the report.

National Medals for Museum and Library Service   This program recognizes outstanding museums and libraries that provide meaningful public service for their communities.

Sources of Information

Electronic Access   Information about IMLS programs, application guidelines, and lists of grantees are available electronically. Internet, www.imls.gov. Email, imlsinfo@imls.gov.

Grants, Contracts, and Cooperative Agreements   For information about applying for IMLS funding, contact the appropriate program office. Museums should contact the Office of Museum Services, Institute of Museum and Library Services, 1800 M Street NW., Ninth Floor, Washington, DC 20036. Phone, 202–653–4798. Libraries should contact the Office of Library Services, Institute of Museum and Library Services, 1800 M Street NW., Washington, DC 20036. Phone, 202–653–4700.

For further information, contact the Office of Public and Legislative Affairs, Institute of Museum and Library Services, 1800 M Street NW., Washington, DC 20036. Phone, 202–653–4757. Email, imlsinfo@imls.gov. Internet, http://www.imls.gov.

NATIONAL LABOR RELATIONS BOARD

1099 Fourteenth Street NW., Washington, DC 20570
Phone, 202–273–1000. TDD, 202–273–4300. Internet, http://www.nlrb.gov.

Chairman	WILMA B. LIEBMAN
Members	PETER C. SCHAUMBER, CRAIG BECKER, MARK G. PEARCE, BRIAN HAYES
Executive Secretary	LESTER A. HELTZER
Director, Representation Appeals	(VACANCY)

Solicitor	WILLIAM B. COWEN
Chief Administrative Law Judge	ROBERT A. GIANNASI
Director, Office of Public Affairs	NANCY CLEELAND
Inspector General	DAVID P. BERRY
General Counsel	LAFE E. SOLOMON, *Acting*
Deputy General Counsel	JOHN E. HIGGINS, JR.
Director, Equal Employment Opportunity	ROBERT J. POINDEXTER
Director, Employee Development	TOM CHRISTMAN
Chief Information Officer	BRYAN BURNETT
Director, Administration	GLORIA J. JOSEPH
Associate General Counsel, Division of Operations-Management	RICHARD A. SIEGEL
Associate General Counsel, Division of Enforcement Litigation	JOHN H. FERGUSON
Associate General Counsel, Division of Advice	BARRY J. KEARNEY

[For the National Labor Relations Board statement of organization, see the Federal Register of June 14, 1979, 44 FR 34215]

The National Labor Relations Board prevents and remedies unfair labor practices committed by private sector employers and unions and safeguards employees' rights to organize and determine whether to have unions as their bargaining representative.

The National Labor Relations Board (NLRB) is an independent agency created by Congress to administer the National Labor Relations Act of 1935 (Wagner Act; 29 U.S.C. 167). The Board is authorized to designate appropriate units for collective bargaining and to conduct secret ballot elections to determine whether employees desire representation by a labor organization.

Activities

The NLRB has two principal functions: preventing and remedying unfair labor practices by employers and labor organizations or their agents and conducting secret ballot elections among employees in appropriate collective-bargaining units to determine whether or not they desire to be represented by a labor organization in bargaining with employers about their wages, hours, and working conditions. The agency also conducts secret ballot elections among employees who have been covered by a union-security agreement to determine whether or not they wish to revoke their union's authority to make such agreements. In jurisdictional disputes between two or more unions, the Board determines which competing group of

workers is entitled to perform the work involved.

The regional directors and their staffs process representation, unfair labor practice, and jurisdictional dispute cases. They issue complaints in unfair labor practice cases, seek settlement of unfair labor practice charges, obtain compliance with Board orders and court judgments, and petition district courts for injunctions to prevent or remedy unfair labor practices. The regional directors conduct hearings in representation cases, hold elections pursuant to the agreement of the parties or the decision-making authority delegated to them by the Board or pursuant to Board directions, and issue certifications of representatives when unions win or certify the results when unions lose employee elections. They process petitions for bargaining unit clarification, for amendment of certification, and for rescission of a labor organization's authority to make a union-shop agreement. They also conduct national emergency employee referendums.

Administrative law judges conduct hearings in unfair labor practice cases, make findings of fact and conclusions of law, and recommend remedies for violations found. Their decisions can be appealed to the Board for a final agency

NATIONAL LABOR RELATIONS BOARD

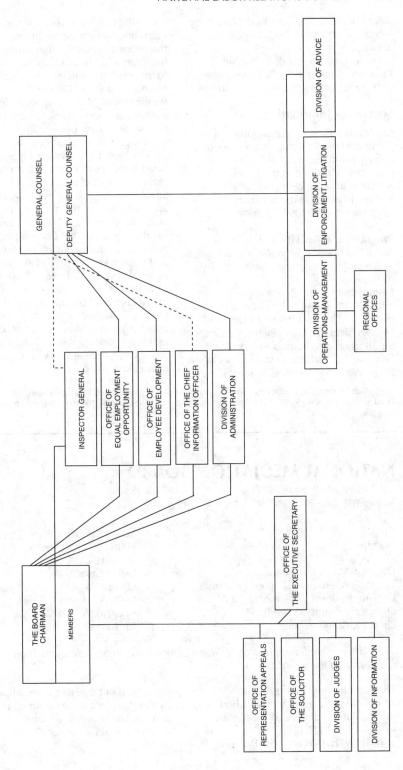

determination. The Board's decisions are subject to review in the U.S. courts of appeals.

For a complete list of National Labor Relations Board field offices, visit www.nlrb.gov/about_us/locating_our_offices.

Sources of Information

Contracts Prospective suppliers of goods and services may inquire about agency procurement and contracting practices by writing to the Chief, Acquisitions Management Branch, National Labor Relations Board, Washington, DC 20570. Phone, 202–273–4047.

Programs and Activities Information about the Board's programs and activities is available through the Internet at www.nlrb.gov.

Employment The Board appoints administrative law judges from a register established by the Office of Personnel Management. The agency hires attorneys for all its offices, field examiners for its field offices, and administrative personnel for its Washington office. Inquiries regarding college and law school recruiting programs should be directed to the nearest regional office. Employment inquiries and applications may be sent to any regional office or the Washington Human Resources Office.

Publications Anyone desiring to inspect formal case documents or read agency publications may use facilities of the Washington or field offices, or search the agency Web site at www.nlrb.gov. The Board's offices offer free informational leaflets in limited quantities.

Speakers To give the public and persons appearing before the agency a better understanding of the National Labor Relations Act and the Board's policies, procedures, and services, Washington and regional office personnel participate as speakers or panel members before bar associations, labor, educational, civic, or management organizations, and other groups. Requests for speakers or panelists may be made to Washington officials or to the appropriate regional director or agency Web site at www.nlrb.gov/about_us/speakers.aspx.

For further information, contact the Office of Public Affairs, National Labor Relations Board, 1099 Fourteenth Street NW., Suite 11550, Washington, DC 20570. Phone, 202–273–1991. Internet, http://www.nlrb.gov.

NATIONAL MEDIATION BOARD

1301 K Street NW., Suite 250 East, Washington, DC 20005
Phone, 202–692–5000. Internet, http://www.nmb.gov.

Chair	ELIZABETH DOUGHERTY
Members	HARRY R. HOGLANDER, LINDA PUCHALA
Director, Mediation Services	LARRY GIBBONS
Director, Alternative Dispute Resolution	DANIEL RAINEY
General Counsel, Office of Legal Affairs	MARY JOHNSON
Director, Arbitration Services	ROLAND WATKINS
Director, Administration	JUNE KING

The National Mediation Board facilitates labor-management relations within the railroads and the airlines.

The National Mediation Board (NMB) is an independent agency established by the 1934 amendments to the Railway Labor Act of 1926 (45 U.S.C. 151–158, 160–162, 1181–1188). The Board is composed of three members, appointed by the President and confirmed by the Senate. The Board designates a Chairman on a yearly basis.

The Agency's dispute-resolution processes are designed to resolve disputes over the negotiation of new or revised collective bargaining agreements and the interpretation or application of existing agreements. It also effectuates employee rights of self-organization where a representation dispute exists.

Activities

Mediation Following receipt of an application for mediation, the NMB assigns a mediator to assist the parties in reaching an agreement. The NMB is obligated to use its best efforts to bring about a peaceful resolution to the dispute. If such efforts do not settle the dispute, the NMB advises the parties and offers interest arbitration as an alternative approach to resolve the remaining issues. If either party rejects this offer, the NMB releases the parties from formal mediation. This release triggers a 30-day cooling off period. During this period, NMB continues to work with the parties to achieve a consensual resolution. If, however, an agreement is not reached by the end of the 30-day period, the parties are free to exercise lawful self-help, such as carrier-imposed working conditions or a strike by the union/organization.

Alternative Dispute Resolution In addition to traditional mediation services, NMB also provides voluntary Alternative Dispute Resolution (ADR) services. ADR services include facilitation, training, grievance mediation, and an Online Dispute Resolution component, which applies technology to the dispute resolution process. The purpose of the ADR program is to assist the parties in learning and applying more effective, less confrontational methods for resolving their disputes and to help them resolve more of their own disputes without outside intervention.

Presidential Emergency Board If NMB determines that a dispute threatens to substantially deprive any section of the country of essential transportation service, it notifies the President. The President may, at his discretion, establish a Presidential Emergency Board (PEB) to investigate and report back within 30 days. After the PEB has been created and

for 30 days after it has made its report to the President, neither party to the dispute may exercise self-help.

There are also special emergency procedures for unresolved disputes affecting publicly funded and operated commuter railroads and their employees. If the mediation procedures are exhausted, the parties to the dispute, or the Governor of any State where the railroad operates, may request that the President establish a PEB. The President is required to establish such a board if requested. If no settlement is reached within 60 days following the creation of the PEB, NMB is required to conduct a public hearing on the dispute. If there is no settlement within 120 days after the creation of the PEB, any party, or the Governor of any affected state, may request a second, final-offer PEB. No self-help is permitted pending the exhaustion of these emergency procedures.

Representation When a labor organization or individual files an application with NMB to represent employees, the Agency assigns an investigator to conduct a representation investigation. Should the applicant meet the requirements, NMB continues the investigation, usually with a secret telephone or Internet election. NMB is responsible for ensuring that the requirements for a fair election process have been maintained. If the employees vote to be represented, NMB issues a certification which commences the carrier's statutory duty to bargain with the certified representative.

Arbitration NMB provides both grievance arbitration and interest arbitration. Grievance arbitration is a process for resolving disputes regarding the interpretation or application of an existing collective bargaining agreement. Grievances must be handled through grievance arbitration if not otherwise resolved, and cannot be used by the parties to trigger self-help actions. NMB has significant administrative responsibilities for grievance arbitration in the railroad industry, which includes those before the National Railroad Adjustment Board (NRAB), as well as the two types of arbitration panels

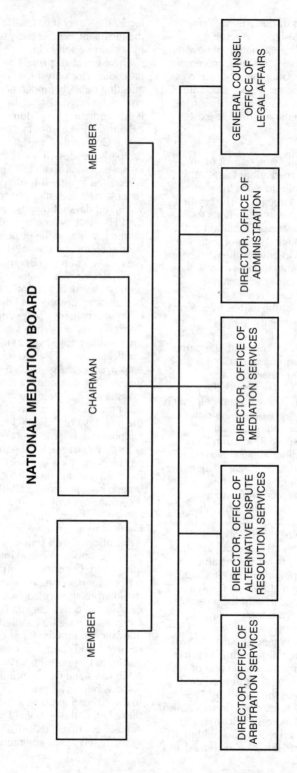

NATIONAL MEDIATION BOARD

MEMBER

CHAIRMAN

MEMBER

GENERAL COUNSEL, OFFICE OF LEGAL AFFAIRS

DIRECTOR, OFFICE OF ADMINISTRATION

DIRECTOR, OFFICE OF MEDIATION SERVICES

DIRECTOR, OFFICE OF ALTERNATIVE DISPUTE RESOLUTION SERVICES

DIRECTOR, OFFICE OF ARBITRATION SERVICES

established by the labor-management parties at each railroad: public law boards (PLBs) and special boards of adjustment (SBAs). Grievance arbitration in the airline industry is accomplished at the various system boards of adjustment created jointly by labor and management at the parties' expense. NMB furnishes panels of prospective arbitrators for the parties' selection in both the airline and railroad industries. NMB also pays the salary and travel expenses of the arbitrators for railroad arbitration proceedings. Grievance arbitration decisions are final and binding with very limited grounds for judicial review.

Interest arbitration is a process to establish the terms of a new or modified collective bargaining agreement through arbitration, rather than through negotiations. Unlike grievance arbitration, its use is not statutorily required. NMB offers the parties the opportunity to use binding interest arbitration when the agency has determined that further mediation efforts will not be successful. In addition, the parties may directly agree to resolve their collective bargaining dispute or portions of their dispute through interest arbitration. NMB generally provides the parties with panels of potential arbitrators from which they choose an individual to resolve their dispute. In some instances,

however, the parties agree to allow NMB to directly appoint an arbitrator. Interest arbitration decisions are final and binding with very narrow grounds for judicial appeal.

Sources of Information

Electronic Access Information pertaining to Board operations, including weekly case activity reports, representation determinations, press releases, and an agency directory, are available at www.nmb.gov.
NMB Knowledge Store The Knowledge Store contains over 100,000 documents in an easily searchable format, including arbitration awards, representation decisions, annual reports, PEB reports, industry contracts, and union constitutions and bylaws.
Publications The Annual Reports of the National Mediation Board are available in the Knowledge Store at www.nmb.gov. A limited supply of paper copies is also available for public distribution by calling 202–692–5031.
Virtual Reading Room Copies of collective-bargaining agreements between labor and management of various rail and air carriers and NMB Determinations (back to at least October 1, 1998) are available in the Knowledge Store at www.nmb.gov.

For further information, contact the Public Information Officer, National Mediation Board, Suite 250 East, 1301 K Street NW., Washington, DC 20005–7011. Phone, 202–692–5050. Internet, http://www.nmb.gov.

NATIONAL RAILROAD PASSENGER CORPORATION (AMTRAK)

60 Massachusetts Avenue NE., Washington, DC 20002
Phone, 202–906–3000. Internet, http://www.amtrak.com.

Chairman	TOM CARPER
Members	RAY LAHOOD, JOSEPH H. BOARDMAN, DONNA MCLEAN, NANCY NAPLES, (4 VACANCIES)

Officers

Vice President, General Counsel and Corporate Secretary	ELEANOR D. ACHESON
Vice President, Government Affairs and Corporate Communications	JOSEPH H. MCHUGH

Officers

Vice President, Human Resources and Diversity Initiatives	LORRAINE A. GREEN
Vice President, Labor Relations	(VACANCY)
Vice President, Marketing and Product Management	EMMETT FREMAUX
Vice President, Security, Strategy, and Special Operations	WILLIAM ROONEY
Vice President, Strategic Partnerships	STEPHEN J. GARDNER
Chief Financial Officer	D. J. STADTLER
Chief Information Officer	ED TRAINOR
Chief Operating Officer	WILLIAM L. CROSBIE
Inspector General	TED ALVES

[For the National Railroad Passenger Corporation statement of organization, see the Code of Federal Regulations, Title 49, Part 700]

The National Railroad Passenger Corporation provides intercity rail passenger service in the United States.

The National Railroad Passenger Corporation (Amtrak) was created by the Rail Passenger Service Act of 1970, as amended (49 U.S.C. 241), and was incorporated under the laws of the District of Columbia to provide a balanced national transportation system by developing, operating, and improving U.S. intercity rail passenger service.

Amtrak operates approximately 300 trains per day, serving over 500 stations in 46 States, over a system of 21,800 route miles. Of this route system, Amtrak owns about 530 route miles in the Northeast and several other small track segments elsewhere in the country.

Amtrak owns or leases its stations and owns its own repair and maintenance facilities. The Corporation employs a total workforce of approximately 19,000 and provides all reservation, station, and onboard service staffs, as well as train and engine operating crews. Outside the Northeast Corridor, Amtrak may enter into contracts with privately or publicly owned railroads to operate over their track. These railroads are responsible for the condition of the roadbed and for coordinating the flow of traffic.

In fiscal year 2009, Amtrak transported over 27 million people with 74,000 passengers traveling on Amtrak per day. Also, Amtrak runs commuter trains under contract with several commuter agencies.

Although Amtrak's basic route system was originally designated by the Secretary of Transportation in 1971, modifications have been made to the Amtrak system and to individual routes that have resulted in more efficient and cost-effective operations. Capital funding has increased in recent years, allowing Amtrak to make progress in bringing its network to a state of good repair and in reducing debt load.

For further information, contact the Government Affairs Department, Amtrak, 60 Massachusetts Avenue NE., Washington, DC 20002. Phone, 202–906–3918. Internet, http://www.amtrak.com.

NATIONAL SCIENCE FOUNDATION

4201 Wilson Boulevard, Arlington, VA 22230
Phone, 703–292–5111. TDD, 800–281–8749. Internet, http://www.nsf.gov. Email, info@nsf.gov.

National Science Board

Chair	RAY M. BOWEN

National Science Board

Vice Chairman	ESIN GULARI
Members	MARK R. ABBOT, DAN E. ARVIZU, BARRY C. BARISH, STEPHEN C. BEERING, CAMILLA P. BENBOW, JOHN T. BRUER, G. WAYNE CLOUGH, FRANCE A. CORDOVA, KELVIN K. DROEGEMEIER, PATRICIA D. GALLOWAY, JOSE-MARIE GRIFFITHS, ELIZABETH HOFFMAN, LOUIS J. LANZEROTTI, ALAN I. LESHNER, GEORGE P. PETERSON, DOUGLAS D. RANDALL, ARTHUR K. REILLY, DIANE L. SOUVAINE, JON C. STRAUSS, KATHRYN D. SULLIVAN, THOMAS N. TAYLOR, RICHARD F. THOMPSON
(Ex officio)	CORA B. MARRETT
Executive Officer/Director	MICHAEL L. VAN WOERT

National Science Foundation

Director	CORA B. MARRETT, *Acting*
Deputy Director	CORA B. MARRETT, *Acting*
General Counsel	LAWRENCE RUDOLPH
Director, Office of Cyberinfrastructure	JOSE MUNOZ, *Acting*
Director, Office of Integrative Activities	W. LANCE HAWORTH
Director, Office of International Science and Engineering	LARRY H. WEBER
Director, Office of Legislative and Public Affairs	JEFF NESBIT
Director, Office of Polar Programs	KARL A. ERB
Director, Office of Equal Opportunity Programs	CLAUDIA POSTELL
Inspector General	ALLISON C. LERNER
Assistant Director for Biological Sciences	JOANN P. ROSKOSKI, *Acting*
Assistant Director for Computer and Information Science and Engineering	JEANNETTE M. WING
Assistant Director for Education and Human Resources	JOAN FERRINI-MUNDY, *Acting*
Assistant Director for Engineering	THOMAS W. PETERSON, *Acting*
Assistant Director for Geosciences	TIMOTHY L. KILLEEN
Assistant Director for Mathematical and Physical Sciences	EDWARD SEIDEL, *Acting*
Assistant Director for Social, Behavioral, and Economic Sciences	MYRON GURMANN
Chief Financial Officer and Director, Office of Budget, Finance, and Award Management	MARTHA A. RUBENSTEIN
Chief Human Capital Officer and Director, Office of Information and Resource Management	ANTHONY A. ARNOLIE

[For the National Science Foundation statement of organization, see the Federal Register of February 8, 1993, 58 FR 7587–7595; May 27, 1993, 58 FR 30819; May 2, 1994, 59 FR 22690; and October 6, 1995, 60 FR 52431]

The National Science Foundation promotes the progress of science and engineering through the support of research and education programs.

The National Science Foundation (NSF) is an independent agency created by the National Science Foundation Act of 1950, as amended (42 U.S.C. 1861–1875).

NSF purposes are to increase the Nation's base of scientific and engineering knowledge; to strengthen its ability to conduct research in all areas of science and engineering; to develop and help implement science and engineering education programs that can better prepare the Nation for meeting the challenges of the future; and to promote international cooperation through science and engineering. In its role as a leading Federal supporter of science and engineering, the agency also has an important role in national policy planning.

The Director and the Deputy Director are appointed by the President, with the advice and consent of the Senate, to a 6-year term and an unspecified term, respectively. The Foundation's activities are guided by the National Science Board (NSB). NSB is composed of a chairman, a vice chairman, 24 board members, and the Director ex officio. Members are appointed by the President with the advice and consent of the Senate for 6-year terms, with one-third appointed every 2 years. They are selected because of their records of distinguished service in science, engineering, education, research management, or public affairs to be broadly representative of the views of national science and engineering leadership. The Board also has a broad national policy responsibility to monitor and make recommendations to promote the health of U.S. science and engineering research and education.

The Foundation's Office of Inspector General is responsible for conducting and supervising audits, inspections, and investigations relating to the programs and operations of the Foundation, including allegations of misconduct in science.

Activities

NSF initiates and supports fundamental, long-term, merit-selected research in all the scientific and engineering disciplines. This support is made through grants, contracts, and other agreements awarded to universities, colleges, academic consortia, and nonprofit and small business institutions. Most of this research is directed toward the resolution of scientific and engineering questions concerning fundamental life processes, natural laws and phenomena, fundamental processes influencing the human environment, and the forces affecting people as members of society as well as the behavior of society as a whole.

The Foundation encourages cooperative efforts by universities, industries, and government. It also promotes the application of research and development for better products and services that improve the quality of life and stimulate economic growth.

The Foundation promotes the development of research talent through support of undergraduate and graduate students, as well as postdoctoral researchers. It administers special programs to identify and encourage participation by groups underrepresented in science and technology and to strengthen research capability at smaller institutions, small businesses, undergraduate colleges, and universities.

The Foundation supports major national and international science and engineering activities, including the U.S. Antarctic Program, the Ocean Drilling Program, global geoscience studies, and others. Cooperative scientific and engineering research activities support exchange programs for American and foreign scientists and engineers, execution of jointly designed research projects, participation in the activities of international science and engineering organizations, and travel to international conferences.

Support is provided through contracts and cooperative agreements with national centers where large facilities are made available for use by qualified scientists and engineers. Among the types of centers supported by the Foundation are astronomy and atmospheric sciences, biological and engineering research,

NATIONAL SCIENCE FOUNDATION

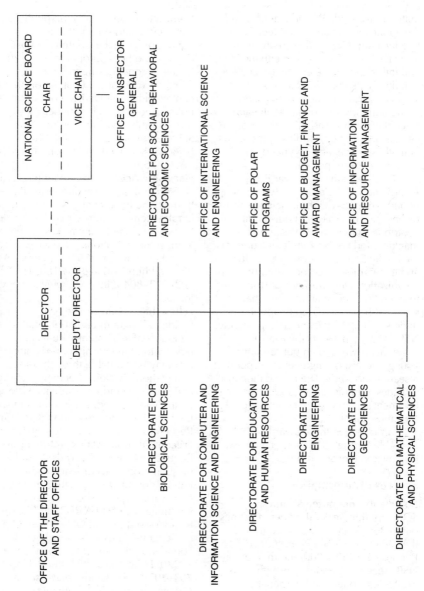

science and technology, supercomputers, and long-term ecological research sites.

The Foundation's science and engineering education activities include grants for research and development activities directed to model instructional materials for students and teachers and the application of advanced technologies to education. Grants also are available for teacher preparation and enhancement and informal science education activities. Funding is also provided for college science instrumentation, course and curriculum improvement, faculty and student activities, and minority resource centers. In addition, studies of the status of math, science, and engineering education are supported.

NSF presents the Vannevar Bush Award annually to a person who, through public service activities in science and technology, has made an outstanding contribution toward the welfare of mankind and the Nation. It also presents the Public Service Award to an individual and to a company, corporation, or organization who, through contributions to public service in areas other than research, have increased the public understanding of science or engineering. NSF annually presents the Alan T. Waterman Award to an outstanding young scientist or engineer for support of research and study. The Foundation also provides administrative support for the President's Committee on the National Medal of Science.

Information on these awards is available through the Internet at www.nsf.gov/home/nsb/start.htm.

Sources of Information

Board and Committee Minutes Summary minutes of the open meetings of the Board may be obtained from the National Science Board Office. Phone, 703–292–7000. Information on NSB meetings, minutes, and reports is available through the Internet at www.nsf.gov/home/nsb/start.htm. Summary minutes of the Foundation's advisory groups may be obtained from the contacts listed in the notice of meetings published in the Federal Register or under "News and Media" on the Foundation's

Web site. General information about the Foundation's advisory groups may be obtained from the Division of Human Resource Management, Room 315, Arlington, VA 22230. Phone, 703–292–8180.

Contracts The Foundation publicizes contracting and subcontracting opportunities in the Commerce Business Daily and other appropriate publications. Organizations seeking to undertake contract work for the Foundation should contact either the Division of Contracts, Policy, and Oversight (phone, 703–292–8240) or the Division of Administrative Services (phone, 703–292–8190), National Science Foundation, Arlington, VA 22230.

Electronic Access Information regarding NSF programs and services is available through the Internet at www.nsf.gov.

Employment Inquiries may be directed to the Division of Human Resource Management, National Science Foundation, Room 315, Arlington, VA 22230. Phone, 703–292–8180. TDD, 703–292–8044. Internet, www.nsf.gov/jobs.

Fellowships Consult the NSF Guide to Programs and appropriate announcements and brochures for postdoctoral fellowship opportunities that may be available through some Foundation divisions. Beginning graduate and minority graduate students wishing to apply for fellowships should contact the Directorate for Education and Human Resources. Phone, 703–292–8601.

Freedom of Information Act Requests Requests for agency records should be submitted in accordance with the Foundation's FOIA regulation at 45 CFR 612. Such requests should be clearly identified with "FOIA REQUEST" and be addressed to the FOIA Officer, Office of General Counsel, National Science Foundation, Room 1265, Arlington, VA 22230. Phone, 703–292–8060. Fax, 703–292–9041. Email, foia@nsf.gov.

Grants Individuals or organizations who plan to submit grant proposals should refer to the NSF Guide to Programs, Grant Proposal Guide (NSF–01–2), and appropriate program brochures and announcements that may be obtained as indicated in the Publications section. Grant information is also available through the Internet at www.nsf.gov.

Office of Inspector General General inquiries may be directed to the Office of Inspector General, National Science Foundation, Room 1135, Arlington, VA 22230. Phone, 703–292–7100.

Privacy Act Requests Requests for personal records should be submitted in accordance with the Foundation's Privacy Act regulation at 45 CFR 613. Such requests should be clearly identified with "PRIVACY ACT REQUEST" and be addressed to the Privacy Act Officer, National Science Foundation, Room 1265, Arlington, VA 22230. Phone, 703–292–8060.

Publications The National Science Board assesses the status and health of science and its various disciplines, including such matters as human and material resources, in reports submitted to the President for submission to the Congress. The National Science Foundation issues publications that announce and describe new programs, critical dates, and application procedures for competitions. Single copies of these publications can be ordered by writing to NSF Clearinghouse, P.O. Box 218, Jessup, MD 20794–0218. Phone, 301–947–2722. Internet, www.nsf.gov. Email, pubinfo@nsf.gov.

Reading Room A collection of Foundation policy documents and staff instructions, as well as current indexes, are available to the public for inspection and copying during regular business hours, 8:30 a.m. to 5 p.m., Monday through Friday, in the National Science Foundation Library, Room 225, Arlington, VA 22230. Phone, 703–292–7830.

Small Business Activities The Office of Small Business Research and Development provides information on opportunities for Foundation support to small businesses with strong research capabilities in science and technology. Phone, 703–292–8330. The Office of Small and Disadvantaged Business Utilization oversees agency compliance with the provisions of the Small Business Act and the Small Business Investment Act of 1958, as amended (15 U.S.C. 631, 661, 683). Phone, 703–292–8330.

For further information, contact the National Science Foundation Information Center, 4201 Wilson Boulevard, Arlington, VA 22230. Phone, 703–292–5111. TDD, 800–281–8749. Email, info@nsf.gov. Internet, http://www.nsf.gov.

NATIONAL TRANSPORATION SAFETY BOARD

490 L'Enfant Plaza SW., Washington, DC 20594
Phone, 202–314–6000. Fax, 202–314–6110. Internet, http://www.ntsb.gov.

Chairman	DEBORAH A.P. HERSMAN
Vice Chairman	CHRISTOPHER A. HART
Members	ROBERT L. SUMWALT, (2 VACANCIES)
Managing Director	DAVID L. MAYER
Chief Financial Officer	STEVEN GOLDBERG
General Counsel	GARY HALBERT
Director, Office of Communications	NANCY LEWIS
Director, Office of Equal Employment Opportunity	FARA GUEST
Director, Office of Research and Engineering	JOESPH M. KOLLY
Director, Office of Railroad, Pipeline, and Hazardous Materials Investigations	(VACANCY)
Director, Office of Aviation Safety	THOMAS E. HAUETER
Director, Office of Highway Safety	BRUCE MAGLADRY
Director, Office of Marine Safety	JOHN SPENCER
Chief Information Officer	ROBERT SCHERER

Director, Office of Administration
Chief Administrative Law Judge

LOLA WARD.
WILLIAM E. FOWLER, JR.

[For the National Transportation Safety Board statement of organization, see the Code of Federal Regulations, Title 49, Part 800]

The National Transportation Safety Board investigates accidents, conducts studies, and makes recommendations to Government agencies, the transportation industry, and others on safety measures and practices.

The National Transportation Safety Board (NTSB) was established in 1967 and became totally independent on April 1, 1975, by the Independent Safety Board Act of 1974 (49 U.S.C. 1111).

NTSB consists of five Members appointed for 5-year terms by the President with the advice and consent of the Senate. The President designates two of these Members as Chairman and Vice Chairman of the Board for 2-year terms. The designation of the Chairman is made with the advice and consent of the Senate.

Activities

Accident Investigation   NTSB is responsible for investigating, determining probable cause, making safety recommendations, and reporting the facts and circumstances of incidents in the following areas: U.S. civil aviation and certain public-use aircraft accidents; railroad accidents in which there is a fatality or substantial property damage, or that involve a passenger train; pipeline accidents in which there is a fatality, substantial property damage, or significant injury to the environment; highway accidents, including railroad grade-crossing accidents, that the Board selects in cooperation with the States; major marine casualties and marine accidents involving a public vessel and a nonpublic vessel, in accordance with regulations prescribed jointly by the Board and the U.S. Coast Guard; certain accidents involving hazardous materials; and other transportation accidents that are catastrophic, involve problems of a recurring character, or otherwise should be investigated in the judgment of the Board.

Safety Problem Identification   NTSB makes recommendations on matters pertaining to transportation safety and is a catalyst for transportation accident prevention by conducting safety studies and special investigations; assessing techniques of accident investigation and publishing recommended procedures; establishing regulatory requirements for reporting accidents; evaluating the transportation safety consciousness and efficacy of other Government agencies in the prevention of accidents; evaluating the adequacy of safeguards and procedures concerning the transportation of hazardous materials and the performance of other Government agencies charged with ensuring the safe transportation of such materials; and reporting annually to the Congress on its activities.

Family Assistance for Aviation Disasters   NTSB coordinates the resources of the Federal Government and other organizations to support the efforts of local and State governments and airlines to meet the needs of aviation disaster victims and their families. It assists in making Federal resources available to local authorities and airlines.

Certificate, Civil Penalty, and License Appeal   On appeal, NTSB reviews the suspension, amendment, modification, revocation, or denial of certain certificates, licenses, and assessments of civil penalties issued by the Secretary of Transportation. NTSB also reviews on appeal from the orders of any administrative law judge, the decisions of the Commandant of the Coast Guard revoking, suspending, or denying certain licenses, certificates, documents, and registers.

For further contact information for National Transportation Safety Board Aviation, Highway, and Railroad Safety regional offices, visit www.ntsb.gov/Abt_NTSB/offices.htm.

NATIONAL TRANSPORTATION SAFETY BOARD

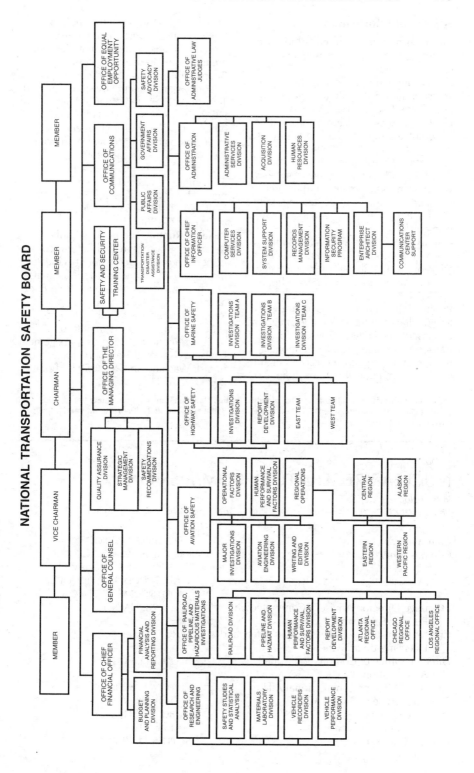

Sources of Information

Contracts and Procurement   Inquiries regarding NTSB's procurement and contracting activities should be addressed to the Contracting Officer, National Transportation Safety Board, Washington, DC 20594. Phone, 202–314–6102.

Electronic Access   Agency information, including aircraft accident data, synopses of aircraft accidents, speeches and congressional testimony given by Board members and staff, press releases, job vacancy announcements, and notices of Board meetings, public hearings, and other agency events, is available at www.ntsb.gov.

Employment   Send applications for employment to the Human Resources Division, National Transportation Safety Board, Washington, DC 20594. Phone, 202–314–6230.

Publications   Publications are provided free of charge to the following categories of subscribers: Federal, State, or local transportation agencies; international transportation organizations or foreign governments; educational institutions

or public libraries; nonprofit public safety organizations; and the news media. Persons in these categories who are interested in receiving copies of Board publications should contact the Records Management Division, National Transportation Safety Board, Washington, DC 20594. Phone, 202–314–6551. All other persons interested in receiving publications must purchase them from the National Technical Information Service, 5285 Port Royal Road, Springfield, VA 22161. Orders may be placed by telephone to the Subscription Unit at 703–487–4630 or the sales desk at 703–487–4768.

Reading Room   The Board's Public Reference Room is available for record inspection or photocopying. It is located in Room 6500 at the Board's Washington, DC, headquarters and is open from 8:45 a.m. to 4:45 p.m. every business day. Requests for access to public records should be made in person at Room 6500 or by writing to the Records Management Division, National Transportation Safety Board, Washington, DC 20594. Phone, 202–314–6551.

For further information, contact the Office of Public Affairs, National Transportation Safety Board, 490 L'Enfant Plaza SW., Washington, DC 20594. Phone, 202–314–6100. Fax, 202–314–6110. Internet, http://www.ntsb.gov.

NUCLEAR REGULATORY COMMISSION

Washington, DC 20555
Phone, 301–415–7000. Internet, http://www.nrc.gov. Email, opa.resource@nrc.gov.

Chairman	GREGORY B. JACZKO
Commissioners	KRISTINE L. SVINICKI, GEORGE APOSTOLAKIS, WILLIAM D. MAGWOOD, IV, WILLIAM C. OSTENDORFF
Executive Director, Advisory Committee on Reactor Safeguards	EDWIN M. HACKETT
Chief Administrative Judge, Atomic Safety and Licensing Board Panel	E. ROY HAWKENS
Director, Office of Commission Appellate Adjudication	BROOKE D. POOLE
Director, Office of Congressional Affairs	REBECCA L. SCHMIDT
Director, Office of Public Affairs	ELIOT B. BRENNER
Inspector General	HUBERT T. BELL, JR.
Chief Financial Officer	JAMES E. DYER
General Counsel	STEPHEN G. BURNS

Director, Office of International Programs	MARGARET M. DOANE
Secretary of the Commission	ANNETTE L. VIETTI-COOK
Executive Director for Operations	R. WILLIAM BORCHARDT
Deputy Executive Director for Reactor and Preparedness Programs	MARTIN J. VIRGILIO
Deputy Executive Director for Materials, Waste, Research, State, Tribal, and Compliance Programs	MICHAEL F. WEBER
Deputy Executive Director for Corporate Management	DARREN B. ASH

[For the Nuclear Regulatory Commission statement of organization, see the Code of Federal Regulations, Title 10, Part I]

The Nuclear Regulatory Commission licenses and regulates civilian use of nuclear energy to protect public health and safety and the environment.

The Nuclear Regulatory Commission (NRC) was established as an independent regulatory agency under the provisions of the Energy Reorganization Act of 1974 (42 U.S.C. 5801 et seq.) and Executive Order 11834 of January 15, 1975. All licensing and related regulatory functions formerly assigned to the Atomic Energy Commission were transferred to the Commission.

The Commission's major program components are the Office of Nuclear Reactor Regulation, the Office of New Reactors, the Office of Nuclear Material Safety and Safeguards, the Office of Federal and State Materials and Environmental Management Programs, and the Office of Nuclear Regulatory Research. Headquarters offices are located in suburban Maryland, and there are four regional offices.

The Commission ensures that the civilian uses of nuclear materials and facilities are conducted in a manner consistent with the public health and safety, environmental quality, national security, and the antitrust laws. Most of the Commission's effort is focused on regulating the use of nuclear energy to generate electric power.

Activities

NRC is primarily responsible for the following functions: licensing the construction, operation, and closure of nuclear reactors and other nuclear facilities, such as nuclear fuel cycle facilities, low-level radioactive waste disposal sites under NRC jurisdiction, the geologic repository for high-level radioactive waste, and nonpower test and research reactors; licensing the possession, use, processing, handling, and export of nuclear material; licensing the operators of nuclear power and nonpower test and research reactors; inspecting licensed facilities and activities; conducting the U.S. Government research program on light-water reactor safety; developing and implementing rules and regulations that govern licensed nuclear activities; investigating nuclear incidents and allegations concerning any matter regulated by NRC; maintaining the NRC Incident Response Program; collecting, analyzing, and disseminating information about the operational safety of commercial nuclear power reactors and certain nonreactor activities; developing effective working relationships with the States regarding reactor operations and the regulation of nuclear material; and assuring that adequate regulatory programs are maintained by those States that exercise regulatory control over certain nuclear materials in the State.

Sources of Information

Freedom of Information Act Requests
Requests for copies of records should be directed to the FOIA/Privacy Act Officer, Mail Stop T–5 F09, Nuclear Regulatory Commission, Washington, DC 20555–0001. Phone, 301–415–7169. Fax, 301–415–5130. Requests may also

NUCLEAR REGULATORY COMMISSION

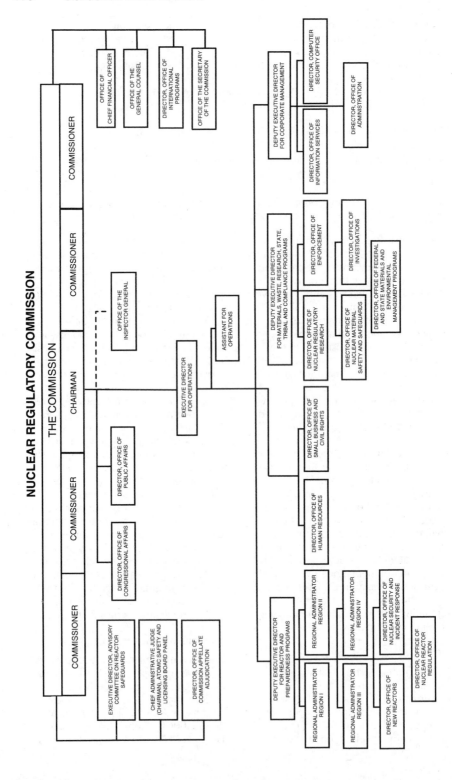

be submitted using the form at www.nrc. gov/reading-rm/foia/foia-submittal-form. html.

Publications NRC publishes scientific, technical, and administrative information dealing with licensing and regulation of civilian nuclear facilities and materials, as well as periodic and annual reports. Some publications and documents are available at www.nrc.gov. The U.S. Government Printing Office (GPO) and the National Technical Information Service (NTIS) sell single copies of, or subscriptions to, NRC publications. To obtain prices and order NRC publications, contact the Superintendent of Documents, GPO, Mail Stop SSOP, Washington, DC 20402–0001 (phone, 202–512–1800; Internet, http://bookstore. gpo.gov) or NTIS, Springfield, VA 22161– 0002 (phone, 703–605–6000; Internet, www.ntis.gov).

Active Regulatory Guides Active Regulatory Guides may be obtained without charge at www.nrc.gov/reading-rm/doc-collections/reg-guides, by faxed request to 301–415–2289, by email request to distribution.resource@nrc.gov, or by written request to the Reproduction and Distribution Services Section, U.S. Nuclear Regulatory Commission, Washington, DC 20555–0001. They may also be purchased, as they are issued, on standing orders from NTIS. These Regulatory Guides are published in 10 subject areas: power reactors, research and test reactors, fuels and materials facilities, environmental and siting, materials and plant protection, products, transportation, occupational health, antitrust and financial review, and general.

Draft Regulatory Guides Draft Regulatory Guides are issued for public comment. These drafts may be downloaded through the Internet at www.nrc.gov/reading-rm/doc-collections/ reg-guides. They may be commented on through the Internet using the form at www.nrc.gov/public-involve/doc-comment/form.html. Comments may also be emailed to nrcrep.resource@nrc. gov. Draft Regulatory Guides may also be obtained, to the extent of supply, by faxed request to 301–415–2289, by email request to distribution.resource@nrc.gov, or by written request to the Reproduction and Distribution Services Section, U.S. Nuclear Regulatory Commission, Washington, DC 20555–0001.

Reading Rooms The headquarters Public Document Room maintains an extensive collection of documents related to NRC licensing proceedings and other significant decisions and actions. Documents issued prior to October 1999 are available in paper or microfiche. Documents issued after October 1999 are also available from NRC's full-text document management system, which is accessible from the NRC Web site at www.nrc.gov/reading-rm/adams.html. The headquarters Public Document Room is located on the first floor at One White Flint North, 11555 Rockville Pike, Rockville, MD, and is open Monday through Friday, from 7:45 a.m. to 4:15 p.m., except on Federal holidays.

Documents Documents from the collection may be reproduced, with some exceptions, on paper, microfiche, or CD–ROM for a nominal fee. For additional information regarding the Public Document Room, go to www.nrc. gov/reading-rm/pdr.html or contact the Nuclear Regulatory Commission, Public Document Room, Washington, DC 20555–0001. Phone, 301–415–4737 or 800–397–4209. Email, pdr.resource@nrc. gov. Fax, 301–415–3548.

Microfiche Collections Selected regional libraries of the Government Printing Office Federal Depository Library Program maintain permanent microfiche collections of NRC documents released between January 1981 and October 1999. For further information, contact the Public Document Room at 301–415–4737 or 800–397–4209.

For further information, contact the Office of Public Affairs, Nuclear Regulatory Commission, Washington, DC 20555–0001. Phone, 301–415–8200. Email, opa.resource@nrc.gov. Internet, http://www.nrc.gov.

OCCUPATIONAL SAFETY AND HEALTH REVIEW COMMISSION

1120 Twentieth Street NW., Washington, DC 20036–3457
Phone, 202–606–5376. Fax, 202–418–3487. Internet, http://www.oshrc.gov.

Chairman	THOMASINA V. ROGERS
Commissioners	HORACE A. THOMPSON, CYNTHIA L. ATTWOOD
General Counsel	NADINE N. MANCINI
Chief Administrative Law Judge	IRVING SOMMER
Executive Director	RICHARD C. LOEB
Deputy Executive Director	DEBRA A. HALL
Executive Secretary	RAY H. DARLING, JR.

The Occupational Safety and Health Review Commission ensures the timely and fair resolution of cases involving the alleged exposure of American workers to unsafe or unhealthy working conditions.

The Occupational Safety and Health Review Commission is an independent, quasi-judicial agency established by the Occupational Safety and Health Act of 1970 (29 U.S.C. 651–678).

The Commission rules on cases when disagreements arise over the results of safety and health inspections performed by the Department of Labor's Occupational Safety and Health Administration (OSHA). Employers have the right to dispute any alleged job safety or health violation found during the inspection by OSHA, the penalties it proposes, and the time given to correct any hazardous situation.

The Occupational Safety and Health Act covers virtually every employer in the country. Its purpose is to reduce the incidence of personal injuries, illness, and deaths among working men and women in the United States that result from their employment. It requires employers to provide a working environment free from recognized hazards that are causing or likely to cause death or serious physical harm to the employees and to comply with occupational safety and health standards promulgated under the act.

Activities

The Commission was created to adjudicate enforcement actions initiated under the act when they are contested by employers, employees, or representatives of employees. A case arises when a citation, issued to an employer as the result of an OSHA inspection, is contested within 15 working days of receipt of the report.

There are two levels of adjudication within the Commission. All cases are first assigned to an administrative law judge. A hearing is generally held in the community or as close as possible to where the alleged violation occurred. After the hearing, the judge issues a decision, based on findings of fact and conclusions of law.

A substantial number of the judge's decisions become final orders of the Commission. However, if a party petitions the Commission members for review of the judge's decision and the petition is granted, Commission members will issue the final order.

After a final order is issued, any party to the case may seek a review of the decision in the United States Courts of Appeals.

The Commission's principal office is in Washington, DC. Administrative law judges are also located in Atlanta and Denver regional offices.

Sources of Information

Publications Copies of the Commission's publications and decisions are available from the Office of the Executive Secretary. Phone, 202–606–

OCCUPATIONAL SAFETY AND HEALTH REVIEW COMMISSION

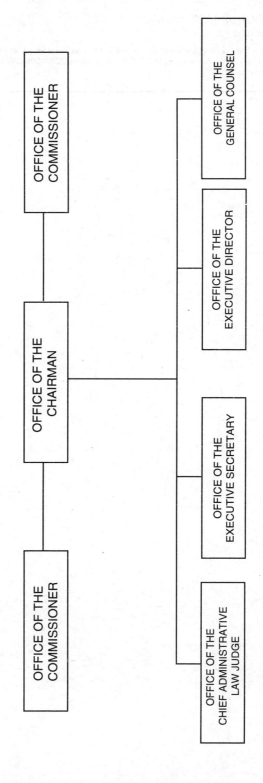

5400. Fax, 202–606–5050. Internet,
www.oshrc.gov.

For further information, contact the Executive Director, Occupational Safety and Health Review Commission, 1120 Twentieth Street NW., Washington, DC 20036–3457. Phone, 202–606–5376. Fax, 202–418–3487. Internet, http://www.oshrc.gov.

EDITORIAL NOTE: The Office of the Director of National Intelligence did not meet the publication deadline for submitting updated information of its activities, functions, and sources of information as required by the automatic disclosure provisions of the Freedom of Information Act (5 U.S.C. 552(a)(1)(A)).

OFFICE OF THE DIRECTOR OF NATIONAL INTELLIGENCE

Washington, DC 20511
Phone, 703–733–8600. Internet, http://www.dni.gov.

Director	JAMES R. CLAPPER, JR.
Director of the Intelligence Staff	LT. GEN. JOHN F. KIMMONS, USA
General Counsel	ROBERT S. LITT
Director, National Counterterrorism Center	MICHAEL LEITER
Deputy Director for Analysis	PETER LAVOY
Deputy Director for Policy, Plans and Requirements	DAVID R. SHEDD
Deputy Director for Collection	GLENN A. GAFFNEY
Deputy Director for Future Capabilities	ALDEN MUNSON
Director, National Counterproliferation Center	KENNETH BRILL
Program Manager, Information Sharing Environment	THOMAS E. MCNAMARA
Intelligence Community Chief Information Officer	PRISCILLA GUTHRIE
Chief Financial Officer	MARILYN A. VACCA
Chief Human Capital Officer	PAULA J. ROBERTS
Director, Intelligence Advanced Research Projects Activity	LISA PORTER
Iran Mission Manager	(VACANCY)
North Korea Mission Manager	SYLVIA COPELAND
Cuba/Venezuela Mission Manager	TIMOTHY LANGFORD
Civil Liberties Protection Officer	ALEXANDER W. JOEL
Inspector General	ROSLYN A. MAZER
Chancellor, National Intelligence University	TERESA DOMZAL
Director, Equal Employment Opportunity and Diversity	PATRICIA T. TAYLOR

The Office of the Director of National Intelligence oversees and coordinates the foreign and domestic activities of the Intelligence Community across the Federal Government.

The Director of National Intelligence (DNI) is a Cabinet-level post established by section 1011 of the Intelligence Reform and Terrorism Prevention Act of 2004 (50 U.S.C. 403). The DNI is responsible for overseeing and coordinating elements of the Intelligence Community and is the principal intelligence adviser to the President. The DNI reports directly to the President of the United States.

The DNI's responsibilities are: coordinating collection, processing, analysis, and dissemination of intelligence information required by the President, the National Security Council, the Secretaries of State and Defense, and other executive branch officials in performing their duties and responsibilities; ensuring the sharing of intelligence information within the Intelligence Community;

OFFICE OF THE DIRECTOR OF NATIONAL INTELLIGENCE

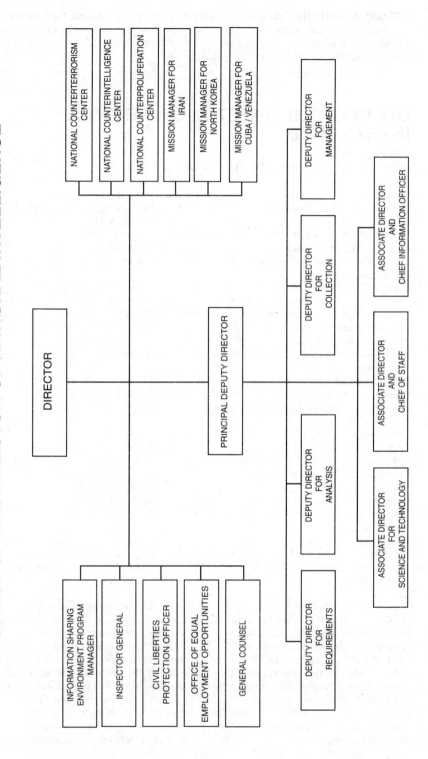

establishing personnel policies and programs applicable to the Intelligence Community; and determining the annual National Intelligence Program budget and directing the expenditure of those funds.

For further information, contact the Office of the Director of National Intelligence, Washington, DC 20511. Phone, 703–733–8600. Internet, http://www.dni.gov.

OFFICE OF GOVERNMENT ETHICS

1201 New York Avenue NW., Suite 500, Washington, DC 20005–3917
Phone, 202–482–9300. TTY, 202–482–9293. Fax, 202–482–9237. Internet, http://www.usoge. gov.

Director	ROBERT I. CUSICK
General Counsel	DON W. FOX
Deputy Director for the Office of International Assistance and Governance Initiatives	JANE S. LEY
Deputy Director for the Office of Agency Programs	JOSEPH E. GANGLOFF
Deputy Director for Administration	BARBARA A. MULLEN-ROTH
Deputy General Counsel	WALTER M. SHAUB, JR.

[For the Office of Government Ethics statement of organization, see the Code of Federal Regulations, Title 5, Part 2600]

The Office of Government Ethics directs executive branch policies related to preventing conflicts of interest on the part of Government employees and resolving those conflicts of interest that do occur.

The Office of Government Ethics (OGE) is an executive branch agency established under the Ethics in Government Act of 1978, as amended (5 U.S.C. app. 401).

The Director of OGE is appointed by the President with the advice and consent of the Senate for a 5-year term.

Activities

The Office of Government Ethics develops appropriate ethics policies for the executive branch through the promulgation of regulations on Standards of Ethical Conduct, public and confidential financial disclosure of executive branch officials, ethics training programs, and the identification and resolution of actual and potential conflicts of interest by reviewing the financial disclosure reports submitted by nominees to, the incumbents of, and those leaving executive branch positions requiring Presidential appointment with Senate confirmation. These financial disclosure reports are also available for public inspection.

The OGE also provides education and training to 6,000 ethics officials through instructor-led and Web-based training programs; assesses the effectiveness of public and confidential financial disclosure systems maintained by over 130 executive branch agencies; maintains an extensive program to provide advice on Standards of Ethical Conduct and conflict of interest laws; conducts onsite reviews of agency ethics programs; orders corrective action on the part of agencies and employees, including orders to establish or modify an agency's ethics program; evaluates the effectiveness of the Ethics Act, the conflict of interest laws, and other related statutes and recommending appropriate changes; and provides technical assistance in support of U.S. international anticorruption and good governance initiatives.

Sources of Information

Electronic Access Information regarding OGE's services and programs is available in electronic format on the Internet at www.usoge.gov.

Publications The Office of Government Ethics periodically updates its publication, The Informal Advisory Letters and Memoranda and Formal Opinions of the United States Office of Government Ethics, available from the Government Printing Office. In addition, OGE has ethics publications and instructional resources available. Upon request, OGE also provides copies of executive branch public financial disclosure reports (SF 278s) in accordance with the Ethics Act and OGE's regulations.

For further information, contact the Office of Government Ethics, Suite 500, 1201 New York Avenue NW., Washington, DC 20005–3917. Phone, 202–482–9300. TTY, 800–877–8339. Fax, 202–482–9237. Email, contactoge@oge.gov. Internet, http://www.usoge.gov.

OFFICE OF PERSONNEL MANAGEMENT

1900 E Street NW., Washington, DC 20415–0001
Phone, 202–606–1800. TTY, 202–606–2532. Internet, http://www.opm.gov.

Director	JOHN BERRY
Deputy Director	CHRISTINE M. GRIFFIN
Chief of Staff	ELIZABETH MONTOYA
Inspector General	PATRICK E. MCFARLAND
General Counsel	ELAINE KAPLAN
Chief Financial Officer	NANCY H. KICHAK, *Acting*
Associate Director and Chief Human Capital Officer	NANCY H. KICHAK
Associate Director, Merit System Audit and Compliance	JEFFERY E. SUMBERG
Associate Director* Retirement and Benefits	BILL ZIELINSKI
Associate Director, Federal Investigative Services	KATHY DILLAMAN
Associate Director, Human Resources Solutions	KAY ELY
Director, Congressional and Legislative Affairs	TANIA SHAND
Director, Communications and Public Liaison	SEDELTA VERBLE
Director, Planning and Policy	JOHN O'BRIEN
Director, Executive Secretariat and Ombudsman	RICHARD B. LOWE
Director, Facilities, Security, and Contracting	TINA B. MCGUIRE
Director, Equal Employment Opportunity	LORNA LEWIS
Director, Internal Oversight and Compliance	MARK LAMBERT
Chief Information Officer	MATTHEW E. PERRY, *Acting*
Executive Director, Chief Human Capital Officer Council	KATHRYN MEDINA

[For the Office of Personnel Management statement of organization, see the Federal Register of Jan. 5, 1979, 44 FR 1501]

The Office of Personnel Management administers a merit system to ensure compliance with personnel laws and regulations and assists agencies in recruiting, examining, and promoting people on the basis of their knowledge and skills, regardless of their race, religion, sex, political influence, or other nonmerit factors.

The Office of Personnel Management (OPM) was created as an independent establishment by Reorganization Plan No. 2 of 1978 (5 U.S.C. app.), pursuant to Executive Order 12107 of December 28, 1978. Many of the functions of the former United States Civil Service Commission were transferred to OPM.

Activities

Employee Benefits OPM manages numerous activities that directly affect the well-being of the Federal employee and indirectly enhance employee effectiveness. These include health benefits, life insurance, and retirement benefits.

Examining and Staffing The Office of Personnel Management is responsible for providing departments and agencies with technical assistance and guidance in examining competitive positions in the Federal civil service for General Schedule grades 1 through 15 and Federal Wage system positions. In addition, OPM is responsible for the following duties: providing testing and examination services, at the request of an agency, on a reimbursable basis; establishing basic qualification standards for all occupations; certifying agency delegated examining units to conduct examining; providing employment information for competitive service positions; and providing policy direction and guidance on promotions, reassignments, appointments in the excepted and competitive services, reinstatements, temporary and term employment, veterans preference, workforce planning and reshaping, organizational design, career transition, and other staffing provisions.

Executive Resources OPM leads in the selection, management, and development of Federal executives. OPM provides policy guidance, consulting services, and technical support on Senior Executive Service (SES) recruitment, selection, succession planning, mobility performance, awards, and removals. It reviews agency nominations for SES career appointments and administers the Qualifications Review Boards that certify candidates' executive qualifications. It manages SES, senior-level, and scientific and professional space allocations to agencies, administers the Presidential Rank Awards program, and

conducts orientation sessions for newly appointed executives. In addition, OPM manages three interagency residential development and training centers for executives and managers.

Investigations The Office of the Inspector General conducts comprehensive and independent audits, investigations, and evaluations relating to OPM programs and operations. It is responsible for administrative actions against health care providers who commit sanctionable offenses with respect to the Federal Employees' Health Benefits Program or other Federal programs.

For further information, contact the Office of the Inspector General. Phone, 202–606–1200.

Personnel Systems OPM provides leadership and guidance to agencies on systems to support the manager's personnel management responsibilities. These include the following: white- and blue-collar pay systems, including SES and special occupational pay systems; geographical adjustments and locality payments; special rates to address recruitment and retention problems; allowances and differentials, including recruitment and relocation bonuses, retention allowances, and hazardous duty/environmental pay; and premium pay; annual and sick leave, court leave, military leave, leave transfer and leave bank programs, family and medical leave, excused absence, holidays, and scheduling of work, including flexible and compressed work schedules; performance management, covering appraisal systems, performance pay and awards, and incentive awards for suggestions, inventions, and special acts; classification policy and standards for agencies to determine the series and grades for Federal jobs; labor-management relations, including collective bargaining, negotiability, unfair labor practices, labor-management cooperation, and consulting with unions on Governmentwide issues; systems and techniques for resolving disputes with employees; quality of worklife initiatives, such as employee health and fitness, work and family, AIDS in the workplace,

OFFICE OF PERSONNEL MANAGEMENT

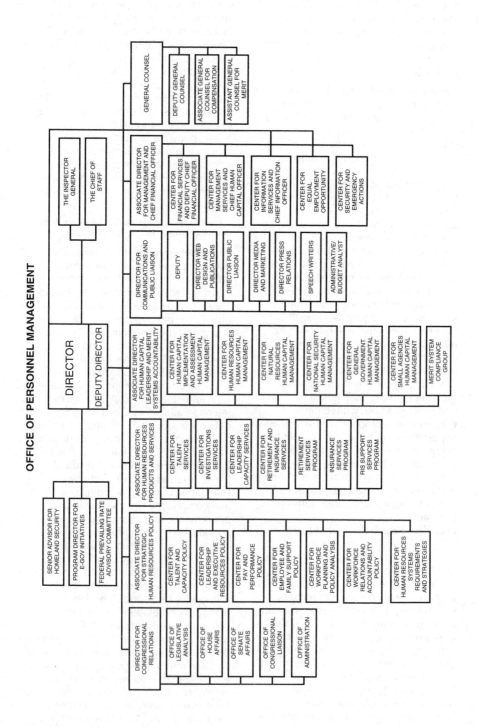

OFFICE OF PERSONNEL MANAGEMENT **429**

and employee assistance programs; human resources development, including leadership and administration of the Human Resources Development Council and the Government Performance and Results Act interest group; the Training and Management Assistance program, to help agencies develop training and human resources management solutions, including workforce planning and succession management strategies, e-learning applications, traditional classroom training materials, compensation and performance management systems, and other customized products; information systems to support and improve Federal personnel management decisionmaking; and Governmentwide instructions for personnel processing and recordkeeping and for release of personnel data under the Freedom of Information Act and the Privacy Act.

OPM also provides administrative support to special advisory bodies, including the Federal Prevailing Rate Advisory Committee, the Federal Salary Council, and the Presidential Advisory Committee on Expanding Training Opportunities.

Oversight OPM assesses human capital management Governmentwide and within agencies to gather information for policy development and program refinement, ensure compliance with law and regulation, and enhance agency capability for human resources management accountability. Agency accountability systems help ensure that human capital decisions are consistent with merit principles and that human capital strategies are aligned with mission accomplishment. OPM also works with agencies to find better and more strategic ways to manage Federal human capital.

Workforce Diversity OPM provides leadership, direction, and policy for Governmentwide affirmative recruiting programs for women, minorities, individuals with disabilities, and veterans. It also provides leadership, guidance, and technical assistance to promote merit and equality in systemic workforce recruitment, employment, training, and retention. In addition, OPM gathers,

analyzes, and maintains statistical data on the diversity of the Federal workforce and prepares evaluation reports for Congress and others on individual agency and Governmentwide progress toward full workforce representation for all Americans in the Federal sector.

Other Personnel Programs OPM coordinates the temporary assignment of employees between Federal agencies and State, local, and Indian tribal governments, institutions of higher education, and other eligible nonprofit organizations for up to 2 years, for work of mutual benefit to the participating organizations. It administers the Presidential Management Intern Program, which provides 2-year, excepted appointments with Federal agencies to recipients of graduate degrees in appropriate disciplines. In addition, the Office of Personnel Management administers the Federal Merit System Standards, which apply to certain grant-aided State and local programs.

Federal Executive Boards Federal Executive Boards (FEBs) were established by Presidential memorandum on November 10, 1961, to improve internal Federal management practices and to provide a central focus for Federal participation in civic affairs in major metropolitan centers of Federal activity. They carry out their functions under OPM supervision and control.

FEBs serve as a means for disseminating information within the Federal Government and for promoting discussion of Federal policies and activities of importance to all Federal executives in the field. Each Board is composed of heads of Federal field offices in the metropolitan area. A chairman is elected annually from among the membership to provide overall leadership to the Board's operations. Committees and task forces carry out interagency projects consistent with the Board's mission.

Federal Executive Boards are located in 28 metropolitan areas that are important centers of Federal activity. These areas are as follows: Albuquerque-Santa Fe, NM; Atlanta, GA; Baltimore, MD; Boston, MA; Buffalo, NY; Chicago, IL; Cincinnati,

OH; Cleveland, OH; Dallas-Fort Worth, TX; Denver, CO; Detroit, MI; Honolulu, HI; Houston, TX; Kansas City, MO; Los Angeles, CA; Miami, FL; Minneapolis-St. Paul, MN; New Orleans, LA; New York, NY; Newark, NJ; Oklahoma City, OK; Philadelphia, PA; Pittsburgh, PA; Portland, OR; St. Louis, MO; San Antonio, TX; San Francisco, CA; and Seattle, WA.

Federal Executive Associations or Councils have been locally organized in approximately 65 other metropolitan areas to perform functions similar to the Federal Executive Boards but on a lesser scale of organization and activity.

For further information, contact the Director for Federal Executive Board Operations, Office of Personnel Management, Room 5524, 1900 E Street NW., Washington, DC 20415–0001. Phone, 202–606–1000.

Sources of Information

Contacts For information, contact the Chief, Contracting Division, Office of Personnel Management, Washington, DC 20415–0071. Phone, 202–606–2240. Internet, www.opm.gov/procure/index.htm.

Employment Information about Federal employment and current job openings is available from USAJobs (phone, 478–757–3000; TTY, 478–744–2299; Internet, www.usajobs.opm.gov). For information about employment opportunities within the Office of Personnel Management, contact the Director of Human Resources. Phone, 202–606–2400.

Publications The Chief, Publications Services Division, can provide information about Federal personnel management publications. Phone, 202–606–1822. Internet, apps.opm.gov/publications.

For further information, contact the Office of Communications, Office of Personnel Management, 1900 E Street NW., Washington, DC 20415–0001. Phone, 202–606–1800. TTY, 202–606–2532. Internet, http://www.opm.gov.

OFFICE OF SPECIAL COUNSEL

1730 M Street NW., Suite 218, Washington, DC 20036–4505
Phone, 202–254–3600; 800–872–9855. Fax, 202–653–5151. Internet, http://www.osc.gov.

Special Counsel	WILLIAM REUKAUF, *Acting*
Deputy Special Counsel	(VACANCY)
Associate Special Counsel/Director of Field Office Operations and Alternative Dispute Resolution	WILLIAM REUKAUF
Associate Special Counsel for Investigation and Prosecution Division	LEONARD DRIBINSKY
Outreach Director	SHIRINE MOAZED, *Acting*
Associate Special Counsel for Legal Counsel and Policy Division	ERIN M. MCDONNELL
Director of Congressional Affairs	PATRICK BOULAY, *Acting*
Director of Public Affairs	DARSHAN SHETH, *Acting*

The Office of Special Counsel investigates allegations of certain activities prohibited by civil service laws, rules, or regulations and litigates before the Merit Systems Protection Board.

The Office of Special Counsel (OSC) was established on January 1, 1979, by Reorganization Plan No. 2 of 1978 (5 U.S.C. app.). The Civil Service Reform Act of 1978 (5 U.S.C. 1101 note),

which became effective on January 11, 1979, enlarged its functions and powers. Pursuant to provisions of the Whistleblower Protection Act of 1989 (5 U.S.C. 1211 et seq.), OSC functions as an independent investigative and prosecutorial agency within the executive branch that litigates before the Merit Systems Protection Board.

Activities

The primary role of OSC is to protect employees, former employees, and applicants for employment from prohibited personnel practices, especially reprisal for whistleblowing. Its basic areas of statutory responsibility are the following: receive and investigate allegations of prohibited personnel practices and other activities prohibited by civil service law, rule, or regulation and, if warranted, initiate corrective or disciplinary action; provide a secure channel through which information substantiating a violation of any law, rule, or regulation, gross mismanagement, gross waste of funds, abuse of authority, or substantial and specific danger to public health or safety may be disclosed without fear of retaliation and without disclosure of identity, except with the employee's consent; and enforce the provisions of the Hatch Act and the Uniformed Services Employment and Reemployment Rights Act.

Sources of Information

For a complete listing of Office of Special Counsel contacts, including field offices, media inquiries, and the whistleblower disclosure hotline, visit www.osc.gov/ contacts.htm.

For further information, contact the Office of Special Counsel, 1730 M Street NW., Suite 218, Washington, DC 20036–4505. Phone, 202–254–3600 or 800–872–9855. Fax, 202- 254–3711. Internet, http://www.osc. gov.

OVERSEAS PRIVATE INVESTMENT CORPORATION

1100 New York Avenue NW., Washington, DC 20527
Phone, 202–336–8400. Fax, 202–336–7949. Internet, http://www.opic.gov.

President and Chief Executive Officer	LAWRENCE SPINELLI, *Acting*
Executive Vice President	(VACANCY)
Vice President of External Affairs	LAWRENCE SPINELLI, *Acting*
Vice President of Investment Policy	BERTA M. HEYBEY, *Acting*
Vice President of Structured Finance	ROBERT B. DRUMHELLER
Vice President of Small and Medium Enterprise Finance	JAMES C. POLAN
Vice President of Insurance	ROD MORRIS
Vice President of Investment Funds	BARBARA K. DAY, *Acting*
Vice President and General Counsel	ROBERT C. O'SULLIVAN, *Acting*
Vice President and Chief Financial Officer	JACQUELINE STRASSER, *Acting*

[For the Overseas Private Investment Corporation statement of organization, see the Code of Federal Regulations, Title 22, Chapter VII]

The Overseas Private Investment Corporation promotes economic growth in developing countries and emerging markets by encouraging U.S. private investment in those nations.

The Overseas Private Investment Corporation (OPIC) was established in 1971 as an independent agency by the Foreign Affairs Reform and Restructuring Act (112 Stat. 2681–790). OPIC helps U.S. businesses invest overseas, fosters economic development in new and emerging markets, complements the private sector in managing risks associated with foreign direct investment, and supports U.S. foreign policy. OPIC charges market-based fees for its products, and it operates on a self-sustaining basis at no net cost to taxpayers.

OPIC helps U.S. businesses compete in emerging markets when private sector support is not available. OPIC offers up to $250 million in long-term financing and/or political risk insurance to U.S. companies investing in over 150 emerging markets and developing countries. Backed by the full faith and credit of the U.S. Government, OPIC advocates for U.S. investment, offers experience in risk management, and draws on an outstanding record of success.

OPIC mobilizes America's private sector to advance U.S. foreign policy and development initiatives. Projects supported by OPIC expand economic development, which encourages political stability and free market reforms. Over the agency's 35 year history, OPIC has supported $177 billion worth of investments that have helped developing countries to generate over $13 billion in host government revenues and over 800,000 host country jobs. OPIC projects have also generated $71 billion in U.S. exports and supported more than 271,000 American jobs. OPIC promotes U.S. best practices by requiring projects to adhere to international standards on the environment, worker rights, and human rights.

Activities

OPIC insures U.S. investors, contractors, exporters, and financial institutions against political violence, expropriation of assets by foreign governments, and the inability to convert local currencies into U.S. dollars. OPIC can insure up to $250 million per project and has no minimum investment size requirements. Insurance is available for investments in new ventures, expansions of existing enterprises, privatizations, and acquisitions with positive developmental benefits.

OPIC provides financing through direct loans and loan guaranties for medium- and long-term private investment. Loans range from $100,000 to $250 million for projects sponsored by U.S. companies, and financing can be provided on a project finance or corporate finance basis. In most cases, the U.S. sponsor is expected to contribute at least 25 percent of the project equity, have a track record in the industry, and have the means to contribute to the financial success of the project.

To address the lack of sufficient equity investment in emerging markets, OPIC has supported the creation of privately owned and managed investment funds that make direct equity and equity-related investments in new, expanding, or privatizing companies. These funds, which have a regional or sectoral focus, provide the long-term growth capital that can serve as a catalyst for private sector economic activity in developing countries and the creation of new markets and opportunities for American companies.

Helping America's small businesses grow through investments in emerging markets is an important OPIC priority. Any small business with annual revenues less than $35 million is eligible for small business center programs. For businesses with annual revenues over $35 million and under $250 million, OPIC's regular small business programs are available. OPIC provides direct loans to U.S. small businesses and offers insurance products to meet the special needs of small businesses. Other client services include streamlined applications and processing procedures and online small business resources available at www.opic.gov.

Sources of Information

General Inquiries   Inquiries should be directed to the Information Officer, Overseas Private Investment Corporation, 1100 New York Avenue NW., Washington, DC 20527. Phone, 202–336–8799. Email, info@opic.gov. Internet, www.opic.gov.
Publications   OPIC programs are further detailed in the Annual Report and the Program Handbook. These publications are available free of charge at www.opic. gov.

For further information, contact the Overseas Private Investment Corporation, 1100 New York Avenue NW., Washington, DC 20527. Phone, 202–336–8400. Fax, 202–336–7949. Internet, http://www.opic.gov.

PEACE CORPS

1111 Twentieth Street NW., Washington, DC 20526
Phone, 202–692–2000; 800–424–8580. Fax, 202–692–2231. Internet, http://www.peacecorps.gov.

Director	AARON S. WILLIAMS
Deputy Director	CAROLYN HESSLER-RADELET
Chief of Staff/Operations	STACY RHODES
Associate Director for Global Operations	ESTHER BENJAMIN
Director of Congressional Relations	SUZIE CARROLL, *Acting*
General Counsel	CARL SOSEBEE
Director of Communications	ALLISON PRICE
Director of Press Relations	JOSHUA FIELD
Director of Office of Strategic Information, Research, and Planning	CATHRYN THORUP
American Diversity Program Manager	SHIRLEY EVEREST
Director of Private Sector Initiatives	JENNIFER CHAVEZ RUBIO
Inspector General	KATHY BULLER
Director of Peace Corps Response	SARAH MORGENTHAU
Regional Director/Africa Operations	LYNN FODEN, *Acting*
Regional Director/Europe, Mediterranean, and Asia Operations	DAVID BURGESS, *Acting*
Regional Director/Inter-American and the Pacific Operations	CARLOS TORRES
Director, Overseas Programming and Training Support	STEPHEN MILLER
Chief Financial Officer	THOMAS BELLAMY, *Acting*
Associate Director for Management	EARL YATES
Associate Director for Volunteer Support	JULES DELAUNE
Chief Information Officer	CHRIS SARANDOS, *Acting*
Associate Director for Volunteer Recruitment and Selection	ROSIE MAUK
Associate Director for Safety and Security	EDWARD HOBSON
Chief Acquisition Officer	CAREY FOUNTAIN
Chief Compliance Officer	(VACANCY)
AIDS Relief Coordinator	(VACANCY)
Director of Intergovernmental Affairs	C.D. GLIN
Director of Public Engagement	DAVID MEDINA
Director of Innovation	(VACANCY)

The Peace Corps helps people of interested countries meet their need for trained men and women and promotes mutual understanding between Americans and citizens of other countries.

The Peace Corps was established by the Peace Corps Act of 1961, as amended (22 U.S.C. 2501), and was made an independent agency by title VI of the International Security and Development Cooperation Act of 1981 (22 U.S.C. 2501–1).

Activities

The Peace Corps consists of a Washington, DC, headquarters, 9 area offices, and overseas operations in 77 countries, utilizing more than 7,600 volunteers.

To fulfill the Peace Corps mandate, men and women are trained for a 9- to 14-week period in the appropriate local language, the technical skills necessary for their particular jobs, and the cross-cultural skills needed to adjust to a society with traditions and attitudes different from their own. Volunteers serve

PEACE CORPS

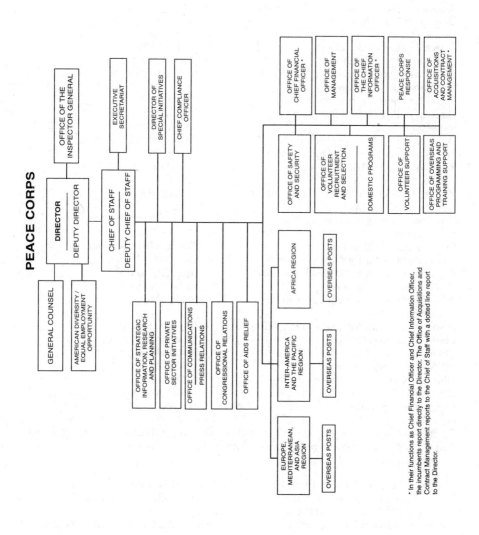

* In their functions as Chief Financial Officer and Chief Information Officer, the incumbents report directly to the Director. The Office of Acquisitions and Contract Management reports to the Chief of Staff with a dotted line report to the Director.

for a period of 2 years, living among the people with whom they work. Volunteers are expected to become a part of the community through their voluntary service.

Thousands of volunteers serve throughout the world, working in six program areas: education, health and HIV/AIDS, environment, youth development, agriculture, and business development. Community-level projects are designed to incorporate the skills of volunteers with the resources of host-country agencies and other international assistance organizations to help solve specific development problems, often in conjunction with private volunteer organizations.

In the United States, the Peace Corps is working to promote an understanding of people in other countries. Through its World Wise Schools Program, volunteers are matched with elementary and junior high schools in the United States to encourage an exchange of letters, pictures, music, and artifacts. Participating students increase their knowledge of geography, languages, and different cultures, while gaining an appreciation for voluntarism.

The Peace Corps offers other domestic programs involving former volunteers, universities, local public school systems, and private businesses and foundations in a partnership to help solve some of the United States most pressing domestic problems.

The Peace Corps Office of Private Sector Initiatives works with schools, civic groups, businesses, and neighborhood and youth organizations in the United States to facilitate their support of Peace Corps initiatives here and abroad.

For a complete listing of Peace Corps area offices, including addresses, telephone numbers, and areas served, visit www.peacecorps.gov.

Sources of Information

Becoming a Peace Corps Volunteer Contact the nearest area office. Phone, 800–424–8580. Internet, www. peacecorps.gov.
Employment Contact the Peace Corps, Office of Human Resource Management, Washington, DC 20526. Phone, 202–692–1200. Internet, www.peacecorps. gov. For recorded employment opportunities, call 800–818–9579.
General Inquiries Information or assistance may be obtained by contacting the Peace Corps Washington, DC, headquarters or any of its area offices.

For further information, contact the Press Office, Peace Corps, 1111 Twentieth Street NW., Washington, DC 20526. Phone, 202–692–2230 or 800–424–8580. Fax, 202–692–2201. Internet, http://www.peacecorps.gov.

PENSION BENEFIT GUARANTY CORPORATION

1200 K Street NW., Washington, DC 20005
Phone, 202–326–4400; 800–736–2444. Internet, http://www.pbgc.gov.

Board of Directors

Chairman (Secretary of Labor)	HILDA L. SOLIS
Member (Secretary of the Treasury)	TIMOTHY F. GEITHNER
Member (Secretary of Commerce)	GARY F. LOCKE

Officials

Director	VINCENT K. SNOWBARGER, *Acting*
Deputy Director, Operations	VINCENT K. SNOWBARGER
Director, Communications and Public Affairs	JEFFREY SPEICHER, *Acting*
Director, Legislative and Regulatory	JOHN HANLEY
Director, Policy, Research, and Analysis	DAVID GUSTAFSON
Deputy Director, Policy	(VACANCY)
Chief of Staff	(VACANCY)

Officials

Chief Information Officer	RICHARD MACY, *Acting*
Chief Management Officer	VINCENT K. SNOWBARGER, *Acting*
Director, Budget and Organizational Performance Department	EDGAR BENNETT
Director, Facilities and Services	PATRICIA DAVIS
Director, Human Resources	ARRIE ETHERIDGE
Director, Procurement	STEVE BLOCK, *Acting*
Chief Operating Officer	RICHARD MACY
Director, Benefits Administration and Payment	BENNIE L. HAGANS
Chief Investment Officer	JOHN H. GREENBERG
Chief Insurance Program Officer	TERRENCE M. DENEEN
Chief Counsel	ISRAEL GOLDOWITZ
Director, Insurance Supervision and Compliance	JOSEPH HOUSE

The Pension Benefit Guaranty Corporation protects the pension benefits of nearly 44 million Americans who participate in defined-benefit pension plans sponsored by private-sector employees.

The Pension Benefit Guaranty Corporation (PBGC) is a self-financing, wholly owned Government corporation subject to the Government Corporation Control Act (31 U.S.C. 9101–9109). The Corporation, established by title IV of the Employee Retirement Income Security Act of 1974 (29 U.S.C. 1301–1461), operates in accordance with policies established by its Board of Directors, which consists of the Secretaries of Labor, Commerce, and the Treasury. The Secretary of Labor is Chairman of the Board. A seven-member Advisory Committee, composed of two labor, two business, and three public members appointed by the President, advises the agency on investment issues.

Activities

Coverage   The Corporation insures most private-sector defined-benefit pension plans, which provide a pension benefit based on factors such as age, years of service, and salary.

The Corporation administers two insurance programs, separately covering single-employer and multiemployer plans. Nearly 44 million workers and retirees participate in more than 29,000 covered plans.

Single-Employer Insurance   Under the single-employer program, the Corporation guarantees payment of basic pension benefits if an insured plan terminates without sufficient assets to pay those benefits. However, the law limits the total monthly benefit that the agency may guarantee for one individual to $4,500 per month for a 65-year-old individual in a pension plan that terminates in 2010. The law also sets other restrictions on PBGC's guarantee, including limits on the insured amount of recent benefit increases. In certain cases, the Corporation may also pay some benefits above the guaranteed amount depending on the funding level of the plan and amounts recovered from employers.

A plan sponsor may terminate a single-employer plan in a standard termination if the plan has sufficient assets to purchase private annuities to cover all benefit liabilities. If a plan does not have sufficient assets, the sponsor may seek to transfer the pension liabilities to the PBGC by demonstrating that it meets the legal criteria for a distress termination. In either termination, the plan administrator must inform participants in writing at least 60 days prior to the date the administrator proposes to terminate the plan. Only a plan that has sufficient assets to pay all benefit liabilities may terminate in a standard termination. The Corporation also may institute termination of underfunded plans in certain specified circumstances.

PENSION BENEFIT GUARANTY CORPORATION

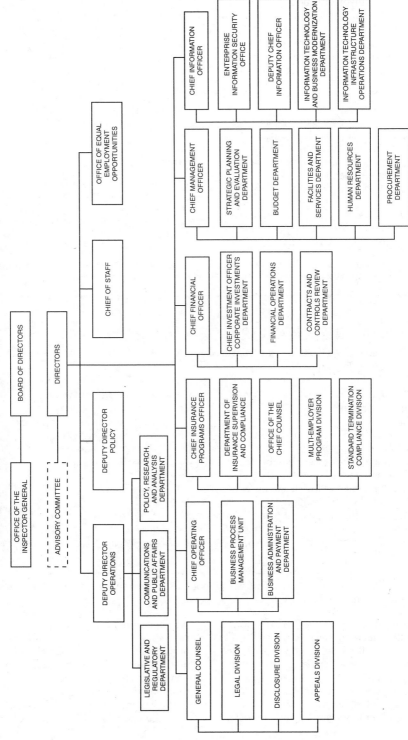

Multiemployer Insurance Under title IV, as revised in 1980 by the Multiemployer Pension Plan Amendments Act (29 U.S.C. 1001 note), which changed the insurable event from plan termination to plan insolvency, the Corporation provides financial assistance to multiemployer plans that are unable to pay nonforfeitable benefits. The plans are obligated to repay such assistance. The act also made employers withdrawing from a plan liable to the plan for a portion of its unfunded vested benefits.

Premium Collections All defined-benefit pension plans insured by PBGC are required to pay premiums to the Corporation according to rates set by Congress. The per-participant flat-rate premium for plan years beginning in 2010 is $35.00 for single-employer plans and $9.00 for multiemployer plans. Underfunded single-employer plans must also pay an additional premium equal to $9 per $1,000 of unfunded vested benefits. A termination premium of $1,250 per participant per year applies to certain distress and involuntary plan terminations occurring on or after January 1, 2006, payable for 3 years after the termination.

Sources of Information

Access to the Pension Benefit Guaranty Corporation is available through the Internet at www.pbgc.gov. TTY/TDD users may call the Federal Relay Service toll free at 800–877–8339 and ask to be connected to 800–736–2444.

For further information, contact the Pension Benefit Guaranty Corporation, 1200 K Street NW., Washington, DC 20005–4026. Phone, 202–326–4400 or 800–736–2444. Internet, http://www.pbgc.gov.

POSTAL REGULATORY COMMISSION

901 New York Avenue NW., Suite 200, Washington, DC 20268–0001
Phone, 202–789–6800. Fax, 202–789–6861. Internet, http://www.prc.gov.

Chairman	RUTH Y. GOLDWAY
Vice Chairman	TONY HAMMOND
Commissioners	MARK ACTON, DAN G. BLAIR, NANCI E. LANGLEY
General Counsel	STEPHEN L. SHARFMAN
Director, Office of Accountability and Compliance	JOHN D. WALLER
Director, Office of Public Affairs and Government Relations	ANN FISHER
Director, Office of Secretary and Administration	SHOSHANA M. GROVE
Inspector General	JOHN F. CALLENDER

[For the Postal Regulatory Commission statement of organization, see the Code of Federal Regulations, Title 39, Part 3002]

The Postal Regulatory Commission develops and implements a modern system of postal rate regulation.

The Postal Regulatory Commission is the successor agency to the Postal Rate Commission, which was created by the Postal Reorganization Act, as amended (39 U.S.C. 101 et seq.). The Commission was established as an independent agency in the executive branch of Government by the Postal Accountability and Enhancement Act (39 U.S.C. 501). It is composed of five Commissioners, appointed by the President with the advice and consent of the Senate, one of whom is designated as Chairman.

The Commission promulgates rules and regulations, establishes procedures, and takes other actions necessary to carry out its obligations. It considers complaints received from interested persons relating to United States Postal Service rates, regulations, and services. The Commission also has certain reporting obligations, including a report on universal postal service and the postal monopoly.

Sources of Information

Employment The Commission's programs require attorneys, economists, statisticians, accountants, industrial engineers, marketing specialists, and administrative and clerical personnel to fulfill its responsibilities. Requests for employment information should be directed to the Personnel Officer.

Electronic Access Electronic access to current docketed case materials is available through the Internet at www.prc.gov. Email can be sent to the Commission at prc-admin@prc.gov and prc-dockets@prc.gov.

Reading Room Facilities for inspection and copying of records, viewing automated daily lists of docketed materials, and accessing the Commission's Internet site are located at Suite 200, 901 New York Avenue NW., Washington, DC. The room is open from 8 a.m. to 4:30 p.m., Monday through Friday, except legal holidays.

Rules of Practice and Procedure The Postal Regulatory Commission's Rules of Practice and Procedure governing the conduct of proceedings before the Commission may be found in parts 3001, 3010, 3015, 3020, 3030, 3031, and 3060 of title 39 of the Code of Federal Regulations.

For further information, contact the Secretary, Postal Regulatory Commission, 901 New York Avenue NW., Suite 200, Washington, DC 20268–0001. Phone, 202–789–6840. Internet, http://www.prc.gov.

RAILROAD RETIREMENT BOARD

844 North Rush Street, Chicago, IL 60611–2092
Phone, 312–751–4777 or 312–751–7154. Internet, http://www.rrb.gov. Email, opa@rrb.gov.

Chairman	MICHAEL S. SCHWARTZ
Labor Member	V.M. SPEAKMAN, JR.
Management Member	JEROME F. KEVER
Inspector General	MARTIN J. DICKMAN
Director, Administration and Senior Executive Officer	HENRY M. VALIULIS
Director, Equal Opportunity	LYNN E. COUSINS
Director, Human Resources	KEITH B. EARLEY
Supervisor, Public Affairs	ANITA J. ROGERS
Supervisor, Acquisition Management	PAUL T. AHERN
Facility Manager	SCOTT L. RUSH
General Counsel	STEVEN A. BARTHOLOW
Director, Legislative Affairs	MARGARET A. LINDSLEY
Director, Hearings and Appeals	KARL T. BLANK
Secretary to the Board	BEATRICE E. EZERSKI
Chief Actuary	FRANK J. BUZZI
Chief Information Officer	TERRI S. MORGAN
Chief Financial Officer	KENNETH P. BOEHNE
Director, Programs	DOROTHY A. ISHERWOOD
Director, Assessment and Training	CATHERINE A. LEYSER
Director, Field Service	MARTHA M. BARRINGER
Director, Operations	ROBERT J. DUDA
Director, Policy and Systems	RONALD RUSSO
Chief, Resource Management Center	JANET M. HALLMAN

[For the Railroad Retirement Board statement of organization, see the Code of Federal Regulations, Title 20, Part 200]

The Railroad Retirement Board administers comprehensive retirement-survivor and unemployment-sickness benefit programs for the Nation's railroad workers and their families.

The Railroad Retirement Board was originally established by the Railroad Retirement Act of 1934, as amended (45 U.S.C. 201—228z-1).

The Board derives statutory authority from the Railroad Retirement Act of 1974 (45 U.S.C. 231–231u) and the Railroad Unemployment Insurance Act (45 U.S.C. 351–369). It administers these acts and participates in the administration of the Social Security Act and the Health Insurance for the Aged Act insofar as they affect railroad retirement beneficiaries.

The Board is composed of three members appointed by the President with the advice and consent of the Senate: one upon recommendations of representatives of employees; one upon recommendations of carriers; and one, the Chairman, as a public member.

Activities   The Railroad Retirement Act provides for the payment of annuities to individuals who have completed at least 10 years of creditable railroad service, or 5 years if performed after 1995, and have ceased compensated service upon their attainment of specified ages, or at any age if permanently disabled for all employment. In some circumstances occupational disability annuities or supplemental annuities are provided for career employees.

A spouse's annuity is provided, under certain conditions, for the wife or husband of an employee annuitant. Divorced spouses may also qualify.

Survivor annuities are awarded to the qualified spouses, children, and parents of deceased career employees, and various lump-sum benefits are also available under certain conditions.

Benefits based upon qualifying railroad earnings in a preceding 1-year period are provided under the Railroad Unemployment Insurance Act to individuals who are unemployed in a benefit year, but who are ready and willing to work, and to individuals who

are unable to work because of sickness or injury.

The Board maintains, through its field offices, a placement service for unemployed railroad personnel.

Sources of Information

Benefit Inquiries   The Board maintains direct contact with railroad employees and railroad retirement beneficiaries through its field offices located across the country. Field personnel explain benefit rights and responsibilities on an individual basis, assist employees in applying for benefits, and answer questions related to the benefit programs. The Board also relies on railroad labor groups and employers for assistance in keeping railroad personnel informed· about its benefit programs. To locate the nearest field office, individuals should check with their rail employer or local union official. Information may also be obtained by calling the Board at 877–772–5772 or by visiting the Board's Web site at www.rrb.gov. Most offices are open to the public from 9 a.m. to 3:30 p.m., Monday through Friday.

Electronic Access   Railroad Retirement Board information is available online at www.rrb.gov.

Employment   Employment inquiries should be directed to the Bureau of Human Resources, Railroad Retirement Board, 844 North Rush Street, Chicago, IL 60611–2092. Phone, 312–751–4580. Email, recruit@rrb.gov.

Congressional and Legislative Assistance Congressional offices making inquiries regarding constituents' claims should contact the Office of Administration, Congressional Inquiry Section. Phone, 312–751–4970. Fax, 312–751–7154. Email, opa@rrb.gov. For information regarding legislative matters, contact the Office of Legislative Affairs, Suite 500, 1310 G Street NW., Washington, DC

RAILROAD RETIREMENT BOARD

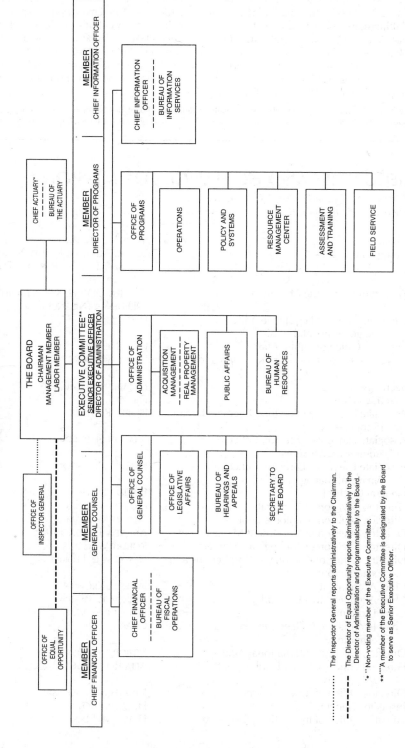

THE BOARD
CHAIRMAN
MANAGEMENT MEMBER
LABOR MEMBER

OFFICE OF INSPECTOR GENERAL

CHIEF ACTUARY*
BUREAU OF THE ACTUARY

EXECUTIVE COMMITTEE**
SENIOR EXECUTIVE OFFICER
DIRECTOR OF ADMINISTRATION

OFFICE OF EQUAL OPPORTUNITY

MEMBER
CHIEF FINANCIAL OFFICER

CHIEF FINANCIAL OFFICER
BUREAU OF FISCAL OPERATIONS

MEMBER
GENERAL COUNSEL

OFFICE OF GENERAL COUNSEL

OFFICE OF LEGISLATIVE AFFAIRS

BUREAU OF HEARINGS AND APPEALS

SECRETARY TO THE BOARD

OFFICE OF ADMINISTRATION

ACQUISITION MANAGEMENT
REAL PROPERTY MANAGEMENT

PUBLIC AFFAIRS

BUREAU OF HUMAN RESOURCES

MEMBER
DIRECTOR OF PROGRAMS

OFFICE OF PROGRAMS

OPERATIONS

POLICY AND SYSTEMS

RESOURCE MANAGEMENT CENTER

ASSESSMENT AND TRAINING

FIELD SERVICE

MEMBER
CHIEF INFORMATION OFFICER

CHIEF INFORMATION OFFICER
BUREAU OF INFORMATION SERVICES

.... The Inspector General reports administratively to the Chairman.

— — — The Director of Equal Opportunity reports administratively to the Director of Administration and programmatically to the Board.

"*" Non-voting member of the Executive Committee.

**"A member of the Executive Committee is designated by the Board to serve as Senior Executive Officer.

20005–3004. Phone, 202–272–7742. Fax, 202–272–7728. Email, ola@rrb.gov. **Publications** General information pamphlets on benefit programs may be obtained from the Board's field offices or Chicago headquarters. Requests for annual reports or statistical data should be directed to Public Affairs at the Chicago headquarters. Phone, 312–751–4777. Fax, 312–751–7154. Email, opa@rrb.gov. **Telecommunications Devices for the Deaf (TDD)** The Board provides TDD services. Phone 312–751–4701 for beneficiary inquiries or 312–751–4334 for equal opportunity inquiries.

For further information, contact Public Affairs, Railroad Retirement Board, 844 North Rush Street, Chicago, IL 60611–2092. Phone, 312–751–4777. Fax, 312–751–7154. Email, opa@rrb.gov. Internet, http://www.rrb.gov.

SECURITIES AND EXCHANGE COMMISSION

100 F Street NE., Washington, DC 20549
Phone, 202–551–7500. Internet, http://www.sec.gov.

Chairman	MARY L. SCHAPIRO
Commissioners	LUIS A. AGUILAR, KATHLEEN L. CASEY, TROY A. PAREDES, ELISSE B. WALTER
Senior Adviser to the Chairmans	STEPHEN DEVINE, KAYLA J. GILLAN, DIDEM A. NISANCI
Director, Office of Legislative and Intergovernmental Affairs	ERIC J. SPITLER
Director, Office of Public Affairs	JOHN NESTER
Director, Office of Investor Education and Advocacy	LORI J. SCHOCK
Secretary	ELIZABETH M. MURPHY
Executive Director	DIEGO T. RUIZ
Associate Executive Director, Office of Human Resources	JEFFREY RISINGER
Associate Executive Director, Office of Administrative Services	SHARON SHEEHAN
Associate Executive Director, Office of Financial Management	KENNETH JOHNSON
General Counsel	DAVID BECKER
Director, Division of Corporation Finance	MEREDITH CROSS
Director, Division of Enforcement	ROBERT KHUZAMI
Director, Division of Investment Management	ANDREW J. DONOHUE
Director, Division of Trading and Markets	ROBERT W. COOK
Director, Office of Compliance Inspections and Examinations	CARLO V. DI FLORIO
Chief Accountant	JAMES L. KROEKER
Chief Administrative Law Judge	BRENDA P. MURRAY
Chief Economist	JAMES OVERDAHL
Chief Information Officer, Office of Information Technology	CHARLES BOUCHER
Director, Office of International Affairs	ETHIOPIS TAFARA
Director, Office of Equal Employment Opportunity	ALTA RODRIGUEZ
Director, Office of Risk Assessment	JONATHAN S. SOKOBIN
Inspector General	H. DAVID KOTZ
Director, Office of Interactive Disclosure	DAVID M. BLASZKOWSKY

[For the Securities and Exchange Commission statement of organization, see the Code of Federal Regulations, Title 17, Part 200]

The Securities and Exchange Commission administers Federal securities laws to provide protection for investors, to ensure that securities markets are fair and honest, and when necessary, to provide the means to enforce securities laws through sanctions.

The Securities and Exchange Commission (SEC) was created under authority of the Securities Exchange Act of 1934 (15 U.S.C. 78a-78jj) and was organized on July 2, 1934. The Commission serves as adviser to United States district courts in connection with reorganization proceedings for debtor corporations in which there is a substantial public interest. The Commission also has certain responsibilities under section 15 of the Bretton Woods Agreements Act of 1945 (22 U.S.C. 286k-1) and section 851(e) of the Internal Revenue Code of 1954 (26 U.S.C. 851(e)).

The Commission is vested with quasi-judicial functions. Persons aggrieved by its decisions in the exercise of those functions have a right of review by the United States courts of appeals.

For a complete listing of SEC regional offices, including addresses, telephone numbers, and key officials, visit www.sec.gov/contact/addresses.htm.

Activities

Full and Fair Disclosure   The Securities Act of 1933 (15 U.S.C. 77a) requires issuers of securities and their controlling persons making public offerings of securities in interstate commerce or via mail to file with the Commission registration statements containing financial and other pertinent data about the issuer and the securities being offered. There are limited exemptions, such as government securities, nonpublic offerings, and intrastate offerings, as well as certain offerings not exceeding $1.5 million. The effectiveness of a registration statement may be refused or suspended after a public hearing if the statement contains material misstatements or omissions, thus barring sale of the securities until it is appropriately amended.

Regulation of Investment Advisers
Persons who, for compensation, engage in the business of advising others with respect to securities must register with the Commission. The Commission is authorized to define what practices are considered fraudulent or deceptive and to prescribe means to prevent those practices.

Regulation of Mutual Funds and Other Investment Companies   The Commission registers investment companies and regulates their activities to protect investors. The regulation covers sales load, management contracts, composition of boards of directors, and capital structure. The Commission must also determine the fairness of various transactions of investment companies before these actually occur.

The Commission may institute court action to enjoin the consummation of mergers and other plans of reorganization of investment companies if such plans are unfair to securities holders. It also may impose sanctions by administrative proceedings against investment company management for violations of the act and other Federal securities laws and file court actions to enjoin acts and practices of management officials involving breaches of fiduciary duty and personal misconduct and to disqualify such officials from office.

Regulation of Securities Markets   The Securities Exchange Act of 1934 assigns to the Commission broad regulatory responsibilities over the securities markets, the self-regulatory organizations within the securities industry, and persons conducting a business in securities. Persons who transactions in securities generally are required to register with the Commission as broker-dealers. Securities exchanges and certain clearing agencies are required to register with the Commission, and associations of brokers

SECURITIES AND EXCHANGE COMMISSION

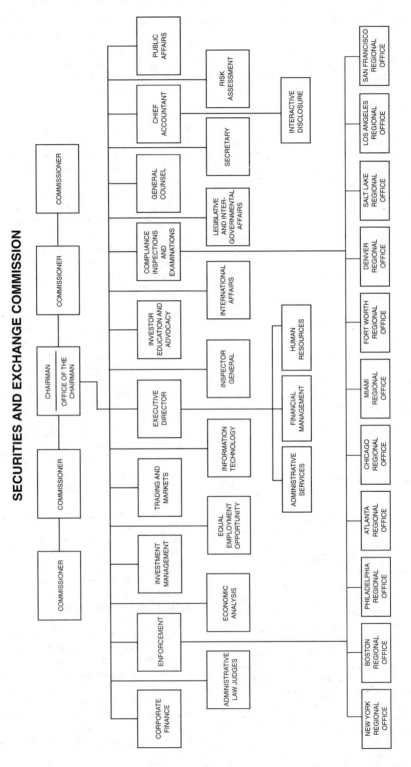

or dealers are permitted to register with the Commission. The Act also provides for the establishment of the Municipal Securities Rulemaking Board to formulate rules for the municipal securities industry.

The Commission oversees the self-regulatory activities of the national securities exchanges and associations, registered clearing agencies, and the Municipal Securities Rulemaking Board. In addition, the Commission regulates industry professionals, such as securities brokers and dealers, certain municipal securities professionals, Government securities brokers and dealers, and transfer agents.

Rehabilitation of Failing Corporations In cases of corporate reorganization proceedings administered in Federal courts, the Commission may participate as a statutory party. The principal functions of the Commission are to protect the interests of public investors involved in such cases through efforts to ensure their adequate representation and to participate in legal and policy issues that are of concern to public investors generally.

Representation of Debt Securities Holders The Commission safeguards the interests of purchasers of publicly offered debt securities issued pursuant to trust indentures.

Enforcement Activities The Commission's enforcement activities are designed to secure compliance with the Federal securities laws administered by the Commission and the rules and regulations adopted thereunder. These activities include measures to do the following: compel compliance with the disclosure requirements of the registration and other provisions of the relevant acts; prevent fraud and deception in the purchase and sale of securities; obtain court orders enjoining acts and practices that operate as a fraud upon investors or otherwise violate the laws; suspend or revoke the registrations of brokers, dealers, investment companies, and investment advisers who willfully engage in such acts and practices; suspend or bar from association persons associated with brokers, dealers, investment companies, and investment advisers who have

violated any provision of the Federal securities laws; and prosecute persons who have engaged in fraudulent activities or other willful violations of those laws.

In addition, attorneys, accountants, and other professionals who violate the securities laws face possible loss of their privilege to practice before the Commission.

To this end, private investigations are conducted into complaints or other indications of securities violations. Evidence thus established of law violations is used in appropriate administrative proceedings to revoke registration or in actions instituted in Federal courts to restrain or enjoin such activities. Where the evidence tends to establish criminal fraud or other willful violation of the securities laws, the facts are referred to the Attorney General for criminal prosecution of the offenders. The Commission may assist in such prosecutions.

Sources of Information

Inquiries regarding the following matters should be directed to the appropriate office, Securities and Exchange Commission, 100 F Street NE., Washington, DC 20549.

Contracts Inquiries regarding SEC procurement and contracting activities should be directed to the Office of Administrative Services. Phone, 202–551–7400.

Electronic Access Information on the Commission is available through the Internet at www.sec.gov.

Employment With the exception of the attorney category, positions are in the competitive civil service, which means applicants must apply for consideration for a particular vacancy and go through competitive selection procedures. The Commission operates a college and law school recruitment program, including on-campus visitations for interview purposes. Inquiries should be directed to the Office of Human Resources. Phone, 202–942–7500. Fax, 703–914–0592.

Investor Assistance and Complaints The Office of Investor Education and Advocacy answers questions from investors, assists investors with specific problems regarding their relations with broker-dealers and companies, and advises the Commission and other offices and divisions regarding problems frequently encountered by investors and possible regulatory solutions to such problems. Phone, 202–551–6551. Consumer information line, 800–732–0330. Fax, 202–772–9295. Complaints and inquiries may also be directed to any regional or district office.

Publications   Blank copies of SEC forms and other publications are available in the Publications Unit. Phone, 202–551–4040.

Reading Rooms   The Commission maintains a public reference room in Washington, DC, where registration statements and other public documents filed with the Commission are available for public inspection Monday through Friday, except on holidays, between the hours of 10:00 a.m. and 3:00 p.m. Phone, 202–551–5850. Copies of public material may be purchased from the Commission's contract copying service at prescribed rates. The Commission also maintains a library where additional information may be obtained. Phone, 202–551–5450. Fax, 202–772–9326.

Small Business Activities   Information on securities laws that pertain to small businesses in relation to securities offerings may be obtained from the Commission. Phone, 202–551–3460.

For further information, contact the Office of Public Affairs, Securities and Exchange Commission, 100 F Street NE., Washington, DC 20549. Phone, 202–551–4120. Fax, 202–777–1026. Internet, http://www.sec. gov.

SELECTIVE SERVICE SYSTEM

National Headquarters, Arlington, VA 22209–2425
Phone, 703–605–4100. Internet, http://www.sss.gov.

Director	LAWRENCE G. ROMO
Deputy Director	EDWARD T. ALLARD, III
Chief of Staff	DEBORAH H. HUBBARD
General Counsel	RUDY G. SANCHEZ, JR.
Associate Director for Public and Intergovernmental Affairs	RICHARD S. FLAHAVAN
Associate Director for Financial Management/ CFO	CARLO VERDINO
Associate Director for Information Management/CIO	JERRY KLOTZ, *Acting*
Associate Director for Operations	ERNEST E. GARCIA
Associate Director for Support Services	EDWARD A. BLACKADAR, JR.

[For the Selective Service System statement of organization, see the Code of Federal Regulations, Title 32, Part 1605]

The Selective Service System provides manpower to the Armed Forces in an emergency and operates an Alternative Service Program during a draft for men classified as conscientious objectors.

The Selective Service System was established by the Military Selective Service Act (50 U.S.C. app. 451–471a). The act requires the registration of male citizens of the United States and all other male persons who are in the United States and who are ages 18 to 25. The act exempts members of the active Armed Forces and nonimmigrant aliens. Proclamation 4771 of July 20, 1980, requires male persons born on or after January 1, 1960, and who have attained age 18 but have not attained age 26 to register. Registration is conducted at post

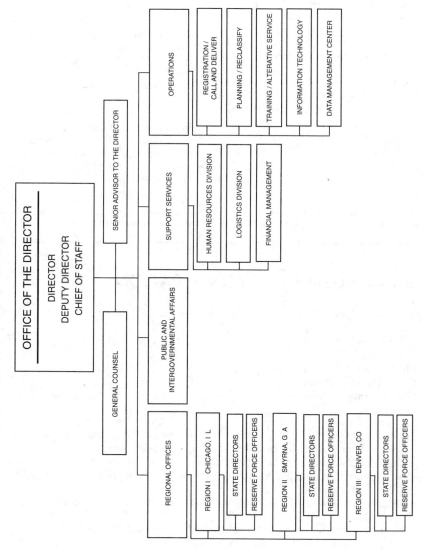

SELECTIVE SERVICE SYSTEM

offices within the United States, at U.S. embassies and consulates outside the United States, and online at www.sss.gov.

The act imposes liability for training and service in the Armed Forces upon registrants who are ages 18 to 26, except those who are exempt or deferred. Persons who have been deferred remain liable for training and service until age 35. Aliens are not liable for training and service until they have remained in the United States for more than 1 year. Conscientious objectors who are found to be opposed to all service in the Armed Forces are required to perform civilian work in lieu of induction into the Armed Forces.

The authority to induct registrants, including doctors and allied medical specialists, expired July 1, 1973.

For a complete listing of the regional offices of the Selective Service System, visit www.sss.gov/regions.

Sources of Information

Employment   Inquiries and applications should be sent to the Director, Selective Service System, Attn: SPT/HR, Arlington, VA 22209–2425. Phone, 703–605–4056.
Procurement   Inquiries should be sent to the Director, Selective Service System, Attn: STP/LO, Arlington, VA 22209–2425. Phone, 703–605–4064.
Publications   Selective Service regulations appear in chapter XVI of title 32 of the Code of Federal Regulations.
Requirements of Law   Persons seeking information concerning the requirements of the Military Selective Service Act should contact the National Headquarters of the Selective Service System. Phone, 703–605–4100.

For further information, contact the Office of Public and Intergovernmental Affairs, Selective Service System, Arlington, VA 22209–2425. Phone, 703–605–4100. Internet, http://www.sss.gov.

SMALL BUSINESS ADMINISTRATION

409 Third Street SW., Washington, DC 20416
Phone, 202–205–6600. Fax, 202–205–7064. Internet, http://www.sba.gov.

Administrator	KAREN MILLS
Deputy Administrator	(VACANCY)
Chief Counsel for Advocacy	SUSAN WALTHALL, *Acting*
Chief Financial Officer	JASON CARVER
Chief Information Officer	PAUL CHRISTY, *Acting*
Chief of Staff	ANA MA
General Counsel	SARA LIPSCOMB
Inspector General	PEGGY GUSTAFSON
Associate Administrator for Disaster Assistance	JAMES RIVERA
Associate Administrator for Field Operations	JESS KNOX
Assistant Administrator for Communications and Public Liaison	JONATHAN SWAIN
Assistant Administrator for Congressional and Legislative Affairs	NICHOLAS COUTSOS
Assistant Administrator for Equal Employment Opportunity and Civil Rights Compliance	MARGARETH BENNETT
Assistant Administrator for Faith Based Community Initiatives	GERALD FLAVIN
Assistant Administrator for Hearings and Appeals	DELORICE FORD
Assistant Administrator for Native American Affairs	CLARA PRATTE

Assistant Administrator for Policy and Strategic Planning	JAMES HAMMERSLEY
Assistant Administrator for Regulatory Enforcement Fairness	ESTHER VASSAR
Assistant Administrator for Veterans Business Development	WILLIAM ELMORE
Associate Administrator for Capital Access	ERIC ZARNIKOW
Director of Credit Risk Management	BRYAN HOOPER
Director of Financial Assistance	GRADY HEDGESPETH
Director of International Trade	LUZ HOPEWELL
Director of Investment	SEAN GREENE
Director of Surety Guarantees	FRANK LALUMIERE
Associate Deputy Administrator for Entrepreneurial Development	PENNY PICKETT
Director of Business and Community Initiatives	ELLEN M. THRASHER
Director of Small Business Development Centers	ANTONIO DOSS
Director of Women's Business Ownership	ANA HARVEY
Associate Administrator for Government Contracting and Business Development	JOSEPH JORDAN
Director of Government Contracting	DEAN KOPPELL
Director of Business Development	(VACANCY)
Director of HUBZone Empowerment Contracting	DEREK BOURCHARD-HALL
Associate Administrator for Management and Administration	DARRYL HAIRSTON
Director of Business Operations	AJOY SINHA, *Acting*
Chief Human Capital Officer	KEVIN MAHONEY
Director of Executive Secretariat	KIM BRADLEY

[For the Small Business Administration statement of organization, see the Code of Federal Regulations, Title 13, Part 101]

The Small Business Administration aids, counsels, assists, and protects the interests of small business; ensures that small-business concerns receive a fair portion of Government purchases, contracts and subcontracts, and sales of Government property; makes loans to small-business concerns, State and local development companies, and the victims of natural disasters or of certain types of economic injury; and licenses, regulates, and makes loans to small-business investment companies.

The Small Business Administration (SBA) was created by the Small Business Act of 1953 and derives its present existence and authority from the Small Business Act (15 U.S.C. 631 et seq.) and the Small Business Investment Act of 1958 (15 U.S.C. 661).

Activities

Advocacy The Office of Advocacy is mandated by Congress to serve as an independent voice within the Federal Government for the approximately 27.2 million small businesses throughout the country. The Office is headed by the Chief Counsel for Advocacy, appointed by the President from the private sector with the advice and consent of the Senate, who advances the views, concerns, and interests of small business before the Congress, the White House, and Federal and State regulatory agencies.

The Office monitors and reports annually on Federal agency compliance with the Regulatory Flexibility Act (RFA), which requires agencies to analyze the impact of their regulations on small businesses and consider less burdensome alternatives. Small entities include small businesses, nonprofit organizations,

and governmental jurisdictions.
Executive Order 13272 requires Federal agencies to take the Office's comments into consideration before proposed regulations are finalized and requires the Office to train Federal agencies on RFA compliance.

The Office is one of the leading national sources for information on the state of small business and the issues that affect small-business success and growth. It conducts economic and statistical research into matters affecting the competitive strength of small business, jobs created by small businesses, and the impact of Federal laws, regulations, and programs on small businesses, making recommendations to policymakers for appropriate adjustments to meet the special needs of small business.

Additionally, regional advocates enhance communication between the small-business community and the Chief Counsel. As the Chief Counsel's direct link to local business owners, State and local government agencies, State legislatures, and small-business organizations, they help identify new issues and problems of small business by monitoring the effect of Federal and State regulations and policies on the local business communities within their regions.

For further information, contact the Office of Advocacy. Phone, 202–205–6533. Email, advocacy@ sba.gov.

Business and Community Initiatives
The Office of Business and Community Initiatives (OBCI) develops and cosponsors counseling, education, training, and information resources for small businesses. It has partnered with the private sector to promote entrepreneurial development. OBCI directs the national program of the Service Corps of Retired Executives (SCORE), a resource partner of SBA. SCORE provides free counseling, mentoring, training seminars, and specialized assistance to veterans and active military personnel. For more information, visit www.score.org. OBCI also offers young entrepreneurs a teen-business site at www.sba.gov/teens.

The Office of International Visitors briefs foreign delegations, business

organizations, and international nongovernmental organizations (NGOs) on the SBA model.

In addition to education and training events, SBA offers an online management series on business growth and sustainability at www.sba.gov/library/pubs.

For further information, contact the Office of Business and Community Initiatives. Phone, 202–205–6665.

Capital Access   The Office of the Associate Administrator for Capital Access provides overall direction for SBA's financial programs. It offers a comprehensive array of debt and equity programs for startup and expanding businesses. In addition to lending to businesses that sell their products and services domestically, the Office provides financial assistance programs for small-business exporters in the form of loan programs and technical assistance. The Office also oversees a surety bond guarantee program for small-business contractors and SBA's lender oversight programs.

For further information, contact the Office of Capital Access. Phone, 202–205–6657.

Disaster Assistance   SBA serves as the Federal disaster bank for nonfarm, private sector losses. It lends money to help the victims of disasters repair or replace most disaster-damaged property. Direct loans with subsidized interest rates are made to assist individuals, homeowners, businesses of all sizes, and nonprofit organizations.

For further information, contact the Office of Disaster Assistance. Phone, 202–205–6734.

Financial Assistance   SBA provides its guarantee to lending institutions and certified development companies that make loans to small-business concerns, which in turn use the loans for working capital and financing the acquisition of land and buildings; the construction, conversion, or expansion of facilities; and the purchase of machinery and equipment.

The Administration also provides small-scale financial and technical assistance to very small businesses through loans and grants to nonprofit organizations that act as intermediaries under SBA's microloan program.

For further information, contact the nearest Small Business Administration district office (see Field Operations below).

Government Contracting SBA helps small businesses, including small disadvantaged businesses, women-owned small businesses, HUBZone-certified firms, and service-disabled veteran-owned small businesses obtain a fair share of Government procurement through a variety of programs and services. The contracting liaison helps small businesses secure an equitable share of natural resources sold by the Federal Government. It works closely with Federal agencies and the Office of Management and Budget to establish policy and regulations concerning small-business access to Government contracts. It assists in the formulation of small-business procurement policies as they relate to size standards, the Small Business Innovation Research Program, and the Small Business Technology Transfer Program.

For further information, contact the nearest Office of Government Contracting. Phone, 202–205–6459. Internet, http://www.sba.gov/GC/indexcontacts. html.

International Trade The Office of International Trade (OIT) supports small-business access to export markets and participates in broader U.S. Government activities related to trade policy and international commercial affairs to encourage an environment of trade and international economic policies favorable to small businesses. These activities are designed to facilitate both entrance and growth into the international marketplace, including educational initiatives, technical assistance programs and services, and risk management and trade finance products.

SBA's export promotion activities for small business combine financial and technical assistance through a nationwide delivery system. Export-finance products include long-term, short-term, and revolving lines of credit through SBA's 7(a) Loan Program, administered by a

staff of field-based export specialists located in U.S. Export Assistance Centers (USEACs). They work with the U.S. Department of Commerce and the Export-Import Bank of the United States, and the effort is leveraged through close collaboration with commercial lenders, Small Business Development Centers, and local business development organizations.

Available financial assistance can provide a business with up to $1.25 million, with terms up to 25 years for real estate and 15 years for equipment. Export Working Capital Program loans generally provide 12 months of renewable financing. For smaller loan amounts, SBA Export Express has a streamlined, quick approval process for businesses needing up to $250,000. Technical assistance includes making available to current and potential small-business exporters export training, export legal assistance, and collaboration with the 30 Small Business Development Centers with international trade expertise and the Government's USA Trade Information Center.

SBA is required to work with the Government's international trade agencies to ensure that small business is adequately represented in bilateral and multilateral trade negotiations. OIT represents SBA and the Government on two official U.S. Government-sponsored multilateral organizations concerned with small business: the Organization for Economic Cooperation and Development and Asia-Pacific Economic Cooperation. SBA's trade policy involvement is carried out with the U.S. Trade Representative and the Commerce Department's International Trade Administration. Private sector input on trade policy is achieved through participation with the small-business Industry Sector Advisory Committee on international trade. OIT also lends support to the Government's key trade initiatives, such as Trade Promotion Authority, the Central American Free Trade Area, and the Free Trade Area of the Americas. The Commerce and State Departments, the Agency for International Development, and the U.S. Trade Representative look to the SBA to share ideas and provide small-business technical expertise to certain countries.

OIT's office in Washington, DC, coordinates SBA's participation/operation of USEACs, including budget, policy, and administration. It participates in a variety of interagency trade efforts and financial programs. OIT provides representations to the Cabinet-level Trade Promotion Coordinating Committee concerning trade and international economic policy. It also participates on the Industry Sector Advisory Council on Small Business International Trade and the congressionally sponsored Task Force on Small Business International Trade. SBA's Administrator is also a sitting member of the President's Export Council.

OIT's field offices provide a nationwide network of service delivery for small-business exporters. Full-time SBA export specialists staff 16 USEACs. Their outreach efforts are supplemented by the 68 SBA district offices staffed by employees with collateral duties as international trade officers.

For further information, contact the Office of International Trade. Phone, 202–205–6720. Internet, http://www.sba.gov/oit.

Venture Capital   The Small Business Investment Company (SBIC) program was created in 1958 to fill the gap between the availability of venture capital and the needs of small businesses in startup and growth situations. The structure of the program is unique in that SBICs are privately owned and managed venture capital funds, licensed and regulated by the SBA, that use their own capital plus funds borrowed with an SBA guarantee to make equity and debt investments in qualifying small businesses. The New Markets Venture Capital (NMVC) program is a sister program focused on low-income areas, which augments the contribution made by SBICs to small businesses in the United States. In addition, NMVC companies may make technical assistance grants to potential portfolio companies.

The Government itself does not make direct investments or target specific industries in the SBIC program. Fund portfolio management and investment decisions are left to qualified private fund managers. To obtain an SBIC license, an experienced team of private equity managers must secure minimum commitments from private investors. SBICs may only invest in small businesses having net worth of less than $18 million and average aftertax income for the previous 2 years of less than $6 million.

For further information, contact the Investment Division. Phone, 202–205–6510. Internet, http://www.sba.gov/inv.

HUBZone Program   The HUBZone Program provides Federal contracting assistance for qualified small businesses located in historically underutilized business zones in an effort to increase employment, capital investment, and economic development in these areas, including Indian reservations. The Office coordinates efforts with other Federal agencies and local municipal governments to leverage resources to assist qualified small businesses located in HUBZone areas. The program provides for set-asides, sole source awards, and price evaluation preferences for HUBZone small businesses and establishes goals for awards to such firms.

For further information, call 202–205–6731. Internet, http://www.sba.gov/hubzone.

Business Development   The Office of Business Development is responsible for the 8(a) Business Development Program. The Office assists small businesses by providing access to capital and credit, business counseling, training workshops, technical guidance, and assistance with contracts and loans. Its primary business development tools are the Mentor-Protege Program and the 7(j) Management and Technical Assistance Program.

For further information, call 202–205–5852. Internet, http://www.sba.gov/8abd.

Native American Affairs   The Office of Native American Affairs was established to assist and encourage the creation, development, and expansion of Native American-owned small businesses by developing and implementing initiatives designed to address those difficulties encountered by Native Americans as they start, develop, and expand small businesses. In addition, in an effort to address the unique conditions encountered by reservation-based entrepreneurs, the Office is developing a Web-based resource entitled the "Tribal Self Assessment Tool." It is intended to allow tribal nations to assess their vision and goals relative to their governance structure, culture, capabilities, and resources. The tool is free and will be available on the Internet.

For further information, contact the Office of Native American Affairs. Phone, 202–205–7364.

Regulatory Fairness Program   Congress established the National Ombudsman and 10 Regulatory Fairness Boards in 1996 as part of the Small Business Regulatory Enforcement Fairness Act (SBREFA). The National Ombudsman's primary mission is to assist small businesses when they experience excessive or unfair federal regulatory enforcement actions, such as repetitive audits or investigations, excessive fines, penalties, threats, retaliation or other unfair enforcement action by a Federal agency. The National Ombudsman receives comments from small-business concerns and acts as a liaison between them and Federal agencies. Comments received from small businesses are forwarded to Federal agencies for review and Federal agencies are requested to consider the fairness of their enforcement action. A copy of the agency's response is sent to the small business owner by the Office of the National Ombudsman. In some cases, fines have been lowered or eliminated and decisions changed in favor of the small-business owner.

Each of the Regulatory Fairness Boards (RegFair) has five volunteer members who are owners, operators, or officers of small-business concerns that are appointed by the SBA Administrator for 3-year terms. Each RegFair Board meets at least annually with the Ombudsman on matters of concern to small businesses relating to the enforcement or compliance activities of Federal agencies; reports to the Ombudsman on substantiated instances of excessive enforcement; and, prior to publication,

provides comment on the annual report to Congress.

For further information, contact the Office of the National Ombudsman. Phone, 202–205–2417 or 888–734–3247. Internet, http://www.sba.gov/ombudsman.

Small Business Development Centers   The Office of Small Business Development Centers (OSBDC) provides counseling and training to existing and prospective small-business owners at more than 950 service locations in every State, Puerto Rico, the U.S. Virgin Islands, Guam, and American Samoa. OSBDC develops national policies and goals, establishes standards for the selection and performance of its Small Business Development Centers (SBDCs), monitors compliance with applicable Office of Management and Budget circulars and laws, and implements new approaches to improve existing centers. OSBDC also oversees 63 lead centers and maintains liaison with other Federal, State, and local agencies and private organizations whose activities relate to its centers. It also assesses how the program is affected by substantive developments and policies in other SBA areas, Government agencies, and the private sector.

The Small Business Development Center Program is a cooperative effort of the private sector, the educational community, and Federal, State, and local governments. The program enhances local economic development by providing small businesses with the management and technical assistance they need to succeed. It also provides services such as development of business plans, manufacturing assistance, financial packages, procurement contracts, and international trade assistance. Special areas include ecommerce; technology transfer; IRS, EPA, and OSHA regulatory compliance; research and development; defense economic transition assistance; disaster recovery assistance; and market research. Based on client need assessments, business trends, and individual business requirements, SBDCs modify their services to meet the evolving needs of the small-business community.

For further information, contact the Office of Small Business Development Centers. Phone, 202–205–6766.

Surety Bonds   Through its Surety Bond Guarantee Program, SBA helps small and emerging contractors to obtain the bonding necessary for them to bid on and receive contracts up to $5 million. SBA guarantees bonds that are issued by participating surety companies and reimburses between 70 percent and 90 percent of losses and expenses incurred should a small business default on the contract. Construction, service, or supply contractors are eligible for the program if they, together with their affiliates, meet the size standard for the primary industry in which the small business is engaged, as defined by the North American Industry Classification System (NAICS).

For further information, contact the Office of Surety Guarantees. Phone, 202–205–6540. Internet, http://www.sba.gov/osg.

Technology   The Office of Technology has authority and responsibility for directing and monitoring the Governmentwide activities of the Small Business Innovation Research Program (SBIR) and the Small Business Technology Transfer Program (STTR). The Office develops and issues policy directives for the general conduct of the programs within the Federal Government and maintains a source file and information program to provide each interested and qualified small-business concern with information on opportunities to compete for SBIR and STTR program awards. The Office also coordinates with each participating Federal agency in developing a master release schedule of all program solicitations; publishes the Presolicitation Announcement quarterly online, which contains pertinent facts on upcoming solicitations; and surveys and monitors program operations within the Federal Government and reports on the progress of the programs each year to Congress.

The Office has four main objectives: to expand and improve SBIR and STTR; to increase private sector commercialization of technology developed through Federal research and development; to increase small-business participation in Federal research and development; and to improve the dissemination of information concerning SBIR and STTR, particularly with regard to participation by women-owned small-business concerns and by socially and economically disadvantaged small-business concerns.

For further information, contact the Office of Technology. Phone, 202–205–6450. Email, technology@sba.gov.

Veterans Affairs   The Office of Veterans Business Development (OVBD) is responsible for the formulation, execution, and promotion of policies and programs that provide assistance to small-business concerns owned and controlled by veterans and service-disabled veterans. This includes reserve component members of the U.S. military. Additionally, OVBD serves as an ombudsman for the full consideration of veterans in all programs of the Administration.

OVBD provides ecounseling and works with every SBA program to ensure that veterans receive special consideration in the operation of that program. OVBD also provides numerous tools, such as the Vet Gazette newsletter, Reserve and Guard business assistance kits, program design assistance, and training events. Additionally, OVBD manages five Veterans Business Outreach Centers to provide outreach, directed referrals, and tailored entrepreneurial development services such as business training, counseling, and mentoring to veterans, including service-disabled veterans, and reservists. These Centers provide an in-depth resource for existing and potential veteran entrepreneurs. The Office also coordinates SBA collaborative efforts with veterans service organizations; the Departments of Defense, Labor, and Veterans Affairs; the National Veterans Business Development Corporation; State departments of veterans affairs; the National Committee for Employer Support of the Guard and Reserve; the Department of Defense Yellow Ribbon Reintegration Program; and other public, civic, and private organizations to ensure that the entrepreneurial needs of veterans, service-disabled veterans, and self-employed members of the Reserve and National Guard are being met.

For further information, contact the Office of Veterans Business Development. Phone, 202–205–6773. Internet, http://www.sba.gov/vets.

Women's Business Ownership

The Office of Women's Business Ownership (OWBO) provides assistance to current and potential women business owners and acts as their advocate in the public and private sectors. OWBO assists women in becoming full partners in economic development by providing business training, counseling, mentoring, and other assistance through representatives in local SBA offices, Women's Business Centers (WBCs), and mentoring roundtables. Each WBC is tailored to meet the needs of its individual community and places a special emphasis on helping women who are socially and economically disadvantaged. Assistance covers every stage of business, from startup to going public. There are WBCs in almost every State and U.S. Territory.

OWBO works with other SBA programs, Federal agencies, and private sector organizations to leverage its resources and improve opportunities for women-owned businesses to access Federal procurement and international trade opportunities. OWBO also works with the National Women's Business Council and the Department of Labor to maintain the most current research on women's business ownership.

SBA has loan guaranty programs to help women access the credit and capital they need to start and grow successful businesses. The 7(a) Loan Guaranty Program offers a number of effective ways to finance business needs, including unsecured smaller loans and revolving lines of credit. The 504 Program provides long-term, fixed-rate financing for major fixed assets, such as land and buildings, through certified development programs. Equity financing is available through the Small Business Investment Company Program. The Microloan Program offers direct small loans, combined with business assistance, through SBA-licensed intermediaries nationwide. The SBA does not offer grants for small businesses.

For further information, contact the Women's Business Ownership representative in your SBA district office. Phone, 202–205–6673. Email, owbo@sba.gov. Internet, http://www.sba.gov/aboutsba/sbaprograms/onlinewbc/index.html.

Field Operations

The Office of Field Operations provides management direction and oversight to SBA's 10 regional and 68 district offices, acting as the liaison between the district offices, the Administration's program delivery system, and the headquarters administrative and program offices.

For a complete listing of the regional, district, and disaster field offices of the SBA, including addresses, telephone numbers, and key officials, visit www.sba.gov/localresources/index.html.

For further information, contact the Office of Field Operations. Phone, 202–205–6808.

Sources of Information

Electronic Access Information on the Small Business Administration is available electronically by various means. Internet, www.sba.gov. FTP, ftp.sbaonline.sba.gov. Access the U.S. Business Adviser through the Internet at www.business.gov. Access the Administration's electronic bulletin board by modem at 800–697–4636 (limited access), 900–463–4636 (full access), or 202–401–9600 (Washington, DC, metropolitan area).

General Information Contact the nearest Small Business Administration field office listed in the preceding text, or call the SBA answer desk. Phone, 800–827–5722. Fax, 202–205–7064. TDD, 704–344–6640.

Public Affairs For public inquiries and small-business advocacy affairs, contact the Office of Public Communications and Public Liaison, 409 Third Street SW., Washington, DC 20416. Phone, 202–205–6740. Internet, www.sba.gov.

Publications A free copy of The Resource Directory for Small Business Management, a listing of for-sale publications and videotapes, is available from any local SBA office or the SBA answer desk.

For further information, contact the Office of Public Communications and Public Liaison, Small Business Administration, 409 Third Street SW., Washington, DC 20416. Phone, 202–205–6740. Internet, http://www.sba.gov.

SOCIAL SECURITY ADMINISTRATION

6401 Security Boulevard, Baltimore, MD 21235
Phone, 410–965–1234. Internet, http://www.socialsecurity.gov.

Commissioner	MICHAEL J. ASTRUE
Deputy Commissioner	(VACANCY)
Executive Secretary	ROBIN F. KAPLAN
International Programs	DIANE K. BRAUNSTEIN
Chief Actuary	STEPHEN C. GOSS
Chief Information Officer	FRANKLIN H. BAITMAN
Deputy Commissioner for Communications	JAMES J. COURTNEY
Deputy Commissioner for Budget, Finance, and Management	MICHAEL G. GALLAGHER
Deputy Commissioner for Disability Adjudication and Review	GLENN E. SKLAR
Deputy Commissioner for Human Resources	REGINALD F. WELLS
Deputy Commissioner for Legislative and Regulatory Affairs	SCOTT L. FREY
Deputy Commissioner for Operations	MARY E. GLENN-CROFT
Deputy Commissioner for Quality Performance	RONALD T. RABORG
Deputy Commissioner for Retirement and Disability Policy	DAVID A. RUST
Deputy Commissioner for Systems	G. KELLY CROFT
General Counsel	DAVID F. BLACK
Inspector General	PATRICK P. O'CARROLL

[For the Social Security Administration statement of organization, see the Code of Federal Regulations, Title 20, Part 422]

The Social Security Administration manages the retirement, survivors, and disability insurance programs commonly known as Social Security; administers the Supplemental Security Income program for the aged, blind, and disabled; assigns Social Security numbers to U.S. citizens; and maintains earnings records for workers under their Social Security numbers.

The Social Security Administration (SSA) was established by Reorganization Plan No. 2 of 1946 (5 U.S.C. app.), effective July 16, 1946. It became an independent agency in the executive branch by the Social Security Independence and Program Improvements Act of 1994 (42 U.S.C. 901), effective March 31, 1995.

The Administration is headed by a Commissioner, appointed by the President with the advice and consent of the Senate.

In administering the programs necessary to carry out the Administration's mission, by law the Commissioner is assisted by a Deputy Commissioner who performs duties assigned or delegated by the Commissioner, a Chief Financial Officer, a Chief Information Officer, a General Counsel, a Chief Actuary, and an Inspector General.

Programs and Activities

Old-Age, Survivors, and Disability Insurance   The agency administers social insurance programs that provide monthly benefits to retired and disabled workers, their spouses and children, and survivors of insured workers. Financing is under a system of contributory social insurance, whereby employees, employers, and the self-employed pay contributions that are pooled in special trust funds. When earnings stop or are reduced because the worker retires, dies, or becomes disabled, monthly cash benefits are paid to partially replace the earnings the family has lost.

Supplemental Security Income The agency administers this needs-based program for the aged, blind, and disabled. A basic Federal monthly payment is financed out of general revenue, rather than a special trust fund. Some States, choosing to provide payments to supplement the benefits, have agreements with the Administration under which it administers the supplemental payments for those States.

Medicare While the administration of Medicare is the responsibility of the Centers for Medicare and Medicaid Services, SSA provides Medicare assistance to the public through SSA field offices and call centers and adjudicates requests for hearings and appeals of Medicare claims.

Black Lung By agreement with the Department of Labor, SSA is involved in certain aspects of the administration of the black lung benefits provisions of the Federal Coal Mine Health and Safety Act of 1969, as amended (30 U.S.C. 901).

Regional Offices Social Security Administration operations are decentralized to provide services at the local level. Each of the 10 SSA regions, under the overall direction of its Regional Commissioner, contains a network of field offices and call centers, which serve as the contacts between SSA and the public. The Administration operates approximately 1300 field offices, 37 call centers, 7 Social Security card centers, and 7 processing centers. These installations are responsible for informing the public of the purposes and provisions of Social Security programs and their rights and responsibilities; assisting with claims filed for retirement, survivors, disability, or health insurance benefits, black lung benefits, or Supplemental Security Income; developing and adjudicating claims; assisting certain beneficiaries in claiming reimbursement for medical expenses; developing cases involving earnings records, coverage, and fraud-related questions; making rehabilitation service referrals; and assisting claimants in filing appeals on SSA determinations of benefit entitlement or amount.

Hearing Offices SSA also administers a nationwide hearings and appeals program which provides a mechanism for individuals dissatisfied with determinations affecting their rights to and amounts of benefits or their participation in programs under the Social Security Act. The act allows for administrative appeals of these determinations in accordance with the requirements of the Administrative Procedure and Social Security Acts. SSA has approximately 140 hearing offices located in the 10 SSA regions.

For further information, contact the Social Security Administration. Phone, 800–772–1213. TTY, 800–325–0778.

Sources of Information

Inquiries on the following subjects may be directed to the appropriate office, Social Security Administration, 6401 Security Boulevard, Baltimore, MD 21235.

Contracts and Small Business Activities Contact the Office of Acquisitions and Grants. Phone, 410–965–7467.

Electronic Access Information regarding the Social Security Administration may be obtained through the Internet at www.socialsecurity.gov.

Employment For information about careers with the Social Security Administration, go to www.socialsecurity.gov/careers. For current vacancies, go to www.usajobs.opm.gov.

General Information The Office of the Deputy Commissioner for Operations manages SSA's toll-free public service telephone. Phone, 800–772–1213. TTY, 800–325–0778.

Inspector General The Office of the Inspector General maintains a toll-free hotline that operates between the hours of 10 a.m. and 4 p.m. eastern standard time (phone, 800–269–0271; TTY, 866–501–2101) to receive allegations of fraud. Persons may submit allegations online at www.socialsecurity.gov/oig, by fax at 410–597–0118, or by mail at Social Security Fraud Hotline, P.O. Box 17768, Baltimore, MD 21235–7768.

Publications The Office of the Deputy Commissioner for Communications publishes numerous pamphlets

SOCIAL SECURITY ADMINISTRATION

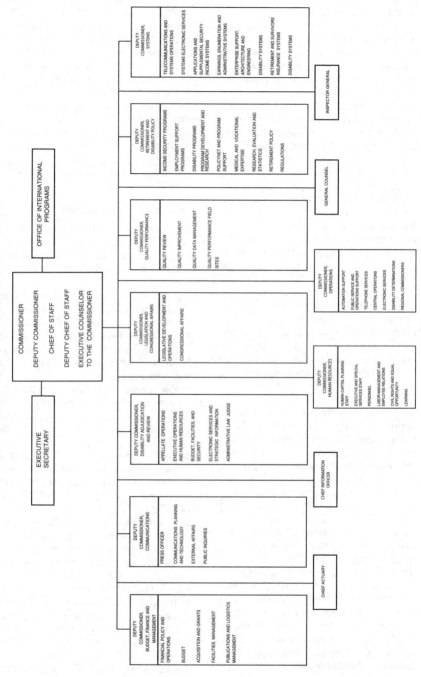

concerning SSA programs. SSA also collects a substantial volume of economic, demographic, and other data in furtherance of its program mission. Basic data on employment, payments, and other items of program interest are published regularly in the Social Security Bulletin, its Annual Statistical Supplement, and in special releases and reports that appear periodically on selected topics of interest to the public. Single copies may be obtained at any local office or by calling 800–772–1213. Requests for bulk orders of publications should be sent to the Social Security Administration, Office of Supply and Warehouse Management, 239 Supply Building, 6301 Security Boulevard, Baltimore, MD 21235. In addition, over 150 publications are available online in English, Spanish, and other languages at www.socialsecurity.gov.

Reading Room   Requests for information, for copies of records, or to inspect records may be made at any local office or the Headquarters Contact Station, Room G–44, Altmeyer Building.

Speakers and Films   SSA makes speakers, films, and exhibits available to public or private organizations, community groups, schools, etc., throughout the Nation. Requests for this service should be directed to the local Social Security office.

For further information, contact the Office of Public Inquiries, Social Security Administration, 6401 Security Boulevard, Windsor Park Building, Baltimore, MD 21235. Phone, 800–772–1213. TTY, 800–325–0778. Internet, http://www.socialsecurity.gov.

TENNESSEE VALLEY AUTHORITY

400 West Summit Hill Drive, Knoxville, TN 37902
Phone, 865–632–2101. Internet, http://www.tva.com.

Chairman	DENNIS BOTTORFF
Directors	ROBERT M. DUNCAN, THOMAS GILLILAND, WILLIAM GRAVES, HOWARD A. THRAILKILL, (4 VACANCIES)
President and Chief Executive Officer	TOM D. KILGORE
Group President of Strategy and External Relations	KIMBERLY S. GREENE
Senior Vice President, Communications, Government and Valley Relations	EMILY J. REYNOLDS
Chief Operating Officer	WILLIAM R. MCCOLLUM, JR.
Chief Nuclear Officer and Executive Vice President, TVA Nuclear	PRESTON D. SWAFFORD
Chief Financial Officer	JOHN M. THOMAS, III

The Tennessee Valley Authority conducts a unified program of resource development for the advancement of economic growth in the Tennessee Valley region.

The Tennessee Valley Authority (TVA) is a wholly owned Government corporation created by the act of May 18, 1933 (16 U.S.C. 831–831dd). All functions of the Authority are vested in its nine-member Board of Directors, the members of which are appointed by the President with the advice and consent of the Senate. The Board designates one member as Chairman.

Programs and Activities

TVA's programs and activities include flood control, navigation, electric power production and transmission, recreation improvement, water supply and water quality management, environmental stewardship, and economic development.

TVA's electric power program is financially self-supporting and operates as part of an independent system with TVA's system of dams on the Tennessee River and its larger tributaries. These dams provide flood regulation on the Tennessee River and contribute to regulation of the lower Ohio and Mississippi Rivers. The system maintains a continuous 9-foot-draft navigation channel for the length of the 650-mile Tennessee River main stream, from Paducah, KY, to Knoxville, TN. The dams harness the power of the rivers to produce electricity. They also provide other benefits, notably outdoor recreation and water supply.

TVA operates the river management system and provides assistance to State and local governments in reducing local flood problems. It also works with other agencies to encourage full and effective use of the navigable waterway by industry and commerce.

TVA is the wholesale power supplier for 155 local municipal and cooperative electric systems serving customers in parts of 7 States. It supplies power to 56 industries and Federal installations whose power requirements are large or unusual. Power to meet these demands is supplied from dams, coal-fired powerplants, nuclear powerplants, combustion turbine and diesel installations, solar energy sites, wind turbines, a methane gas facility, and a pumped-storage hydroelectric plant; U.S. Corps of Engineers dams in the Cumberland Valley; and Aluminum Company of America dams, whose operation is coordinated with TVA's system.

Economic development is at the heart of TVA's mission of making the Tennessee Valley a better place to live. A healthy economy means quality jobs, more investment in the region, sustainable growth, and opportunities for residents in the southeastern region to build more prosperous lives. TVA Economic Development takes a regional approach to economic growth by partnering with power distributors and both public and private organizations to attract new investments and quality jobs, supporting retention and growth of existing businesses and industries, preparing communities for leadership and economic growth, and providing financial and technical services.

Sources of Information

Citizen Participation   TVA Communications, 400 West Summit Hill Drive, Knoxville, TN 37902–1499. Phone, 865–632–2101.

Contracts   Purchasing, WT 3A, 400 West Summit Hill Drive, Knoxville, TN 37902–1499. Phone, 865–632–4796. This office will direct inquiries to the appropriate procurement officer.

Economic Development   OCP 2A–NST, One Century Place, 26 Century Boulevard, Suite 100, Nashville, TN 37214. Mailing address: P.O. Box 292409, Nashville, TN 37229–2409. Phone, 615–232–6051.

Electric Power Supply   1101 Market Street, Chattanooga, TN 37402. Phone, 423–751–6000.

Electric Rates   One Century Plaza, 26 Century Boulevard, Suite 100, Nashville, TN 37214–3685.

Employment   For employment inquiries, visit www.tva.gov.

Library Services   TVA Research Library, 400 West Summit Hill Drive, Knoxville, TN 37902–1499. Phone, 865–632–3464. Chattanooga Office Complex, LP4A–C, 1101 Market Street, Chattanooga, TN 37402–2791. Phone, 423–751–4913. P.O. Box 1010, CTR 1E–M, Muscle Shoals, AL 35662. Phone, 256–386–2872.

Maps   Maps Information and Photo Records, HV 1C–C, 2837 Hickory Valley Road, Chattanooga, TN 37421. Phone, 423–499–6285 or 800–627–7882.

Publications   TVA Communications, WT 7D, 400 West Summit Hill Drive, Knoxville, TN 37902–1499. Phone, 865–632–6000.

For further information, contact the Tennessee Valley Authority at either 400 West Summit Hill Drive, Knoxville, TN 37902–1499, phone, 865–632–3199 or One Massachusetts Avenue NW., Washington, DC 20044, phone, 202–898–2999. Internet, http://www.tva.gov.

TRADE AND DEVELOPMENT AGENCY

1000 Wilson Boulevard, Suite 1600, Arlington, VA 22209–3901
Phone, 703–875–4357. Fax, 703–875–4009. Internet, http://www.ustda.gov. Email, info@ustda.
gov.

Director	LEOCADIA I. ZAK
General Counsel	JAMES A. WILDEROTTER
Chief of Staff	CHRIS H. WYANT
Director for Policy and Programs	GEOFFREY JACKSON
Resource Advisor	MICHAEL HILLIER
Director, Congressional Affairs and Public Relations	THOMAS R. HARDY
Economist/Evaluation Officer	DAVID DENNY
Financial Manager	NOREEN ST. LOUIS
Contracts Manager	WALTER KNOTT
Administrative Officer	CAROLYN HUM
Grants Administrator	PATRICIA DAUGHETEE
Regional Directors	
East Asia	GEOFFREY JACKSON
Europe and Eurasia	DAVID HESTER, *Acting*
Latin America and Caribbean	NATHAN YOUNGE
Middle East and North Africa	CARL B. KRESS
South and Southeast Asia	HENRY D. STEINGASS
Sub-Saharan Africa	PAUL MARIN

The Trade and Development Agency advances economic development and U.S.
commercial interest in developing and middle-income countries.

The Trade and Development Program was established on July 1, 1980, as a component organization of the International Development Cooperation Agency. Section 2204 of the Omnibus Trade and Competitiveness Act of 1988 (22 U.S.C. 2421) made it a separate component agency. The organization was renamed the Trade and Development Agency (USTDA) and made an independent agency within the executive branch of the Federal Government on October 28, 1992, by the Jobs Through Exports Act of 1992 (22 U.S.C. 2421).

USTDA is a foreign assistance agency that delivers its program commitments through overseas grants, contracts with U.S. firms, and the use of trust funds at several multilateral development bank groups. The projects supported by USTDA activities represent strong and measurable development priorities in host countries and offer opportunities for commercial participation by U.S. firms. Public and private sector project sponsors in developing and middle-income

countries request USTDA support to assist them in implementing their development priorities.

USTDA helps countries establish a favorable trading environment and a modern infrastructure that promotes sustainable economic development. To this end, USTDA funds overseas projects and sponsors access to U.S. private sector expertise in project definition and investment analysis and in trade capacity building and sector development. Project definition and investment analysis involves activities that support large capital investments that contribute to overseas infrastructure development. Trade capacity building and sector development supports the establishment of industry standards, rules and regulations, trade agreements, market liberalization, and other policy reform.

USTDA works with other U.S. Government agencies to bring their particular expertise and resources to a development objective. These agencies include the Departments of State, the

TRADE AND DEVELOPMENT AGENCY

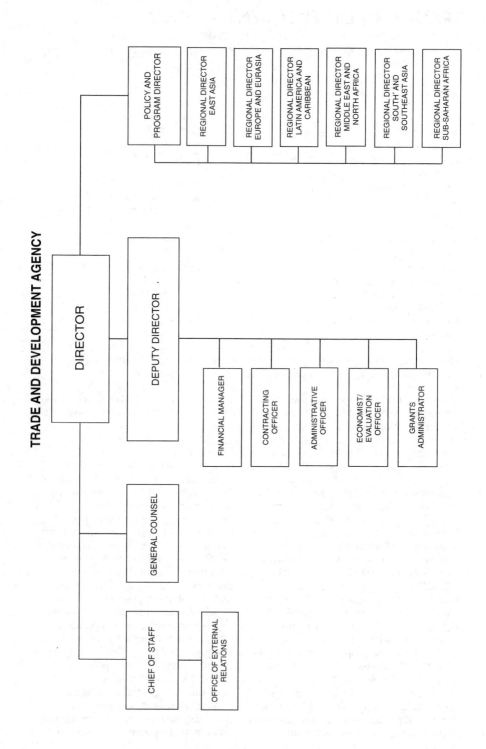

DIRECTOR

CHIEF OF STAFF

OFFICE OF EXTERNAL RELATIONS

GENERAL COUNSEL

DEPUTY DIRECTOR

FINANCIAL MANAGER

CONTRACTING OFFICER

ADMINISTRATIVE OFFICER

ECONOMIST/ EVALUATION OFFICER

GRANTS ADMINISTRATOR

POLICY AND PROGRAM DIRECTOR

REGIONAL DIRECTOR EAST ASIA

REGIONAL DIRECTOR EUROPE AND EURASIA

REGIONAL DIRECTOR LATIN AMERICA AND CARIBBEAN

REGIONAL DIRECTOR MIDDLE EAST AND NORTH AFRICA

REGIONAL DIRECTOR SOUTH' AND SOUTHEAST ASIA

REGIONAL DIRECTOR SUB-SAHARAN AFRICA

Treasury, Commerce, Transportation, Energy, Agriculture, and Homeland Security; the Office of the U.S. Trade Representative; the Export-Import Bank of the United States; and the Overseas Private Investment Corporation.

Activities USTDA funds various forms of technical assistance, training, early investment analysis, orientation visits, and business workshops that support the development of a modern infrastructure and a fair and open trading environment. Working closely with a foreign project sponsor, USTDA makes its funds available on the condition that the foreign entity contract with a U.S. firm to perform the activity funded. This affords American firms market entry, exposure, and information, thus helping them to establish a position in markets that are otherwise difficult to penetrate. USTDA is involved in many sectors, including transportation, energy, telecommunications, environment, health care, mining and minerals development, biotechnology, and agriculture.

USTDA-funded studies evaluate the technical, economic, and financial aspects of a development project. They also advise the host nation about the availability of U.S. goods and services and can be used by financial institutions in assessing the creditworthiness of an undertaking. Grants are based on an official request for assistance made by the sponsoring government or private sector organization of a developing or middle-income nation. Study costs typically are shared by USTDA and the U.S. firm developing the project.

The Agency makes decisions on funding requests based on the recommendations contained in definitional mission or desk study reports, the advice of the U.S. Embassy, and its own internal analysis.

Sources of Information

Requests for proposals to conduct USTDA-funded technical assistance and feasibility studies or definitional missions involving review of projects under consideration for USTDA support are listed on the Federal Business Opportunities (FBO) Web site. Links to the FBO postings can be found at www.ustda.gov.

Small and minority U.S. firms that wish to be considered for future USTDA desk study solicitations should register with the Agency's online Consultant Database at www.ustda.gov/consultantdb.

In an effort to provide timely information on Agency-supported activities, USTDA sends out an electronic newsletter with current business opportunities and a calendar of events on a biweekly basis. A free email subscription is available at www.ustda.gov. The Agency's printed newsletter, USTDA Update, contains current items of interest on a variety of program activities. Region- or sector-specific factsheets and case studies are also available. An annual report summarizes the Agency's activities.

Agency news, reports, and lists of current business opportunities and upcoming events are available at www.ustda.gov.

USTDA's library maintains final reports on the Agency's activities. The reports are available for public review Monday through Friday, from 8:30 a.m. to 5:30 p.m. Copies of completed studies may be purchased through the Department of Commerce's National Technical Information Service at www.ntis.gov.

Regional program inquiries should be directed to the assigned Country Manager. Phone, 703–875–4357. Fax, 703–875–4009. Email, info@ustda.gov.

For further information, contact the Trade and Development Agency, Suite 1600, 1000 Wilson Boulevard, Arlington, VA 22209–3901. Phone, 703–875–4357. Fax, 703–875–4009. Email, info@ustda.gov. Internet, http://www.ustda.gov.

UNITED STATES AGENCY FOR INTERNATIONAL DEVELOPMENT

1300 Pennsylvania Avenue NW., Washington, DC 20523
Phone, 202–712–0000. Internet, http://www.usaid.gov.

Administrator	RAJIV SHAH
Deputy Administrator	(VACANCY)
Counselor	JAMES MICHEL
Chief Operating Officer	ALONZO FULGHAM
Assistant Administrator for Africa	EARL GAST
Assistant Administrator for Asia	MARGOT ELLIS, *Acting*
Assistant Administrator for Middle East	GEORGE LAUDATO, *Acting*
Assistant Administrator for Democracy, Conflict and Humanitarian Assistance	SHARON L. CROMER, *Acting*
Assistant Administrator for Economic Growth, Agriculture and Trade	MICHAEL YATES, *Acting*
Assistant Administrator for Europe and Eurasia	KEN YAMASHITA, *Acting*
Assistant Administrator for Global Health	GLORIA STEEL, *Acting*
Assistant Administrator for Latin America and the Caribbean	MARK FEINSTEIN
Assistant Administrator for Legislative and Public Affairs	CHRIS MILLIGAN, *Acting*
Assistant Administrator for Management	DREW LUTEN, *Acting*
Director of Office of Development Partners	KAREN TURNER
Director of Security	RANDY STREUFERT
Director of Equal Opportunity Programs	JESSALYN L. PENDARVIS
Director of Small and Disadvantaged Business Utilization/Minority Resource Center	MAURICIO VERA
General Counsel	LISA GOMER
Inspector General	DONALD A. GAMBATESA

[For the Agency for International Development statement of organization, see the Federal Register of Aug. 26, 1987, 52 FR 32174]

The United States Agency for International Development administers U.S. foreign economic and humanitarian assistance programs worldwide in the developing world, Central and Eastern Europe, and Eurasia.

The United States Agency for International Development (USAID) is an independent Federal agency established by 22 U.S.C. 6563. Its principal statutory authority is the Foreign Assistance Act of 1961, as amended (22 U.S.C. 2151 et seq.). USAID serves as the focal point within the Government for economic matters affecting U.S. relations with developing countries. USAID administers international economic and humanitarian assistance programs. The Administrator is under the direct authority and foreign policy guidance of the Secretary of State.

Programs

The Agency meets its post-cold war era challenges by utilizing its strategy for achieving sustainable development in developing countries. It supports programs in four areas: population and health, broad-based economic growth, environment, and democracy. It also provides humanitarian assistance and aid to countries in crisis and transition.

Population and Health The Agency contributes to a cooperative global effort to stabilize world population growth and support women's reproductive rights. The types of population and health programs supported vary with the particular needs of individual countries and the kinds of approaches that local communities initiate and support. Most USAID resources are directed to the following areas: support for voluntary family planning systems, reproductive health care, needs of adolescents and young adults, infant and child health, and education for girls and women.

Economic Growth   The Agency promotes broad-based economic growth by addressing the factors that enhance the capacity for growth and by working to remove the obstacles that stand in the way of individual opportunity. In this context, programs concentrate on strengthening market economies, expanding economic opportunities for the disadvantaged in developing countries, and building human skills and capacities to facilitate broad-based participation.

Environment   The Agency's environmental programs support two strategic goals: reducing long-term threats to the global environment, particularly loss of biodiversity and climate change; and promoting sustainable economic growth locally, nationally, and regionally by addressing environmental, economic, and developmental practices that impede development and are unsustainable. Globally, Agency programs focus on reducing sources and enhancing sinks of greenhouse gas emissions and on promoting innovative approaches to the conservation and sustainable use of the planet's biological diversity. The approach to national environmental problems differs on a country-by-country basis, depending on a particular country's environmental priorities. Country strategies may include improving agricultural, industrial, and natural resource management practices that play a central role in environmental degradation; strengthening public policies and institutions to protect the environment; holding dialogues with country governments on environmental issues and with international agencies on the environmental impact of lending practices and the design . and implementation of innovative mechanisms to support environmental

work; and environmental research and education.

Democracy   The Agency's strategic objective in the democracy area is the transition to and consolidation of democratic regimes throughout the world. Programs focus on such problems as human rights abuses; misperceptions about democracy and free-market capitalism; lack of experience with democratic institutions; the absence or weakness of intermediary organizations; nonexistent, ineffectual, or undemocratic political parties; disenfranchisement of women, indigenous peoples, and minorities; failure to implement national charter documents; powerless or poorly defined democratic institutions; tainted elections; and the inability to resolve conflicts peacefully.

Humanitarian Assistance and Post-Crisis Transitions   The Agency provides humanitarian assistance that saves lives, reduces suffering, helps victims return to self-sufficiency, and reinforces democracy. Programs focus on disaster prevention, preparedness, and mitigation; timely delivery of disaster relief and short-term rehabilitation supplies and services; preservation of basic institutions of civil governance during disaster crisis; support for democratic institutions during periods of national transition; and building and reinforcement of local capacity to anticipate and handle disasters and their aftermath.

Overseas Organizations   U.S. Agency for International Development country organizations are located in countries where a bilateral program is being implemented. The in-country organizations are subject to the direction and guidance of the chief U.S. diplomatic representative in the country, usually the Ambassador. The organizations report to the Agency's Assistant Administrators for the four geographic bureaus: the Bureaus for Africa, Asia and Near East, Europe and the New Independent States, and Latin America and the Caribbean.

The overseas program activities that involve more than one country are administered by regional offices. These offices may also perform country organizational responsibilities for

UNITED STATES AGENCY FOR INTERNATIONAL DEVELOPMENT

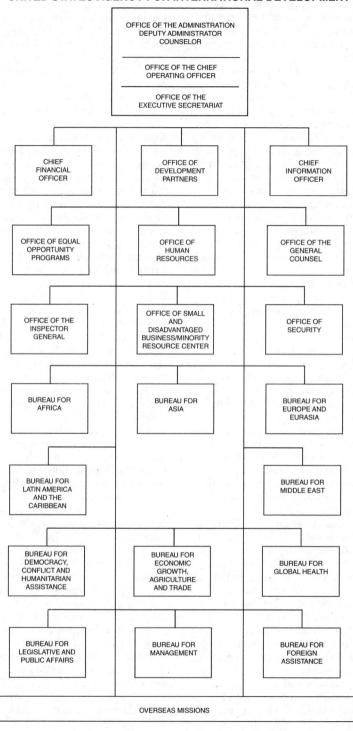

assigned countries. Generally, the offices are headed by a regional development officer.

Development Assistance Coordination and Representative Offices provide liaison with various international organizations and represent U.S. interests in development assistance matters. Such offices may be only partially staffed by Agency personnel and may be headed by employees of other U.S. Government agencies.

For a complete listing of USAID overseas missions and links to mission Web sites, visit www.usaid.gov/locations/missiondirectory.html.

Sources of Information

General Inquiries Inquiries may be directed to the Bureau for Legislative and Public Affairs, USAID/LPA, Washington, DC 20523–0001. Phone, 202–712–4810. Fax, 202–216–3524.

Congressional Affairs Congressional inquiries may be directed to the Bureau for Legislative and Public Affairs, USAID/LPA, Washington, DC 20523–0001. Phone, 202–712–4810.

Contracting and Small Business Inquiries For information regarding contracting opportunities, contact the Office of Small and Disadvantaged Business Utilization, U.S. Agency for International Development, Washington, DC 20523–0001. Phone, 202–712–1500. Fax, 202–216–3056.

Employment For information regarding employment opportunities, contact the Workforce Planning, Recruitment, and Personnel Systems Division, Office of Human Resources, U.S. Agency for International Development, Washington, DC 20523–0001. Internet, www.usaid.gov.

News Media Inquiries from the media only should be directed to the Press Relations Division, Bureau for Legislative and Public Affairs, USAID/LPA, Washington, DC 20523–0001. Phone, 202–712–4320.

For further information, contact the United States Agency for International Development, 1300 Pennsylvania Avenue NW., Washington, DC 20523–0001. Phone, 202–712–0000. Internet, http://www.usaid.gov.

UNITED STATES COMMISSION ON CIVIL RIGHTS

624 Ninth Street NW., Washington, DC 20425
Phone, 202–376–8177. TTY, 202–376–8116. Internet, http://www.usccr.gov.

Chairman	GERALD A. REYNOLDS
Vice Chair	ABIGAIL THERNSTROM
Commissioners	TODD GAZIANO, GAIL HERIOT, PETER KIRSANOW, ARLAN D. MELENDEZ, ASHLEY TAYLOR, MICHAEL YAKI
Staff Director	MARTIN DANNENFELSER
Deputy Staff Director	(VACANCY)
Associate Deputy Staff Director	DEBRA A. CARR
General Counsel	DAVID BLACKWOOD
Assistant Staff Director for Civil Rights Evaluation	CHRISTOPHER BYRNES, *Acting*
Assistant Staff Director for Congressional Affairs	(VACANCY)
Assistant Staff Director for Management	TINALOUISE MARTIN
Chief, Public Affairs Unit	LENORE OSTROWSKY, *Acting*
Chief, Regional Programs Coordination	PETER MINARIK, *Acting*

[For the Commission on Civil Rights statement of organization, see the Code of Federal Regulations, Title 45, Part 701]

UNITED STATES COMMISSION ON CIVIL RIGHTS

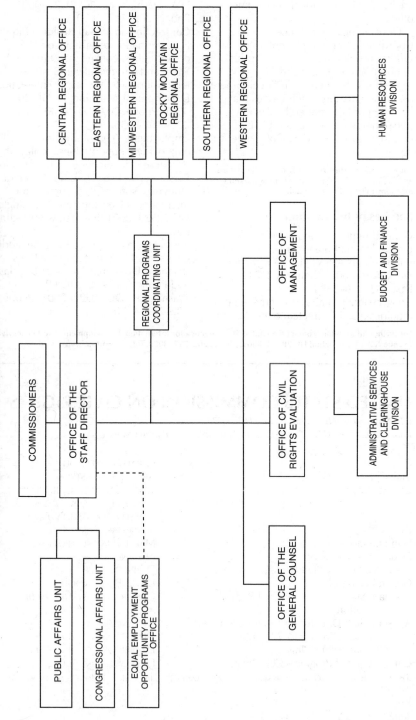

The Commission on Civil Rights collects and studies information on discrimination or denials of equal protection of the laws because of race, color, religion, sex, age, disability, national origin, or in the administration of justice in such areas as voting rights, enforcement of Federal civil rights laws, and equal opportunity in education, employment, and housing.

The Commission on Civil Rights was first created by the Civil Rights Act of 1957, as amended, and reestablished by the United States Commission on Civil Rights Act of 1994, as amended (42 U.S.C. 1975).

Activities

The Commission makes findings of fact but has no enforcement authority. Findings and recommendations are submitted to the President and Congress, and many of the Commission's recommendations have been enacted, either by statute, Executive order, or regulation. The Commission evaluates Federal laws and the effectiveness of Government equal opportunity programs. It also serves as a national clearinghouse for civil rights information.

Regional Programs   The Commission maintains six regional divisions. For a complete listing of the regional divisions of the U.S. Commission on Civil Rights, including addresses, telephone numbers, and areas served visit www.usccr.gov/regofc/rondx.htm.

Sources of Information

Complaints   Complaints alleging denials of civil rights may be reported to Complaints Referral, 624 Ninth Street NW., Washington, DC 20425. Phone, 202–376–8513 or 800–552–6843. Internet, www.usccr.gov.

Employment   Human Resources Office, Room 510, 624 Ninth Street NW., Washington, DC 20425. Phone, 202–376–8364.

Publications   Commission publications are made available upon request from the Administrative Services and Clearinghouse Division, Room 550, 624 Ninth Street NW., Washington, DC 20425. Phone, 202–376–8105. A catalog of publications may be obtained from this office.

Reading Room   The National Civil Rights Clearinghouse Library is located in Room 602, 624 Ninth Street NW., Washington, DC 20425. Phone, 202–376–8110.

For further information, contact the Office of the Staff Director, United States Commission on Civil Rights, Room 730, 624 Ninth Street NW., Washington, DC 20425. Phone, 202–376–7700. TTY, 202–376–8116. Internet, http://www.usccr.gov.

UNITED STATES INTERNATIONAL TRADE COMMISSION

500 E Street SW., Washington, DC 20436
Phone, 202–205–2000. Internet, http://www.usitc.gov.

Chairman	SHARA L. ARANOFF
Vice Chairman	DANIEL R. PEARSON
Commissioners	CHARLOTTE R. LANE, DEANNA TANNER OKUN, DEAN A. PINKERT, IRVING A. WILLIAMSON
Chief Administrative Law Judge	PAUL J. LUCKERN
Director of Operations	(VACANCY)
Director, Office of Economics	ROBERT B. KOOPMAN
Director, Office of Industries	KAREN LANEY
Director, Office of Investigations	CATHERINE B. DEFILIPPO

Director, Office of Tariff Affairs and Trade Agreements	DAVID BECK
Director, Office of Unfair Import Investigations	LYNN LEVINE
General Counsel	JAMES M. LYONS
Director, Office of External Relations	LYN M. SCHLITT
Chief Information Officer	STEPHEN MCLAUGHLIN
Director, Office of Information Technology Services	ROBERT N. RIESS
Director of Administration	STEPHEN MCLAUGHLIN
Director, Office of Human Resources	CYNTHIA A. ROSCOE
Director, Office of Facilities Management	JONATHAN BROWN
Director, Office of Finance	PATRICIA KATSOUROS
Secretary	MARILYN R. ABBOTT
Inspector General	PHILIP M. HENEGHAN
Director, Office of Equal Employment Opportunity	JACQUELINE A. WATERS

The United States International Trade Commission furnishes studies, reports, and recommendations involving international trade and tariffs to the President, the U.S. Trade Representative, and congressional committees. The Commission also conducts a variety of investigations pertaining to international trade relief.

The United States International Trade Commission (USITC) is an independent agency created by the Revenue Act (39 Stat. 795) and originally named the United States Tariff Commission. The name was changed to the United States International Trade Commission by section 171 of the Trade Act of 1974 (19 U.S.C. 2231).

Six Commissioners are appointed by the President with the advice and consent of the Senate for 9-year terms, unless appointed to fill an unexpired term. The Chairman and Vice Chairman are designated by the President for 2-year terms, and succeeding Chairmen may not be of the same political party. The Chairman generally is responsible for the administration of the Commission. Not more than three Commissioners may be members of the same political party (19 U.S.C. 1330).

Activities

The Commission performs a number of functions pursuant to the statutes referred to above. Under the Tariff Act of 1930, the Commission is given broad powers of investigation relating to the customs laws of the United States and foreign countries; the volume of importation in comparison with domestic production

and consumption; the conditions, causes, and effects relating to competition of foreign industries with those of the United States; and all other factors affecting competition between articles of the United States and imported articles. The Commission is required, whenever requested, to make available to the President, the House Committee on Ways and Means, and the Senate Committee on Finance all information at its command and is directed to make such investigations and reports as may be requested by the President, Congress, or the committees mentioned above.

In order to carry out these responsibilities, the Commission is required to engage in extensive research, conduct specialized studies, and maintain a high degree of expertise in all matters relating to the commercial and international trade policies of the United States.

Imported Articles Subsidized or Sold at Less Than Fair Value The Commission conducts preliminary-phase investigations to determine whether imports of foreign merchandise allegedly being subsidized or sold at less than fair value injure or threaten to injure an industry in the United States. If the Commission's determination is affirmative and the Secretary of Commerce determines

UNITED STATES INTERNATIONAL TRADE COMMISSION

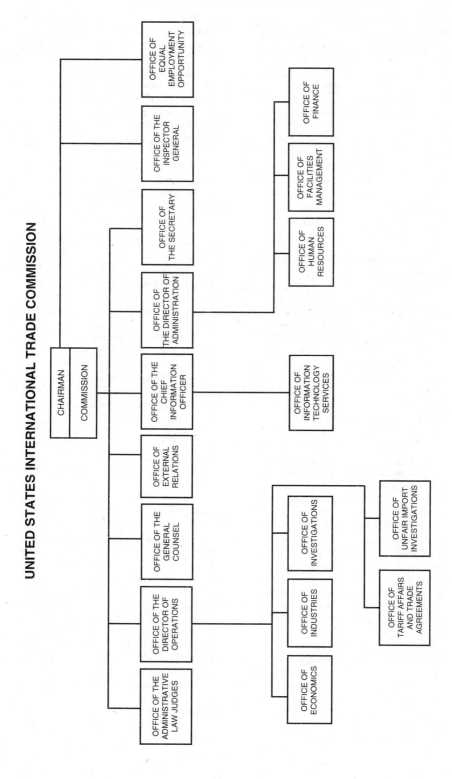

there is reason to believe or suspect such unfair practices are occurring, then the Commission conducts final-phase investigations to determine the injury or threat of injury to an industry because of such imports.

Under the Uruguay Round Agreements Act, the Commission also conducts sunset reviews. In these reviews, the Commission evaluates whether material injury to a U.S. industry would continue or recur if the antidumping duty or countervailing duty order under review were revoked. Such injury reviews must be conducted on all antidumping duty and countervailing duty orders every 5 years for as long as the orders remain in effect.

Unfair Practices in Import Trade   The Commission applies U.S. statutory and common law of unfair competition to the importation of products into the United States and their sale. If the Commission determines that there is a violation of law, it is to direct that the articles involved be excluded from entry into the United States, or it may issue cease-and-desist orders directing the person engaged in such violation to cease and desist from engaging in such unfair methods or acts.

Trade Negotiations   The Commission advises the President as to the probable economic effect on the domestic industry and on consumers of modification of duties and other barriers to trade that may be considered for inclusion in any proposed trade agreement with foreign countries.

Generalized System of Preferences   With respect to articles that may be considered for preferential removal of the duty on imports from designated developing countries, the Commission advises the President as to the probable economic effect such removal will have on the domestic industry and on consumers.

Industry Adjustment to Import Competition (Global Safeguard Actions)   The Commission conducts investigations upon petition on behalf of an industry, a firm, a group of workers, or other entity representative of an industry to determine whether an article is being imported in such increased quantities as to injure or

threaten to injure the domestic industry producing an article like or directly competitive with the imported article. If the Commission's finding is affirmative, it recommends to the President the action that would address such a threat and be most effective in facilitating positive adjustment by the industry to import competition. The President determines if import relief is appropriate.

The Commission reports on developments within an industry that has been granted import relief and advises the President of the probable economic effect of the reduction or elimination of the tariff increase that has been granted. The President may continue, modify, or terminate the import relief previously granted.

Imports From NAFTA Countries (Bilateral Safeguard Actions)   The Commission conducts investigations to determine whether, as a result of the reduction or elimination of a duty provided for under the North American Free Trade Agreement (NAFTA), a Canadian article or a Mexican article, as the case may be, is being imported into the United States in such increased quantities and under such conditions so that imports of the article constitute a substantial cause of serious injury or (except in the case of a Canadian article) a threat of serious injury to the domestic industry producing an article that is like or directly competitive with the imported article. If the Commission's determination is in the affirmative, the Commission recommends to the President the relief that is necessary to prevent or remedy serious injury. Commission investigations under these provisions are similar procedurally to those conducted under the global safeguard action provisions.

Imports from China (Bilateral Safeguard Actions)   The Commission conducts investigations to determine whether products from China are being imported into the United States in such increased quantities or under such conditions as to cause or threaten to cause market disruption to the domestic producers of like or directly competitive products. If the Commission makes an affirmative determination, it proposes a remedy.

The Commission sends its reports to the President and the U.S. Trade Representative. The President makes the final remedy decision.

Market Disruption From Communist Countries The Commission conducts investigations to determine whether increased imports of an article produced in a Communist country are causing market disruption in the United States. If the Commission's determination is in the affirmative, the President may take the same action as in the case of serious injury to an industry, except that the action would apply only to imports of the article from the Communist country. Commission investigations conducted under this provision are similar procedurally to those conducted under the global safeguard action provisions.

Import Interference With Agricultural Programs The Commission conducts investigations, at the direction of the President, to determine whether imports or potential imports may interfere with the Department of Agriculture's agricultural programs or reduce the amount of any product processed in the United States. After investigating, the Commission discloses findings and makes recommendations. The President may then restrict the imports in question by imposing import fees or quotas. Such fees or quotas may be applied only against countries that are not members of the World Trade Organization.

Uniform Statistical Data The Commission, in cooperation with the Secretary of the Treasury and the Secretary of Commerce, establishes for statistical purposes an enumeration of articles imported into the United States and exported from the United States and seeks to establish comparability of such statistics with statistical programs for domestic production.

Harmonized Tariff Schedule of the United States, Annotated The Commission issues a publication containing the U.S. tariff schedules and related matters and considers questions concerning the arrangement of such schedules and the classification of articles.

International Trade Studies The Commission conducts studies, investigations, and research projects on a broad range of topics relating to international trade, pursuant to requests of the President, the House Ways and Means Committee, the Senate Finance Committee, either branch of the Congress, or on its own motion. Public reports of these studies, investigations, and research projects are issued in most cases.

The Commission also keeps informed of the operation and effect of provisions relating to duties or other import restrictions of the United States contained in various trade agreements. Occasionally, the Commission is required by statute to perform specific trade-related studies.

Sources of Information

Inquiries should be directed to the specific organizational unit or to the Secretary, United States International Trade Commission, 500 E Street SW., Washington, DC 20436. Phone, 202–205–2000.

Contracts Procurement inquiries should be directed to the Procurement Executive. Phone, 202–205–2745.

Electronic Access Commission publications, news releases, Federal Register notices, scheduling information, the Commission's interactive Trade and Tariff DataWeb, and general information about USITC are available for electronic access. Investigation-related public inspection files are available through the Electronic Document Imaging System (EDIS). Internet, www.usitc.gov.

Employment Information on employment can be obtained from the Director, Office of Human Resources. The Agency employs international economists, attorneys, accountants, commodity and industry specialists and analysts, and clerical and other support personnel. Phone, 202–205–2651.

Publications The Commission publishes results of investigations concerning various commodities and subjects. Other publications include an annual report to the Congress on the operation of the trade agreements program and an annual review of Commission activities. Specific information regarding these publications can be obtained from the Office of the Secretary.

Reading Rooms Reading rooms are open to the public in the Office of the Secretary and the USITC Main Library. The USITC Law Library is available to individuals who make prior arrangements by calling 202–205–3287.

For further information, contact the Secretary, United States International Trade Commission, 500 E Street SW., Washington, DC 20436. Phone, 202–205–2000. Internet, http://www.usitc.gov.

UNITED STATES POSTAL SERVICE

475 L'Enfant Plaza SW., Washington, DC 20260
Phone, 202–268–2000. Internet, http://www.usps.gov.

Board of Governors

Chairman	LOUIS J. GIULIANO
Vice Chairman	THURGOOD MARSHALL, JR.
Governors	MICKEY D. BARNETT, JAMES H. BILBRAY, CAROLYN LEWIS GALLAGHER, ALAN C. KESSLER, JAMES C. MILLER, III, ELLEN C. WILLIAMS, (VACANCY)
Postmaster General, Chief Executive Officer	JOHN E. POTTER
Deputy Postmaster General, Chief Operating Officer	PATRICK R. DONAHOE
Secretary	JULIE S. MOORE
Inspector General	DAVID C. WILLIAMS

Officers

Postmaster General, Chief Executive Officer	JOHN E. POTTER
Deputy Postmaster General, Chief Operating Officer	PATRICK R. DONAHOE
Managing Director and Vice President, Global Business	PRANAB SHAH
Operations Senior Vice President	STEVEN FORTE
Network Operations Vice President	JORDAN SMALL
Delivery and Post Office Operations Vice President	DEAN GRANHOLM
Engineering Vice President	DAVID E. WILLIAMS
Facilities Vice President	TOM SAMRA
Sustainability Vice President	SAM PULCRANO
Intelligent Mail and Address Quality Senior Vice President	THOMAS DAY
Business Mail Entry and Payment Technologies Vice President	PRITHA MEHRA
Capitol Metro Area Vice President	(VACANCY)
Eastern Area Vice President	MEGAN BRENNAN
Great Lakes Area Vice President	JO ANN FEINDT
Northeast Area Vice President	TIM HANEY
Pacific Area Vice President	DREW ALIPARTO
Southeast Area Vice President	LINDA WELCH
Southwest Area Vice President	ELLIS BURGOYNE
Western Area Vice President	SYLVESTER BLACK
Mailing and Shipping Services President	SUSAN PLONKEY, *Acting*

Officers

Expedited Shipping Vice President	GARY REBLIN
Ground Shipping Vice President	JAMES COCHRANE
Sales Vice President	SUSAN PLONKEY
Retail Products and Services Vice President	TIMOTHY C. HEALY
Chief Information Officer and Executive Vice President	ROSS PHILO
Information Technology Solutions Vice President	JOHN EDGAR
Chief Financial Officer, Executive Vice President	JOSEPH CORBETT
Supply Management Vice President	SUSAN BROWNELL
Controller Vice President	VINCENT DEVITO
Finance and Planning Vice President	STEPHEN MASSE
Chief Human Resources Officer, Executive Vice President	ANTHONY VEGLIANTE
Employee Resource Management Vice President	DEBORAH GIANNONI-JACKSON
Employee Development and Diversity Vice President	SUSAN LACHANCE
Labor Relations Vice President	DOUG TULINO
Customer Relations Senior Vice President	STEPHEN KEARNEY
Consumer Advocate Vice President	DELORES KILLETTE
Corporate Communications Vice President	MITZI BETMAN
Pricing Vice President	MAURA ROBINSON
General Counsel	MARY ANNE GIBBONS
Strategy and Transition Senior Vice President	LINDA KINGSLEY
Government Relations and Public Policy Vice President	MARIE THERESE DOMINGUEZ
Chief Postal Inspector	WILLIAM GILLIGAN
Judicial Officer	WILLIAM CAMPBELL

[For the United States Postal Service statement of organization, see the Code of Federal Regulations, Title 39, Part 221]

The United States Postal Service provides mail processing and delivery services to individuals and businesses within the United States.

The Postal Service was created as an independent establishment of the executive branch by the Postal Reorganization Act (39 U.S.C. 101 et seq.), approved August 12, 1970. The present United States Postal Service commenced operations on July 1, 1971.

In 2009, the Postal Service had approximately 623,000 career employees and handled about 177 billion pieces of mail. The chief executive officer of the Postal Service, the Postmaster General, is appointed by the nine Governors of the Postal Service, who are appointed by the President with the advice and consent of the Senate. The Governors and the Postmaster General appoint the Deputy Postmaster General, and these 11 people constitute the Board of Governors.

In addition to the national headquarters, there are area and district offices supervising more than 36,000 post offices, branches, stations, and community post offices throughout the United States.

Activities

In order to expand and improve service to the public, the Postal Service is engaged in customer cooperation activities, including the development of programs for both the general public and major customers. The Consumer Advocate, a postal ombudsman, represents the interest of the individual

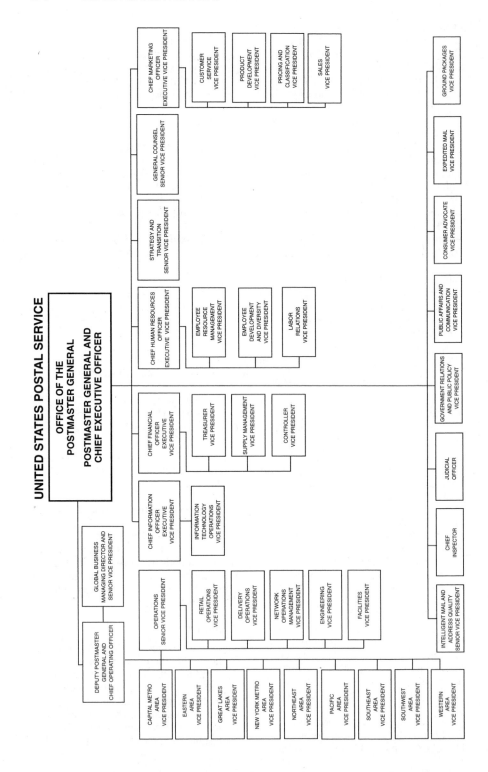

UNITED STATES POSTAL SERVICE

mail customer in matters involving the Postal Service by bringing complaints and suggestions to the attention of top postal management and solving the problems of individual customers. To provide postal services responsive to public needs, the Postal Service operates its own planning, research, engineering, real estate, and procurement programs specially adapted to postal requirements and maintains close ties with international postal organizations.

The Postal Service is the only Federal agency whose employment policies are governed by a process of collective bargaining under the National Labor Relations Act. Labor contract negotiations, affecting all bargaining unit personnel, as well as personnel matters involving employees not covered by collective bargaining agreements, are administered by Labor Relations or Human Resources.

The U.S. Postal Inspection Service is the Federal law enforcement agency which has jurisdiction in criminal matters affecting the integrity and security of the mail. Postal Inspectors enforce more than 200 Federal statutes involving mail fraud, mail bombs, child pornography, illegal drugs, mail theft, and other postal crimes, as well as being responsible for the protection of all postal employees.

Postal Service customers and employees can file mail fraud complaints, find local Postal Inspection Service offices, and receive helpful preventative tips at 1–877–876–2455 or at https://postalinspectors.uspis.gov.

Sources of Information

Consumer Information   Customers may check shipping rates, buy stamps, print postage, track packages, locate ZIP Codes, shop at the Postal Store, change addresses, or obtain answers to frequently asked questions by visiting www.usps.com. For general information 24 hours a day, call 1–800–ASK–USPS (1–800–275–8777). For the Express Mail, Priority Mail, and Package Support Line, call 1–800–222–1811. Information on past and present schemes used to defraud the public is available at https://postalinspectors.uspis.gov. Reports of

fraudulent activity involving the mail may be made to the Mail Fraud Hotline, 1–800–372–8347.

Contracts and Small Business Activities Contact Supplier Diversity. Phone, 202–268–4633.

Employment   General information about jobs such as clerk, letter carrier, etc., including information about programs for veterans, may be obtained by contacting the nearest post office or from the Postal Service Web site at www.usps.gov/employment. Information about U.S. Postal Inspector Service employment may be obtained online at www.usps.com/postalinspectors.

Inspector General   The Office of Inspector General maintains a toll-free hotline as a means for individuals to report activities involving fraud, waste, or mismanagement. Such reports may be made by email to hotline@uspsoig.gov, by telephone at 1–888–USPS–OIG (1–888–877–7644), by fax at 1–866–756–6741, or by mail to the United States Postal Service, Office of Inspector General Hotline, 10th Floor, 1735 North Lynn Street, Arlington, VA 22209–2020. Publicly available documents and information on the Office of Inspector General and some Freedom of Information Act documents are available at www.uspsoig.gov.

Philatelic Sales   Contact Stamp Fulfillment Services, Kansas City, MO 64179–1009. Phone, 800–782–6724.

Publications   Pamphlets on mailability, postage rates and fees, and many other topics may be obtained free of charge from the nearest post office. Most postal regulations are contained in Postal Service manuals covering domestic and international mail, postal operations, administrative support, and employee and labor relations. These manuals and other publications, including the National Five-Digit ZIP Code and Post Office Directory (Publication 65), may be purchased from the Superintendent of Documents, Government Printing Office, Washington, DC 20402–0001. The National Five-Digit ZIP Code and Post Office Directory is also available through local post offices.

Reading Rooms   Reading rooms are located at USPS Headquarters on the 11th Floor North, Library. Phone, 202–268–2900.

For further information, contact the United States Postal Service, 475 L'Enfant Plaza SW., Washington, DC 20260. Phone, 202–268–2000. Internet, http://www.usps.gov.

Boards, Commissions, and Committees

Below is a list of Federal boards, commissions, councils, etc., not listed elsewhere in the Manual, which were established by congressional or Presidential action, whose functions are not strictly limited to the internal operations of a parent department or agency and which are authorized to publish documents in the Federal Register. While the editors have attempted to compile a complete and accurate listing, suggestions for improving coverage of this guide are welcome. Please address your comments to the Office of the Federal Register, National Archives and Records Administration, Washington, DC 20408. Phone, 202–741–6040. E-mail, fedreg.info@nara.gov. Internet, www.ofr.gov.

Federal advisory committees, as defined by the Federal Advisory Committee Act, as amended (5 U.S.C. app.), have not been included here. Information on Federal advisory committees may be obtained from the Committee Management Secretariat, General Services Administration, General Services Building (MC), Room G–230, Washington, DC 20405. Phone, 202–273–3556. Internet, www.gsa.gov/committeemanagement.

Administrative Committee of the Federal Register

Office of the Federal Register, National Archives and Records Administration, 8601 Adelphi Road, College Park, MD 20740–6001. Phone, 202–741–6010. E-mail, fedreg.info@nara.gov. Internet, www.ofr.gov.

Advisory Council on Historic Preservation

1100 Pennsylvania Avenue NW., Suite 803, Washington, DC 20004. Phone, 202–606–8503. E-mail, achp@achp.gov. Internet, www.achp.gov.

American Battle Monuments Commission

2300 Clarendon Boulevard, Court House Plaza 2, Suite 500, Arlington, VA 22201. Phone, 703–696–6900. E-mail, info@abmc.gov. Internet, www.abmc.gov.

Appalachian Regional Commission

1666 Connecticut Avenue NW., Suite 700, Washington, DC 20009–1068. Phone, 202–884–7700. E-mail, info@arc.gov. Internet, www.arc.gov.

Architectural and Transportation Barriers Compliance Board[1]

1331 F Street NW., Suite 1000, Washington, DC 20004–1111. Phone, 202–272–0080 or TTY, 202–272–0082. Fax, 202–272–0081. Internet, www.access-board.gov.

Arctic Research Commission

4350 North Fairfax Drive, Suite 510, Arlington, VA 22203. Phone, 703–525–0111. E-mail, info@arctic.gov. Internet, www.arctic.gov.

---

[1]Also known as the Access Board.

Arthritis and Musculoskeletal Interagency Coordinating Committee

National Institutes of Health/NIAMS, Building 31—MSC 2350, Room 4C02, 31 Center Drive, Bethesda, MD 20892–2350. Phone, 301–496–8190. Fax, 301–480–2814. Internet, www.niams.nih.gov.

Barry M. Goldwater Scholarship and Excellence in Education Program

Phone, 319–341–2333. Internet, www.act.org/goldwater.

Chemical Safety and Hazard Investigation Board

2175 K Street NW., Suite 400, Washington, DC 20037–1809. Phone, 202–261–7600. Fax, 202–261–7650. Internet, www.csb.gov.

Citizens' Stamp Advisory Committee

United States Postal Service c/o Stamp Development, 1735 N. Lynn Street, Suite 5013, Arlington, VA 22209–6432. Phone, 703–292–3810. Fax, 703–292–3634. Internet, www.usps.com/communications/organization/csac.htm.

U.S. Commission of Fine Arts

National Building Museum, 401 F Street NW., Suite 312, Washington, DC 20001–2728. E-mail, cfastaff@cfa.gov. Internet, www.cfa.gov.

Committee on Foreign Investment in the United States

Department of the Treasury, 1500 Pennsylvania Avenue NW., Washington, DC 20220. Phone, 202–622–1860. E-mail, CFIUS@do.treas.gov. Internet, www.treas.gov/offices/international-affairs/cfius/.

Committee for the Implementation of Textile Agreements

Office of Textiles and Apparel, U.S. Department of Commerce, DC 20230. Phone, 202–482–5078. Fax, 202–482–2331. E-mail, OTEXA@mail.doc.gov. Internet, http://otexa.ita.doc.gov/cita.htm.

Committee for Purchase From People Who Are Blind or Severely Disabled

1421 Jefferson Davis Highway, Jefferson Plaza 2, Suite 10800, Arlington, VA 22202–3259. Phone, 703–603–7740. Fax, 703–608–0655. E-mail, info@abilityone.gov. Internet, www.abilityone.gov.

Coordinating Council on Juvenile Justice and Delinquency Prevention

Department of Justice, Office of Juvenile Justice and Delinquency Prevention, 810 7th Street NW., Washington, DC 20531. Phone, 202–307–9963. Fax, 202–307–9093. E-mail, robin.delany-shabazz@usdoj.gov. Internet, www.juvenilecouncil.gov.

Delaware River Basin Commission

25 State Police Drive, P.O. Box 7360, West Trenton, NJ 08628–0360. Phone, 609–883–9500. Fax, 609–883–9522. E-mail, clarke.rupert@drbc.state.nj.us. Internet, www.state.nj.us/drbc.

Endangered Species Program

4401 N. Fairfax Drive, Room 420, Arlington, VA 22203. Phone, 703–358–1985. Internet, http://endangered.fws.gov.

Export Administration Operating Committee

Department of Commerce, Bureau of Industry and Security, 14th Street and Constitution Avenue, NW., Washington, DC 20230. Phone, 202–482–4811. Internet, http://www.bis.doc.gov/index.htm.

Federal Financial Institutions Examination Council

3501 Fairfax Drive, D8073a, Arlington, VA 22226. Phone, 703–516–5590. Internet, www.ffiec.gov.

Federal Financing Bank

Department of the Treasury, 1500 Pennsylvania Avenue NW., Washington, DC 20220. Phone, 202–622–2470. Fax,

202–622–0707. E-mail, ffb@do.treas.gov. Internet, www.ustreas.gov/ffb.

Federal Interagency Committee on Education

Department of Education, 400 Maryland Avenue SW., Washington, DC 20202. Phone, 202–401–3673.

Federal Laboratory Consortium for Technology Transfer

Washington, DC Liaison Office. Phone, 202–296–7201. Internet, www. federallabs.org.

Federal Library and Information Center Committee

Library of Congress, 101 Independence Avenue SE., Washington, DC 20540–4935. Phone, 202–707–4800. Internet, www.loc.gov/flicc/.

Harry S. Truman Scholarship Foundation

712 Jackson Place NW., Washington, DC 20006. Phone, 202–395–4831. Fax, 202–395–6995. E-mail, office@truman. gov. Internet, www.truman.gov.

Indian Arts and Crafts Board

U.S. Department of the Interior, Room MS 2528–MIB, 1849 C Street NW., Washington, DC 20240. Phone, 202–208–3773. E-mail, iacb@ios.doi.gov. Internet, www.iacb.doi.gov.

J. William Fulbright Foreign Scholarship Board

Department of State, Bureau of Educational and Cultural Affairs, 2200 C Street NW., Washington, DC 20522. Phone, 202–632–6170. E-mail, fulbright@state.gov. Internet, http://fulbright.state.gov.

James Madison Memorial Fellowship Foundation

2000 K Street NW., Suite 303, Washington, DC 20006–1809. Phone, 202–653–8700. Fax, 202–653–6045. Internet, www.jamesmadison.com.

Japan-US Conference on Cultural and Educational Interchange (CULCON)

1201 15th Street NW., Suite 330, Washington, DC 20005. Phone, 202–653–9800. Fax, 202–653–9802. E-mail, culcon@jusfc.gov. Internet, www.jusfc.gov.

Joint Board for the Enrollment of Actuaries

Internal Revenue Service, SE: OPR, 1111 Constitution Avenue NW., Washington, DC 20224. Phone, 202–622–8229. Fax, 202–622–8300. E-mail, nhqjbea@irs.gov. Internet, www.irs.gov/taxpros/actuaries/index.html.

Marine Mammal Commission

4340 East-West Highway, Suite 700, Bethesda, MD 20814. Phone, 301–504–0087. Fax, 301–504–0099. E-mail, mmc@mmc.gov. Internet, www.mmc.gov.

Medicare Payment Advisory Commission

601 New Jersey Avenue NW., Suite 9000, Washington, DC 20001. Phone, 202–220–3700. Internet, www.medpac.gov.

Migratory Bird Conservation Commission

Secetary, Migration Bird Conservation Commission, Mail Code: ARLSQ–622, 4401 North Fairfax Drive, Arlington, VA 22203–1610. Phone, 703–358–1716. Fax, 703–358–2223. Internet, www.fws.gov/refuges/realty/mbcc.html.

Mississippi River Commission

Mississippi River Commission, 1400 Walnut Street, Vicksburg, MS 39180–0080. Phone, 601–634–5760. E-mail, cenvd-ex@usace.army.mil. Internet, www.mvd.usace.army.mil/mrc/.

Morris K. and Stewart L. Udall Foundation

130 South Scott Avenue, Tucson, AZ 85701–1922. Phone, 520–901–8500. Fax, 520–670–5530. Internet, www.udall.gov.

National Council on Disability

1331 F Street NW., Suite 850,
Washington, DC 20004. Phone, 202–
272–2004. TTY, 202–272–2074. Fax,
202–272–2022. E-mail, ncd@ncd.gov.
Internet, www.ncd.gov.

National Indian Gaming Commission

1441 L Street NW., Suite 9100,
Washington, DC 20005. Phone, 202–
632–7003. Fax, 202–632–7066. E-mail,
contactus@nigc.gov. Internet,
www.nigc.gov.

National Park Foundation

1201 I Street NW., Suite 550B,
Washington, DC 20005. Phone, 202–
354–6460. Fax, 202–371–2066. E-mail,
ask-npf@nationalparks.org. Internet,
www.nationalparks.org.

Northwest Power and Conservation Council

851 SW. Sixth Avenue, Suite 1100,
Portland, OR 97204. Phone, 503–222–
5161 or 800–452–5161. Fax, 503–820–
2370. Internet, www.nwcouncil.org.

Office of Navajo and Hopi Indian Relocation

201 East Birch Avenue, Flagstaff, AZ
86001. Phone, 928–779–2721 or
800–321–3114. Fax, 928–774–1977.
E-mail, webmaster@onhir.gov. Internet,
http://onhir.gov

Permanent Committee for the Oliver Wendell Holmes Devise

Library of Congress, Manuscript Division,
Washington, DC 20540–4680. Phone,
202–707–1082.

President's Intelligence Advisory Board

New Executive Office Building, Room
5020, Washington, DC 20502. Phone,
202–456–2352. Fax, 202–395–3403.

Presidio Trust

34 Graham Street, P.O. Box 29052, San
Francisco, CA 94129–0052. Phone,

415–561–5300. TTY, 415–561–5301.
Fax, 415–561–5315. E-mail, presidio@
presidiotrust.gov. Internet,
www.presidio.gov.

Social Security Advisory Board

400 Virginia Avenue SW., Suite 625,
Washington, DC 20024. Phone, 202–
475–7700. Fax, 202–475–7715. E-mail,
info@ssab.gov. Internet, www.ssab.gov.

Susquehanna River Basin Commission

1721 North Front Street, Harrisburg,
PA 17102. Phone, 717–238–0423. Fax,
717–238–2436. E-mail, msrbc@srbc.net.
Internet, www.srbc.net.

Trade Policy Staff Committee

Office of the United States Trade
Representative, 600 17th Street
NW.,Washington, DC 20508. Phone,
202–395–3475. Fax, 202–395–5141.
Internet, www.ustr.gov.

United States Holocaust Memorial Museum

100 Raoul Wallenberg Place SW.,
Washington, DC 20024–2126. Phone,
202–488–0400. TTY, 202–488–0406.
Internet, www.ushmm.org.

United States Nuclear Waste Technical Review Board

2300 Clarendon Boulevard, Suite 1300,
Arlington, VA 22201. Phone, 703–235–
4473. Fax, 703–235–4495. Internet,
www.nwtrb.gov.

Veterans Day National Committee

Department of Veterans Affairs, 810
Vermont Avenue NW., Mail Code
002C, Washington, DC 20420. Phone,
202–461–7449. E-mail, vetsday@va.gov.
Internet, www1.va.gov/opa/vetsday.

White House Commission on Presidential Scholars

Department of Education, 400 Maryland
Avenue SW., Washington, DC 20202–

8173. Phone, 202–401–0961. Fax, 202–260–7464. E-mail, presidential.scholars@ed.gov. Internet, www.ed.gov/programs/psp/index.html.

QUASI-OFFICIAL AGENCIES

LEGAL SERVICES CORPORATION

3333 K Street NW., Washington, DC 20007
Phone, 202–295–1500. Fax, 202–337–6797. Internet, http://www.lsc.gov.

President	VICTOR M. FORTUNO
General Counsel	VICTOR M. FORTUNO
Corporate Secretary	PATRICIA BATIE, *Acting*
Vice President for Programs and Compliance	KAREN SARJEANT
Director, Office of Compliance and Enforcement	DANILO CARDONA
Director, Office of Information Management	JOHN MEYER
Director, Office of Program Performance	JANET LABELLA
Chief Administrative Officer	CHARLES JEFFRESS
Comptroller/Treasurer	DAVID L. RICHARDSON
Director, Office of Human Resources	ALICE DICKERSON
Director, Office of Information Technology	JEFF MORNINGSTAR
Director, Government Relations and Public Affairs	JOHN CONSTANCE
Inspector General	JEFFREY E. SCHANZ

[For the Legal Services Corporation statement of organization, see the Code of Federal Regulations, Title 45, Part 1601]

The Legal Services Corporation's mission is to promote equal access to justice in our Nation and to provide high-quality civil legal assistance to low-income persons.

The Legal Services Corporation (LSC) is a private, nonprofit corporation established by the Legal Services Act of 1974, as amended (42 U.S.C. 2996), to promote equal access to justice under the law for all Americans.

LSC is headed by an 11-member Board of Directors, appointed by the President and confirmed by the Senate. By law, the Board is bipartisan and no more than six members may be of the same political party.

LSC is funded by congressional appropriations and provides legal services through grants to independent local legal services provider programs selected through a system of competition. In 2009, LSC funded 136 programs. Together, they serve every county and congressional district in the Nation, as well as the U.S. territories. Programs are also funded to address the needs of Native Americans and migrant farmworkers.

The legal services delivery system is based on several principles: local priorities, national accountability, competition for grants, and a strong public-private partnership. Local programs are governed by their own boards of directors, which set priorities and determine the types of cases that will be handled subject to restrictions set by Congress. A majority of each local board is appointed by local bar associations and one-third of each local board is composed of client representatives appointed by client groups. Each board hires its own executive director. Programs

485

LEGAL SERVICES CORPORATION

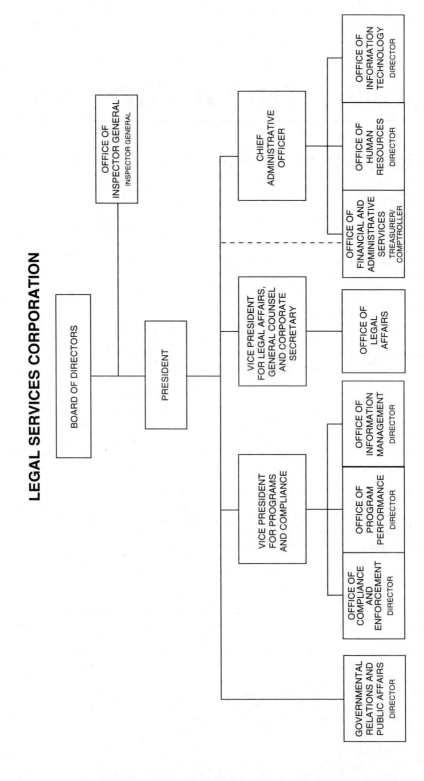

may supplement their LSC grants with additional funds from State and local governments and other sources. They further leverage Federal funds by involving private attorneys in the delivery of legal services for the poor, mostly through volunteer pro bono work.

LSC-funded programs do not handle criminal cases, nor do they accept fee-generating cases that private attorneys are willing to accept on a contingency basis. In addition, in 1996 a series of new limitations were placed upon activities in which LSC-funded programs may engage on behalf of their clients, even with non-LSC funds. All Legal Services programs must comply with laws enacted by Congress and the implementing regulations promulgated by the Legal Services Corporation.

For further information, contact the Office of Government Relations and Public Affairs, Legal Services Corporation, 3333 K Street NW., Washington, DC 20007–3522. Phone, 202–295–1500. Fax, 202–337–6797. Internet, http://www.lsc.gov.

SMITHSONIAN INSTITUTION

1000 Jefferson Drive SW., Washington, DC 20560
Phone, 202–633–1000. TDD, 202–357–1729. Internet, http://www.smithsonian.org.

Board of Regents

The Chief Justice of the United States (Chancellor)	JOHN G. ROBERTS, JR.
The Vice President of the United States	JOSEPH R. BIDEN, JR.
Member of the Senates	THAD COCHRAN, CHRISTOPHER DODD, PATRICK J. LEAHY
Member of the House of Representatives	XAVIER BECERRA, SAMUEL JOHNSON, DORIS MATSUI
Citizen Members	ELI BROAD, FRANCE A. CORDOVA, PHILLIP FROST, SHIRLEY ANN JACKSON, ROBERT P. KOGOD, JOHN W. MCCARTER, ROGER SANT, ALAN G. SPOON, PATTY STONESIFER

Officials

Secretary	G. WAYNE CLOUGH
Inspector General	A. SPRIGHTLEY RYAN
Director, Communications and Public Affairs	EVELYN LIEBERMAN
Director, Equal Employment and Minority Affairs	ERA MARSHALL
Director, External Affairs	VIRGINIA B. CLARK
Director, Government Relations	NELL PAYNE
General Counsel	JUDITH E. LEONARD
Undersecretary for Finance and Administration	ALISON MCNALLY
Chief Financial Officer	ALICE C. MARONI
Chief Information Officer	ANN SPEYER
Director, Accessibility Program	ELIZABETH ZIEBARTH
Director, Exhibits Central	MICHAEL HEADLEY
Director, Facilities Engineering and Operations	BRUCE KENDALL
Director, Human Resources	JAMES DOUGLAS
Director, Investments	AMY CHEN
Director, National Collections	BILL THOMPKINS
Director, Policy and Analysis	CAROLE P. NEVES
Director, Smithsonian Institution Archives	ANNE VAN CAMP
Director, Special Events and Protocol	NICOLE L. KRAKORA

Officials

Ombudsman	CHANDRA HEILMAN
Undersecretary for History, Art, and Culture	RICHARD KURIN
Director, Anacostia Community Museum	CAMILE AKEJU
Director, Archives of American Art	JOHN W. SMITH
Director, Asian Pacific American Program	FRANKLIN ODO
Director, Center for Folklife and Cultural Heritage	DANIEL SHEEHY, *Acting*
Director, Cooper-Hewitt National Design Museum	WILLIAM MOGGRIDGE
Director, Freer Gallery of Art and Arthur M. Sackler Gallery	JULIAN RABY
Director, Hirshhorn Museum and Sculpture Garden	RICHARD KOSHALEK
Director, National Museum of African American History and Culture	LONNIE BUNCH
Director, National Museum of African Art	JOHNNETTA B. COLE
Director, National Museum of American History	BRENT GLASS
Director, National Museum of the American Indian	KEVIN GOVER
Director, National Portrait Gallery	MARTIN SULLIVAN
Director, National Postal Museum	ALLEN KANE
Director, National Programs	RICHARD KURIN, *Acting*
Director, Smithsonian Affiliations	HAROLD CLOSTER
Director, Smithsonian Associates	BARBARA TUCELING
Director, Smithsonian American Art Museum and Renwick Gallery	ELIZABETH BROUN
Director, Smithsonian Center for Education and Museum Studies	STEPHANIE NORBY
Director, Smithsonian Latino Center	EDUARDO DIAZ
Director, Smithsonian Institution Traveling Exhibition Service	ANNA R. COHN
Undersecretary for Science	EVA PELL
Director, National Air and Space Museum	JOHN R. DAILEY
Director, National Museum of Natural History	CRISTIAN SAMPER
Director, National Science Resources Center	SALLY SHULER
Director, Fellowships	CATHERINE HARRIS
Director, National Zoological Park	DENNIS KELLY
Director, Smithsonian Astrophysical Observatory	CHARLES ALCOCK
Director, Smithsonian Environmental Research Center	ANSON H. HINES
Director, Smithsonian Institution Libraries	NANCY E. GWINN
Director, Smithsonian Marine Station	VALERIE PAUL
Director, Smithsonian Museum Conservation Institute	ROBERT KOESTLER
Director, Smithsonian Tropical Research Institute	ELDREDGE BERMINGHAM
President, Smithsonian Enterprises and Smithsonian Publishing	TOM OTT
Editor-in-Chief, Smithsonian Magazine	CAREY WINFREY
Group Publisher, Smithsonian Publishing	KERRY BIANCHI

The Smithsonian Institution is an independent trust instrumentality of the United States which comprises the world's largest museum and research complex; includes 19 museums and galleries, the National Zoo, and research facilities in several States and the Republic of Panama; and is dedicated to public education, national service, and scholarship in the arts, sciences, history, and culture.

The Smithsonian Institution was created by an act of Congress on August 10, 1846 (20 U.S.C. 41 et seq.), to carry out the terms of the will of British scientist James Smithson (1765–1829), who in 1826 had bequeathed his entire estate to the United States "to found at Washington, under the name of the Smithsonian Institution, an establishment for the increase and diffusion of knowledge among men." On July 1, 1836, Congress accepted the legacy and pledged the faith of the United States to the charitable trust.

In September 1838, Smithson's legacy, which amounted to more than 100,000 gold sovereigns, was delivered to the mint at Philadelphia. Congress vested responsibility for administering the trust in the Secretary of the Smithsonian and the Smithsonian Board of Regents, composed of the Chief Justice, the Vice President, three Members of the Senate, three Members of the House of Representatives, and nine citizen members appointed by joint resolution of Congress. To carry out Smithson's mandate, the Institution executes the following functions: conducts scientific and scholarly research; publishes the results of studies, explorations, and investigations; preserves for study and reference more than 136 million artifacts, works of art, and scientific specimens; organizes exhibits representative of the arts, the sciences, and American history and culture; shares Smithsonian resources and collections with communities throughout the Nation; and engages in educational programming and national and international cooperative research.

Smithsonian activities are supported by its trust endowments and revenues; gifts, grants, and contracts; and funds appropriated to it by Congress. Admission to the museums in Washington, DC, is free.

Activities

Anacostia Community Museum   The Museum, located in the historic Fort Stanton neighborhood of southeast Washington, serves as a national resource for exhibitions, historical documentation, and interpretive and educational programs relating to African American history and culture.

For further information, contact the Anacostia Community Museum, 1901 Fort Place SE., Washington, DC 20020. Phone, 202–633–1000. Internet, http://www.si.edu/anacostia.

Archives of American Art   The Archives contains the Nation's largest collection of documentary materials reflecting the history of visual arts in the United States. On the subject of art in America, it is the largest archives in the world, holding more than 16 million documents. The Archives gathers, preserves, and microfilms the papers of artists, craftsmen, collectors, dealers, critics, and art societies. These papers include manuscripts, letters, diaries, notebooks, sketchbooks, business records, clippings, exhibition catalogs, transcripts of tape-recorded interviews, and photographs of artists and their work. The Archives is located at 750 Ninth Street NW., in Washington, DC.

For further information, contact the Archives of American Art, Smithsonian Institution, Washington, DC 20560. Phone, 202–275–2156. Internet, http://archivesofamericanart.si.edu.

Cooper-Hewitt National Design Museum   The Museum is the only museum in the country devoted exclusively to historical and contemporary design. Collections include objects in such areas as applied arts and industrial design, drawings and prints, glass, metalwork, wallcoverings, and textiles. Changing exhibits and public programs seek to educate by exploring the role of design in daily life. The Museum is open daily, except Mondays and holidays. The general admission fee is $12, $7 for students and senior citizens with ID, and free for members and children under 12.

For further information, contact Cooper-Hewitt National Design Museum, 2 East Ninety-First Street, New York, NY 10128. Phone, 212–849–8400. Internet, http://www.si.edu/ndm.

SMITHSONIAN INSTITUTION

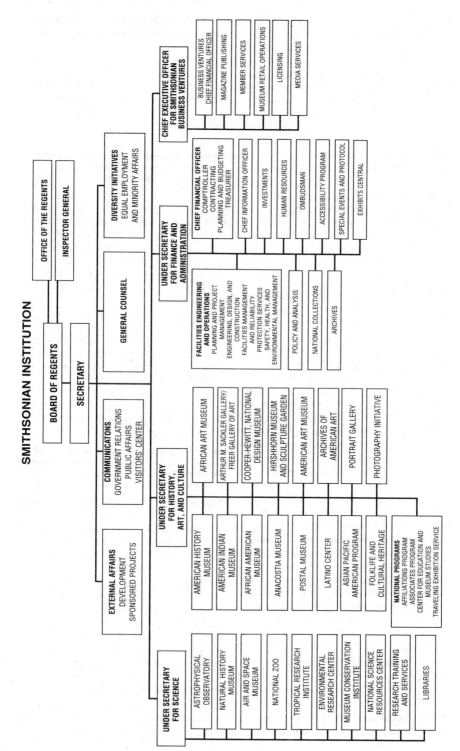

Freer Gallery of Art   The building, the original collection, and an endowment were the gift of Charles Lang Freer (1854–1919). The Gallery houses one of the world's most renowned collections of Asian art, an important group of ancient Egyptian glass, early Christian manuscripts, and works by 19th- and early 20th-century American artists. The objects in the Asian collection represent the arts of East Asia, the Near East, and South and Southeast Asia, including paintings, manuscripts, scrolls, screens, ceramics, metalwork, glass, jade, lacquer, and sculpture. Members of the staff conduct research on objects in the collection and publish results in scholarly journals and books for general and scholarly audiences.

For further information, contact the Freer Gallery of Art, Jefferson Drive at Twelfth Street SW., Washington, DC 20560. Phone, 202–633–1000. Internet, http://www.asia.si.edu.

Hirshhorn Museum and Sculpture Garden   From cubism to minimalism, the Museum houses major collections of modern and contemporary art. The nucleus of the collection is the gift and bequest of Joseph H. Hirshhorn (1899–1981). Supplementing the permanent collection are loan exhibitions. The Museum houses a collection research facility, a specialized art library, and a photographic archive, available for consultation by prior appointment. The outdoor sculpture garden is located nearby on the National Mall. There is an active program of public service and education, including docent tours, lectures on contemporary art and artists, and films of historic and artistic interest.

For further information, contact the Hirshhorn Museum and Sculpture Garden, Seventh Street and Independence Avenue SW., Washington, DC 20560. Phone, 202–633–1000. Internet, http://www.hirshhorn.si.edu.

National Air and Space Museum Created to memorialize the development and achievements of aviation and space flight, the Museum collects, displays, and preserves aeronautical and space flight artifacts of historical significance as well as documentary and artistic materials related to air and space. Among its artifacts are full-size planes, models, and instruments. Highlights of the collection include the Wright brothers' Flyer, Charles Lindbergh's Spirit of St. Louis, a Moon rock, and Apollo spacecraft. The exhibitions and study collections record the human conquest of the air from its beginnings to recent achievements. The principal areas in which work is concentrated include flight craft of all types, space flight vehicles, and propulsion systems. The Museum's IMAX Theater and the 70-foot domed Einstein Planetarium are popular attractions. The Museum's Steven F. Udvar-Hazy Center, at Washington Dulles International Airport, opened in December 2003. Its featured artifacts include a space shuttle and the Enola Gay B–29 World War II bomber.

For further information, contact the National Air and Space Museum, Sixth Street and Independence Avenue SW., Washington, DC 20560. Phone, 202–633–1000. Internet, http://www.nasm.si.edu.

National Museum of African Art   This is the only art museum in the United States dedicated exclusively to portraying the creative visual traditions of Africa. Its research components, collection, exhibitions, and public programs establish the Museum as a primary source for the examination and discovery of the arts and culture of Africa. The collection includes works in wood, metal, fired clay, ivory, and fiber. The Eliot Elisofon Photographic Archives includes slides, photos, and film segments on Africa. There is also a specialized library.

For further information, contact the National Museum of African Art, 950 Independence Avenue SW., Washington, DC 20560. Phone, 202–633–1000. Internet, http://www.nmafa.si.edu.

National Museum of African American History and Culture   The Museum was established in 2003 and will be the only national museum devoted exclusively to the documentation of African American life, art, history, and culture.

For further information, contact the National Museum of African American History and Culture, 470 L'Enfant Plaza SW., Washington, DC 20560. Phone, 202–633–1000. Internet, http://www.nmaahc.si.edu.

Smithsonian American Art Museum The Museum's art collection spans centuries of American painting, sculpture, folk art, photography, and graphic art. A major center for research in American art, the Museum has contributed to such resources as the Inventory of American Paintings Executed Before 1914; the Smithsonian Art Index; and the Inventory of American Sculpture. The library, shared with the National Portrait Gallery, contains volumes on art, history, and biography, with special emphasis on the United States. The Donald W. Reynolds Center for American Art and Portraiture is home to both the Smithsonian American Art Museum and the National Portrait Gallery. Hundreds of images from the collection and extensive information on its collections, publications, and activities are available electronically at www.saam. si.edu.

For further information, contact the Smithsonian American Art Museum, Eighth and G Streets NW., Washington, DC 20560. Phone, 202–633–1000. Internet, http://www.americanart.si.edu.

Renwick Gallery   The Gallery is dedicated to exhibiting crafts of all periods and to collecting 20th-century American crafts. It offers changing exhibitions of American crafts and decorative arts, both historical and contemporary, and a rotating selection from its permanent collection. The Gallery's grand salon is elegantly furnished in the Victorian style of the 1860s and 1870s.

For further information, contact the Renwick Gallery, Seventeenth Street and Pennsylvania Avenue NW., Washington, DC 20560. Phone, 202–633–1000. Internet, http://www.americanart. si.edu/renwick.

National Museum of American History In pursuit of its fundamental mission to inspire a broader understanding of the United States and its people, the Museum provides learning opportunities, stimulates the imagination of visitors, and presents challenging ideas about the Nation's past. The Museum's exhibits provide a unique view of the American experience. Emphasis is placed upon innovative individuals representing a wide range of cultures, who have shaped our heritage, and upon science and the remaking of our world through technology. Exhibits draw upon strong collections in the sciences and

engineering, agriculture, manufacturing, transportation, political memorabilia, costumes, musical instruments, coins, Armed Forces history, photography, computers, ceramics, and glass. Classic cars, icons of the American Presidency, First Ladies' gowns, the Star-Spangled Banner flag, Whitney's cotton gin, Morse's telegraph, the John Bull locomotive, Dorothy's ruby slippers from "The Wizard of Oz," and other American icons are highlights of the collection.

For further information, contact the National Museum of American History, Fourteenth Street and Constitution Avenue NW., Washington, DC 20560. Phone, 202–633–1000. Internet, http://www. americanhistory.si.edu.

National Museum of the American Indian   The Museum was established in 1989, and the building on the National Mall opened September 2004. Much of the collection of the Museum is comprised of the collection of the former Heye Foundation in New York City. It is an institution of living cultures dedicated to the collection, preservation, study, and exhibition of the life, languages, literature, history, and arts of the Native peoples of the Americas. Highlights include Northwest Coast carvings; dance masks; pottery and weaving from the Southwest; painted hides and garments from the North American Plains; goldwork of the Aztecs, Incas, and Maya; and Amazonian featherwork. The National Museum of the American Indian also operates the George Gustav Heye Center at the Alexander Hamilton U.S. Custom House in New York City.

For further information, contact the National Museum of the American Indian, Fourth Street and Independence Avenue SW., Washington, DC 20560. Phone, 202–633–1000. Internet, http://www.nmai. si.edu.

National Museum of Natural History Dedicated to understanding the natural world and the place of humans in it, the Museum's permanent exhibitions focus on human cultures, Earth sciences, biology, and anthropology, with the most popular displays featuring gem stones such as the Hope Diamond, dinosaurs, insects, marine ecosystems, birds, and mammals. In 2010, the Museum celebrated its 100th anniversary with the opening of a new permanent exhibition, the David H. Koch Hall of Human Origins. An IMAX theater offers large-format films. The Museum's encyclopedic collections comprise more than 126 million specimens, making the Museum one of the world's foremost facilities for natural history research. The Museum's seven departments are anthropology, botany, entomology, invertebrate zoology, mineral sciences, paleobiology, and vertebrate zoology. Doctorate-level staff researchers ensure the continued growth and value of the collection by conducting studies in the field and laboratory.

For further information, contact the National Museum of Natural History, Tenth Street and Constitution Avenue NW., Washington, DC 20013. Phone, 202–633–1000. Internet, http://www.mnh.si.edu.

National Portrait Gallery   The Gallery was established in 1962 for the exhibition and study of portraiture depicting men and women who have made significant contributions to the history, development, and culture of the United States. The Gallery contains more than 19,000 works, including photographs and glass negatives. The first floor of the Gallery is devoted to changing exhibitions from the Gallery's collection of paintings, sculpture, prints, photographs, and drawings as well as to special portrait collections. Featured on the second floor are the permanent collection of portraits of eminent Americans and the Hall of Presidents, including the famous Gilbert Stuart portrait-from-life of George Washington. The two-story American Victorian Renaissance Great Hall on the third floor of the Gallery houses a Civil War exhibit and is used for special events and public programs. The Gallery shares a large library with the Smithsonian American Art Museum and the Archives of American Art. The education department offers public programs; outreach programs for schools, senior adults, hospitals, and nursing homes; and walk-in and group tours.

For further information, contact the National Portrait Gallery, Eighth and F Streets NW., Washington, DC 20560. Phone, 202–633–1000. Internet, http://www.npg.si.edu.

National Postal Museum   The Museum houses the Nation's postal history and philatelic collection, the largest of its kind in the world, with more than 13 million objects. The Museum is devoted to the history of America's mail service, and major galleries include exhibits on mail service in colonial times and during the Civil War, the Pony Express, modern mail service, automation, mail transportation, and the art of letters, as well as displays of the Museum's priceless stamp collection. Highlights include three mail planes, a replica of a railway mail car, displays of historic letters, handcrafted mail boxes, and rare U.S. and foreign-issue stamps and covers.

For further information, contact the National Postal Museum, 2 Massachusetts Avenue NE., Washington, DC 20560. Phone, 202–633–1000. Internet, http://www.si.edu/postal.

National Zoological Park   The National Zoo is an international leader in wildlife conservation, education, and research. Home to more than 2,000 animals, the Zoo encompasses 163 acres along Rock Creek Park in Northwest Washington, DC. Exhibits include the Fujifilm Giant Panda Habitat, where the giant pandas Mei Xiang and Tian Tian can be found. Built to mimic the animals' natural habitat in China, it is part of the Zoo's Asia Trail, which also takes visitors through the habitats of red pandas, Asian small-clawed otters, fishing cats, sloth bears, and clouded leopards. Other highlights include the Elephant House, home to the Asian elephant Kandula, who was born at the Zoo in 2001; Amazonia, a 15,000-square-foot rain forest habitat; the Reptile Discovery Center, featuring African pancake tortoises and the world's largest lizards, Komodo dragons; and the Great Ape House, home to gorillas, orangutans, and other primates.

For further information, contact the National Zoo, 3000 Connecticut Avenue NW., Washington, DC 20008. Phone, 202–673–4717. Internet, http://www.si.edu/natzoo.

Center for Folklife and Cultural Heritage
The Center is responsible for research, documentation, and presentation of grassroots cultural traditions. It maintains a documentary collection and produces Smithsonian Folkways Recordings,

educational materials, documentary films, publications, and traveling exhibits, as well as the annual Smithsonian Folklife Festival on the National Mall. Recent Folklife festivals have featured a range of American music styles, a number of State tributes, and performers from around the world. Admission to the festival is free. The 2-week program includes Fourth of July activities on the National Mall.

For further information, contact the Center for Folklife and Cultural Heritage, Suite 4100, 750 Ninth Street NW., Washington, DC 20560. Phone, 202–633–1000. Internet, http://www.folklife.si.edu.

International Center   The International Center supports Smithsonian activities abroad and serves as liaison for the Smithsonian's international interests. The Smithsonian seeks to encourage a broadening of public understanding of the histories, cultures, and natural environments of regions throughout the world. The International Center provides a meeting place and an organizational channel to bring together the world's scholars, museum professionals, and the general public to attend and participate in conferences, public forums, lectures, and workshops.

For further information, contact the Office of International Relations, MRC 705, 1100 Jefferson Drive SW., Washington, DC 20560. Phone, 202–633–1000.

Arthur M. Sackler Gallery   This Asian art museum opened in 1987 on the National Mall. Changing exhibitions drawn from major collections in the United States and abroad, as well as from the permanent holdings of the Sackler Gallery, are displayed in the distinctive below-ground museum. The Gallery's growing permanent collection is founded on a group of art objects from China, South and Southeast Asia, and the ancient Near East that were given to the Smithsonian by Arthur M. Sackler (1913–1987). The Museum's current collection features Persian manuscripts; Japanese paintings; ceramics, prints, and textiles; sculptures from India; and paintings and metalware from China, Korea, Japan, and Southeast Asia. The Sackler Gallery is connected by an underground exhibition space to the neighboring Freer Gallery.

For further information, contact the Arthur M. Sackler Gallery, 1050 Independence Avenue SW., Washington, DC 20560. Phone, 202–633–1000. Internet, http://www.asia.si.edu.

Smithsonian Institution Archives   The Smithsonian Institution Archives acquires, preserves, and makes available for research the official records of the Smithsonian Institution and the papers of individuals and organizations associated with the Institution or with its work. These holdings document the growth of the Smithsonian and the development of American science, history, and art.

For further information, contact the Smithsonian Institution Archives, MRC 414, 900 Jefferson Drive SW., Washington, DC 20560. Phone, 202–633–1000.

Smithsonian Astrophysical Observatory   The Smithsonian Astrophysical Observatory and the Harvard College Observatory have coordinated research activities under a single director in a cooperative venture, Harvard-Smithsonian Center for Astrophysics. The Center's research activities are organized in the following areas of study: atomic and molecular physics, radio and geoastronomy, high-energy astrophysics, optical and infrared astronomy, planetary sciences, solar and stellar physics, and theoretical astrophysics. Research results are published in the Center Preprint Series and other technical and nontechnical bulletins and distributed to scientific and educational institutions around the world.

For more information, contact the Smithsonian Astrophysical Observatory, 60 Garden Street, Cambridge, MA 02138. Phone, 617–495–7461. Internet, http://www.cfa.harvard.edu/sao.

Smithsonian Museum Conservation Institute   The Institute researches preservation, conservation, and technical study and analysis of collection materials. Its researchers investigate the chemical and physical processes that are involved in the care of art, artifacts, and specimens and attempt to formulate conditions and procedures for storage, exhibit, and stabilization that optimize the preservation of these objects. In interdisciplinary collaborations with archeologists, anthropologists, and art historians, natural and physical scientists study and analyze objects from the collections and related materials to expand knowledge and understanding of their historical and scientific context.

For further information, contact the Museum Conservation Institute, Museum Support Center, Suitland, MD 20746. Phone, 301–238–1240.

Smithsonian Environmental Research Center (SERC)

The Center is the leading national research center for understanding environmental issues in the coastal zone. SERC is dedicated to increasing knowledge of the biological and physical processes that sustain life on Earth. The Center, located near the Chesapeake Bay, trains future generations of scientists to address ecological questions of the Nation and the globe.

For further information, contact the Smithsonian Environmental Research Center, 647 Contees Wharf Road, Edgewater, MD 21037. Phone, 443–482–2200. Internet, http://www.serc.si.edu.

Smithsonian Institution Libraries

The Smithsonian Institution Libraries include more than 1 million volumes (among them, 40,000 rare books) with strengths in natural history, art, science, humanities, and museology. Many volumes are available through interlibrary loan.

For further information, contact the Smithsonian Institution Libraries, Tenth Street and Constitution Avenue NW., Washington, DC 20560. Phone, 202–633–2240. Email, libhelp@sil.si.edu. Internet, http://www.sil.si.edu.

Smithsonian Institution Traveling Exhibition Service (SITES)

Since 1952, SITES has been committed to making Smithsonian exhibitions available to millions of people who cannot view them firsthand at the Smithsonian museums. Exhibitions on art, history, and science travel to more than 250 locations each year.

For further information, contact the Smithsonian Institution Traveling Exhibition Service, Suite 7103, 470 L'Enfant Plaza SW., Washington, DC 20024. Phone, 202–633–1000.

Smithsonian Marine Station

The research institute features a state-of-the-art laboratory where Station scientists catalog species and study marine plants and animals. Among the most important projects being pursued at the site is the search for possible causes of fishkills, including Pfiesteria and other organisms.

For further information, contact the Smithsonian Marine Station, 701 Seaway Drive, Fort Pierce, FL 34946. Phone, 772–465–6630. Internet, http://www.sms.si.edu.

Smithsonian Tropical Research Institute (STRI)

The Institute is a research organization for advanced studies of tropical ecosystems. Headquartered in the Republic of Panama, STRI maintains extensive facilities in the Western Hemisphere tropics. It is the base of a corps of tropical researchers who study the evolution, behavior, ecology, and history of tropical species of systems ranging from coral reefs to rain forests.

For further information, contact the Smithsonian Tropical Research Institute, 1100 Jefferson Drive SW., Suite 3123, Washington, DC 20560. Phone, 202–633–4016. Phone (Panama), 011–507–212–8000. Internet, http://www.stri.org.

Sources of Information

Contracts and Small Business Activities Information regarding procurement of supplies, property management and utilization services for Smithsonian Institution organizations, and contracts for construction, services, etc., may be obtained from the Director, Office of Contracting, Smithsonian Institution, Washington, DC 20560. Phone, 202–275–1600.

Education and Research Write to the Directors of the following offices at the Smithsonian Institution, Washington, DC 20560: Office of Fellowships and Grants, Center for Folklife and Cultural Heritage, National Science Resources Center, and Smithsonian Center for Education and Museum Studies.

Electronic Access Information about the Smithsonian Institution is available electronically through the Internet at www.si.edu or www.smithsonian.org.

Employment Employment information for the Smithsonian is available from the Office of Human Resources, Smithsonian Institution, Suite 6100, 750 Ninth Street NW., Washington, DC 20560. Phone, 202–275–1102. Recorded message, 202–287–3102.

Media Affairs Members of the press may contact the Smithsonian Office of Public Affairs, 1000 Jefferson Drive SW., Washington, DC 20560. Phone, 202–633–2400. Internet, http://newsdesk. si.edu.

Memberships   For information about Smithsonian membership (Resident Program), write to the Smithsonian Associates, MRC 701, 1100 Jefferson Drive SW., Washington, DC 20560. Phone, 202–357–3030. For information about Smithsonian membership (National Program), call 202–357–4800. For information about the Contributing Membership, call 202–357–1699. For information about the Young Benefactors, call 202–786–9049. Information about activities of the Friends of the National Zoo and their magazine, The Zoogoer, is available by writing to FONZ, National Zoological Park, Washington, DC 20008. Phone, 202–673–4950.

Photographs   Color and black-and-white photographs and slides are available to Government agencies, research and educational institutions, publishers, and the general public from the Smithsonian photographic archives. A searchable database of images is available through the Internet. Information, order forms, and price lists may be obtained from the Office of Imaging, Printing, and Photographic Services, MAH CB–054, Smithsonian Institution, Washington, DC 20560. Internet, http://photos.si.edu. Email, psdmx@sivm.si.edu.

Publications   To purchase the Smithsonian Institution's annual report, Smithsonian Year, call 202–633–2400. The Smithsonian Institution Press publishes a range of books and studies related to the sciences, technology, history, culture, air and space, and the arts. A book catalog is available from Publications Sales. Phone, 800–782–4612. To purchase a recording of the Smithsonian Folkways Recordings, call 800–410–9815. Internet, www.si.edu/folkways.

A free brochure providing a brief guide to the Smithsonian Institution is published in English and several foreign languages. For a copy, call Visitor Information at 202–633–1000 or pick up a copy at the information desks in the museums.

A visitor's guide for individuals with disabilities is also available.

Smithsonian Institution Research Reports, containing news of current research projects in the arts, sciences, and history that are being conducted by Smithsonian staff, is produced by the Smithsonian Office of Public Affairs, Smithsonian Institution Building, 1000 Jefferson Drive SW., Washington, DC 20560. Phone, 202–633–2400.

To request a copy of Smithsonian Runner, a newsletter about Native American-related activities at the Smithsonian, contact the National Museum of the American Indian, Smithsonian Institution, Washington, DC 20560. Phone, 800–242–NMAI.

For the newsletter Art to Zoo for teachers of fourth through eighth graders, write to the Smithsonian Center for Education and Museum Studies, Room 1163, MRC 402, Arts and Industries Building, Washington, DC 20560. Phone, 202–357–2425.

Telephone   Dial-A–Museum, 202–633–1000, provides a taped message with daily announcements on new exhibits and special events. For a Spanish listing of Smithsonian events, call 202–633–9126.

Tours   For information about museum and gallery tours, contact the Smithsonian Information Center, 1000 Jefferson Drive SW., Washington, DC 20560. Phone, 202–633–1000. School groups are welcome. Special behind-the-scenes tours are offered through the various memberships.

Visitor Information   The Smithsonian Information Center, located in the original Smithsonian building, commonly known as The Castle, provides general orientation through films, computer interactive programs, and visitor information specialists to help members and the public learn about the national collections, museum events, exhibitions, and special programs. Write to the Smithsonian Information Center, 1000 Jefferson Drive SW., Washington, DC 20560. Phone, 202–633–1000. TTY, 202–633–5285.

Volunteer Service Opportunities   The Smithsonian Institution welcomes volunteers and offers a variety of interesting service opportunities. For information, write to the Visitor Information and Associates' Reception Center, 1000 Jefferson Drive SW., Washington, DC 20560. Phone, 202–633–1000. TTY, 202–633–5285.

For further information, contact the Smithsonian Information Center, 1000 Jefferson Drive SW., Washington, DC 20560. Phone, 202–633–1000. TDD, 202–357–1729. Internet, http://www.smithsonian.org.

John F. Kennedy Center for the Performing Arts

John F. Kennedy Center for the Performing Arts, Washington, DC 20566
Phone, 202–467–4600. Internet, http://www.kennedy-center.org.

Chairman	DAVID M. RUBENSTEIN
President	MICHAEL M. KAISER

The Kennedy Center is the only official memorial to President Kennedy in Washington, DC. Since its opening in 1971, the Center has presented a year-round program of the finest in music, dance, and drama from the United States and abroad. The Kennedy Center box offices are open daily, and general information and tickets may be obtained by calling 202–467–4600 or 202–416–8524 (TTY). A limited number of half-price tickets are available for some attractions to eligible patrons, including senior citizens over the age of 65, enlisted military personnel of grade E–1 through E–4, persons with fixed low incomes, full-time students, and persons with permanent disabilities. Visitor services are provided by the Friends of the Kennedy Center volunteers. Tours are available free of charge between 10 a.m. and 5 p.m. on weekdays and between 10 a.m. and 1 p.m. on weekends. Free performances are given every day at 6 p.m. on the Millennium Stage in the Grand Foyer.

Sources of Information

Contracts and Small Business Activities Contact the John F. Kennedy Center for the Performing Arts, Washington, DC 20566.

Education and Research For information regarding Kennedy Center education programs, contact the John F. Kennedy Center for the Performing Arts, Washington, DC 20566. Phone, 202–416–8000.

Electronic Access Information on the John F. Kennedy Center for the Performing Arts is available through the Internet at www.kennedy-center.org.

Employment For information on employment opportunities at the John F. Kennedy Center for the Performing Arts, contact the Human Resources Department, Washington, DC 20566. Phone, 202–416–8610.

Memberships Information about the national and local activities of Friends of the Kennedy Center (including the bimonthly Kennedy Center News for members) is available at the information desks within the Center or by writing to Friends of the Kennedy Center, Washington, DC 20566.

Special Functions Inquiries regarding the use of Kennedy Center facilities for special functions may be directed to the Office of Special Events, John F. Kennedy Center for the Performing Arts, Washington, DC 20566. Phone, 202–416–8000.

Theater Operations Inquiries regarding the use of the Kennedy Center's theaters may be addressed to the Booking Coordinator, John F. Kennedy Center for the Performing Arts, Washington, DC 20566. Phone, 202–416–8000.

Volunteer Service Opportunities For information about volunteer opportunities at the Kennedy Center, write to Friends of the Kennedy Center, Washington, DC 20566. Phone, 202–416–8000.

For further information, contact the Kennedy Center. Phone, 202–467–4600. Internet, http://www.kennedy-center.org.

National Gallery of Art

4th and Constitution Avenue NW., Washington, DC 20565
Phone, 202–737–4215. Internet, http://www.nga.gov.

President   VICTORIA P. SANT
Director   EARL A. POWELL, III

Activities

The Gallery houses one of the finest
collections in the world, illustrating
Western man's achievements in painting,
sculpture, and the graphic arts. The West
Building includes European (13th through
early 20th century) and American (18th
through early 20th century) works. An
extensive survey of Italian painting and
sculpture, including the only painting
by Leonardo da Vinci in the Americas,
is presented here. Rich in Dutch masters
and French impressionists, the collection
offers superb surveys of American,
British, Flemish, Spanish, and 15th
and 16th century German art, as well
as Renaissance medals and bronzes,
Chinese porcelains, and about 97,000
works of graphic art from the 12th to
the 20th centuries. The East Building
collections and Sculpture Garden
include important works by major 20th-
century artists. The Gallery represents
a partnership of Federal and private
resources. Its operations and maintenance
are supported through Federal
appropriations, and all of its acquisitions
of works of art, as well as numerous
special programs, are made possible
through private donations and funds.
Graduate and postgraduate research is
conducted under a fellowship program;
education programs for schoolchildren
and the general public are conducted
daily; and an extension service provides
slide teaching and multimedia programs,
videocassettes, CD–ROMs, DVDs, and
videodiscs to millions of people each
year.

Sources of Information

Calendar of Events   The Calendar of
Events is available through the Internet
at www.nga.gov/programs/calendar/.
To receive email notices when new
calendars go online, send your name,
street address, and email address to
calendar@nga.gov.
Concerts   Concerts by world-renowned
musicians are presented Sunday
evenings from October through June.
For information, call the Concert Line at
202–842–6941. Internet, www.nga.gov/
programs/music.
Contracts and Small Business Activities
Contact National Gallery of Art, Office
of Procurement and Contracts, 2000B
South Club Drive, Landover, MD 20785.
Phone, 202–842–6745.
Educational Resources   The National
Gallery of Art provides slide teaching and
multimedia programs, videocassettes,
CD–ROMs, DVDs, and videodiscs at no
charge to individuals, schools, and civic
organizations throughout the country.
Contact the Department of Education
Resources, National Gallery of Art,
2000B South Club Drive, Landover, MD
20785. Phone, 202–842–6273. Internet,
www.nga.gov/education/classroom/
loanfinder. Please write or email
EdResources@nga.gov to request a free
catalog of programs.
Electronic Access   Information on
the National Gallery of Art is available
through the Internet at www.nga.gov.
NGAkids (www.nga.gov/kids) includes
interactive activities and adventures with
works of art in the Gallery's collection
and an animated tale set in the Gallery's
Sculpture Garden.
Employment   For information on
employment opportunities at the National
Gallery, contact the Personnel Office,
National Gallery of Art, 601 Pennsylvania
Avenue South NW., Second Floor,
Washington, DC 20004. Phone, 202–
842–6282. TDD, 202–842–6176. Internet,
www.nga.gov/resources/employ.htm.
Family Programs   The Gallery offers a
full range of free family programs suitable
for children ages 4 and up, including
workshops, children's films, music

performances, and storytelling. Phone, 202–789–3030. Internet, www.nga.gov/kids.

Fellowships   For information about research fellowship programs, contact the Center for Advanced Study in the Visual Arts. Phone, 202–842–6482. Fax, 202–842–6733. Internet, www.nga.gov/resources/casva.htm.

Films   An ongoing free program of independent films, major retrospectives, classic cinema, and area premieres are presented. Visiting filmmakers and scholars are often invited to discuss films with the audience following screenings. The auditorium is equipped with an FM wireless listening system for the hearing impaired. Receivers, earphones, and neck loops are available at the East Building Art Information Desk near the main entrance. Phone, 202–842–6799. Internet, www.nga.gov/programs/film. htm.

Internships   For information about National Gallery internship programs for college graduates, master's degree students, and Ph.D. candidates, contact the Department of Academic Programs, National Gallery of Art, 2000B South Club Drive, Landover, MD 20785. Email, intern@nga.gov. Phone, 202–842–6257. Fax, 202–842–6935.

Lectures   An ongoing schedule of lectures, symposia, and works in progress are free and open to the public on a first-come, first-serve basis. Internet, www. nga.gov/programs/lectures.

Library   The Gallery's collection of more than 330,000 books and periodicals on the history, theory, and criticism of art and architecture emphasizes Western art from the Middle Ages to the present and American art from the colonial era to the present. The library is open by appointment on Mondays from 12 noon to 4:30 p.m. and Tuesday through Friday from 10 a.m. to 4:30 p.m. It is closed on all Federal holidays. Adult researchers may gain access to the library by calling 202–842–6511. Internet, www.nga.gov/resources/dldesc.shtm.

Library Image Collections   The Department of Image Collections is the study and research center for images of Western art and architecture

at the National Gallery of Art. The collection now numbers nearly 10 million photographs, slides, negatives, and microform images, making it one of the largest resources of its kind. The Department serves the Gallery's staff, members of the Center for Advanced Study in the Visual Arts, visiting scholars, and serious adult researchers. The library is open by appointment on Mondays from 12 noon to 4:30 p.m. and Tuesday through Friday from 10 a.m. to 4:30 p.m. It is closed on all Federal holidays. Phone, 202–842–6026. Internet, www. nga.gov/resources/dlidesc.shtm.

Memberships   The Circle of the National Gallery of Art is a membership program which provides support for special projects for which Federal funds are not available. For more information about membership in The Circle of the National Gallery of Art, please write to The Circle, National Gallery of Art, 2000B South Club Drive, Landover, MD 20785. Phone, 202–842–6450. Internet, www. nga.gov/support.

Publications   The National Gallery shop makes available quality reproductions and publications about the Gallery's collections. To order, call 202–842–6002. Items are also available for sale online at shop.nga.gov. The Office of Press and Public Information offers a free bimonthly calendar of events, which can be ordered by calling 202–842–6662 or through email at calendar@nga.gov. The calendar and Brief Guide to the National Gallery of Art are also available at art information desks throughout the Gallery or by calling Visitor Services at 202–842–6691.

Tours   The Education Division of the National Gallery of Art offers gallery talks and lectures. Phone, 202–842–6247. Internet, www.nga.gov/education/school. htm or www.nga.gov/programs/tours.htm.

Visitor Services   The Visitor Services Office of the National Gallery of Art provides individual assistance to those with special needs, responds to written and telephone requests, and provides information to those planning to visit the Washington, DC, area. For more information, write to the National Gallery of Art, Office of Visitor Services, 2000B South Club Drive, Landover, MD 20785.

Phone, 202–842–6691. Internet, www. nga.gov/ginfo/index.shtm.
Volunteer Opportunities   For information about volunteering as a docent or as an Art Information Desk volunteer, please call Volunteer Opportunities at 202–789–3013. Internet, www.nga.gov/education/volunteer.htm.
Library Volunteering   Phone, 202–842–6510. Internet, www.nga.gov/education/volunteer.shtm.

Horticulture Volunteers   Phone, 202–842–6844. Email, gardens@nga.gov.
Works on Paper   Works of art on paper that are not on view may be seen by appointment on weekdays by calling 202–842–6380. The Matisse cutouts are on view in the East Building Concourse from 10 a.m. to 3 p.m., Monday through Saturday, and from 11 a.m. to 4 p.m. on Sunday.

For further information, contact the National Gallery of Art. Phone, 202–737–4215. TTY, 202–842–6176. Internet, http://www.nga.gov.

Woodrow Wilson International Center for Scholars

Scholar Administration Office, Woodrow Wilson Center, One Woodrow Wilson Plaza, 1300 Pennsylvania Avenue NW., Washington, DC 20004–3027
Phone, 202–691–4000. Fax, 202–691–4001. Internet, http://www.wilsoncenter.org.

Director	LEE H. HAMILTON
Deputy Director	MICHAËL VAN DUSEN
Chairman, Board of Trustees	JOSEPH B. GILDENHORN

Activities

The Center was established by Congress in 1968 as the Nation's official memorial to its 28th President. The Center is a nonpartisan institution of advanced study that promotes scholarship in public affairs, supported by both public and private funds. The Center convenes scholars and policymakers, businesspeople, and journalists in a neutral forum for open, serious, and informed dialogue. The Center supports research in social sciences and humanities, with an emphasis on history, political science, and international relations. In providing an essential link between the worlds of ideas and public policy, the Center addresses current and emerging challenges confronting the United States and the world.

Sources of Information

Electronic Access   Information on the Woodrow Wilson Center is available through the Internet at www.wilsoncenter.org.
Employment   For information on employment opportunities at the Woodrow Wilson Center, contact the Office of Human Resources, One Woodrow Wilson Plaza, 1300 Pennsylvania Avenue NW., Washington, DC 20004–3027. Internet, www.wilsoncenter.org/hr/index.htm.
Fellowships and Internship   The Woodrow Wilson Center offers residential fellowships that allow academics, public officials, journalists, business professionals, and others to pursue their research and writing at the Center while interacting with policymakers in Washington. The Center also invites public policy scholars and senior scholars from a variety of disciplines to conduct research for varying lengths of time in residence. For more information, call 202–691–4213. The Center also has a year-round need for interns to assist the program and projects staff and to act as research assistants for scholars and fellows. For more information, call 202–691–4053.
Media Affairs   Members of the press may contact the Woodrow Wilson Center at 202–691–4016.
Presidential Memorial Exhibit   The Woodrow Wilson Center houses the Woodrow Wilson Presidential Memorial Exhibit which includes memorabilia,

historical information, photographs, several short films, and a memorial hall with quotations. The exhibit is open Monday through Friday, 8:30 a.m. to 5 p.m. Admission is free.

Publications The Woodrow Wilson Center publishes the monthly newsletter Centerpoint, as well as books written by staff and visiting scholars and fellows, through the Wilson Center Press. It also produces Dialogue, a weekly radio and television program about national and international affairs, history, and culture. For more information, call 202–691–4016.

Visitor Services To hear a listing of events at the Woodrow Wilson Center, call 202–691–4188. All events, unless otherwise noted, are free and open to the public. Please note that photo identification is required for entry.

For further information, contact the Scholar Administration Office, Woodrow Wilson International Center for Scholars, One Woodrow Wilson Plaza, 1300 Pennsylvania Avenue NW., Washington, DC 20004–3027. Phone, 202–691–4000. Fax, 202–691–4001. Internet, http://www.wilsoncenter.org.

STATE JUSTICE INSTITUTE

Suite 600, 1650 King Street, Alexandria, VA 22314
Phone, 703–684–6100. Internet, http://www.sji.gov.

Board of Directors

Chairman	ROBERT A. MILLER
Vice Chairman	JOSEPH F. BACA
Secretary	SANDRA A. O'CONNOR
Executive Committee Member	KEITH MCNAMARA
Members	TERRENCE B. ADAMSON, ROBERT N. BALDWIN, SOPHIA H. HALL, TOMMY JEWELL, ARTHUR MCGIVERIN, (2 VACANCIES)

Officers

Executive Director	JANICE MUNSTERMAN
Deputy Director	JONATHAN MATTIELLO

The State Justice Institute awards grants to improve judicial administration in the State courts of the United States.

The State Justice Institute (SJI) was created by the State Justice Institute Act of 1984 (42 U.S.C. 10701) as a private, nonprofit corporation to further the development and improvement of judicial administration in the State courts.

SJI is supervised by a Board of Directors consisting of 11 members appointed by the President with the advice and consent of the Senate. The Board is statutorily composed of six judges, a State court administrator, and four members of the public, of whom no more than two can be of the same political party.

In carrying out its mission, SJI develops solutions to common issues faced by State and Federal courts; provides practical products to judges and court staff; ensures that effective approaches in one State are quickly and economically shared with other courts nationwide; supports national, regional, and in-State educational programs to speed the transfer of solutions; and delivers targeted technical assistance to specific jurisdictions.

To accomplish these broad objectives, SJI is authorized to provide funds, through grants, cooperative agreements, and contracts, to State courts and organizations that can assist in the improvement of judicial administration in the State courts.

Sources of Information

Inquiries concerning grants, publications, speakers, or Privacy Act/Freedom of Information Act requests should be directed to the Executive Director or Deputy Director, State Justice Institute, Suite 600, 1650 King Street, Alexandria, VA 22314. Phone, 703–684–6100.

Information regarding the programs and services of the State Justice Institute is also available through the Internet at www.sji.gov.

For further information, contact the State Justice Institute, Suite 600, 1650 King Street, Alexandria, VA 22314. Phone, 703–684–6100. Internet, www.sji.gov.

UNITED STATES INSTITUTE OF PEACE

Suite 200, 1200 Seventeenth Street NW., Washington, DC 20036
Phone, 202–457–1700. Fax, 202–429–6063. Internet, http://www.usip.org.

Board of Directors

Chairman	J. ROBINSON WEST
Vice Chairman	GEORGE E. MOOSE
Members	ANNE H. CAHN, CHESTER A. CROCKER, KERRY KENNEDY, IKRAM U. KHAN, KATHLEEN MARTINEZ, JEREMY A. RABKIN, JUDY VAN REST, NANCY ZIRKIN

Ex Officio Members

Secretary of State	HILLARY RODHAM CLINTON
Secretary of Defense	ROBERT M. GATES
President, National Defense University	ANN E. RONDEAU
President, U.S. Institute of Peace	RICHARD H. SOLOMON

Officials

President	RICHARD H. SOLOMON
Executive Vice President	TARA D. SONENSHINE
Vice President, Headquarters Project	CHARLES E. NELSON
Vice President for Management & CFO	MICHAEL B. GRAHAM
Vice President, Grants and Fellowships	STEVEN HEYDEMANN
Vice President, Center for Conflict Analysis and Prevention	ABIODUN WILLIAMS
Vice President, Domestic Education and Training Center	PAMELA AALL
Vice President, International Education and Training Center	MICHAEL LEKSON
Vice President, Center for Post-conflict Peace and Stability Operations	WILLIAM TAYLOR
Vice President, Center for Mediation and Conflict Resolution	DAVID R. SMOCK
Vice President, External Affairs	(VACANCY)
Vice President, Innovation	DANIEL SERWER
Chief Administrative Officer	CHRISTOPHER DE PAOLA
Chief Information Officer	DOUGLAS LEINS
Chief Human Resource Officer	PAULA KING
Director, Congressional Relations	LAURIE SCHULTZ-HEIM
Director, Public Affairs and Communications	LAUREN SUCHER
Director, Publications	VALERIE NORVILLE
Director, Intergovernmental Affairs	BETH ELLEN COLE

The United States Institute of Peace promotes research, policy analysis, education, and training on international peace and conflict resolution.

The United States Institute of Peace (USIP) is an independent institution, established by Congress pursuant to title XVII of the Defense Authorization Act of 1985, as amended (22 U.S.C. 4601–4611). USIP's mission is to help prevent and resolve violent conflicts, promote postconflict stability and development, and increase peacebuilding capacity, tools, and intellectual capital worldwide. The Institute achieves this by empowering others with knowledge, skills, and resources, as well as by directly engaging in peacebuilding throughout the world.

The Institute is governed by a bipartisan Board of Directors appointed by the President and confirmed by the Senate. The Board is comprised of members from outside the Federal service, the Secretary of State, the Secretary of Defense, and the President of the National Defense University. The Board appoints the President of the Institute.

Programs and Activities

In carrying out its mission, USIP operates on the ground in conflict zones, providing services that include mediating and facilitating dialogue among parties in conflict; building conflict management skills and capacity; indentifying and disseminating best practices in conflict management; promoting the rule of law; reforming and strengthening education systems; strengthening civil society and statebuilding; and educating the public through film, radio, the Internet, special events, and other outreach activities.

The Institute conducts and sponsors relevant research on causes of and solutions to violent conflict. Drawing on this intellectual capital, USIP identifies promising models and innovative approaches and practices. USIP shares these tools with others through its publications, the Internet, and training programs. Tools developed by USIP include book series on international mediation and cultural negotiating behavior, resources on religious peacemaking, a toolkit for promoting the rule of law in fragile states, guidelines for civilian and military interactions in hostile environments, and textbooks on international conflict management.

USIP works in partnership with nongovernmental organizations, higher and secondary educational institutions, and international organizations to promote collaborative problemsolving through conferences, standing working groups, Track II diplomacy, and special events. The Institute offers training on conflict management, including mediation and negotiating skills, for government and military personnel, civil society leaders, and staff of nongovernmental and international organizations.

As part of its efforts to strengthen and professionalize the field of international conflict management, the Institute shares the resources and tools it has developed with the larger conflict management community. In conflict zones, USIP works in partnership with local organizations to build their capacity and promote sustainability. The Institute also extends its reach by investing in nonprofit organizations in the United States and overseas.

Strategic Centers   The Institute includes strategic centers focused on preventing violent conflicts before they occur, mediating and resolving conflicts when they do occur, and promoting postconflict stability once the fighting ends. The efforts of these cross-disciplinary centers focus on education, training, grantmaking, fellowships, scholarships, and innovation in peacebuilding.

Sources of Information

Electronic access to the Institute is available through the Internet at www. usip.org. For further information, contact the Office of Public Affairs and Communications, United States Institute of Peace, 1200 Seventeenth Street NW., Suite 200, Washington, DC 20036–3011. Phone, 202–457–1700. Fax, 202–429–6063. Internet, http://www.usip.org.

INTERNATIONAL ORGANIZATIONS

AFRICAN DEVELOPMENT BANK

Headquarters (temporary): Angle des Trois Rues, Avenue Du Ghana, Rue Pierre De Coubertin, Rue Hedi Nouira, BP. 323, 1002 Tunis Belvedere, Tunisia
Internet, http://www.afdb.org. Email, afdb@afdb.org.

President   DONALD KABERUKA

The African Development Bank (AFDB) was established in 1964 and, by charter amendment, opened its membership to non-African countries in 1982. Its mandate is to contribute to the economic development and social progress of its regional members. Bank members total 77, including 53 African countries and 24 nonregional countries. With the September 1999 ratification of the agreement on the fifth general capital increase, Bank ownership is 60 percent African and 40 percent nonregional.

The African Development Fund (AFDF), the concessional lending affiliate, was established in 1972 to complement AFDB operations by providing concessional financing for high-priority development projects in the poorest African countries. The Fund's membership consists of 25 nonregional member countries, South Africa, and AFDB, which represents its African members and is allocated half of the votes.

In February 2003, security concerns resulted in AFDB headquarters temporarily relocating to Tunis, Tunisia.

ASIAN DEVELOPMENT BANK

Headquarters: 6 ADB Avenue, Mandaluyong City, 1550 Metro Manila, Philippines
Phone, 632–632–4444. Fax, 632–636–2444. Internet, http://www.adb.org. Email, information@adb.org.

President   HARUHIKO KURODA

The Asian Development Bank commenced operations on December 19, 1966. It now has 67 member countries: 48 from Asia and 19 from outside the region.

The purpose of the Bank is to foster sustainable economic development, poverty alleviation, and cooperation among its developing member countries in the Asia/Pacific region.

For further information, contact the Asian Development Bank, P.O. Box 789, 0980 Manila, Philippines, or the ADB North American Representative Office, 815 Connecticut Avenue NW., Washington, DC 20006. Phone, 202–728–1500. Email, adbnaro@adb.org.

EUROPEAN BANK FOR RECONSTRUCTION AND DEVELOPMENT

U.S. Executive Director's Office, European Bank for Reconstruction and Development, One Exchange Square, London EC2A 2JN, United Kingdom
Phone, +44 20 7338 6000. Internet, http://www.ebrd.com.

President	THOMAS MIROW

The EBRD is an international financial institution that supports projects in 29 countries from central Europe to central Asia. Investing primarily in private sector clients whose needs cannot be fully met by the market, the Bank promotes entrepreneurship and fosters transition towards open and democratic market economies.

The Bank, which is owned by 61 countries and two intergovernmental institutions, is based in London.

INTER-AMERICAN DEFENSE BOARD

2600 Sixteenth Street NW., Washington, DC 20441
Phone, 202–939–6041. Fax, 202–387–2880. Internet, http://www.jid.org. Email, protocol1@jid.org.

Chairman	LT. GEN. JOSE ROBERTO MACHADO E SILVA

The Inter-American Defense Board is the oldest permanently constituted, international military organization in the world. It was founded by Resolution XXXIX of the Meeting of Foreign Ministers at Rio de Janeiro in January 1942. Senior army, navy, and air force officers from the member nations staff the various agencies of the Board. Its four major components are the Council of Delegates, the decisionmaking body; the International Staff; the Inter-American Defense College; and the Secretariat, which provides administrative and logistical support.

The Board studies and recommends to member governments measures it feels are necessary for the safety and security of the hemisphere. It also acts as a technical military adviser for the Organization of American States and is involved in projects such as disaster preparedness and humanitarian demining programs in Central and South America.

The Inter-American Defense College, founded in 1962, prepares senior military officers and civilian functionaries for positions in their respective governments. The College's multidisciplinary program uses four annual seminars to focus on the Western Hemisphere's most pressing defense and security issues.

INTER-AMERICAN DEVELOPMENT BANK

Headquarters: 1300 New York Avenue NW., Washington, DC 20577
Phone, 202–623–1000. Internet, http://www.iadb.org.

President	LUIS ALBERTO MORENO

The Inter-American Development Bank (IDB) was established in 1959 to help accelerate economic and social development in Latin America and the Caribbean. It is based in Washington, DC.

The Bank has 47 member countries, 26 of which are borrowing members in Latin America and the Caribbean.

INTER-AMERICAN INVESTMENT CORPORATION

Headquarters: 1350 New York Avenue NW., Washington, DC 20577
Phone, 202–623–3900. Internet, http://www.iic.int.

Chairman, Board of Directors	LUIS ALBERTO MORENO
General Manager	JACQUES ROGOZINSKI

The Inter-American Investment Corporation (IIC), an affiliate of the Inter-American Development Bank based in Washington, DC, began operations in 1989 to promote the economic development of its Latin American and Caribbean members by financing small- and medium-size private enterprises. IIC provides project financing in the form of direct loans and equity investments, lines of credit to local financial intermediaries, and investments in local and regional investment funds.

IIC has 43 member countries, of which 27 are in the Western Hemisphere, including the United States, and 16 are outside the region.

INTERNATIONAL BANK FOR RECONSTRUCTION AND DEVELOPMENT

Headquarters: 1818 H Street NW., Washington, DC 20433
Phone, 202–473–1000. Internet, http://www.worldbank.org.

President	ROBERT ZOELLICK

The International Bank for Reconstruction and Development (IBRD), also known as the World Bank, officially came into existence on December 27, 1945.

The Bank's purpose is to promote economic, social, and environmental progress in developing nations by reducing poverty so that their people may live better and fuller lives. The Bank lends funds at market-determined interest rates, provides advice, and serves as a catalyst to stimulate outside investments. Its resources come primarily from funds raised in the world capital markets, its retained earnings, and repayments on its loans.

International Development Association The International Development Association (IDA) came into existence on September 24, 1960, as an affiliate of IBRD. The Association's resources consist of subscriptions and supplementary resources in the form of general replenishments, mostly from its more industrialized and developed members; special contributions by its richer members; repayments on earlier credits; and transfers from IBRD's net earnings.

The Association promotes economic development, reduces poverty, and raises the standard of living in the least developed areas of the world. It does this by financing their developmental requirements on concessionary terms, which are more flexible and bear less heavily on the balance of payments than those of conventional loans, thereby furthering the objectives of IBRD and supplementing its activities.

INTERNATIONAL FINANCE CORPORATION

Headquarters: 2121 Pennsylvania Avenue NW., Washington, DC 20433
Phone, 202–473–3800. Internet, http://www.ifc.org.

President	ROBERT ZOELLICK
Executive Vice President	LARS THUNELL

The International Finance Corporation (IFC), an affiliate of the World Bank, was established in July 1956 to promote productive private enterprise in developing member countries.

The Corporation pursues its objective principally through direct debt and equity investments in projects that establish new businesses or expand, modify, or diversify existing businesses. It also encourages cofinancing by other investors and lenders.

Additionally, advisory services and technical assistance are provided by IFC to developing member countries in areas such as capital market development, privatization, corporate restructuring, and foreign investment.

INTERNATIONAL MONETARY FUND

700 Nineteenth Street NW., Washington, DC 20431
Phone, 202–623–7000. Fax, 202–623–4661. Internet, http://www.imf.org.

Managing Director and Chairman of the Executive Board	DOMINIQUE STRAUSS-KAHN
First Deputy Managing Director	JOHN LIPSKY
Deputy Managing Directors	MURILO PORTUGAL, NAOYUKI SHINOHARA

The Final Act of the United Nations Monetary and Financial Conference, signed at Bretton Woods, NH, on July 22, 1944, set forth the original Articles of Agreement of the International Monetary Fund (IMF). The Agreement became effective on December 27, 1945, when the President, authorized by the Bretton Woods Agreements Act (22 U.S.C. 286), accepted membership for the United States in IMF, the Agreement having thus been accepted by countries whose combined financial commitments (quotas) equaled approximately 80 percent of IMF's original quotas. The inaugural meeting of the Board of Governors was held in March 1946, and the first meeting of the Executive Directors was held May 6, 1946.

On May 31, 1968, the Board of Governors approved an amendment to the Articles of Agreement for the establishment of a facility based on Special Drawing Rights (SDR) in IMF and for modification of certain IMF rules and practices. The amendment became effective on July 28, 1969, and the Special Drawing Account became operative on August 6, 1969. United States acceptance of the amendment and participation in the Special Drawing Account were authorized by the Special Drawing Rights Act (22 U.S.C. 286 et seq.).

On April 30, 1976, the Board of Governors approved a second amendment to the Articles of Agreement, which entered into force on April 1, 1978. This amendment gave members the right to adopt exchange arrangements of their choice while placing certain obligations on them regarding their exchange rate policies, over which IMF was to exercise firm surveillance. The official price of gold was abolished, and the SDR account was promoted as the principal reserve asset of the international monetary system. United States acceptance of this amendment was authorized by the Bretton Woods

Agreements Act Amendments (22 U.S.C. 286e-5).

On June 28, 1990, the Board of Governors approved a third amendment to the Articles of Agreement, which became effective on November 11, 1992. Under this amendment, a member's voting rights and certain related rights may be suspended by a 70-percent majority of the executive board if the member, having been declared ineligible to use the general resources of the Fund, persists in its failure to fulfill any of its obligations under the Articles.

As of December 31, 2007, IMF had 185 member countries. Total quotas at the end of December 2007 were SDR 217.3 billion (about $343.4 billion).

The IMF promotes international monetary cooperation through a permanent forum for consultation and collaboration on international monetary problems; facilitates the expansion and balanced growth of international trade; promotes exchange rate stability; assists in the establishment of an open multilateral system of payments for current transactions among members; and gives confidence to members by making

IMF resources temporarily available to them under adequate safeguards.

IMF helps its members correct imbalances in their international balances of payments. It periodically examines the economic developments and policies of its member countries, offers policy advice, and at member's request and upon executive board approval, provides financial assistance through a variety of financial facilities designed to address specific problems. These financing mechanisms provide access to the Fund's general resources to offer short-term assistance during crises of market confidence, compensatory financing to countries suffering declines in export earnings, emergency assistance for countries recovering from natural disasters or armed conflict, and low-interest rate resources to support structural adjustment and promote growth in the poorest countries. IMF also provides technical assistance and training to its members. As of December 31, 2007, IMF usable resources were SDR 165.4 billion ($261.4 billion), and 1-year forward commitment capacity was SDR 127.7 billion ($201.7 billion).

For further information, contact the Chief, Public Affairs Division, External Relations Department, International Monetary Fund, 700 Nineteenth Street NW., Washington, DC 20431. Phone, 202–623–7300. Fax, 202–623–6278. Email, publicaffairs@imf.org. Internet, http://www.imf.org.

INTERNATIONAL ORGANIZATION FOR MIGRATION

Headquarters: 17 Route des Morillons, Grand-Saconnex, Geneva. Mailing address, P.O. Box 71, CH–1211, Geneva 19, Switzerland
Phone, 011–41–22–717–9111. Fax, 011–41–22–798–6150. Internet, http://www.iom.int.

Washington Office: Suite 700, 1752 N Street NW., Washington, DC 20036
Phone, 202–862–1826. Fax, 202–862–1879. Email, MRFWashington@iom.int.

New York Office: Suite 1610, 122 E. 42d Street, New York, NY 10168
Phone, 212–681–7000. Fax, 212–867–5887. Email, newyork@iom.int.

Director General	WILLIAM LACY SWING
Deputy Director General	LAURA THOMPSON
Regional Representative (Washington, DC)	RICHARD SCOTT
Chief of Mission	MICHAEL GRAY
Permanent Observer to the United Nations	LUCA DALL'OGLIO

The International Organization for Migration (IOM) was formed in 1951 as the Intergovernmental Committee for European Migration (ICEM) to help

solve the postwar problems of refugees and displaced persons in Europe and to assist in orderly transatlantic migration. It adopted its current name in 1989 to reflect its progressively global outreach. Since its creation, IOM has assisted more than 12 million refugees and migrants in over 125 countries. As of April 2010, 127 governments are members of IOM, and 17 others have observer status. IOM has observer status at the United Nations.

IOM's guiding principle is that humane and orderly migration benefits migrants and societies. In carrying out its mandate, IOM helps migrants, governments, and civil society plan and operate international and national migration programs at the request of its member states and in cooperation with other international organizations. Its major objectives are the processing and movement of migrants and refugees to countries offering them permanent resettlement opportunities; the promotion of orderly migration to meet the needs of both emigration and immigration communities; counter-trafficking activities; the transfer of technology through migration in order to promote the economic, educational, and social advancement of developing countries; the provision of a forum for states and other partners to exchange views; the promotion of cooperation and coordination on migration issues; and technical cooperation and advisory services on migration policies and legislation.

MULTILATERAL INVESTMENT GUARANTEE AGENCY

Headquarters: 1818 H Street NW., Washington, DC 20433
Phone, 202–458–9292. Internet, http://www.miga.org.

President	ROBERT ZOELLICK
Executive Vice President	IZUMI KOBAYASHI

The Multilateral Investment Guarantee Agency (MIGA), an affiliate of the World Bank, was formally constituted in April 1988.

Its basic purpose is to facilitate the flow of foreign private investment for productive purposes to developing member countries by offering long-term political risk insurance in the areas of expropriation, transfer restriction, breach of contract, and war and civil disturbance; and by providing advisory and consultative services. The Agency cooperates with national investment insurance schemes, such as OPIC, and with private insurers.

EDITORIAL NOTE: The Organization of American States did not meet the publication deadline for submitted updated information of activities, functions, and sources of information.

ORGANIZATION OF AMERICAN STATES

Seventeenth Street and Constitution Avenue NW., Washington, DC 20006
Phone, 202–458–3000. Fax, 202–458–3967. Internet, http://www.oas.org.

Secretary General	JOSE MIGUEL INSULZA
Assistant Secretary General	ALBERT R. RAMDIN

The Organization of American States (OAS) brings together the countries of the Western Hemisphere to strengthen cooperation and advance common interests. At the core of the OAS mission is a commitment to democracy. Building on this foundation, OAS works to promote good governance, strengthen human rights, foster peace and security, expand trade, and address the complex problems caused by poverty, drugs, and corruption. Through decisions made by its political bodies and programs carried out by its General Secretariat, OAS promotes greater inter-American cooperation and understanding.

OAS member states have intensified their cooperation since the end of the cold war, taking on new and important challenges. In 1994, the region's 34 democratically elected presidents and prime ministers met in Miami for the First Summit of the Americas, where they established broad political, economic, and social development goals. They have continued to meet periodically since then to examine common interests and priorities. Through the ongoing Summits of the Americas process, the region's leaders have entrusted the OAS with a growing number of responsibilities to help advance the countries' shared vision.

With four official languages—English, Spanish, Portuguese, and French— the OAS reflects the rich diversity of peoples and cultures across the Americas. The OAS has 35 member states: the independent nations of North, Central, and South America, and of the Caribbean. Since 1962, Cuba has been barred from participation by resolution

of the Eighth Meeting of Consultation of Ministers of Foreign Affairs. Countries from all around the world are permanent observers, closely following the issues that are critical to the Americas and often providing key financial support for OAS programs.

Member states set major policies and goals through the General Assembly, which gathers the hemisphere's foreign ministers once a year in regular session. The Permanent Council, made up of ambassadors appointed by member states, meets regularly at OAS headquarters in Washington, DC, to guide ongoing policies and actions. The chairmanship of the Permanent Council rotates every 3 months, in alphabetical order of countries. Each member state has an equal voice, and most decisions are made through consensus.

Also under the OAS umbrella are several specialized agencies that have considerable autonomy: the Pan American Health Organization in Washington, DC; the Inter-American Children's Institute in Montevideo, Uruguay; the Inter-American Institute for Cooperation on Agriculture in San Jose, Costa Rica; and the Pan American Institute of Geography and History and the Inter-American Indian Institute, both in Mexico City.

In 1948, 21 nations of the hemisphere signed the OAS Charter at the Ninth International Conference of American States. They were Argentina, Bolivia, Brazil, Chile, Colombia, Costa Rica, Cuba (barred from participation), Dominican Republic, Ecuador, El Salvador, Guatemala, Haiti, Honduras,

Mexico, Nicaragua, Panama, Paraguay, Peru, United States of America, Uruguay, and Venezuela.

Subsequently, 14 other countries joined the OAS by signing and ratifying the Charter. They were Barbados, Trinidad and Tobago, Jamaica, Grenada, Suriname, Dominica, Saint Lucia, Antigua and Barbuda, Saint Vincent and the Grenadines, the Bahamas, Saint Kitts and Nevis, Canada, Belize, and Guyana. This brings the number of member states to 35.

For further information, contact the Organization of American States, Seventeenth Street and Connecticut Avenue NW., Washington, DC 20006. Phone, 202–458–3000. Fax, 202–458–3967.

UNITED NATIONS

United Nations, New York, NY 10017
Phone, 212–963–1234. Internet, http://www.un.org.

United Nations Office at Geneva: Palais des Nations, 1211 Geneva 10, Switzerland

United Nations Office at Vienna: Vienna International Centre, P.O. Box 500, A–1400, Vienna, Austria

Washington, DC: U.N. Information Centre, Suite 400, 1775 K Street NW., Washington, DC 20006
Phone, 202–331–8670. Fax, 202–331–9191. Internet, http://www.unicwash.org. Email, unicdc@ unicwas.org.

Secretary-General	BAN KI-MOON
Director-General, U.N. Office at Geneva	SERGEI ORDZHONIKIDZE
Director-General, U.N. Office at Vienna	ANTONIO MARIA COSTA
Director, Washington DC Information Centre	WILL DAVIS

The United Nations is an international organization that was set up in accordance with the Charter drafted by governments represented at the Conference on International Organization meeting at San Francisco. The Charter was signed on June 26, 1945, and came into force on October 24, 1945, when the required number of ratifications and accessions had been made by the signatories. Amendments increasing membership of the Security Council and the Economic and Social Council came into effect on August 31, 1965.

The United Nations now consists of 191 member states, of which 51 are founding members.

The purposes of the United Nations set out in the Charter are to maintain international peace and security; to develop friendly relations among nations; to achieve international cooperation in solving international problems of an economic, social, cultural, or humanitarian character and in promoting respect for human rights; and to be a center for harmonizing the actions of nations in the attainment of these common ends.

The principal organs of the United Nations are as follows:

General Assembly   All states that are members of the United Nations are members of the General Assembly. Its functions are to consider and discuss any matter within the scope of the Charter of the United Nations and to make recommendations to the members of the United Nations and other organs. It approves the budget of the organization, the expenses of which are borne by the members as apportioned by the General Assembly.

The General Assembly may call the attention of the Security Council to situations likely to endanger international peace and security, may initiate studies, and may receive and consider reports from other organs of the United Nations. Under the "Uniting for Peace" resolution adopted by the General Assembly in November 1950, if the Security Council fails to act on an apparent threat to or breach of the peace or act of aggression because of lack of unanimity of its five permanent members, the Assembly itself may take up the matter within 24 hours—in emergency special session—and recommend collective measures, including, in case of a breach of the peace or act of aggression, use of armed force when necessary to maintain or restore international peace and security.

The General Assembly normally meets in regular annual session from September through December. It also has met in special sessions and emergency special sessions.

Security Council   The Security Council consists of 15 members, of which 5—the People's Republic of China, France, Russia, the United Kingdom, and the United States of America—are permanent members. The 10 nonpermanent members are elected for 2-year terms by the General Assembly. The primary responsibility of the Security Council is to act on behalf of the members of the United Nations in maintenance of international peace and security. Measures that may be employed by the Security Council are outlined in the Charter.

The Security Council, together with the General Assembly, also elects the judges of the International Court of Justice and makes a recommendation to the General Assembly on the appointment of the Secretary-General of the organization.

The Security Council first met in London on January 17, 1946, and is so organized as to be able to function continuously.

Economic and Social Council   This organ is responsible, under the authority of the General Assembly, for the economic and social programs of the United Nations. Its functions include making or initiating studies, reports, and recommendations on international economic, social, cultural, educational, health, and related matters; promoting respect for and observance of human rights and fundamental freedoms for all; calling international conferences and preparing draft conventions for submission to the General Assembly

on matters within its competence; negotiating agreements with the specialized agencies and defining their relationship with the United Nations; coordinating the activities of the specialized agencies; and consulting with nongovernmental organizations concerned with matters within its competence. The Council consists of 54 members of the United Nations elected by the General Assembly for 3-year terms; 18 are elected each year.

The Council usually holds two regular sessions a year. It has also held a number of special sessions.

Trusteeship Council   The Trusteeship Council was initially established to consist of any member states that administered trust territories, permanent members of the Security Council that did not administer trust territories, and enough other nonadministering countries elected by the General Assembly for 3-year terms to ensure that membership would be equally divided between administering and nonadministering members. Under authority of the General Assembly, the Council considered reports from members administering trust territories, examined petitions from trust territory inhabitants, and provided for periodic inspection visits to trust territories.

With the independence of Palau, the last remaining U.N. trust territory, the Trusteeship Council formally suspended operations after nearly half a century. The Council will henceforth meet only on an extraordinary basis, as the need may arise.

International Court of Justice   The International Court of Justice is the principal judicial organ of the United Nations. It has its seat at The Hague, the Netherlands. All members of the United Nations are ipso facto parties to the Statute of the Court. Nonmembers of the United Nations may become parties to the Statute of the Court on conditions prescribed by the General Assembly on the recommendation of the Security Council.

The jurisdiction of the Court comprises all cases that the parties refer to it and all matters specially provided for in the Charter of the United Nations or in treaties and conventions in force.

The Court consists of 15 judges known as "members" of the Court. They are elected for 9-year terms by the General Assembly and the Security Council, voting independently, and may be reelected.

Secretariat The Secretariat consists of a Secretary-General and "such staff as the Organization may require." The Secretary-General, who is appointed by the General Assembly on the recommendation of the Security Council, is the chief administrative officer of the United Nations. He acts in that capacity for the General Assembly, the Security Council, the Economic and Social Council, and the Trusteeship Council. Under the Charter, the Secretary-General "may bring to the attention of the Security Council any matter that in his opinion may threaten the maintenance of international peace and security."

OTHER INTERNATIONAL ORGANIZATIONS

Below is a list of other international organizations in which the United States participates, but do not have separate entries elsewhere in the Manual. The United States participates in these organizations in accordance with the provisions of treaties, other international agreements, congressional legislation, or executive arrangements. In some cases, no financial contribution is involved.

Various commissions, councils, or committees subsidiary to the organizations listed here are not named separately on this list. These include the international bodies for drugs and crime, which are subsidiary to the United Nations.

This listing is provided for reference purposes and should not be considered exhaustive. For more information on international organizations and United States participation in them, contact the State Department's Bureau of International Organizations. Phone, 202–647–9326. Internet, www.state.gov/p/io.

I. United Nations (UN) and Specialized Agencies of the UN

United Nations
Food and Agricultural Organization
International Atomic Energy Agency
International Civil Aviation Organization
International Fund for Agriculture Development
International Labor Organization
International Maritime Organization
United Nations Educational, Scientific and Cultural Organization
United Nations International Telecommunication Union
Universal Postal Union
World Health Organization
World Intellectual Property Organization
World Meteorological Organization

II. Peacekeeping and Political Missions Administered by UN Department of Peacekeeping Operations

Africa

United Nations Mission in the Central African Republic and Chad (MINURCAT)
African Union-United Nations Hybrid Operation in Darfur (UNAMID)
United Nations Mission in the Sudan (UNMIS)
United Nations Operation in Côte d'Ivoire (UNOCI)
United Nations Mission in Liberia (UNMIL)
United Nations Organization Mission in the Democratic Republic of the Congo (MONUC)
United Nations Mission for the Referendum in Western Sahara (MINURSO)
United Nations Stabilization Mission in the Democratic Republic of the Congo (MONUSCO)

Americas

United Nations Stabilization Mission in Haiti (MINUSTAH)

Middle East

United Nations Truce Supervision Organization (UNTSO)

III. Inter-American Organizations

Caribbean Postal Union

515

Inter-American Center of Tax Administrators
Inter-American Indian Institute
Inter-American Institute for Cooperation on Agriculture
Inter-American Institute for Global Change Research
Inter-American Tropical Tuna Commission
Organization of American States
Pan American Health Organization
Pan American Institute of Geography and History
Pan American Railway Congress Association
Postal Union of the Americas and Spain and Portugal
Organization of American States
Inter-American Commission of Women
Inter-American Children's Institute
Inter-American Commission of Human Rights
Inter-American Council for Integral Development
Inter-American Telecommunications Commission
Inter-American Defense Board
Inter-American Drug Abuse Control Commission
Inter-American Committee Against Terrorism
Inter-American Committee on Natural Disaster Reduction

IV. Regional Organizations

Antarctic Treaty System
Arctic Council
Asia-Pacific Economic Cooperation
Asia Pacific Energy Research Center
Colombo Plan for Cooperative Economic and Social Development in Asia and the Pacific
Commission for Environmental Cooperation
Commission for Labor Cooperation
International Commission for the Conservation of Atlantic Tunas
North Atlantic Assembly
North Atlantic Treaty Organization
North Atlantic Salmon Conservation Organization
North Pacific Anadromous Fish Commission
North Pacific Coast Guard Forum

North Pacific Marine Science Organization
Northwest Atlantic Fisheries Organization
Secretariat of the Pacific Community
South Pacific Regional Environment Program
Western and Central Pacific Fisheries Commission

V. Other International Organizations

Center for International Forestry Research
Commission for the Conservation of Antarctic Marine Living Resources
Community of Democracies
Comprehensive Nuclear Test Ban Treaty Organization
Consultative Group on International Agricultural Research
COPAS–SARSAT (Search and Rescue Satellite System)
Global Biodiversity Information Facility
Integrated Ocean Drilling Program Council
International Agency for Research on Cancer (IARC)
International Bureau for the Permanent Court of Arbitration
International Bureau of Weights and Measures
International Center for Agricultural Research in the Dry Areas
International Center for Migration Policy Development
International Center for the Study of the Preservation and the Restoration of Cultural Property
International Coffee Organization
International Committee of the Red Cross
International Cotton Advisory Committee
International Council for the Exploration of the Seas
International Criminal Police Organization (INTERPOL)
International Customs Tariffs Bureau
International Development Law Organization
International Energy Agency
International Energy Forum Secretariat
International Fertilizer Development Center
International Grains Council
International Hydrographic Organization
International Institute for Applied Systems Analysis

International Institute for Cotton

International Institute for the Unification of Private Law

International Mobile Satellite Organization

International North Pacific Fisheries Commission

International Organization for Legal Metrology

International Organization for Migration (IOM)

International Organization for Migration

International Organization of Supreme Audit Institutions

International Rubber Study Group

International Science and Technology Center

International Seed Testing Association

International Service for National Agriculture Research

International Sugar Council

International Telecommunications Satellite Organization

International Tropical Timber Organization

International Union of Credit and Investment Insurers (Berne Union)

International Whaling Commission

Inter-Parliamentary Union

Iran-United States Claims Tribunal

Hague Conference on Private International Law

Human Frontier Science Program Organization

Multinational Force and Observers

Nuclear Energy Agency

Organization for Economic Cooperation and Development

Organization for the Prohibition of Chemical Weapons

Pacific Aviation Safety Office

Permanent International Association of Navigation Congresses

Preparatory Commission for the Comprehensive Nuclear Test-Ban Treaty

Regional Environmental Center for Central and Eastern Europe

Science and Technology Center in Ukraine

Sierra Leone Special Court

Wassenaar Arrangement

World Heritage Fund

World Customs Organization

World Organization for Animal Health

World Trade Organization

VI. Special Voluntary Programs

Asian Vegetable Research and Development Center

Convention on International Trade in Endangered Species of Wild Fauna and Flora (CITES)

Global Fund to Fight HIV/AIDS, Tuberculosis, and Malaria

International Council for Science

International Crop Research Institute for Semi-Arid Tropics

International Federation of the Red Cross and Red Crescent Societies

International Food Policy Research Institute

International Fund for Agricultural Development

International Institute of Tropical Agriculture

International Strategy for Disaster Reduction

Joint United Nations Program on HIV/AIDS (UNAIDS)

Korean Peninsula Energy Development Organization

Multilateral Fund for the Implementation of the Montreal Protocol

Permanent Interstate Committee for Drought Control in the Sahel

Ramsar Convention on Wetlands

United Nations Children's Fund (UNICEF)

United Nations Democracy Fund

United Nations Development Fund for Women (UNIFEM)

United Nations Development Program

United Nations Environment Program

United Nations Framework Convention on Climate Change

United Nations Convention to Combat Desertification

United Nations High Commissioner for Human Rights Programs

United Nations High Commissioner for Refugees Programs

United Nations Human Settlements Program (UN HABITAT)

United Nations Population Fund (UNFPA)

United Nations Relief and Works Agency

United Nations Voluntary Fund for Technical Cooperation in the Field of Human Rights

United Nations Voluntary Fund for the
   Victims of Torture
United Nations World Food Program

World Agroforestry Center
World Health Organization Special
   Programs

SELECTED BILATERAL ORGANIZATIONS

Below is a list of bilateral organizations in which the United States participates with its two neighbors, Mexico and Canada. This listing is for reference purposes only and should not be considered exhaustive.

Border Environment Cooperation Commission

United States Section: P.O. Box 221648, El Paso, TX 79913. Phone, 877–277–17030. Fax, 915–975–82800 Internet, www.becc.org.

Mexican Section: Bulevar Tomas Fernadez 8069, Ciudad Juarez, Chihuahua, 32470. Phone, 011–52–656–688–4600. Fax, 011–52–656–625–6180. Internet, www.cocef.org.

Great Lakes Fishery Commission

2100 Commonwealth Boulevard, Suite 100, Ann Arbor, MI 48105. Telephone, 734–662–3209. Fax, 734–741–2010. Email, info@glfc.org. Internet, www.glfc.org.

International Boundary Commission, United States and Canada

United States Section: 2000 L Street NW., Suite 615, Washington, DC 20036. Phone, 202–736–9102. Fax, 202–632–2008. Internet, http://www. internationalboundarycommission.org.

Canadian Section: 575–615 Booth Street, Ottawa, Ontario K1A 0E9 Canada. Phone, 780–495–2519. Fax, 780–495–2725.

International Boundary and Water Commission,United States and Mexico

United States Section: Building C, Suite 100, 4171 North Mesa Street, El Paso, TX 79902. Phone, 800–262–8857. Internet, www.ibwc.state.gov.

Mexican Section: Avenue Universidad 2180, Zona Chamizal, Ciudad Juarez, Chihuahua, 32310. Phone, 011–52–656–613–7311 or 011–52–656–613–7363. Fax, 011–52–656–613–9943. Internet, www.sre.gob.mx/cila.

International Joint Commission—United States and Canada

United States Section: 2000 L Street NW., Washington, DC 20036. Phone, 202–736–9024. Fax, 202–632–2006. Internet, www.ijc.org.

Canadian Section: 234 Laurier Avenue West, 22d Floor, Ottawa, Ontario K1P 6K6. Phone, 613–947–1420. Fax, 613–993–5583.

Great Lakes Regional Office: 100 Ouellette Avenue, 8th Floor, Windsor, Ontario N9A 6T3. Phone, 519–257–6714. Fax, 519–257–6740.

Joint Mexican-United States Defense Commission

United States Section: Room 2E773, The Pentagon, Washington, DC 20318. Phone, 703–695–8164.

Mexican Section: 6th Floor, 1911 Pennsylvania Avenue NW., Mexican Embassy, Washington, DC 20006. Phone, 202–728–1748.

519

Permanent Joint Board on Defense—United States and Canada

United States Section: Room 2E773, The Pentagon, Washington, DC 20318. Phone, 703–695–8164.

Canadian Section: Director of Western Hemisphere, 101 Colonel By Drive, Ottawa, ON K1A 0K2. Phone, 613–992–4423.

COMMONLY USED AGENCY ACRONYMS

ABMC AMERICAN BATTLE MONUMENTS COMMISSION

ACF ADMINISTRATION OF CHILDREN AND FAMILIES

ACFR ADMINISTRATIVE COMMITTEE OF THE FEDERAL REGISTER

ADF AFRICAN DEVELOPMENT FOUNDATION

AFRH ARMED FORCES RETIREMENT HOME

AHRQ AGENCY FOR HEALTHCARE RESEARCH AND QUALITY

AID AGENCY FOR INTERNATIONAL DEVELOPMENT

AMC ANTITRUST MODERNIZATION COMMISSION

AMS AGRICULTURAL MARKETING SERVICE

AOA ADMINISTRATION ON AGING

APHIS ANIMAL AND PLANT HEALTH INSPECTION SERVICE

APPAL APPALACHIAN STATES LOW LEVEL RADIOACTIVE WASTE COMMISSION

ARCTIC ARCTIC RESEARCH COMMISSION

ARS AGRICULTURAL RESEARCH SERVICE

ARTS NATIONAL FOUNDATION ON THE ARTS AND THE HUMANITIES

ATBCB ARCHITECTURAL AND TRANSPORTATION BARRIERS COMPLIANCE BOARD

ATF ALCOHOL, TOBACCO, FIREARMS, AND EXPLOSIVES BUREAU

ATSDR AGENCY FOR TOXIC SUBSTANCES AND DISEASE REGISTRY

BBG BROADCASTING BOARD OF GOVERNORS

BGSEEF BARRY M. GOLDWATER SCHOLARSHIP AND EXCELLENCE IN EDUCATION FOUNDATION

BIA BUREAU OF INDIAN AFFAIRS

BIS BUREAU OF INDUSTRY AND SECURITY

BLM BUREAU OF LAND MANAGEMENT

BLS BUREAU OF LABOR STATISTICS

BOP FEDERAL PRISONS BUREAU

BOR BUREAU OF RECLAMATION

BPA BONNEVILLE POWER ADMINISTRATION

BPD BUREAU OF PUBLIC DEBT

CCC COMMODITY CREDIT CORPORATION

CCJJDP COORDINATING COUNCIL ON JUVENILE JUSTICE AND DELINQUENCY PREVENTION

CDC CENTERS FOR DISEASE CONTROL AND PREVENTION

CDFI   COMMUNITY DEVELOPMENT FINANCIAL INSTITUTIONS FUND

CEQ   COUNCIL ON ENVIRONMENTAL QUALITY

CFTC   COMMODITY FUTURES TRADING COMMISSION

CIA   CENTRAL INTELLIGENCE AGENCY

CITA   COMMITTEE FOR THE IMPLEMENTATION OF TEXTILE AGREEMENTS

CMS   CENTERS FOR MEDICARE & MEDICAID SERVICES

CNCS   CORPORATION FOR NATIONAL AND COMMUNITY SERVICE

COE   CORPS OF ENGINEERS

COFA   COMMISSION OF FINE ARTS

COLC   COPYRIGHT OFFICE, LIBRARY OF CONGRESS

CORP   CORPORATION FOR NATIONAL AND COMMUNITY SERVICE

CPPBSD   COMMITTEE FOR PURCHASE FROM PEOPLE WHO ARE BLIND OR SEVERELY DISABLED

CPSC   CONSUMER PRODUCT SAFETY COMMISSION

CRB   COPYRIGHT ROYALTY BOARD, LIBRARY OF CONGRESS

CRC   CIVIL RIGHTS COMMISSION

CSB   CHEMICAL SAFETY AND HAZARD INVESTIGATION BOARD

CSEO   CHILD SUPPORT ENFORCEMENT OFFICE

CSOSA   COURT SERVICES AND OFFENDER SUPERVISION AGENCY FOR THE DISTRICT OF COLUMBIA

CSREES   COOPERATIVE STATE RESEARCH, EDUCATION, AND EXTENSION SERVICE

DARS   DEFENSE ACQUISITION REGULATIONS SYSTEM

DC   DENALI COMMISSION

DEA   DRUG ENFORCEMENT ADMINISTRATION

DEPO   DISABILITY EMPLOYMENT POLICY OFFICE

DHS   DEPARTMENT OF HOMELAND SECURITY

DIA   DEFENSE INTELLIGENCE AGENCY

DISA   DEFENSE INFORMATION SYSTEMS AGENCY

DLA   DEFENSE LOGISTICS AGENCY

DNFSB   DEFENSE NUCLEAR FACILITIES SAFETY BOARD

DOC   DEPARTMENT OF COMMERCE

DOD   DEPARTMENT OF DEFENSE

DOE   DEPARTMENT OF ENERGY

DOI   DEPARTMENT OF THE INTERIOR

DOJ   DEPARTMENT OF JUSTICE

DOL   DEPARTMENT OF LABOR

DOS   DEPARTMENT OF STATE

DOT   DEPARTMENT OF TRANSPORTATION

DRBC   DELAWARE RIVER BASIN COMMISSION

EAB   BUREAU OF ECONOMIC ANALYSIS

EAC   ELECTION ASSISTANCE COMMISSION

EBSA   EMPLOYEE BENEFITS SECURITY ADMINISTRATION

ECAB   EMPLOYEES' COMPENSATION APPEALS BOARD

ECSA   ECONOMICS AND STATISTICS ADMINISTRATION

ED   DEPARTMENT OF EDUCATION

EDA   ECONOMIC DEVELOPMENT ADMINISTRATION

EEOC   EQUAL EMPLOYMENT OPPORTUNITY COMMISSION

EERE   ENERGY EFFICIENCY AND RENEWABLE ENERGY OFFICE

EIA	ENERGY INFORMATION ADMINISTRATION
EIB	EXPORT IMPORT BANK OF THE UNITED STATES
EOA	ENERGY OFFICE, AGRICULTURE DEPARTMENT
EOIR	EXECUTIVE OFFICE FOR IMMIGRATION REVIEW
EOP	EXECUTIVE OFFICE OF THE PRESIDENT
EPA	ENVIRONMENTAL PROTECTION AGENCY
ERS	ECONOMIC RESEARCH SERVICE
ESA	EMPLOYMENT STANDARDS ADMINISTRATION
ETA	EMPLOYMENT AND TRAINING ADMINISTRATION
FAA	FEDERAL AVIATION ADMINISTRATION
FAR	FEDERAL ACQUISITION REGULATION
FAS	FOREIGN AGRICULTURAL SERVICE
FASAB	FEDERAL ACCOUNTING STANDARDS ADVISORY BOARD
FBI	FEDERAL BUREAU OF INVESTIGATION
FCA	FARM CREDIT ADMINISTRATION
FCC	FEDERAL COMMUNICATIONS COMMISSION
FCIC	FEDERAL CROP INSURANCE CORPORATION
FCSIC	FARM CREDIT SYSTEM INSURANCE CORPORATION
FDA	FOOD AND DRUG ADMINISTRATION
FDIC	FEDERAL DEPOSIT INSURANCE CORPORATION
FEC	FEDERAL ELECTION COMMISSION
FEMA	FEDERAL EMERGENCY MANAGEMENT AGENCY
FERC	FEDERAL ENERGY REGULATORY COMMISSION
FFIEC	FEDERAL FINANCIAL INSTITUTIONS EXAMINATION COUNCIL
FHFA	FEDERAL HOUSING FINANCE AGENCY
FHFB	FEDERAL HOUSING FINANCE BOARD
FHWA	FEDERAL HIGHWAY ADMINISTRATION
FINCEN	FINANCIAL CRIMES ENFORCEMENT NETWORK
FINCIC	FINANCIAL CRISIS INQUIRY COMMISSION
FISCAL	FISCAL SERVICE
FLETC	FEDERAL LAW ENFORCEMENT TRAINING CENTER
FLRA	FEDERAL LABOR RELATIONS AUTHORITY
FMC	FEDERAL MARITIME COMMISSION
FMCS	FEDERAL MEDIATION AND CONCILIATION SERVICE
FMCSA	FEDERAL MOTOR CARRIER SAFETY ADMINISTRATION
FNS	FOOD AND NUTRITION SERVICE
FPPO	FEDERAL PROCUREMENT POLICY OFFICE
FR	OFFICE OF THE FEDERAL REGISTER
FRA	FEDERAL RAILROAD ADMINISTRATION
FRS	FEDERAL RESERVE SYSTEM
FRTIB	FEDERAL RETIREMENT THRIFT INVESTMENT BOARD
FS	FOREST SERVICE
FSA	FARM SERVICE AGENCY
FSIS	FOOD SAFETY AND INSPECTION SERVICE
FTA	FEDERAL TRANSIT ADMINISTRATION
FTC	FEDERAL TRADE COMMISSION
FTZB	FOREIGN TRADE ZONES BOARD
FWS	FISH AND WILDLIFE SERVICE
GAO	GOVERNMENT ACCOUNTABILITY OFFICE

GEO   GOVERNMENT ETHICS OFFICE

GIPSA   GRAIN INSPECTION, PACKERS AND STOCKYARDS ADMINISTRATION

GPO   GOVERNMENT PRINTING OFFICE

GSA   GENERAL SERVICES ADMINISTRATION

HHS   DEPARTMENT OF HEALTH AND HUMAN SERVICES

HHSIG   INSPECTOR GENERAL OFFICE, HEALTH AND HUMAN SERVICES DEPARTMENT

HOPE   BOARD OF DIRECTORS OF THE HOPE FOR HOMEOWNERS PROGRAM

HPAC   HISTORIC PRESERVATION, ADVISORY COUNCIL

HRSA   HEALTH RESOURCES AND SERVICES ADMINISTRATION

HST   HARRY S. TRUMAN SCHOLARSHIP FOUNDATION

HUD   DEPARTMENT OF HOUSING AND URBAN DEVELOPMENT

IAF   INTER AMERICAN FOUNDATION

ICEB   IMMIGRATION AND CUSTOMS ENFORCEMENT BUREAU

IHS   INDIAN HEALTH SERVICE

IIO   INTERNATIONAL INVESTMENT OFFICE

IRS   INTERNAL REVENUE SERVICE

ISOO   INFORMATION SECURITY OVERSIGHT OFFICE

ITA   INTERNATIONAL TRADE ADMINISTRATION

ITC   INTERNATIONAL TRADE COMMISSION

JBEA   JOINT BOARD FOR ENROLLMENT OF ACTUARIES

LMSO   LABOR MANAGEMENT STANDARDS OFFICE

LOC   LIBRARY OF CONGRESS

LSC   LEGAL SERVICES CORPORATION

MARAD   MARITIME ADMINISTRATION

MBDA   MINORITY BUSINESS DEVELOPMENT AGENCY

MCC   MILLENNIUM CHALLENGE CORPORATION

MISS   MISSISSIPPI RIVER COMMISSION

MKU   MORRIS K. UDALL SCHOLARSHIP AND EXCELLENCE IN NATIONAL ENVIRONMENTAL POLICY FOUNDATION

MMC   MARINE MAMMAL COMMISSION

MMS   MINERALS MANAGEMENT SERVICE

MSHA   MINE SAFETY AND HEALTH ADMINISTRATION

MSHFRC   FEDERAL MINE SAFETY AND HEALTH REVIEW COMMISSION

MSPB   MERIT SYSTEMS PROTECTION BOARD

NARA   NATIONAL ARCHIVES AND RECORDS ADMINISTRATION

NASA   NATIONAL AERONAUTICS AND SPACE ADMINISTRATION

NASS   NATIONAL AGRICULTURAL STATISTICS SERVICE

NCD   NATIONAL COUNCIL ON DISABILITY

NCLIS   NATIONAL COMMISSION ON LIBRARIES AND INFORMATION SCIENCE

NCPPCC   NATIONAL CRIME PREVENTION AND PRIVACY COMPACT COUNCIL

NCS   NATIONAL COMMUNICATIONS SYSTEM

NCUA   NATIONAL CREDIT UNION ADMINISTRATION

NEC   NATIONAL ECONOMIC COUNCIL

NEIGHBOR   NEIGHBORHOOD REINVESTMENT CORPORATION

NHTSA   NATIONAL HIGHWAY TRAFFIC SAFETY ADMINISTRATION

NIFA   NATIONAL INSTITUTE OF FOOD AND AGRICULTURE

NIGC   NATIONAL INDIAN GAMING COMMISSION

NIH   NATIONAL INSTITUTES OF HEALTH

NIL   NATIONAL INSTITUTE FOR LITERACY

NIST   NATIONAL INSTITUTE OF STANDARDS AND TECHNOLOGY

NLRB   NATIONAL LABOR RELATIONS BOARD

NMB   NATIONAL MEDIATION BOARD

NNSA   NATIONAL NUCLEAR SECURITY ADMINISTRATION

NOAA   NATIONAL OCEANIC AND ATMOSPHERIC ADMINISTRATION

NPREC   NATIONAL PRISON RAPE ELIMINATION COMMISSION

NPS   NATIONAL PARK SERVICE

NRC   NUCLEAR REGULATORY COMMISSION

NRCS   NATURAL RESOURCES CONSERVATION SERVICE

NSA   NATIONAL SECURITY AGENCY/CENTRAL SECURITY SERVICE

NSF   NATIONAL SCIENCE FOUNDATION

NTIA   NATIONAL TELECOMMUNICATIONS AND INFORMATION ADMINISTRATION

NTSB   NATIONAL TRANSPORTATION SAFETY BOARD

NWTRB   NUCLEAR WASTE TECHNICAL REVIEW BOARD

OCC   COMPTROLLER OF THE CURRENCY

ODNI   OFFICE OF THE DIRECTOR OF NATIONAL INTELLIGENCE

OEPNU   OFFICE OF ENERGY POLICY AND NEW USES

OFAC   OFFICE OF FOREIGN ASSETS CONTROL

OFCCP   OFFICE OF FEDERAL CONTRACT COMPLIANCE PROGRAMS

OFHEO   FEDERAL HOUSING ENTERPRISE OVERSIGHT OFFICE

OFPP   OFFICE OF FEDERAL PROCUREMENT POLICY

OJJDP   JUVENILE JUSTICE AND DELINQUENCY PREVENTION OFFICE

OJP   JUSTICE PROGRAMS OFFICE

OMB   OFFICE OF MANAGEMENT AND BUDGET

ONDCP   OFFICE OF NATIONAL DRUG CONTROL POLICY

ONHIR   OFFICE OF NAVAJO AND HOPI INDIAN RELOCATION

OPIC   OVERSEAS PRIVATE INVESTMENT CORPORATION

OPM   OFFICE OF PERSONNEL MANAGEMENT

OPPM   OFFICE OF PROCUREMENT AND POLICY MANAGEMENT

OSC   OFFICE OF SPECIAL COUNSEL

OSHA   OCCUPATIONAL SAFETY AND HEALTH ADMINISTRATION

OSHRC   OCCUPATIONAL SAFETY AND HEALTH REVIEW COMMISSION

OSM   OFFICE OF SURFACE MINING RECLAMATION AND ENFORCEMENT

OSTP   OFFICE OF SCIENCE AND TECHNOLOGY POLICY

OTS   OFFICE OF THRIFT SUPERVISION

PACIFIC   PACIFIC NORTHWEST ELECTRIC POWER

	AND CONSERVATION PLANNING COUNCIL
PBGC	PENSION BENEFIT GUARANTY CORPORATION
PC	PEACE CORPS
PHMSA	PIPELINE AND HAZARDOUS MATERIALS SAFETY ADMINISTRATION
PHS	PUBLIC HEALTH SERVICE
PRC	POSTAL RATE COMMISSION
PRES	PRESIDENTIAL DOCUMENTS
PT	PRESIDIO TRUST
PTO	PATENT AND TRADEMARK OFFICE
RATB	RECOVERY ACCOUNTABILITY AND TRANSPARENCY BOARD
RBS	RURAL BUSINESS COOPERATIVE SERVICE
RHS	RURAL HOUSING SERVICE
RISC	REGULATORY INFORMATION SERVICE CENTER
RITA	RESEARCH AND INNOVATIVE TECHNOLOGY ADMINISTRATION
RMA	RISK MANAGEMENT AGENCY
RRB	RAILROAD RETIREMENT BOARD
RTB	RURAL TELEPHONE BANK
RUS	RURAL UTILITIES SERVICE
SAMHSA	SUBSTANCE ABUSE AND MENTAL HEALTH SERVICES ADMINISTRATION
SBA	SMALL BUSINESS ADMINISTRATION
SEC	SECURITIES AND EXCHANGE COMMISSION
SIGIR	SPECIAL INSPECTOR GENERAL FOR IRAQ RECONSTRUCTION
SJI	STATE JUSTICE INSTITUTE
SLSDC	SAINT LAWRENCE SEAWAY DEVELOPMENT CORPORATION

SRBC	SUSQUEHANNA RIVER BASIN COMMISSION
SSA	SOCIAL SECURITY ADMINISTRATION
SSS	SELECTIVE SERVICE SYSTEM
STB	SURFACE TRANSPORTATION BOARD
SWPA	SOUTHWESTERN POWER ADMINISTRATION
TA	TECHNOLOGY ADMINISTRATION
TREAS	DEPARTMENT OF THE TREASURY
TSA	TRANSPORTATION SECURITY ADMINISTRATION
TTB	ALCOHOL AND TOBACCO TAX AND TRADE BUREAU
TVA	TENNESSEE VALLEY AUTHORITY
URMCC	UTAH RECLAMATION MITIGATION AND CONSERVATION COMMISSION
USA	ARMY DEPARTMENT
USAF	AIR FORCE DEPARTMENT
USBC	BUREAU OF THE CENSUS
USCBP	CUSTOMS AND BORDER PROTECTION BUREAU
USCC	U.S. CHINA ECONOMIC AND SECURITY REVIEW COMMISSION
USCG	COAST GUARD
USCIS	U.S. CITIZENSHIP AND IMMIGRATION SERVICES
USDA	DEPARTMENT OF AGRICULTURE
USEIB	EXPORT IMPORT BANK
USIP	UNITED STATES INSTITUTE OF PEACE
USJC	JUDICIAL CONFERENCE OF THE UNITED STATES
USMINT	UNITED STATES MINT
USN	NAVY DEPARTMENT
USPC	PAROLE COMMISSION
USPS	POSTAL SERVICE
USSC	UNITED STATES SENTENCING COMMISSION
USSS	SECRET SERVICE

USTR	OFFICE OF UNITED STATES TRADE REPRESENTATIVE	VETS	VETERANS EMPLOYMENT AND TRAINING SERVICE
USUHS	UNIFORMED SERVICES UNIVERSITY OF THE HEALTH SCIENCES	WAPA	WESTERN AREA POWER ADMINISTRATION
VA	DEPARTMENT OF VETERANS AFFAIRS	WCPO	WORKERS COMPENSATION PROGRAMS OFFICE
VCNP	VALLES CALDERA TRUST	WHD	WAGE AND HOUR DIVISION

HISTORY OF AGENCY ORGANIZATIONAL CHANGES

NOTE: Italicized terms indicate obsolete agencies, organizations, and programs. Refer to the name of the obsolete entity in this index for more explanation. Some dates prior to March 4, 1933 are included to provide additional information.

Entries are indexed using the most significant term in their titles, or when there is more than one significant term, the entry uses the first significant term. Thus, **Bureau of the Budget** is found at **Budget, Bureau of the,** and **Annual Assay Commission** is found at **Assay Commission, Annual.**

Accounting Office, General   Established by act of June 10, 1921 (42 Stat. 20). Renamed Government Accountability Office by act of July 7, 2004 (118 Stat. 814).

ACTION   Established by Reorg. Plan No. 1 of 1971 (5 U.S.C. app.), effective July 1, 1971. Reorganized by act of Oct. 1, 1973 (87 Stat. 405). Functions relating to SCORE and ACT programs transferred to Small Business Administration by EO 11871 of July 18, 1975 (40 FR 30915). Functions exercised by the Director of ACTION prior to Mar. 31, 1995, transferred to the Corporation for National and Community Service (107 Stat. 888 and Proclamation 6662 of Apr. 4, 1994 (57 FR 16507)).

Acts of Congress   *See* **State, Department of**

Administrative Conference of the United States   Established by act of Aug. 30, 1964 (78 Stat. 615). Terminated by act of Nov. 19, 1995 (109 Stat. 480). Reauthorized in 2004, 2008, and 2009 by acts of Oct. 30, 2004 (118 Stat. 2255), July 30, 2008 (122 Stat. 2914), and March 11, 2009 (123 Stat. 656). Reestablished by Congress on Mar. 3, 2010 upon confirmation of chairman.

Advanced Research Projects Agency   *See* Defense Advanced Research Projects Agency

Advisory Board.   *See other part of title*

Aeronautical Board   Organized in 1916 by agreement of *War* and Navy Secretaries. Placed under supervision of President by military order of July 5, 1939. Dissolved by Secretary of Defense letter of July 27, 1948, and functions transferred to *Munitions Board* and *Research and Development Board.* Military order of July 5, 1939, revoked by military order of Oct. 18, 1948.

Aeronautics, Bureau of   Established in the Department of the Navy by act of July 12, 1921 (42 Stat. 140). Abolished by act of Aug. 18, 1959 (73 Stat. 395) and functions transferred to *Bureau of Naval Weapons.*

Aeronautics, National Advisory Committee for   Established by act of Mar. 3, 1915 (38 Stat. 930). Terminated by act of July 29, 1958 (72 Stat. 432), and functions transferred to National Aeronautics and Space Administration, established by same act.

Aeronautics Administration, Civil   *See* **Aeronautics Authority, Civil**

Aeronautics Authority, Civil   Established under act of June 23, 1938 (52 Stat. 973). Renamed *Civil Aeronautics Board* and Administrator transferred to the Department of Commerce by Reorg. Plan Nos. III and IV of 1940, effective June 30, 1940. Office of Administrator designated *Civil Aeronautics Administration* by Department Order 52 of Aug. 29, 1940. *Administration* transferred to *Federal Aviation Agency* by act of Aug. 23, 1958 (72 Stat. 810). Functions of *Board* under act of Aug. 23, 1958 (72 Stat. 775), transferred to National Transportation Safety Board by act of Oct. 15, 1966 (80 Stat. 931). Functions of *Board* terminated or transferred—effective in part Dec. 31, 1981; in part Jan. 1, 1983; and in part Jan. 1, 1985—by act of Aug. 23, 1958 (92 Stat. 1744). Most remaining functions transferred to Secretary of Transportation, remainder to U.S. Postal Service. Termination of *Board* finalized by act of Oct. 4, 1984 (98 Stat. 1703).

Aeronautics Board, Civil   *See* **Aeronautics Authority, Civil**

Aeronautics Branch   Established in the Department of Commerce to carry out provisions of act of May 20, 1926 (44 Stat. 568). Renamed *Bureau of Air Commerce* by Secretary's administrative order of July 1, 1934. Personnel and property transferred to *Civil Aeronautics Authority* by EO 7959 of Aug. 22, 1938.

Aeronautics and Space Council, National   Established by act of July 29, 1958 (72 Stat. 427). Abolished by Reorg. Plan No. 1 of 1973, effective June 30, 1973.

Aging, Administration on   Established by *Secretary of Health, Education, and Welfare* on Oct. 1, 1965, to carry out provisions of act of July 14, 1965 (79 Stat. 218). Reassigned to *Social and Rehabilitation Service* by Department reorganization of Aug. 15, 1967. Transferred to Office of Assistant Secretary for Human Development by Secretary's order of June 15, 1973. Transferred to the Office of the Secretary of Health and Human Services by Secretary's reorganization notice dated Apr. 15, 1991.

Aging, Federal Council on   Established by Presidential memorandum of Apr. 2, 1956. Reconstituted at Federal level by Presidential letter of Mar. 7, 1959, to *Secretary of Health, Education, and Welfare.* Abolished by EO 11022 of May 15, 1962, which established *President's Council on Aging.*

Aging, Office of   Established by *Secretary of Health, Education, and Welfare* June 2, 1955, as *Special Staff on Aging.* Terminated Sept. 30, 1965, and functions assumed by Administration on Aging.

Aging, President's Council on   Established by EO 11022 of May 14, 1962. Terminated by EO 11022, which was revoked by EO 12379 of Aug. 17, 1982.

Agricultural Adjustment Administration   Established by act of May 12, 1933 (48 Stat. 31). Consolidated into *Agricultural Conservation and Adjustment Administration* as *Agricultural Adjustment Agency,* Department of Agriculture, by EO 9069 of Feb. 23, 1942. Grouped with other agencies to form *Food Production Administration* by EO 9280 of Dec. 5, 1942. Transferred to *War Food Administration* by EO 9322 of Mar. 26, 1943. Administration terminated by EO 9577 of June 29, 1945, and functions transferred to Secretary of Agriculture. Transfer made permanent by Reorg. Plan No. 3 of 1946, effective July 16, 1946. Functions of *Agricultural Adjustment Agency* consolidated with *Production and Marketing Administration* by Secretary's Memorandum 1118 of Aug. 18, 1945.

Agricultural Adjustment Agency   *See* **Agricultural Adjustment Administration**

Agricultural Advisory Commission, National   Established by EO 10472 of July 20, 1953. Terminated Feb. 4, 1965, on resignation of members.

Agricultural Chemistry and Engineering, Bureau of   *See* **Agricultural Engineering, Bureau of**

Agricultural Conservation and Adjustment Administration   Established by EO 9069 of Feb. 23, 1942, consolidating *Agricultural Adjustment Agency, Sugar Agency, Federal Crop Insurance Corporation,* and *Soil Conservation Service.* Consolidated into *Food Production Administration* by EO 9280 of Dec. 5, 1942.

Agricultural Conservation Program Service   Established by Secretary of Agriculture Jan. 21, 1953, from part of *Production and Marketing Administration.* Merged with *Commodity Stabilization Service* by Secretary's Memorandum 1446, supp. 2, of Apr. 19, 1961.

Agricultural Developmental Service, International   Established by Secretary of Agriculture memorandum of July 12, 1963. Functions and delegations of authority transferred to Foreign Agricultural Service by Secretary's memorandum of Mar. 28, 1969. Functions transferred by Secretary to *Foreign Economic Development Service* Nov. 8, 1969.

Agricultural Economics, Bureau of   Established by act of May 11, 1931 (42 Stat. 532). Functions transferred to other units of the Department of Agriculture, including *Consumer and Marketing Service* and Agricultural Research Service, under Secretary's Memorandum 1320, supp. 4, of Nov. 2, 1953.

Agricultural Engineering, Bureau of   Established by act of Feb. 23, 1931 (46 Stat. 1266). Merged with *Bureau of Chemistry and Soils* by Secretarial order of Oct. 16, 1938, to form *Bureau of Agricultural Chemistry and Engineering.*

Agricultural and Industrial Chemistry, Bureau of   *Bureau of Chemistry* and *Bureau of Soils,* created in 1901, combined into *Bureau of Chemistry and Soils* by act of Jan. 18, 1927 (44 Stat. 976). Soils units transferred to other agencies of the Department of Agriculture and remaining units of *Bureau of Chemistry and Soils* and *Bureau of Agricultural Engineering* consolidated with *Bureau of Agricultural Chemistry and Engineering* by Secretary's order of Oct. 16, 1938. In February 1943 agricultural engineering research made part of *Bureau of Plant Industry, Soils, and Agricultural Engineering,* and organization for continuing agricultural chemistry research relating to crop utilization named *Bureau of Agricultural and Industrial Chemistry,* in accordance with *Research Administration* Memorandum 5 issued pursuant to EO 9069 of Feb. 23, 1942, and in conformity with Secretary's Memorandums 960 and 986. Functions transferred to *Agricultural Research Service* under Secretary's Memorandum 1320, supp. 4, of Nov. 2, 1953.

Agricultural Library, National   Established by Secretary of Agriculture Memorandum 1496 of Mar. 23, 1962. Consolidated into *Science and Education Administration* by Secretary's order of Jan. 24, 1978. Reestablished as National Agricultural Library by Secretary's order of June 16, 1981. Became part of Agricultural Research Service in 1994 under Department of Agriculture reorganization.

Agricultural Marketing Administration   Established by EO 9069 of Feb. 23, 1942, consolidating *Surplus Marketing Administration, Agricultural Marketing Service,* and *Commodity Exchange Administration. Division of Consumers' Counsel* transferred to *Administration* by Secretary's memorandum of Feb. 28, 1942. Consolidated into *Food Distribution Administration* in the Department of Agriculture by EO 9280 of Dec. 5, 1942.

Agricultural Marketing Service   Established by the Secretary of Agriculture pursuant to act of June 30, 1939 (53 Stat. 939). Merged into *Agricultural Marketing Administration* by EO 9069 of Feb. 23, 1942. Renamed *Consumer and Marketing Service* by

Secretary's Memorandum 1567, supp. 1, of Feb. 8, 1965. Reestablished as Agricultural Marketing Service by the Secretary of Agriculture on Apr. 2, 1972, under authority of Reorg. Plan No. 2 of 1953 (67 Stat. 633).

Agricultural Relations, Office of Foreign See **Agricultural Service, Foreign**

Agricultural Research Administration Established by EO 9069 of Feb. 23, 1942. Superseded by Agricultural Research Service.

Agricultural Research Service Established by Secretary of Agriculture Memorandum 1320, supp. 4, of Nov. 2, 1953. Consolidated into *Science and Education Administration* by Secretary's order of Jan. 24, 1978. Reestablished as Agricultural Research Service by Secretarial order of June 16, 1981.

Agricultural Service, Foreign Established by act of June 5, 1930 (46 Stat. 497). Economic research and agricultural attaché activities administered by *Foreign Agricultural Service Division, Bureau of Agricultural Economics,* until June 29, 1939. Transferred by Reorg. Plan No. II of 1939, effective July 1, 1939, from the Department of Agriculture to the Department of State. Economic research functions of *Division* transferred to *Office of Foreign Agricultural Relations* June 30, 1939. Functions of *Office* transferred to Foreign Agricultural Service Mar. 10, 1953. Agricultural attachés placed in the Department of Agriculture by act of Aug. 28, 1954 (68 Stat. 908).

Agricultural Stabilization and Conservation Service Established June 5, 1961, by the Secretary of Agriculture under authority of revised statutes (5U.S.C. 301) and Reorg. Plan No. 2 of 1953 (5 U.S.C. app.). Abolished and functions assumed by the *Farm Service Agency* by Secretary's Memorandum 1010–1 dated Oct. 20, 1994 (59 FR 60297, 60299).

Agricultural Statistics Division Transferred to *Bureau of Agricultural Economics* by EO 9069 of Feb. 23, 1942.

Agriculture, Division of See **Farm Products, Division of**

Air Commerce, Bureau of See **Aeronautics Branch**

Air Coordinating Committee Established Mar. 27, 1945, by interdepartmental memorandum; formally established by EO 9781 of Sept. 19, 1946. Terminated by EO 10883 of Aug. 11, 1960, and functions transferred for liquidation to *Federal Aviation Agency.*

Air Force Management Engineering Agency Established in 1975 in Air Force as separate operating unit. Made subordinate unit of Air Force Military Personnel Center (formerly Air Force Manpower and Personnel Center) in 1978. Reestablished as separate operating unit of Air Force, effective Mar. 1, 1985, by Secretarial order.

Air Force Manpower and Personnel Center Certain functions transferred on activation of Air Force Management Engineering Agency, which was made separate operating unit from Air Force Manpower and

Personnel Center (later Air Force Military Personnel Center) in April 1985 by general order of Chief of Staff.

Air Force Medical Service Center Renamed Air Force Office of Medical Support by Program Action Directive 85–1 of Mar. 6, 1985, approved by Air Force Vice Chief of Staff.

Air Mail, Bureau of Established in Interstate Commerce Commission to carry out provisions of act of June 12, 1934 (48 Stat. 933). Personnel and property transferred to *Civil Aeronautics Authority* by EO 7959 of Aug. 22, 1938.

Air Patrol, Civil Established in *Civilian Defense Office* by Administrative Order 9 of Dec. 8, 1941. Transferred to *Department of War* as auxiliary of Army Air Forces by EO 9339 of Apr. 29, 1943. Transferred to the Department of the Air Force by Secretary of Defense order of May 21, 1948. Established as civilian auxiliary of U.S. Air Force by act of May 26, 1948 (62 Stat. 274).

Air Safety Board Established by act of June 23, 1938 (52 Stat. 973). Functions transferred to *Civil Aeronautics Board* by Reorg. Plan No. IV of 1940, effective June 30, 1940.

Airways Modernization Board Established by act of Aug. 14, 1957 (71 Stat. 349). Transferred to *Federal Aviation Agency* by EO 10786 of Nov. 1, 1958.

Alaska, Board of Road Commissioners for Established in *Department of War* by act of Jan. 27, 1905 (33 Stat. 616). Functions transferred to the Department of Interior by act of June 30, 1932 (47 Stat. 446), and delegated to *Alaska Road Commission.* Functions transferred to the Department of Commerce by act of June 29, 1956 (70 Stat. 377), and terminated by act of June 25, 1959 (73 Stat. 145).

Alaska, Federal Field Committee for Development Planning in Established by EO 11182 of Oct. 2, 1964. Abolished by EO 11608 of July 19, 1971.

Alaska, Federal Reconstruction and Development Planning Commission for Established by EO 11150 of Apr. 2, 1964. Abolished by EO 11182 of Oct. 2, 1964, which established *President's Review Committee for Development Planning in Alaska* and *Federal Field Committee for Development Planning in Alaska.*

Alaska, President's Review Committee for Development Planning in Established by EO 11182 of Oct. 2, 1964. Superseded by *Federal Advisory Council on Regional Economic Development* etablished by EO 11386 of Dec. 28, 1967. EO 11386 revoked by EO 12553 f Feb. 25, 1986.

Alaska Communication System Operational responsibility vested in Secretary of the Army by act of May 26, 1900 (31 Stat. 206). Transferred to Secretary of the Air Force by Secretary of Defense reorganization order of May 24, 1962.

Alaska Engineering Commission See **Alaska Railroad**

Alaska Game Commission   Established by act of Jan. 13, 1925 (43 Stat. 740). Expired Dec. 31, 1959, pursuant to act of July 7, 1958 (72 Stat. 339).

Alaska International Rail and Highway Commission   Established by act of Aug. 1, 1956 (70 Stat. 888). Terminated June 30, 1961, under terms of act.

Alaska Natural Gas Transportation System, Office of Federal Inspector of Construction for the   Established by Reorg. Plan No. 1 of 1979 (5 U.S.C. app.), effective July 1, 1979. Abolished by act of Oct. 24, 1992 (106 Stat. 3128) and functions and authority vested in the Inspector transferred to the Secretary of Energy. Functions vested in the Secretary of Energy transferred to the Federal Coordinator, Office of the Federal Coordinator for Alaska Natural Gas Transportation Projects by act of Oct. 13, 2004 (118 Stat. 1261).

Alaska Power Administration   Established by the Secretary of the Interior in 1967. Transferred to the Department of Energy by act of Aug. 4, 1977 (91 Stat. 578).

Alaska Railroad   Built pursuant to act of Mar. 12, 1914 (38 Stat. 305), which created *Alaska Engineering Commission.* Placed under the Secretary of the Interior by EO 2129 of Jan. 26, 1915, and renamed Alaska Railroad by EO 3861 of June 8, 1923. Authority to regulate tariffs granted to Interstate Commerce Commission by EO 11107 of Apr. 25, 1963. Authority to operate Railroad transferred to the Secretary of Transportation by act of Oct. 15, 1966 (80 Stat. 941), effective Apr. 1, 1967. Railroad purchased by State of Alaska, effective Jan. 5, 1985.

Alaska Road Commission   *See* **Alaska, Board of Road Commissioners for**

Alcohol, Bureau of Industrial   Established by act of May 27, 1930 (46 Stat. 427). Consolidated into *Bureau of Internal Revenue* by EO 6166 of June 10, 1933. Consolidation deferred until May 11, 1934, by EO 6639 of Mar. 10, 1934. Order also transferred to Internal Revenue Commissioner certain functions imposed on Attorney General by act of May 27, 1930, with relation to enforcement of criminal laws concerning intoxicating liquors remaining in effect after repeal of 18th amendment; personnel of, and appropriations for, *Bureau of Industrial Alcohol;* and necessary personnel and appropriations of *Bureau of Prohibition,* Department of Justice.

Alcohol, Drug Abuse, and Mental Health Administration   Established by the *Secretary of Health, Education, and Welfare* by act of May 21, 1972 (88 Stat. 134). Redesignated as an agency of the Public Health Service from the *National Institute of Mental Health* Sept. 25, 1973, by the Secretary of Health, Education, and Welfare. Functions transferred to the Department of Health and Human Services by act of Oct. 17, 1979 (93 Stat. 695). Established as an agency of the Public Health Service by act of Oct. 27, 1986 (100 Stat. 3207–106). Renamed Substance Abuse and Mental Health Services Administration by act of July 10, 1992 (106 Stat. 325).

Alcohol Abuse and Alcoholism, National Institute on   Established within the National Institute of Mental Health, *Department of Health, Education, and Welfare* by act of Dec. 31, 1970 (84 Stat. 1848). Removed from within the National Institute of Mental Health and made an entity within the Alcohol, Drug Abuse, and Mental Health Administration by act of May 14, 1974 (88 Stat. 1356). Functions transferred to the Department of Health and Human Services by act of Oct. 17, 1979 (93 Stat. 695). (*See also* act of Oct. 27, 1986; 100 Stat. 3207–106.) Abolished by act of July 10, 1992 (106 Stat. 331). Reestablished by act of July 10, 1992 (106 Stat. 359).

Alcohol Administration, Federal   *See* **Alcohol Control Administration, Federal**

Alcohol Control Administration, Federal   Established by EO 6474 of Dec. 4, 1933. Abolished Sept. 24, 1935, on induction into office of Administrator, *Federal Alcohol Administration,* as provided in act of Aug. 29, 1935 (49 Stat. 977). Abolished by Reorg. Plan No. III of 1940, effective June 30, 1940, and functions consolidated with activities of Internal Revenue Service.

Alcohol, Tobacco, and Firearms, Bureau of   Established within Treasury Department by Treasury Order No. 221, eff. July 1, 1972. Transferred to Bureau of Alcohol, Tobacco, Firearms, and Explosives in Justice Department by act of Nov. 25, 2002, except some authorities, functions, personnel, and assets relating to administration and enforcement of certain provisions of the Internal Revenue Code of 1986 and title 27 of the U.S. Code (116 Stat. 2275).

Alexander Hamilton Bicentennial Commission   Established by act of Aug. 20, 1954 (68 Stat. 746). Terminated Apr. 30, 1958.

Alien Property Custodian   Appointed by President Oct. 22, 1917, under authority of act of Oct. 6, 1917 (40 Stat. 415). Office transferred to *Alien Property Division,* Department of Justice, by EO 6694 of May 1, 1934. Powers vested in President by act delegated to Attorney General by EO 8136 of May 15, 1939. Authority vested in Attorney General by EO's 6694 and 8136 transferred by EO 9142 of Apr. 21, 1942, to *Office of Alien Property Custodian, Office for Emergency Management,* as provided for by EO 9095 of Mar. 11, 1942.

American Forces Information Service   Established by Secretary of Defense Directive 5122.10 of March 13, 1989. Dissolved by Secretary's Directive 5105.74 of Dec. 18, 2007 and functions transferred to Defense Media Activity effective Oct. 1, 2008.

American Republics, Office for Coordination of Commercial and Cultural Relations between the   Established by *Council of National Defense* order approved by President Aug. 16, 1940. Succeeded by *Office of the Coordinator of Inter-American Affairs, Office for Emergency Management,* established by EO 8840 of July 30, 1941. Renamed *Office of Inter-American Affairs* by EO 9532 of Mar. 23, 1945. Information functions transferred to the Department of State by EO 9608 of Aug. 31, 1945. Terminated by EO 9710 of Apr. 10, 1946, and functions transferred to the Department of State, functioning as *Institute of Inter-American Affairs.* Transferred to *Foreign Operations Administration* by Reorg. Plan No. 7, effective Aug. 1, 1953.

American Revolution Bicentennial Administration *See* **American Revolution Bicentennial Commission**

American Revolution Bicentennial Commission Established by act of July 4, 1966 (80 Stat. 259). *American Revolution Bicentennial Administration*

established by act of Dec. 11, 1973 (87 Stat. 697), to replace *Commission. Administration* terminated June 30, 1977, pursuant to terms of act. Certain continuing functions transferred to the Secretary of the Interior by EO 12001 of June 29, 1977.

Anacostia Neighborhood Museum Renamed Anacostia Museum by Smithsonian Institution announcement of Apr. 3, 1987.

Animal Industry, Bureau of Established in the Department of Agriculture by act of May 29, 1884 (23 Stat. 31). Functions transferred to Agricultural Research Service by Secretary's Memorandum 1320, supp. 4, of Nov. 2, 1953.

Apprenticeship, Federal Committee on Previously known as *Federal Committee on Apprentice Training,* established by EO 6750–C of June 27, 1934. Functioned as part of *Division of Labor Standards,* Department of Labor, pursuant to act of Aug. 16, 1937 (50 Stat. 664). Transferred to *Office of Administrator, Federal Security Agency,* by EO 9139 of Apr. 18, 1942. Transferred to *Bureau of Training, War Manpower Commission,* by EO 9247 of Sept. 17, 1942. Returned to the Department of Labor by EO 9617 of Sept. 19, 1945.

Archive of Folksong Renamed Archive of Folk Culture by administrative order of Deputy Librarian of Congress, effective Sept. 21, 1981.

Archives Council, National Established by act of June 19, 1934 (48 Stat. 1122). Transferred to General Services Administration by act of June 30, 1949 (63 Stat. 378). Terminated on establishment of Federal Records Council by act of Sept. 5, 1950 (64 Stat. 583).

Archives Establishment, National *Office of Archivist of the U.S.* and *National Archives* created by act of June 19, 1934 (48 Stat. 1122). Transferred to General Services Administration by act of June 30, 1949 (63 Stat. 381), and incorporated as *National Archives and Records Service* by order of General Services Administrator, together with functions of *Division of the Federal Register, National Archives Council, National Historical Publications Commission,* National Archives Trust Fund Board, *Trustees of the Franklin D. Roosevelt Library,* and Administrative Committee of the Federal Register. Transferred from General Services Administration to National Archives and Records Administration by act of Oct. 19, 1984 (98 Stat. 2283), along with certain functions of Administrator of General Services

transferred to Archivist of the United States, effective Apr. 1, 1985.

Archives and Records Service, National *See* **Archives Establishment, National**

Archives Trust Fund Board, National *See* **Archives Establishment, National**

Area Redevelopment Administration Established May 8, 1961, by the Secretary of Commerce pursuant to act of May 1, 1961 (75 Stat. 47) and Reorg. Plan No. 5 of 1950, effective May 24, 1950. Terminated Aug. 31, 1965, by act of June 30, 1965 (79 Stat. 195). Functions transferred to Economic Development Administration in the Department of Commerce by Department Order 4–A, effective Sept. 1, 1965.

Arlington Memorial Amphitheater Commission Established by act of Mar. 4, 1921 (41 Stat. 1440). Abolished by act of Sept. 2, 1960 (74 Stat. 739), and functions transferred to the Secretary of Defense.

Arlington Memorial Bridge Commission Established by act of Mar. 4, 1913 (37 Stat. 885; D.C. Code (1951 ed.) 8–158). Abolished by EO 6166 of June 10, 1933, and functions transferred to *Office of National Parks, Buildings, and Reservations.*

Armed Forces, U.S. Court of Appeals for the *See* **Military Appeals, United States Court of**

Armed Forces Medical Library Founded in 1836 as *Library of the Surgeon General's Office,* U.S. Army. Later known as *Army Medical Library,* then *Armed Forces Medical Library* in 1952. Personnel and property transferred to National Library of Medicine established in Public Health Service by act of Aug. 3, 1956 (70 Stat. 960).

Armed Forces Museum Advisory Board, National Established by act of Aug. 30, 1961 (75 Stat. 414). Functions discontinued due to lack of funding.

Armed Forces Staff College Renamed Joint Forces Staff College by act of Oct. 30, 2000 (144 Stat. 165A–230).

Armed Services Renegotiation Board Established by Secretary of Defense directive of July 19, 1948. Abolished by Secretary's letter of Jan. 18, 1952, and functions transferred to *Renegotiation Board.*

Arms Control and Disarmament Agency, U.S. Established by act of Sept. 26, 1961 (75 Stat. 631). Abolished by act of Oct. 21, 1998 (112 Stat. 2681–767) and functions transferred to the Secretary of State.

Army Communications Command, U.S. Renamed U.S. Army Information Systems Command by Department General Order No. 26 of July 25, 1984.

Army Materiel Development and Readiness Command, U.S. Renamed U.S. Army Materiel Command by Department General Order No. 28 of Aug. 15, 1984.

Army and Navy, Joint Board   Placed under direction of President by military order of July 5, 1939. Abolished Sept. 1, 1947, by joint letter of Aug. 20, 1947, to President from Secretaries of *War* and Navy.

Army and Navy Staff College   Established Apr. 23, 1943, and operated under Joint Chiefs of Staff. Redesignated the National War College, effective July 1, 1946.

Army Specialist Corps   Established in *Department of War* by EO 9078 of Feb. 26, 1942. Abolished by the *Secretary of War* Oct. 31, 1942, and functions merged into central *Officer Procurement Service.*

Arts, National Collection of Fine   Established within Smithsonian Institution by act of Mar. 24, 1937 (50 Stat. 51). Renamed *National Museum of American Art* in Smithsonian Institution by act of Oct. 13, 1980 (94 Stat. 1884).

Arthritis, Diabetes, and Digestive and Kidney Diseases, National Institute of   See **Arthritis, Metabolism, and Digestive Diseases, National Institute of**

Arthritis, Metabolism, and Digestive Diseases, National Institute of   Renamed *National Institute of Arthritis, Diabetes, and Digestive and Kidney Diseases* by Secretary's order of June 15, 1981, pursuant to act of Dec. 19, 1980 (94 Stat. 3184). Renamed National Institute of Diabetes and Digestive and Kidney Diseases and National Institute of Arthritis and Musculoskeletal and Skin Diseases by act of Nov. 20, 1985 (99 Stat. 820).

Arts, Advisory Committee on the   Established under authority of act of Sept. 20, 1961 (75 Stat. 527). Terminated July 1973 by act of Oct. 6, 1972. Formally abolished by Reorg. Plan No. 2 of 1977, effective Apr. 1, 1978.

Arts, National Council on the   Established in Executive Office of the President by act of Sept. 3, 1964 (78 Stat. 905). Transferred to National Foundation on the Arts and the Humanities by act of Sept. 29, 1965 (79 Stat. 845).

Assay Commission, Annual   Established initially by act of Apr. 2, 1792 (1 Stat. 250) and by act of Feb. 12, 1873 (Revised Statute sec. 3647; 17 Stat. 432). Terminated and functions transferred to the Secretary of the Treasury by act of Mar. 14, 1980 (94 Stat. 98).

Assistance, Bureau of Public   Renamed *Bureau of Family Services* by order of the *Secretary of Health, Education, and Welfare,* effective Jan. 1, 1962. Functions redelegated to *Social and Rehabilitation Service* by Secretary's reorganization of Aug. 15, 1967.

Assistance Coordinating Committee, Adjustment   Established by act of Jan. 3, 1975 (88 Stat. 2040). Inactive since 1981.

Assistance Payments Administration   Established by *Secretary of Health, Education, and Welfare* reorganization of Aug. 15, 1967. Transferred by *Secretary's* reorganization of Mar. 8, 1977 (42 FR

13262), from *Social and Rehabilitation Service* to Social Security Administration.

Athletics, Interagency Committee on International   Established by EO 11117 of Aug. 13, 1963. Terminated by EO 11515 of Mar. 13, 1970.

Atlantic-Pacific Interoceanic Canal Study Commission   Established by act of Sept. 22, 1964 (78 Stat. 990). Terminated Dec. 1, 1970, pursuant to terms of act.

Atomic Energy Commission   Established by act of Aug. 1, 1946 (60 Stat. 755). Abolished by act of Oct. 11, 1974 (88 Stat. 1237) and functions transferred to *Energy Research and Development Administration* and Nuclear Regulatory Commission.

Aviation, Interdepartmental Committee on Civil International   Established by Presidential letter of June 20, 1935. Terminated on organization of *Civil Aeronautics Authority.*

Aviation Agency, Federal   Established by act of Aug. 23, 1958 (72 Stat. 731). Transferred to Secretary of Transportation by act of Oct. 15, 1966 (80 Stat. 931). *Agency* reestablished as Federal Aviation Administration by act of Jan 12, 1983 (96 Stat. 2416).

Aviation Commission, Federal   Established by act of June 12, 1934 (48 Stat. 938). Terminated Feb. 1, 1935, under provisions of act.

Beltsville Research Center   Established to operate with other agencies of the Department of Agriculture under *Agricultural Research Administration.* Consolidated into *Agricultural Research Administration,* the Department of Agriculture, by EO 9069 of Feb. 23, 1942.

Bilingual Education and Minority Languages Affairs, Office of   Renamed Office of English Language Acquisition, Language Enhancement, and Academic Achievement for Limited English Proficient Students by act of Jan. 8, 2002 (115 Stat. 2089).

Biobased Products and Bioenergy, Advisory Committee on   Established by EO 13134 of June 3, 1999. Abolished by EO 13423 of Jan. 24, 2007.

Biobased Products and Bioenergy, Interagency Council on   Established by EO 13134 of June 3, 1999. Abolished by EO 13423 of Jan. 24, 2007.

Biobased Products and Bioenergy Coordination Office, National   Established by EO 13134 of June 3, 1999. Abolished by EO 13423 of Jan. 24, 2007.

Biological Service, National   Established in the Department of the Interior in 1995 by Secretarial order. Transferred to U.S. Geological Survey as new Biological Resources Division by Secretarial Order No. 3202, Sept. 30, 1996.

Biological Survey, Bureau of   Established by Secretary's order July 1, 1885, as part of *Division of Entomology,* Department of Agriculture. Made separate bureau by act of Apr. 23, 1904 (33 Stat. 276). Transferred to the Department of the Interior by Reorg. Plan No. II of 1939, effective July 1, 1939. Consolidated with *Bureau of Fisheries* into *Fish and Wildlife Service* by Reorg. Plan No. III of 1940, effective June 30, 1940.

Biological Survey, National Established in the the Department of the Interior by Secretarial Order 3173 of Sept. 29, 1993. Renamed *National Biological Service* by Secretarial order in 1995.

Blind, Inc., American Printing House for the Established in 1858 as privately owned institution in Louisville, KY. Functions of the Secretary of the Treasury, except that relating to perpetual trust funds, transferred to *Federal Security Agency* by Reorg. Plan No. II of 1939, effective July 1, 1939. Functions performed by *Department of Health, Education, and Welfare* transferred to the Department of Education.

Blind-made Products, Committee on Purchases of Established by act of June 25, 1938 (52 Stat. 1196). Renamed *Committee for Purchase of Products and Services of the Blind and Other Severely Handicapped* by act of June 23, 1971 (85 Stat. 77). Renamed *Committee for Purchase from the Blind and Other Severely Handicapped* by act of July 25, 1974 (88 Stat. 392). Renamed Committee for Purchase From People Who Are Blind or Severely Disabled by act of Oct. 29, 1992 (106 Stat. 4486).

Blind and Other Severely Handicapped, Committee for Purchase of Products and Services of the *See* **Blind-made Products, Committee on Purchases of**

Board. *See other part of title*

Bond and Spirits Division Established as *Taxes and Penalties Unit,* as announced by Assistant to Attorney General in departmental circular of May 25, 1934, pursuant to EO 6639 of May 10, 1934. Abolished by administrative order of October 1942, and functions transferred to Tax, Claims, and Criminal Divisions, Department of Justice.

Bonneville Power Administration Established by the Secretary of the Interior pursuant to act of Aug. 20, 1937 (50 Stat. 731). Transferred to the Department of Energy by act of Aug. 4, 1977 (91 Stat. 578).

Boston National Historic Sites Commission Established by joint resolution of June 16, 1955 (69 Stat. 137). Terminated June 16, 1960, by act of Feb. 19, 1957 (71 Stat. 4).

Brazil-U.S. Defense Commission, Joint Established in May 1942 by agreement between the U.S. and Brazil. Terminated in September 1977 at direction of Brazilian Government.

Broadcast Bureau Merged with *Cable Television Bureau* to form Mass Media Bureau by Federal Communications Commission order, effective Nov. 30, 1982.

Broadcast Intelligence Service, Foreign *See* **Broadcast Monitoring Service, Foreign**

Broadcast Monitoring Service, Foreign Established

in Federal Communications Commission by Presidential directive of Feb. 26, 1941. Renamed *Foreign Broadcast Intelligence Service* by FCC order of July 28, 1942. Transferred to *Department of War* by Secretarial order of Dec. 30, 1945. Act of May 3, 1945 (59 Stat. 110), provided for liquidation 60 days after Japanese armistice. Transferred to *Central Intelligence Group* Aug. 5, 1946, and renamed *Foreign Broadcast Information Service.*

Budget, Bureau of the Established by act of June 10, 1921 (42 Stat. 20), in the Department of the Treasury under immediate direction of President. Transferred to Executive Office of the President by Reorg. Plan No. I of 1939, effective July 1, 1939. Reorganized by Reorg. Plan No. 2 of 1970, effective July 1, 1970, and renamed Office of Management and Budget.

Buildings Administration, Public Established as part of *Federal Works Agency* by Reorg. Plan No. I of 1939, effective July 1, 1939. Abolished by act of June 30, 1949 (63 Stat. 380), and functions transferred to General Services Administration.

Buildings Branch, Public Organized in *Procurement Division,* established in the Department of the Treasury by EO 6166 of June 10, 1933. Consolidated with *Branch of Buildings Management,* National Park Service, to form *Public Buildings Administration, Federal Works Agency,* under Reorg. Plan No. I of 1939, effective July 1, 1939.

Buildings Commission, Public Established by act of July 1, 1916 (39 Stat. 328). Abolished by EO 6166 of June 10, 1933, and functions transferred to *Office of National Parks, Buildings, and Reservations,* Department of the Interior. Functions transferred to *Public Buildings Administration, Federal Works Agency,* under Reorg. Plan No. I of 1939, effective July 1, 1939.

Buildings Management, Branch of Functions of National Park Service (except those relating to monuments and memorials) consolidated with *Public Buildings Branch, Procurement Division,* Department of the Treasury, to form *Public Buildings Administration, Federal Works Agency,* in accordance with Reorg. Plan No. I of 1939, effective July 1, 1939.

Buildings and Public Parks of the National Capital, Office of Public Established by act of Feb. 26, 1925 (43 Stat. 983), by consolidation of *Office of Public Buildings and Grounds* under Chief of Engineers, U.S. Army, and *Office of Superintendent of State, War, and Navy Department Buildings.* Abolished by EO 6166 of June 10, 1933, and functions transferred to *Office of National Parks, Buildings, and Reservations,* Department of the Interior.

Bureau. *See other part of title*

Business, Cabinet Committee on Small Established by Presidential letter of May 31, 1956. Dissolved January 1961.

Business Administration, Domestic and International *See* **Business and Defense Services Administration**

**Business and Defense Services
Administration**   Established by the Secretary of
Commerce Oct. 1, 1953, and operated under
Department Organization Order 40–1. Abolished
by Department Organization Order 40–1A of Sept.
15, 1970, and functions transferred to *Bureau of
Domestic Commerce*. Functions transferred to
Domestic and International Business Administration,
effective Nov. 17, 1972. *Administration* terminated
by Secretary's order of Dec. 4, 1977, and functions
assumed by *Industry and Trade Administration*.

Business Economics, Office of   Established by the
Secretary of Commerce Jan. 17, 1946. Renamed
Office of Economic Analysis Dec. 1, 1953.
Transferred to the *Administration of Social and
Economic Statistics* along with Bureau of the Census
and renamed Bureau of Economic Analysis on Jan.
1, 1972.

**Business Operations, Bureau of
International**   Established by the Secretary of
Commerce Aug. 8, 1961, by Departmental Orders
173 and 174. Abolished by Departmental Order
182 of Feb. 1, 1963, which established *Bureau of
International Commerce*. Functions transferred to
Domestic and International Business Administration,
effective Nov. 17, 1972.

Cable Television Bureau   Merged with *Broadcast
Bureau* by Federal Communications Commission
order to form Mass Media Bureau, effective Nov. 30,
1982.

California Debris Commission   Established by act of
Mar. 1, 1893 (27 Stat. 507). Abolished by act of Nov.
17, 1986 (100 Stat. 4229), and functions transferred
to the Secretary of the Interior.

Canal Zone Government   Established by act of Aug.
24, 1912 (37 Stat. 561). Abolished by act of Sept. 27,
1979 (93 Stat. 454).

Capital Housing Authority, National   Established by
act of June 12, 1934 (48 Stat. 930). Made agency of
District of Columbia government by act of Dec. 24,
1973 (87 Stat. 779), effective July 1, 1974.

Capital Park Commission, National   Established by
act of June 6, 1924 (43 Stat. 463). *National Capital
Park and Planning Commission* named successor
by act of Apr. 30, 1926 (44 Stat. 374). Functions
transferred to National Capital Planning Commission
by act of July 19, 1952 (66 Stat. 781).

**Capital Park and Planning Commission,
National**   *See* **Capital Park Commission, National**

**Capital Regional Planning Council,
National**   Established by act of July 19, 1952 (66
Stat. 785). Terminated by Reorg. Plan No. 5 of 1966,
effective Sept. 8, 1966.

**Capital Transportation Agency,
National**   Established by act of July 14, 1960 (74 Stat
537). Authorized to establish rapid rail transit system
by act of Sept. 8, 1965 (79 Stat. 663). Functions
transferred to Washington Metropolitan Area Transit
Authority by EO 11373 of Sept. 20, 1967.

Career Executive Board   Established by EO 10758
of Mar. 4, 1958. Terminated July 1, 1959, and EO
10758 revoked by EO 10859 of Feb. 5, 1960.

Caribbean Organization   Act of June 30, 1961 (75
Stat. 194), provided for acceptance by President of
Agreement for the Establishment of the Caribbean
Organization, signed at Washington, June 21, 1960.
Article III of Agreement provided for termination of
Caribbean Commission, authorized by Agreement
signed Oct. 30, 1946, on first meeting of Caribbean
Council, governing body of *Organization*.
Terminated, effective Dec. 31, 1965, by resolution
adopted by Council.

**Cemeteries and Memorials in Europe,
National**   Supervision transferred from *Department
of War* to American Battle Monuments Commission
by EO 6614 of Feb. 26, 1934, which transfer was
deferred to May 21, 1934, by EO 6690 of Apr. 25,
1934.

Cemeteries and Parks, National   *Department of
War* functions regarding National Cemeteries and
Parks located in continental U.S. transferred to *Office
of National Parks, Buildings, and Reservations,*
Department of the Interior, by EO 6166 of June 10,
1933.

Cemetery System, National   Established in the
Veterans' Administration by act of June 18, 1973 (87
Stat. 75). Redesignated as the National Cemetery
Administration by act of Nov. 11, 1998 (112 Stat.
3337).

Censorship, Office of   Established by EO 8985 of
Dec. 19, 1941. Terminated by EO 9631 of Sept. 28,
1945.

Censorship Policy Board   Established by EO 8985
of Dec. 19, 1941. Terminated by EO 9631 of Sept.
28, 1945.

Census, Bureau of the   *See* **Census Office**

Census Office   Established temporarily within the
Department of the Interior in accordance with act of
Mar. 3, 1899. Established as a permanent office by
act of Mar. 6, 1902. Transferred from the Department
of the Interior to *Department of Commerce and Labor*
by act of Feb. 14, 1903. Remained in the Department
of Commerce under provisions of Reorganization
Plan No. 5 of May 24, 1950, effective May 24, 1950.

Center.   *See other part of title*

Central.   *See other part of title*

Chemistry and Soils, Bureau of   *See* **Agricultural
and Industrial Chemistry, Bureau of**

**Chesapeake Bay Center for Environmental
Studies**   Established in 1965 in Annapolis, MD,
as part of Smithsonian Institution by Secretarial
order. Merged with *Radiation Biology Laboratory* by
Secretarial Order July 1, 1983, to form Smithsonian
Environmental Research Center.

Chief Information Officers Council   Established by EO 13011 of July 16, 1996. Abolished by EO 13403 of May 12, 2006.

Chief People Officer, Office of the   Renamed Office of the Chief Human Capital Officer by administrative order 5440.597 of June 16, 2006.

Chief Strategic Officer, Office of the   Established by the Commissioner of Social Security Dec. 20, 2002. Abolished by Commissioner's memorandum of Jan. 14, 2008, and functions transferred to the Office of the Deputy Commissioner for Budget, Finance, and Management.

Child Development, Office of   *See* **Children's Bureau**

Children's Bureau   Established by act of Apr. 9, 1912 (37 Stat. 79). Placed in the Department of Labor by act of Mar. 4, 1913 (37 Stat. 737). Transferred, with exception of child labor functions, to *Social Security Administration, Federal Security Agency,* by Reorg. Plan No. 2 of 1946, effective July 16, 1946. Continued under *Administration* when *Agency* functions assumed by the *Department of Health, Education, and Welfare.* Reassigned to *Welfare Administration* by Department reorganization of Jan. 28, 1963. Reassigned to *Social and Rehabilitation Service* by Department reorganization of Aug. 15, 1967. Reassigned to *Office of Child Development* by Department reorganization order of Sept. 17, 1969.

Child Health and Human Development, National Institute of   Established by act of Oct. 17, 1962 (76 Stat. 1072). Renamed Eunice Kennedy Shriver National Institute of Child Health and Human Development by act of Dec. 21, 2007 (121 Stat. 1826).

China, U.S. Court for   Established by act of June 30, 1906 (34 Stat. 814). Transferred to the Department of Justice by EO 6166 of June 10, 1933, effective Mar. 2, 1934. Act of June 30, 1906, repealed effective Sept. 1, 1948 (62 Stat. 992).

Christopher Columbus Quincentenary Jubilee Commission   Established by act of Aug. 7, 1984 (98 Stat. 1257). Terminated pursuant to terms of act.

Civil defense.   *See* **Defense**

Civil Rights, Commission on   Established by act of Sept. 9, 1957 (71 Stat. 634). Terminated in 1983 and reestablished by act of Nov. 30, 1983 (97 Stat. 1301). Renamed United States Commission on Civil Rights by act of Nov. 2, 1994 (108 Stat. 4683).

Civil Service Commission, U.S.   Established by act of Jan. 16, 1883 (22 Stat. 403). Redesignated as Merit Systems Protection Board and functions transferred to Board and Office of Personnel Management by Reorg. Plan No. 2 of 1978, effective Jan. 1, 1979.

Civil War Centennial Commission   Established by act of Sept. 7, 1957 (71 Stat. 626). Terminated May 1, 1966, pursuant to terms of act.

Civilian Conservation Corps   Established by act of

June 28, 1937 (50 Stat. 319). Made part of *Federal Security Agency* by Reorg. Plan No. I of 1939, effective July 1, 1939. Liquidation provided for by act of July 2, 1942 (56 Stat. 569), not later than June 30, 1943.

Civilian Health and Medical Program of the United States, Office of   Established as field activity in the Department of Defense in 1974. Functions consolidated into the TRICARE Management Activity in November 1997 by Defense Reform Initiative.

Civilian Production Administration   Established by EO 9638 of Oct. 4, 1945. Consolidated with other agencies to form *Office of Temporary Controls, Office for Emergency Management,* by EO 9809 of Dec. 12, 1946.

Civilian Service Awards Board, Distinguished   Established by EO 10717 of June 27, 1957. Terminated by EO 12014 of Oct. 19, 1977, and functions transferred to U.S. *Civil Service Commission.*

Claims, U.S. Court of   Established Feb. 25, 1855 (10 Stat. 612). Abolished by act of Apr. 2, 1982 (96 Stat. 26) and trial jurisdiction transferred to U.S. *Claims Court* and appellate functions merged with those of U.S. *Court of Customs and Patent Appeals* to form U.S. Court of Appeals for the Federal Circuit. U.S. *Claims Court* renamed U.S. Court of Federal Claims by act of Oct. 29, 1992 (106 Stat. 4516).

Claims Commission of the United States, International   Established in the Department of State by act of Mar. 10, 1950 (64 Stat. 12). Abolished by Reorg. Plan No. 1 of 1954, effective July 1, 1954, and functions transferred to Foreign Claims Settlement Commission of the United States.

Claims Settlement Commission of the United States, Foreign   Established by Reorg. Plan No. 1 of 1954, effective July 1, 1954. Transferred to the Department of Justice by act of Mar. 14, 1980 (94 Stat. 96).

Clark Sesquicentennial Commission, George Rogers   Established by Public Resolution 51 (45 Stat. 723). Expenditures ordered administered by the Department of the Interior by EO 6166 of June 10, 1933.

Classification Review Committee, Interagency   Established by EO 11652 of Mar. 8, 1972. Abolished by EO 12065 of June 28, 1978.

Clemency Board, Presidential   Established in Executive Office of the President by EO 11803 of Sept. 16, 1974. Final recommendations submitted to President Sept. 15, 1975, and *Board* terminated by EO 11878 of Sept. 10, 1975.

Coal Commission, National Bituminous   Established under authority of act of Aug. 30, 1935 (49 Stat. 992). Abolished by Reorg. Plan No. II of 1939, effective July 1, 1939, and functions transferred to *Bituminous Coal Division,* Department of the Interior.

Coal Consumers' Counsel, Office of the Bituminous   Established by act of Apr. 11, 1941

(55 Stat. 134), renewing provisions of act of Apr. 23, 1937 (50 Stat. 72) for 2 years to continue functions of *Consumers' Counsel Division,* Department of the Interior. Functions continued by acts of Apr. 24, 1943 (57 Stat. 68), and May 21, 1943 (57 Stat. 82). Terminated Aug. 24, 1943.

Coal Division, Bituminous   Established July 1, 1939, by Secretary of the Interior Order 1394 of June 16, 1939, as amended by Order 1399, of July 5, 1939, pursuant to act of Apr. 3, 1939 (53 Stat. 562) and Reorg. Plan No. II of 1939, effective July 1, 1939. Administered functions vested in *National Bituminous Coal Commission* by act of Apr. 23, 1937 (50 Stat. 72). Act extended to Aug. 24, 1943, on which date it expired.

Coal Labor Board, Bituminous   Established by act of July 12, 1921 (42 Stat. 140). Abolished as result of U.S. Supreme Court decision, May 18, 1936, in case of *Carter* v. *Carter Coal Company et al.*

Coal Mine Safety Board of Review, Federal   Established by act of July 16, 1952 (66 Stat. 697). Inactive after Mar. 30, 1970, pursuant to act of Dec. 30, 1969 (83 Stat. 803).

Coal Mines Administration   Established by the Secretary of the Interior July 1, 1943. Abolished by Secretary's Order 1977 of Aug. 16, 1944, as amended by Order 1982 of Aug. 31, 1944, and functions assumed by *Solid Fuels Administration for War. Administration* reestablished in the Department of the Interior by EO 9728 of May 21, 1946. Terminated June 30, 1947, by act of Mar. 27, 1942 (56 Stat. 176).

Coal Research, Office of   Established in the Department of the Interior by act of July 7, 1960 (74 Stat. 336). Functions transferred to *Energy Research and Development Administration* by act of Oct. 11, 1974 (88 Stat. 1237).

Coalition Provisional Authority, Inspector General of the   Established by act of Nov. 6, 2003 (117 Stat. 1234). Renamed Special Inspector General for Iraq Reconstruction by act of Oct. 28, 2004 (118 Stat. 2078.)

Coalition Provisional Authority, Office of the Inspector General of the   Established by act of Nov. 6, 2003 (117 Stat. 1234). Renamed Office of the Special Inspector General for Iraq Reconstruction by act of Oct. 28, 2004 (118 Stat. 2078).

Coast and Geodetic Survey   *See* **Coast Survey**

Coast Guard, U.S.   Established by act of Jan. 28, 1915 (38 Stat. 800) as a military service and branch of the U.S. Armed Forces at all times and as a service in Treasury Department, except when operating as a service in the Navy. Transferred from the Department of the Treasury to the Department of the Navy by EO 8929 of Nov. 1, 1941. Returned to the Department of the Treasury by EO 9666 of Dec. 28, 1945. Transferred to the Department of Transportation by act of Oct. 15, 1966 (80 Stat. 931). Transferred to Homeland Security Department by act of Nov. 25, 2002 (116 Stat. 2249) with related authorities and functions of the Secretary of Transportation.

Coast Survey   Established by act of Feb. 10, 1807 (2 Stat. 413). Redesignated as *Coast and Geodetic Survey* by act of June 20, 1878 (20 Stat. 206). Transferred to *Environmental Science Services Administration* by Reorg. Plan No. 2 of 1965, effective July 13, 1965.

Codification Board   Established by act of June 19, 1937 (50 Stat. 304). Abolished by Reorg. Plan No. II of 1939, effective July 1, 1939, and functions transferred to *Division of the Federal Register.*

Coinage, Joint Commission on the   Established by act of July 23, 1965 (79 Stat. 258). Expired Jan. 4, 1975, pursuant to act of Oct. 6, 1972 (88 Stat. 776).

Columbia Institution for the Instruction of the Deaf and Dumb, and the Blind   Established by act of Feb. 16, 1857 (11 Stat. 161). Renamed *Columbia Institution for the Instruction of the Deaf and Dumb* by act of Feb. 23, 1865 (13 Stat. 436). Renamed *Columbia Institution for the Deaf* by act of Mar. 4, 1911 (36 Stat. 1422). Renamed *Gallaudet College* by act of June 18, 1954 (68 Stat. 265). Functions of the *Department of Health, Education, and Welfare* transferred to the Department of Education by act of Oct. 17, 1979 (93 Stat. 695). Renamed Gallaudet University by act of Aug. 4, 1986 (100 Stat. 781).

Commander in Chief, U.S. Fleet, and Chief of Naval Operations   Duties of two positions prescribed by EO 8984 of Dec. 18, 1941. Combined under one officer by EO 9096 of Mar. 12, 1942.

Commerce, Bureau of Domestic   *See* **Business and Defense Services Administration**

Commerce, Bureau of Foreign   Established by the Secretary of Commerce Oct. 12, 1953, by Reorg. Plan No. 5 of 1950, effective May 24, 1950. Abolished by department order of Aug. 7, 1961, and functions vested in *Bureau of International Programs* and *Bureau of International Business Operations.*

Commerce, Bureau of Foreign and Domestic   Established by act of Aug. 23, 1912 (37 Stat. 407). Functions reassigned to other offices of the Department of Commerce due to internal reorganizations.

Commerce, Bureau of International   *See* **Business Operations, Bureau of International**

Commerce Service, Foreign   Established in *Bureau of Foreign and Domestic Commerce,* Department of Commerce, by act of Mar. 3, 1927 (44 Stat. 1394). Transferred to the Department of State as part of Foreign Service by Reorg. Plan No. II of 1939, effective July 1, 1939.

Commercial Company, U.S.   Established Mar. 27, 1942, as subsidiary of *Reconstruction Finance Corporation.* Transferred to *Office of Economic Warfare* by EO 9361 of July 15, 1943. *Office* consolidated into *Foreign Economic Administration* by EO 9380 of Sept. 25, 1943. Functions returned to *Corporation* by EO 9630 of Sept. 27, 1945, until June 30, 1948.

Commercial Policy, Executive Committee on   Established by Presidential letter of Nov. 11, 1933, to Secretary of State. Abolished by EO 9461 of Aug. 7, 1944.

Commercial Standards Division   Transferred with *Division of Simplified Trade Practice* from *National Bureau of Standards* to the Secretary of Commerce by Reorg. Plan No. 3 of 1946, effective July 16, 1946, to permit reassignment to *Office of Domestic Commerce.* Functions transferred to *National Bureau of Standards* by the Department of Commerce Order 90, June 7, 1963, pursuant to Reorg. Plan No. 5 of 1950, effective May 24, 1950.

Commission.   *See other part of title*

Committee.   *See also other part of title*

Committee Management Secretariat   Established in the Office of Management and Budget Jan. 5, 1973, by act of Oct. 6, 1972 (86 Stat. 772). Functions transferred to General Services Administrator by Reorg. Plan No. 1 of 1977, effective Apr. 1, 1978. Reassigned to the *National Archives and Records Service* by GSA order of Feb. 22, 1979. Transferred in Archives to Office of the Federal Register by GSA order of Oct. 14, 1980. Transferred to Office of the Archivist of the United States by GSA order of Sept. 24, 1982. Reassigned to Office of Program Initiatives, GSA, by GSA order of May 18, 1984. Transferred to Office of Management Services, GSA, by GSA order of Apr. 7, 1986.

Commodities Corporation, Federal Surplus   *See* **Relief Corporation, Federal Surplus**

Commodity Credit Corporation   Organized by EO 6340 of Oct. 16, 1933, and managed in close affiliation with *Reconstruction Finance Corporation.* Transferred to the Department of Agriculture by Reorg. Plan No. I of 1939, effective July 1, 1939.

Commodity Exchange Administration   *See* **Grain Futures Administration**

Commodity Exchange Authority   *See* **Grain Futures Administration**

Commodity Exchange Commission   Established by act of Sept. 21, 1922 (42 Stat. 998). Functions transferred to Commodity Futures Trading Commission by act of Oct. 23, 1974 (88 Stat. 1414).

Commodity Stabilization Service   Established in the Department of Agriculture Nov. 2, 1953, by Secretary's Memorandum 1320, supp. 4. Renamed Agricultural Stabilization and Conservation Service by Secretary's Memorandum 1458 of June 14, 1961, effective June 5, 1961.

Communication Agency, International   *See Information Agency, U.S.*

Communications Program, Joint Tactical   Combined with *Joint Interoperability of the Tactical Command and Control Systems Programs* to form Joint Tactical Command, Control, and Communications Agency in July 1984, pursuant to DOD Directive 5154.28.

Community Development Corporation   Established in the Department of Housing and Urban Development by act of Dec. 31, 1970 (84 Stat. 1791). Renamed *New Community Development Corporation* by act of Aug. 22, 1974 (88 Stat. 725). Abolished Nov. 30, 1983, by act of Nov. 30, 1983 (97 Stat. 1238), and functions transferred to Assistant Secretary for Community Planning and Development, Department of Housing and Urban Development.

Community Development Corporation, New   *See* **Community Development Corporation**

Community Facilities, Bureau of   Established in 1945 by *Federal Works Administrator.* Transferred by act of June 30, 1949 (63 Stat. 380), to General Services Administration, functioning as *Community Facilities Service.* Certain functions transferred to various agencies, including the Department of the Interior, *Housing and Home Finance Agency,* and *Federal Security Agency* by Reorg. Plans Nos. 15, 16, and 17 of 1950, effective May 24, 1950.

Community Facilities Administration   Established in *Housing and Home Finance Agency* by Administrator's Organizational Order 1 of Dec. 23, 1954. Terminated by act of Sept. 9, 1965 (79 Stat. 667), and functions transferred to the Department of Housing and Urban Development.

Community Organization, Committee on   Established in *Office of Defense Health and Welfare Services* Sept. 10, 1941. Functions transferred to *Federal Security Agency* by EO 9338 of Apr. 29, 1943.

Community Relations Service   Established in the Department of Commerce by act of July 2, 1964 (78 Stat. 241). Transferred to the Department of Justice by Reorg. Plan No. 1 of 1966, effective Apr. 22, 1966.

Community Service, Commission on National and   Established by act of Nov. 16, 1990 (104 Stat. 3168). Abolished by act of Sept. 21, 1993, and functions vested in the Board of Directors or the Executive Director prior to Oct. 1, 1993, transferred to the Corporation for National and Community Service (107 Stat. 873, 888).

Community Services Administration   Established by act of Jan. 4, 1975 (88 Stat. 2291) as successor to *Office of Economic Opportunity.* Abolished as independent agency through repeal of act of Aug. 20, 1964 (except titles VIII and X of such act) by act of Aug. 13, 1981 (95 Stat. 519).

Community Services Administration   Functions concerning Legal Services Program transferred to Legal Services Corporation by act of July 25, 1974 (88 Stat. 389). Renamed *Public Services Administration* by *Health, Education, and Welfare* departmental notice of Nov. 3, 1976. Transferred to *Office of Human Development* by Secretary's reorganization of Mar. 8, 1977 (42 FR 13262).

Community War Services   Established in *Office of the Administrator* under EO 9338 of Apr. 29, 1943, and *Federal Security Agency* order. Terminated Dec. 31, 1946, by act of July 26, 1946 (60 Stat. 695).

Conciliation Service, U.S.   Established by act of Mar. 4, 1913 (37 Stat. 738). Functions transferred to Federal Mediation and Conciliation Service, established by act of June 23, 1947 (61 Stat. 153).

Conference on Security and Cooperation in Europe   Renamed Organization for Security and Cooperation in Europe by EO 13029, Dec. 3, 1996 (61 FR 64591).

Consolidated Farm Service Agency   Established by act of Oct. 13, 1994 (108 Stat. 3214). Renamed Farm Service Agency (61 FR 1109), effective Jan. 16, 1996.

Constitution, Commission on the Bicentennial of the United States   Established by act of Sept. 29, 1983, as amended (97 Stat. 722). Terminated by act of Dec. 3, 1991 (105 Stat. 1232).

Constitution, transfer of functions   *See* Statutes at Large and other matters

Construction, Collective Bargaining Committee in   Established by EO 11849 of Apr. 1, 1975. Inactive since Jan. 7, 1976. Formally abolished by EO 12110 of Dec. 28, 1978.

Construction, Equipment and Repairs, Bureau of   Established in the Department of the Navy by act of Aug. 31, 1842 (5 Stat. 579). Abolished by act of July 5, 1862 (12 Stat. 510), and functions distributed among *Bureau of Equipment and Recruiting, Bureau of Construction and Repair,* and *Bureau of Steam Engineering.*

Construction Branch   Established in the Department of the Treasury in 1853 and designated *Bureau of Construction* under control of *Office of Supervising Architect* by Sept. 30, 1855. *Office* incorporated into *Public Buildings Branch, Procurement Division,* by EO 6166 of June 10, 1933. Transferred to *Federal Works Agency* by Reorg. Plan No. I of 1939, effective July 1, 1939, when *Public Buildings Branch* of *Procurement Division, Bureau of Buildings Management,* National Park Service, Department of the Interior—so far as latter concerned with operation of public buildings for other departments or agencies—and *U.S. Housing Corporation* consolidated with *Public Buildings Administration, Federal Works Agency.*

Construction Industry Stabilization Committee   Established by EO 11588 of Mar. 29, 1971. Abolished by EO 11788 of June 18, 1974.

Construction and Repair, Bureau of   Established by act of July 5, 1862 (12 Stat. 510), replacing *Bureau of Construction, Equipment and Repairs.* Abolished by act of June 20, 1940 (54 Stat. 492), and functions transferred to *Bureau of Ships.*

Consumer Advisory Council   Established by EO 11136 of Jan. 3, 1964. *Office of Consumer Affairs* established in Executive Office of the President by EO 11583 of Feb. 24, 1971, and Council reestablished in *Office.*

Consumer Affairs, Office of   Established by EO 11583 of Feb. 24, 1971. Transferred to the *Department of Health, Education, and Welfare* by EO 11702 of Jan. 25, 1973.

Consumer Affairs Staff, National Business Council for   Established in the Department of Commerce by departmental organization order of Dec. 16, 1971. Terminated by departmental order of Dec. 6, 1973, due to lack of funding.

Consumer agencies   Consumer agencies of *National Emergency Council* and *National Recovery Administration* reorganized and functions transferred, together with those of *Consumers' Advisory Board, NRA,* and *Cabinet Committee on Price Policy,* to *Consumers' Division, NRA,* by EO 7120 of July 30, 1935. *Division* transferred to the Department of Labor by EO 7252 of Dec. 21, 1935. Transferred to *Division of Consumers' Counsel, Agricultural Adjustment Administration,* Department of Agriculture, by Secretary of Labor letter of Aug. 30, 1938, to the Secretary of Agriculture. Continued as *Consumer Standards Project* until June 30, 1941. Research on consumer standards continued by *Consumer Standards Section, Consumers' Counsel Division,* transferred to *Agricultural Marketing Administration* by administrative order of Feb. 28, 1942. Other project activities discontinued.

Consumer Cooperative Bank, National   Established by act of Aug. 20, 1978 (92 Stat. 499). Removed from mixed-ownership, Government corporation status by acts of Sept. 13, 1982 (96 Stat. 1062) and Jan. 12, 1983 (96 Stat. 2478).

Consumer Interests, President's Committee on   Established by EO 11136 of Jan. 3, 1964. Abolished by EO 11583 of Feb. 24, 1971.

Consumer and Marketing Service   Established by the Secretary of Agriculture Feb. 2, 1965. Renamed Agricultural Marketing Service Apr. 2, 1972, by Secretary's order and certain functions transferred to Animal and Plant Health Inspection Service.

Consumers' Counsel   Established in *National Bituminous Coal Commission* by act of Aug. 30, 1935 (49 Stat. 993). Office abolished by Reorg. Plan No. II of 1939, effective July 1, 1939, and functions transferred to Office of Solicitor, Department of the Interior, to function as *Consumers' Counsel Division* under direction of the Secretary of the Interior. Functions transferred to *Office of the Bituminous Coal Consumers' Counsel* June 1941 by act of Apr. 11, 1941 (55 Stat. 134).

Consumers' Counsel Division   *See* Consumers' Counsel

Consumers' Counsel, Division of   Established by act of May 12, 1933 (48 Stat. 31). Transferred by order of the Secretary of Agriculture from *Agricultural Adjustment Administration* to supervision of *Director of Marketing,* effective Feb. 1, 1940. Transferred to *Agricultural Marketing Administration* by administrative order of Feb. 28, 1942.

Consumers' Problems, Adviser on   *See* Consumer agencies

Contract Committee Government *See* **Contract Compliance, Committee on Government**

Contract Compliance, Committee on Government Established by EO 10308 of Dec. 3, 1951. Abolished by EO 10479 of Aug. 13, 1953, which established successor *Government Contract Committee.* Abolished by EO 10925 of Mar. 6, 1961, and records and property transferred to *President's Committee on Equal Employment Opportunity.*

Contract Settlement, Office of Established by act of July 1, 1944 (58 Stat. 651). Transferred to *Office of War Mobilization and Reconversion* by act of Oct. 3, 1944 (58 Stat. 785). Terminated by EO 9809 of Dec. 12, 1946, and Reorg. Plan No. 1 of 1947, effective July 1, 1947, and functions transferred to the Department of the Treasury. Functions transferred to General Services Administration by act of June 30, 1949 (63 Stat. 380).

Contract Settlement Advisory Board Established by act of July 1, 1944 (58 Stat. 651). Transferred to the Department of the Treasury by EO 9809 of Dec. 12, 1946, and by Reorg. Plan No. 1 of 1947, effective July 1, 1947. Transferred to General Services Administration by act of June 30, 1949 (63 Stat. 380) and established as *Contract Review Board.* Renamed Board of Contract Appeals in 1961 by Administrator's order. Board established as independent entity within General Services Administration Feb. 27, 1979, pursuant to act of Nov. 1, 1978 (92 Stat. 2383).

Contract Settlement Appeal Board, Office of Established by act of July 1, 1944 (58 Stat. 651). Transferred to the Department of the Treasury by EO 9809 of Dec. 12, 1946, and by Reorg. Plan No. 1 of 1947, effective July 1, 1947. Functions transferred to General Services Administration by act of June 30, 1949 (63 Stat. 380). Abolished by act of July 14, 1952 (66 Stat. 627).

Contract Termination Board, Joint Established Nov. 12, 1943, by *Director of War Mobilization.* Functions assumed by *Office of Contract Settlement.*

Contracts Division, Public Established in the Department of Labor to administer act of June 30, 1936 (49 Stat. 2036). Consolidated with Wage and Hour Division by Secretarial order of Aug. 21, 1942. Absorbed by Wage and Hour Division by Secretarial order of May 1971.

Cooperation Administration, International Established by Department of State Delegation of Authority 85 of June 30, 1955, pursuant to EO 10610 of May 9, 1955. Abolished by act of Sept. 4, 1961 (75 Stat. 446), and functions redelegated to Agency for International Development pursuant to Presidential letter of Sept. 30, 1961, and EO 10973 of Nov. 3, 1961.

Cooperative State Research, Education, and Extension Service Established by act of Oct. 13, 1994 (108 Stat. 3178). Reorganized into the National Institute of Food and Agriculture by Secretary's Memorandum 1062–001 of Sept. 17, 2009.

Cooperative State Research Service Established in the Department of Agriculture. Incorporated into Cooperative State, Research, Education, and Extension Service under Department of Agriculture reorganization in 1995.

Coordinating Council for Comparative Effectiveness Research, Federal Terminated by act of Mar. 23, 2010 (124 Stat. 747).

Coordinating Service, Federal *Office of Chief Coordinator* created by Executive order promulgated in *Bureau of the Budget* Circular 15, July 27, 1921, and duties enlarged by other *Bureau* circulars. Abolished by EO 6166 of June 10, 1933. Contract form, Federal traffic, and surplus property functions transferred to *Procurement Division* by order of the Secretary of the Treasury, approved by President Oct. 9, 1933, issued pursuant to EO's 6166 of June 10, 1933, and 6224 of July 27, 1933.

Copyright Arbitration Royalty Panels Established by act of Dec. 17, 1993 (107 Stat. 2304). Replaced by Copyright Royalty Judges under act of Nov. 30, 2004 (118 Stat. 2351).

Copyright Royalty Tribunal Established as an independent entity within the legislative branch by act of Oct. 19, 1976 (90 Stat. 2594). Abolished by act of Dec. 17, 1993 (107 Stat. 2304), and functions transferred to copyright arbitration royalty panels.

Copyrighted Works, National Commission on New Technological Uses of Established by act of Dec. 31, 1974 (88 Stat. 1873). Terminated Sept. 29, 1978, pursuant to terms of act.

Corporate Payments Abroad, Task Force on Questionable Established by Presidential memorandum of Mar. 31, 1976. Terminated Dec. 31, 1976, pursuant to terms of memorandum.

Corporation, Federal Facilities Established in the Department of the Treasury by EO 10539 of June 22, 1954. Placed under supervision of Director appointed by General Services Administrator by EO 10720 of July 11, 1957. Dissolved by act of Aug. 30, 1961 (75 Stat. 418), and functions transferred to Administrator of General Services.

Corregidor-Bataan Memorial Commission Established by act of Aug. 5, 1953 (67 Stat. 366). Terminated May 6, 1967, by act of Dec. 23, 1963 (77 Stat. 477).

Cost Accounting Standards Board Established by act of Aug. 15, 1970 (84 Stat. 796). Terminated Sept. 30, 1980, due to lack of funding. Reestablished by act of Nov. 17, 1988 (102 Stat. 4059).

Cost of Living Council Established by EO 11615 of Aug. 15, 1971. Abolished by EO 11788 of June 18, 1974.

Cotton Stabilization Corporation Organized June 1930 under laws of Delaware by *Federal Farm Board* pursuant to act of June 15, 1929 (46 Stat. 11). Certificate of dissolution filed with Corporation Commission of Delaware Dec. 27, 1934.

Cotton Textile Industry, Board of Inquiry for the   Established by EO 6840 of Sept. 5, 1934. Abolished by EO 6858 of Sept. 26, 1934.

Council.   *See other part of title*

Counterespionage Section   Transferred from the Criminal Division to the National Security Division by act of Mar. 9, 2006 (120 Stat. 249).

Counterintelligence, Office of   Established within the Department of Energy by Public Law 106–65 of Oct. 5, 1999 (113 Stat. 955). Merged with *Office of Intelligence* to form *Office of Intelligence and Counterintelligence* by memorandum of March 9, 2006 of the Secretary of Energy.

Counterterrorism Section   Transferred from the Criminal Division to the National Security Division by act of Mar. 9, 2006 (120 Stat. 249).

Courts   Under act of Aug. 7, 1939 (53 Stat. 1223), and revised June 25, 1948 (62 Stat. 913), to provide for administration of U.S. courts, administrative jurisdiction over all continental and territorial courts transferred to Administrative Office of the U.S. Courts, including U.S. courts of appeals and district courts, District Court for the Territory of Alaska, U.S. District Court for the District of the Canal Zone, District Court of Guam, District Court of the Virgin Islands, Court of Claims, Court of Customs and Patent Appeals, and Customs Courts.

Credit Unions, Bureau of Federal   *See* **Credit Union System, Federal**

Credit Union System, Federal   Established by act of June 26, 1934 (48 Stat. 1216), to be administered by *Farm Credit Administration.* Transferred to Federal Deposit Insurance Corporation by EO 9148 of Apr. 27, 1942, and Reorg. Plan No. 1 of 1947, effective July 1, 1947. Functions transferred to *Bureau of Federal Credit Unions, Federal Security Agency,* established by act of June 29, 1948 (62 Stat. 1091). Functions transferred to the *Department of Health, Education, and Welfare* by Reorg. Plan No. 1 of 1953, effective Apr. 11, 1953. Functions transferred to National Credit Union Administration by act of Mar. 10, 1970 (84 Stat. 49).

Crime, National Council on Organized   Established by EO 11534 of June 4, 1970. Terminated by EO 12110 of Dec. 28, 1978.

Critical Materials Council, National   Established within Executive Office of the President by act of July 31, 1984 (98 Stat. 1250). *Office* abolished in September 1993 due to lack of funding and functions transferred to the Office of Science and Technology Policy.

Crop Insurance Corporation, Federal   Established by act of Feb. 16, 1938. Consolidated with the *Agricultural Stabilization and Conservation Service* and *Farmers' Home Administration* in 1995 to form the *Farm Service Agency* pursuant to act of Oct. 13, 1994 (108 Stat. 3178).

Crop Production Loan Office   Authorized by Presidential letters of July 26, 1918, and July 26, 1919, to the Secretary of Agriculture. Further authorized by act of Mar. 3, 1921 (41 Stat. 1347). Transferred to Farm Credit Administration by EO 6084 of Mar. 27, 1933.

Cultural Center, National   Established in Smithsonian Institution by act of Sept. 2, 1958 (72 Stat. 1698). Renamed John F. Kennedy Center for the Performing Arts by act of Jan. 23, 1964 (78 Stat. 4).

Customs, Bureau of   Established under sec. 1 of act of Mar. 3, 1927 (19 U.S.C. 2071) in Treasury Department. Functions relating to award of numbers to undocumented vessels, vested in *Collectors of Customs,* transferred to Commandant of Coast Guard by EO 9083 of Feb. 27, 1942. Transfer made permanent by Reorg. Plan No. 3 of 1946, effective July 16, 1946. Redesignated U.S. Customs Service by the Department of the Treasury Order 165–23 of Apr. 4, 1973. Functions transferred to and agency established within Homeland Security Department by act of Nov. 25, 2002 (116 Stat. 2178).

Customs Court, U.S.   Formerly established as Board of General Appraisers by act of June 10, 1890 (26 Stat. 136). Renamed *U.S. Customs Court* by act of May 26, 1926 (44 Stat. 669). Renamed U.S. Court of International Trade by act of Oct. 10, 1980 (94 Stat. 1727).

Customs and Patent Appeals, U.S. Court of   Established by act of Mar. 2, 1929 (45 Stat. 1475). Abolished by act of Apr. 2, 1982 (96 Stat. 28) and functions merged with appellate functions of *U.S. Court of Claims* to form U.S. Court of Appeals for the Federal Circuit.

Dairy Industry, Bureau of   *Bureau of Dairying* established in the Department of Agriculture by act of May 29, 1924 (43 Stat. 243). *Bureau of Dairy Industry* designation first appeared in act of May 11, 1926 (44 Stat. 499). Functions transferred to Agricultural Research Service by Secretary's Memorandum 1320, supp. 4, of Nov. 2, 1953.

Defense, Advisory Commission to the Council of National   *See* **Defense, Council of National**

Defense, Council of National   Established by act of Aug. 29, 1916 (39 Stat. 649). *Advisory Commission*—composed of Advisers on Industrial Production, Industrial Materials, Employment, Farm Products, Price Stabilization, Transportation, and Consumer Protection—established by *Council* pursuant to act and approved by President May 29, 1940. *Commission* decentralized by merging divisions with newly created national defense units. Agencies evolved from *Commission,* except *Office of Agricultural War Relations* and *Office of Price Administration,* made units of *Office for Emergency Management. Council* inactive.

Defense, Office of Civilian   Established in *Office for Emergency Management* by EO 8757 of May 20, 1941. Terminated by EO 9562 of June 4, 1945.

Defense Administration, Federal Civil   Established in *Office for Emergency Management* by EO 10186 of Dec. 1, 1950; subsequently established as independent agency by act of Jan. 12, 1951 (64 Stat. 1245). Functions transferred to *Office of Defense and Civilian Mobilization* by Reorg. Plan No. 1 of 1958, effective July 1, 1958.

Defense Advanced Research Projects Agency   Established as a separate agency of the Department of Defense by DOD Directive 5105.41 dated July 25, 1978. Renamed *Advanced Research Projects Agency* by order of the Secretary of Defense dated July 13, 1993. Reestablished by act of Feb. 10, 1996 (110 Stat. 406).

Defense Advisory Council, Civil   Established by act of Jan. 12, 1951 (64 Stat. 1245). Transferred to *Office of Defense and Civilian Mobilization* by Reorg. Plan No. 1 of 1958, effective July 1, 1958.

Defense Aid Reports, Division of   Established in *Office for Emergency Management* by EO 8751 of May 2, 1941. Abolished by EO 8926 of Oct. 28, 1941, which created *Office of Lend-Lease Administration.*

Defense Air Transportation Administration   Established Nov. 12, 1951, by Department of Commerce Order 137. Abolished by Amendment 3 of Sept. 13, 1962, to Department Order 128 (revised) and functions transferred to *Office of the Under Secretary of Commerce for Transportation.*

Defense Atomic Support Agency   Renamed *Defense Nuclear Agency* by General Order No. 1 of July 1, 1971.

Defense Audiovisual Agency   Established by DOD Directive 5040.1 of June 12, 1979. Abolished by Secretary's memorandum of Apr. 19, 1985, and functions assigned to the military departments.

Defense Audit Service   Established by DOD Directive of Oct. 14, 1976. Abolished by Deputy Secretary's memorandum of Nov. 2, 1982, and functions transferred to Office of the Inspector General.

Defense Civil Preparedness Agency   Functions transferred from the Department of Defense to the Federal Emergency Management Agency by EO 12148 of July 20, 1979.

Defense and Civilian Mobilization Board   Established by EO 10773 of July 1, 1938. Redesignated *Civil and Defense Mobilization Board* by act of Aug. 26, 1958 (72 Stat. 861). Abolished by *Office of Emergency Preparedness* Circular 1200.1 of Oct. 31, 1962.

Defense Communications Agency   Established by direction of the Secretary of Defense on May 12, 1960. Renamed Defense Information Systems Agency by DOD Directive 5105.19 dated June 25, 1991.

Defense Communications Board   Established by EO 8546 of Sept. 24, 1940. Renamed *Board*

of War Communications by EO 9183 of June 15, 1942. Abolished by EO 9831 of Feb. 24, 1947, and property transferred to Federal Communications Commission.

Defense Coordinating Board, Civil   Established by EO 10611 of May 11, 1955. EO 10611 revoked by EO 10773 of July 1, 1958.

Defense Electric Power Administration   Established by Order 2605 of Dec. 4, 1950 of the Secretary of the Interior. Abolished June 30, 1953, by Secretary's Order 2721 of May 7, 1953. Reestablished by Departmental Manual Release No. 253 of Aug. 6, 1959. Terminated by Departmental Manual Release No. 1050 of Jan. 10, 1977.

Defense Fisheries Administration   Established by Order 2605 of Dec. 4, 1950 of the Secretary of the Interior. Abolished June 30, 1953, by Secretary's Order 2722 of May 13, 1953.

Defense Health and Welfare Services, Office of   Established by EO 8890 of Sept. 3, 1941. Terminated by EO 9338 of Apr. 29, 1943, and functions transferred to *Federal Security Agency.*

Defense Homes Corporation   Incorporated pursuant to President's letter to the Secretary of the Treasury of Oct. 18, 1940. Transferred to *Federal Public Housing Authority* by EO 9070 of Feb. 24, 1942.

Defense Housing Coordinator   Office established July 21, 1940, by *Advisory Commission to Council of National Defense.* Functions transferred to *Division of Defense Housing Coordination, Office for Emergency Management,* by EO 8632 of Jan. 11, 1941.

Defense Housing Division, Mutual Ownership   Established by Administrator of *Federal Works Agency* under provisions of act of June 28, 1941 (55 Stat. 361). Functions transferred to *Federal Public Housing Authority, National Housing Agency,* by EO 9070 of Feb. 24, 1942.

Defense Intelligence College   Established by DOD Directive 3305.1 of January 28, 1983. Renamed Joint Military Intelligence College by DOD Directive 3305.1 of January 14, 1998. *See also Defense Intelligence School.*

Defense Intelligence School   Established by DOD Directive 5105.25 of November 2, 1962. Renamed Defense Intelligence College by DOD Directive 3305.1 of January 28, 1983.

Defense Investigative Service   Established by the Secretary of Defense Jan. 1, 1972. Renamed Defense Security Service in November 1997 by Defense Reform Initiative.

Defense Manpower Administration   Established by the Secretary of Labor by General Order 48, pursuant to EO 10161 of Sept. 9, 1950, and Reorg. Plan No. 6 of 1950, effective May 24, 1950. General Order 48 revoked by General Order 63 of Aug. 25, 1953, which established *Office of Manpower Administration* in Department.

Defense Mapping Agency   Established as a the Department of Defense agency in 1972. Functions transferred to the National Imagery and Mapping Agency by act of Sept. 23, 1996 (110 Stat. 2677).

Defense Materials Procurement Agency   Established by EO 10281 of Aug. 28, 1951. Abolished by EO 10480 of Aug. 14, 1953, and functions transferred to General Services Administration.

Defense Materials Service   *See* **Emergency Procurement Service**

Defense Mediation Board, National   Established by EO 8716 of Mar. 19, 1941. Terminated on creation of *National War Labor Board, Office for Emergency Management* by EO 9017 of Jan. 12, 1942. Transferred to the Department of Labor by EO 9617 of Sept. 19, 1945. *Board* terminated by EO 9672 of Dec. 31, 1945, which established *National Wage Stabilization Board* in the Department of Labor. Terminated by EO 9809 of Dec. 12, 1946, and functions transferred to the Secretary of Labor and the Department of the Treasury, effective Feb. 24, 1947.

Defense Medical Programs Activity   Functions consolidated into the TRICARE Management Activity in November 1997 by Defense Reform Initiative.

Defense Minerals Administration   Established by Order 2605 of Dec. 4, 1950 of the Secretary of the Interior. Functions assigned to *Defense Materials Procurement Agency.* Functions of exploration for critical and strategic minerals redelegated to the Secretary of the Interior and administered by *Defense Minerals Exploration Administration* by Secretary's Order 2726 of June 30, 1953. Termination of program announced by Secretary June 6, 1958. Certain activities continued in *Office of Minerals Exploration, Department of the Interior.*

Defense Minerals Exploration Administration   *See* **Defense Minerals Administration**

Defense Mobilization, Office of   Established in Executive Office of the President by EO 10193 of Dec. 16, 1950. Superseded by *Office of Defense Mobilization* established by Reorg. Plan No. 3 of 1953, effective June 12, 1953, which assumed functions of former *Office, National Security Resources Board,* and critical materials stockpiling functions of Army, Navy, Air Force, and Interior Secretaries and of *Army and Navy Munitions Board.* Consolidated with *Federal Civil Defense Administration* into *Office of Defense and Civilian Mobilization* by Reorg. Plan No. 1 of 1958, effective July 1, 1958, and offices of Director and Deputy Director terminated.

Defense Mobilization Board   Established by EO 10200 of Jan. 3, 1951, and restated in EO 10480 of Aug. 14, 1953. Terminated by EO 10773 of July 1, 1958.

Defense Nuclear Agency   Established in 1971. Renamed *Defense Special Weapons Agency* by DOD Directive 5105.31 of June 14, 1995.

Defense Nuclear Counterintelligence, Office of   Established by act of Oct. 5, 1999 (113 Stat. 960). Abolished by act of Oct. 17, 2006 (120 Stat. 2507) and functions transferred to the Secretary of Energy.

Defense Plant Corporation   Established by act of June 25, 1940 (54 Stat. 572). Transferred from *Federal Loan Agency* to the Department of Commerce by EO 9071 of Feb. 24, 1942. Returned to *Federal Loan Agency* pursuant to act of Feb. 24, 1945 (59 Stat. 5). Dissolved by act of June 30, 1945 (59 Stat. 310), and functions transferred to *Reconstruction Finance Corporation.*

Defense Plants Administration, Small   Established by act of July 31, 1951 (65 Stat. 131). Terminated July 31, 1953, by act of June 30, 1953 (67 Stat. 131). Functions relating to liquidation transferred to Small Business Administration by EO 10504 of Dec. 1, 1953.

Defense Production Administration   Established by EO 10200 of Jan. 3, 1951. Terminated by EO 10433 of Feb. 4, 1953, and functions transferred to *Office of Defense Mobilization.*

Defense Property Disposal Service   Renamed Defense Reutilization and Marketing Service by Defense Logistics Agency General Order 10–85, effective July 1, 1985.

Defense Prisoner of War/Missing in Action Office   Established by DOD Directive 5110.10, July 16, 1993. Renamed Defense Prisoner of War/Missing Personnel Office by Secretary of Defense memorandum of May 30, 1996.

Defense Public Works Division   Established in *Public Works Administration.* Transferred to *Office of Federal Works Administrator* by administrative order of July 16, 1941. Abolished by administrative order of Mar. 6, 1942, and functions transferred to *Office of Chief Engineer, Federal Works Agency.*

Defense Purchases, Office for the Coordination of National   Established by order of *Council of National Defense,* approved June 27, 1940. Order revoked Jan. 7, 1941, and records transferred to Executive Office of the President.

Defense Research Committee, National   Established June 27, 1940, by order of *Council of National Defense.* Abolished by order of *Council* June 28, 1941, and reestablished in *Office of Scientific Research and Development* by EO 8807 of June 28, 1941. *Office* terminated by EO 9913 of Dec. 26, 1947, and property and records transferred to *National Military Establishment.*

Defense Resources Committee   Established by Administrative Order 1496 of June 15, 1940. Replaced by *War Resources Council* by Administrative Order 1636 of Jan. 14, 1942. Inactive.

Defense Security Assistance Agency   Established on Sept. 1, 1971. Renamed the Defense Security Cooperation Agency by DOD Directive 5105.38.

Defense Solid Fuels Administration Established by Order 2605 of Dec. 4, 1950 of the Secretary of the Interior. Abolished June 29, 1954, by Secretary's Order 2764.

Defense Special Weapons Agency Established by General Order No. 1 of July 1, 1971. Functions transferred to the Defense Threat Reduction Agency by DOD Directive 5105.62 of Sept. 30, 1998.

Defense Stockpile Manager, National Established by act of Nov. 14, 1986 (100 Stat. 4067). Functions transferred from the Administrator of General Services to the Secretary of Defense by EO 12626 of Feb. 25, 1988.

Defense Supplies Corporation Established under act of June 25, 1940 (54 Stat. 572). Transferred from *Federal Loan Agency* to the Department of Commerce by EO 9071 of Feb. 24, 1942. Returned to *Federal Loan Agency* by act of Feb. 24, 1945 (59 Stat. 5). Dissolved by act of June 30, 1945 (59 Stat. 310), and functions transferred to *Reconstruction Finance Corporation.*

Defense Supply Agency Renamed Defense Logistics Agency by DOD Directive 5105.22 of Jan. 22, 1977.

Defense Supply Management Agency Established in the Department of Defense by act of July 1, 1952 (66 Stat. 318). Abolished by Reorg. Plan No. 6 of 1953, effective June 30, 1953, and functions transferred to the Secretary of Defense.

Defense Technology Security Administration Established on May 10, 1985. Functions transferred to the Defense Threat Reduction Agency by DOD Directive 5105.62 of Sept. 30, 1998.

Defense Transport Administration Established Oct. 4, 1950, by order of Commissioner of *Interstate Commerce Commission* in charge of *Bureau of Service,* pursuant to EO 10161 of Sept. 9, 1950. Terminated by DTA Commissioner's order, effective July 1, 1955, and functions transferred to *Bureau of Safety and Service, Interstate Commerce Commission.*

Defense Transportation, Office of Established in *Office for Emergency Management* by EO 8989 of Dec. 18, 1941. Terminated by EO 10065 of July 6, 1949.

Director. *See other part of title*

Disarmament Administration, U.S. Established in the Department of State. Functions transferred to *U.S. Arms Control and Disarmament Agency* by act of Sept. 26, 1961 (75 Stat. 638).

Disarmament Problems, President's Special Committee on Established by President Aug. 5, 1955. Dissolved in February 1958.

Disaster Assistance Administration, Federal Functions transferred from the Department of Housing and Urban Development to the Federal Emergency Management Agency by EO 12148 of July 20, 1979.

Disaster Loan Corporation Grouped with other agencies to form *Federal Loan Agency* by Reorg. Plan No. I of 1939, effective July 1, 1939. Transferred to the Department of Commerce by EO 9071 of Feb. 24, 1942. Returned to *Federal Loan Agency* by act of Feb. 24, 1945 (59 Stat. 5). Dissolved by act of June 30, 1945 (59 Stat. 310), and functions transferred to *Reconstruction Finance Corporation.*

Disease Control, Center for Established within the Public Health Service by the *Secretary of Health, Education, and Welfare* on July 1, 1973. Renamed *Centers for Disease Control* by Health and Human Services Secretary's notice of Oct. 1, 1980 (45 FR 67772). Renamed Centers for Disease Control and Prevention by act of Oct. 27, 1992 (106 Stat. 3504).

Displaced Persons Commission Established by act of June 25, 1948 (62 Stat. 1009). Terminated Aug. 31, 1952, pursuant to terms of act.

District of Columbia Established by acts of July 16, 1790 (1 Stat. 130), and Mar. 3, 1791. *Corporations of Washington and Georgetown* and *levy court of Washington County* abolished in favor of territorial form of government in 1871. Permanent commission government established July 1, 1878. District Government created as municipal corporation by act of June 11, 1878 (20 Stat. 102). Treated as branch of U.S. Government by various statutory enactments of Congress. District Government altered by Reorg. Plan No. 3 of 1967, effective Nov. 3, 1967. Charter for local government in District of Columbia provided by act of Dec. 24, 1973 (87 Stat. 774).

District of Columbia, Highway Commission of the Established by act of Mar. 2, 1893 (27 Stat 532). *National Capital Park and Planning Commission* named successor by act of Apr. 30, 1926 (44 Stat. 374). Functions transferred to National Capital Planning Commission by act of July 19, 1952 (66 Stat. 781).

District of Columbia, Reform-School of the Established by act of May 3, 1876 (19 Stat. 49). Renamed *National Training School for Boys* by act of May 27, 1908 (35 Stat. 380). Transferred to the Department of Justice by Reorg. Plan No. II of 1939, effective July 1, 1939, to be administered by Director of Bureau of Prisons.

District of Columbia Auditorium Commission Established by act of July 1, 1955 (69 Stat. 243). Final report submitted to Congress Jan. 31, 1957, pursuant to act of Apr. 27, 1956 (70 Stat. 115).

District of Columbia Redevelopment Land Agency Established by act of Aug. 2, 1946 (60 Stat. 790). Agency established as instrumentality of District Government by act of Dec. 24, 1973 (87 Stat. 774), effective July 1, 1974.

District of Columbia-Virginia Boundary Commission Established by act of Mar. 21, 1934 (48 Stat. 453). Terminated Dec. 1, 1935, to which date it had been extended by Public Resolution 9 (49 Stat. 67).

Division. *See other part of title*

Domestic Council Established in Executive Office of the President by Reorg. Plan No. 2 of 1970, effective July 1, 1970. Abolished by Reorg. Plan No. 1 of 1977, effective Mar. 26, 1978, and functions transferred to President and staff designated as *Domestic Policy Staff.* Pursuant to EO 12045 of Mar. 27, 1978, *Staff* assisted President in performance of transferred functions. Renamed Office of Policy Development in 1981. Abolished in February 1992 by President's reorganizational statement, effective May 1992.

Domestic Policy Staff *See* **Domestic Council**

Dominican Customs Receivership Transferred from *Division of Territories and Island Possessions,* Department of the Interior, to the Department of State by Reorg. Plan No. IV of 1940, effective June 30, 1940.

Drug Abuse, National Institute on Established within the National Institute of Mental Health, *Department of Health, Education, and Welfare* by act of Mar. 21, 1972 (86 Stat. 85). Removed from within the National Institute of Mental Health and made an entity within the Alcohol, Drug Abuse, and Mental Health Administration by act of May 14, 1974 (88 Stat. 136). Functions transferred to the Department of Health and Human Services by act of Oct. 17, 1979 (93 Stat. 695). (*See also* act of Oct. 27, 1986; 100 Stat. 3207–106.) Abolished by act of July 10, 1992 (106 Stat. 331). Reestablished by act of July 10, 1992 (106 Stat. 361).

Drug Abuse, President's Advisory Commission on Narcotic and Established by EO 11076 of Jan. 15, 1963. Terminated November 1963 under terms of order.

Drug Abuse Control, Bureau of Established in Food and Drug Administration, Department of Health and Human Services, to carry out functions of act of July 15, 1965 (79 Stat. 226). Functions transferred to *Bureau of Narcotics and Dangerous Drugs,* Department of Justice, by Reorg. Plan No. 1 of 1968, effective Apr. 8, 1968. Abolished by Reorg. Plan No. 2 of 1973, effective July 1, 1973, and functions transferred to Drug Enforcement Administration.

Drug Abuse Law Enforcement, Office of Established by EO 11641 of Jan. 28, 1972. Terminated by EO 11727 of July 6, 1973, and functions transferred to Drug Enforcement Administration.

Drug Abuse Policy, Office of Established in Executive Office of the President by act of Mar. 19, 1976 (90 Stat. 242). Abolished by Reorg. Plan No. 1 of 1977, effective Mar. 26, 1978, and functions transferred to President.

Drug Abuse Prevention, Special Action Office for Established by EO 11599 of June 17, 1971, and act of Mar. 21, 1972 (86 Stat. 65). Terminated June 30, 1975, pursuant to terms of act.

Drug Abuse Prevention, Treatment, and Rehabilitation, Cabinet Committee on Established Apr. 27, 1976, by Presidential announcement. Terminated by Presidential memorandum of Mar. 14, 1977.

Drug Law Enforcement, Cabinet Committee for Established Apr. 27, 1976, pursuant to Presidential message to Congress of Apr. 27, 1976. Abolished by Presidential memorandum of Mar. 14, 1977.

Drugs, Bureau of Narcotics and Dangerous *See* **Drug Abuse Control, Bureau of**

Drugs and Biologics, National Center for Renamed *Center for Drugs and Biologics* by Food and Drug Administration notice of Mar. 9, 1984 (49 FR 10166). Reestablished as Center for Drug Evaluation and Research and Center for Biologics Evaluation and Research by Secretary's notice of Oct. 6, 1987 (52 FR 38275).

Drunk Driving, Presidential Commission on Established by EO 12358 of Apr. 14, 1982. Terminated Dec. 31, 1983, by EO 12415 of Apr. 5, 1983.

Dryden Research Center, Hugh L. Formerly separate field installation of National Aeronautics and Space Administration. Made component of Ames Research Center by NASA Management Instruction 1107.5A of Sept. 3, 1981.

Economic Administration, Foreign Established in *Office for Emergency Management* by EO 9380 of Sept. 25, 1943. Functions of *Office of Lend-Lease Administration, Office of Foreign Relief and Rehabilitation Operations, Office of Economic Warfare* (together with *U.S. Commercial Company, Rubber Development Corporation, Petroleum Reserves Corporation,* and *Export-Import Bank of Washington* and functions transferred thereto by EO 9361 of July 15, 1943), and foreign economic operations of *Office of Foreign Economic Coordination* transferred to *Administration.* Foreign procurement activities of *War Food Administration* and Commodity Credit Corporation transferred by EO 9385 of Oct. 6, 1943. Terminated by EO 9630 of Sept. 27, 1945, and functions redistributed to the Departments of State, Commerce, and Agriculture and the *Reconstruction Finance Corporation.*

Economic Analysis, Office of *See* **Business Economics, Office of**

Economic Cooperation Administration Established by act of Apr. 3, 1948 (62 Stat. 138). Abolished by act of Oct. 10, 1951 (65 Stat. 373), and functions transferred to *Mutual Security Agency* pursuant to EO 10300 of Nov. 1, 1951.

Economic Coordination, Office of Foreign *See* **Board of Economic Operations**

Economic Defense Board Established by EO 8839 of July 30, 1941. Renamed *Board of Economic Warfare* by EO 8982 of Dec. 17, 1941. *Board* terminated by EO 9361 of July 15, 1943, and *Office* of Economic Warfare established in *Office for Emergency Management. Office of Economic Warfare* consolidated with *Foreign Economic Administration* by EO 9380 of Sept. 25, 1943.

Economic Development, Office of
Regional   Established by the Secretary of Commerce Jan. 6, 1966, pursuant to act of Aug. 26, 1965 (79 Stat. 552). Abolished by Department Order 5A, Dec. 22, 1966, and functions vested in Economic Development Administration.

Economic Development Service,
Foreign   Established by order of the Secretary of Agriculture Nov. 8, 1969. Abolished by order of Secretary Feb. 6, 1972, and functions transferred to Economic Research Service.

Economic Growth and Stability, Advisory Board
on   Established by Presidential letter to Congress of June 1, 1953. Superseded by *National Advisory Board on Economic Policy* by Presidential direction Mar. 12, 1961. *Cabinet Committee on Economic Growth* established by President Aug. 21, 1962, to succeed *Board.*

Economic Management Support Center   Established by Secretary of Agriculture Memorandum 1836 of Jan. 9, 1974. Consolidated with other Department units into *Economics, Statistics, and Cooperatives Service* by Secretary's Memorandum 1927, effective Dec. 23, 1977.

Economic Operations, Board of   Established by Department of State order of Oct. 7, 1941. Abolished by departmental order of June 24, 1943, and functions transferred to *Office of Foreign Economic Coordination* established by same order. *Office* abolished by departmental order of Nov. 6, 1943, pursuant to EO 9380 of Sept. 25, 1943.

Economic Opportunity, Office of   Established in Executive Office of the President by act of Aug. 20, 1964 (78 Stat. 508). All OEO programs except three transferred by administrative action to the Departments of *Health, Education, and Welfare,* Labor, and Housing and Urban Development July 6, 1973. Community Action, Economic Development, and Legal Services Programs transferred to *Community Services Administration* by act of Jan. 4, 1975 (88 Stat. 2310).

Economic Policy, Council on   Established by Presidential memorandum of Feb. 2, 1973. Functions absorbed by *Economic Policy Board* Sept. 30, 1974.

Economic Policy, Council on Foreign   Established Dec. 22, 1954, by Presidential letter of Dec. 11, 1954. Abolished by President Mar. 12, 1961, and functions transferred to Secretary of State.

Economic Policy, Council on
International   Established in Executive Office of the President by Presidential memorandum of January 1971. Reestablished by act of Aug. 29, 1972 (86 Stat. 646). Terminated Sept. 30, 1977, on expiration of statutory authority.

Economic Policy, National Advisory Board on   *See* **Economic Growth and Stability, Advisory Board on**

Economic Policy Board, President's   Established by EO 11808 of Sept. 30, 1974. Terminated by EO 11975 of Mar. 7, 1977.

Economic Research Service   Established by Secretary of Agriculture Memorandum 1446, supp. 1, of Apr. 3, 1961. Consolidated with other Department of Agriculture units into *Economics, Statistics, and Cooperatives Service* by Secretary's Memorandum 1927, effective Dec. 23, 1977. Redesignated as Economic Research Service by Secretarial order of Oct. 1, 1981.

Economic Security, Advisory Council
on   Established by EO 6757 of June 29, 1934. Terminated on approval of act of Aug. 14, 1935 (49 Stat. 620) Aug. 14, 1935.

Economic Security, Committee on   Established by EO 6757 of June 29, 1934. Terminated as formal agency in April 1936, as provided in act, but continued informally for some time thereafter.

Economic Stabilization, Office of   Established in *Office for Emergency Management* by EO 9250 of Oct. 3, 1942. Terminated by EO 9620 of Sept. 20, 1945, and functions transferred to *Office of War Mobilization and Reconversion.* Reestablished in *Office for Emergency Management* by EO 9699 of Feb. 21, 1946. Transferred by EO 9762 of July 25, 1946, to *Office of War Mobilization and Reconversion.* Consolidated with other agencies to form *Office of Temporary Controls* by EO 9809 of Dec. 12, 1946.

Economic Stabilization Agency   Established by EO 10161 of Sept. 9, 1950, and EO 10276 of July 31, 1951. Terminated, except for liquidation purposes, by EO 10434 of Feb. 6, 1953. Liquidation completed Oct. 31, 1953, pursuant to EO 10480 of Aug. 14, 1953.

Economic Stabilization Board   Established by EO 9250 of Oct. 3, 1942. Transferred to *Office of War Mobilization and Reconversion* by EO 9620 of Sept. 20, 1945. Returned to *Office of Economic Stabilization* on reestablishment by EO 9699 of Feb. 21, 1946. *Board* returned to *Office of War Mobilization and Reconversion* by EO 9762 of July 25, 1946. Functions terminated by EO 9809 of Dec. 12, 1946.

Economic Warfare, Board of   *See* **Economic Defense Board**

Economic Warfare, Office of   *See* **Economic Defense Board**

Economics, Bureau of Industrial   Established by the Secretary of Commerce Jan. 2, 1980, in conjunction with Reorg. Plan No. 3 of 1979, effective Oct. 1, 1980, and operated under Department Organization Order 35–5B. Abolished at bureau level by Secretarial order, effective Jan. 22, 1984 (49 FR 4538). Industry-related functions realigned and transferred from Under Secretary for Economic

Affairs to Under Secretary for International Trade. Under Secretary for Economic Affairs retained units to support domestic macroeconomic policy functions.

Economics, Statistics, and Cooperatives Service   Renamed *Economics and Statistics Service* by Secretary of Agriculture Memorandum 2025 of Sept. 17, 1980. Redesignated as Economic Research Service and *Statistical Reporting Service* by Secretarial order of Oct. 1, 1981.

Economy Board, Joint   Placed under direction of President by military order of July 5, 1939. Abolished Sept. 1, 1947, by joint letter of Aug. 20, 1947, from Secretaries of *War* and Navy to President.

Education, Federal Board for Vocational   Established by act of Feb. 23, 1917 (39 Stat. 929). Functions transferred to the Department of the Interior by EO 6166 of June 10, 1933. Functions assigned to *Commissioner of Education* Oct. 10, 1933. *Office of Education* transferred from the Department of the Interior to the *Federal Security Agency* by Reorg. Plan No. I of 1939, effective July 1, 1939. *Board* abolished by Reorg. Plan No. 2 of 1946, effective July 16, 1946.

Education, National Institute of   Established by act of June 23, 1972 (86 Stat. 327). Transferred to Office of Educational Research and Improvement, Department of Education, by act of Oct. 17, 1979 (93 Stat. 678), effective May 4, 1980.

Education, Office of   Established as independent agency by act of Mar. 2, 1867 (14 Stat. 434). Transferred to the Department of the Interior by act of July 20, 1868 (15 Stat. 106). Transferred to *Federal Security Agency* by Reorg. Plan No. I of 1939, effective July 1, 1939. Functions of *Federal Security Administrator* administered by *Office of Education* relating to student loans and defense-related education transferred to *War Manpower Commission* by EO 9247 of Sept. 17, 1942.

Education, Office of Bilingual   Abolished by act of Oct. 17, 1979 (93 Stat. 675), and functions transferred to Office of Bilingual Education and Minority Languages Affairs, Department of Education.

Education Beyond the High School, President's Committee on   Established by act of July 26, 1956 (70 Stat. 676). Terminated Dec. 31, 1957. Certain activities continued by *Bureau of Higher Education, Office of Education.*

Education Division   Established in the *Department of Health, Education, and Welfare* by act of June 23, 1972 (86 Stat. 327). Functions transferred to the Department of Education by act of Oct. 17, 1979 (93 Stat. 677).

Education Goals Panel, National   Terminated by Congressional mandate, March 15, 2002.

Education Statistics, National Center for   Established in the Office of the Assistant Secretary, Department of Health and Human Services, by act of Aug. 21, 1974 (88 Stat. 556). Transferred to the Office of Educational Research and

Improvement, Department of Education, by act of Oct. 17, 1979 (93 Stat. 678), effective May 4, 1980. Renamed *Center for Education Statistics* by act of Oct. 17, 1986 (100 Stat. 1579). Renamed National Center for Education Statistics by act of Apr. 28, 1988 (102 Stat. 331).

Educational and Cultural Affairs, Bureau of   Established by Secretary of State in 1960. Terminated by Reorg. Plan No. 2 of 1977, effective July 1, 1978, and functions transferred to *International Communication Agency,* effective Apr. 1, 1978.

Educational and Cultural Affairs, Interagency Council on International   Established Jan. 20, 1964, by Foreign Affairs Manual Circular, under authority of act of Sept. 21, 1961 (75 Stat. 527). Terminated Oct. 1973 following creation of Subcommittee on International Exchanges by National Security Council directive.

Educational Exchange, U.S. Advisory Commission on   Established by act of Jan. 27, 1948 (62 Stat. 10). Abolished by act of Sept. 21, 1961 (75 Stat. 538), and superseded by U.S. Advisory Commission on International Educational and Cultural Affairs.

Efficiency, Bureau of   Organized under act of Feb. 28, 1916 (39 Stat. 15). Abolished by act of Mar. 3, 1933 (47 Stat. 1519), and records transferred to *Bureau of the Budget.*

Elderly, Committee on Mental Health and Illness of the   Established by act of July 29, 1975 (89 Stat. 347). Terminated Sept. 30, 1977.

Electoral votes for President and Vice President, transfer of functions   *See* **State, Department of**

Electric Home and Farm Authority   Incorporated Aug. 1, 1935, under laws of District of Columbia. Designated as U.S. agency by EO 7139 of Aug. 12, 1935. Continued by act of June 10, 1941 (55 Stat. 248). Grouped with other agencies in *Federal Loan Agency* by Reorg. Plan. No. I of 1939, effective July 1, 1939. Functions transferred to the Department of Commerce by EO 9071 of Feb. 24, 1942. Terminated by EO 9256 of Oct. 13, 1942.

Electric Home and Farm Authority, Inc.   Organized Jan. 17, 1934, under laws of State of Delaware by EO 6514 of Dec. 19, 1933. Dissolved Aug. 1, 1935, and succeeded by *Electric Home and Farm Authority.*

Electricity Delivery and Energy Reliability, Office of   Established by Secretary of Energy announcement of June 9, 2005. Position of director elevated to Assistant Secretary of Electricity Delivery and Energy Reliability by Secretary's memorandum EXEC–2007–010607 of Oct. 24, 2007.

Electricity Transmission and Distribution, Office of   Renamed *Office of Electricity Delivery and Energy Reliability* by the Secretary of Energy's memo of Feb. 15, 2005.

Emergency Administration of Public Works, Federal   Established by act of June 16, 1933 (48

Stat. 200). Operation continued by subsequent legislation, including act of June 21, 1938 (52 Stat. 816). Consolidated with *Federal Works Agency* as *Public Works Administration* by Reorg. Plan No. I of 1939, effective July 1, 1939. Functions transferred to *Office of Federal Works Administrator* by EO 9357 of June 30, 1943.

Emergency Conservation Work   Established by EO 6101 of Apr. 5, 1933. Succeeded by *Civilian Conservation Corps.*

Emergency Council, National   Established by EO 6433–A of Nov. 17, 1933. Consolidated with *Executive Council* by EO 6889–A of Oct. 29, 1934. Abolished by Reorg. Plan No. II of 1939, effective July 1, 1939, and functions (except those relating to *Radio Division* and *Film Service*) transferred to Executive Office of the President.

Emergency Council, Office of Economic Adviser to National   Established by EO 6240 of Aug. 3, 1933, in connection with *Executive Council,* which later consolidated with *National Emergency Council.* Records and property used in preparation of statistical and economic summaries transferred to *Central Statistical Board* by EO 7003 of Apr. 8, 1935.

Emergency Management, Office for   Established in Executive Office of the President by administrative order of May 25, 1940, in accordance with EO 8248 of Sept. 8, 1939. Inactive.

Emergency Management Agency, Federal   Established in EO 12127 of Mar. 31, 1979. Functions transferred to Department of Homeland Security by act of Nov. 25, 2002 (116 Stat. 2213). Established as a distinct entity with the Department of Homeland Security by act of Oct. 4, 2006 (120 Stat. 1400).

Emergency Mobilization Preparedness Board   Established Dec. 17, 1981, by the President. Abolished by Presidential directive of Sept. 16, 1985.

Emergency Planning, Office of   Established as successor to *Office of Civil and Defense Mobilization* by act of Sept. 22, 1961 (75 Stat. 630). Renamed *Office of Emergency Preparedness* by act of Oct. 21, 1968 (82 Stat. 1194). Terminated by Reorg. Plan No. 2 of 1973, effective July 1, 1973, and functions transferred to the the the Departments of the Treasury and Housing and Urban Development and the General Services Administration.

Emergency Preparedness, Office of   *See* **Emergency Planning, Office of**

Emergency Procurement Service   Established Sept. 1, 1950, by Administrator of General Services. Renamed *Defense Materials Service* Sept. 7, 1956. Functions transferred to *Property Management and Disposal Service* July 29, 1966. *Service* abolished July 1, 1973, and functions transferred to Federal Supply Service, Public Buildings Service, and Federal Property Resources Service.

Emergency Relief Administration, Federal   Established by act of May 12, 1933 (48 Stat. 55). Expired June 30, 1938, having been liquidated by *Works Progress Administrator* pursuant to act of May 28, 1937 (50 Stat. 352).

Employee-Management Relations Program, President's Committee on the Implementation of the Federal   Established by EO 10988 of Jan. 17, 1962. Terminated upon submission of report to President June 21, 1963.

Employees' Compensation, Bureau of   Transferred from *Federal Security Agency* to the Department of Labor by Reorg. Plan No. 19 of 1950, effective May 24, 1950. Functions absorbed by Employment Standards Administration Mar. 13, 1972.

Employees' Compensation Appeals Board   Transferred from *Federal Security Agency* to the Department of Labor by Reorg. Plan No. 19 of 1950, effective May 24, 1950.

Employees' Compensation Commission, U.S.   Established by act of Sept. 7, 1916 (39 Stat. 742). Abolished by Reorg. Plan No. 2 of 1946, effective July 16, 1946, and functions transferred to *Federal Security Administrator.*

Employment Board, Fair   Established by *U.S. Civil Service Commission* pursuant to EO 9980 of July 26, 1948. Abolished by EO 10590 of Jan. 18, 1955.

Employment of People With Disabilities, President's Committee on   Created by EO 12640 of May 10, 1988. Duties subsumed by the Office of Disability Employment within the Department of Labor as directed by Public Law 106–554 of Dec. 21, 2000.

Employment of the Physically Handicapped, President's Committee on   Established by EO 10640 of Oct. 10, 1955, continuing *Committee* established by act of July 11, 1949 (63 Stat. 409). Superseded by President's Committee on Employment of the Handicapped established by EO 10994 of Feb. 14, 1962.

Employment Policy, President's Committee on Government   Established by EO 10590 of Jan. 18, 1955. Abolished by EO 10925 of Mar. 6, 1961, and functions transferred to *President's Committee on Equal Employment Opportunity.*

Employment Practice, Committee on Fair   Established in *Office of Production Management* by EO 8802 of June 25, 1941. Transferred to *War Manpower Commission* by Presidential letter effective July 30, 1942. Committee terminated on establishment of *Committee on Fair Employment Practice, Office for Emergency Management,* by EO 9346 of May 27, 1943. Terminated June 30, 1946, by act of July 17, 1945 (59 Stat. 743).

Employment Security, Bureau of   Transferred from *Federal Security Agency* to the Department of Labor by Reorg. Plan No. 2 of 1949, effective Aug. 20, 1949. Abolished by order of Mar. 14, 1969 of the Secretary of Labor, and functions transferred to *Manpower Administration.*

Employment Service, U.S.   Established in the Department of Labor in 1918 by departmental order. Abolished by act of June 6, 1933 (48 Stat. 113), and created as bureau with same name. Functions consolidated with unemployment compensation functions of *Social Security Board, Bureau of Employment Security,* and transferred to *Federal Security Agency* by Reorg. Plan No. I of 1939, effective July 1, 1939. *Service* transferred to *Bureau of Placement, War Manpower Commission,* by EO 9247 of Sept. 17, 1942. Returned to the Department of Labor by EO 9617 of Sept. 19, 1945. Transferred to *Federal Security Agency* by act of June 16, 1948 (62 Stat. 443), to function as part of *Bureau of Employment Security,* Social Security Administration. *Bureau,* including *U.S. Employment Service,* transferred to the Department of Labor by Reorg. Plan No. 2 of 1949, effective Aug. 20, 1949. Abolished by reorganization of *Manpower Administration,* effective Mar. 17, 1969, and functions assigned to *U.S. Training and Employment Service.*

Employment Stabilization Board, Federal   Established by act of Feb. 10, 1931 (46 Stat. 1085). Abolished by EO 6166 of June 10, 1933. Abolition deferred by EO 6623 of Mar. 1, 1934, until functions of *Board* transferred to *Federal Employment Stabilization Office,* established in the Department of Commerce by same order. *Office* abolished by Reorg. Plan No. I of 1939, effective July 1, 1939, and functions transferred from the Department of Commerce to *National Resources Planning Board,* Executive Office of the President.

Employment Stabilization Office, Federal.   *See* **Employment Stabilization Board, Federal**

Employment and Training, Office of Comprehensive   Established in the Department of Labor. Terminated due to expiration of authority for appropriations after fiscal year 1982. Replaced by *Office of Employment and Training Programs.*

Employment and Training Programs, Office of   Renamed Office of Job Training Programs by Employment and Training Administration reorganization in the Department of Labor, effective June 1984.

Endangered Species Scientific Authority   Established by EO 11911 of Apr. 13, 1976. Terminated by act of Dec. 28, 1979 (93 Stat. 1228), and functions transferred to the Secretary of the Interior.

Energy Administration, Federal   Established by act of May 7, 1974 (88 Stat. 96). Assigned additional responsibilities by acts of June 22, 1974 (88 Stat. 246), Dec. 22, 1975 (89 Stat. 871), and Aug. 14, 1976 (90 Stat. 1125). Terminated by act of Aug. 4, 1977 (91 Stat. 577), and functions transferred to the Department of Energy.Energy Advisory Support Office, Secretary of Abolished by secretarial decision of Feb. 6, 2006.

Energy Assurance, Office of   Abolished pursuant to Conference Report No. 108–729 on H.R. 4818, Consolidated Appropriations Act. Functions merged with *Office of Electricity Delivery and Energy Reliability.*

Energy Conservation, Office of   Established by Interior Secretarial Order 2953 May 7, 1973. Functions transferred to *Federal Energy Administration* by act of May 7, 1974 (88 Stat. 100).

Energy Data and Analysis, Office of   Established by Interior Secretarial Order 2953 of May 7, 1973. Functions transferred to *Federal Energy Administration* by act of May 7, 1974 (88 Stat. 100).

Energy Policy Office   Established in Executive Office of the President by EO 11726 of June 29, 1973. Abolished by EO 11775 of Mar. 26, 1974.

Energy Programs, Office of   Established by Department of Commerce Organization Order 25–7A, effective Sept. 24, 1975. Terminated by act of Aug. 4, 1977 (91 Stat. 581), and functions transferred to the Department of Energy.

Energy Research and Development Administration   Established by act of Oct. 11, 1974 (88 Stat. 1234). Assigned responsibilities by acts of Sept. 3, 1974 (88 Stat. 1069, 1079), Oct. 26, 1974 (88 Stat. 1431), and Dec. 31, 1974 (88 Stat. 1887). Terminated by act of Aug. 4, 1977 (91 Stat. 577), and functions transferred to the Department of Energy.

Energy Resources Council   Established in Executive Office of the President by act of Oct. 11, 1974 (88 Stat. 1233). Establishing authority repealed by act of Aug. 4, 1977 (91 Stat. 608), and *Council* terminated.

Energy Supplies and Resources Policy, Presidential Advisory Committee on   Established July 30, 1954, by President. Abolished Mar. 12, 1961, by President and functions transferred to the Secretary of the Interior.

Enforcement Commission, National   Established by General Order 18 of *Economic Stabilization Administrator,* effective July 30, 1952. Functions transferred to Director, *Office of Defense Mobilization,* and Attorney General by EO 10494 of Oct. 14, 1953.

Engineering, Bureau of   *See* **Steam Engineering, Bureau of**

Entomology, Bureau of   *See* **Entomology and Plant Quarantine, Bureau of**

Entomology and Plant Quarantine, Bureau of   *Bureau of Entomology* and *Bureau of Plant Quarantine* created by acts of Apr. 23, 1904 (33 Stat. 276), and July 7, 1932 (47 Stat. 640), respectively. Consolidated with disease control and eradication functions of *Bureau of Plant Industry* into *Bureau of Entomology and Plant Quarantine* by act of Mar. 23, 1934 (48 Stat. 467). Functions transferred to Agricultural Research Service by Secretary's Memorandum 1320, supp. 4, of Nov. 2, 1953.

Environment, Cabinet Committee on the   *See* **Environmental Quality Council**

Environmental Financing Authority   Established by act of Oct. 18, 1972 (86 Stat. 899). Expired June 30, 1975, pursuant to terms of act.

Environmental Quality Council Established by EO 11472 of May 29, 1969. Renamed *Cabinet Committee on the Environment* by EO 11514 of Mar. 5, 1970. EO 11514 terminated by EO 11541 of July 1, 1970.

Environment, Safety, and Health, Office of Established by act of Aug. 4, 1977 (91 Stat. 570). Abolished by Secretary of Energy memorandum 2006–007929 of Aug. 30, 2006, and functions transferred to Office of Health, Safety, and Security.

Environmental Science Services Administration Established in the Department of Commerce by Reorg. Plan No. 2 of 1965, effective July 13, 1965, by consolidating *Weather Bureau* and *Coast and Geodetic Survey.* Abolished by Reorg. Plan No. 4 of 1970, effective Oct. 3, 1970, and functions transferred to National Oceanic and Atmospheric Administration.

Equal Employment Opportunity, President's Committee on Established by EO 10925 of Mar. 6, 1961. Abolished by EO 11246 of Sept. 24, 1965, and functions transferred to the Department of Labor and *U.S. Civil Service Commission.*

Equal Opportunity, President's Council on Established by EO 11197 of Feb. 5, 1965. Abolished by EO 11247 of Sept. 24, 1965, and functions transferred to the Department of Justice.

Equipment, Bureau of Established as *Bureau of Equipment and Recruiting* by act of July 5, 1862 (12 Stat. 510), replacing *Bureau of Construction, Equipment and Repairs.* Designated as *Bureau of Equipment* in annual appropriation acts commencing with fiscal year 1892 (26 Stat. 192) after cognizance over enlisted personnel matters transferred, effective July 1, 1889, to *Bureau of Navigation.* Functions distributed among bureaus and offices in the Department of the Navy by act of June 24, 1910 (61 Stat. 613). Abolished by act of June 30, 1914 (38 Stat. 408).

Ethics, Office of Government Established in the Office of Personnel Management by act of Oct. 26, 1978 (92 Stat. 1862). Became a separate executive agency status by act of Nov. 3, 1988 (102 Stat. 3031).

European Migration, Intergovernmental Committee for Renamed Intergovernmental Committee for Migration by Resolution 624, passed by Intergovernmental Committee for European Migration Council, effective Nov. 11, 1980.

Evacuation, Joint Committee on *See* **Health and Welfare Aspects of Evacuation of Civilians, Joint Committee on**

Exchange Service, International Established in 1849 in Smithsonian Institution. Renamed Office of Publications Exchange by Secretary's internal directive of Jan. 11, 1985.

Executive Branch of the Government, Commission on Organization of the Established by act of July 7, 1947 (61 Stat. 246). Terminated June 12, 1949, pursuant to terms of act. Second *Commission*

on Organization of the Executive Branch of the Government established by act of July 10, 1953 (67 Stat. 142). Terminated June 30, 1955, pursuant to terms of act.

Executive Council Established by EO 6202–A of July 11, 1933. Consolidated with *National Emergency Council* by EO 6889–A of Oct. 29, 1934.

Executive Exchange, President's Commission on *See* **Personnel Interchange, President's Commission on**

Executive orders *See* **State, Department of**

Executive Organization, President's Advisory Council on Established by President Apr. 5, 1969. Terminated May 7, 1971.

Executive Protective Service *See* **Secret Service Division**

Executives, Active Corps of Established in ACTION by act of Oct. 1, 1973 (87 Stat. 404). Transferred to Small Business Administration by EO 11871 of July 18, 1975.

Export Administration, Bureau of Established as a separate agency within the Department of Commerce on Oct. 1, 1987 (50 USC app. 2401 *et seq.*). Renamed Bureau of Industry and Security by Department of Commerce internal organization order of Apr. 18, 2002 (67 FR 20630).

Export Control, Administrator of Functions delegated to Administrator by Proc. 2413 of July 2, 1940, transferred to *Office of Export Control, Economic Defense Board,* by EO 8900 of Sept. 15, 1941. Renamed *Board of Economic Warfare* by EO 8982 of Dec. 17, 1941. *Board* terminated by EO 9361 of July 15, 1943.

Export Control, Office of *See* **Export Control, Administrator of**

Export-Import Bank of Washington Organization of District of Columbia banking corporation directed by EO 6581 of Feb. 2, 1934. Certificate of incorporation filed Feb. 12, 1934. Grouped with other agencies to form *Federal Loan Agency* by Reorg. Plan No. I of 1939, effective July 1, 1939. Transferred to the Department of Commerce by EO 9071 of Feb. 24, 1942. Functions transferred to *Office of Economic Warfare* by EO 9361 of July 15, 1943. Established as permanent independent agency by act of July 31, 1945 (59 Stat. 526). Renamed Export-Import Bank of the U.S. by act of Mar. 13, 1968 (82 Stat. 47).

Export-Import Bank of Washington, DC, Second Authorized by EO 6638 of Mar. 9, 1934. Abolished by EO 7365 of May 7, 1936, and records transferred to *Export-Import Bank of Washington,* effective June 30, 1936.

Export Marketing Service Established by the Secretary of Agriculture Mar. 28, 1969. Merged with Foreign Agricultural Service by Secretary's memorandum of Dec. 7, 1973, effective Feb. 3, 1974.

Exports and Requirements, Division of Established in *Office of Foreign Economic Coordination* by the Department of State order of Feb. 1, 1943. Abolished by departmental order of Nov. 6, 1943, pursuant to EO 9380 of Sept. 25, 1943.

Extension Service Established by act of May 14, 1914 (38 Stat. 372). Consolidated into *Science and Education Administration* by Secretary's order of Jan. 24, 1978. Reestablished as *Extension Service* by Secretarial order of June 16, 1981. Became part of Cooperative State, Research, Education, and Extension Service under Department of Agriculture's reorganization in 1995.

Facts and Figures, Office of Established in *Office for Emergency Management* by EO 8922 of Oct. 24, 1941. Consolidated with *Office of War Information* in *Office for Emergency Management* by EO 9182 of June 13, 1942.

Family Security Committee Established in *Office of Defense Health and Welfare Services* Feb. 12, 1941, by administrative order. Terminated Dec. 17, 1942.

Family Services, Bureau of *See* **Assistance, Bureau of Public**

Family Support Administration Established on Apr. 4, 1986, in the Department of Health and Human Services under authority of section 6 of Reorganization Plan No. 1 of 1953, effective Apr. 11, 1953 (*see also* 51 FR 11641). Merged into Administration for Children and Families by Secretary's reorganization notice dated Apr. 15, 1991.

Farm Board, Federal Established by act of June 15, 1929 (46 Stat. 11). Renamed Farm Credit Administration and certain functions abolished by EO 6084 of Mar. 27, 1933. Administration placed under the Department of Agriculture by Reorg. Plan No. I of 1939, effective July 1, 1939. Made independent agency in the executive branch of the Government, to be housed in the Department of Agriculture, by act of Aug. 6, 1953 (67 Stat. 390). Removed from the Department of Agriculture by act of Dec. 10, 1971 (85 Stat. 617).

Farm Credit Administration *See* **Farm Board, Federal**

Farm Loan Board, Federal Established in the Department of the Treasury to administer act of July 17, 1916 (39 Stat. 360). Offices of appointed members of *Board,* except member designated as *Farm Loan Commissioner,* abolished by EO 6084 of Mar. 27, 1933, and *Board* functions transferred to *Farm Loan Commissioner,* subject to jurisdiction and control of Farm Credit Administration. Title changed to *Land Bank Commissioner* by act of June 16, 1933. Abolished by act of Aug. 6, 1953 (67 Stat. 393).

Farm Loan Bureau, Federal Established in the Department of the Treasury under supervision of *Federal Farm Loan Board* and charged with execution of act of July 17, 1916 (39 Stat. 360). Transferred to *Farm Credit Administration* by EO 6084 of Mar. 27, 1933.

Farm Loan Commissioner *See* **Farm Loan Board, Federal**

Farm Mortgage Corporation, Federal Established by act of Jan. 31, 1934 (48 Stat. 344). Transferred to the Department of Agriculture by Reorg. Plan No. I of 1939, effective July 1, 1939, to operate under supervision of Farm Credit Administration. Abolished by act of Oct. 4, 1961 (75 Stat. 773).

Farm Products, Division of (Also known as *Division of Agriculture*)

Established by *Advisory Commission to Council of National Defense* pursuant to act of Aug. 29, 1916 (39 Stat. 649). *Office of Agricultural Defense Relations* (later known as *Office for Agricultural War Relations*) established in the Department of Agriculture by Presidential letter of May 5, 1941, which transferred to the Secretary of Agriculture functions previously assigned to *Division of Agriculture.* Functions concerned with food production transferred to *Food Production Administration* and functions concerned with food distribution transferred to *Food Distribution Administration* by EO 9280 of Dec. 5, 1942.

Farm Security Administration *See* **Resettlement Administration**

Farm Service Agency Established by Secretary's Memorandum 1010–1 dated Oct. 20, 1994, under authority of the act of Oct. 13, 1994 (7 U.S.C. 6901), and assumed certain functions of the *Agricultural Stabilization and Conservation Service,* the *Farmers' Home Administration,* and the *Federal Crop Insurance Corporation.* Renamed *Consolidated Farm Service Agency* by Acting Administrator on Dec. 19, 1994.

Farmer Cooperative Service Established by Secretary of Agriculture Memorandum 1320, supp. 4, of Dec. 4, 1953. Consolidated with other Department of Agriculture units into *Economics, Statistics, and Cooperatives Service* by Secretary's Memorandum 1927, effective Dec. 23, 1977.

Farmers' Home Administration. *See* **Resettlement Administration**

Federal. *See also other part of title*

Federal Advisory Council Established in *Federal Security Agency* by act of June 6, 1933 (48 Stat. 116). Transferred to the Department of Labor by Reorg. Plan No. 2 of 1949, effective Aug. 20, 1949.

Federal Register, Administrative Committee of the *See* **Archives Establishment, National**

Federal Register, Division of the Established by act of July 26, 1935 (49 Stat. 500). Transferred to General Services Administration as part of *National Archives and Records Service* by act of June 30, 1949 (63 Stat. 381). Renamed Office of the Federal Register by order of General Services Administrator, Feb. 6, 1959. Transferred to National Archives and Records Administration by act of Oct. 19, 1984 (98 Stat. 2283).

Federal Register, Office of the   *See* **Federal Register, Division of the**

Federal Reserve Board   Renamed Board of Governors of the Federal Reserve System, and Governor and Vice Governor designated as Chairman and Vice Chairman, respectively, of Board by act of Aug. 23, 1935 (49 Stat. 704).

Federal Tax Reform, President's Advisory Panel on   Established by EO 13369 of Jan. 7, 2005. Abolished by EO 13446 of Sept. 28, 2007.

Field Services, Office of   Established by the Secretary of Commerce Feb. 1, 1963, by Department Organization Order 40–3. Terminated by Department Organization Order 40–1A of Sept. 15, 1970, and functions transferred to *Bureau of Domestic Commerce.*

Filipino Rehabilitation Commission   Established by act of June 29, 1944 (58 Stat. 626). Inactive pursuant to terms of act.

Film Service, U.S.   Established by *National Emergency Council* in September 1938. Transferred to *Office of Education, Federal Security Agency,* by Reorg. Plan No. II of 1939, effective July 1, 1939. Terminated June 30, 1940.

Films, Coordinator of Government   Director of *Office of Government Reports* designated *Coordinator of Government Films* by Presidential letter of Dec. 18, 1941. Functions transferred to *Office of War Information* by EO 9182 of June 13, 1942.

Financial Operations, Bureau of Government   Renamed Financial Management Service by Order 145–21 of the Secretary of the Treasury, effective Oct. 10, 1984.

Fire Administration, U.S.   *See* **Fire Prevention and Control Administration, National**

Fire Council, Federal   Established by EO 7397 of June 20, 1936. Transferred July 1, 1939, to *Federal Works Agency* by EO 8194 of July 6, 1939, with functions under direction of *Federal Works Administrator.* Transferred with *Federal Works Agency* to General Services Administration by act of June 30, 1949 (63 Stat. 380). Transferred to the Department of Commerce by EO 11654 of Mar. 13, 1972.

Fire Prevention and Control, National Academy for   Established in the Department of Commerce by act of Oct. 29, 1974 (88 Stat. 1537). Transferred to Federal Emergency Management Agency by Reorg. Plan No. 3 of 1978, effective Apr. 1, 1979.

Fire Prevention and Control Administration, National   Renamed U.S. Fire Administration by act of Oct. 5, 1978 (92 Stat. 932). Transferred to Federal Emergency Management Agency by Reorg. Plan No. 3 of 1978, effective Apr. 1, 1979.

Fish Commission, U.S.   *Commissioner of Fish and Fisheries* established as head of *U.S. Fish Commission* by joint resolution of Feb. 9, 1871 (16 Stat. 594).

Commission established as *Bureau of Fisheries* in *Department of Commerce and Labor* by act of Feb. 14, 1903 (32 Stat. 827). Department of Labor created by act of Mar. 4, 1913 (37 Stat. 736), and *Bureau* remained in the Department of Commerce. Transferred to the Department of the Interior by Reorg. Plan No. II of 1939, effective July 1, 1939. Consolidated with *Bureau of Biological Survey* into *Fish and Wildlife Service* by Reorg. Plan No. III of 1940, effective June 30, 1940.

Fish and Wildlife Service   Established by Reorg. Plan No. III of 1940, effective June 30, 1940, consolidating *Bureau of Fisheries* and *Bureau of Biological Survey.* Succeeded by U.S. Fish and Wildlife Service.

Fisheries, Bureau of   *See* **Fish Commission, U.S.**

Fisheries, Bureau of Commercial   Organized in 1959 under U.S. Fish and Wildlife Service, the Department of the Interior. Abolished by Reorg. Plan No. 4 of 1970, effective Oct. 3, 1970, and functions transferred to National Oceanic and Atmospheric Administration.

Fishery Coordination, Office of   Established in the Department of the Interior by EO 9204 of July 21, 1942. Terminated by EO 9649 of Oct. 29, 1945.

Flood Indemnity Administration, Federal   Established in *Housing and Home Finance Agency* by Administrator's Organizational Order 1, effective Sept. 28, 1956, redesignated as Administrator's Organizational Order 2 on Dec. 7, 1956, pursuant to act of Aug. 7, 1956 (70 Stat. 1078). Abolished by Administrator's Organizational Order 3, effective July 1, 1957, due to lack of funding.

Food, Cost of Living Council Committee on   Established by EO 11695 of Jan. 11, 1973. Abolished by EO 11788 of June 18, 1974.

Food, Drug, and Insecticide Administration   Established by act of Jan. 18, 1927 (44 Stat. 1002). Renamed Food and Drug Administration by act of May 27, 1930 (46 Stat. 422). Transferred from the Department of Agriculture to *Federal Security Agency* by Reorg. Plan No. IV of 1940, effective June 30, 1940. Transferred to *Department of Health, Education, and Welfare* by Reorg. Plan No. 1 of 1953, effective Apr. 11, 1953.

Food Distribution Administration   Established in the Department of Agriculture by EO 9280 of Dec. 5, 1942, consolidating *Agricultural Marketing Administration, Sugar Agency,* distribution functions of *Office for Agricultural War Relations,* regulatory work of *Bureau of Animal Industry,* and food units of *War Production Board.* Consolidated with other agencies by EO 9322 of Mar. 26, 1943, to form *Administration of Food Production and Distribution.*

Food and Drug Administration   *See* **Food, Drug, and Insecticide Administration**

Food Industry Advisory Committee   Established by EO 11627 of Oct. 15, 1971. Abolished by EO 11781 of May 1, 1974.

Food and Nutrition Service   Established Aug. 8, 1969, by Secretary of Agriculture under authority of 5 U.S.C. 301 and Reorg. Plan No. 2 of 1953 (5 U.S.C. app.). Abolished by Secretary's Memorandum 1010–1 dated Oct. 20, 1994. Functions assumed by Food and Consumer Service.

Food Production Administration   Established in the Department of Agriculture by EO 9280 of Dec. 5, 1942, which consolidated *Agricultural Adjustment Agency,* Farm Credit Administration, *Farm Security Administration,* Federal Crop Insurance Corporation, Soil Conservation Service, and food production activities of *War Production Board, Office of Agricultural War Relations,* and *Division of Farm Management and Costs, Bureau of Agricultural Economics.* Consolidated with other agencies by EO 9322 of Mar. 26, 1943, to form *Administration of Food Production and Distribution.*

Food Production and Distribution, Administration of   Established by consolidation of *Food Production Administration, Food Distribution Administration,* Commodity Credit Corporation, and Extension Service, Department of Agriculture, by EO 9322 of Mar. 26, 1943, under direction of Administrator, directly responsible to President. Renamed *War Food Administration* by EO 9334 of Apr. 19, 1943. Terminated by EO 9577 of June 29, 1945, and functions transferred to the Secretary of Agriculture. Transfer made permanent by Reorg. Plan No. 3 of 1946, effective July 16, 1946.

Food Safety and Quality Service   Renamed Food Safety and Inspection Service by Agriculture Secretary's memorandum of June 19, 1981.

Foods, Bureau of   Renamed Center for Food Safety and Applied Nutrition by Food and Drug Administration notice of Mar. 9, 1984 (49 FR 10166).

Foreign.   *See also other part of title*

Foreign Aid, Advisory Committee on Voluntary   Established by President May 14, 1946. Transferred from the Department of State to the Director of the *Mutual Security Agency,* and later to Director of the *Foreign Operations Administration,* by Presidential letter of June 1, 1953.

Foreign Intelligence Advisory Board, President's   Established by EO 12863 of Sept. 13, 1993. Abolished by EO 13462 of Feb. 29, 2008.

Foreign Operations Administration   Established by Reorg. Plan No. 7 of 1953, effective Aug. 1, 1953, and functions transferred from *Office of Director of Mutual Security, Mutual Security Agency, Technical Cooperation Administration, Institute of Inter-American Affairs.* Abolished by EO 10610 of May 9, 1955, and functions and offices transferred to the Departments of State and Defense.

Foreign Scholarships, Board of   Renamed J. William Fulbright Foreign Scholarship Board by act of Feb. 16, 1990 (104 Stat. 49).

Forest Reservation Commission, National   Established by act of Mar. 1, 1911 (36 Stat.

962). Terminated by act of Oct. 22, 1976 (90 Stat. 2961), and functions transferred to the Secretary of Agriculture.

Forests, Director of   Established by Administrative Order 1283 of May 18, 1938. Made part of *Office of Land Utilization,* Department of the Interior, by Administrative Order 1466 of Apr. 15, 1940.

Freedmen's Hospital   Established by act of Mar. 3, 1871 (16 Stat. 506; T. 32 of D.C. Code). Transferred from the Department of the Interior to *Federal Security Agency* by Reorg. Plan No. IV of 1940, effective June 30, 1940.

Fuel Yards   Established by act of July 1, 1918 (40 Stat. 672). Transferred from *Bureau of Mines,* Department of Commerce, to *Procurement Division,* Department of the Treasury, by EO 6166 of June 10, 1933, effective Mar. 2, 1934.

Fuels Coordinator for War, Office of Solid   *See* **Fuels Administration for War, Solid**

Fuels Corporation, U.S. Synthetic   Established by act of June 30, 1980 (94 Stat. 636). Terminated Apr. 18, 1986, by act of Dec. 19, 1985 (99 Stat. 1249), and functions transferred to the Secretary of the Treasury.

Fund-Raising Within the Federal Service, President's Committee on   Established by EO 10728 of Sept. 6, 1957. Abolished by EO 10927 of Mar. 18, 1961, and functions transferred to *U.S. Civil Service Commission.*

Gallaudet College   *See* **Columbia Institution for the Instruction of the Deaf and Dumb, and the Blind**

General Programs, Office of   Renamed Office of Public Programs by the Chairman, National Endowment for the Humanities, in January 1991.

Geographic Board, U.S.   Established by EO 27–A of Sept. 4, 1890. Abolished by EO 6680 of Apr. 17, 1935, and duties transferred to *U.S. Board on Geographical Names,* Department of the Interior, effective June 17, 1934. *Board* abolished by act of July 25, 1947 (61 Stat. 457), and duties assumed by *Board on Geographic Names.*

Geographical Names, U.S. Board on   *See* **Geographic Board, U.S.**

Geography, Office of   Function of standardizing foreign place names placed in the Department of the Interior conjointly with the *Board on Geographic Names* by act of July 25, 1947 (61 Stat. 456). Functions transferred to the Department of Defense by memorandum of understanding by the Departments of the Interior and Defense and the *Bureau of the Budget* Mar. 9, 1968.

Geological Survey   Established in the the Department of the Interior by act of Mar. 3, 1879 (20 Stat. 394). Renamed United States Geological Survey by acts of Nov. 13, 1991 (105 Stat. 1000) and May 18, 1992 (106 Stat. 172).

Germany, Mixed Claims Commission, U.S. and Established by agreement of Aug. 10, 1922, between U.S. and Germany. Duties extended by agreement of Dec. 31, 1928. Time limit for filing claims expired June 30, 1928. All claims disposed of by Oct. 30, 1939. Terminated June 30, 1941.

Global Communications, Office of Established within the White House Office by EO 13283 of Jan. 21, 2003. Abolished by EO 13385 of Sept.

Goethals Memorial Commission Established by act of Aug. 4, 1935 (49 Stat. 743). Placed under jurisdiction of *Department of War* by EO 8191 of July 5, 1939.

Government. *See other part of title*

Grain Futures Administration Established in the Department of Agriculture under provisions of act of Sept. 21, 1922 (42 Stat. 998). Superseded by *Commodity Exchange Administration* by order of Secretary, effective July 1, 1936. Consolidated with other agencies into *Commodity Exchange Branch, Agricultural Marketing Administration,* by EO 9069 of Feb. 23, 1942. Functions transferred to the Secretary of Agriculture by EO 9577 of June 29, 1945. Transfer made permanent by Reorg. Plan No. 3 of 1946, effective July 16, 1946. Functions transferred to *Commodity Exchange Authority* by Secretary's Memorandum 1185 of Jan. 21, 1947. Functions transferred to Commodity Futures Trading Commission by act of Oct. 23, 1974 (88 Stat. 1414).

Grain Inspection Service, Federal Established in the Department of Agriculture by act of Oct. 21, 1976 (90 Stat. 2868). Abolished by Secretary's Memorandum 1010–1 dated Oct. 20, 1994, and program authority and functions transferred to the Grain Inspection, Packers and Stockyards Administration.

Grain Stabilization Corporation Organized as Delaware corporation to operate in connection with *Federal Farm Board* pursuant to act of June 15, 1929 (46 Stat. 11). Terminated by filing of certificate of dissolution with Corporation Commission of State of Delaware Dec. 14, 1935.

Grant Administration, Office of Transferred from the Office of the General Council to the Deputy Director, U.S. Trade and Development Agency by administrative order of Apr. 25, 2007.

Grants and Program Systems, Office of Abolished and functions transferred to Cooperative State Research Service, Department of Agriculture, by Secretarial Memorandum 1020–26 of July 1, 1986.

Grazing Service Consolidated with *General Land Office* into Bureau of Land Management, Department of the Interior, by Reorg. Plan No. 3 of 1946, effective July 16, 1946.

Great Lakes Basin Commission Established by EO 11345 of Apr. 20, 1967. Terminated by EO 12319 of Sept. 9, 1981.

Great Lakes Pilotage Administration Established in the Department of Commerce to administer act of June 30, 1960 (74 Stat. 259). Administration of act transferred to the Secretary of Transportation by act of Oct. 15, 1966 (80 Stat. 931).

Greening the Government through Waste Prevention and Recycling, Steering Committee Established by EO 13101 of Sept. 14, 1998. Abolished by EO 13423 of Jan. 24, 2007.

Handicapped, National Center on Education Media and Materials for the Established by agreement between the *Secretary of Health, Education, and Welfare* and Ohio State University, pursuant to acts of Aug. 20, 1969 (83 Stat. 102) and Apr. 13, 1970 (84 Stat. 187). Authorization deleted by act of Nov. 29, 1975 (89 Stat. 795), and the Secretary was authorized to enter into agreements with non-Federal organizations to establish and operate centers for handicapped.

Handicapped, National Council on the Established in the *Department of Health, Education, and Welfare* by act of Nov. 6, 1978 (92 Stat. 2977). Transferred to the Department of Education by act of Oct. 17, 1979 (93 Stat. 677). Reorganized as independent agency by act of Feb. 22, 1984 (98 Stat. 26).

Handicapped Employees, Interagency Committee on Alternately renamed Interagency Committee on Employment of People with Disabilities by EO 12704 of Feb. 26, 1990.

Handicapped Individuals, White House Conference on Established by act of Dec. 7, 1974 (88 Stat. 1617). Terminated Dec. 30, 1977, pursuant to terms of act.

Handicapped Research, National Institute of Renamed National Institute on Disability and Rehabilitation Research by act of Oct. 21, 1986 (100 Stat. 1820).

Health, Cost of Living Council Committee on Established by EO 11695 of Jan. 11, 1973. Abolished by EO 11788 of June 18, 1974.

Health, Education, and Welfare, Department of Established by Reorganization Plan No. 1 of 1953 (5 U.S.C. app.), effective Apr. 11, 1953. Renamed Department of Health and Human Services by act of Oct. 17, 1979 (93 Stat. 695).

Health, Welfare, and Related Defense Activities, Office of the Coordinator of *Federal Security Administrator* designated as Coordinator of health, welfare, and related fields of activity affecting national defense, including aspects of education under *Federal Security Agency,* by *Council of National Defense,* with approval of President, Nov. 28, 1940. Office of Coordinator superseded by *Office of Defense Health and Welfare Services,* established in *Office for Emergency Services* by EO 8890 of Sept. 3, 1941.

Health Care Technology, National Council on Established by act of July 1, 1944, as amended (92 Stat. 3447). Renamed *Council on Health Care*

Technology by act of Oct. 30, 1984 (98 Stat. 2820). Name lowercased by act of Oct. 7, 1985 (99 Stat. 493). Terminated by act of Dec. 19, 1989 (103 Stat. 2205).

Health Facilities, Financing, Compliance, and Conversion, Bureau of   Renamed Bureau of Health Facilities by Department of Health and Human Services Secretarial order of Mar. 12, 1980 (45 FR 17207).

Health Industry Advisory Committee   Established by EO 11695 of Jan. 11, 1973. Abolished by EO 11781 of May 1, 1974.

Health Manpower, Bureau of   Renamed Bureau of Health Professions by Department of Health and Human Services Secretarial order of Mar. 12, 1980 (45 FR 17207).

Health and Medical Committee   Established by *Council of National Defense* order of Sept. 19, 1940. Transferred to *Federal Security Agency* by *Council* order approved by President Nov. 28, 1940. Reestablished in *Office of Defense Health and Welfare Services, Office for Emergency Management,* by EO 8890 of Sept. 3, 1941. *Committee* transferred to *Federal Security Agency* by EO 9338 of Apr. 29, 1943.

Health Resources Administration   Established in Public Health Service. Abolished by Department of Health and Human Services Secretarial reorganization of Aug. 20, 1982 (47 FR 38409), and functions transferred to Health Resources and Services Administration.

Health Service, Public   Originated by act of July 16, 1798 (1 Stat. 605). Transferred from the Department of the Treasury to the *Federal Security Agency* by Reorg. Plan No. I of 1939, effective July 1, 1939.

Health Services Administration   Established in Public Health Service. Abolished by Department of Health and Human Services Secretarial reorganization of Aug. 20, 1982 (47 FR 38409), and functions transferred to Health Resources and Services Administration.

Health Services Industry, Committee on the   Established by EO 11627 of Oct. 15, 1971. Abolished by EO 11695 of Jan. 11, 1973.

Health Services and Mental Health Administration   Established in Public Health Service Apr. 1, 1968. Abolished by *Department of Health, Education, and Welfare* reorganization order and functions transferred to *Centers for Disease Control, Health Resources Administration,* and *Health Services Administration,* effective July 1, 1973.

Health Services Research, National Center for   Established by act of July 23, 1974 (88 Stat. 363). Transferred from *Health Resources Administration* to Office of the Assistant Secretary for Health by *Department of Health, Education, and Welfare* reorganization, effective Dec. 2, 1977. Renamed *National Center for Health Services Research and Health Care Technology Assessment* by

Secretary's order, pursuant to act of Oct. 30, 1984 (98 Stat. 2817). Terminated by act of Dec. 19, 1989 (103 Stat. 2205).

Health Statistics, National Center for   Established by act of July 23, 1974 (88 Stat. 363). Transferred from *Health Resources Administration* to Office of the Assistant Secretary for Health by the *Department of Health, Education, and Welfare* reorganization, effective Dec. 2, 1977. Transferred to *Centers for Disease Control* by Secretary's notice of Apr. 2, 1987 (52 FR 13318).

Health and Welfare Activities, Interdepartmental Committee to Coordinate   Appointed by President Aug. 15, 1935, and reestablished by EO 7481 of Oct. 27, 1936. Terminated in 1939.

Health and Welfare Aspects of Evacuation of Civilians, Joint Committee on   Established August 1941 as joint committee of *Office of Defense Health and Welfare Services* and *Office of Civilian Defense.* Reorganized in June 1942 and renamed *Joint Committee on Evacuation. Office of Defense Health and Welfare Services* abolished by EO 9388 of Apr. 29, 1943, and functions transferred to *Federal Security Agency. Committee* terminated.

Heart and Lung Institute, National   Renamed National Heart, Lung, and Blood Institute by act of Apr. 22, 1976 (90 Stat. 402).

Heritage Conservation and Recreation Service   Established by the Secretary of the Interior Jan. 25, 1978. Abolished by Secretarial Order 3060 of Feb. 19, 1981, and functions transferred to National Park Service.

Hemispheric Defense Studies, Center for   Established by Department of Defense Directive 3200.12 of Sept. 3, 1997. Abolished by act of Oct. 17, 2006 (120 Stat. 2353).

Highway Safety Agency, National   Established in the Department of Commerce by act of Sept. 9, 1966 (80 Stat. 731). Functions transferred to the Department of Transportation by act of Oct. 15, 1966 (80 Stat. 931). Functions transferred to *National Highway Safety Bureau* by EO 11357 of June 6, 1967. *Bureau* renamed National Highway Traffic Safety Administration by act of Dec. 31, 1970 (84 Stat. 1739).

Highway Safety Bureau, National   *See* **Highway Safety Agency, National**

Home Economics, Bureau of Human Nutrition and   *See* **Home Economics, Office of**

Home Economics, Office of   Renamed *Bureau of Home Economics* by Secretary's Memorandum 436, effective July 1, 1923, pursuant to act of Feb. 26, 1923 (42 Stat. 1289). Redesignated *Bureau of Human Nutrition and Home Economics* February 1943 in accordance with *Research Administration* Memorandum 5 issued pursuant to EO 9069 of Feb. 23, 1942, and in conformity with Secretary's Memorandums 960 and 986. Functions transferred to Agricultural

Research Service by Secretary's Memorandum 1320, supp. 4, of Nov. 2, 1953.

Home Loan Bank Administration, Federal See **Home Loan Bank Board, Federal**

Home Loan Bank Board See **Home Loan Bank Board, Federal**

Home Loan Bank Board, Federal Established by acts of July 22, 1932 (47 Stat. 725), June 13, 1933 (48 Stat. 128), and June 27, 1934 (48 Stat. 1246). Grouped with other agencies to form *Federal Loan Agency* by Reorg. Plan No. I of 1939, effective July 1, 1939. Functions transferred to *Federal Home Loan Bank Administration, National Housing Agency,* by EO 9070 of Feb. 24, 1942. Abolished by Reorg. Plan No. 3, effective July 27, 1947, and functions transferred to *Home Loan Bank Board, Housing and Home Finance Agency.* Renamed *Federal Home Loan Bank Board* and made independent agency by act of Aug. 11, 1955 (69 Stat. 640). Abolished by act of Aug. 9, 1989 (103 Stat. 354, 415), and functions transferred to Office of Thrift Supervision, Resolution Trust Corporation, Federal Deposit Insurance Corporation, and Federal Housing Finance Board.

Home Loan Bank System, Federal Grouped with other agencies to form *Federal Loan Agency* by Reorg. Plan No. I of 1939, effective July 1, 1939. Functions transferred to *Federal Home Loan Bank Administration, National Housing Agency,* by EO 9070 of Feb. 24, 1942. Transferred to *Housing and Home Finance Agency* by Reorg. Plan No. 3 of 1947, effective July 27, 1947.

Home Mortgage Credit Extension Committee, National Voluntary Established by act of Aug. 2, 1954 (68 Stat 638). Terminated Oct. 1, 1965, pursuant to terms of act.

Home Owners' Loan Corporation Established by act of June 13, 1933 (48 Stat. 128), under supervision of *Federal Home Loan Bank Board.* Grouped with other agencies to form *Federal Loan Agency* by Reorg. Plan No. I of 1939, effective July 1, 1939. Transferred to *Federal Home Loan Bank Administration, National Housing Agency,* by EO 9070 of Feb. 24, 1942. Board of Directors abolished by Reorg. Plan No. 3 of 1947, effective July 27, 1947, and functions transferred, for liquidation of assets, to *Home Loan Bank Board, Housing and Home Finance Agency.* Terminated by order of *Secretary of the Home Loan Bank Board,* effective Feb. 3, 1954, pursuant to act of June 30, 1953 (67 Stat. 121).

Homesteads, Division of Subsistence Established by act of June 16, 1933 (48 Stat. 205). Secretary of the Interior authorized to administer section 208 of act by EO 6209 of July 21, 1933. *Federal Subsistence Homesteads Corporation* created by Secretary's order of Dec. 2, 1933, and organization incorporated under laws of Delaware. Transferred to *Resettlement Administration* by EO 7041 of May 15, 1935.

Homesteads Corporation, Federal Subsistence See **Homesteads, Division of Subsistence**

Hospitalization, Board of Federal Organized Nov. 1, 1921. Designated as advisory agency to *Bureau of the Budget* May 7, 1943. Terminated June 30, 1948, by Director's letter of May 28, 1948.

Housing, President's Committee on Equal Opportunity in Established by EO 11063 of Nov. 20, 1962. Inactive as of June 30, 1968.

Housing Administration, Federal Established by act of June 27, 1934 (48 Stat. 1246). Grouped with other agencies to form *Federal Loan Agency* by Reorg. Plan No. I of 1939, effective July 1, 1939. Functions transferred to *Federal Housing Administration, National Housing Agency,* by EO 9070 of Feb. 24, 1942. Transferred to *Housing and Home Finance Agency* by Reorg. Plan No. 3, effective July 27, 1947. Functions transferred to the Department of Housing and Urban Development by act of Sept. 9, 1965 (79 Stat. 667).

Housing Administration, Public Established as constituent agency of *Housing and Home Finance Agency* by Reorg. Plan No. 3 of 1947, effective July 27, 1947. Functions transferred to the Department of Housing and Urban Development by act of Sept. 9, 1965 (79 Stat. 667).

Housing Agency, National Established by EO 9070 of Feb. 24, 1942, to consolidate housing functions relating to *Federal Home Loan Bank Board, Federal Home Loan Bank System, Federal Savings and Loan Insurance Corporation, Home Owners' Loan Corporation, U.S. Housing Corporation, Federal Housing Administration, U.S. Housing Authority, Defense Homes Corporation, Division of Defense Housing Coordination, Central Housing Committee, Farm Security Administration* with respect to nonfarm housing, *Public Buildings Administration, Division of Defense Housing, Mutual Ownership Defense Housing Division, Office of Administrator of Federal Works Agency,* and the Departments of *War* and the Navy with respect to housing located off military installations. Agency dissolved on creation of *Housing and Home Finance Agency* by Reorg. Plan No. 3 of 1947, effective July 27, 1947.

Housing Authority, Federal Public Established by EO 9070 of Feb. 24, 1942. Public housing functions of *Federal Works Agency, the Departments of War* and the Navy (except housing located on military installations), and *Farm Security Administration* (nonfarm housing) transferred to *Authority,* and *Defense Homes Corporation* administered by the Commissioner of the *Authority'.* Functions transferred to *Public Housing Administration, Housing and Home Finance Agency,* by Reorg. Plan No. 3 of 1947, effective July 27, 1947.

Housing Authority, U.S. Established in the Department of the Interior by act of Sept. 1, 1937 (50 Stat. 888). Transferred to *Federal Works Agency* by Reorg. Plan No. I of 1939, effective July 1, 1939. Transferred to *Federal Public Housing Authority, National Housing Agency,* by EO 9070 of Feb. 24, 1942. Office of Administrator abolished by Reorg. Plan No. 3 of 1947, effective July 27, 1947, and functions transferred to *Public Housing Administration, Housing and Home Finance Agency.*

Housing Corporation, U.S. Incorporated July 10, 1918, under laws of New York. Transferred from the Department of Labor to the Department of the Treasury by EO 7641 of June 22, 1937. Transferred from the Department of the Treasury to the *Public Buildings Administration, Federal Works Agency,* by EO 8186 of June 29, 1939. Functions transferred for liquidation to *Federal Home Loan Bank Administration, National Housing Agency,* by EO 9070 of Feb. 24, 1942. Terminated Sept. 8, 1952, by the *Secretary of the Home Loan Bank Board.*

Housing Council, National Established in *Housing and Home Finance Agency* by Reorg. Plan No. 3 of 1947, effective July 27, 1947. Terminated by Reorg. Plan No. 4 of 1965, effective July 27, 1965, and functions transferred to President.

Housing Division Established in *Public Works Administration* by act of June 16, 1933 (48 Stat. 195). Functions transferred to *U.S. Housing Authority* by EO 7732 of Oct. 27, 1937.

Housing Enterprise Oversight, Office of Federal Office and positions of Director and Deputy Director established within the Department of Housing and Urban Development by the act of October 28, 1992 (106 Stat. 3944). Abolished by the act of July 30, 2008 (122 Stat. 2794), and functions, personnel, and property transferred to Federal Housing Finance Agency.

Housing Expediter, Office of the Established in *Office of War Mobilization and Reconversion* by Presidential letter of Dec. 12, 1945, to *Housing Expediter.* Functions of *Housing Expediter* defined by EO 9686 of Jan. 26, 1946. *Housing Expediter* confirmed in position of *National Housing Administrator* Feb. 6, 1946. *Office of the Housing Expediter* established by act of May 22, 1946 (60 Stat. 208). Functions of *Office* and *National Housing Administrator* segregated by EO 9820 of Jan. 11, 1947. Housing functions of *Civilian Production Administration* transferred to *Office* by EO 9836 of Mar. 22, 1947, effective Apr. 1, 1947. Rent control functions of *Office of Temporary Controls* transferred to *Office* by EO 9841 of Apr. 23, 1947. *Office* terminated by EO 10276 of July 31, 1951, and functions transferred to *Economic Stabilization Agency.*

Housing Finance Board, Federal Established by the act of August 9, 1989 (103 Stat. 354, 415), and certain functions transferred from Federal Home Loan Bank Board. Abolished by the act of July 30, 2008 (122 Stat. 2797), and functions, personnel, and property transferred to Federal Housing Finance Agency.

Housing and Home Finance Agency Established by Reorg. Plan No. 3 of 1947, effective July 27, 1947. Terminated by act of Sept. 9, 1965 (79 Stat. 667), and functions transferred to the Department of Housing and Urban Development.

Howard University Established by act of Mar. 2, 1867 (14 Stat. 438). Functions of the Department of the Interior transferred to *Federal Security Agency* by Reorg. Plan No. IV of 1940, effective June 30, 1940.

Functions of the *Department of Health, Education, and Welfare* transferred to the Department of Education by act of Oct. 17, 1979 (93 Stat. 678).

Human Development, Office of Established in *Department of Health, Education, and Welfare.* Renamed Office of Human Development Services and component units transferred to or reorganized under new administrations in Office by Secretary's reorganization order of July 26, 1977. Merged into the Administration for Children and Families by Secretary of Health and Human Services reorganization notice dated Apr. 15, 1991.

Human Development Services, Office of *See* **Human Development, Office of**

Human Embryo Stem Cell Registry Approved by Presidential announcement of Aug. 9, 2001 and established through National Institute of Health's Departmental Notice NOT–OD–01–058 of Aug. 27, 2001. Renamed Human Pluripotent Stem Cell Registry by EO 13435 of June 20, 2007.

Hydrographic Office Jurisdiction transferred from *Bureau of Navigation* to Chief of Naval Operations by EO 9126 of Apr. 8, 1942, and by Reorg. Plan No. 3 of 1946, effective July 16, 1946. Renamed U.S. Naval Oceanographic Office by act of July 10, 1962 (76 Stat. 154).

Imagery and Mapping Agency, National Established by act of Sept. 23, 1996 (110 Stat. 2677). Renamed National Geospatial-Intelligence Agency by act of Nov. 24, 2003 (117 Stat. 1568).

Imagery Office, Central Established as a Department of Defense agency on May 6, 1992. Functions transferred to National Imagery and Mapping Agency by act of Sept. 23, 1996 (110 Stat. 2677).

Immigration, Bureau of Established as branch of the Department of the Treasury by act of Mar. 3, 1891 (26 Stat. 1085). Transferred to *Department of Commerce and Labor* by act of Feb. 14, 1903 (34 Stat. 596). Made *Bureau of Immigration and Naturalization* by act of June 29, 1906 (37 Stat. 736). Made separate division after the Department of Labor created by act of Mar. 4, 1913 (37 Stat. 736). Consolidated into Immigration and Naturalization Service, Department of Labor, by EO 6166 of June 10, 1933. Transferred to the Department of Justice by Reorg. Plan No. V of 1940, effective June 14, 1940. Abolished by act of Nov. 25, 2002 (116 Stat. 2205) and functions transferred to Homeland Security Department.

Immigration, Commissioners of Offices of commissioners of immigration of the several ports created by act of Aug. 18, 1894 (28 Stat. 391). Abolished by Reorg. Plan No. III of 1940, effective June 30, 1940, and functions transferred to *Bureau of Immigration and Naturalization,* Department of Labor.

Immigration and Naturalization, Bureau of *See* **Immigration, Bureau of**

Immigration and Naturalization, District Commissioner of Created by act of Aug. 18, 1894 (28 Stat. 391). Abolished by Reorg. Plan No. III of 1940, effective June 30, 1940. Functions administered by the Commissioner of Immigration and Naturalization, Department of Justice, through district immigration and naturalization directors.

Immigration and Naturalization Service *See* **Immigration, Bureau of**

Import Programs, Office of Established by the Secretary of Commerce Feb. 14, 1971. Functions transferred to *Domestic and International Business Administration,* effective Nov. 17, 1972.

Indian Claims Commission Established by act of Aug. 13, 1946 (60 Stat. 1049). Terminated by act of Oct. 8, 1976 (90 Stat. 1990), and pending cases transferred to *U.S. Court of Claims* Sept. 30, 1978.

Indian Commissioners, Board of Established by section 2039, Revised Statutes. Abolished by EO 6145 of May 25, 1933.

Indian Education Programs, Office of Established within the Bureau of Indian Affairs, Department of the Interior, by act of June 23, 1972 (86 Stat. 343). Renamed Bureau of Indian Education by Departmental Manual Release No. 3721 of Aug. 29, 2006.

Indian Medical Facilities Functions transferred from the Department of the Interior to the *Department of Health, Education, and Welfare,* to be administered by the Surgeon General of Public Health Service, by act of Aug. 5, 1954 (68 Stat. 674).

Indian Opportunity, National Council on Established by EO 11399 of Mar. 6, 1968. Terminated Nov. 26, 1974, by act of Nov. 26, 1969 (83 Stat. 220).

Indian Policy Review Commission, American Established by act of Jan. 2, 1975 (88 Stat. 1910). Terminated June 30, 1977, pursuant to terms of act.

Industrial Analysis, Committee of Established by EO 7323 of Mar. 21, 1936. Terminated Feb. 17, 1937.

Industrial Cooperation, Coordinator for Established by EO 7193 of Sept. 26, 1935. Continued by EO 7324 of Mar. 30, 1936. Terminated June 30, 1937.

Industrial Emergency Committee Established by EO 6770 of June 30, 1934. Consolidated with *National Emergency Council* by EO 6889–A of Oct. 29, 1934.

Industrial Pollution Control Council Staff, National Established by Department of Commerce Organization Order 35–3 of June 17, 1970. *Staff* abolished by departmental organization order of Sept. 10, 1973. Council inactive.

Industrial Recovery Board, National Established by EO 6859 of Sept. 27, 1934. Terminated by EO 7075 of June 15, 1935.

Industrial Recovery Board, Special Established by EO 6173 of June 16, 1933. Functions absorbed by *National Emergency Council* under terms of EO 6513 of Dec. 18, 1933.

Industrial Relations, Office of Activated in the Department of the Navy Sept. 14, 1945. Superseded June 22, 1966, by creation of *Office of Civilian Manpower Management.*

Industry and Trade Administration *See* **Business and Defense Services Administration**

Information, Committee for Reciprocity Established by EO 6750 of June 27, 1934; reestablished by EO 10004 of Oct. 5, 1948, which revoked EO 6750. Superseded by EO 10082 of Oct. 5, 1949; abolished by EO 11075 of Jan. 15, 1963, which revoked EO 10082.

Information, Coordinator of Established by Presidential order of July 11, 1941. Functions exclusive of foreign information activities transferred by military order of June 13, 1942, to jurisdiction of Joint Chiefs of Staff, *War Department,* as *Office of Strategic Services.* Foreign information functions transferred to *Office of War Information* by EO 9182 of June 13, 1942.

Information, Division of Established pursuant to Presidential letter of Feb. 28, 1941, to *Liaison Officer, Office of Emergency Management.* Abolished by EO 9182 of June 13, 1942. Functions relating to public information on war effort transferred and consolidated with *Office of War Information,* and publication services relating to specific agencies of OEM transferred to those agencies.

Information, Office of Coordinator of Transferred, exclusive of foreign information activities, to *Office of War Information* by EO 9182 of June 13, 1942. Designated *Office of Strategic Services* and transferred to jurisdiction of Joint Chiefs of Staff by military order of June 13, 1942. Terminated by EO 9621 of Sept. 20, 1945, and functions distributed to the Departments of State and *War.*

Information Administration, International Transferred from the Department of State to the *U.S. Information Agency* by Reorg. Plan No. 8 of 1953, effective Aug. 1, 1953.

Information Agency, U.S. Established by Reorg. Plan No. 8 of 1953, effective Aug. 1, 1953. Abolished by Reorg. Plan No. 2 of 1977, effective Apr. 1, 1978; replaced by and functions transferred to *International Communication Agency.* Redesignated *U.S. Information Agency* by act of Aug. 24, 1982 (96 Stat. 291). Abolished by act of Oct. 21, 1998 (112 Stat. 2681–761), and functions transferred to the Department of State, effective Oct. 1, 1999.

Information and Public Affairs, Office of Merged with *Office of Intergovernmental Affairs* to form Office of Public and Intergovernmental Affairs by Order 1–85 of June 5, 1985 of the Secretary of Labor.

Information Resources Management, Office of *See* **Telecommunications Service, Automated Data**

**Information Resources Management
Service**   Established in the General Services
Administration. Renamed Information Technology
Service in 1995.

**Information Security Committee,
Interagency**   Established by EO 12065 of June 28,
1978. Abolished by EO 12356 of Apr. 2, 1982.

Information Security Oversight Office   Established
in General Services Administration by EO 12065 of
June 28, 1978. EO 12065 revoked by EO 12356 of
Apr. 2, 1982, which provided for continuation of
Office.

Information Service, Government   *See* **Information
Service, U.S.**

**Information Service, Interim
International**   Established in the Department of State
by EO 9608 of Aug. 31, 1945. Abolished Dec. 31,
1945, pursuant to terms of order.

Information Service, U.S.   Established in March
1934 as division of *National Emergency Council.*
Transferred to *Office of Government Reports* by
Reorg. Plan No. II of 1939, effective July 1, 1939.
Consolidated, along with other functions of *Office,*
into *Division of Public Inquiries, Bureau of Special
Services, Office of War Information,* by EO 9182 of
June 13, 1942. *Bureau of Special Services* renamed
Government Information Service and transferred to
Bureau of the Budget by EO 9608 of Aug. 31, 1945.
Service transferred to *Office of Government Reports*
by EO 9809 of Dec. 12, 1946.

Information Systems Council   Established by EO
13356 of Aug. 27, 2004. Abolished by EO 13388 of
Oct. 25, 2005 (70 FR 62025).

Information Technology Service   Established in
General Services Administration. Abolished by
General Services Administrative Order No. 5440.492,
Aug. 21, 1996, and functions transferred to Federal
Telecommunications Service.

Insane, Government Hospital for the   Established
by act of Mar. 3, 1855 (10 Stat. 682). Renamed
Saint Elizabeth's Hospital by act of July 1, 1916 (39
Stat. 309). Transferred from the Department of the
Interior to *Federal Security Agency* by Reorg. Plan
No. IV of 1940, effective June 30, 1940. Transferred
to *Department of Health, Education, and Welfare* by
Reorg. Plan No. 1 of 1953, effective Apr. 11, 1953.
Functions redelegated to National Institute of Mental
Health by Secretary's reorganization order of Aug.
9, 1967. Property and administration transferred to
District of Columbia Government by act of Nov. 8,
1984 (98 Stat. 3369).

Installations, Director of   Established in the
Department of Defense by act of July 14, 1952 (66
Stat. 625). Abolished by Reorg. Plan No. 6 of 1953,
effective June 30, 1953, and functions transferred to
the Secretary of Defense.

Insular Affairs, Bureau of   Transferred from
Department of War to *Division of Territories and
Island Possessions,* the Department of the Interior,

by Reorg. Plan No. II of 1939, effective July 1,
1939.

Insurance Administrator, Federal   Established by act
of Aug. 1, 1968 (82 Stat. 567). Functions transferred
to Federal Emergency Management Agency by Reorg.
Plan No. 3 of 1978, effective Apr. 1, 1979.

**Integrity and Efficiency, President's Council
on**   Established by EO 12301 of Mar. 26, 1981
(46 FR 19211). Abolished and reestablished by EO
12625 of Jan 27, 1988 (53 FR 2812). Abolished and
reestablished by EO 12805 of May 11, 1992 (57 FR
20627).

Intelligence, Office of   Established within the
Department of Energy by Public Law 106–65 of
Oct. 5, 1999 (113 Stat. 955). Merged with *Office of
Counterintelligence* to form Office of Intelligence and
Counterintelligence by memorandum of March 9,
2006 of the Secretary of Energy.

**Intelligence Activities, President's Board of
Consultants on Foreign**   Established by EO 10656
of Feb. 6, 1956. EO 10656 revoked by EO 10938
of May 4, 1961, and *Board* terminated. Functions
transferred to President's Foreign Intelligence
Advisory Board.

**Intelligence Advisory Board, President's
Foreign**   Established by EO 11460 of Mar. 20,
1969. Abolished by EO 11984 of May 4, 1977.
Reestablished by EO 12331 of Oct. 20, 1981.

Intelligence Authority, National   Established by
Presidential directive of Jan. 22, 1946. Terminated
on creation of Central Intelligence Agency under
National Security Council by act of July 26, 1947 (61
Stat. 497).

Intelligence Group, Central   Terminated on creation
of Central Intelligence Agency by act of July 26, 1947
(61 Stat. 497).

**Intelligence Policy and Review, Office
of**   Transferred from the Criminal Division to the
National Security Division by act of Mar. 9, 2006
(120 Stat. 249).

Inter-American Affairs, Institute of   *See* **American
Republics, Office for Coordination of Commercial
and Cultural Relations between the**

Inter-American Affairs, Office of   *See* **American
Republics, Office for Coordination of Commercial
and Cultural Relations between the**

**Inter-American Affairs, Office of the Coordinator
of**   *See* **American Republics, Office for
Coordination of Commercial and Cultural Relations
between the**

Interagency.   *See other part of title*

Interdepartmental.   *See also other part of title*

Interdepartmental Advisory Council   Established
January 1941 to advise *Coordinator of Health,
Welfare, and Related Defense Activities.* Terminated

on creation of *Office of Defense Health and Welfare Service* Sept. 3, 1941.

Interest and Dividends, Committee on   Established by EO 11695 of Jan. 11, 1973. Abolished by EO 11781 of May 1, 1974.

Intergovernmental Affairs, Office of   Merged with *Office of Information and Public Affairs* to form Office of Public and Intergovernmental Affairs by Order 1–85 of June 5, 1985 of the Secretary of Labor.

Intergovernmental and Interagency Affairs, Office of   Abolished by decision of March 21, 2005 of the Secretary of Education under authority of section 413 of the Department of Education Organization Act.

Intergovernmental Relations, Advisory Commission on   Established by act of Sept. 24, 1959 (73 Stat. 703). Terminated pursuant to act of Nov. 19, 1995 (109 Stat. 480). Continued in existence by act of Oct. 19, 1996 (110 Stat. 4004).

Intergovernmental Relations, Commission on   Established by act of July 10, 1953 (67 Stat. 145). Final report submitted to Congress by June 30, 1955, pursuant to act of Feb. 7, 1955 (69 Stat. 7).

Intergovernmental Relations, Office of   Established by EO 11455 of Feb. 14, 1969. Functions transferred to *Domestic Council* by EO 11690 of Dec. 14, 1972.

Interim Compliance Panel   Established by Dec. 30, 1969 (83 Stat. 774). Terminated June 30, 1976, pursuant to terms of act.

Internal Revenue Service   Functions relating to alcohol, tobacco, firearms, and explosives transferred to Bureau of Alcohol, Tobacco, and Firearms by Department of Treasury order of July 1, 1972.

Internal Security Division   Established July 9, 1945, by transfer of functions from Criminal Division. Abolished Mar. 22, 1973, and functions transferred to Criminal Division, Department of Justice.

International   *See also other part of title*

International Activities, Office of   Renamed *Office of Service and Protocol* by Secretary of the Smithsonian Institution internal directive of Jan. 11, 1985.

International Development, Agency for   Transferred from the Department of State to *U.S. International Development Cooperation Agency* by Reorg. Plan No. 2 of 1979, effective Oct. 1, 1979. Continued as agency within *IDCA* by IDCA Delegation of Authority No. 1 of Oct. 1, 1979. By act of Oct. 21, 1998 (112 Stat. 2681–790), became independent agency.

International Development Cooperation Agency, U.S.   Established by Reorg. Plan No. 2 of 1979, effective Oct. 1, 1979. Abolished by act of Oct. 21, 1998 (112 Stat. 2681–790) and functions transferred to the Department of State, U.S. Agency for International Development, and Overseas Private Investment Corporation.

Interstate Commerce Commission   Created by act of Feb. 4, 1887 (24 Stat. 379). Certain functions as cited in act of Oct. 15, 1966 (80 Stat. 931) transferred to the Secretary of Commerce. Functions relating to railroad and pipeline safety transferred to Federal Railroad Administrator and motor carrier safety to Federal Highway Administrator by act. Abolished by act of Dec. 29, 1995 (109 Stat. 932) and many functions transferred to the newly created Surface Transportation Board within the Department of Transportation.

Investigation, Bureau of   Established by act of May 22, 1908 (35 Stat. 235). Functions consolidated with investigative functions of *Bureau of Prohibition, Division of Investigation,* Department of Justice, by EO 6166 of June 10, 1933, effective Mar. 2, 1934.

Investigation, Division of   Designated as Federal Bureau of Investigation in the Department of Justice by act of Mar. 22, 1935 (49 Stat. 77).

Investigation and Research, Board of   Established by act of Sept. 18, 1940 (54 Stat. 952). Extended to Sept. 18, 1944, by Proc. 2559 of June 26, 1942.

Investigations, Division of   Established by administrative order of Apr. 27, 1933. Abolished Jan. 17, 1942, by administrative order and functions transferred to *Branch of Field Examination, General Land Office,* Department of the Interior.

Investments, Office of Foreign Direct   Established in the Department of Commerce Jan. 2, 1968, by Departmental Organization Order 25–3 to carry out provisions of EO 11387 of Jan. 1, 1968. Controls on foreign investments terminated Jan. 29, 1974.

Iraq Reconstruction, Office of the Inspector General for   Established by act of Nov. 6, 2003 (117 Stat. 1234). Abolished by act of Oct. 17, 2006 (120 Stat. 2397).

Jamestown-Williamsburg-Yorktown National Celebration Commission   Established by act of Aug. 13, 1953 (67 Stat. 576). Terminated upon submission of final report to Congress Mar. 1, 1958.

Job Corps, Office of   Transferred from the Employment and Training Administration to the Office of the Secretary, U.S. Department of Labor by act of Dec. 30, 2005 (119 Stat. 2842). Transferred from the Office of the Secretary, U.S. Department of Labor to the Employment and Training Administration by act of Dec. 16, 2009 (123 Stat. 3238).

Joint.   *See also other part of title*

Joint Resolutions of Congress   *See* **State, Department of**

Judicial Procedure, Commission on International Rules of   Established by act of Sept. 2, 1958 (72 Stat. 1743). Terminated Dec. 31, 1966, by act of Aug. 30, 1964 (78 Stat. 700).

Justice Assistance, Research, and Statistics, Office of   Established in the Department of Justice by act of Dec. 27, 1979 (93 Stat. 1201).

Abolished by act of Oct. 12, 1984 (98 Stat. 2091).

Kennedy, Commission To Report Upon the Assassination of President John F. Established by EO 11130 of Nov. 29, 1963. Report submitted Sept. 24, 1964, and *Commission* discharged by Presidential letter of same date.

Labor, President's Committee on Migratory Appointed by Presidential letter of Aug. 26, 1954. Formally established by EO 10894 of Nov. 15, 1960. Terminated Jan. 6, 1964, by the Secretary of Labor in letter to members, with approval of President.

Labor and Commerce, Department of Established by act of Feb. 14, 1903 (32 Stat. 825). Reorganized into separate Departments of Labor and Commerce by act of Mar. 4, 1913 (37 Stat. 736).

Labor Department, Solicitor for Transferred from the Department of Justice to the Department of Labor by EO 6166 of June 10, 1933.

Labor-Management Advisory Committee Established by EO 11695 of Jan. 11, 1973. Abolished by EO 11788 of June 18, 1974.

Labor-Management Policy, President's Advisory Committee on Established by EO 10918 of Feb. 16, 1961. Abolished by EO 11710 of Apr. 4, 1973.

Labor-Management Relations Services, Office of Established by Order 3–84 of May 3, 1984 of the Secretary of Labor. Renamed Bureau of Labor-Management Relations and Cooperative Programs by Secretarial Order 7–84 of Sept. 20, 1984 (49 FR 38374).

Labor-Management Services Administration *Office of Pension and Welfare Benefit Programs* transferred from *Administration* and constituted as separate unit by Order 1–84 of Jan. 20, 1984 of the Secretary of Labor (49 FR 4269). Remaining labor-management relations functions reassigned by Secretarial Order 3–84 of May 3, 1984.

Labor Organization, International Established in 1919 by Treaty of Versailles with U.S. joining in 1934. U.S. membership terminated Nov. 1, 1977, at President's direction. The U.S. rejoined the organization in February 1980.

Labor Relations Council, Federal Established by EO 11491 of Oct. 29, 1969. Abolished by Reorg. Plan No. 2 of 1978, effective Jan. 1, 1979, and functions transferred to Federal Labor Relations Authority.

Labor Standards, Apprenticeship Section, Division of Transferred to *Federal Security Agency* by EO 9139 of Apr. 18, 1942, functioning as *Apprentice Training Service.* Transferred to *War Manpower Commission* by EO 9247 of Sept. 17, 1942, functioning in *Bureau of Training.* Returned to the Department of Labor by EO 9617 of Sept. 19, 1945.

Labor Standards, Bureau of Established by Labor departmental order in 1934. Functions absorbed by

Occupational Safety and Health Administration in May 1971.

Land Bank Commissioner *See* **Farm Loan Board, Federal**

Land Law Review Commission, Public Established by act of Sept. 19, 1964 (78 Stat. 982). Terminated Dec. 31, 1970, pursuant to terms of act.

Land Office, General Consolidated with *Grazing Service* into Bureau of Land Management, Department of the Interior, by Reorg. Plan No. 3 of 1946, effective July 16, 1946.

Land Office, Office of Recorder of the General Created in the Department of the Interior by act of July 4, 1836 (5 Stat. 111). Abolished by Reorg. Plan No. III of 1940, effective June 30, 1940, and functions transferred to *General Land Office.*

Land Policy Section Established in 1934 as part of *Program Planning Division, Agricultural Adjustment Administration.* Personnel taken over by *Resettlement Administration* in 1935.

Land Problems, Committee on National Established by EO 6693 of Apr. 28, 1934. Abolished by EO 6777 of June 30, 1934.

Land Program, Director of Basis of program found in act of June 16, 1933 (48 Stat. 200). *Special Board of Public Works* established by EO 6174 of June 16, 1933. Land Program established by *Board* by resolution passed Dec. 28, 1933, and amended July 18, 1934. *Federal Emergency Relief Administration* designated to administer program Feb. 28, 1934. Land Program transferred to *Resettlement Administration* by EO 7028 of Apr. 30, 1935. Functions of *Administration* transferred to the Secretary of Agriculture by EO 7530 of Dec. 31, 1936. Land conservation and land-utilization programs administered by *Administration* transferred to *Bureau of Agricultural Economics* by Secretary's Memorandum 733. Administration of land programs placed under Soil Conservation Service by Secretary's Memorandum 785 of Oct. 6, 1938.

Land Use Coordination, Office of Established by Secretary of Agriculture Memorandum 725 of July 12, 1937. Abolished Jan. 1, 1944, by General Departmental Circular 21 and functions administered by *Land Use Coordinator.*

Land Use and Water Planning, Office of Established in the Department of the Interior by Secretarial Order No. 2953 of May 7, 1973. Abolished by Secretarial Order No. 2988 of Mar. 11, 1976.

Law Enforcement Assistance Administration Established by act of June 19, 1968 (82 Stat. 197). Operations closed out by the Department of Justice due to lack of appropriations and remaining functions transferred to *Office of Justice Assistance, Research, and Statistics.*

Law Enforcement Training Center, Federal *See* **Law Enforcement Training Center, Consolidated Federal**

Law Enforcement Training Center, Consolidated Federal Established by Treasury Order No. 217, Mar. 2, 1970. Renamed Federal Law Enforcement Training Center by Amendment No. 1 to Treasury Order No. 217 on Aug. 14, 1975. Transferred to Department of Homeland Security by act of Nov. 25, 2002 (116 Stat. 2178).

Legislative Affairs, Office of Renamed Office of Intergovernmental and Legislative Affairs Feb. 24, 1984, by Attorney General's Order 1054–84 (49 FR 10177).

Lend-Lease Administration, Office of Established by EO 8926 of Oct. 28, 1941, to replace *Division of Defense Aid Reports*. Consolidated with *Foreign Economic Administration* by EO 9380 of Sept. 25, 1943.

Lewis and Clark Trail Commission Established by act of Oct. 6, 1964 (78 Stat. 1005). Terminated October 1969 by terms of act.

Libraries and Information Science, National Commission on Established by act of July 20, 1970 (84 Stat. 440). As per close out activities, the Commission was abolished by act of Dec. 26, 2007 (121 Stat. 2204), and functions transferred to the Institute of Museum and Library Services pursuant to instructions set forth in House Report 110–231 and Senate Report 110–107.

Library of Congress Police Established by act of Aug. 4, 1950 (64 Stat. 411). Personnel transferred to United States Capitol Police by acts of Dec. 26, 2007 (121 Stat. 2228) and Jan. 7, 2008 (121 Stat. 2546).

Lighthouses, Bureau of Established in the Department of Commerce by act of Aug. 7, 1789 (1 Stat. 53). Consolidated with U.S. Coast Guard by Reorg. Plan No. II of 1939, effective July 1, 1939.

Lincoln Sesquicentennial Commission Established by joint resolution of Sept. 2, 1957 (71 Stat. 587). Terminated Mar. 1, 1960, pursuant to terms of joint resolution.

Liquidation, Director of Established in *Office for Emergency Management* by EO 9674 of Jan. 4, 1946. Terminated by EO 9744 of June 27, 1946.

Liquidation Advisory Committee Established by EO 9674 of Jan. 4, 1946. Terminated by EO 9744 of June 27, 1946.

Literacy, National Institute for Established by act of July 25, 1991 (105 Stat. 333). Abolished by act of Dec. 16, 2009 (123 Stat. 3267).

Loan Agency, Federal Established by Reorg. Plan No. I of 1939, effective July 1, 1939, by consolidating *Reconstruction Finance Corporation*—including subordinate units of *RFC Mortgage Company, Disaster Loan Corporation, Federal National Mortgage Association, Defense Plant Corporation, Defense Homes Corporation, Defense Supplies Corporation, Rubber Reserve Company, Metals Reserve Company,* and *War Insurance Corporation* (later known as *War Damage Corporation*)—

with *Federal Home Loan Bank Board, Home Owners' Loan Corporation, Federal Savings and Loan Insurance Corporation, Federal Housing Administration, Electric Home and Farm Authority,* and *Export-Import Bank of Washington. Federal Home Loan Bank Board, Federal Savings and Loan Insurance Corporation, Home Owners' Loan Corporation, Federal Housing Administration,* and *Defense Homes Corporation* transferred to *National Housing Agency* by EO 9070 of Feb. 24, 1942. *Reconstruction Finance Corporation* and its units (except *Defense Homes Corporation*), *Electric Home and Farm Authority,* and *Export-Import Bank of Washington* transferred to the Department of Commerce by EO 9071 of Feb. 24, 1942. *RFC* and units returned to *Federal Loan Agency* by act of Feb. 24, 1945 (59 Stat. 5). *Agency* abolished by act of June 30, 1947 (61 Stat. 202), and all property and records transferred to *Reconstruction Finance Corporation.*

Loan Fund, Development Established in *International Cooperation Administration* by act of Aug. 14, 1957 (71 Stat. 355). Created as independent corporate agency by act of June 30, 1958 (72 Stat. 261). Abolished by act of Sept. 4, 1961 (75 Stat. 445), and functions redelegated to Agency for International Development.

Loan Policy Board Established by act of July 18, 1958 (72 Stat. 385). Abolished by Reorg. Plan No. 4 of 1965, effective July 27, 1965, and functions transferred to Small Business Administration.

Longshoremen's Labor Board, National Established in the Department of Labor by EO 6748 of June 26, 1934. Terminated by Proc. 2120 of Mar. 11, 1935.

Low-Emission Vehicle Certification Board Established by act of Dec. 31, 1970 (84 Stat. 1701). Terminated by act of Mar. 14, 1980 (94 Stat. 98).

Lowell Historic Canal District Commission Established by act of Jan. 4, 1975 (88 Stat. 2330). Expired January 1977 pursuant to terms of act.

Loyalty Review Board Established Nov. 10, 1947, by *U.S. Civil Service Commission,* pursuant to EO 9835 of Mar. 21, 1947. Abolished by EO 10450 of Apr. 27, 1953.

Management, Budget and Evaluation, Office of Established within the Department of Energy pursuant to the Conference Report No. 107–258 on H.R. 2311, Energy and Water Development Appropriations Act, 2002. Abolished by memorandum of July 28, 2005 of the Secretary of Energy, and various functions transferred within the Department of Energy to the *Office of Management, Office of Chief Financial Officer,* and *Office of Human Capital Management.*

Management Improvement, Advisory Committee on Established by EO 10072 of July 29, 1949. Abolished by EO 10917 of Feb. 10, 1961, and functions transferred to *Bureau of the Budget.*

Management Improvement, President's Advisory Council on   Established by EO 11509 of Feb. 11, 1970. Inactive as of June 30, 1973.

Manpower, President's Committee on   Established by EO 11152 of Apr. 15, 1964. Terminated by EO 11515 of Mar. 13, 1970.

Manpower Administration   Renamed Employment and Training Administration by Order 14–75 of Nov. 12, 1975 of the Secretary of Labor.

Manpower Management, Office of Civilian   Renamed Office of Civilian Personnel by Notice 5430 of Oct. 1, 1976 of the Secretary of the Navy.

Marine Affairs, Office of   Established by the Secretary of the Interior Apr. 30, 1970, to replace *Office of Marine Resources,* created by Secretary Oct. 22, 1968. Abolished by Secretary Dec. 4, 1970.

Marine Corps Memorial Commission, U.S.   Established by act of Aug. 24, 1947 (61 Stat. 724). Terminated by act of Mar. 14, 1980 (94 Stat. 98).

Marine Debris Coordinating Committee   Renamed Interagency Marine Debris Coordinating Committee by act of Dec. 22, 2006 (120 Stat. 3337).

Marine Inspection and Navigation, Bureau of   *See* **Navigation and Steamboat Inspection, Bureau of**

Marine Resources and Engineering Development, National Council on   Established in Executive Office of the President by act of June 17, 1966 (80 Stat. 203). Terminated Apr. 30, 1971, due to lack of funding.

Maritime Administration   Established in the Department of Commerce by Reorg. Plan No. 21 of 1950, effective May 24, 1950. Transferred to the Department of Transportation by act of Aug. 6, 1981 (95 Stat. 151).

Maritime Advisory Committee   Established by EO 11156 of June 17, 1964. Terminated by EO 11427 of Sept. 4, 1968.

Maritime Board, Federal   *See* **Maritime Commission, U.S.**

Maritime Commission, U.S.   Established by act of June 29, 1936 (49 Stat. 1985), as successor agency to *U.S. Shipping Board* and *U.S. Shipping Board Merchant Fleet Corporation.* Training functions transferred to Commandant of Coast Guard by EO 9083 of Feb. 27, 1942. Functions further transferred to *War Shipping Administration* by EO 9198 of July 11, 1942. Abolished by Reorg. Plan No. 21 of 1950, effective May 24, 1950, which established *Federal Maritime Board* and *Maritime Administration* as successor agencies. *Board* abolished, regulatory functions transferred to Federal Maritime Commission, and functions relating to subsidization of merchant marine transferred to the Secretary of Commerce by Reorg. Plan No. 7 of 1961, effective Aug. 12, 1961.

Maritime Labor Board   Authorized by act of June 23, 1938 (52 Stat. 968). Mediatory duties abolished by act of June 23, 1941 (55 Stat. 259); title expired June 22, 1942.

Marketing Administration, Surplus   Established by Reorg. Plan No. III of 1940, effective June 30, 1940, consolidating functions vested in *Federal Surplus Commodities Corporation* and *Division of Marketing and Marketing Agreements, Agricultural Adjustment Administration.* Consolidated with other agencies into *Agricultural Marketing Administration* by EO 9069 of Feb. 23, 1942.

Marketing and Marketing Agreements, Division of   Established in the Department of Agriculture by act of June 3, 1937 (50 Stat. 246). Consolidated with *Federal Surplus Commodities Corporation* into *Surplus Marketing Administration* by Reorg. Plan No. III of 1940, effective June 30, 1940.

Mediation, U.S. Board of   Established by act of May 20, 1926 (44 Stat. 577). Abolished by act of June 21, 1934 (48 Stat. 1193), and superseded by National Mediation Board, July 21, 1934.

Medical Information Systems Program Office, Tri-Service   Renamed Defense Medical Systems Support Center by memorandum of the Assistant Secretary of Defense (Health Affairs) May 3, 1985.

Medical Services Administration   Established by the *Secretary of Health, Education, and Welfare* reorganization of Aug. 15, 1967. Transferred from *Social and Rehabilitation Service* to Health Care Financing Administration by Secretary's reorganization of Mar. 8, 1977 (42 FR 13262).

Medicine and Surgery, Department of   Established in the *Veterans Administration* by act of Sept. 2, 1958 (72 Stat. 1243). Renamed *Veterans Health Services and Research Administration* in the the Department of Veterans Affairs by act of Oct. 25, 1988 (102 Stat. 2640). Renamed Veterans Health Administration by act of May 7, 1991 (105 Stat. 187).

Memorial Commission, National   Established by Public Resolution 107 of Mar. 4, 1929 (45 Stat. 1699). Terminated by EO 6166 of June 10, 1933, and functions transferred to *Office of National Parks, Buildings, and Reservations,* Department of the Interior.

Mental Health, National Institute of   Established by act of July 3, 1946 (60 Stat. 425). Made entity within the Alcohol, Drug Abuse, and Mental Health Administration by act of May 14, 1974 (88 Stat. 135). Functions transferred to the Department of Health and Human Services by act of Oct. 17, 1979 (93 Stat. 695). (*See also* act of Oct. 27, 1986; 100 Stat. 3207–106.) Abolished by act of July 10, 1992 (106 Stat. 331). Reestablished by act of July 10, 1992 (106 Stat. 364).

Metals Reserve Company   Established June 28, 1940, by act of Jan. 22, 1932 (47 Stat. 5). Transferred from *Federal Loan Agency* to the Department of Commerce by EO 9071 of Feb. 24, 1942. Returned to *Federal Loan Agency* by act of Feb. 24, 1945 (59 Stat.

5). Dissolved by act of June 30, 1945 (59 Stat. 310), and functions transferred to *Reconstruction Finance Corporation.*

Metric Board, U.S. Established by act of Dec. 23, 1975 (89 Stat. 1007). Terminated Oct. 1, 1982, due to lack of funding.

Mexican-American Affairs, Interagency Committee on Established by Presidential memorandum of June 9, 1967. Renamed *Cabinet Committee on Opportunities for Spanish-Speaking People* by act of Dec. 30, 1969 (83 Stat. 838). Terminated Dec. 30, 1974, pursuant to terms of act.

Mexican Claims Commission, American Established by act of Dec. 18, 1942 (56 Stat. 1058). Terminated Apr. 4, 1947, by act of Apr. 3, 1945 (59 Stat. 59).

Mexican Claims Commission, Special Established by act of Apr. 10, 1935 (49 Stat. 149). Terminated by EO 7909 of June 15, 1938.

Mexico Commission for Border Development and Friendship, U.S. Established through exchange of notes of Nov. 30 and Dec. 3, 1966, between U.S. and Mexico. Terminated Nov. 5, 1969.

Micronesian Claims Commission Established by act of July 1, 1971 (85 Stat. 92). Terminated Aug. 3, 1976, pursuant to terms of act.

Migration, Intergovernmental Committee for European Renamed Intergovernmental Committee for Migration by Resolution 624, passed by *Intergovernmental Committee for European Migration Council,* effective Nov. 11, 1980.

Migration, International Committee for Created in 1951. Renamed International Organization for Migration pursuant to article 29, paragraph 2, of the ICM constitution, effective Nov. 14, 1989.

Migratory Bird Conservation Commission Chairmanship transferred from the Secretary of Agriculture to the Secretary of the Interior by Reorg. Plan No. II of 1939, effective July 1, 1939.

Military Air Transport Service Renamed *Military Airlift Command* in U.S. Air Force by HQ MATS/MAC Special Order G–164 of Jan. 1, 1966.

Military Airlift Command Inactivated June 1, 1992.

Military Appeals, United States Court of Established under Article I of the Constitution of the United States pursuant to act of May 5, 1950, as amended. Renamed United States Court of Appeals for the Armed Forces by act of Oct. 5, 1994 (108 Stat. 2831).

Military Establishment, National Established as executive department of the Government by act of July 26, 1947 (61 Stat. 495). Designated Department of Defense by act of Aug. 10, 1949 (63 Stat. 579).

Military Intelligence College, Joint Established by DoD Directive 3305.1 of January 14, 1998. Renamed *National Defense Intelligence College by*

DOD Instruction 3305.01 of Dec. 22, 2006. See also Defense Intelligence College.

Military Purchases, Interdepartmental Committee for Coordination of Foreign and Domestic Informal liaison committee created on Presidential notification of Dec. 6, 1939, to the Secretaries of the Treasury and *War* and the Acting Secretary of the Navy. Committee dissolved in accordance with Presidential letter to the Secretary of the Treasury Apr. 14, 1941, following approval of act of Mar. 11, 1941 (55 Stat. 31).

Military Renegotiation Policy and Review Board Established by directive of the Secretary of Defense July 19, 1948. Abolished by Secretary's letter of Jan. 18, 1952, which transferred functions to *Renegotiation Board.*

Military Sea Transportation Service Renamed Military Sealift Command in U.S. Navy by COMSC notice of Aug. 1, 1970.

Militia Bureau Established in 1908 as *Division of Militia Affairs, Office of the Secretary of War.* Superseded in 1933 by National Guard Bureau.

Mine Health and Safety Academy, National Transferred from the Department of the Interior to the Department of Labor by act of July 25, 1979 (93 Stat. 111).

Minerals Exploration, Office of Established by act of Aug. 21, 1958 (72 Stat. 700). Functions transferred to *Geological Survey* by Order 2886 of Feb. 26, 1965 of the Secretary of the Interior.

Minerals Management Service Established on Jan. 19, 1982 by Secretarial order. Renamed as the Bureau of Ocean Energy Management, Regulation and Enforcement on June 18, 2010 by Secretarial order 3302.

Minerals Mobilization, Office of Established by the Secretary of the Interior pursuant to act of Sept. 8, 1950 (64 Stat. 798) and EO 10574 of Nov. 5, 1954, and by order of *Office of Defense Mobilization.* Succeeded by *Office of Minerals and Solid Fuels* Nov. 2, 1962. *Office of Minerals Policy Development* combined with *Office of Research and Development* in the Department of the Interior May 21, 1976, under authority of Reorg. Plan No. 3 of 1950, to form *Office of Minerals Policy and Research Analysis.* Abolished Sept. 30, 1981, by Secretarial Order 3070 and functions transferred to Bureau of Mines.

Minerals Policy and Research Analysis, Office of *See* **Minerals Mobilization, Office of**

Minerals and Solid Fuels, Office of Established by the Secretary of the Interior Oct. 26, 1962. Abolished and functions assigned to Deputy Assistant Secretary—Minerals and Energy Policy, Office of the Assistant Secretary—Mineral Resources, effective Oct. 22, 1971.

Mines, Bureau of Established in the Department of the Interior by act of May 16, 1910 (36 Stat. 369). Transferred to the Department of Commerce by EO 4239 of June 4, 1925. Transferred to the Department of the Interior by EO 6611 of Feb. 22, 1934. Renamed United States Bureau of Mines by act of May 18, 1992 (106 Stat. 172). Terminated pursuant to act of Jan. 26, 1996 (110 Stat. 32). Certain functions transferred to Secretary of Energy by act of Apr. 26, 1996 (110 Stat. 1321–167).

Mining Enforcement and Safety Administration   Established by Order 2953 of May 7, 1973 of the Secretary of the Interior. Terminated by departmental directive Mar. 9, 1978, and functions transferred to Mine Safety and Health Administration, Department of Labor, established by act of Nov. 9, 1977 (91 Stat. 1319).

Minority Business Enterprise, Office of   Renamed Minority Business Development Agency by Commerce Secretarial Order DOO–254A of Nov. 1, 1979.

Mint, Bureau of the   Renamed U.S. Mint by Treasury Secretarial order of Jan. 9, 1984 (49 FR 5020).

Missile Sites Labor Commission   Established by EO 10946 of May 26, 1961. Abolished by EO 11374 of Oct. 11, 1967, and functions transferred to Federal Mediation and Conciliation Service.

Missouri Basin Survey Commission   Established by EO 10318 of Jan. 3, 1952. Final report of *Commission* submitted to President Jan. 12, 1953, pursuant to EO 10329 of Feb. 25, 1952.

Missouri River Basin Commission   Established by EO 11658 of Mar. 22, 1972. Terminated by EO 12319 of Sept. 9, 1981.

Mobilization, Office of Civil and Defense   See **Mobilization, Office of Defense and Civilian**

Mobilization, Office of Defense and Civilian   Established by Reorg. Plan No. 1 of 1958, effective July 1, 1958. Redesignated as *Office of Civil and Defense Mobilization* by act of Aug. 26, 1958 (72 Stat. 861), consolidating functions of *Office of Defense Mobilization* and *Federal Civil Defense Administration.* Civil defense functions transferred to the Secretary of Defense by EO 10952 of July 20, 1961, and remaining organization redesignated *Office of Emergency Planning* by act of Sept. 22, 1961 (75 Stat. 630).

Mobilization Policy, National Advisory Board on   Established by EO 10224 of Mar. 15, 1951. EO 10224 revoked by EO 10773 of July 1, 1958.

Monetary and Financial Problems, National Advisory Council on International   Established by act of July 31, 1945 (59 Stat. 512). Abolished by Reorg. Plan No. 4 of 1965, effective July 27, 1965, and functions transferred to President. Functions assumed by National Advisory Council on International Monetary and Financial Policies, established by EO 11269 of Feb. 14, 1966.

Monument Commission, National   Established by act of Aug. 31, 1954 (68 Stat. 1029). Final report submitted in 1957, and audit of business completed September 1964.

Monuments in War Areas, American Commission for the Protection and Salvage of Artistic and Historic   Established by President June 23, 1943; announced by Secretary of State Aug. 20, 1943. Activities assumed by the Department of State Aug. 16, 1946.

Morris K. Udall Foundation   Established by act of Mar. 19, 1992 (106 Stat 79). Renamed Morris K. Udall and Stewart L. Udall Foundation by act of Nov. 3, 2009 (123 Stat. 2977).

Mortgage Association, Federal National   Chartered Feb. 10, 1938, by act of June 27, 1934 (48 Stat. 1246). Grouped with other agencies to form *Federal Loan Agency* by Reorg. Plan No. I of 1939, effective July 1, 1939. Transferred to the Department of Commerce by EO 9071 of Feb. 24, 1942. Returned to *Federal Loan Agency* by act of Feb. 24, 1945 (59 Stat. 5). Transferred to *Housing and Home Finance Agency* by Reorg. Plan No. 22 of 1950, effective July 10, 1950. Rechartered by act of Aug. 2, 1954 (68 Stat. 590) and made constituent agency of *Housing and Home Finance Agency.* Transferred with functions of *Housing and Home Finance Agency* to the Department of Housing and Urban Development by act of Sept. 9, 1965 (79 Stat. 667). Made Government-sponsored, private corporation by act of Aug. 1, 1968 (82 Stat. 536).

Motor Carrier Claims Commission   Established by act of July 2, 1948 (62 Stat. 1222). Terminated Dec. 31, 1952, by acts of July 11, 1951 (65 Stat. 116), and Mar. 14, 1952 (66 Stat. 25).

Mount Rushmore National Memorial Commission   Established by act of Feb. 25, 1929 (45 Stat. 1300). Expenditures ordered administered by the Department of the Interior by EO 6166 of June 10, 1933. Transferred to National Park Service, Department of the Interior, by Reorg. Plan No. II of 1939, effective July 1, 1939.

Mounted Horse Unit   Transferred from the United States Capitol Police to the United States Park Police by Public Law 109–55 of Aug. 2, 2005 (119 Stat. 572).

Munitions Board   Established in the Department of Defense by act of July 26, 1947 (61 Stat. 499). Abolished by Reorg. Plan No. 6 of 1953, effective June 30, 1953, and functions vested in the Secretary of Defense.

Munitions Board, Joint Army and Navy   Organized in 1922. Placed under direction of President by military order of July 5, 1939. Reconstituted Aug. 18, 1945, by order approved by President. Terminated on establishment of *Munitions Board* by act of July 26, 1947 (61 Stat. 505).

Museum of American Art, National   Renamed Smithsonian American Art Museum by Act of October 27, 2000 (114 Stat. 1463).

Museum of History and Technology, National   Renamed National Museum of American History in Smithsonian Institution by act of Oct. 13, 1980 (94 Stat. 1884).

Museum Services, Institute of Established by act of June 23, 1972 (86 Stat. 327). Transferred to Office of Educational Research and Improvement, Department of Education, by act of Oct. 17, 1979 (93 Stat. 678), effective May 4, 1980. Transferred to National Foundation on the Arts and the Humanities by act of Dec. 23, 1981 (95 Stat. 1414). Functions transferred to the Institute of Museum and Library Services by act of Sept. 30, 1996 (110 Stat. 3009–307).

Narcotics, Bureau of Established in the Department of the Treasury by act of June 14, 1930 (46 Stat. 585). Abolished by Reorg. Plan No. 1 of 1968, effective Apr. 8, 1968, and functions transferred to *Bureau of Narcotics and Dangerous Drugs,* Department of Justice.

Narcotics, President's Council on Counter- Renamed President's Drug Policy Council by EO 13023, Nov. 6, 1996 (61 FR 57767).

Narcotics Control, Cabinet Committee on International Established by Presidential memorandum of Aug. 17, 1971. Terminated by Presidential memorandum of Mar. 14, 1977.

National. *See other part of title*

Naval Material, Office of Established by act of Mar. 5, 1948 (62 Stat. 68). Abolished by the Department of Defense reorg. order of Mar. 9, 1966, and functions transferred to the Secretary of the Navy (31 FR 7188).

Naval Material Command *See* **Naval Material Support Establishment**

Naval Material Support Establishment Established by Department of the Navy General Order 5 of July 1, 1963 (28 FR 7037). Replaced by *Naval Material Command* pursuant to General Order 5 of Apr. 29, 1966 (31 FR 7188). Functions realigned to form Office of Naval Acquisition Support, and termination of *Command* effective May 6, 1985.

Naval Observatory Jurisdiction transferred from *Bureau of Navigation* to Chief of Naval Operations by EO 9126 of Apr. 8, 1942, and by Reorg. Plan No. 3 of 1946, effective July 16, 1946.

Naval Oceanography Command Renamed Naval Meteorology and Oceanography Command in 1995.

Naval Petroleum and Oil Shale Reserves, Office of Established by the Secretary of the Navy, as required by law (70A Stat. 457). Jurisdiction transferred to the Department of Energy by act of Aug. 4, 1977 (91 Stat. 581).

Naval Reserve Established by act of Mar. 3, 1915 (38 Stat. 940). Redesignated Navy Reserve by Public Law 109–163 of Jan. 6, 2006 (119 Stat. 3233).

Naval Weapons, Bureau of Established by act of Aug. 18, 1959 (73 Stat. 395), to replace *Bureau of Ordnance and Aeronautics.* Abolished by Department of Defense reorg. order of Mar. 9, 1966, and functions transferred to the Secretary of the Navy (31 FR 7188), effective May 1, 1966.

Navigation, Bureau of Created by act of July 5, 1884 (23 Stat. 118), as special service under the Department of the Treasury. Transferred to the *Department of Commerce and Labor* by act of Feb. 4, 1903 (32 Stat. 825). Consolidated with *Bureau of Navigation and Steamboat Inspection* by act of June 30, 1932 (47 Stat. 415).

Navigation, Bureau of Renamed Bureau of Naval Personnel by act of May 13, 1942 (56 Stat. 276).

Navigation and Steamboat Inspection, Bureau of Renamed *Bureau of Marine Inspection and Navigation* by act of May 27, 1936 (49 Stat. 1380). Functions transferred to *Bureau of Customs,* Department of the Treasury, and U.S. Coast Guard by EO 9083 of Feb. 28, 1942. Transfer made permanent and *Bureau* abolished by Reorg. Plan. No. 3 of 1946, effective July 16, 1946.

Navy, Department of Defense housing functions transferred to *Federal Public Housing Authority, National Housing Agency,* by EO 9070 of Feb. 24, 1942.

Navy Bureau of Medicine and Surgery, Dental Division of Renamed Dental Corps of the Navy Bureau of Medicine and Surgery by act of Oct. 17, 2006 (120 Stat. 2234).

Navy Commissioners, Board of Established by act of Feb. 7, 1815 (3 Stat. 202). Abolished by act of Aug. 31, 1842 (5 Stat. 579).

Neighborhoods, National Commission on Established by act of Apr. 30, 1977 (91 Stat. 56). Terminated May 4, 1979, pursuant to terms of act.

Neighborhoods, Voluntary Associations and Consumer Protection, Office of Abolished and certain functions transferred to Office of the Assistant Secretary for Housing—Federal Housing Commissioner and Office of the Assistant Secretary for Community Planning and Development. Primary enabling legislation, act of Oct. 31, 1978 (92 Stat. 2119), repealed by act of Aug. 13, 1981 (95 Stat. 398). Abolishment of *Office* and transfer of functions carried out by Housing and Urban Development Secretarial order.

New England River Basins Commission Established by EO 11371 of Sept. 6, 1967. Terminated by EO 12319 of Sept. 9, 1981.

Nicaro Project Responsibility for management of Nicaro nickel producing facilities in Oriente Province, Cuba, transferred from *Office of Special Assistant to the Administrator (Nicaro Project)* to *Defense Materials Service* by General Services Administrator, effective July 7, 1959. Facilities expropriated by Cuban Government and nationalized Oct. 26, 1960.

Noble Training Center Transferred from Public Health Service to the Center for Domestic Preparedness, Department of Homeland Security by act of Oct. 4, 2006 (120 Stat. 1433).

Northern Mariana Islands Commission on Federal Laws Created by joint resolution of Mar. 24, 1976 (90 Stat. 263). Terminated upon submission of final report in August 1985.

Nursing Research, National Center for Renamed National Institute of Nursing Research by act of June 10, 1993 (107 Stat. 178).

Nutrition Division Functions transferred from *Department of Health, Education, and Welfare* to the Department of Agriculture by EO 9310 of Mar. 3, 1943.

Ocean Mining Administration Established by Interior Secretarial Order 2971 of Feb. 24, 1975. Abolished by Department Manual Release 2273 of June 13, 1980.

Oceanography, Interagency Committee on Established by *Federal Council for Science and Technology* pursuant to EO 10807 of Mar. 13, 1959. Absorbed by *National Council on Marine Resources and Engineering Development* pursuant to Vice Presidential letter of July 21, 1967.

Office *See also other part of title*

Office Space, President's Advisory Commission on Presidential Established by act of Aug. 3, 1956 (70 Stat. 979). Terminated June 30, 1957, by act of Jan. 25, 1957 (71 Stat. 4).

Official Register Function of preparing *Official Register* vested in Director of the Census by act of Mar. 3, 1925 (43 Stat. 1105). Function transferred to *U.S. Civil Service Commission* by EO 6166 of June 10, 1933. Yearly compilation and publication required by act of Aug. 28, 1935 (49 Stat. 956). Act repealed by act of July 12, 1960 (74 Stat. 427), and last *Register* published in 1959.

Ohio River Basin Commission Established by EO 11578 of Jan. 13, 1971. Terminated by EO 12319 of Sept. 9, 1981.

Oil and Gas, Office of Established by the Secretary of the Interior May 6, 1946, in response to Presidential letter of May 3, 1946. Transferred to *Federal Energy Administration* by act of May 7, 1974 (88 Stat. 100).

Oil Import Administration Established in the Department of the Interior by Proc. 3279 of Mar. 10, 1959. Merged into *Office of Oil and Gas* Oct. 22, 1971.

Oil Import Appeals Board Established by the Secretary of Commerce Mar. 13, 1959, and made part of Office of Hearings and Appeals Dec. 23, 1971.

On-Site Inspection Agency Established on Jan. 26, 1988. Functions transferred to the Defense Threat Reduction Agency by DOD Directive 5105.62 of Sept. 30, 1998.

Operations Advisory Group Established by EO 11905 of Feb. 18, 1976. Abolished by Presidential Directive No. 2 of Jan. 20, 1977.

Operations Coordinating Board Established by EO 10483 of Sept. 2, 1953, which was superseded by EO 10700 of Feb. 25, 1957. EO 10700 revoked by EO 10920 of Feb. 18, 1961, and *Board* terminated.

Ordnance, Bureau of *See* **Ordnance and Hydrography, Bureau of**

Ordnance and Hydrography, Bureau of Established in the Department of the Navy by act of Aug. 31, 1842 (5 Stat. 579). Replaced under act of July 5, 1862 (12 Stat. 510), by *Bureau of Ordnance* and *Bureau of Navigation*. Abolished by act of Aug. 18, 1959 (73 Stat. 395), and functions transferred to *Bureau of Naval Weapons*.

Organization, President's Advisory Committee on Government Established by EO 10432 of Jan. 24, 1953. Abolished by EO 10917 of Feb. 10, 1961, and functions transferred to *Bureau of the Budget* for termination.

Organizations Staff, International Functions merged with Foreign Agricultural Service by memorandum of Dec. 7, 1973 of , effective Feb. 3, 1974.

Overseas Private Investment Corporation Transferred as separate agency to *U.S. International Development Cooperation Agency* by Reorg. Plan No. 2 of 1979, effective Oct. 1, 1979. Became an independent agency following the abolition of *IDCA* by act of Oct. 21, 1998 (112 Stat. 2681–790).

Oversight Board (for the Resolution Trust Corporation) Established by act of Aug. 9, 1989 (103 Stat. 363). Renamed *Thrift Depositor Protection Oversight Board* by act of Dec. 12, 1991 (105 Stat. 1767). Abolished by act of July 29, 1998 (112 Stat. 908). Authority and duties transferred to the Secretary of the Treasury.

Pacific Northwest River Basins Commission Established by EO 11331 of Mar. 6, 1967. Terminated by EO 12319 of Sept. 9, 1981.

Packers and Stockyards Administration Established by Memorandum 1613, supp. 1, of May 8, 1967 of the Secretary of Agriculture. Certain functions consolidated into Agricultural Marketing Service by Secretary's Memorandum 1927 of Jan. 15, 1978. Remaining functions incorporated into the Grain Inspection, Packers and Stockyards Administration by Secretary's Memorandum 1010–1 dated Oct. 20, 1994.

Panama Canal Operation of piers at Atlantic and Pacific terminals transferred to *Panama Railroad Company* by EO 7021 of Apr. 19, 1935. Panama Canal reestablished as *Canal Zone Government* by act of Sept. 26, 1950 (64 Stat. 1038).

Panama Canal Commission Established by act of Oct. 1, 1979, as amended (22 U.S.C. 3611). U.S.

responsibility terminated by stipulation of the Panama Canal Treaty of 1977, which transferred responsibility for the Panama Canal to the Republic of Panama, effective Dec. 31, 1999. Commission terminated by act of Sept. 30, 2004 (118 Stat. 1140).

Panama Canal Company Established by act of June 29, 1948 (62 Stat. 1076). Abolished and superseded by *Panama Canal Commission* (93 Stat. 454).

Panama Railroad Company Incorporated Apr. 7, 1849, by New York State Legislature. Operated under private control until 1881, when original *French Canal Company* acquired most of its stock. *Company* and its successor, *New Panama Canal Company,* operated railroad as common carrier and also as adjunct in attempts to construct canal. In 1904 their shares of stock in *Panama Railroad Company* passed to ownership of U.S. as part of assets of *New Panama Canal Company* purchased under act of June 28, 1902 (34 Stat. 481). Remaining shares purchased from private owners in 1905. *Panama Railroad Company* reincorporated by act of June 29, 1948 (62 Stat. 1075) pursuant to requirements of act of Dec. 6, 1945 (59 Stat. 597). Reestablished as *Panama Canal Company* by act of Sept. 26, 1950 (64 Stat. 1038). The Secretary of the Army was directed to discontinue commercial operations of *Company* by Presidential letter of Mar. 29, 1961.

Paperwork, Commission on Federal Established by act of Dec. 27, 1974 (88 Stat. 1789). Terminated January 1978 pursuant to terms of act.

Park Service, National Functions in District of Columbia relating to space assignment, site selection for public buildings, and determination of priority in construction transferred to *Public Buildings Administration, Federal Works Agency,* under Reorg. Plan No. I of 1939, effective July 1, 1939.

Park Trust Fund Board, National Established by act of July 10, 1935 (49 Stat. 477). Terminated by act of Dec. 18, 1967 (81 Stat. 656), and functions transferred to National Park Foundation.

Parks, Buildings, and Reservations, Office of National Established in the Department of the Interior by EO 6166 of June 10, 1933. Renamed National Park Service by act of Mar. 2, 1934 (48 Stat. 362).

Parole, Board of Established by act of June 25, 1948 (62 Stat. 854). Abolished by act of Mar. 15, 1976 (90 Stat. 219), and functions transferred to U.S. Parole Commission.

Patent Office Provisions of first patent act administered by the Department of State, with authority for granting patents vested in board comprising Secretaries of State and *War* and Attorney General. Board abolished, authority transferred to Secretary of State, and registration system established by act of Feb. 21, 1793 (1 Stat. 318). *Office* made bureau in the Department of State in October 1802, headed by *Superintendent of Patents. Office* reorganized in 1836 by act of June 4, 1836 (5 Stat. 117) under *Commissioner of Patents. Office* transferred to the Department of the Interior in 1849.

Office transferred to the Department of Commerce by EO 4175 of Mar. 17, 1925.

Patents Board, Government Established by EO 10096 of Jan. 23, 1950. Abolished by EO 10930 of Mar. 24, 1961, and functions transferred to the Secretary of Commerce.

Pay Board Established by EO 11627 of Oct. 15, 1971. Abolished by EO 11695 of Jan. 11, 1973.

Payment Limitations, Commission on Application of Established by act of May 13, 2002 (116 Stat. 216). Abolished by acts of May 22, 2008 (122 Stat. 1025) and June 18, 2008 (122 Stat. 1753).

Peace Corps Established in the Department of State by EO 10924 of Mar. 1, 1961, and continued by act of Sept. 22, 1961 (75 Stat. 612), and EO 11041 of Aug. 6, 1962. Functions transferred to ACTION by Reorg. Plan No. 1 of 1971, effective July 1, 1971. Made independent agency in executive branch by act of Dec. 29, 1981 (95 Stat. 1540).

Pennsylvania Avenue, Temporary Commission on Established by EO 11210 of Mar. 25, 1956. Inactive as of Nov. 15, 1969, due to lack of funding.

Pennsylvania Avenue Development Corporation Established by act of Oct. 27, 1972 (86 Stat. 1266). Terminated pursuant to act of Jan. 26, 1996 (110 Stat. 32) and act of Apr. 26, 1996 (110 Stat. 1321–198). Functions transferred to General Services Administration, National Capital Planning Commission, and National Park Service (61 FR 11308), effective Apr. 1, 1996.

Pension and Welfare Benefit Programs, Office of *See* **Labor-Management Services Administration**

Pensions, Commissioner of Provided for by act of Mar. 2, 1833 (4 Stat. 668). Continued by act of Mar. 3, 1835 (4 Stat. 779), and other acts as *Office of the Commissioner of Pensions.* Transferred to the Department of the Interior as bureau by act of Mar. 3, 1849 (9 Stat. 395). Consolidated with other bureaus and agencies into *Veterans Administration* by EO 5398 of July 21, 1930.

Pensions, Office of the Commissioner of *See* **Pensions, Commissioner of**

Perry's Victory Memorial Commission Created by act of Mar. 3, 1919 (40 Stat. 1322). Administration of Memorial transferred to National Park Service by act of June 2, 1936 (49 Stat. 1393). *Commission* terminated by terms of act and membership reconstituted as advisory board to the Secretary of Interior.

Personal Property, Office of *See* **Supply Service, Federal**

Personnel, National Roster of Scientific and Specialized Established by *National Resources Planning Board* pursuant to Presidential letter of June 18, 1940, to the Secretary of the Treasury. After Aug. 15, 1940, administered jointly by *Board* and *U.S. Civil Service Commission.* Transferred to *War*

Manpower Commission by EO 9139 of Apr. 18, 1942. Transferred to the Department of Labor by EO 9617 of Sept. 19, 1945. Transferred with *Bureau of Employment Security* to *Federal Security Agency* by act of June 16, 1948 (62 Stat. 443). Transferred to the Department of Labor by Reorg. Plan No. 2 of 1949, effective Aug. 20, 1949, and became inactive. Roster functions transferred to National Science Foundation by act of May 10, 1950 (64 Stat. 154). Reactivated in 1950 as *National Scientific Register* by *Office of Education, Federal Security Agency,* through *National Security Resources Board* grant of funds, and continued by National Science Foundation funds until December 1952, when *Register* integrated into Foundation's National Register of Scientific and Technical Personnel project in Division of Scientific Personnel and Education.

Personnel Administration, Council of   Established by EO 7916 of June 24, 1938, effective Feb. 1, 1939. Made unit in *U.S. Civil Service Commission* by EO 8467 of July 1, 1940. Renamed *Federal Personnel Council* by EO 9830 of Feb. 24, 1947. Abolished by act of July 31, 1953 (67 Stat. 300), and personnel and records transferred to *Office of Executive Director, U.S. Civil Service Commission.*

Personnel Council, Federal   *See* **Personnel Administration, Council of**

Personnel Interchange, President's Commission on   Established by EO 11451 of Jan. 19, 1969. Continued by EO 12136 of May 15, 1979, and renamed *President's Commission on Executive Exchange.* Continued by EO 12493 of Dec. 5, 1984. Abolished by EO 12760 of May 2, 1991.

Personnel Management, Liaison Office for   Established by EO 8248 of Sept. 8, 1939. Abolished by EO 10452 of May 1, 1953, and functions transferred to *U.S. Civil Service Commission.*

Petroleum Administration for Defense   Established under act of Sept. 8, 1950 (64 Stat. 798) by Order 2591 of Oct. 3, 1950 of the Secretary of the Interior, pursuant to EO 10161 of Sept. 9, 1950. Continued by Secretary's Order 2614 of Jan. 25, 1951, pursuant to EO 10200 of Jan. 3, 1951, and PAD Delegation 1 of Jan. 24, 1951. Abolished by Secretary's Order 2755 of Apr. 23, 1954.

Petroleum Administration for War   *See* **Petroleum Coordinator for War, Office of**

Petroleum Administrative Board   Established Sept. 11, 1933, by the Secretary of the Interior. Terminated Mar. 31, 1936, by EO 7076 of June 15, 1935. The Secretary of the Interior was authorized to execute functions vested in President by act of Feb. 22, 1935 (49 Stat. 30) by EO 7756 of Dec. 1, 1937. Secretary also authorized to establish *Petroleum Conservation Division* to assist in administering act. Records of ● , *Petroleum Administrative Board* and *Petroleum Labor Policy Board* housed with *Petroleum Conservation Division, Office of Oil and Gas,* acting as custodian for the Secretary of the Interior.

Petroleum Coordinator for War, Office of   Secretary of the Interior designated *Petroleum Coordinator for National Defense* pursuant to Presidential letter of May 28, 1941, and approved *Petroleum Coordinator for War* pursuant to Presidential letter of Apr. 20, 1942. *Office* abolished by EO 9276 of Dec. 2, 1942, and functions transferred to *Petroleum Administration for War,* established by same EO. *Administration* terminated by EO 9718 of May 3, 1946.

Petroleum Labor Policy Board   Established by the Secretary of the Interior, as *Administrator of Code of Fair Competition for Petroleum Industry,* on recommendation of Planning and Coordination Committee Oct. 10, 1933. Reorganized by Secretary Dec. 19, 1933, and reorganization confirmed by order of Mar. 8, 1935. Terminated Mar. 31, 1936, when *Petroleum Administrative Board* abolished by EO 7076 of June 15, 1935.

Petroleum Reserves Corporation   Established June 30, 1943, by *Reconstruction Finance Corporation.* Transferred to *Office of Economic Warfare* by EO 9360 of July 15, 1943. *Office* consolidated into *Foreign Economic Administration* by EO 9380 of Sept. 25, 1943. Functions transferred to *Reconstruction Finance Corporation* by EO 9630 of Sept. 27, 1945. *RFC's* charter amended Nov. 9, 1945, to change name to *War Assets Corporation. Corporation* designated by *Surplus Property Administrator* as disposal agency for all types of property for which *Reconstruction Finance Corporation* formerly disposal agency. Domestic surplus property functions of *Corporation* transferred to *War Assets Administration* by EO 9689 of Jan. 31, 1946. *Reconstruction Finance Corporation Board of Directors* ordered by President to dissolve *War Assets Corporation* as soon after Mar. 25, 1946, as practicable.

Philippine Alien Property Administration   Established in *Office for Emergency Management* by EO 9789 of Oct. 14, 1946. Abolished by EO 10254 of June 15, 1951, and functions transferred to the Department of Justice.

Philippine War Damage Commission   Established by act of Apr. 30, 1946 (60 Stat. 128). Terminated Mar. 31, 1951, by act of Sept. 6, 1950 (64 Stat. 712).

Photographic Interpretation Center, National   Functions transferred to the National Imagery and Mapping Agency by act of Sept. 23, 1996 (110 Stat. 2677).

Physical Fitness, Committee on   Established in *Office of Federal Security Administrator* by EO 9338 of Apr. 29, 1943. Terminated June 30, 1945.

Physical Fitness, President's Council on   *See* **Youth Fitness, President's Council on**

Physician Payment Review Commission   Established by act of Apr. 7, 1986 (100 Stat. 190). Terminated by act of Aug. 5, 1997 (111 Stat. 354). Assets, staff, and continuing responsibility for reports transferred to the Medicare Payment Advisory Commission.

Planning Board, National   Established by *Administrator of Public Works* July 30, 1933. Terminated by EO 6777 of June 30, 1934.

Plant Industry, Bureau of   Established by act of Mar. 2, 1902 (31 Stat. 922). Soil fertility and soil microbiology work of *Bureau of Chemistry and Soils* transferred to *Bureau* by act of May 17, 1935. Soil chemistry and physics and soil survey work of *Bureau of Chemistry and Soils* transferred to *Bureau* by Secretary's Memorandum 784 of Oct. 6, 1938. In February 1943 engineering research of *Bureau of Agricultural Chemistry and Engineering* transferred to *Bureau of Plant Industry, Soils, and Agricultural Engineering* by Research Administration Memorandum 5 issued pursuant to EO 9069 of Feb. 23, 1942, and in conformity with Secretary's Memorandums 960 and 986. Functions transferred to Agricultural Research Service by Secretary's Memorandum 1320, supp. 4, of Nov. 2, 1953.

Plant Industry, Soils, and Agricultural Engineering, Bureau of   *See* **Plant Industry, Bureau of**

Plant Quarantine, Bureau of   *See* **Entomology and Plant Quarantine, Bureau of**

Policy Development, Office of   *See* **Domestic Council**

Post Office, Department of   *See* **Postal Service**

Postal Rate Commission   Renamed Postal Regulatory Commission by act of Dec. 20, 2006 (120 Stat. 3241).

Postal Savings System   Established by act of June 25, 1910 (36 Stat. 814). System closed by act of Mar. 28, 1966 (80 Stat. 92).

Postal Service   Created July 26, 1775, by Continental Congress. Temporarily established by Congress by act of Sept. 22, 1789 (1 Stat. 70), and continued by subsequent acts. *Department of Post Office* made executive department under act of June 8, 1872 (17 Stat. 283). Offices of First, Second, Third, and Fourth Assistant Postmasters General abolished and Deputy Postmaster General and four Assistant Postmasters General established by Reorg. Plan No. 3 of 1949, effective Aug. 20, 1949. Reorganized as U.S. Postal Service in executive branch by act of Aug. 12, 1970 (84 Stat. 719), effective July 1, 1971.

Power Commission, Federal   Established by act of June 10, 1920 (41 Stat. 1063). Terminated by act of Aug. 4, 1977 (91 Stat. 578), and functions transferred to the Department of Energy.

Preparedness, Office of   Renamed *Federal Preparedness Agency* by General Services Administrator's order of June 26, 1975.

Preparedness Agency, Federal   Functions transferred from General Services Administration to Federal Emergency Management Agency by EO 12148 of July 20, 1979.

Presidential.   *See other part of title*

President's.   *See other part of title*

Press Intelligence, Division of   Established in August 1933. Made division of *National Emergency Council* July 10, 1935. Continued in *Office of Government Reports* by Reorg. Plan No. II of 1939, effective July 1, 1939. Transferred to *Office of War Information* by EO 9182 of June 13, 1942, functioning in *Bureau of Special Services. Office* abolished by EO 9608 of Aug. 31, 1945, and *Bureau* transferred to *Bureau of the Budget.* Upon reestablishment of *Office of Government Reports,* by EO 9809 of Dec. 12, 1946, *Division of Press Intelligence* made unit of *Office.*

Price Administration, Office of   Established by EO 8734 of Apr. 11, 1941, combining *Price Division* and *Consumer Division* of *National Defense Advisory Commission.* Renamed *Office of Price Administration* by EO 8875 of Aug. 28, 1941, which transferred *Civilian Allocation Division* to *Office of Production Management.* Consolidated with other agencies into *Office of Temporary Controls* by EO 9809 of Dec. 12, 1946, except *Financial Reporting Division,* transferred to Federal Trade Commission.

Price Commission   Established by EO 11627 of Oct. 15, 1971. Abolished by EO 11695 of Jan. 11, 1973.

Price Decontrol Board   Established by act of July 25, 1946 (60 Stat. 669). Effective period of act of Jan. 30, 1942 (56 Stat. 23), extended to June 30, 1947, by joint resolution of June 25, 1946 (60 Stat. 664).

Price Stability for Economic Growth, Cabinet Committee on   Established by Presidential letter of Jan. 28, 1959. Abolished by Presidential direction Mar. 12, 1961.

Price Stabilization, Office of   Established by General Order 2 of *Economic Stabilization Administrator* Jan. 24, 1951. *Director of Price Stabilization* provided for in EO 10161 of Sept. 9, 1950. Terminated Apr. 30, 1953, by EO 10434 of Feb. 6, 1953, and provisions of acts of June 30, 1952 (66 Stat. 296) and June 30, 1953 (67 Stat. 131).

Prices and Costs, Committee on Government Activities Affecting   Established by EO 10802 of Jan. 23, 1959. Abolished by EO 10928 of Mar. 23, 1961.

Priorities Board   Established by order of *Council of National Defense,* approved Oct. 18, 1940, and by EO 8572 of Oct. 21, 1940. EO 8572 revoked by EO 8629 of Jan. 7, 1941.

Prison Industries, Inc., Federal   Established by EO 6917 of Dec. 11, 1934. Transferred to the Department of Justice by Reorg. Plan No. II of 1939, effective July 1, 1939.

Prison Industries Reorganization Administration   Functioned from Sept. 26, 1935, to Sept. 30, 1940, under authority of act of Apr. 8, 1935 (49 Stat. 115), and of EO's 7194 of Sept. 26, 1935, 7202 of Sept. 28, 1935, and 7649 of June 29, 1937. Terminated due to lack of funding.

Private Sector Programs, Office of   Functions transferred to the Office of Citizen Exchanges within the Bureau of Educational and Cultural Affairs, USIA, by act of Feb. 16, 1990 (104 Stat. 56).

Processing tax   *Agricultural Adjustment Administration's* function of collecting taxes declared unconstitutional by U.S. Supreme Court Jan. 6, 1936. Functions under acts of June 28, 1934 (48 Stat. 1275), Apr. 21, 1934 (48 Stat. 598), and Aug. 24, 1935 (49 Stat. 750) discontinued by repeal of these laws by act of Feb. 10, 1936 (49 Stat. 1106).

Processing Tax Board of Review   Established in the Department of the Treasury by act of June 22, (49 Stat. 1652). Abolished by act of Oct. 21, 1942 (56 Stat. 967).

Proclamations   *See* **State, Department of**

Procurement, Commission on Government   Established by act of Nov. 26, 1969 (83 Stat. 269). Terminated Apr. 30, 1973, due to expiration of statutory authority.

Procurement and Assignment Service   Established by President Oct. 30, 1941. Transferred from *Office of Defense Health and Welfare Services* to *War Manpower Commission* by EO 9139 of Apr. 18, 1942. Transferred to *Federal Security Agency* by EO 9617 of Sept. 19, 1945, which terminated *Commission.*

Procurement Division   Established in the Department of the Treasury by EO 6166 of June 10, 1933. Renamed *Bureau of Federal Supply* by Department of the Treasury Order 73 of Nov. 19, 1946, effective Jan. 1, 1947. Transferred to General Services Administration as Federal Supply Service by act of June 30, 1949 (63 Stat. 380).

Procurement Policy, Office of Federal   Established within Office of Management and Budget by act of Aug. 30, 1974 (88 Stat. 97). Abolished due to lack of funding and functions transferred to Office of Management and Budget by act of Oct 28, 1993 (107 Stat. 1236).

Product Standards Policy, Office of   Formerly separate operating unit under Assistant Secretary for Productivity, Technology, and Innovation, Department of Commerce. Transferred to *National Bureau of*

Production Areas, Committee for Congested   Established in Executive Office of the President by EO 9327 of Apr. 7, 1943. Terminated Dec. 31, 1944, by act of June 28, 1944 (58 Stat. 535).

Production Authority, National   Established in the Department of Commerce Sept. 11, 1950, by EO's 10161 of Sept. 9, 1950, 10193 of Dec. 16, 1950, and 10200 of Jan. 3, 1951. Abolished by order of Oct. 1, 1953 of the Secretary of Commerce, and functions merged into *Business and Defense Services Administration.*

Production Management, Office of   Established in *Office for Emergency Management* by EO 8629 of Jan. 7, 1941. Abolished by EO 9040 of Jan. 24, 1942, and personnel and property transferred to *War Production Board.*

Production and Marketing Administration   Established by Secretary of Agriculture Memorandum 1118 of Aug. 18, 1945. Functions transferred under Department reorganization by Secretary's Memorandum 1320, supp. 4, of Nov. 2, 1953.

Productivity Council, National   Established by EO 12089 of Oct. 23, 1978. EO 12089 revoked by EO 12379 of Aug. 17, 1982.

Programs, Bureau of International   Established by the Secretary of Commerce Aug. 8, 1961, by Departmental Orders 173 and 174. Abolished by Departmental Order 182 of Feb. 1, 1963, which established *Bureau of International Commerce.* Functions transferred to *Domestic and International Business Administration,* effective Nov. 17, 1972.

Programs, Office of Public   Established in the National Archives and Records Administration. Reorganized by Archivist under Notice 96–260, Sept. 23, 1996, effective Jan. 6, 1997. Functions restructured and transferred to Office of Records Services—Washington, DC.

Prohibition, Bureau of   Established by act of May 27, 1930 (46 Stat. 427). Investigative functions consolidated with functions of *Bureau of Investigation* into *Division of Investigation,* Department of Justice. by EO 6166 of June 10, 1933, which set as effective date Mar. 2, 1934, or such later date as fixed by President. All other functions performed by *Bureau of Prohibition* ordered transferred to such division in the Department of Justice as deemed desirable by Attorney General.

Property, Office of Surplus   Established in *Procurement Division,* Department of the Treasury, by EO 9425 of Feb. 19, 1944, and act of Oct. 3, 1944 (58 Stat. 765), under general direction of *Surplus Property Board* established by same legislation. Transferred to the Department of Commerce by EO 9541 of Apr. 19, 1945. Terminated by EO 9643 of Oct. 19, 1945, and activities and personnel transferred to *Reconstruction Finance Corporation.*

Property Administration, Surplus   *See* **War Property Administration, Surplus**

Property Board, Surplus   *See* **War Property Administration, Surplus**

Property Council, Federal   Established by EO 11724 of June 25, 1973, and reconstituted by EO 11954 of Jan. 7, 1977. Terminated by EO 12030 of Dec. 15, 1977.

Property Management and Disposal Service   *See* **Emergency Procurement Service**

Property Office, Surplus   Established in *Division of Territories and Island Possessions,* Department of the Interior, under Regulation 1 of *Surplus Property Board,* Apr. 2, 1945. Transferred to *War Assets Administration* by EO 9828 of Feb. 21, 1947.

Property Review Board   Established by EO 12348 of Feb. 25, 1982. EO 12348 revoked by EO 12512 of Apr. 29, 1985.

Protective Service, Federal Functions established in the *Federal Works* Agency by act of June 1, 1948 (62 Stat. 281). Functions transferred to General Services Administrator by act of June 30, 1949 (63 Stat. 380). Established as an agency within General Services Administration by GSA Administrator on Jan. 11, 1971 (ADM. 5440.46). Transferred to Homeland Security.Department by act of Nov. 25, 2002 (116 Stat. 2178).

Prospective Payment Assessment Commission Established by act of Apr. 20, 1983 (97 Stat. 159). Terminated by act of Aug. 5, 1997 (111 Stat. 354). Assets, staff, and continuing responsibility for reports transferred to the Medicare Payment Advisory Commission.

Provisions and Clothing, Bureau of Established by acts of Aug. 31, 1842 (5 Stat. 579), and July 5, 1862 (12 Stat. 510). Designated *Bureau of Supplies and Accounts* by act of July 19, 1892 (27 Stat. 243). Abolished by Department of Defense reorg. order of Mar. 9, 1966, and functions transferred to the Secretary of the Navy (31 FR 7188).

Public *See other part of title*

Publications Commission, National Historical Established by act of Oct. 22, 1968 (82 Stat. 1293). Renamed National Historical Publications and Records Commission by act of Dec. 22, 1974 (88 Stat. 1734).

Puerto Rican Hurricane Relief Commission Established by act of Dec. 21, 1928 (45 Stat. 1067). No loans made after June 30, 1934, and *Commission* abolished June 3, 1935, by Public Resolution 22 (49 Stat. 320). Functions transferred to *Division of Territories and Island Possessions,* Department of the Interior. After June 30, 1946, collection work performed in *Puerto Rico Reconstruction Administration.* Following termination of *Administration,* remaining collection functions transferred to the Secretary of Agriculture by act of July 11, 1956 (70 Stat. 525).

Puerto Rico, U.S.-Puerto Rico Commission on the Status of Established by act of Feb. 20, 1964 (78 Stat. 17). Terminated by terms of act.

Puerto Rico Reconstruction Administration Established in the Department of the Interior by EO 7057 of May 28, 1935. Terminated Feb. 15, 1955, by act of Aug. 15, 1953 (67 Stat. 584).

Radiation Biology Laboratory *See* **Radiation and Organisms, Division of**

Radiation Council, Federal Established by EO 10831 of Aug. 14, 1959, and act of Sept. 23, 1959 (73 Stat. 688). Abolished by Reorg. Plan No. 3 of 1970, effective Dec. 2, 1970, and functions transferred to Environmental Protection Agency.

Radiation and Organisms, Division of Established by Secretarial order of May 1, 1929, as part of Smithsonian Astrophysical Observatory. Renamed *Radiation Biology Laboratory* by Secretarial order of Feb. 16, 1965. Merged with *Chesapeake Center for*

Environmental Studies by Secretarial order of July 1, 1983, to form Smithsonian Environmental Research Center.

Radio Commission, Federal Established by act of Feb. 23, 1927 (44 Stat. 1162). Abolished by act of June 19, 1934 (48 Stat. 1102), and functions transferred to Federal Communications Commission.

Radio Division Established by *National Emergency Council* July 1, 1938. Transferred to *Office of Education, Federal Security Agency,* by Reorg. Plan No. II of 1939, effective July 1, 1939. Terminated June 30, 1940, by terms of act of June 30, 1939 (53 Stat. 927).

Radio Propagation Laboratory, Central Transferred from *National Bureau of Standards* to *Environmental Science Services Administration* by the Department of Commerce Order 2–A, effective July 13, 1965.

Radiological Health, National Center for Devices and Renamed Center for Devices and Radiological Health by Food and Drug Administration notice of Mar. 9, 1984 (49 FR 10166).

Rail Public Counsel, Office of Established by act of Feb. 5, 1976 (90 Stat. 51). Terminated Dec. 1, 1979, due to lack of funding.

Railroad Administration, U.S. *See* **Railroads, Director General of**

Railroad and Airline Wage Board Established by *Economic Stabilization Administrator's* General Order 7 of Sept. 27, 1951, pursuant to act of Sept. 8, 1950 (64 Stat. 816). Terminated Apr. 30, 1953, by EO 10434 of Feb. 6, 1953, and acts of June 30, 1952 (66 Stat. 296), and June 30, 1953 (67 Stat. 131).

Railroads, Director General of Established under authority of act of Aug. 29, 1916 (39 Stat. 645). Organization of *U.S. Railroad Administration* announced Feb. 9, 1918. Office abolished by Reorg. Plan No. II of 1939, effective July 1, 1939, and functions transferred to the Secretary of the Treasury.

Railway Association, U.S. Established by act of Jan. 2, 1974 (87 Stat. 985). Terminated Apr. 1, 1987, by act of Oct. 21, 1986 (100 Stat. 1906).

Railway Labor Panel, National Established by EO 9172 of May 22, 1942. EO 9172 revoked by EO 9883 of Aug. 11, 1947.

Real Estate Board, Federal Established by EO 8034 of Jan. 14, 1939. Abolished by EO 10287 of Sept. 6, 1951.

Reclamation, Bureau of *See* **Reclamation Service**

Reclamation Service Established July 1902 in *Geological Survey* by the Secretary of the Interior, pursuant to act of June 17, 1902 (32 Stat. 388). Separated from Survey in 1907 and renamed *Bureau of Reclamation* June 1923. Power marketing functions transferred to the Department of Energy by act of Aug. 4, 1977 (91 Stat. 578). *Bureau* renamed *Water and Power Resources Service* by Secretarial

Order 3042 of Nov. 6, 1979. Renamed Bureau of Reclamation by Secretarial Order 3064 of May 18, 1981.

Reconciliation Service   Established by Director of Selective Service pursuant to EO 11804 of Sept. 16, 1974. Program terminated Apr. 2, 1980.

Reconstruction Finance Corporation   Established Feb. 2, 1932, by act of Jan. 22, 1932 (47 Stat. 5). Grouped with other agencies to form *Federal Loan Agency* by Reorg. Plan No. I of 1939, effective July 1, 1939. Transferred to the Department of Commerce by EO 9071 of Feb. 24, 1942. Returned to *Federal Loan Agency* by act of Feb. 24, 1945 (59 Stat. 5). *Agency* abolished by act of June 30, 1947 (61 Stat. 202), and functions assumed by *Corporation.* Functions relating to financing houses or site improvements, authorized by act of Aug. 10, 1948 (61 Stat. 1275), transferred to *Housing and Home Finance Agency* by Reorg. Plan No. 23 of 1950, effective July 10, 1950. *Corporation* Board of Directors, established by act of Jan. 22, 1932 (47 Stat. 5), abolished by Reorg. Plan No. 1 of 1951, effective May 1, 1951, and functions transferred to Administrator and *Loan Policy Board* established by same plan, effective Apr. 30, 1951. Act of July 30, 1953 (67 Stat. 230), provided for *RFC* succession until June 30, 1954, and for termination of its lending powers Sept. 28, 1953. Certain functions assigned to appropriate agencies for liquidation by Reorg. Plan No. 2 of 1954, effective July 1, 1954. *Corporation* abolished by Reorg. Plan No. 1 of 1957, effective June 30, 1957, and functions transferred to *Housing and Home Finance Agency,* General Services Administration, Small Business Administration, and the Department of the Treasury.

Records Administration, Office of   Established in the National Archives and Records Administration. Reorganized by Archivist under Notice 96–260, Sept. 23, 1996, effective Jan. 6, 1997. Functions restructured and transferred to Office of Records Services—Washington, DC.

Records Centers, Office of Federal   Established in the National Archives and Records Administration. Reorganized by Archivist under Notice 96–260, Sept. 23, 1996, effective Jan. 6, 1997. Functions restructured and transferred to Office of Regional Records Services.

Records and Information Management, Office of   Functions transferred from *National Archives and Records Service* to *Automated Data and Telecommunications Service* by General Services Administrator's decision, effective Jan. 10, 1982, regionally and Apr. 1, 1982, in Washington, DC.

Recovery Administration, Advisory Council, National   Established by EO 7075 of June 15, 1935. Transferred to the Department of Commerce by EO 7252 of Dec. 21, 1935, and functions ordered terminated not later than Apr. 1, 1936, by same order. *Committee of Industrial Analysis* created by EO 7323 of Mar. 21, 1936, to complete work of *Council.*

Recovery Administration, National   Established by President pursuant to act of June 16, 1933 (48 Stat. 194). Provisions of title I of act repealed by Public

Resolution 26 of June 14, 1935 (49 Stat. 375), and extension of *Administration* in skeletonized form authorized until Apr. 1, 1936. *Office of Administrator, National Recovery Administration,* created by EO 7075 of June 15, 1935. *Administration* terminated by EO 7252 of Dec. 21, 1935, which transferred *Division of Review, Division of Business Corporation,* and *Advisory Council* to the Department of Commerce for termination of functions by Apr. 1, 1936. *Consumers' Division* transferred to the Department of Labor by same order.

Recovery Review Board, National   Established by EO 6632 of Mar. 7, 1934. Abolished by EO 6771 of June 30, 1934.

Recreation, Bureau of Outdoor   Established in the Department of the Interior by act of May 28, 1963 (77 Stat. 49). Terminated by Secretary's order of Jan. 25, 1978, and functions assumed by *Heritage Conservation and Recreation Service.*

Recreation and Natural Beauty, Citizens' Advisory Committee on   Established by EO 11278 of May 4, 1966. Terminated by EO 11472 of May 29, 1969.

Recreation and Natural Beauty, President's Council on   Established by EO 11278 of May 4, 1966. Terminated by EO 11472 of May 29, 1969.

Recreation Resources Review Commission, Outdoor   Established by act of June 28, 1958 (72 Stat. 238). Final report submitted to President January 1962 and terminated Sept. 1, 1962.

Regional Action Planning Commissions   Authorized by act of Aug. 26, 1965 (79 Stat. 552). Federal role abolished through repeal by act of Aug. 13, 1981 (95 Stat. 766). At time of repeal, eight commissions— Coastal Plains, Four Corners, New England, Old West Ozarks, Pacific Northwest, Southwest Border, Southwest Border Region, and Upper Great Lakes— affected.

Regional Archives, Office of Special and   Established in the National Archives and Records Administration. Reorganized by Archivist under Notice 96–260, Sept. 23, 1996, effective Jan. 6, 1997. Functions restructured and transferred between Office of Records Services—Washington, DC and Office of Regional Records Services.

Regional Councils, Federal   Established by EO 12314 of July 22, 1981. Abolished by EO 12407 of Feb. 22, 1983.

Regional Operations, Executive Director of   Established in Food and Drug Administration by order of May 20, 1971 of the *Secretary of Health, Education, and Welfare.* Merged into Office of Regulatory Affairs by order of Nov. 5, 1984 of the Secretary of Health and Human Services.

Regulations and Rulings, Office of   Established in the U.S. Customs and Border Protection. Abolished by act of Oct. 13, 2006 (120 Stat. 1924) and functions transferred to the Office of International Trade.

Regulatory Council, U.S. Disbanded by Vice Presidential memorandum of Mar. 25, 1981. Certain functions continued in Regulatory Information Service Center.

Regulatory Relief, Presidential Task Force on Establishment announced in President's remarks Jan. 22, 1981. Disbanded and functions transferred to Office of Management and Budget in August 1983.

Rehabilitation Services Administration Functions transferred from *Department of Health, Education, and Welfare* to Office of Special Education and Rehabilitative Services, Department of Education, by act of Oct. 17, 1979 (93 Stat. 678), effective May 4, 1980.

Relief Corporation, Federal Surplus Organized under powers granted to President by act of June 16, 1933 (48 Stat. 195). Charter granted by State of Delaware Oct. 4, 1933, and amended Nov. 18, 1935, changing name to *Federal Surplus Commodities Corporation* and naming the Secretary of Agriculture, *Administrator of Agricultural Adjustment Administration,* and *Governor of Farm Credit Administration* as Board of Directors. Continued as agency under the Secretary of Agriculture by acts of June 28, 1937 (50 Stat. 323) and Feb. 16, 1938 (52 Stat. 38). Consolidated with *Division of Marketing and Marketing Agreements* into *Surplus Marketing Administration* by Reorg. Plan No. III of 1940, effective June 30, 1940. Merged into *Agricultural Marketing Administration* by EO 9069 of Feb. 23, 1942.

Relief and Rehabilitation Operations, Office of Foreign Established in the Department of State as announced by White House Nov. 21, 1942. Consolidated with *Foreign Economic Administration* by EO 9380 of Sept. 25, 1943.

Renegotiation Board Established by act of Mar. 23, 1951 (65 Stat. 7). Terminated Mar. 31, 1979, by act of Oct. 10, 1978 (92 Stat. 1043).

Rent Advisory Board Established by EO 11632 of Nov. 22, 1971. Abolished by EO 11695 of Jan. 11, 1973.

Rent Stabilization, Office of Established by General Order 9 of *Economic Stabilization Administrator* July 31, 1951, pursuant to act of June 30, 1947 (61 Stat. 193), and EO's 10161 of Sept. 9, 1950, and 10276 of July 31, 1951. Abolished by EO 10475 of July 31, 1953, and functions transferred to *Office of Defense Mobilization. Office of Research and Development* combined with *Office of Minerals Policy Development* in the Department of the Interior May 21, 1976, under authority of Reorg. Plan No. 3 of 1950, effective May 24, 1950, to form *Office of Minerals Policy and Research Analysis.* Abolished Sept. 30, 1981, by Secretarial Order 3070 and functions transferred to *Bureau of Mines.*

Reports, Office of Government Established July 1, 1939, to perform functions of *National Emergency Council* abolished by Reorg. Plan No. II of 1939, effective July 1, 1939. Established as administrative unit of Executive Office of the President by EO 8248

of Sept. 8, 1939. Consolidated with *Office of War Information, Office for Emergency Management,* by EO 9182 of June 13, 1942. Reestablished in Executive Office of the President by EO 9809 of Dec. 12, 1946, which transferred to it functions of *Media Programming Division* and *Motion Picture Division, Office of War Mobilization and Reconversion,* and functions transferred from *Bureau of Special Services, Office of War Information,* to *Bureau of the Budget* by EO 9608 of Aug. 31, 1945. Subsequent to enactment of act of July 30, 1947 (61 Stat. 588), functions of *Office* restricted to advertising and motion picture liaison and operation of library. Terminated June 30, 1948.

Research, Office of University Transferred from *Office of Program Management and Administration,* Research and Special Programs Administration, to Office of Economics, Office of the Assistant Secretary for Policy and International Affairs, under authority of the Department of Transportation appropriation request for FY 1985, effective Oct. 1, 1984.

Research and Development Board Established in the Department of Defense by act of July 26, 1947 (61 Stat. 499). Abolished by Reorg. Plan No. 6 of 1953, effective June 30, 1953, and functions vested in the Secretary of Defense.

Research and Development Board, Joint Established June 6, 1946, by charter of Secretaries of *War* and Navy. Terminated on creation of *Research and Development Board* by act of July 26, 1947 (61 Stat. 506).

Research and Intelligence Service, Interim Established in the Department of State by EO 9621 of Sept. 20, 1945. Abolished Dec. 31, 1945, pursuant to terms of order.

Research and Special Programs Administration Established by act of Oct. 24, 1992 (106 Stat. 3310). Abolished and certain duties and powers transferred to both the Pipeline Hazardous Materials Safety Administration and the Administrator of the Research and Innovative Technology Administration, Department of Transportation, by act of Nov. 30, 2004 (118 Stat. 2424–2426).

Research Resources, Division of Established in National Institutes of Health, Department of Health and Human Services. Renamed National Center for Research Resources by Secretarial notice of Feb. 23, 1990 (55 FR 6455) and act of June 10, 1993 (107 Stat. 178).

Research Service, Cooperative State Established by Secretary of Agriculture Memorandum 1462, supp. 1, of Aug. 31, 1961. Consolidated into *Science and Education Administration* by Secretary's order of Jan. 24, 1978. Reestablished as Cooperative State Research Service by Secretarial order of June 16, 1981.

Research and Service Division, Cooperative Functions transferred to the Secretary of Agriculture in *Farmer Cooperative Service* by act of Aug. 6, 1953 (67 Stat. 390).

Resettlement Administration Established by EO 7027 of Apr. 30, 1935. Functions transferred to the Department of Agriculture by EO 7530 of Dec. 31, 1936. Renamed *Farm Security Administration* by Secretary's Memorandum 732 of Sept. 1, 1937. Abolished by act of Aug. 14, 1946 (60 Stat. 1062) and functions incorporated into the *Farmers' Home Administration,* effective Jan. 1, 1947. *Farmers' Home Administration* abolished, effective Dec. 27, 1994, under authority of Secretary's Memorandum 1010–1 dated Oct. 20, 1994 (59 FR 66441). Functions assumed by the *Consolidated Farm Service Agency* and the *Rural Housing and Community Development Service.*

Resolution Trust Corporation Established by act of Aug. 9, 1989 (103 Stat. 369). Board of Directors of the Corporation abolished by act of Dec. 12, 1991 (105 Stat. 1769). Corporation functions terminated pursuant to act of Dec. 17, 1993 (107 Stat. 2369).

Resources Board and Advisory Committee, National Established by EO 6777 of June 30, 1934. Abolished by EO 7065 of June 7, 1935, and functions transferred to *National Resources Committee.*

Resources Committee, National Established by EO 7065 of June 7, 1935. Abolished by Reorg. Plan No. I of 1939, effective July 1, 1939, and functions transferred to *National Resources Planning Board* in Executive Office of the President. *Board* terminated by act of June 26, 1943 (57 Stat. 169).

Resources Planning Board, National *See* **Resources Committee, National**

Retired Executives, Service Corps of Established in ACTION by act of Oct. 1, 1973 (87 Stat. 404). Transferred to Small Business Administration by EO 11871 of July 18, 1975.

Retraining and Reemployment Administration Established by EO 9427 of Feb. 24, 1944, and act of Oct. 3, 1944 (58 Stat. 788). Transferred from *Office of War Mobilization and Reconversion* to the Department of Labor by EO 9617 of Sept. 19, 1945. Terminated pursuant to terms of act.

Revenue Sharing, Office of Established by the Secretary of the Treasury to administer programs authorized by acts of Oct. 20, 1972 (86 Stat. 919), and July 22, 1976 (90 Stat. 999). Transferred from the Office of the Secretary to Assistant Secretary (Domestic Finance) by Department of the Treasury Order 242, rev. 1, of May 17, 1976.

Review, Division of Established in *National Recovery Administration* by EO 7075 of June 15, 1935. Transferred to the Department of Commerce by EO 7252 of Dec. 21, 1935, and functions terminated Apr. 1, 1936. *Committee of Industrial Analysis* created by EO 7323 of Mar. 21, 1936, to complete work of *Division.*

RFC Mortgage Company Organized under laws of Maryland Mar. 14, 1935, pursuant to act of Jan. 22, 1932 (47 Stat. 5). Grouped with other agencies to form *Federal Loan Agency* by Reorg. Plan No. I

of 1939, effective July 1, 1939. Transferred to the Department of Commerce by EO 9071 of Feb. 24, 1942. Returned to *Federal Loan Agency* by act of Feb. 24, 1945 (59 Stat. 5). Assets and liabilities transferred to *Reconstruction Finance Corporation* by act of June 30, 1947 (61 Stat. 207).

River Basins, Neches, Trinity, Brazos, Colorado, Guadalupe, San Antonio, Nueces, and San Jacinto, and Intervening Areas, U.S. Study Commission on Established by act of Aug. 28, 1958 (72 Stat. 1058). Terminated June 30, 1962.

River Basins, Savannah, Altamaha, Saint Marys, Apalachicola-Chattahoochee, and Perdido-Escambia, and Intervening Areas, U.S. Study Commission on Established by act of Aug. 28, 1958 (72 Stat. 1090). Terminated Dec. 23, 1962.

Road Inquiry, Office of Established by the Secretary of Agriculture under authority of act of Aug. 8, 1894 (28 Stat. 264). Federal aid for highways to be administered by the Secretary of Agriculture through *Office of Public Roads and Rural Engineering* authorized by act of July 11, 1916 (39 Stat. 355), known as *Bureau of Public Roads* after July 1918. Transferred to *Federal Works Agency* by Reorg. Plan No. I of 1939, effective July 1, 1939, and renamed *Public Roads Administration.* Transferred to General Services Administration as *Bureau of Public Roads* by act of June 30, 1949 (63 Stat. 380). Transferred to the Department of Commerce by Reorg. Plan No. 7 of 1949, effective Aug. 20, 1949. Transferred to the Secretary of Transportation by act of Oct. 15, 1966 (80 Stat. 931), and functions assigned to Federal Highway Administration.

Roads, Bureau of Public *See* **Road Inquiry, Office of**

Roads Administration, Public *See* **Road Inquiry, Office of**

Roads and Rural Engineering, Office of Public *See* **Road Inquiry, Office of**

Rock Creek and Potomac Parkway Commission Established by act of Mar. 14, 1913 (37 Stat. 885). Abolished by EO 6166 of June 10, 1933, and functions transferred to *Office of National Parks, Buildings, and Reservations,* Department of the Interior.

Roosevelt Centennial Commission, Theodore Established by joint resolution of July 28, 1955 (69 Stat. 383). Terminated Oct. 27, 1959, pursuant to terms of act.

Roosevelt Library, Franklin D. Functions assigned to National Park Service by Reorg. Plan No. 3 of 1946, effective July 16, 1946, transferred to General Services Administration by Reorg. Plan No. 1 of 1963, effective July 27, 1963.

Roosevelt Library, Trustees of the Franklin D. Established by joint resolution of July 18, 1939 (53 Stat. 1063). Transferred to General Services Administration by act of June 30, 1949 (63 Stat. 381). Abolished by act of Mar. 5, 1958 (72 Stat. 34), and

Library operated by *National Archives and Records Service,* General Services Administration.

Roosevelt Memorial Commission, Franklin Delano Established by joint resolution of Aug. 11, 1955 (69 Stat. 694). Terminated by act of Nov. 14, 1997 (111 Stat. 1601).

Rubber Development Corporation Establishment announced Feb. 20, 1943, by the Secretary of Commerce. Organized under laws of Delaware as subsidiary of *Reconstruction Finance Corporation.* Assumed all activities of *Rubber Reserve Company* relating to development of foreign rubber sources and procurement of rubber therefrom. Functions transferred to *Office of Economic Warfare* by EO 9361 of July 15, 1943. *Office* consolidated into *Foreign Economic Administration* by EO 9380 of Sept. 25, 1943. *Office* returned to *Reconstruction Finance Corporation* by EO 9630 of Sept. 27, 1945. Certificate of incorporation expired June 30, 1947.

Rubber Producing Facilities Disposal Commission Established by act of Aug. 7, 1953 (67 Stat. 408). Functions transferred to *Federal Facilities Corporation* by EO 10678 of Sept. 20, 1956.

Rubber Reserve Company Established June 28, 1940, under act of Jan. 22, 1932 (47 Stat. 5). Transferred from *Federal Loan Agency* to the Department of Commerce by EO 9071 of Feb. 24, 1942. Returned to *Federal Loan Agency* by act of Feb. 24, 1945 (59 Stat. 5). Dissolved by act of June 30, 1945 (59 Stat. 310), and functions transferred to *Reconstruction Finance Corporation.*

Rural Areas Development, Office of Established by Secretary of Agriculture memorandum in 1961 (revised Sept. 21, 1962). Renamed *Rural Community Development Service* by Secretary's Memorandum 1570 of Feb. 24, 1965.

Rural Business and Cooperative Development Service Established within the Department of Agriculture by Secretary's Memorandum 1020–34 dated Dec. 31, 1991. Renamed Rural Business-Cooperative Service (61 FR 2899), effective Jan. 30, 1996.

Rural Community Development Service Established by Secretary of Agriculture Memorandum 1570 of Feb. 25, 1965, to supersede *Office of Rural Areas Development.* Abolished Feb. 2, 1970, by Secretary's Memorandum 1670 of Jan. 30, 1970, and functions transferred to other agencies in the Department of Agriculture.

Rural Development Administration Established within the Department of Agriculture by Secretary's Memorandum 1020–34 dated Dec. 31, 1991. Abolished Dec. 27, 1994 (59 FR 66441) under authority of Secretary's Memorandum 1010–1 dated Oct. 20, 1994. Functions assumed by the Rural Business and Cooperative Development Service.

Rural Development Committee *See* **Rural Development Program, Committee for**

Rural Development Policy, Office of Established initially as *Office of Rural Development Policy Management and Coordination,* Farmers Home Administration, by Secretary of Agriculture Memorandum 1020–3 of Oct. 26, 1981. Abolished in 1986 due to lack of funding.

Rural Development Program, Committee for Established by EO 10847 of Oct. 12, 1959. Abolished by EO 11122 of Oct. 16, 1963, which established *Rural Development Committee. Committee* superseded by EO 11307 of Sept. 30, 1966, and functions assumed by the Secretary of Agriculture.

Rural Development Service Established by Agriculture Secretarial order in 1973. Functions transferred to *Office of Rural Development Coordination and Planning, Farmers Home Administration,* by Secretarial order in 1978.

Rural Electrification Administration Established by EO 7037 of May 11, 1935. Functions transferred by EO 7458 of Sept. 26, 1936, to *Rural Electrification Administration* established by act of May 20, 1936 (49 Stat. 1363). Transferred to the Department of Agriculture by Reorg. Plan No. II of 1939, effective July 1, 1939. Abolished by Secretary's Memorandum 1010–1 dated Oct. 20, 1994, and functions assumed by Rural Utilities Service.

Rural Housing and Community Development Service Established by act of Oct. 13, 1994 (108 Stat. 3219). Renamed Rural Housing Service (61 FR 2899), effective Jan. 30, 1996.

Rural Rehabilitation Division Established April 1934 by act of May 12, 1933 (48 Stat. 55). Functions transferred to *Resettlement Administration* by *Federal Emergency Relief Administrator's* order of June 19, 1935.

Saint Elizabeth's Hospital *See* **Insane, Government Hospital for the**

Saint Lawrence Seaway Development Corporation Established by act of May 13, 1954 (68 Stat. 92). Secretary of Commerce given direction of general policies of *Corporation* by EO 10771 of June 20, 1958. Transferred to the Department of Transportation by act of Oct. 15, 1966 (80 Stat. 931).

Salary Stabilization, Office of *See* **Salary Stabilization Board**

Salary Stabilization Board Established May 10, 1951, by *Economic Stabilization Administrator's* General Order 8. Stabilization program administered by *Office of Salary Stabilization.* Terminated Apr. 30, 1953, by EO 10434 of Feb. 6, 1953, and acts of June 30, 1952 (66 Stat. 296), and June 30, 1953 (67 Stat. 131).

Sales Manager, Office of the General Established by the Secretary of Agriculture Feb. 29, 1976. Consolidated with Foreign Agricultural Service by Secretary's Memorandum 2001 of Nov. 29, 1979.

Savings Bonds, Interdepartmental Committee for the Voluntary Payroll Savings Plan for the Purchase of U.S.   Established by EO 11532 of June 2, 1970. Superseded by EO 11981 of Mar. 29, 1977, which established Interagency Committee for the Purchase of U.S. Savings Bonds.

Savings and Loan Advisory Council, Federal   Established by act of Oct. 6, 1972 (86 Stat. 770). Continued by act of Dec. 26, 1974 (88 Stat. 1739). Terminated by act of Aug. 9, 1989 (103 Stat. 422).

Savings and Loan Insurance Corporation, Federal   Established by act of June 27, 1934 (48 Stat. 1246). Grouped with other agencies to form *Federal Loan Agency* by Reorg. Plan No. I of 1939, effective July 1, 1939. Transferred to *Federal Home Loan Bank Administration, National Housing Agency,* by EO 9070 of Feb. 24, 1942. Board of Trustees abolished by Reorg. Plan No. 3 of 1947, effective July 27, 1947, and functions transferred to *Home Loan Bank Board.* Abolished by act of Aug. 9, 1989 (103 Stat. 354).

Savings Bonds Division, United States   Established by Departmental Order 62 of Dec. 26, 1945, as successor to the War and Finance Division, War Savings Staff, and Defense Savings Staff. Functions transferred to Bureau of Public Debt by Departmental Order 101–05 of May 11, 1994, and *Division* renamed Savings Bond Marketing Office.

Science, Engineering, and Technology, Federal Coordinating Council for   Established by act of May 11, 1976 (90 Stat. 471). Abolished by Reorg. Plan No. 1 of 1977, effective Feb. 26, 1978, and functions transferred to President. Functions redelegated to Director of the Office of Science and Technology Policy and Federal Coordinating Council for Science, Engineering, and Technology, established by EO 12039 of Feb. 24, 1978.

Science, Engineering, and Technology Panel, Intergovernmental   Established by act of May 11, 1976 (90 Stat. 465). Abolished by Reorg. Plan No. 1 of 1977, effective Feb. 26, 1978, and functions transferred to President. Functions redelegated to Director of Office of Science and Technology Policy by EO 12039 of Feb. 24, 1978, which established Intergovernmental Science, Engineering, and Technology Advisory Panel.

Science Advisory Committee, President's   Established by President Apr. 20, 1951, and reconstituted Nov. 22, 1957. Terminated with *Office of Science and Technology,* effective July 1, 1973.

Science Exhibit-Century 21 Exposition, U.S.   Established Jan. 20, 1960, by Department of Commerce Order 167. Abolished by revocation of order on June 5, 1963.

Science and Technology, Federal Council for   *See* **Scientific Research and Development, Interdepartmental Committee on**

Science and Technology, Office of   Established by Reorg. Plan No. 2 of 1962, effective June 8, 1962.

Office abolished by Reorg. Plan No. 1 of 1973, effective June 30, 1973, and functions transferred to National Science Foundation.

Science and Technology, President's Committee on   Established by act of May 11, 1976 (90 Stat. 468). Abolished by Reorg. Plan No. 1 of 1977, effective Feb. 26, 1978, and functions transferred to President.

Scientific and Policy Advisory Committee   Established by act of Sept. 26, 1961 (75 Stat. 631). Terminated Apr. 30, 1996 under terms of act.

Scientific Research and Development, Interdepartmental Committee on   Established by EO 9912 of Dec. 24, 1947. EO 9912 revoked by EO 10807 of Mar. 13, 1959, which established *Federal Council for Science and Technology.* Abolished by act of May 11, 1976 (90 Stat. 472).

Scientific Research and Development, Office of   Established in *Office for Emergency Management* by EO 8807 of June 28, 1941. Terminated by EO 9913 of Dec. 26, 1947, and property transferred to *National Military Establishment* for liquidation.

Scientists and Engineers, National Committee for the Development of   Established by President Apr. 3, 1956. Renamed *President's Committee on Scientists and Engineers* May 7, 1957. Final report submitted Dec. 17, 1958, and expired Dec. 31, 1958.

Scientists and Engineers, President's Committee on   *See* **Scientists and Engineers, National Committee for the Development of**

Screw Thread Commission, National   Established by act of July 18, 1918 (40 Stat. 912). Terminated by EO 6166 of June 10, 1933, and records transferred to the Department of Commerce, effective Mar. 2, 1934. Informal Interdepartmental Screw Thread Committee established on Sept. 14, 1939, consisting of representatives of the Departments of *War,* the Navy, and Commerce.

Sea Grant Review Panel, National   Established by act of Oct. 8, 1976 (90 Stat. 1967). Renamed National Sea Grant Advisory Board by act of Oct. 13, 2008 (122 Stat. 4207.

Secret Service, United States   *See* **Secret Service Division**

Secret Service Division   Established July 5, 1865, as a Bureau under Treasury Department. Acknowledged as distinct agency within Treasury Department in 1883. *White House Police Force* created on October 1, 1922, and placed under supervision of *Secret Service Division* in 1930. *White House Police Force* renamed *Executive Protective Service* by act of June 30, 1970 (84 Stat. 358). *Executive Protective Service* renamed U.S. Secret Service Uniformed Division by act of Nov. 15, 1977 (91 Stat. 1371). *Treasury Police Force* merged into Secret Service on Oct. 5, 1986. U.S. Secret Service transferred to Homeland Security Department by act of Nov. 25, 2002 (116 Stat. 2224).

Security and Safety Performance Assurance, Office of Established by Secretary of Energy memorandum of December 2, 2003. Abolished by Secretary's Memorandum 2006–007929 of Aug. 30, 2006 and functions transferred to the Office of Health, Safety and Security.

Security, Commission on Government Established by act of Aug. 9, 1955 (69 Stat. 595). Terminated Sept. 22, 1957, pursuant to terms of act.

Security, Office of the Director for Mutual *See* Security Agency, Mutual

Security Agency, Federal Established by Reorg. Plan No. I of 1939, effective July 1, 1939, grouping under one administration *Office of Education, Public Health Service, Social Security Board, U.S. Employment Service, Civilian Conservation Corps,* and *National Youth Administration.* Abolished by Reorg. Plan No. 1 of 1953, effective Apr. 11, 1953, and functions and units transferred to *Department of Health, Education, and Welfare.*

Security Agency, Mutual Established and continued by acts of Oct. 10, 1951 (65 Stat. 373) and June 20, 1952 (66 Stat. 141). *Agency* and *Office of Director for Mutual Security* abolished by Reorg. Plan No. 7 of 1953, effective Aug. 1, 1953, and functions transferred to *Foreign Operations Administration,* established by same plan.

Security and Individual Rights, President's Commission on Internal Established by EO 10207 of Jan. 23, 1951. Terminated by EO 10305 of Nov. 14, 1951.

Security Resources Board, National Established by act of July 26, 1947 (61 Stat. 499). Transferred to Executive Office of the President by Reorg. Plan No. 4 of 1949, effective Aug. 20, 1949. Functions of *Board* transferred to Chairman and *Board* made advisory to him by Reorg. Plan No. 25 of 1950, effective July 10, 1950. Functions delegated by Executive order transferred to *Office of Defense Mobilization* by EO 10438 of Mar. 13, 1953. *Board* abolished by Reorg. Plan No. 3 of 1953, effective June 12, 1953, and remaining functions transferred to *Office of Defense Mobilization.*

Security Training Commission, National Established by act of June 19, 1951 (65 Stat. 75). Expired June 30, 1957, pursuant to Presidential letter of Mar. 25, 1957.

Seed Loan Office Authorized by Presidential letters of July 26, 1918, and July 26, 1919, to the Secretary of Agriculture. Further authorized by act of Mar. 3, 1921 (41 Stat. 1347). Office transferred to Farm Credit Administration by EO 6084 of Mar. 27, 1933.

Selective Service Appeal Board, National Established by EO 9988 of Aug. 20, 1948. Inactive as of Apr. 11, 1975.

Selective Service Records, Office of *See* Selective Service System

Selective Service System Established by act of Sept. 16, 1940 (54 Stat. 885). Placed under jurisdiction of

War Manpower Commission by EO 9279 of Dec. 5, 1942, and designated *Bureau of Selective Service.* Designated Selective Service System, separate agency, by EO 9410 of Dec. 23, 1943. Transferred for liquidation to *Office of Selective Service Records* established by act of Mar. 31, 1947 (61 Stat. 31). Transferred to Selective Service System by act of June 24, 1948 (62 Stat. 604).

Self-Help Development and Technical Development, Office of Established in *National Consumer Cooperative Bank* by act of Aug. 20, 1978 (92 Stat. 499). Abolished by act of Aug. 13, 1981 (95 Stat. 437), and assets transferred to Consumer Cooperative Development Corporation, Department of Commerce, Dec. 30, 1982.

Services, Bureau of Special *See* Office of War Information

Services, Division of Central Administrative Established by *Liaison Officer for Emergency Management* pursuant to Presidential letter of Feb. 28, 1941. Terminated by EO 9471 of Aug. 25, 1944, and functions discontinued or transferred to constituent agencies of *Office for Emergency Management* and other agencies.

Shipbuilding Stabilization Committee Originally organized by *National Defense Advisory Commission* in 1940. Established August 1942 by *War Production Board.* Transferred to the Department of Labor from *Civilian Production Administration,* successor agency to *Board,* by EO 9656 of Nov. 15, 1945. Terminated June 30, 1947.

Shipping Board, U.S. Established by act of Sept. 7, 1916 (39 Stat. 729). Abolished by EO 6166 of June 10, 1933, and functions, including those with respect to *U.S. Shipping Board Merchant Fleet Corporation,* transferred to *U.S. Shipping Board Bureau,* Department of Commerce, effective Mar. 2, 1934. Separation of employees deferred until Sept. 30, 1933, by EO 6245 of Aug. 9, 1933. Functions assumed by *U.S. Maritime Commission* Oct. 26, 1936, pursuant to act of June 29, 1936 (49 Stat. 1985).

Shipping Board Bureau, U.S. *See* Shipping Board, U.S.

Shipping Board Emergency Fleet Corporation, U.S. Established Apr. 16, 1917, under authority of act of Sept. 7, 1916 (39 Stat. 729). Renamed *U.S. Shipping Board Merchant Fleet Corporation* by act of Feb. 11, 1927 (44 Stat. 1083). Terminated Oct. 26, 1936, under provisions of act of June 29, 1936 (49 Stat. 1985), and functions transferred to *U.S. Maritime Commission.*

Shipping Board Merchant Fleet Corporation, U.S. *See* Shipping Board Emergency Fleet Corporation, U.S.

Ships, Bureau of Established by act of June 20, 1940 (54 Stat. 493), to replace *Bureau of Engineering* and *Bureau of Construction and Repair.* Abolished by Department of Defense reorg. order of Mar. 9, 1966, and functions transferred to the Secretary of the Navy (31 FR 7188).

Simpson Historical Research Center, Albert F.   Renamed Headquarters USAF Historical Research Center by special order of Dec. 16, 1983 of the Secretary of Defense.

Small and Disadvantaged Business Utilization, Office of   Established within certain Defense Departments by act of Oct. 24, 1978 (92 Stat. 1770). Renamed Office of Small Business Programs by Public Law 109–163 of Jan. 6, 2006 (119 Stat. 3399).

Smithsonian Symposia and Seminars, Office of   Renamed Office of Interdisciplinary Studies by Smithsonian Institution announcement of Mar. 16, 1987.

Social Development Institute, Inter-American   Established by act of Dec. 30, 1969 (83 Stat. 821). Renamed Inter-American Foundation by act of Feb. 7, 1972 (86 Stat. 34).

Social Protection, Committee on   Established in *Office of Defense Health and Welfare Services* by administrative order June 14, 1941. Functions transferred to *Federal Security Agency* by EO 9338 of Apr. 29, 1943.

Social and Rehabilitation Service   Established by the *Secretary of Health, Education, and Welfare* reorganization of Aug. 15, 1967. Abolished by Secretary's reorganization of Mar. 8, 1977 (42 FR 13262), and constituent units—*Medical Services Administration, Assistance Payments Administration, Office of Child Support Enforcement, and Public Services Administration*—transferred.

Social Security Administration   *See* **Social Security Board**

Social Security Board   Established by act of Aug. 14, 1935 (49 Stat. 620). Incorporated into *Federal Security Agency* by Reorg. Plan No. I of 1939, effective July 1, 1939. *Social Security Board* abolished and Social Security Administration established by Reorg. Plan No. 2 of 1946 (5 U.S.C. app.), effective July 16, 1946, and functions of the *Board* transferred to *Federal Security Administrator*. Social Security Administration transferred from the *Federal Security Agency* by Reorganization Plan No. 1 of 1953 (5 U.S.C. app.), effective Apr. 11, 1953, to the *Department of Health, Education, and Welfare*. Social Security Administration became an independent agency in the executive branch by act of Aug. 15, 1994 (108 Stat. 1464), effective Mar. 31, 1995.

Soil Conservation Service   *See* **Soil Erosion Service**

Soil Erosion Service   Established in the Department of the Interior following allotment made Aug. 25, 1933. Transferred to the Department of Agriculture by Secretary of Interior administrative order of Mar. 25, 1935. Made *Soil Conservation Service* by order of the Secretary of Agriculture, Apr. 27, 1935, pursuant to provisions of act of Apr. 27, 1935 (49 Stat. 163). Certain functions of *Soil Conservation Service* under jurisdiction of the Department of the Interior transferred from the Department of Agriculture to the Department of the Interior by Reorg. Plan No. IV

of 1940, effective June 30, 1940. *Soil Conservation Service* abolished by act of Oct. 13, 1994 (108 Stat. 3225) and functions assumed by the Natural Resources Conservation Service.

Soils, Bureau of   *See* **Agricultural and Industrial Chemistry, Bureau of** and **Plant Industry, Bureau of**

Solicitor General, Office of Assistant   Established in the Department of Justice by act of June 16, 1933 (48 Stat. 307). Terminated by Reorg. Plan No. 2 of 1950, effective May 24, 1950.

Southeastern Power Administration   Established by the Secretary of the Interior in 1943 to carry out functions under act of Dec. 22, 1944 (58 Stat. 890). Transferred to the Department of Energy by act of Aug. 4, 1977 (91 Stat. 578).

Southwestern Power Administration   Established by the Secretary of the Interior in 1943 to carry out functions under act of Dec. 22, 1944 (58 Stat. 890). Transferred to the Department of Energy by act of Aug. 4, 1977 (91 Stat. 578).

Space Access and Technology, Office of   Established in the National Aeronautics and Space Administration. Abolished by Administrator's order of Feb. 24, 1997.

Space Communications, Office of   Established in the National Aeronautics and Space Administration. Abolished by Administrator's order of Feb. 24, 1997.

Space Payload Technology Organization, Joint Operationally Responsive   Established by act of Jan. 6, 2006 (119 Stat. 3408). Abolished by acts of Oct. 17, 2006 (120 Stat. 2358) and Dec. 20, 2006 (120 Stat. 3286).

Space Science, Office of   *See* **Space and Terrestrial Applications, Office of**

Space Science Board   Renamed Space Studies Board by authority of the National Research Council, National Academy of Sciences, effective May 8, 1989.

Space Station, Office of   Established in the National Aeronautics and Space Administration. Abolished in 1990 and remaining functions transferred to the Office of Space Flight.

Space Technology Laboratories, National   Renamed John C. Stennis Space Center by EO 12641 of May 20, 1988.

Space and Terrestrial Applications, Office of   Combined with *Office of Space Science* to form Office of Space Science and Applications by National Aeronautics and Space Administrator's announcement of Sept. 29, 1981.

Space Tracking and Data Systems, Office of   Renamed Office of Space Operations by National Aeronautics and Space Administrator's announcement of Jan. 9, 1987.

Space Transportation Operations, Office of Combined with *Office of Space Transportation Systems* to form Office of Space Transportation Systems, National Aeronautics and Space Administration, effective July 1982.

Space Transportation Systems, Office of *See* **Space Transportation Operations, Office of**

Spanish-Speaking People, Cabinet Committee on Opportunities for *See* **Mexican-American Affairs, Interagency Committee on**

Special. *See other part of title*

Specifications Board, Federal Established by *Bureau of the Budget* Circular 42 of Oct. 10, 1921. Transferred from *Federal Coordinating Service* to *Procurement Division* by order of Oct. 9, 1933 of the Secretary of the Treasury. *Board* superseded by *Federal Specifications Executive Committee*, set up by *Director of Procurement* under Circular Letter 106 of July 16, 1935.

Sport Fisheries and Wildlife, Bureau of Established in the Department of the Interior by act of Aug. 8, 1956 (70 Stat. 1119). *Bureau* replaced by U.S. Fish and Wildlife Service pursuant to act of Apr. 22, 1974 (88 Stat. 92).

Standards, National Bureau of *See* **Weights and Measures, Office of Standard**

State, Department of Duty of Secretary of State of procuring copies of all statutes of the States, as provided for in act of Sept. 28, 1789 (R.S. 206), abolished by Reorg. Plan No. 20 of 1950, effective May 24, 1950. Functions of numbering, editing, and distributing proclamations and Executive orders transferred from the Department of State to the *Division of the Federal Register, National Archives,* by EO 7298 of Feb. 18, 1936. Duty of Secretary of State of publishing Executive proclamations and treaties in newspapers in District of Columbia, provided for in act of July 31, 1876 (19 Stat. 105), abolished by Reorg. Plan No. 20 of 1950, effective May 24, 1950. Functions concerning publication of U.S. Statutes at Large, acts and joint resolutions in pamphlet form known as slip laws, and amendments to the Constitution; electoral votes for President and Vice President; and Territorial papers transferred from the Department of State to the Administrator of the General Services Administration by Reorg. Plan No. 20 of 1950. (*See also* **Archives Establishment, National**)

State and Local Cooperation, Division of Established by *Advisory Commission to Council of National Defense* Aug. 5, 1940. Transferred to *Office of Civilian Defense.*

State and Local Government Cooperation, Committee on Established by EO 11627 of Oct 15, 1971. Abolished by EO 11695 of Jan. 11, 1973.

State Technical Services, Office of Established by the Secretary of Commerce Nov. 19, 1965, pursuant to act of Sept. 14, 1965 (79 Stat. 697). Abolished by Secretary, effective June 30, 1970.

Statistical Board, Central Organized Aug. 9, 1933, by EO 6225 of July 27, 1933. Transferred to *Bureau of the Budget* by Reorg. Plan No. I of 1939, effective July 1, 1939. Expired July 25, 1940, and functions taken over by *Division of Statistical Standards, Bureau of the Budget.*

Statistical Committee, Central Established by act of July 25, 1935 (49 Stat. 498). Abolished by Reorg. Plan No. I of 1939, effective July 1, 1939, and functions transferred to *Bureau of the Budget.*

Statistical Policy Coordination Committee Established by EO 12013 of Oct. 7, 1977. Abolished by EO 12318 of Aug. 21, 1981.

Statistical Reporting Service Established by Memorandum 1446, supp. 1, part 3, of 1961 of the Secretary of Agriculture. Consolidated with other departmental units into *Economics, Statistics, and Cooperatives Service* by Secretary's Memorandum 1927, effective Dec. 23, 1977. Redesignated as *Statistical Reporting Service* by Secretary's order of Oct. 1, 1981. Renamed National Agricultural Statistics Service.

Statistics Administration, Social and Economic Established Jan. 1, 1972, by the Secretary of Commerce. Terminated by Department of Commerce Organization Order 10–2, effective Aug. 4, 1975 (40 FR 42765). Bureau of Economic Analysis and Bureau of the Census restored as primary operating units of the Department of Commerce by Organization Orders 35–1A and 2A, effective Aug. 4, 1975.

Statutes at Large *See* **State, Department of**

Statutes of the States *See* **State, Department of**

Steam Engineering, Bureau of Established in the Department of the Navy by act of July 5, 1862 (12 Stat. 510). Redesignated as *Bureau of Engineering* by act of June 4, 1920 (41 Stat. 828). Abolished by act of June 20, 1940 (54 Stat. 492), and functions transferred to *Bureau of Ships.*

Steamboat Inspection Service President authorized to appoint *Service* by act of June 28, 1838 (5 Stat. 252). Secretary of Treasury authorized to establish boards of local inspectors at enumerated ports throughout the U.S. by act of Feb. 28, 1871 (16 Stat. 440). Authority to appoint boards of local inspectors delegated to *Secretary of Commerce and Labor* by act of Mar. 4, 1905 (33 Stat. 1026). Consolidated with *Bureau of Navigation and Steamboat Inspection* by act of June 30, 1932 (47 Stat. 415).

Stock Catalog Board, Federal Standard Originated by act of Mar. 2, 1929 (45 Stat. 1461). Transferred from *Federal Coordinating Service* to *Procurement Division* by order of Oct. 9, 1933 of the Secretary of the Treasury.

Strategic Defense Initiative Organization Established in 1986 as a separate agency of the Department of Defense. Renamed Ballistic Missile Defense Organization by Deputy Secretary's memorandum in May 1993.

Strategic Posture of the United States, Commission on the Implementation of the New   Established by act of Jan. 6, 2006 (119 Stat. 3431). Terminated by act of Jan. 28, 2009 (122 Stat. 328)

Strategic Services, Office of   *See* **Information, Office of Coordinator of**

Strategic Trade, Office of   Established in the U.S. Customs and Border Protection pursuant to Customs Service Reorganization plan, effective Sept. 30, 1995. Abolished by act of Oct. 13, 2006 (120 Stat. 1924) and functions transferred to the Office of International Trade.

Subversive Activities Control Board   Established by act of Sept. 23, 1950 (64 Stat. 987). Terminated June 30, 1973, due to lack of funding.

Sugar Division   Created by act of May 12, 1933 (48 Stat. 31), authorized by act of Sept. 1, 1937 (50 Stat. 903). Taken from *Agricultural Adjustment Administration* and made independent division of the Department of Agriculture by Secretary's Memorandum 783, effective Oct. 16, 1938. Placed under *Agricultural Conservation and Adjustment Administration* by EO 9069 of Feb. 23, 1942, functioning as *Sugar Agency.* Functions transferred to *Food Distribution Administration* by EO 9280 of Dec. 5, 1942.

Sugar Rationing Administration   Established by Memorandum 1190 of Mar. 31, 1947, of the Secretary of Agriculture under authority of act of Mar. 31, 1947 (61 Stat. 35). Terminated Mar. 31, 1948, on expiration of authority.

Supplies and Accounts, Bureau of   *See* **Provisions and Clothing, Bureau of**

Supplies and Shortages, National Commission on   Established by act of Sept. 30, 1974 (88 Stat. 1168). Terminated Mar. 31, 1977, pursuant to terms of act.

Supply, Bureau of Federal   *See* **Procurement Division**

Supply, Office of   Renamed Office of Procurement and Property by Smithsonian Institution announcement of Nov. 4, 1986.

Supply Committee, General   Established by act of June 17, 1910 (36 Stat. 531). Abolished by EO 6166 of June 10, 1933, effective Mar. 2, 1934, and functions transferred to *Procurement Division,* the Department of the Treasury.

Supply Priorities and Allocations Board   Established in *Office for Emergency Management* by EO 8875 of Aug. 28, 1941. Abolished by EO 9024 of Jan. 16, 1942, and functions transferred to *War Production Board.*

Supply Service, Federal   Renamed *Office of Personal Property* by General Services Administration (GSA) order, effective Sept. 28, 1982; later renamed *Office of Federal Supply and Services* by GSA order of Jan. 22, 1983; then redesignated *Federal Supply*

Service. Merged with *Federal Technology Service* to form Federal Acquisition Service by GSA Order No. 5440.591 of Sept. 9, 2005. *See also* act of Oct. 6, 2006 (120 Stat. 1735).

Surveys and Maps, Federal Board of   *See* **Surveys and Maps of the Federal Government, Board of**

Surveys and Maps of the Federal Government, Board of   Established by EO 3206 of Dec. 30, 1919. Renamed *Federal Board of Surveys and Maps* by EO 7262 of Jan. 4, 1936. Abolished by EO 9094 of Mar. 10, 1942, and functions transferred to Director, *Bureau of the Budget.*

Space System Development, Office of   Established in the National Aeronautics and Space Administration. Renamed Office of Space Access and Technology in 1995.

Tariff Commission, U.S.   Established by act of Sept. 8, 1916 (39 Stat. 795). Renamed U.S. International Trade Commission by act of Jan. 3, 1975 (88 Stat. 2009).

Tax Appeals, Board of   Established as an independent agency within the executive branch by act of June 2, 1924 (43 Stat. 336). Continued by acts of Feb. 26, 1926 (44 Stat. 105) and Feb. 10, 1939 (53 Stat. 158). Renamed *Tax Court of the United States* by act of Aug. 16, 1954 (68A Stat. 879). Renamed United States Tax Court by act of Dec. 30, 1969 (83 Stat. 730).

Technical Cooperation Administration   Transferred from the Department of State to *Mutual Security Agency* by EO 10458 of June 1, 1953. Transferred to *Foreign Operations Administration* by Reorg. Plan No. 7 of 1953, effective Aug. 1, 1953.

Technical Services, Office of   Designated unit of Office of the Secretary of Commerce by Department Order 179, July 23, 1962. Functions transferred to *National Bureau of Standards* by Order 90 of Jan. 30, 1964.

Technology Administration   Established by act of Oct. 24, 1988 (102 Stat. 2593). Abolished by act of Aug. 9, 2007 (121 Stat. 587) and functions absorbed by National Institute of Standards and Technology, Department of Commerce.

Technology Assessment, Office of   Created by act of Oct. 13, 1972 (86 Stat. 797). Office inactive as of Sept. 30, 1995.

Technology, Automation, and Economic Progress, National Commission on   Established by act of Aug. 19, 1964 (78 Stat. 463). Terminated January 1966 pursuant to terms of act.

Technology Service, Federal   Merged with *Federal Supply Service* to form Federal Acquisition Service by General Services Administration Order No. 5440.591 of Sept. 9, 2005. *See also* act of Oct. 6, 2006 (120 Stat. 1735).

Telecommunications Adviser to the President   Established in Executive Office of the President by EO 10297 of Oct. 9, 1951. EO 10297 revoked by EO 10460 of June 16, 1953, and functions transferred to Director of *Office of Defense Mobilization.*

Telecommunications Management, Director of Established in *Office of Emergency Planning* by EO 10995 of Feb. 16, 1962. Assignment of radio frequencies delegated to Government agencies and foreign diplomatic establishments by EO 11084 of Feb. 16, 1963. Abolished by Reorg. Plan No. 1 of 1970, effective Apr. 20, 1970.

Telecommunications Policy, Office of Established in Executive Office of the President by Reorg. Plan No. 1 of 1970, effective Apr. 20, 1970. Abolished by Reorg. Plan No. 1 of 1977, effective Mar. 26, 1978, and certain functions transferred to President with all other functions transferred to the Department of Commerce.

Telecommunications Service, Automated Data Renamed *Office of Information Resources Management* by General Services Administration order of Aug. 17, 1982. Later renamed Information Resources Management Service.

Temporary Controls, Office of Established in *Office for Emergency Management* by EO 9809 of Dec. 12, 1946, consolidating *Office of War Mobilization and Reconversion, Office of Economic Stabilization, Office of Price Administration,* and *Civilian Production Administration.* Functions with respect to Veterans' Emergency Housing Program transferred to *Housing Expediter* by EO 9836 of Mar. 22, 1947. Functions with respect to distribution and price of sugar products transferred to the Secretary of Agriculture by act of Mar. 31, 1947 (61 Stat. 36). Office terminated by EO 9841 of Apr. 23, 1947, and remaining functions redistributed.

Temporary Emergency Court of Appeals Established by act of Dec. 22, 1971 (85 Stat. 749). Abolished by act of Oct. 29, 1992, effective Apr. 30, 1993 (106 Stat. 4507). Court's jurisdiction and pending cases transferred to the United States Court of Appeals for the Federal Circuit.

Territorial Affairs, Office of Established by Interior Secretarial Order 2951 of Feb. 6, 1973. Abolished by Departmental Manual Release 2270 of June 6, 1980, and functions transferred to Office of Assistant Secretary for Territorial and International Affairs.

Territorial papers *See* **State, Department of**

Territories, Office of Established by the Secretary of the Interior July 28, 1950. Functions reassigned to *Deputy Assistant Secretary for Territorial Affairs* in *Office of the Assistant Secretary—Public Land Management,* Department of the Interior, by Secretarial Order 2942, effective July 1, 1971.

Terrorism, Cabinet Committee To Combat Established by Presidential memorandum of Sept. 25, 1972. Terminated by National Security Council memorandum of Sept. 16, 1977.

Terrorist Threat Integration Center Established on May 1, 2003, pursuant to Presidential initiative. Transferred to the National Counterterrorism Center by act of Dec. 17, 2004 (118 Stat. 3697).

Textile National Industrial Relations Board Established by administrative order of June 28, 1934. Abolished by EO 6858 of Sept. 26, 1934, which created *Textile Labor Relations Board* in connection with the Department of Labor. *Board* terminated July 1, 1937, and functions absorbed by *U.S. Conciliation Service,* Department of Labor.

Textile National Industrial Relations Board, Cotton Established by original Code of Fair Competition for the Cotton Textile Industry, as amended July 10, 1934. Abolished by EO 6858 of Sept. 26, 1934.

Textile Work Assignment Board, Cotton Amendments to Code of Fair Competition for Cotton Textile Industry approved by EO 6876 of Oct. 16, 1934, and *Cotton Textile Work Assignment Board* appointed by *Textile Labor Relations Board. Board* expired June 15, 1935.

Textile Work Assignment Board, Silk Appointed by *Textile Labor Relations Board* following President's approval of amendments to Code of Fair Competition for Silk Textile Industry by EO 6875 of Oct. 16, 1934. Terminated June 15, 1935.

Textile Work Assignment Board, Wool Established by EO 6877 of Oct. 16, 1934. Terminated June 15, 1935.

Textiles, Office of Established by the Secretary of Commerce Feb. 14, 1971. Functions transferred to *Domestic and International Business Administration,* effective Nov. 17, 1972.

Thrift Depositor Protection Oversight Board. *See* **Oversight Board (of the Resolution Trust Corporation).**

Trade, Special Adviser to the President on Foreign Established by EO 6651 of Mar. 23, 1934. Terminated on expiration of *National Recovery Administration.*

Trade Administration, International *See* **Business and Defense Services Administration**

Trade Agreements, Interdepartmental Committee on Established by Secretary of State in 1934 and reestablished by EO 9832 of Feb. 25, 1947. Abolished by EO 11075 of Jan. 15, 1963.

Trade and Development Program Established by act of Sept. 4, 1961, as amended (88 Stat. 1804). Designated separate entity within the *U.S. International Development Cooperation Agency* by act of Sept. 4, 1961, as amended (102 Stat. 1329). Renamed Trade and Development Agency by act of Oct. 28, 1992 (106 Stat. 3657).

Trade Expansion Act Advisory Committee Established by EO 11075 of Jan. 15, 1963. Abolished by EO 11846 of Mar. 27, 1975, and records transferred to Trade Policy Committee established by same EO.

Trade Negotiations, Office of the Special Representative for   Renamed Office of the U.S. Trade Representative by EO 12188 of Jan. 4, 1980.

Trade Policy Committee   Established by EO 10741 of Nov. 25, 1957. Abolished by EO 11075 of Jan. 15, 1963.

Traffic Safety, President's Committee for   Established by Presidential letter of Apr. 14, 1954. Continued by EO 10858 of Jan. 13, 1960. Abolished by EO 11382 of Nov. 28, 1967.

Traffic Safety Agency, National   Established in the Department of Commerce by act of Sept. 9, 1966 (80 Stat. 718). Activity transferred to the Department of Transportation by act of Oct. 15, 1966 (80 Stat. 931). Responsibility placed in *National Highway Safety Bureau* by EO 11357 of June 6, 1967.

Training and Employment Service, U.S.   Established in *Manpower Administration,* Department of Labor, Mar. 17, 1969. Abolished by Secretary's letter of Dec. 6, 1971, and functions assigned to *Office of Employment Development Programs* and *U.S. Employment Service.*

Training School for Boys, National   See **District of Columbia, Reform-School of the**

Transportation, Federal Coordinator of   Established by act of June 16, 1933 (48 Stat. 211). Expired June 16, 1936, under provisions of Public Resolution 27 (49 Stat. 376).

Transportation, Office of   Established in the Department of Agriculture by Secretary's Memorandum 1966 dated Dec. 12, 1978. Abolished by Secretary's Memorandum 1030–25 dated Dec. 28, 1990.

Transportation and Communications Service   Established by General Services Administrator Oct. 19, 1961. Abolished by Administrator's order, effective July 15, 1972. Motor equipment, transportation, and public utilities responsibilities assigned to Federal Supply Service; telecommunications function assigned to *Automated Data Telecommunications Service.*

Transportation and Public Utilities Service   Abolished by General Services Administration order of Aug. 17, 1982. Functions transferred to various GSA organizations.

Transportation Safety Board, National   Established in the Department of Transportation by act of Oct. 15, 1966 (80 Stat. 935). Abolished by act of Jan. 3, 1975 (88 Stat. 2156), which established independent National Transportation Safety Board.

Transportation Security Administration   Established by act of Nov. 19, 2001 (115 Stat. 597). Functions transferred from Department of Transportation to Department of Homeland Security by act of Nov. 25, 2002 (116 Stat. 2178).

Transportation Statistics, Bureau of   Established by act of Dec. 18, 1991 (105 Stat. 2172). Transferred to Research and Innovative Technology Administration, Transportation Department, by act of Nov. 30, 2004 (118 Stat. 2424).

Travel Service, U.S.   Replaced by *U.S. Travel and Tourism Administration,* Department of Commerce, pursuant to act of Oct. 16, 1981 (95 Stat. 1014).

Travel and Tourism Administration, U.S.   Established by act of Oct. 16, 1981 (95 Stat. 1014). Abolished by act of Oct. 11, 1996 (110 Stat. 3407).

Travel and Tourism Advisory Board   Established by act of Oct. 16, 1981 (95 Stat. 1017). Abolished by act of Oct. 11, 1996 (110 Stat. 3407).

Treasury, Office of the Assistant Secretary of the— Electronics and Information Technology   Established by Secretary's Order 114–1 of Mar. 14, 1983. Abolished by Secretary's Order 114–3 of May 17, 1985, and functions transferred to Office of the Assistant Secretary for Management. Certain provisions effective Aug. 31, 1985 (50 FR 23573).

Treasury, Solicitor of the   Position established when certain functions of *Solicitor of the Treasury* transferred to the Department of Justice by EO 6166 of June 10, 1933. *Solicitor of the Treasury* transferred from the Department of Justice to the Department of the Treasury by same order. *Office of Solicitor of the Treasury* abolished by act of May 10, 1934 (48 Stat. 758), and functions transferred to General Counsel, the Department of the Treasury.

Treasury Police Force   See **Secret Service Division**

Treasury Secretary, Assistant   Office abolished by Reorg. Plan No. III of 1940, effective June 30, 1940, and functions transferred to Fiscal Assistant Secretary, Department of the Treasury.

Treasury Under Secretary for Enforcement, Office of   Established by act of Oct. 28, 1993 (107 Stat. 1234). Office abolished by act of Dec. 8, 2004 (118 Stat. 3245), and functions transferred to the Office of the Under Secretary for Terrorism and Financial Crimes, Department of the Treasury.

Treaties   See **State, Department of**

Typhus Commission, U.S. of America   Established in *Department of War* by EO 9285 of Dec. 24, 1942. Abolished June 30, 1946, by EO 9680 of Jan. 17, 1946.

U.S.   *See other part of title*

Uniformed Services University of the Health Sciences, School of Medicine of the   Renamed F. Edward Hébert School of Medicine by act of Sept. 24, 1983 (97 Stat. 704).

United Nations Educational, Scientific and Cultural Organization   U.S. membership in UNESCO authorized by act of July 30, 1946 (60 Stat. 712). Announcement of U.S. intention to withdraw made Dec. 28, 1983, in accordance with UNESCO constitution. Official U.S. withdrawal effective Dec. 31, 1984, by Secretary of State's letter of Dec. 19, 1984. The U.S. maintained status as an observer mission in UNESCO from 1984–2003, and rejoined the organization in October 2003.

Upper Mississippi River Basin Commission   Established by EO 11659 of Mar. 22, 1972. Terminated by EO 12319 of Sept. 9, 1981.

Urban Affairs, Council for   Established in Executive Office of the President by EO 11452 of Jan. 23, 1969. Terminated by EO 11541 of July 1, 1970.

Urban Mass Transportation Administration   Functions regarding urban mass transportation established in the Department of Housing and Urban Development by act of July 9, 1964 (78 Stat. 302). Most functions transferred to the Department of Transportation by Reorg. Plan No. 2 of 1968, effective June 30, 1968 (82 Stat. 1369), and joint responsibility assigned to the Departments of Transportation and Housing and Urban Development for functions relating to research, technical studies, and training. Transportation and Housing and Urban Development Under Secretaries agreed in November 1969 that the Department of Transportation should be focal point for urban mass transportation grant administration; at which time functions transferred to the Department of Transportation. Renamed Federal Transit Administration by act of Dec. 18, 1991 (105 Stat. 2088).

Urban Renewal Administration   Established in *Housing and Home Finance Agency* by Administrator's Organizational Order 1 of Dec. 23, 1954. Functions transferred to the Department of Housing and Urban Development by act of Sept. 9, 1965 (78 Stat. 667), and *Administration* terminated.

Utilization and Disposal Service   Established July 1, 1961, by Administrator of General Services and assigned functions of Federal Supply Service and Public Buildings Service. Functions transferred to *Property Management and Disposal Service* July 29, 1966.

Veterans Administration   Legal work in defense of suits against the U.S. arising under act of June 7, 1924 (43 Stat. 607), transferred to the Department of Justice by EO 6166 of June 10, 1933. Transfer deferred to Sept. 10, 1933, by EO 6222 of July 27, 1933. Established as an independent agency under the President by Executive Order 5398 of July 21, 1930, in accordance with the act of July 3, 1930 (46 Stat. 1016) and the act of Sept. 2, 1958 (72 Stat. 1114). Made an executive department in the executive branch and redesignated the Department of Veterans Affairs by act of Oct. 25, 1988 (102 Stat. 2635).

Veterans Appeals, U.S. Court of   Established by act of Nov. 18, 1988 (102 Stat. 4113). Renamed U.S. Court of Appeals for Veterans Claims by act of Nov. 11, 1998 (112 Stat. 3341).

Veterans Education Appeals Board   *See* **Veterans Tuition Appeals Board**

Veterans Employment and Training, Advisory Committee on   Renamed Advisory Committee on Veterans Employment, Training, and Employer Outreach by act of June 15, 2006 (120 Stat. 403).

Veterans Employment Service   Renamed Veterans' Employment and Training Service by Order 4–83 of Mar. 24, 1983 of the Secretary of Labor (48 FR 14092).

Veterans Health Administration   *See* **Medicine and Surgery, Department of**

Veterans Health Services and Research Administration   *See* **Medicine and Surgery, Department of**

Veterans Placement Service Board   Established by act of June 22, 1944 (58 Stat. 293). Abolished by Reorg. Plan No. 2 of 1949, effective Aug. 20, 1949, and functions transferred to the Secretary of Labor.

Veterans Tuition Appeals Board   Established by act of Aug. 24, 1949 (63 Stat. 654). Functions assumed by *Veterans Education Appeals Board* established by act of July 13, 1950 (64 Stat. 336). *Board* terminated by act of Aug. 28, 1957 (71 Stat. 474).

Veterinary Medicine, Bureau of   Established in Food and Drug Administration, *Department of Health, Education, and Welfare*. Renamed Center for Veterinary Medicine by FDA notice of Mar. 9, 1984 (49 FR 10166).

Virgin Islands   Public works programs under act of Dec. 20, 1944 (58 Stat. 827), transferred from General Services Administrator to the Secretary of the Interior by Reorg. Plan No. 15 of 1950, effective May 24, 1950.

Virgin Islands Company   Established in 1934. Reincorporated as Government corporation by act of June 30, 1949 (63 Stat. 350). Program terminated June 30, 1965, and *Corporation* dissolved July 1, 1966.

Virgin Islands Corporation   *See* **Virgin Islands Company**

Visitor Facilities Advisory Commission, National   Established by act of Mar. 12, 1968 (82 Stat. 45). Expired Jan. 5, 1975, pursuant to act of Oct. 6, 1972 (86 Stat. 776).

Vocational Rehabilitation, Office of   Established to administer provisions of act of July 6, 1943 (57 Stat. 374). Other duties delegated by acts of Aug. 3, 1954 (68 Stat. 652), Nov. 8, 1965 (79 Stat. 1282), July 12, 1960 (74 Stat. 364), and July 10, 1954 (68 Stat. 454). Redesignated *Vocational Rehabilitation Administration* Jan. 28, 1963. Made component of newly created *Social and Rehabilitation Service* as *Rehabilitation Services Administration* by *Department of Health, Education, and Welfare* reorganization of Aug. 15, 1967.

Vocational Rehabilitation Administration   *See* **Vocational Rehabilitation, Office of**

Voluntary Citizen Participation, State Office of Renamed State Office of Volunteerism in ACTION by notice of Apr. 18, 1986 (51 FR 13265), effective May 18, 1986.

Volunteer Service, International, Secretariat for Established in 1962 by International Conference on Middle Level Manpower called by President. Terminated Mar. 31, 1976, due to insufficient funding.

Volunteers in Service to America Established by act of Nov. 8, 1966 (80 Stat. 1472). *Service* administered by *Office of Economic Opportunity* and functions transferred to ACTION by Reorg. Plan No. 1 of 1971, effective July 1, 1971.

Wage Adjustment Board Established May 29, 1942, by the Secretary of Labor at Presidential direction of May 14, 1942, to accomplish purpose of act of Mar. 3, 1931 (46 Stat. 1494), as amended by acts of Aug. 30, 1935 (49 Stat. 1011), and Jan. 30, 1942 (56 Stat. 23). Disbanded on termination of *National Wage Stabilization Board.*

Wage and Price Stability, Council on Established in Executive Office of the President by act of Aug. 24, 1974 (88 Stat. 750). Abolished by EO 12288 of Jan. 29, 1981. Funding ceased beyond June 5, 1981, by act of June 5, 1981 (95 Stat. 74), and authorization for appropriations repealed by act of Aug. 13, 1981 (95 Stat. 432).

Wage and Price Stability Program *See* **Wage and Price Stability, Council on**

Wage Stabilization Board Established by EO 10161 of Sept. 9, 1950. Reconstituted by EO 10377 of July 25, 1952. Terminated Apr. 30, 1953, by EO 10434 of Feb. 6, 1953, and acts of June 30, 1952 (66 Stat. 296), and June 30, 1953 (67 Stat. 131).

Wage Stabilization Board, National *See* **Defense Mediation Board, National**

Wallops Flight Center, Wallops Island, VA Formerly separate field installation of National Aeronautics and Space Administration. Made component of Goddard Space Flight Center by NASA Management Instruction 1107.10A of Sept. 3, 1981.

War, Solid Fuels Administration for Established in the Department of the Interior by EO 9332 of Apr. 19, 1943. Absorbed *Office of Solid Fuels Coordinator for War* (originally established as *Office of Solid Fuels Coordinator for National Defense*) pursuant to Presidential letter of Nov. 5, 1941; later changed by Presidential letter of May 25, 1942. Terminated by EO 9847 of May 6, 1947.

War Assets Administration Established in *Office for Emergency Management* by EO 9689 of Jan. 31, 1946. Functions transferred to *Surplus Property Administration* by Reorg. Plan No. 1 of 1947, effective July 1, 1947, and agency renamed *War Assets Administration.* Abolished by act of June 30, 1949 (63 Stat. 738), and functions transferred for liquidation to General Services Administration.

War Assets Corporation *See* **Petroleum Reserves Corporation**

War Claims Commission Established by act of July 3, 1948 (62 Stat. 1240). Abolished by Reorg. Plan No. 1 of 1954, effective July 1, 1954, and functions transferred to Foreign Claims Settlement Commission of the U.S.

War Commodities Division Established in *Office of Foreign Economic Coordination* by Department of State Order of Aug. 27, 1943. *Office* abolished by departmental order of Nov. 6, 1943, pursuant to EO 9380 of Sept. 25, 1943, which established *Foreign Economic Administration* in *Office for Emergency Management.*

War Communications, Board of *See* **Defense Communications Board**

War Contracts Price Adjustment Board Established by act of Feb. 25, 1944 (58 Stat. 85). Abolished by act of Mar. 23, 1951 (65 Stat. 7), and functions transferred to *Renegotiation Board,* established by same act, and General Services Administrator.

War Damage Corporation *See* **War Insurance Corporation**

War, Department of Established by act of Aug. 7, 1789 (1 Stat. 49), succeeding similar department established prior to adoption of the Constitution. Three military departments—Army; Navy, including naval aviation and U.S. Marine Corps; and Air Force—reorganized under *National Military Establishment* by act of July 26, 1947 (61 Stat. 495).

War Finance Corporation Established by act of Apr. 5, 1918 (40 Stat. 506). Functions and obligations transferred by Reorg. Plan No. II of 1939, effective July 1, 1939, to the Secretary of the Treasury for liquidation not later than Dec. 31, 1939.

War Food Administration *See* **Food Production and Distribution, Administration of**

War Information, Office of Established in *Office of Emergency Management* by EO 9182 of June 13, 1942, consolidating *Office of Facts and Figures; Office of Government Reports; Division of Information, Office for Emergency Management;* and *Foreign Information Service—Outpost, Publications, and Pictorial Branches, Coordinator of Information.* Abolished by EO 9608 of Aug. 31, 1945. *Bureau of Special Services* and functions with respect to review of publications of Federal agencies transferred to *Bureau of the Budget.* Foreign information activities transferred to the Department of State.

War Insurance Corporation Established Dec. 13, 1941, by act of June 10, 1941 (55 Stat. 249). Charter filed Mar. 31, 1942. Renamed *War Damage Corporation* by act of Mar. 27, 1942 (56 Stat. 175). Transferred from *Federal Loan Agency* to the Department of Commerce by EO 9071 of Feb. 24, 1942. Returned to *Federal Loan Agency* by act of Feb. 24, 1945 (59 Stat. 5). *Agency* abolished by act of June 30, 1947 (61 Stat. 202), and functions assumed by *Reconstruction Finance Corporation.* Powers of

War Damage Corporation, except for purposes of liquidation, terminated as of Jan. 22, 1947.

War Labor Board, National *See* **Defense Mediation Board, National**

War Manpower Commission Established in *Office for Emergency Management* by EO 9139 of Apr. 18, 1942. Terminated by EO 9617 of Sept. 19, 1945, and functions, except *Procurement and Assignment Service,* transferred to the Department of Labor.

War Mobilization, Office of Established by EO 9347 of May 27, 1943. Transferred to *Office of War Mobilization and Reconversion* by EO 9488 of Oct. 3, 1944.

War Mobilization and Reconversion, Office of Established by act of Oct. 3, 1944 (58 Stat. 785). Consolidated with other agencies by EO 9809 of Dec. 12, 1946, to form *Office of Temporary Controls. Media Programming Division* and *Motion Picture Division* transferred to *Office of Government Reports,* reestablished by same order. Certain other functions transferred to President and the Secretary of Commerce.

War Mobilization and Reconversion Advisory Board, Office of Established by act of Oct. 3, 1944 (58 Stat. 788). Transferred to *Office of Temporary Controls* by EO 9809 of Dec. 12, 1946.

War Plants Corporation, Smaller Established by act of June 11, 1942 (56 Stat. 351). Functions transferred by EO 9665 of Dec. 27, 1945, to *Reconstruction Finance Corporation* and the Department of Commerce. Abolished by act of June 30, 1947 (61 Stat. 202), and functions transferred for liquidation to General Services Administration by Reorg. Plan No. 1 of 1957, effective July 1, 1957.

War and Post War Adjustment Policies, Advisory Unit on Established in *Office of War Mobilization* by Presidential direction Nov. 6, 1943. Report submitted Feb. 15, 1944, and Unit Director and Assistant Director submitted letter to Director of *War Mobilization* ending their work May 12, 1944.

War Production Board Established in *Office for Emergency Management* by EO 9024 of Jan. 16, 1942. *Board* terminated and successor agency, *Civilian Production Administration,* established by EO 9638 of Oct. 4, 1945.

War Property Administration, Surplus Established in *Office of War Mobilization* by EO 9425 of Feb. 19, 1944. Terminated on establishment of *Surplus Property Board* by act of Oct. 3, 1944 (58 Stat. 768). *Surplus Property Administration* established in *Office of War Mobilization and Reconversion* by act of Sept. 18, 1945 (59 Stat. 533), and *Board* abolished. Domestic functions of *Administration* merged into *War Assets Corporation, Reconstruction Finance Corporation,* by EO 9689 of Jan. 31, 1946. Foreign functions transferred to the Department of State by same order. Transfers made permanent by Reorg. Plan No. 1 of 1947, effective July 1, 1947.

War Refugee Board Established in Executive Office of the President by EO 9417 of Jan. 22, 1944. Terminated by EO 9614 of Sept. 14, 1945.

War Relations, Agricultural, Office for *See* **Farm Products, Division of**

War Relief Agencies, President's Committee on Established by Presidential letter of Mar. 13, 1941. *President's War Relief Control Board* established by EO 9205 of July 25, 1942, to succeed *Committee. Board* terminated by EO 9723 of May 14, 1946, and functions transferred to the Department of State.

War Relief Control Board, President's *See* **President's Committee on War Relief Agencies**

War Relocation Authority Established in *Office for Emergency Management* by EO 9102 of Mar. 18, 1942. Transferred to the Department of the Interior by EO 9423 of Feb. 16, 1944. Terminated by EO 9742 of June 25, 1946.

War Resources Board Established in August 1939 as advisory committee to work with *Joint Army and Navy Munitions Board.* Terminated by President Nov. 24, 1939.

War Resources Council *See* **Defense Resources Committee**

War Shipping Administration Established in *Office for Emergency Management* by EO 9054 Feb. 7, 1942. Terminated by act of July 8, 1946 (60 Stat. 501), and functions transferred to *U.S. Maritime Commission,* effective Sept. 1, 1946.

Water, Office of Saline Established to perform functions vested in the Secretary of the Interior by act of July 29, 1971 (85 Stat. 159). Merged with *Office of Water Resources Research* to form *Office of Water Research and Technology* by Secretary's Order 2966 of July 26, 1974.

Water Commission, National Established by act of Sept. 26, 1968 (82 Stat. 868). Terminated Sept. 25, 1973, pursuant to terms of act.

Water Policy, Office of Established by Department of the Interior Manual Release 2374 of Dec. 29, 1981, under authority of Assistant Secretary. Abolished by Secretarial Order No. 3096 of Oct. 19, 1983, and functions transferred to *Geological Survey* and *Office of Policy Analysis.*

Water Pollution Control Administration, Federal Established under the *Secretary of Health, Education, and Welfare* by act of Oct. 2, 1965 (79 Stat. 903). Transferred to the Department of the Interior by Reorg. Plan No. 2 of 1966, effective May 10, 1966. Renamed *Federal Water Quality Administration* by act of Apr. 3, 1970. Abolished by Reorg. Plan No. 3 of 1970, effective Dec. 2, 1970, and functions transferred to Environmental Protection Agency.

Water and Power Resources Service   Renamed Bureau of Reclamation May 18, 1981, by Interior Secretarial Order 3064.

Water Quality Administration, Federal   *See* **Water Pollution Control Administration, Federal Water**

Research and Technology, Office of   Established by Interior Secretarial Order 2966 of July 26, 1974. Abolished by Secretarial order of Aug. 25, 1982, and functions transferred to Bureau of Reclamation, Geological Survey, and *Office of Water Policy.*

Water Resources Council   Established by act of July 22, 1965 (89 Stat 575). Inactive as of Oct. 1, 1982.

Water Resources Research, Office of   Established to perform functions vested in the Secretary of the Interior by act of July 17, 1964 (78 Stat. 329). Merged with *Office of Saline Water* to form *Office of Water Research and Technology* by Secretary's Order 2966 of July 26, 1974.

Watergate Special Prosecution Force   Established by Attorney General order, effective May 25, 1973. Terminated by Attorney General order, effective June 20, 1977.

Waterways Corporation, Inland   Incorporated under act of June 3, 1924 (43 Stat. 360). Transferred from the *Department of War* to the Department of Commerce by Reorg. Plan No. II of 1939, effective July 1, 1939. *Corporation* sold to *Federal Waterways Corporation* under contract of July 24, 1953. Renamed *Federal Barge Lines, Inc.* Liquidated by act of July 19, 1963 (77 Stat. 81).

Weather Bureau   Established in the Department of Agriculture by act of Oct. 1, 1890 (26 Stat. 653). Transferred to the Department of Commerce by Reorg. Plan No. IV of 1940, effective June 30, 1940. Functions transferred to *Environmental Science Services Administration* by Reorg. Plan No. 2 of 1965, effective July 13, 1965.

Weather Control, Advisory Committee on   Established by act of Aug. 13, 1953 (67 Stat. 559). Act of Aug. 28, 1957 (71 Stat. 426), provided for termination by Dec. 31, 1957.

Weed and Seed, Executive Office of   Abolished by Public Law 109–162 of Jan. 5, 2006 (119 Stat. 3107). Functions transferred to Office of Weed and Seed Strategies, Office of Justice Programs, within the Department of Justice.

Weights and Measures, Office of Standard   Renamed *National Bureau of Standards* by act of Mar. 3, 1901 (31 Stat. 1449). *Bureau* transferred from the Department of the Treasury to the *Department of Commerce and Labor* by act of Feb. 14, 1903 (32 Stat. 825). *Bureau* established within the Department of Commerce by act of Mar. 4, 1913 (37 Stat. 736). Renamed National Institute of Standards and Technology by act of Aug. 23, 1988 (102 Stat. 1827).

Welfare Administration   Established by the *Secretary of Health, Education, and Welfare* reorganization of Jan. 28, 1963. Components consisted of *Bureau of Family Services, Children's Bureau, Office of Juvenile Delinquency and Youth Development,* and *Cuban Refugee Staff.* These functions reassigned to *Social and Rehabilitation Service* by Department reorganization of Aug. 15, 1967.

White House Police Force   *See* **Secret Service Division**

Wilson Memorial Commission, Woodrow   Established by act of Oct. 4, 1961 (75 Stat. 783). Terminated on submittal of final report to President and Congress Sept. 29, 1966.

Women, Interdepartmental Committee on the Status of   Established by EO 11126 of Nov. 1, 1963. Terminated by EO 12050 of Apr. 4, 1978.

Women, President's Commission on the Status of   Established by EO 10980 of Dec. 14, 1961. Submitted final report to President Oct. 11, 1963.

Women's Army Auxiliary Corps   Established by act of May 14, 1942 (56 Stat. 278). Repealed in part and superseded by act of July 1, 1943 (57 Stat. 371), which established *Women's Army Corps. Corps* abolished by the Secretary of Defense Apr. 24, 1978, pursuant to provisions of 10 U.S.C. 125A.

Women's Business Enterprise Division   Renamed *Office of Women's Business Enterprise* by Small Business Administrator's reorganization, effective Aug. 19, 1981. Renamed Office of Women's Business Ownership Aug. 19, 1982.

Women's Reserve   Established in U.S. Coast Guard by act of Nov. 23, 1942 (56 Stat. 1020).

Women's Year, 1975, National Commission on the Observance of International   Established by EO 11832 of Jan. 9, 1975. Continued by act of Dec. 23, 1975 (89 Stat. 1003). Terminated Mar. 31, 1978, pursuant to terms of act.

Wood Utilization, National Committee on   Established by Presidential direction in 1925. Abolished by EO 6179–B of June 16, 1933.

Work Projects Administration   *See* **Works Progress Administration**

Work-Training Programs, Bureau of   Abolished by reorganization of *Manpower Administration* and functions assigned to *U.S. Training and Employment Service,* effective Mar. 17, 1969.

Working Life, Productivity and Quality of, National Center for   Established by act of Nov. 28, 1975 (89 Stat. 935). Authorized appropriations expired Sept. 30, 1978, and functions assumed by *National Productivity Council.*

Works, Advisory Committee on Federal Public   Established by President Oct. 5, 1955. Abolished by President Mar. 12, 1961, and functions assigned to *Bureau of the Budget.*

Works Administration, Federal Civil   Established by EO 6420–B of Nov. 9, 1933. Function of employment expired March 1934. Function of settling claims continued under *Works Progress Administration.*

Works Administration, Public   *See* **Emergency Administration of Public Works, Federal**

Works Agency, Federal   Established by Reorg. Plan No. I of 1939, effective July 1, 1939. Functions relating to defense housing transferred to *Federal .Public Housing Authority, National Housing Agency,* by EO 9070 of Feb. 24, 1942. Abolished by act of June 30, 1949 (63 Stat. 380), and functions transferred to General Services Administration.

Works Emergency Housing Corporation, Public   Established by EO 6470 of Nov. 29, 1933. Incorporated under laws of State of Delaware. Abolished and liquidated as of Aug. 14, 1935, by filing of certificate of surrender of corporate rights.

Works Emergency Leasing Corporation, Public   Incorporated Jan. 3, 1934, under laws of Delaware by direction of Administrator of Public Works. Terminated with filed certificate of dissolution with secretary of state of Delaware Jan. 2, 1935.

Works Progress Administration   Established by EO 7034 of May 6, 1935, and continued by subsequent yearly emergency relief appropriation acts. Renamed *Work Projects Administration* by Reorg. Plan No. I of 1939, effective July 1, 1939, which provided for consolidation of *Works Progress Administration* into *Federal Works Agency.* Transferred by President to *Federal Works Administrator* Dec. 4, 1942.

Works, Special Board of Public   *See* **Land Program, Director of**

Yards and Docks, Bureau of   Established by acts of Aug. 31, 1842 (5 Stat. 579), and July 5, 1862 (12 Stat. 510). Abolished by Department of Defense reorg. order of Mar. 9, 1966, and functions transferred to the Secretary of the Navy (31 FR 7188).

Youth Administration, National   Established in *Works Progress Administration* by EO 7086 of June 26, 1935. Transferred to *Federal Security Agency* by Reorg. Plan No. I of 1939, effective July 1, 1939. Transferred to *Bureau of Training, War Manpower Commission,* by EO 9247 of Sept. 17, 1942. Terminated by act of July 12, 1943 (57 Stat. 539).

Youth Crime, President's Committee on Juvenile Delinquency and   Established by EO 10940 of May 11, 1961. Terminated by EO 11529 of Apr. 24, 1970.

Youth Fitness, President's Council on   Established by EO 10673 of July 16, 1956. Renamed *President's Council on Physical Fitness* by EO 11074 of Jan. 8, 1963. Renamed President's Council on Physical Fitness and Sports by EO 11398 of Mar. 4, 1968. Abolished and reestablished by EO 13265 of June 6, 2002.

Youth Opportunity, President's Council on   Established by EO 11330 of Mar. 5, 1967. Inactive as of June 30, 1971; EO 11330 revoked by EO 12379 of Aug. 17, 1982.

Youth Programs, Office of   Established in the Department of the Interior by Secretarial Order No. 2985 of Jan. 7, 1965. Functions moved to Office of Historically Black College and University Programs and Job Corps, Office of the Secretary, by Departmental Manual Release 2788 of Mar. 22, 1988.

NAME INDEX

NOTE: Separate listings of Senators and Representatives can be found beginning on pages 30 and 35, respectively. Any other references to said persons can be found in this index.

THE LAWS OF PLATO